Nanogenerators

Nanogenerators

Basic Concepts, Design Strategies, and Applications

Edited by
Inamuddin
Mohd Imran Ahamed
Rajender Boddula
Tariq Altalhi

CRC Press is an imprint of the
Taylor & Francis Group, an **informa** business

First edition published 2023
by CRC Press
6000 Broken Sound Parkway NW, Suite 300, Boca Raton, FL 33487-2742

and by CRC Press
4 Park Square, Milton Park, Abingdon, Oxon, OX14 4RN

CRC Press is an imprint of Taylor & Francis Group, LLC

Library of Congress Cataloging-in-Publication Data

Names: Inamuddin, 1980- editor. | Ahamed, Mohd Imran, editor. | Boddula, Rajender, editor. | Altalhi, Tariq, editor.
Title: Nanogenerators : basic concepts, design strategies, and applications / edited by Inamuddin, Mohd Imran Ahamed, Rajender Boddula, Tariq A. Altalhi.
Other titles: Nanogenerators (CRC Press)
Description: Boca Raton, FL : CRC Press, 2022. | Includes bibliographical references and index.
Identifiers: LCCN 2021060066 | ISBN 9781032034911 (hbk) | ISBN 9781032034973 (pbk) | ISBN 9781003187615 (ebk)
Subjects: LCSH: Nanogenerators.
Classification: LCC TK2897 .N36 2022 | DDC 621.31/24--dc23/eng/20220311
LC record available at https://lccn.loc.gov/2021060066

ISBN: 978-1-032-03491-1 (hbk)
ISBN: 978-1-032-03497-3 (pbk)
ISBN: 978-1-003-18761-5 (ebk)

DOI: 10.1201/9781003187615

Typeset in Times
by KnowledgeWorks Global Ltd.

Contents

Preface

Nanogenerators are an emerging technology that has potential applications in energy science to convert waste mechanical energy into electricity by applying the small-scale physical change/displacement current in ambient conditions. Basically nanogenerators are classified into piezoelectric, pyroelectric, and triboelectric nanogenerators. Nanogenerators act as sustainable power sources due to their unique characteristics such as mobility, accessibility, and sustainability. They involve only tiny power in microwatt to milliwatts to overcome the actual change. It is an emerging technology utilized in various applications including intelligent electronics, wireless transmitters, the Internet of things, self-powered systems, sensors, and actuators. The understanding of nanogenerators is essential for the industrial and research community.

This book covers an in-depth overview of the history, theory, experimental, and design strategies for nanogenerators and their applications. It also discusses mechanisms, characterizations, device fabrications, and utilization in sustainable power applications. The chapters in this book are contributed by top researchers working in the area of sustainable technologies. It is a good source for readers from many scientific and engineering research fields from undergraduate students to R&D and academic professionals. It will appeal to readers wishing to know how to do research and utilize nanogenerators for modern world applications.

Chapter 1 details the concept of nanogenerators. It enumerated in detail the different types of nanogenerators in addition to energy sources and energy storage devices. The applications of nanotechnology to energy storage are discussed. The integration of nanogenerators in energy storage devices is also highlighted.

Chapter 2 aims to provide a clear idea of the basic working principle and mechanisms of different types of nanogenerators. It details about the basic principles to working principles going through the physics behind nanogenerators.

Chapter 3 discusses about the nanogenerators of the various forms. Furthermore, it discusses the basic theory for their functioning with its working mechanism and applications in various fields. Nanogenerators are a new path for scientists to provide energy from waste resources which will be an efficient source of energy in the near future.

Chapter 4 highlights the role of different 2D materials such as Mxene, graphene, metal-organic frameworks, black phosphorus, and covalent organic frameworks for nanogenerator applications. The major focus is given to the triboelectric nanogenerators for such applications.

Chapter 5 gives information about the carbon materials used in nanogenerator production. Detailed information about the nanogenerators, carbon materials used in this field, and the production methods of these materials are given. Also, recent studies in this field are elaborated.

Chapter 6 provides an in-depth analysis of piezoelectric nanogenerators (PENGs), their applications, and challenges. Different organic, inorganic, and metal halide perovskites-based piezoelectric materials are discussed in detail. Additionally, a thorough overview of different fabrication techniques along with future scope of these nanogenerators is also illustrated.

Chapter 7 discusses the several applications of triboelectric nanogenerators in harvesting energy. Their construction and modes of operation are discussed. The roles of triboelectric nanogenerators in a range of diverse fields such as medicine, textile, and electrochemistry are discussed in detail. The challenges and areas for improvement are also presented.

Chapter 8 explains the potential of hybrid nanogenerators based on different properties such as piezoelectricity, pyroelectricity, and triboelectricity. Additionally, it details their structure, synthesis, and characterization techniques. The promising applications in several fields such as wearable devices, sensors, textiles, biomedical devices, antimicrobial adherence are also presented.

Chapter 9 portrays an overview of the pyroelectric nanogenerators (PNs) as energy harvesting system using the thermal fluctuations from the green and mechanical energy sources. It discusses the different synthesis approaches used to fabricate polymer-based PNs (PPNs) and ceramic-based PNs (CPNs), and their advanced applications in energy technology.

Chapter 10 discusses wearable nanogenerators (WNGs) capable of harnessing different kinds of energy respective to the human body. WNGs based on textiles and flexible sheet configurations and their mechanisms are reported in detail. The outside and the inside body applications of such WNGs are reported along with existing limitations and challenges.

Chapter 11 provides an overview of the different nanogenerators based on piezoelectric or triboelectric materials. In this context, different piezoelectric and triboelectric materials are discussed. It also gives an idea of the wide area of applications it beholds in the field of electronics.

Chapter 12 describes the types, design, constitution, working principle, and use of nanogenerators in devices for sustainable energy. It also focuses on the applications of nanogenerators in medicine, industry, environment monitoring, and technology.

Chapter 13 surveys the basics, implementation, and recent advances of nanogenerators. The applications of nanogenerators in sensors, photovoltaic panels, harvesting of wind, and blue energy are also discussed.

Chapter 14 summarizes different types of nanogenerators based on piezoelectric, triboelectric, and pyroelectric effects. Their large-scale applications in various fields are covered. Following the trends in actual applications, triboelectric nanogenerators are given more thrust than the other two types of nanogenerators.

Chapter 15 discusses the synthesis, mechanical, thermal, and solar energy harvesting applications of electroactive polymer-carbon nanostructures and nanocomposites. The importance of the generation of the electroactive phase of the insulating polymer for nanogenerator application is highlighted. Inherently conducting polymer-based nanocomposites for thermal and solar energy harvesting applications are also discussed.

Chapter 16 provides an overview of various synthesis procedures to fabricate different types of polymer-based nanogenerators. In addition, the biomedical applications of polymer-based nanogenerators, such as wearable smart devices, micro or nano transparent devices, as well as applications as energy sources to power implantable biomedical appliances, are discussed.

Editors

Inamuddin, PhD, is an Assistant Professor in the Department of Applied Chemistry, Aligarh Muslim University, Aligarh, India. He earned his M.Sc. in organic chemistry at Chaudhary Charan Singh (CCS) University, Meerut, India, in 2002. He earned his MPhil and PhD in applied chemistry at Aligarh Muslim University (AMU), India, in 2004 and 2007, respectively. He has extensive research experience in the multidisciplinary fields of analytical chemistry, materials chemistry, electrochemistry, and, more specifically, renewable energy and environment. He has worked on different research projects as project fellow and senior research fellow funded by the University Grants Commission (UGC), Government of India, and the Council of Scientific and Industrial Research (CSIR), Government of India. He has received the Fast Track Young Scientist Award from the Department of Science and Technology, India, to work in the area of bending actuators and artificial muscles. He has completed four major research projects sanctioned by the University Grant Commission, Department of Science and Technology, the Council of Scientific and Industrial Research, and the Council of Science and Technology, India. He has published 196 research articles in international journals of repute and 19 book chapters in knowledge-based book editions published by renowned international publishers. He has published 150 edited books with Springer (UK), Elsevier, Nova Science Publishers, Inc. (USA), CRC Press – Taylor & Francis Asia Pacific, Trans Tech Publications Ltd. (Switzerland), IntechOpen Limited (UK), Wiley-Scrivener (USA) and Materials Research Forum LLC (USA). He is a member of various journals' editorial boards. He is an Associate Editor for several journals (*Environmental Chemistry Letter, Applied Water Science*, and *Euro-Mediterranean Journal for Environmental Integration*, Springer-Nature), Frontiers Section Editor (*Current Analytical Chemistry*, Bentham Science Publishers), Editorial Board Member (*Scientific Reports*, Nature), Editor (*Eurasian Journal of Analytical Chemistry*), and Review Editor (*Frontiers in Chemistry*, Frontiers, UK). He has also guest-edited various thematic special issues to the journals of Elsevier, Bentham Science Publishers, and John Wiley & Sons, Inc. He has attended as well as chaired sessions at various international and national conferences. He has worked as a Postdoctoral Fellow, leading a research team at the Creative Research Initiative Center for Bio-Artificial Muscle, Hanyang University, South Korea, in the field of renewable energy, especially biofuel cells. He has also worked as a Postdoctoral Fellow at the Center of Research Excellence in Renewable Energy, King Fahd University of Petroleum and Minerals, Saudi Arabia, in the field of polymer electrolyte membrane fuel cells and computational fluid dynamics of polymer electrolyte membrane fuel cells. He is a life member of the *Journal of the Indian Chemical Society*. His research interests include ion exchange materials, a sensor for heavy metal ions, biofuel cells, supercapacitors, and bending actuators.

Mohd Imran Ahamed, PhD, is working as a Research Associate at the Department of Chemistry, Aligarh Muslim University (AMU), Aligarh, India. He earned a BSc (Hons) in chemistry and a PhD in chemistry at AMU. He earned an MSc in organic chemistry at Dr. Bhimrao Ambedkar University, Agra, India. Dr. Ahamed has published several research and review articles in various international scientific journals. He has co-edited 57 books with Springer (UK), Elsevier, CRC Press - Taylor & Francis Asia Pacific, Materials Research Forum LLC (USA) and Wiley-Scrivener (USA). His research work includes ion-exchange chromatography, wastewater treatment, and analysis, bending actuator and electrospinning.

Rajender Boddula, PhD, works with the Chinese Academy of Sciences – President's International Fellowship Initiative (CAS-PIFI) at the National Center for Nanoscience and Technology (NCNST, Beijing). He earned his MS in organic chemistry at Kakatiya University, Warangal, India, in 2008.

He earned a PhD in chemistry with highest honors in 2014 for the work titled "Synthesis and Characterization of Polyanilines for Supercapacitor and Catalytic Applications" at the CSIR-Indian Institute of Chemical Technology (CSIR-IICT) and Kakatiya University (India). Before joining the National Center for Nanoscience and Technology (NCNST) as CAS-PIFI research fellow, China, Dr. Boddula worked as a senior research associate and postdoc at National Tsing-Hua University (NTHU, Taiwan), respectively, in the fields of biofuel and CO_2 reduction applications. His academic honors include a University Grants Commission National Fellowship and many merit scholarships, study-abroad fellowships from Australian Endeavour Research Fellowship, and CAS-PIFI. He has published many scientific articles in peer-reviewed international journals, authored around 20 book chapters, and also served as an editorial board member and a referee for reputed international peer-reviewed journals. He has published edited books with Springer (UK), Elsevier, Materials Science Forum LLC (USA), Wiley-Scrivener (USA), and CRC Press – Taylor & Francis Group. His specialized areas of research are energy conversion and storage, which include sustainable nanomaterials, graphene, polymer composites, heterogeneous catalysis for organic transformations, environmental remediation technologies, photoelectrochemical water-splitting devices, biofuel cells, batteries, and supercapacitors.

Tariq Altalhi, PhD, joined the Department of Chemistry at Taif University, Saudi Arabia, as Assistant Professor in 2014. He earned his doctorate degree from University of Adelaide, Australia, in the year 2014 with Dean's Commendation for Doctoral Thesis Excellence. He was promoted to the position of the Head of Chemistry Department at Taif University in 2017 and Vice Dean of Science College in 2019 till now. In 2015, one of his works was nominated for Green Tech Awards from Germany, Europe's largest environmental and business prize, amongst top ten entries. He has co-edited various scientific books. His group is involved in fundamental multidisciplinary research in nanomaterial synthesis and engineering, characterization, and their application in molecular separation, desalination, membrane systems, drug delivery, and biosensing. In addition, he has established key contacts with major industries in Kingdom of Saudi Arabia.

Contributors

Ayooluwa P. Adeagbo
Department of Electrical and Electronics Engineering
Adeleke University
Ede, Osun State, Nigeria

Mohd Imran Ahamed
Department of Chemistry
Aligarh Muslim University
Aligarh, India

Vivian C. Akubude
Department of Agricultural and Bioresource Engineering
Federal University of Technology
Owerri, Nigeria

Muhanna K. Al-Muhanna
The National Center for Composite and High Performances Materials
King Abdulaziz City for Science and Technology (KACST)
Riyadh, Saudi Arabia

Syed Wazed Ali
Department of Textile and Fibre Engineering
Indian Institute of Technology Delhi
New Delhi, India

Pratheep K Annamalai
Australian Institute for Bioengineering and Nanotechnology
The University of Queensland
St. Lucia, Queensland, Australia

Naushad Anwar
Department of Chemistry
Aligarh Muslim University
Aligarh, India

Sandeep Arya
Department of Physics
University of Jammu
Jammu, Jammu and Kashmir, India

Muhammad Shahbaz Aslam
Institute of Biochemistry and Biotechnology
University of the Punjab
Lahore, Pakistan

Satyaranjan Bairagi
Department of Textile and Fibre Engineering
Indian Institute of Technology Delhi
New Delhi, India

Prasun Banerjee
Multiferroic and Magnetic Material Research Laboratory (MMMRL)
Gandhi Institute of Technology and Management (GITAM) University
Bengaluru, Karnataka, India

Swagata Banerjee
Department of Textile and Fibre Engineering
Indian Institute of Technology Delhi
New Delhi, India

Syqa Banoo
Department of Chemistry
Mangalayatan University
Beswan, Aligarh, India

Mehmet Bugdayci
Chemical Engineering Department
Yalova University
Yalova, Turkey
and
Construction Technology Department
Istanbul Medipol University
Istanbul, Turkey

Moises Bustamante-Torres
Biomedical Engineering Department
School of Biological and Engineering
Yachay Tech University
Urcuqui City, Ecuador

Jorge Cárdenas-Gamboa
Departament de Ciència de Materials i Química Física
Institut de Química Teòrica i Computacional (IQTCUB)
Universitat de Barcelona
Barcelona, Spain

Emilio Bucio Carrillo
Department of Radiation Chemistry and Radiochemistry
Institute of Nuclear Sciences
National Autonomous University of Mexico
Mexico City, Mexico

Michael K. Danquah
Chemical Engineering Department
University of Tennessee
Chattanooga, Tennessee, USA

Kannan Deepa
Department of Chemical Engineering
SRM Institute of Science and Technology
Kattankulathur, Tamil Nadu, India

Tanvir Mahady Dip
Department of Yarn Engineering
Bangladesh University of Textiles
Dhaka, Bangladesh

Deepak Dubal
Centre for Materials Science
School of Chemistry and Physics
Queensland University of Technology
Brisbane, Queensland, Australia

Jocelyne Estrella-Nuñez
Chemistry Department
School of Chemical and Engineering
Yachay Tech University
Urcuqui City, Ecuador

Francisco García-Salinas
Centro de Investigación y de Estudios Avanzados Unidad Querétaro
Cinvestav, Santiago de Querétaro, Mexico

Reshma Haridass
Department of Sciences
Amrita School of Engineering
Amrita Vishwa Vidyapeetham
Coimbatore, Tamil Nadu, India

Muhammad Tahir Haseeb
Department of Pharmacy
University of Sargodha
Sargodha, Pakistan

Muhammad Ajaz Hussain
Department of Chemistry
University of Sargodha
Sargodha, Pakistan

Shah Imtiaz
Department of Chemistry
Aligarh Muslim University
Aligarh, India

Muhammad Mudassir Iqbal
Institute of Biochemistry and Biotechnology
University of the Punjab
Lahore, Pakistan

Md Rabiul Islam
Department of Chemistry
Aligarh Muslim University
Aligarh, India

Pallavi Jain
Department of Chemistry
SRM Institute of Science and Technology Delhi, NCR Campus
Ghaziabad, Uttar Pradesh, India

Ampattu R. Jayakrishnan
Department of Physics
School of Basic and Applied Sciences
Central University of Tamil Nadu
Thiruvarur, India

Jaison Jeevanandam
Centro de Química da Madeira (CQM)
Molecular Material Research Group
Universidade da Madeira, Campus da Penteada
Funchal, Portugal

Rita Joshi
Centre of Excellence: Nanotechnology
Indian Institute of Technology Roorkee
Roorkee, India

Koppole Kamakshi
Department of Physics
Indian Institute of Information Technology
Thiruchirapalli, Tamil Nadu, India

Ashish Kapoor
Department of Chemical Engineering
SRM Institute of Science and Technology
Kattankulathur, Tamil Nadu, India

Bhargavi Koneru
Multiferroic and Magnetic Material Research Laboratory (MMMRL)
Gandhi Institute of Technology and Management (GITAM) University
Bengaluru, Karnataka, India

Indranil Lahiri
Centre of Excellence: Nanotechnology
and
Department of Metallurgical and Materials Engineering
Indian Institute of Technology Roorkee
Roorkee, India

Gulzar Muhammad
Department of Chemistry
Government College University Lahore
Lahore, Pakistan

Pranesh Muralidharan
Department of Sciences
Amrita School of Engineering
Amrita Vishwa Vidyapeetham
Coimbatore, Tamil Nadu, India

Ashok Kumar Nanjundan
School of Chemical Engineering
The University of Queensland
St. Lucia, Queensland, Australia

Manikanta P. Narayanaswamy
Department of Applied Sciences
Centre for Post Graduate Studies
Visvesvaraya Technological University
Bengaluru, Karnataka, India

Srinivasan Natarajan
Solid State and Structural Chemistry Unit
Indian Institute of Science
Bangalore, India

Kevin N. Nwaigwe
Department of Mechanical Engineering
University of Botswana
Gaborone, Botswana

Victor C. Okafor
Department of Agricultural and Bioresource Engineering
Federal University of Technology
Owerri, Nigeria

Levent Oncel
Metallurgical and Materials Engineering Department
Faculty of Engineering and Architecture
Sinop University
Sinop, Turkey

Jelili A. Oyedokun
Engineering and Scientific Services Department
National Centre for Agricultural Mechanization
Ilorin, Kwara State, Nigeria

Bhavya Padha
Department of Physics
University of Jammu
Jammu, Jammu and Kashmir, India

Sharadwata Pan
School of Life Sciences
Technical University of Munich
Freising, Germany

Ritvik B. Panicker
Department of Chemical Engineering
SRM Institute of Science and Technology
Kattankulathur, Tamil Nadu, India

Vishal Panwar
Nanomaterials and Applications Laboratory
Department of Metallurgical and Materials Engineering
Indian Institute of Technology Roorkee
Roorkee, India

Sapna Raghav
Department of Chemistry
SRM Institute of Science and Technology Delhi
NCR Campus
Ghaziabad, Uttar Pradesh, India

Dinesh Rangappa
Department of Applied Sciences
Centre for Post Graduate Studies
Visvesvaraya Technological University
Bengaluru, Karnataka, India

Muhammad Arshad Raza
Department of Chemistry
Government College University Lahore
Lahore, Pakistan

Ramanujam B.T.S.
Department of Sciences
Amrita School of Engineering
Amrita Vishwa Vidyapeetham
Coimbatore, Tamil Nadu, India

Tania Saif
Department of Chemistry
Government College University Lahore
Lahore, Pakistan

Sugumaran Sathish
Department of Physics
School of Basic and Applied Sciences
Central University of Tamil Nadu
Thiruvarur, India

Koppole C. Sekhar
Department of Physics
School of Basic and Applied Sciences
Central University of Tamil Nadu
Thiruvarur, India

Nimra Shakeel
Department of Chemistry
Aligarh Muslim University
Aligarh, India

Suraj Sharma
Department of Textiles, Merchandising, and Interiors
University of Georgia
Athens, Georgia, USA

José P. B. da Silva
Centro de Fısica das Universidades do Minho e do Porto (CF-UM-UP)
Campus de Gualtar
Braga, Portugal

Madhur Babu Singh
Department of Chemistry
SRM Institute of Science and Technology Delhi, NCR Campus
Ghaziabad, Uttar Pradesh, India

Prabhakar Sivaraman
Department of Chemical Engineering
SRM Institute of Science and Technology
Kattankulathur, Tamil Nadu, India

Atif Suhail
Nanomaterials and Applications Laboratory
Department of Metallurgical and Materials Engineering
Indian Institute of Technology Roorkee
Roorkee, India

Jhilmil Swapnalin
Multiferroic and Magnetic Material Research Laboratory (MMMRL)
Gandhi Institute of Technology and Management (GITAM) University
Bengaluru, Karnataka, India

Md Mazbah Uddin
Department of Textiles, Merchandising, and Interiors
University of Georgia
Athens, Georgia, USA

Sonali Verma
Department of Physics
University of Jammu
Jammu, Jammu and Kashmir, India

1 Nanogenerators-Based Energy Storage Devices

Vivian C. Akubude, Ayooluwa P. Adeagbo, Jelili A. Oyedokun, Victor C. Okafor, and Kevin N. Nwaigwe

CONTENTS

1.1 INTRODUCTION

With the serious problems of environmental pollution and the possibility of an energy crisis, the desire for green and renewable energy sources for practical applications has gained increasing importance. Energy collection methods including solar energy, wave, wind or mechanical energy have attracted widespread attention for small self-powered electronic devices with low power consumption, such as sensors, wearable devices, electronic skin and implantable devices. One significant challenge for electronic devices is that the energy storage devices are unable to provide sufficient energy for continuous and long-time operation, leading to frequent recharging or inconvenient battery replacement. To satisfy the needs of next-generation electronic devices for sustainable working, conspicuous progress has been achieved regarding the development of nanogenerator-based self-charging energy storage devices. A nanogenerator is a type of technology that converts

DOI: 10.1201/9781003187615-1

mechanical/thermal energy as produced by small-scale physical change into electricity. And, this chapter discusses the various types of nanogenerators, applications of nanotechnology in energy and integration of nanogenerator in energy storage devices.

1.2 CONCEPT OF NANOGENERATORS

Nanogenerator is simply a technology or nanoscale device capable of generating electricity from conversion of mechanical/thermal energy. Basically, there are three techniques of generating electricity: piezoelectricity, triboelectricity and pyroelectricity. The first two generate electrical current from mechanical energy while the third technique produces electricity via thermal energy conversion.

The first principle of nanogenerators stems from the fact that they use displacement current as driving force in effective generation of electrical power from mechanical energy. Maxwell equation shows the governing principle of nanogenerator as given below:

$$\nabla . D = \rho \tag{1.1}$$

$$\nabla . \mathrm{B} = 0 \tag{1.2}$$

$$\nabla * E = -\frac{\partial D}{\partial t} \tag{1.3}$$

$$\nabla * H = J + \frac{\partial D}{\partial t} \tag{1.4}$$

$$D = \varepsilon_0 E + P \tag{1.5}$$

$$P = (\varepsilon - \varepsilon_0) E \tag{1.6}$$

$$D = \varepsilon E \tag{1.7}$$

$$J_D = \frac{\partial D}{\partial t} = \varepsilon \frac{\partial E}{\partial t} \tag{1.8}$$

The Maxwell equation has further been expanded to estimate the power output of nanogenerators. A polarization term was introduced into the equation to stand for the polarization generated by the electrostatic surface charges; therefore, Equation (1.5) can be written as:

$$D = \varepsilon_0 E + P + P_s$$

1.2.1 Types of Nanogenerators

There are different classes of nanogenerators as described in Figure 1.1. They are classified based on the type of energy that is harvested and used for electricity generation. Basically, they are mechanical energy and thermal energy harvesting type. Also, ion stream effect type is gradually springing up.

1.2.1.1 Triboelectric Nanogenerator (TENG)

This is energy harvesting device that converts mechanical energy into electricity based on triboelectric effect. It comprises two unlike films facing opposite to each other with electrodes fixed at the top and bottom of each film. It can be used as mechanical energy harvester (mechanical energy like vibration, wave energy and biomechanical energy), power source for some electronic devices

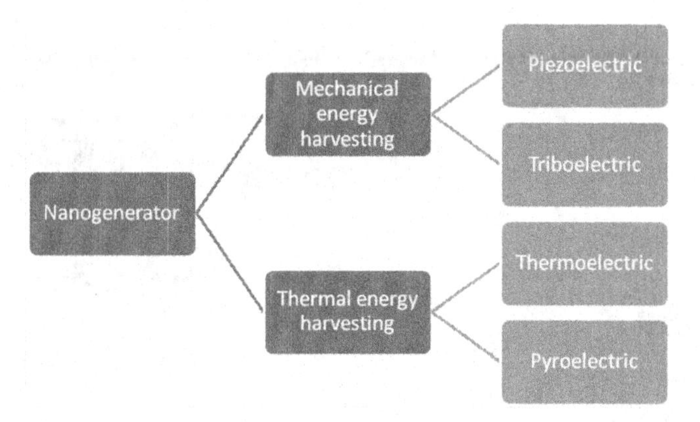

FIGURE 1.1 Classification of nanogenerator.

and self-powered sensors for pressure, flow rate, respiration and motion (Pu et al. 2015; Song et al. 2016; Sridhar et al. 2019). The four basic working modes of this type of nanogenerators are vertical contact separation mode, linear sliding mode, single-electrode mode and free-standing mode (see Figure 1.2).

1.2.1.2 Piezoelectric Nanogenerator (PENG)

This is the first nanogenerator to be discovered in 2006, which is capable of harvesting electric current from mechanical energy via the action of nanostructured piezoelectric material. Examples of piezoelectric materials include polyvinylidene fluoride (PVDF), poly(vinylidene fluoride-co-trifluoroethylene) [P(VDF-TrFE)], BaTiO3(BTO), ZnO, $Pb(Zr_xTi_{1-x})O_3$(PMN-and $(1 - x)$Pb (MgNb2/3)O3 – xPbTiO3(PMN-PT). It converts kinetic energy due to crystal deformation into electricity. The mechanism of operation of this class of nanogenerator is based on piezoelectricity or piezoelectric effect as shown in Figure 1.3. This piezoelectric effect is a change in electric polarization that is generated in piezoelectric materials when subjected to

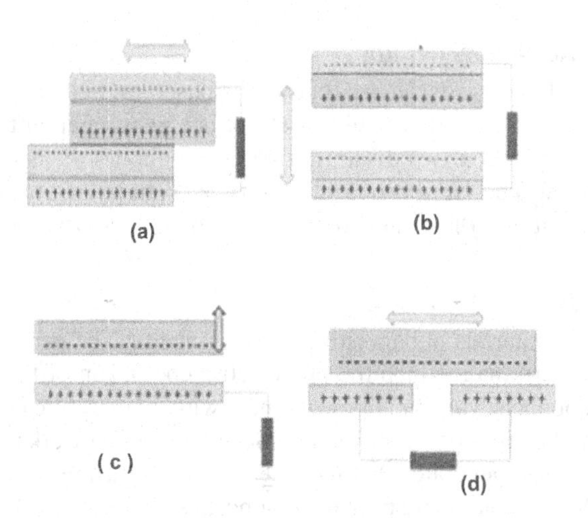

FIGURE 1.2 Four basic working modes of triboelectric nanogenerator: (a) in-plane sliding mode, (b) vertical separation mode, (c) single electrode mode and (d) free standing mode (Sridhar et al. 2019).

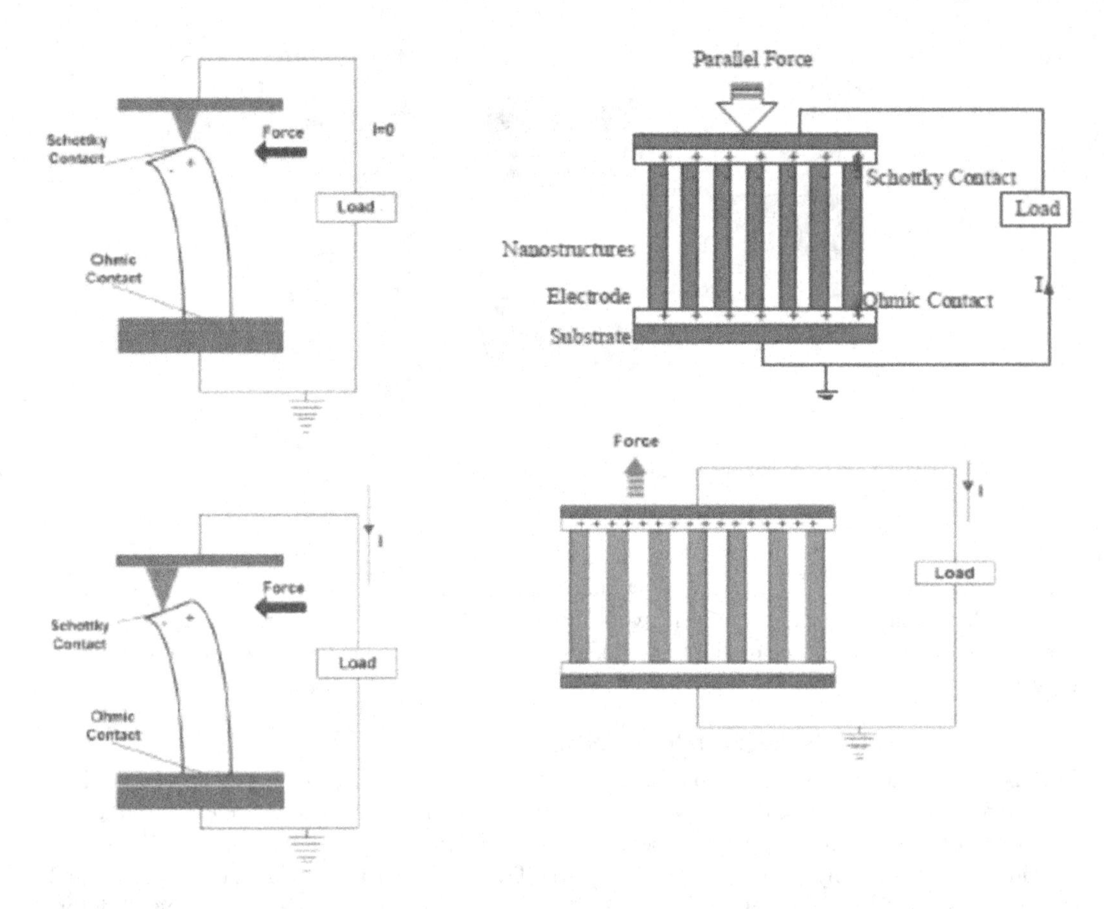

FIGURE 1.3 Piezoelectric nanogenerator under applied force (Sridhar et al. 2019).

mechanical stress or pressure. It can be direct (as used in microphone, pressure sensors or hydrophones) or indirect (as used in speakers or buzzers), see Figure 1.4.

1.2.1.3 Thermoelectric Nanogenerator

This is a class of thermal energy harvester that works based on Seebeck effect as described in Figure 1.5. It is important to note that Seebeck effect is a phenomenon that occurs when electric current is generated between two dissimilar electric conductors as a result of temperature difference between them. The limitations of this type of nanogenerators are their need for high-temperature gradient and the required materials to be used are costly.

1.2.1.4 Pyroelectric Nanogenerator

This applies pyroelectric materials in harvesting thermal energy. The performance of the material is dependent on a key determinant which the pyroelectric coefficient of the material. Pyroelectric coefficient has two components: varying dipole moment with temperature change and piezoelectric effect induced by thermal expansion in the crystal. Several research works have been documented on different pyroelectric nanogenerator that has been fabricated using different materials such as ZnO nanowire arrays, $KNbO_3$ nanowire polymer composites, $Pb(Zr_xTi_{1-x})O_3$(PZT) film (Yang et al. 2012a, 2012b). They can be used to power liquid-crystal display (LCD), light emitting diode (LED), wireless sensors, etc.

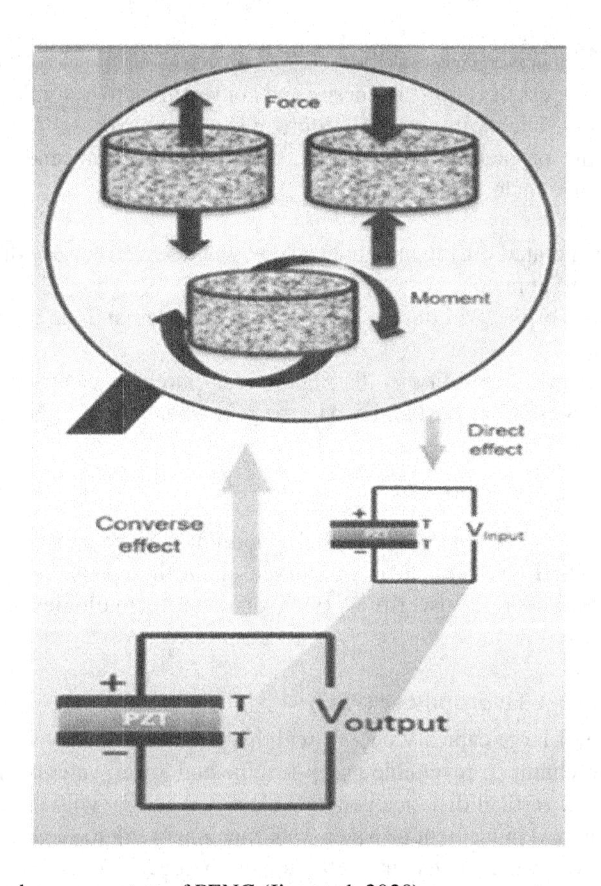

FIGURE 1.4 Direct and converse state of PENG (Jiao et al. 2020).

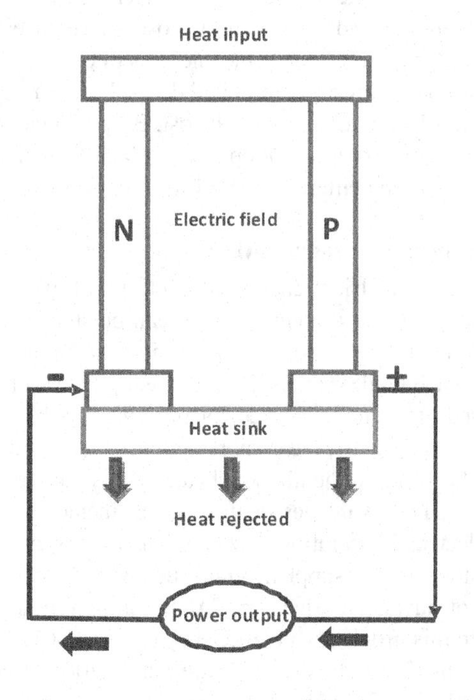

FIGURE 1.5 Seebeck effect.

1.3 ENERGY STORAGE

This is defined as the process of capturing energy and converting it to a storable state that is cheaper to use during its demand. The main aim of the storage is to store excesses for later use. This solves the instability/variability problem associated with energy output from renewable energy systems. Benefits of energy storage include the following:

a. Excess energy generated during abundance supply is reserved for use during limited supply to meet up with demand.
b. It ensures stability of the grid during seasons of rise in demand via its rapid discharge of power to the grid.
c. It gives room for grid flexibility as distributors can purchase electricity during off-peak times when energy is at low cost and sell to the grid when it is in high demand.

1.3.1 Energy Storage Technologies

Energy can be stored in diverse forms using small, medium or large capacity technologies. These technologies are classified based on form of storage or on their expected function (i.e. energy density, power density and response time). Each of these technologies has its strengths and weaknesses.

1.3.1.1 Pumped-Storage Hydropower (PSH)

This is the oldest type of large capacity energy with hydroelectric energy storage configuration. It comprises upper water chamber, reversible pump-turbine and lower water chamber. The two water chambers separated by a vertical distance generate electrical energy with the help of turbine and a generator during downward movement of water from the upper tank to the lower tank. The water is pumped back via upward water movement to the upper chamber where it is stored usually during off-peak periods using excess electricity, thereby storing for later use. It is an advanced storage system that is characterized by long storage periods, high efficiency, relatively low capital cost per unit of energy, no emissions of pollutants and low operating cost as it requires no fuel. It offers long-term solution and facilitates the integration of renewable energies into the system. Its limitations include lack of large space to accommodate big reservoirs and dams, long lead time, disruption of land and wildlife, destruction of trees and vegetation and high cost. Based on its operational design, PSH can be classified as open-loop system and closed-loop system. The former is continuously linked to a natural flowing water source, and the latter is isolated from naturally flowing water source.

1.3.1.2 Compressed Air Energy Storage (CAES)

This technology converts excess electrical energy from electrical grid or renewable energy source to high-pressure compressed gas using a compressor. It can be stored underground (for large-scale purpose) or in specialized tanks (for small-scale purpose) and can be used during the periods of high electricity demand. These storage facilities must be strong enough to withstand the high pressure and air density of the compressed air, hence the need for utilizing durable construction materials with great strength (like carbon fiber). Usually, the compressed air is retrieved back to the surface via diabetic approach that uses fuel to heat up the air. The heated air is then employed in turning a turbine to generate electrical energy. The advantages of this system include high energy saving, better air quality, reliable and stable electrical generation, low maintenance cost and improved service life of air compressor. Its drawback is the need for supplementary fuel, which is detrimental to the environment. However, the development of advanced adiabatic CAES configuration – which is emission-free – offers a promising solution to this problem with efficiency of about 50%, notwithstanding its need for advanced thermal storage method, which is rare (Pickard et al. 2009). Recently, research is focusing on developing isothermal CAES configurations for more efficient energy system.

1.3.1.3 Thermal Energy Storage (TES)

This type of energy storage system stores up energy by heating or cooling a storage medium and the stored energy is later deployed in heating or cooling application and power generation. It is achieved by employing diverse material properties using different mechanisms resulting in three categories, namely sensible heat storage, latent energy storage and thermal-chemical storage. The first category utilizes temperature difference (either increase or decrease), the second category uses change in phase of a material and the third category employs chemical reaction or sorption on the surface of a material. The advantages of TES are increase in energy security and reliability, reduction in CO_2 emissions, cost effectiveness and high system efficiency. Several research works in relation to this type of storage system have been documented in detail by different researchers (Sharma et al. 2009; Chidambaram et al. 2011; Liu et al. 2012; Oró et al. 2012; Zhou et al. 2012; Tian and Zhao 2013; Zhai et al. 2013; Moreno et al. 2014; Joybari et al. 2015; Pintaldi et al. 2015).

1.3.1.4 Batteries

These are electrochemical storage devices that store energy in chemical form and release it for use when needed in electrical form. They follow a cycle of charging and discharging. When charging, the electrical energy is stored as chemical energy; while during the discharge, the chemical energy is converted back to electrical energy. This system comprises battery banks that consist of multiple cells arranged in parallel and/or series form to produce the expected output. A cell is made up of electrodes that are negatively charged anode and positively charged cathode, both immersed in an electrolyte. There are several classifications of battery storage based on the characteristics of the electrode and electrolyte materials. These include lead-acid batteries, nickel-based batteries, lithium-ion batteries, flow batteries, metal-air batteries and lead-acid batteries (Scrosati and Garche 2010; Xia et al. 2012; Li et al. 2015; Hannan et al. 2017; Kurzweil and Garche 2017; Leung et al. 2017; Untereker et al. 2017; Kwon et al. 2018; Vangapally et al. 2018; Zhang et al. 2018; Trocino et al. 2019; XAKTY 2019; Yang et al. 2019) etc. as shown in Figure 1.6.

1.3.1.5 Flywheels Energy Storage

Flywheels are mechanical device that is used to store up rotational energy in form of kinetic energy. It comprises a spinning mass that is driven by electric motor. Usually, they are charged via the rotation of the motor, thereby increasing their rotational speed afterwards. The heavier the flywheel, the more energy it can store and the faster the flywheel mass spin the more energy there is stored. They are characterized by high efficiency, high energy and power density, low maintenance cost, short recharge time, reliable and safe, large energy storage capacity, very high charging and discharging rate and long lifetime. However, it is limited by its high cost, short discharge time and need for larger space in comparison with batteries.

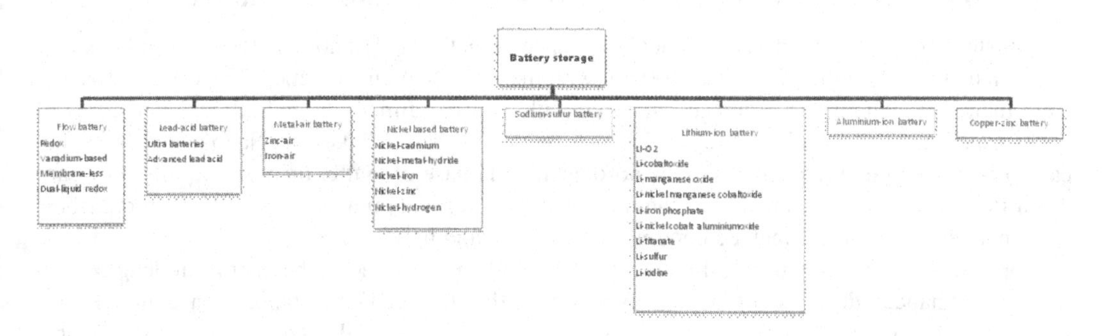

FIGURE 1.6 Classification of battery.

1.3.1.6 Supercapacitors

This is short-term energy storage device that stores electrical energy at an electrode-electrolyte interface. It comprises two electrodes with electrolyte in between them. Its properties include high energy and power density, fast charging rate, long lifetime, reliable, light weight in comparison to battery and rapid charge/discharge cycles. However, it faces some bottleneck such as high self-discharge rate, low energy storage capacity when compared to battery. They can be used in cars for dynamic breaking and in most cases complement batteries in applications.

1.3.1.7 Superconducting Magnetic Energy Storage

This technology generates energy stored in form of magnetic field by circulating direct current through a superconducting coil. The magnetic field form gives room for more quantity of energy to be stored over a short duration with high efficiency. The system comprises superconducting coil, refrigeration system and power conducting system (Liu et al. 2015; Salih et al. 2015). It is characterized by high instantaneous efficiency, rapid response time, ability to release almost all the stored energy and high number of complete charge-discharge cycles. This system is suitable for regulation of network stability and continuous operation application.

1.3.1.8 Fuel Cell-Hydrogen Energy Storage

Hydrogen is a clean energy source with high density that offers lots of potential as alternative source of fuel. Energy is needed for producing hydrogen as well as in removing hydrogen from the carbon portion of fossil fuel and also for the splitting of hydrogen from oxygen in water. Fuel cell is primarily designed to regenerate spent energy used in generating hydrogen via water electrolysis. The storage system comprises electrolysis, fuel cell, and hydrogen buffer tank. During electrolysis, off-peak electricity is utilized in generating hydrogen while the fuel cell uses the generated hydrogen and oxygen from the atmospheric air to produce electric current for offsetting on-peak electricity demand. Also, the buffer tank ensures the accessibility of enough quantity of resources in time of need. The hydrogen-based fuel cell operates based on the oxygen-reduction oxidation-reduction reaction between the hydrogen and oxygen within the fuel cell system comprising of two electrodes separated by an electrolyte medium which permits the transfer of ions or charged particles. The fuel cell can also be operated reversible by using external electric current to produce or split hydrogen and oxygen in the presence of water. There are different configurations of fuel cell such as polymer electrolyte fuel cell (PEFCs), alkaline fuel cell (AFCs), phosphoric acid fuel cell (PAFCs), solid oxide fuel cell (SOFCs), molten carbonate fuel cell (MCFCs), direct methanol fuel cell (DMFCs), direct ammonia fuel cell (DAFCs), direct carbon fuel cell (DCFCS) etc. The different modes of hydrogen storage are compressed, liquefied, metal hydride etc. This storage modes have been improved using nanostructured materials like carbon nanotube, nanocubes of metallo-organic compounds etc. (Christodoulou et al. 2007; Radgen 2007; Winter 2009; Seitz et al. 2013).

1.4 APPLICATIONS OF NANOTECHNOLOGY TO ENERGY STORAGE

Nanomaterial can be defined as a chemical substance or material of nanoscale level or smaller sizes or few nanometers ranging from 1 to 100 nm with their physical and chemical properties differing from their bulk counterpart. They are material with external dimension and internal structure on nanoscale. Examples include titanium oxide, iron oxide, fullerenes, flakes etc. Nanomaterials can be categorized based on their dimension (nano-object) and phase of matter (nanostructure). They can occur in natural form and can also be engineered using several approaches. The various approaches for nanomaterial synthesis include bottom-up, top-down and hybrid method.

Application of nanomaterials in energy storage devices reduces the diffusion length and therefore enhances the kinetics of the operation of the device. For instance, conventional battery energy storage faces certain limitations because of the material used in its design and fabrication. Nanomaterials offer a better solution to these convectional battery materials proffering remedy to challenges, such as electrode volume change, solid-electrolyte interphase formation, electron and ion transport and atom and molecule diffusion. Hence, they can promote the charge-discharge rate as well as the electron transport within the electrode. Research has documented several works with respect to the use of nanomaterials in energy storage systems (Wang et al. 2002; Bruce et al. 2008; Chan et al. 2008; Cheng and Verbrugge 2008; Li et al. 2008; Banerjee et al. 2009; Jayaprakash et al. 2011; Ji et al. 2011; Kalnaus et al. 2011; Sun et al. 2012; Liu et al. 2013; Yang et al. 2014) for the enhancement of the overall performance of the storage system. Nanostructure-based lithium-sulfur battery comprising of grapheme, sulfur and carbon nanocomposite multilayer structures was designed as a cathode material to promote charge-discharge performance (Jin et al. 2015).

Nanomaterials have been used to improve the performance of solar cells via current collection enhancement in amorphous silicon devices (Johlin et al. 2016), plamonic improvement in dye-sensitized solar cells (Sheehan et al. 2013), promotion of light trapping in crystalline silicon.

1.5 INTEGRATION OF NANOGENERATORS IN ENERGY STORAGE DEVICES

So far coupling of nanogenerator with energy storage devices has been proven possible with batteries particularly lithium ion type and supercapacitors. Lithium ion battery is the most common energy storage device that is commercially available. Over the decades, several investigations revolve around it and presently it has been quite developed for diverse applications in electronic system. Also, supercapacitor is another common energy storage device with notable attributes: long lifetime, excellent charging/discharging performance and power density (Liu et al. 2010; Zhu et al. 2011; Chou et al. 2013). Research has captured majorly the integration of nanogenerator with lithium ion batteries and supercapacity using different materials, structure and design.

1.5.1 Triboelectric Nanogenerator-Based Energy Storage Devices

The triboelectric nanogenerators show obvious advantages in collecting low-frequency and irregular mechanical energies such as human motions, wind, water waves and vibration. Under various periodic mechanical motions such as human walking, vibrations and ocean waves, a pulsating alternating current (AC) output is delivered through the triboelectric nanogenerators to an external circuit. It is not possible for triboelectric nano generators to be directly used to drive most electronic devices as a result of the varying frequency and the unpredictable amplitude of the pulsating output. It is important to store the energy harvested by nanogenerators in an energy storage unit and to provide a regulated output. Consequently, self-charging power systems have been developed by hybridizing a nanogenerator with an energy storage unit and since most energy storage units are in DC form, the energy storage unit is charged by the nanogenerator through a full-wave bridge rectifier. In this hybrid system, the energy storage can be directly charged by mechanical motions. Lithium ion battery-based energy storage can be considered as an example. The integration of nanogenerator in the storage can be realized through developing a flexible lithium battery on an arch-shaped triboelectric nanogenerator structure. This will in turn generate an AC in response to the external triggering when the surrounding mechanical energy is applied to the intended power unit which can be stored as electrical energy after rectification in the lithium ion battery. The lithium ion battery does not serve only as energy storage but also as a power regulator and management for the entire system by utilizing the stable electrode-potential difference. Since mechanical motion is available almost everywhere and at any time to replenish the energy draining in the battery, a large capacity battery with a high density may not be necessary to ensure a long operating lifetime. Due to advancement in technology and rapid development recorded in miniaturized electronic devices, micro-supercapacitors have been proved as complements to or replacements for batteries and electrolytic capacitors in many applications because of its remarkable advantages

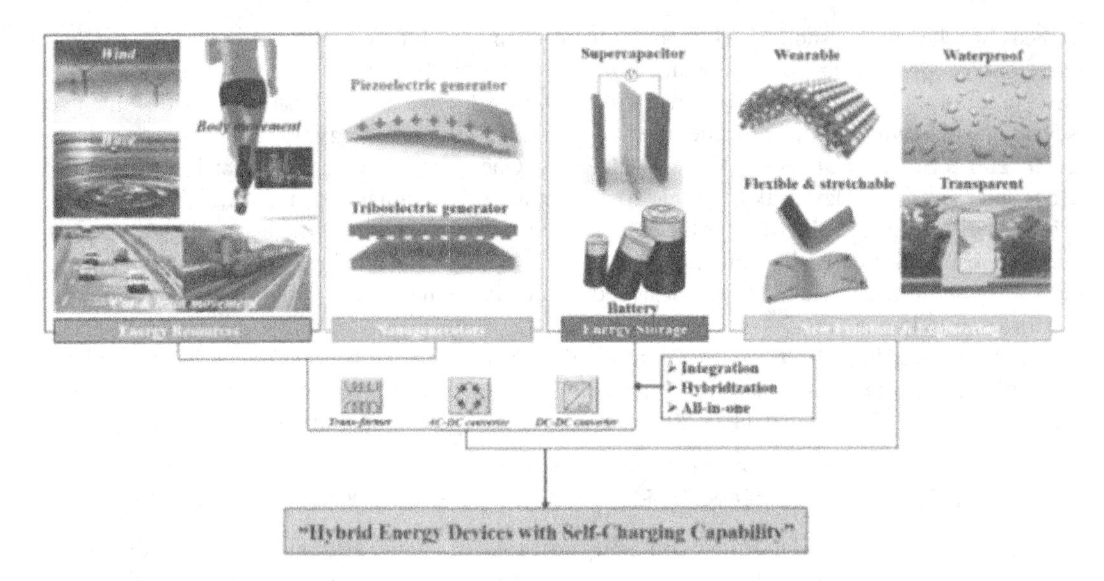

FIGURE 1.7 Nanogenerator-based hybrid energy storage system.

such as: high rate capabilities, high power densities, long life cycles and environmental friendliness. As a result of this, it is possible to develop a triboelectric nanogenerator-based energy harvester and a micro-supercapacitor array storage unit. A laser-induced graphene is a vital material in developing this array due to the excellent electrical conductivity property of graphene and its tremendous potential for energy storage by making them to store more charge and also charge faster. Also, in

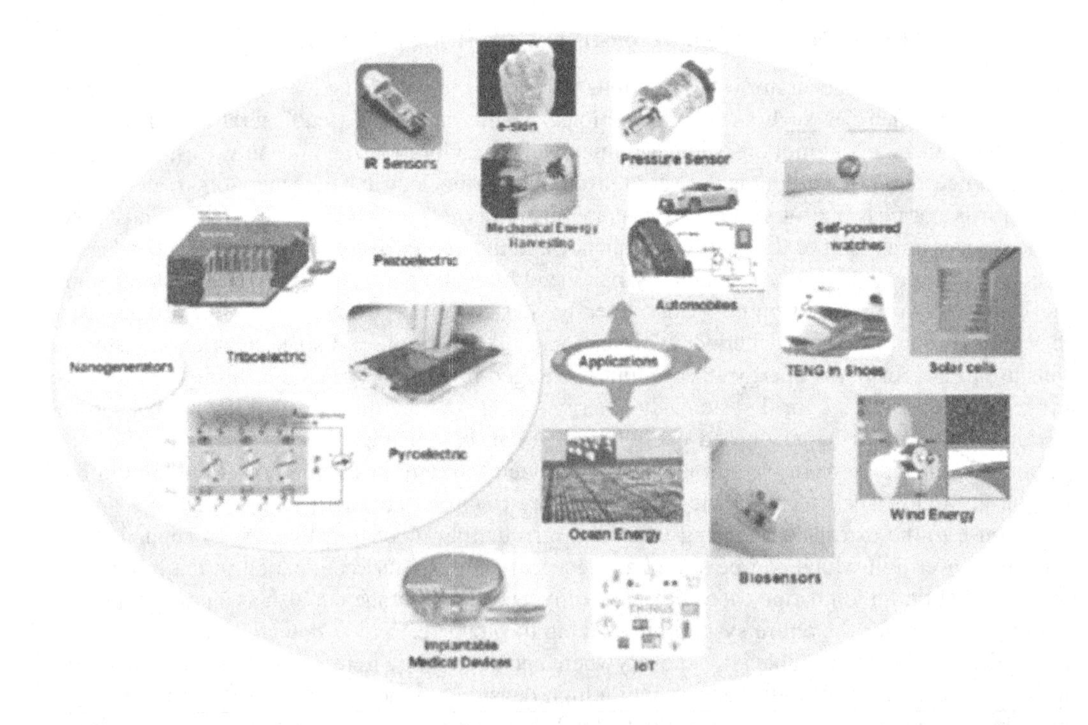

FIGURE 1.8 Applications of nanogenerators (Sridhar et al. 2019).

TABLE 1.1
Characteristics of Different Classes of Battery

S. No	Lead-Acid	Nickel-Based	Li-Ion	Flow	Reference
1	Low specific energy and power	High energy density	High energy and power density	Stores energy in the electrolytes	Scrosati and Garche (2010); Xia et al. (2012); Li et al. (2015); Yang et al. (2019)
2	Periodic maintenance	More charge/discharge cycle	High open circuit voltage	Capacity of the battery is determined by the volume of the electrolytes	Kurzweil and Garche (2017); Untereker et al. (2017)
3	Limited charging/discharging cycle	Less need for maintenance and less expensive	Fast charging and discharging	The membrane-less flow battery type has stable performance with longer lifetime and durable	Hannan et al. (2017); Leung et al. (2017); Kwon et al. (2018)
4	Environmental pollution issues	Nickel hydroxide and cadmium hydroxide are used as positive and negative electrode respectively.	Low self-discharging rate	Vanadium-based flow battery is associated with high power, long life, high efficiency, high safety and high capacity	Zhang et al. (2018); XAKTY (2019)
5	Low cycle life	Safety issues because of the toxicity of cadmium	Long lifetime		Vangapally et al. (2018)
6	High self-discharge rate	Wide temperature range and easy recycling			Trocino et al. (2019)

triboelectric nanogenerator coupled to battery for effective charging for use in hybrid devices, the conversion and storage efficiency and charge retention capacity were studied for different electrode materials. The outcome shows that effective charging process is dependent on the electrode material used (Pu et al. 2016).

1.5.2 Piezoelectric Nanogenerator-Based Energy Storage Devices

Studies show that lithium ion batteries can be integrated with piezoelectric nanogenerator for energy harvesting and storage. Piezoelectric material (piezoelectric PVDF-PZT film) with high piezoelectric performance was designed for self-charging of lithium ion battery. Also, piezoceramic material has been used for energy generation for storage in thin film batteries which has the ability to support structural load (Anton et al. 2010).

Charging of supercapacitor using piezoelectric effect has been investigated, where piezoelectric material can be utilized as separator in an electrochemical storage unit. The nanogenerator depends on the potential difference produced by the material when subjected to mechanical stress and energy generated can be stored in electrochemical storage unit in absence of a rectifier (Ramadoss et al. 2015; Pan et al. 2016). Past studies also showed how lead zirconite-titanate piezoelectric material can employ vibrations in generating energy to charge a capacitor of capacitance 22 microfarad producing 0.42 volts, which was used to power a laser diode. Furthermore, this concept of integration has been applied in wearable devices, all-in-one devices and multifunctional hybrid devices as detailed in the following research report (Pu et al. 2015, 2016; Ramadoss et al. 2015; Guo et al. 2016; Luo et al. 2016; Song et al. 2016; Wang et al. 2016; Wen et al. 2016; Yi et al. 2016).

1.6 CONCLUSION

Coupling of nanogenerators with energy storage has become a necessity with the advent of mobile and portable electronics. These electronics require continuous power to function properly, but nanogenerator is limited in this aspect as it can only produce instantaneous power via mechanical energy source. And this is contrary to the traditional energy storage system like batteries and supercapacitors that can generate continuous power. Hence, it is imperative to integrate nanogenerator with energy storage system for use in electronic systems. Results show that this can be the future next generation energy solution because of their practical applications in our electronic systems. However, studies on suitable material selection are imperative for fabrication of materials with high performance, eco-friendly and cheap. Investigation should be carried out on device life and more research in area of combining nanogenerator with other types of high-performing batteries like metal-air, lithium-sulfur, lithium-silicon etc. Also, their use in wearable and implantable devices should be well studied in terms of safety, device cost and compatibility.

REFERENCES

Anton, S. R., Erturk, A. & Inman, D. J. Multifunctional self-charging structures using piezoceramics and thin-film batteries *Smart Mater. Struct.* 19, 115021–115115 (2010). DOI: 10.1088/0964-1726/19/11/115021.

Banerjee, P., Perez, I., Henn-Lecordier, L., Lee, S. B. & Rubloff, G. W. Nanotubular metal–insulator–metal capacitor arrays for energy storage. *Nat. Nanotechnol.* 4, 292–296 (2009). DOI: 10.1038/nnano.2009.37.

Bruce, P. G., Scrosati, B. & Tarascon, J.-M. Nanomaterials for rechargeable lithium batteries. *Angew. Chem. Int. Ed.* 47, 2930–2946 (2008).

Chan, C. K., Zhang, X. F. & Cui, Y. High capacity Li ion battery anodes using Ge nanowires. *Nano Lett.* 8, 307–309 (2008).

Cheng, Y. T. & Verbrugge, M. W. The influence of surface mechanics on diffusion induced stresses within spherical nanoparticles. *J. Appl. Phys.* 104, 083521 (2008).

Chidambaram, L. A., Ramana, A. S., Kamaraj, G. & Velraj, R. Review of solar cooling methods and thermal storage options. *Renew. Sustain. Energy Rev.* 15, 3220–3228 (2011).

Chou, J. C., Chen, Y. L., Yang, M. H., Chen, Y. Z., Lai, C., Chiu, H. T., Lee, C. Y., Chueh, Y. L. & Gan, J. Y. RuO_2/MnO_2 core-shell nanorods for supercapacitors. *J. Mater. Chem. A* 1, 8753 (2013).

Christodoulou, C., Karagiorgis, G., Poullikkas, A., Karagiorgis, N. & Hadjiargyriou, N. Green electricity production by a grid connected H_2/fuel cell in Cyprus. In: Proceedings of the renewable energy sources and energy efficiency (2007).

Guo, H., Yeh, M. H., Lai, Y. C., Zi, Y., Wu, C., Wen, Z., Hu, C. & Wang, Z. L. All-in-one shape-adaptive self-charging power package for wearable electronics. *ACS Nano* 10, 10580 (2016).

Hannan, M. A., Hoque, M. M., Mohamed, A. & Ayob, A. Review of energy storage systems for electric vehicle applications: issues and challenges. *Renew. Sustain. Energy Rev.* 69, 771–789 (2017). DOI: 10.1016/j.rser.2016.11.171.

Jayaprakash, N., Shen, J., Moganty, S. S., Corona, A. & Archer, L. A. Porous hollow carbon@sulfur composites for high-power lithium–sulfur batteries. *Angew. Chem. Int. Ed.* 50, 5904–5908 (2011).

Ji, L. W., Lin, Z., Alcoutlabi, M. & Zhang, X. W. Recent developments in nanostructured anode materials for rechargeable lithium-ion batteries. *Energy Environ. Sci.* 4, 2682–2699 (2011).

Jiao, P., Egbe, K. I., Xie, Y., Nazar, A. M. & Alavi, A. H. Piezoelectric sensing techniques in structural health monitoring: a state-of-the art review. *Sensors* 20(13), 3730 (2020).

Jin, K., Zhou, X. & Liu, Z. Graphene/sulfur/carbon nanocomposites for high performance lithium sulfur batteries. *Nanomaterials* 5(3): 1481–1489 (2015).

Johlin, E., Al-Obeidi, A., Nogay, G., Stuckelberger, M., Buonassis, T. & Grossman, J. C. Nanohole structuring for improved performance of hydrogenated amorphous silicon photovoltaics. *ACS Appl. Mater. Interfaces* 8(24): 15169–15176 (2016).

Joybari, M., Haghighat, F., Moffat, J. & Sra, P. (2015). Heat and cold storage using phase change materials in domestic refrigeration systems: the state-of-the-art review. *Energy Build.* 106, 111–124.

Kalnaus, S., Rhodes, K. & Daniel, C. A study of lithium ion intercalation induced fracture of silicon particles used as anode material in Li-ion battery. *J. Power Sources* 196, 8116–8124 (2011).

Kurzweil, P. & Garche, J. (2017). 2 Overview of batteries for future automobiles. In J. Garche, E. Karden, P. T. Moseley & D. A. J. Rand (Eds.), *Lead-acid batteries for future automobiles* (p. 2796). Amsterdam: Elsevier.

Kwon, G., Lee, S., Hwang, J., Shim, H.-S., Lee, B., Lee, M. H., ... Kang, K. Multi-redox molecule for high-energy redox flow batteries. *Joule* 2(9), 1771–1782 (2018). DOI: 10.1016/j.joule.2018.05.014.

Leung, P., Martin, T., Liras, M., Berenguer, A., Marcilla, R., Shah, A., ... Palma, J. Cyclohexanedione as the negative electrode reaction for aqueous organic redox flow batteries. *Appl. Energy* 197, 318–326 (2017).

Li, Y. G., Tan, B. & Wu, Y. Y. Mesoporous Co_3O_4 nanowire arrays for lithium ion batteries with high capacity and rate capability. *Nano Lett.* 8, 265–270 (2008).

Li, W., Yang, Y., Zhang, G. & Zhang, Y.-W. Ultrafast and directional diffusion of lithium in phosphorene for high-performance lithium-ion battery. *Nano Lett.* 15(3), 1691–1697 (2015).

Liu, C., Yu, Z., Neff, D., Zhamu A. & Jang B. Z. Graphene-based supercapacitor with an ultrahigh energy density. *Nano Lett.* 10, 4863 (2010).

Liu, M., Saman, W. & Bruno, F. Review on storage materials and thermal performance enhancement techniques for high temperature phase change thermal storage systems. *Renew. Sustain. Energy Rev.* 16, 2118–2132 (2012).

Liu, N., Huo, K. F., McDowell, M. T., Zhao, J. & Cui, Y. Rice husks as a sustainable source of nanostructured silicon for high performance Li-ion battery anodes. *Sci. Rep.* 3, 1919 (2013).

Liu Y., Tang Y., Shi, J., Shi, X., Deng, J. & Gong, K. Application of small-sized SMES in an EV charging station with DC bus and PV system, *IEEE Trans. Appl. Supercond.* 25, 5700406 (2015).

Luo, J., Tang, W., Fan, F. R., Liu, C., Pang, Y., Cao, G. & Wang, Z. L. Transparent and flexible self-charging power film and its application in a sliding unlock system in touchpad technology. *ACS Nano* 10, 8078 (2016).

Moreno, P., Solé, C., Castell, A. & Cabeza, L. F. The use of phase change materials in domestic heat pump and air-conditioning systems for short term storage: a review. *Renew. Sustain. Energy Rev.* 39, 1–13 (2014).

Oró, E., de Gracia, A., Castell, A., Farid, M. M. & Cabeza, L. F. Review on phase change materials (PCMs) for cold thermal energy storage applications. *Appl. Energy* 99, 513–533 (2012).

Pickard, W. F., Hansing, N. J. & Shen, A. Q. Can large scale advanced-adiabatic compressed air storage be justified economically in an age of sustainable energy? *J. Renew. Sustain. Energy* 1(033102), 1–10 (2009). DOI: 10.1063/1.3139449.

Pintaldi, S., Perfumo, C., Sethuvenkatraman, S., White, S. & Rosengarten, G. A review of thermal energy storage technologies and control approaches for solar cooling. *Renew. Sustain. Energy Rev.* 41, 975–995 (2015).

Pu, X., Li, L., Song, H., Du, C., Zhao, Z., Jiang, C., Cao, G., Hu, W. & Wang, Z. L. A self-charging power unit by integration of a textile triboelectric nanogenerator and a flexible lithium-ion battery for wearable electronics. *Adv. Mater.* 27, 2472 (2015).

Pu, X., Li, L., Liu, M., Jiang, C., Du, C., Zhao, Z., Hu, W. & Wang, Z. L. Wearable self-charging power textile based on flexible yarn supercapacitors and fabric nanogenerators. *Adv. Mater.* 28, 98 (2016).

Radgen, P. (2007): Zukunftsmarkt elektrische Energiespeicherung. Umwelt, innovation, Beschäftigung 05/07 (Hrsg. Umweltbundesamt und Bundesministerium für Umwelt, Naturschutz und Reaktorsicherheit). http://www.umweltbundesamt.de/publikationen/zukunftsmarkt-elektrische-energiespeicherung.

Ramadoss, A., Saravanakumar, B., Lee, S. W., Kim, Y.-S., Kim, S. J. & Wang, Z. L. Piezoelectric-driven self-charging supercapacitor power cell. *ACS Nano* 9, 4337 (2015).

Salih E., Lachowicz S., Bass O. & Habibi D. Superconducting magnetic energy storage unit for damping enhancement of a wind farm generation system. *J. Clean Energy Technol.* 3, 398–405 (2015).

Scrosati, B. & Garche, J. (2010). Lithium batteries: status, prospects and future. *J. Power Sources* 195(9), 2419–2430. DOI: 10.1016/j.jpowsour.2009.11.048.

Seitz, S., Moller, B. P., Thielmann, A., Sauer, A., Meister, M., Pero, M., Kleine, O., Rohde, C., Bierwisch, A., de Vries, M. & Kayser, V. (2013). Nanotechnology in the sectors of solar energy and energy storage. Technology Report by the International Electrotechnical Commission (IEC). https://webstore.iec.ch/publication/61191.

Sharma, A., Tyagi, V. V., Chen, C. R. & Buddhi, D. Review on thermal energy storage with phase change materials and applications. *Renew. Sustain. Energy Rev.* 13, 318–345 (2009).

Sheehan, S. W., Noh, H., Brudvig, G. W., Cao, H. & Schmuttenmaer, C. A. Plasmonic enhancement of dye-sensitized solar cells using core-shell-shell nanostructures. *J. Phys. Chem. C* 117(2), 927–934 (2013).

Song, Y., Cheng, X., Chen, H., Huang, J., Chen, X., Han, M., Su, Z., Meng, B., Song, Z. & Zhang, H. Integrated self-charging power unit with flexible supercapacitor and triboelectric nanogenerator. *J. Mater. Chem. A* 4, 14298 (2016).

Sridhar, S. I., Chockalingam, A. V., Kameswara, S. P. O., Faizal, M. & Saidur, R. Nanogenerators as a sustainable power source: state of art. *Appl. Challenges Nanomater.* 9, 773 (2019).

Sun, Y. M., Hu, X. L., Luo, W., Xia, F. F. & Huang, Y. H. Reconstruction of conformal nanoscale MnO on graphene as a high-capacity and long-life anode material for lithium ion batteries. *Adv. Funct. Mater.* 23, 2436–2444 (2012).

Tian, Y. & Zhao, C. Y. A review of solar collectors and thermal energy storage in solar thermal applications. *Appl. Energy* 104, 538–553 (2013).

Trocino, S., Lo Faro, M., Zignani, S. C., Antonucci, V. & Arico, A. S. High performance solid-state iron-air rechargeable ceramic battery operating at intermediate temperatures (500e–50C). *Appl. Energy* 233e234–386e394 (2019). DOI: 10.1016/j.apenergy.2018.10.022.

Untereker, D. F., Schmidt, C. L., Jain, G., Tamirisa, P. A., Hossick-Schott, J. & Viste, M. (2017). 8epower sources and capacitors for pacemakers and implantable cardioverter-defibrillators. In K. A. Ellenbogen, B. L.Wilkoff, G. N. Kay, C.-P. Lau & A. Auricchio (Eds.), *Clinical cardiac pacing, defibrillation and resynchronization therapy* (5th ed., pp. 251–269). Elsevier. https://www.sciencedirect.com/book/9780323378048/clinical-cardiac-pacing-defibrillation-and-resynchronization-therapy

Vangapally, N., Jindal, S., Gaffoor, S. A. & Martha, S. K. Titanium dioxide-reduced graphene oxide hybrid as negative electrode additive for high performance lead-acid batteries. *J. Energy Storage* 20, 204–212 (2018). DOI: 10.1016/j.est.2018.09.015.

Wang, J., Yang, J., Xie, J. & Xu, N. A novel conductive polymer–sulfur composite cathode material for rechargeable lithium batteries. *Adv. Mater.* 14, 963–965 (2002).

Wang, F., Jiang, C., Tang, C., Bi, S., Wang, Q., Du, D. & Song, J. High output nano-energy cell with piezoelectric nanogenerator and porous supercapacitor dual function – a technique to provide sustaining power by harvesting intermittent mechanical energy from surroundings. *Nano Energy* 21, 209 (2016).

Wen, Z., Yeh, M.H., Guo, H., Wang, J., Zi, Y., Xu, W., Deng, J., Zhu, L., Wang, X., Hu, C., Zhu, L., Sun, X. & Wang, Z. L. Self powered textile for wearable electronics by hybridizing fiber-shaped nanogenerators, solar cells and supercapacitors. *Sci. Adv.* 2, e1600097 (2016).

Winter, C. J. Hydrogen energy—abundant, efficient, clean: a debate over the energy system of change. *Int. J. Hydro. Energy* 34:1–52 (2009).

XAKTY. Electrochemistry: Copper zinc battery (2019). Available at http://xaktly.com/Electrochemistry.html.

Xia, H., Luo, Z. & Xie, J. Nanostructured $LiMn_2O_4$ and their composites as high performance cathodes for lithium-ion batteries. *Prog. Nat. Sci. Mater. Int.* 22(6), 572–584 (2012). DOI: 10.1016/j.pnsc.2012.11.014.

Yang, Y., Jung, H. J., Yun, B. K., Zhang, F., Pradel, K. C., Guo, W. & Wang, Z. L. Flexible pyroelectric nanogenerators using composite structure of lead-free $KNbO_3$ nanowires. *Adv. Mater.* 24(39), 5357–5362 (2012a).

Yang, Y., Wang, S., Zhang, Y. & Wang, Z. Pyroelectric nanogenerators for driving wireless sensors. *Nano Lett.* 12(12), 6408–6413 (2012b).

Yang, Y., Ruan, G, Xiang, C., Wang, G. & Tour, J. M. Flexible three-dimensional nanoporous metal-based energy devices. *J. Am. Chem. Soc.* 136 (17), 6187–6190 (2014); DOI: 10.1021/ja501247f.

Yang, N., Fu, Y., Yue, H., Zheng, J., Zhang, X., Yang, C. & Wang, J. An improved semi empirical model for thermal analysis of lithium-ion batteries. *Electrochim. Acta*, 311, 8–20 (2019). DOI: 10.1016/j.electacta.2019.04.129.

Yi, F., Wang, J., Wang, X., Niu, S., Li, S., Liao, Q., Xu, Y., You, Z., Zhang, Y. & Wang, Z. L. Stretcheable and waterproof self-charging power system for harvesting energy from diverse deformation and powering wearable electronics. *ACS Nano* 10, 6519 (2016).

Zhai, X. Q., Wang, X. L., Wang, T. & Wang, R. Z. A review on phase change cold storage in air-conditioning system: materials and applications. *Renew. Sustain. Energy Rev.* 22, 108–120 (2013).

Zhang, C., Wei, Y.-L., Cao, P.-F. & Lin, M.-C. Energy storage system: current studies on batteries and power condition system. *Renew. Sustain. Energy Rev.* 82, 3091–3106 (2018).
Zhou, D., Zhao, C. Y. & Tian, Y. Review on thermal energy storage with phase change materials (PCMs) in building applications. *Appl. Energy* 92, 593–605 (2012).
Zhu, Y., Murali, S., Stoller, M. D. Ganesh, K. J., Cai, W., Ferreira, P. J., Pirkle, A., Wallace, R. M., Cychosz, K. A., Thommes, M., Su, D., Stach, E. A. & Ruoff, R. S. Carbon-based supercapacitors produced by activation of graphene. *Science* 332, 1537 (2011).

2 Working Principles and Mechanisms in Nanogenerators

Rita Joshi and Indranil Lahiri

CONTENTS

2.1 INTRODUCTION

As the present world is migrating towards miniaturization and portability, harvesting energy using nanogenerators has become a demanding field of research. Nanoenergy is a field of science that deals with small-scale energy harvesting from day-to-day activities by using nanomaterials and nanodevices (Menéndez-Manjón et al. 2011; Shi, Liu, and Mei 2020). The energy harvesters in the nano-regime are essential factors in reducing the size of power sources in the applications for sensors (Tat et al. 2021), resonators (Chen et al. 2013), optoelectronics (Grutzmacher, Mikulics, and Hardtdegen 2016), and many more. This can be accepted as one of the best solutions to the energy crisis being faced worldwide. Since 2006 many nanogenerators working on the principle of piezoelectric effect, triboelectric effect and pyroelectric effect for harvesting mechanical and thermal energies have been explored (Wang and Song 2006). Piezoelectric nanogenerator (PENG), triboelectric nanogenerator (TENG) and pyroelectric nanogenerator (PYNG) are the three prominent members of the nanogenerator family used for energy harvesting. This chapter aims to provide a clear idea of basic working principle and mechanisms of the types of nanogenerators. It is divided

DOI: 10.1201/9781003187615-2

into various subsections which include a brief introduction to the kinds of nanogenerators. The following two subsections include the basic principles and working of nanogenerators. The last section concludes by providing a brief summary.

2.2 INTRODUCTION TO PENG

A PENG has been used since 2006 to convert mechanical energy into an electrical form of energy using piezoelectric nanostructured materials (Wang and Song 2006). This effect is popularly called the piezoelectric effect (Manbachi and Cobbold 2011). The word "piezo" is taken from the Greek word *piezein* which means "to press". In 1880, Jacques Curie and Pierre Curie discovered the piezoelectric effect for the first time. Piezoelectric materials are the special class of materials that can produce electric charges on applying external mechanical pressure. The special class of piezoelectric materials includes ZnO, PZT and $BaTiO_3$. Among all these materials, ZnO has attracted most of the attention of the research community due to high transparency, chemical stability and ease of nanostructure design (Li et al. 2015; Zhu et al. 2010). The emergence of PENGs is a basic element in the progress of new energy harvesting sources, as well as a proper utilization of abundant mechanical energy.

2.2.1 Principle of PENG

The basic principle behind the PENGs is the well-known piezoelectric effect. A PENG consists of two electrodes having Fermi levels, precisely balanced. In Figure 2.1(a), when a force or strain is exerted externally in a piezoelectric material, a piezoelectric potential difference is developed between both the Fermi levels. To balance the Fermi levels, the electrons start flowing through the external circuit. This phenomenon is described as the direct piezoelectric effect, as expressed in Equation (2.1).

By the expression,

$$P = d \times \sigma \tag{2.1}$$

where P denotes polarization, d denotes piezoelectric coefficient and σ denotes the applied stress.

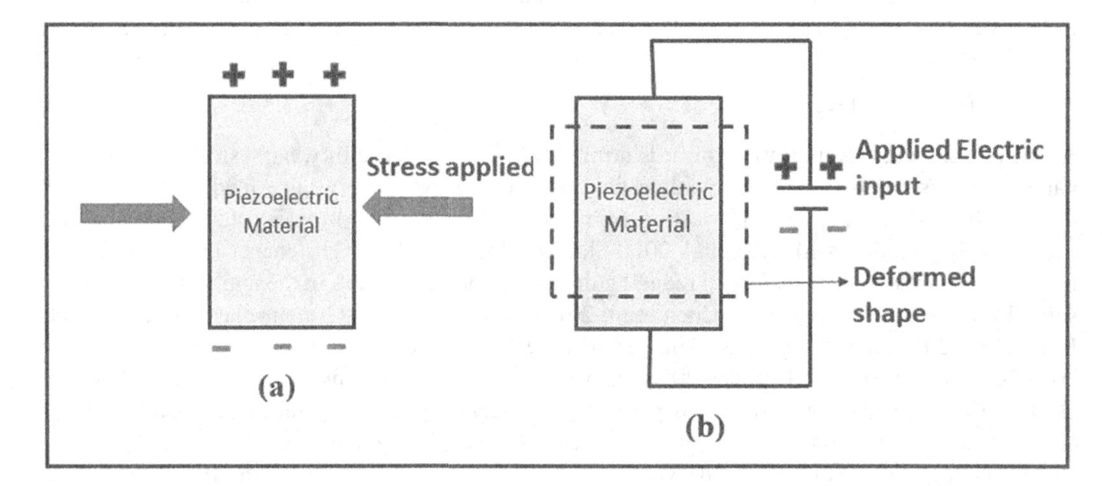

FIGURE 2.1 The basic principle behind piezoelectric nanogenerators: (a) the schematic representation of direct piezoelectric effect and (b) the schematic representation of converse piezoelectric effect.

This equation clearly shows that the change in electric polarization is directly proportional to the externally applied stress. Opposite of the direct piezoelectric effect occurs when an electric field is exerted externally on any piezoelectric material, stress or strain is induced as shown in Figure 2.1(b). It is popularly recognized as the converse piezoelectric effect as described by the expression in Equation (2.2),

$$\varepsilon = \mathrm{d} \times \mathrm{E} \tag{2.2}$$

where ε denotes the induced strain, d denotes the piezoelectric coefficient and E denotes the applied electric field.

The PENGs are mostly preferred due to their easier manufacturing and high output power as compared to other traditional energy harvesting systems. Their basic mechanism relies on converting various forms of mechanical energy into electric energy with the help of piezoelectric materials.

2.2.2 Working and Mechanism of PENG

The piezoelectric materials used for energy harvesting exhibit a different notation for axis orientations as demonstrated in Figure 2.2(a). The directions X, Y and Z are represented as the subscripts 1, 2 and 3, respectively, whereas 4, 5 and 6 denote the rotations about 1, 2 and 3, respectively. Due to the anisotropic nature of piezoelectric materials, these are used as subscripts in the material constants, as "x_{ij}", where "x" denotes a constant value, "i" is the direction of response to the direction of applied stimulus "j" (Damjanovic et al. 2006). The piezoelectric constants that are widely used include the piezoelectric charge constant "d_{ij}", piezoelectric voltage constant "g_{ij}", piezoelectric stress constant "e_{ij}" and piezoelectric strain constant "h_{ij}". The d constant is the charge developed per unit mechanical stress having the units of Coulomb per Newton (C/N). The g constant is the electric field generated per unit of mechanical stress in the units of volt-meter per Newton (V m/N). The e constant is the stress developed per unit of applied electric field in the units of Newton per volt-meter (N/(V m)). The h constant is the strain developed per unit of charge displacement in the units of (N/C) (Briscoe 2014).

The polar axis orientation is denoted as the "3" direction, and the direction perpendicular to the polar axis is denoted as the "1" direction (Figure 2.2(a)). The direction of the applied force relative

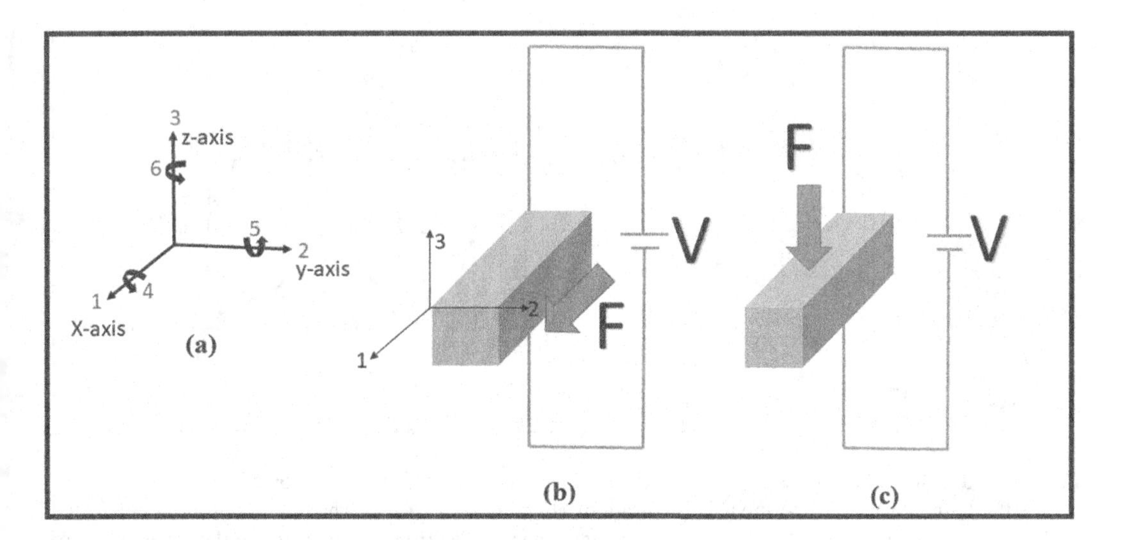

FIGURE 2.2 Schematic representation of (a) axis orientation of a piezoelectric material, (b) 3-1 mode of configuration and (c) 3-3 mode of configuration.

to the polar axis determines the performance of the nanogenerator. The applied force can be either parallel to the direction of the polar axis, known as 3-3 mode configuration, or at right angles to the polar axis, known as 3-1 mode configuration. The schematic of the force applied in 3-1 mode configuration and 3-3 mode configuration is shown in Figure 2.2(b and c), respectively.

2.2.2.1 Direction of Exerted Force Perpendicular to the Polar Axis (3-1 Mode Configuration)

When a parallel force is exerted on the polar axis of the piezoelectric nanowires, a relative compression and stretching is experienced. This is shown schematically in Figure 2.3(a). The portion being compressed experiences a negative strain and a negative potential, whereas the portion of nanostructure being stretched experiences a positive strain and positive potential due to a relative displacement of the charges. To prevent the flow of electrons via an external circuit, a Schottky contact is formed between the top metal electrode and the tip of the nanostructure. On the other hand, an ohmic contact is made between the nanostructure and the bottom electrode that neutralize the electric field generated. First the electrode contacts the stretched positive potential surface, which develops a negative bias voltage at the interface forming a Schottky diode with little current (Figure 2.3(b)). No current is detected when the electrode is connected to the positive potential in the structure. When the electrode is in contact with the negative potential, the electrons pass from

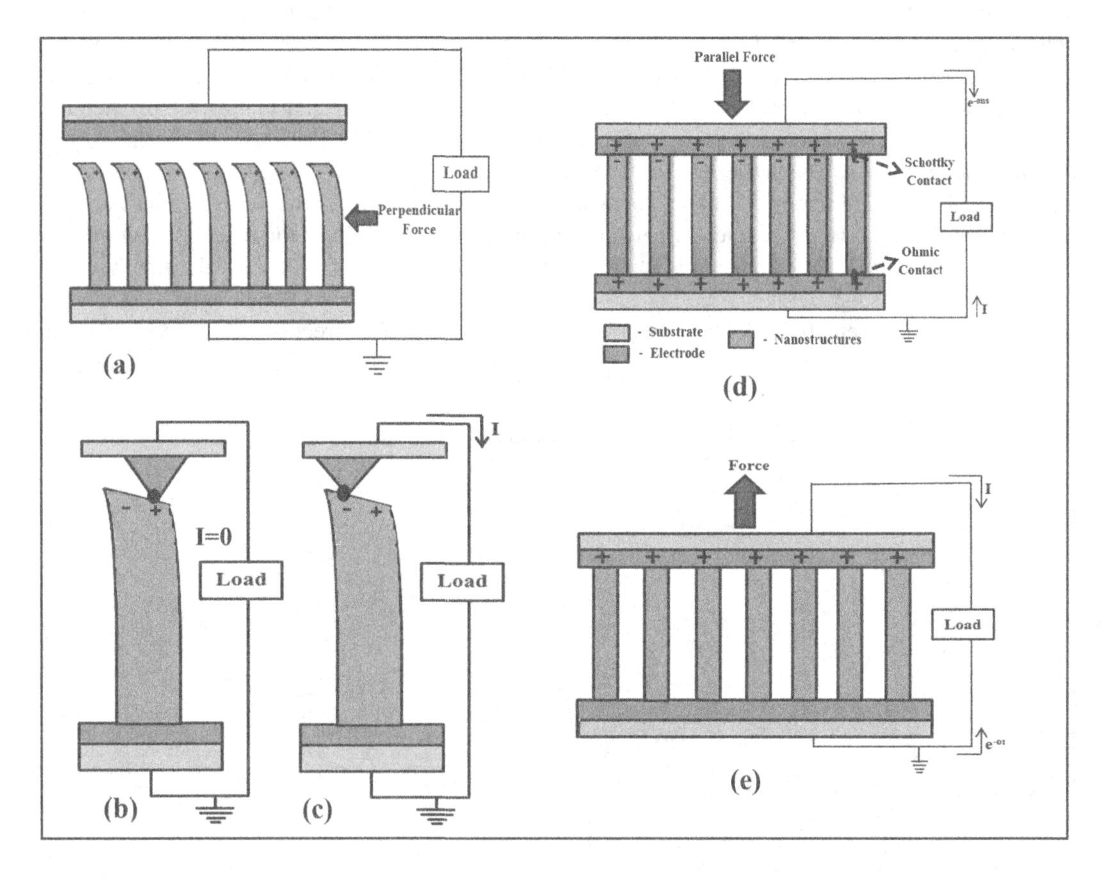

FIGURE 2.3 Operation modes of piezoelectric nanogenerator: (a) when the direction of exerted force is perpendicular to the polar axis (3-1 mode configuration), (b) no flow of current when the electrode is connected to the positive end of nanostructure, (c) flow of current when the electrode is connected to the negative end of nanostructure and (d, e) when the direction of exerted force is parallel to the polar axis (3-3 mode configuration). (Reprinted with permission from Roji, Jiji, and Raj (2017).)

the top electrode to the bottom via Schottky contact and ohmic contact (Figure 2.3(c)). This results in the generation of DC output, hence the DC power nanogenerator.

The maximum amount of voltage that can be generated using nanowires is calculated using the following equation (Gao and Wang 2007; Wang 2008):

$$V_{max} = \pm\frac{3}{4\left(k_0 + k\right)}\left[e_{33} - 2(1+v)e_{15} - 2ve_{31}\right]\frac{a^3}{l^3}v_{max} \quad (2.3)$$

where k_0 denotes the permittivity in a vacuum, k denotes the dielectric constant, v denotes the Poisson ratio, a denotes the radius of the nanostructure, l denotes the length of the nanostructure, V_{max} denotes the maximum deflection of the nanostructure tip, and e_{33}, e_{15} and e_{31} denote the piezoelectric strain constants as defined in Section 2.2.

2.2.2.2 Direction of Exerted Force Parallel to the Polar Axis (3-3 Mode Configuration)

When a parallel force is exerted to the polar axis of the piezoelectric nanowires, a uniaxial compression is created throughout the nanostructure. Figure 2.3(d) shows that a negative potential is created at the upper tip of the nanowires and positive at the other end due to the piezoelectric property of the nanowires. The nanowires between both the electrodes act as a capacitor and drive the electrons to the external circuit. The Schottky contact acts as a gate and restricts the motion of electrons through the nanowires. The movement of the electrons is towards the bottom electrode via an external circuit which develops a positive potential at the top of the nanowire. In Figure 2.3(e), when the force is withdrawn, the electrons migrate back to the top electrode due to the decline of the piezoelectric effect. The transfer of electrons to the top neutralizes the tip of the nanowire that results in voltage in the opposite direction. Hence, alternating current (AC) is generated.

The maximum amount of voltage (V_{max}) that can be generated using nanowires is calculated using the following equation (Choi et al. 2009):

$$V_{max} = Fg_{33}L/A \quad (2.4)$$

where F denotes the force applied on the top electrode, g_{33} denotes the piezoelectric voltage constant as described in Section 2.2, L denotes the length of the nanostructure and A denotes the contact area of the nanostructure.

In the last decade, many successful PENGs were invented, which serve as extraordinary energy harvesting devices from various mechanical energy sources. Numerous ways were studied by the chemists and material scientists to enhance the performance of PENGs and to utilize them in practical applications. ZnO nanowires are the most explored candidate for PENGs. Various ZnO nanowire PENGs with different designs were developed since 2006 (Qin, Wang, and Wang 2008; Yang et al. 2009). Due to the arrangement of zigzag electrodes above the nanowire array, the existing devices resulted in increased contact resistance and instability.

In 2009, Yang et al. (2009) developed a nanogenerator with new design that consisted of a single ZnO nanowire held horizontal on the flexible substrate and electrodes fixed to both ends. When the flexible substrate bends, the nanowire stretches resulting in a tensile strain induced in the nanowire. This tensile strain causes a change in piezoelectric potential and causes a flow of electrons.

Later in 2012, nanogenerators based on vertically grown integrated ZnO nanowires were developed that converted biomechanical energy into electrical energy. The observed output voltage of 58 V and current of 134 μA were appreciably higher than previously reported ones. This is attributed to the insulating layer of Poly methyl methacrylate (PMMA) between the metal electrodes and the nanowire that prevents the leakage current. Further exploring nanogenerators in different ways, multiple PENGs can also be connected parallelly to enhance the power output up to a power density of 0.78 W/cm (Zhu, Wang et al. 2012).

2.3 INTRODUCTION TO TENG

In the previous decade, a wide range of new energy harvesting technologies came into existence as the conventional energy sources are depleting gradually. Among them, a TENG has emerged to harness various mechanical motions into energy. It was accidentally invented in 2012 when Zhong and co-workers fabricated a new type of PENG having a little gap (Fan, Tian, and Wang 2012). It showed a comparatively higher output voltage, which was later related to the triboelectric effect. This special class of nanogenerator works on the principle of contact charging and electrostatic induction combined. It has several advantages over pre-existing energy harvesting technologies, including low cost, lightweight, small size, high power density and flexibility (Wang, Yang, and Wang 2017).

2.3.1 PRINCIPLE OF TENG

The basic principle behind the working of TENG can be explained by a combined effect of electrostatic induction and triboelectrification. In triboelectrification or contact electrification, two materials having different triboelectric polarities come in contact and transfer the charges to develop a potential difference after the separation, as shown in Figure 2.4(c) (Lin et al. 2014). The transferred charges also induce the charges of opposite polarity on the metal electrodes by the phenomenon of electrostatic induction, as shown in Figure 2.4(d) (Nanogenerators, Triboelectric, and Green Energy 2016).

As piezoelectric effect, the triboelectric effect is also used to generate electric energy from mechanical energy; however, the mechanisms of both processes differ. When a force is exerted on a TENG device, two triboelectric materials present come in contact and the charges are transferred. The material that has a strong electron attracting ability produces a negative charge and the other material acquires a positive charge. When this external force is removed, a potential difference is developed between both the materials that can be expressed as follows (Zhu, Pan et al. 2012):

$$V_{oc} = -\frac{\sigma_d}{\varepsilon_0} \tag{2.5}$$

where σ denotes the charge density, d denotes the separation distance and ε_0 denotes the permittivity in a vacuum.

Due to the potential difference created between both the materials, the charged particles flow from one material to the other through the external circuit, as shown in Figure 2.4(c), and produce

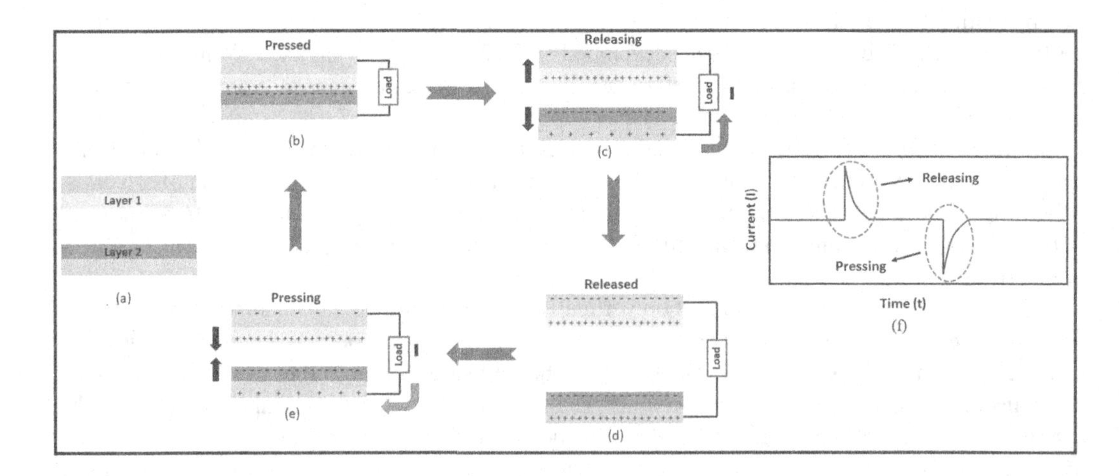

FIGURE 2.4 The basic principle of triboelectric nanogenerator.

a positive short circuit current, as shown in Figure 2.4(f). When the force is applied again, the interlayer distance decreases and hence decreases the potential difference as shown in Figure 2.4(e) and a negative short circuit current is produced as shown in Figure 2.4(f). This causes the flow of current in the opposite direction. Thus, TENG produces alternating current (AC).

2.3.2 Working and Mechanism of TENG

TENG is differentiated into four fundamental modes, on the basis of working. These include vertical contact separation mode, linear sliding mode, single-electrode mode and free-standing mode. The fundamental principle involved here is that due to the displacement of any triboelectric layer, the movement of electrostatic charges takes place. Due to this, a potential difference is developed between both the electrodes and causes the flow of electrons in the external circuit.

2.3.2.1 Vertical Contact Separation Mode

It is the first developed most basic TENG mode of operation. The simple structure and facile fabrication make it to be widely used for the application in low-power electronics (Fan et al. 2012; Wang, Lin, and Wang 2012). In this mode, the triboelectric behaviour is observed when an external force or stress is exerted vertically over the triboelectric material, as shown in Figure 2.5(a). Due to this, charges with different polarities are developed on both the surfaces of the layers and as the distance between them is changed, a potential difference is developed between both the electrodes. As a result, the electrons flow when the plates are in contact and charges accumulate when no contact is present. This process of triboelectrification can take place between dielectric and metallic layers or among two different dielectric materials. This model has a simple design, low cost and easy fabrication (Ahmed et al. 2017; Hou et al. 2013; Zhao et al. 2019). This mode of operation is used for the mechanical energy obtained by deformations, stress, shocks and vibrations.

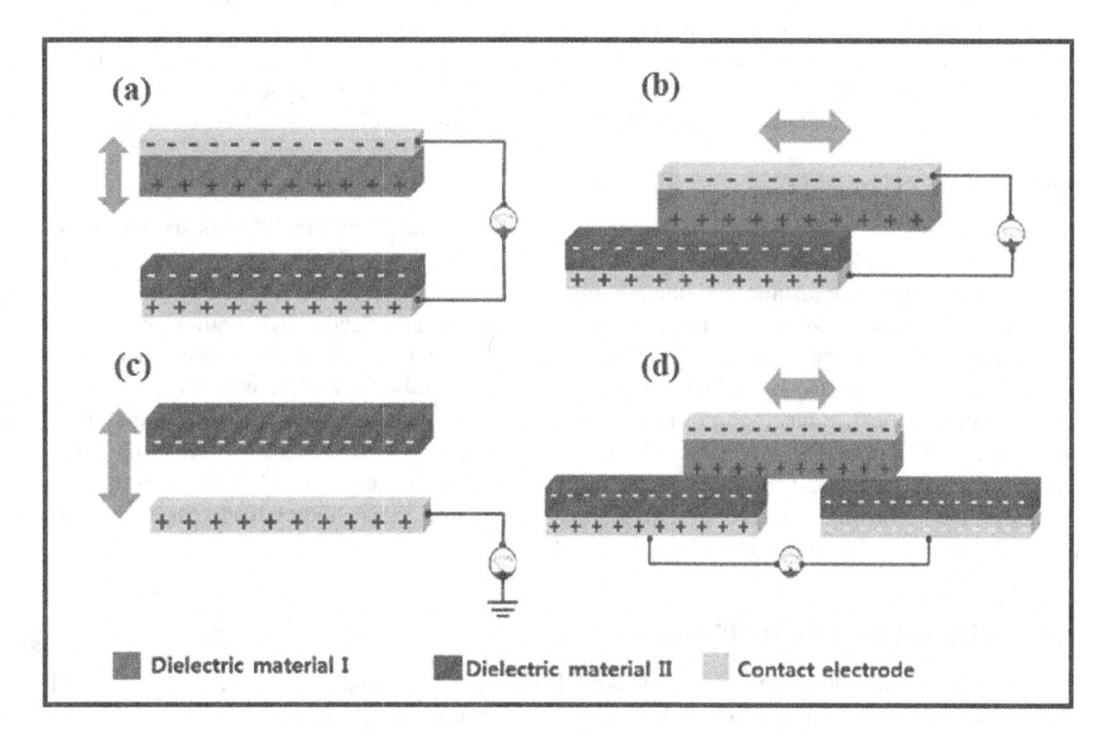

FIGURE 2.5 Modes of operation of triboelectric nanogenerators: (a) vertical contact separation mode, (b) linear sliding mode, (c) single-electrode mode and (d) free-standing mode. (Reprinted with permission from Chandrasekaran et al. (2019).)

2.3.2.2 Linear Sliding Mode

In the linear sliding mode or in-plane sliding mode, the charges are generated due to the in-plane sliding or the to-and-fro sliding motion between both the layers (Wang et al. 2013; Zhu et al. 2013) (Figure 2.5(b)). This periodic sliding and closing motion generate alternating power. The electrodes are connected to the back of the layer, as shown in Figure 2.5(b). To balance the induced potential difference, the lateral polarization is developed between the layers. It drives the electrons from one electrode to the other. The sliding motion and the friction generated due to motion between the surfaces help to improve the generation of triboelectric charges as compared to vertical contact mode. It has low efficiency as compared with contact separation mode (Chen et al. 2018; Yuan et al. 2020). However, the high contact area while sliding produces more charge density than that by merely contacting. This mode of operation allows many new techniques and ways to harvest energy using different types of motion. These sliding motions can be obtained using planar motion, disc rotation or cylindrical motion.

2.3.2.3 Single-Electrode Mode

Both vertical contact mode and lateral sliding mode of operation have a common disadvantage that both the triboelectric layer and back electrode should be connected by a wire. This configuration limits their application in many fields of energy harvesting. This issue is resolved by removing the back electrode to form single-electrode mode TENG (Yang, Zhang et al. 2013; Yang, Zhou et al. 2013). This is the most significant advantage of single-electrode TENG over other modes. A single-electrode mode of TENG is the simplest structure used in some special applications such as fingertip-driven TENG (Yang, Zhou, et al. 2013). In single-electrode mode TENG, energy is harvested using randomly moving objects, unlike other modes of operation. In this mode of TENG, the triboelectric layer and single electrode interact directly with the object in motion. It can be seen in Figure 2.5(c) that it consists of another electrode such as ground or a large conductor that acts as a reference electrode to serve as an electron source. The contact or separation with the electrode changes the electric field distribution and causes the exchange of electrons between both the electrodes to balance the electric potential (Wang et al. 2018). Various applications using single-electrode mode TENG include our day-to-day objects and human skin as triboelectric objects.

2.3.2.4 Free-Standing Mode

Free-standing mode TENG is used to extract energy from different moving objects such as automobiles, airflow and human walking. Two alike electrodes are kept under a dielectric triboelectric layer maintaining a little distance as shown in Figure 2.5(d) (Wang et al. 2014). The asymmetrically distributed charge is produced by contact and separation of the triboelectric layer with the electrode. The electrons start flowing between the electrodes and balance the potential difference. There is no contact between the electrodes and triboelectric layer, so the probability of damage decreases. Due to no screening effect, it has high power conversion efficiency and electrical power conversion as compared to other operating modes. Similar to single-electrode mode TENGs, free-standing mode TENGs also harvest mechanical energy from diverse movements from day-to-day activities that are not accessible to the other two modes. Fabrication of these devices is comparatively easier and can be applied to real-time applications (Nie et al. 2018).

2.4 INTRODUCTION TO PYNG

Transforming the harmful thermal energy using PYNG can be a potential application in the field of environmental monitoring, medical diagnostics, temperature imaging and electronics. A PYNG uses the pyroelectric effect to transform the thermal energy into electrical energy in pyroelectric materials (Ehre, Lavert, and Lahav 2010; Ye, Tamagawa, and Polla 1991). These are the special class of materials, which change their atomic structure (crystal lattice structure) when heated.

2.4.1 Principle of PYNG

The pyroelectric effect can be defined as the temporary variation in the polarization of a material as a result of fluctuating temperature (Thomson 1878).When an external electric field is absent, pyroelectric materials possess an internal spontaneous polarization (P_s), i.e. the alignment of all the dipoles is in the same direction without any wiggling ($\theta = 0$). The spontaneous polarization can also be defined as "the dipole moment per unit volume of the material". If the temperature of the material is kept constant, its polarization also remains constant. When the temperature is increased ($dT/dt > 0$), polarization (P_s) decreases due to thermal vibration. The dipoles wiggle around and the wiggling angle enlarges, which leads to a decrease of induced surface charges. The flow of electrons takes place due to the decrease of surface charges. When the temperature is decreased ($dT/dt < 0$), the wiggling angle becomes smaller than the initial. Polarization (P_s) increases due to reduced thermal vibrations and finally increases the induced surface charges. By the flow of surface charges, a current is generated in the external circuit. When no temperature change occurs ($dT/dt = 0$), thermal equilibrium is attained and the alignment of dipoles is recovered. The localized charges in the material are shielded by equal and opposite free charges, which leads to no flow of electricity (Leng et al. 2014). The process of generating electricity by pyroelectric effect is schematically presented in Figure 2.6(a).

The dependence of change of spontaneous polarization (P_s) and dP_s/dT as a function of temperature in a pyroelectric material is shown in Figure 2.6(b). It can be seen that P_s decreases to zero with the increase in temperature till Curie temperature (T_C) is reached, whereas the change in polarization with increasing temperature gradually increases till the T_C is attained.

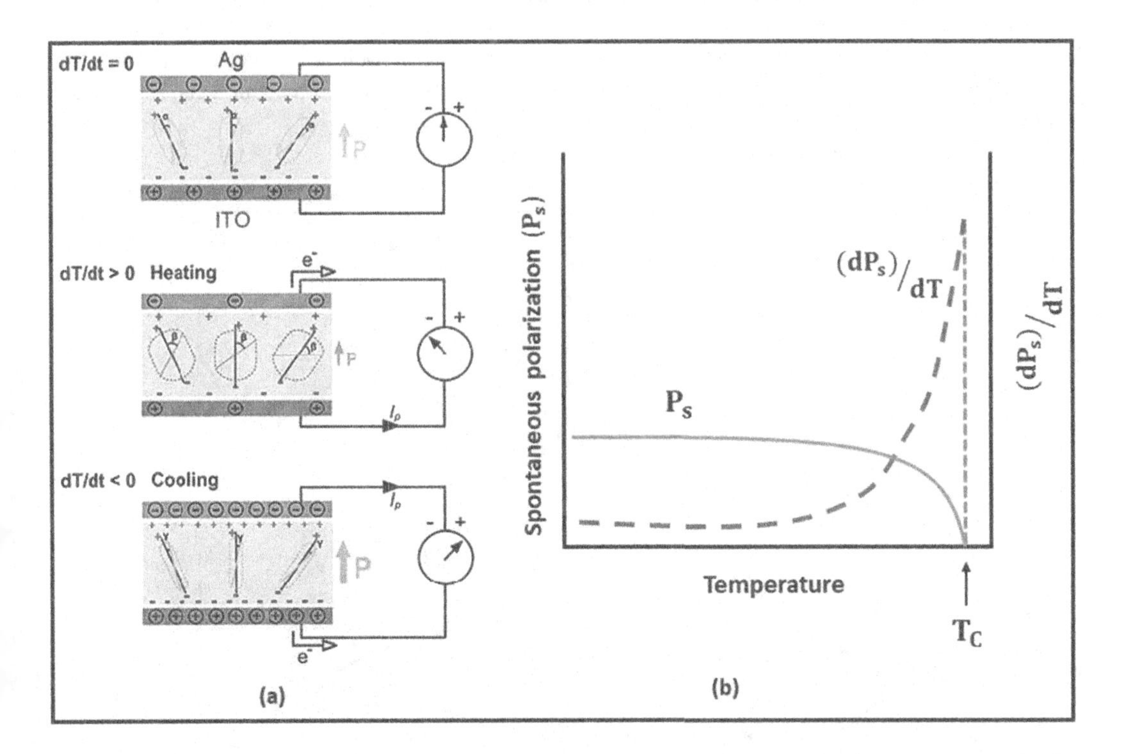

FIGURE 2.6 (a) Schematic representation of generation of pyroelectricity by the action of change of temperature. (Reprinted with permission from Yang, Jung et al. (2012).) (b) The dependence of change of spontaneous polarization (P_s) and dP_s/dT as a function of temperature in a pyroelectric material (Nalwa 2001).

The direction of the electric current is changed when the temperature of a pyroelectric material is altered, i.e. heated or cooled. The pyroelectric efficiency of the material is defined by an important factor defined by the pyroelectric coefficient:

$$\Delta P_s = p^{\sigma,E} \times \Delta T \tag{2.6}$$

or

$$p^{\sigma,E} = \frac{dP}{dT} \tag{2.7}$$

where P_s denotes the polarization, $p^{\sigma,E}$ denotes the pyroelectric coefficient, the subscripts σ and E represent the stress and electric field, respectively and ΔT denotes the temperature variation.

The unit of the pyroelectric coefficient is C m^{-1} K^{-1}. The three nonzero vector components of the pyroelectric coefficient can be expressed as

$$p_m = \frac{\partial P_m}{\partial T} \tag{2.8}$$

where m = 1,2,3….

The pyroelectric coefficient is a sign-dependent quantity that attains a positive value when a pyroelectric material is heated. Further, it attains a negative value when the pyroelectric material is cooled.

PYNGs are one of the most tremendous energy harvesters having the capability to transform heat form of energy to the electric form of energy with the help of pyroelectric materials in nano size. Usually a PYNG consists of three layers as shown in Figure 2.7(a). The bottom layer receives the heat energy and acts as the bottom electrode. The pyroelectric layer is in the middle which transforms thermal form of energy into electricity, while the top layer acts as another electrode.

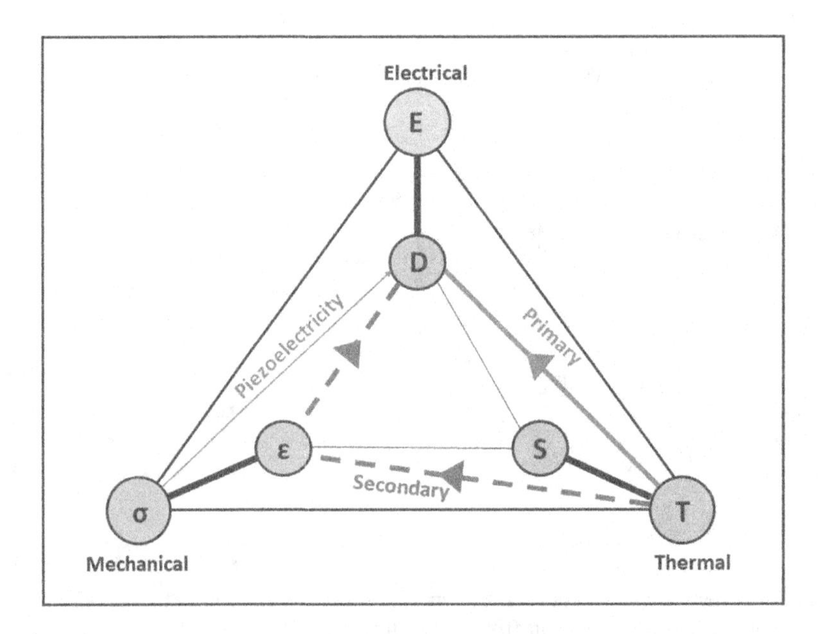

FIGURE 2.7 Heckmann diagram showing the thermodynamically reversible interactions that may occur during the transformation of thermal, mechanical or electrical forms of energies (Lang 2005).

The working and mechanism of PYNGs can be explained in two cases as explained in the following, namely the primary pyroelectric effect and the secondary pyroelectric effect. Figure 2.7 shows a triangular schematic diagram (Heckmann diagram) to illustrate the thermodynamically reversible conversions among mechanical, thermal and electrical energies (Lang 2005). The lines joining each of the circle represent that the variation in one variable causes the change in the other. The outermost black lines joining thermal, mechanical and electrical energies signify the properties of heat capacity, elasticity and electrical permittivity, respectively. The variables E, T and σ represent the electric field, temperature and elastic stress, respectively. The variables D, S and ε denote the electric displacement, entropy and strain, respectively.

The bold green line and the dotted red lines indicate the two types of pyroelectric effects. The first case is primary pyroelectric effect, which is denoted by green bold line. The variation in temperature leads to the variation in electric displacement without experiencing any strain ε. The second case is secondary pyroelectric effect, which is denoted by dotted red lines. The change in temperature causes a strain that finally changes the electric displacement (D) by a piezoelectric process.

2.4.1.1 Primary Pyroelectric Effect

This effect is described as the production of charges in a strain-free case (clamping of the sample) in the materials like barium titanate (BTO), lead zirconate titanate (PZT) and some of the ferroelectric materials. Its mechanism depends on the thermally induced random vibration of the electric dipoles around its own axis, whose magnitude depends on the increasing temperature. The thermal variations cause the free oscillation of electric dipoles from their respective axis. At a certain temperature, the average strength of electric dipole polarization is fixed. If the temperature increases, it causes the electric dipoles to oscillate more around their respective axes. This causes a decrease in average spontaneous polarization due to an increase in oscillation angles. As a result, the quantity of induced charges is reduced and the motion of electrons takes place through the external circuit.

If the temperature decreases, because of the lower thermal variations, the electric dipoles oscillate with smaller angles. This increases the total polarization and the number of induced charges increases. This causes the flow of electrons in the opposite direction. A lead-free $KNbO_3$-nanowire/PDMS-polymer PYNG is shown in Figure 2.8(a) (Yang, Jung et al. 2012). The device comprises three layers, namely an Ag and an indium tin oxide (ITO) layers that act as top and bottom electrodes, respectively, and a nanowire composite layer. Figure 2.8(b) shows the output current of 20 pA and the output voltage of 2 mV corresponding to the cyclic temperature variation in the range of 295–298K.

2.4.1.2 Secondary Pyroelectric Effect

The secondary pyroelectric effect is explained as the production of charges by the anisotropic thermal deformations induced by the strain. The thermal deformation causes the piezoelectric potential difference. It allows the flow of electrons via an external circuit. CdS, ZnO and other wurtzite-type materials having non-centrosymmetric structure are the special class of materials that show the secondary pyroelectric effect. Their working mechanism can be explained by the structure described in Figure 2.9 (Yang, Guo, et al. 2012). It consists of the Ag layer and ITO layers as the top and bottom electrodes, and ZnO nanowire as the middle layer. Pyroelectric potential is developed in the middle layer (ZnO nanowire) when the temperature is varied in the range of 295–304K. Positive potential is developed at the Ag electrode and negative at the ITO electrode. The electrons start to flow from negative potential to positive, i.e. ITO to Ag electrode. Due to the change of temperature to 295K, the electrons from the Ag electrode are ejected and return to the bottom electrode and hence the pyroelectric potential disappears. The output voltage of 5.8 mV and output current of 108.5 pA were obtained corresponding to the temperature variation from 295 to 304K.

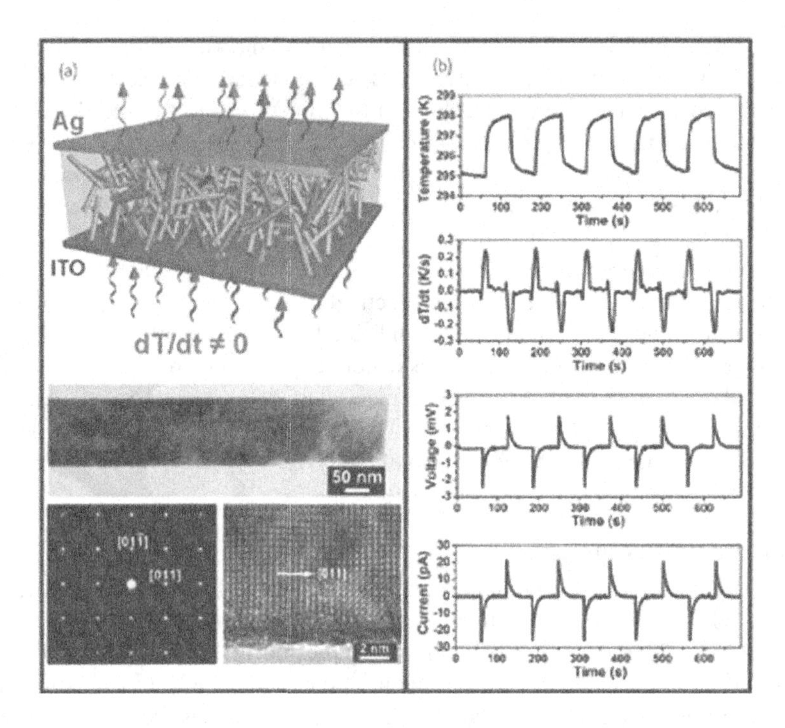

FIGURE 2.8 The working mechanism of the primary pyroelectric effect-based PYNG: (a) schematic structure of PYNG and TEM images, SAED pattern and HRTEM of a single $KNbO_3$ nanowire and (b) the output current and voltage change on the variation of temperature in a PYNG device. (Reprinted with permission from Yang, Jung et al. (2012).)

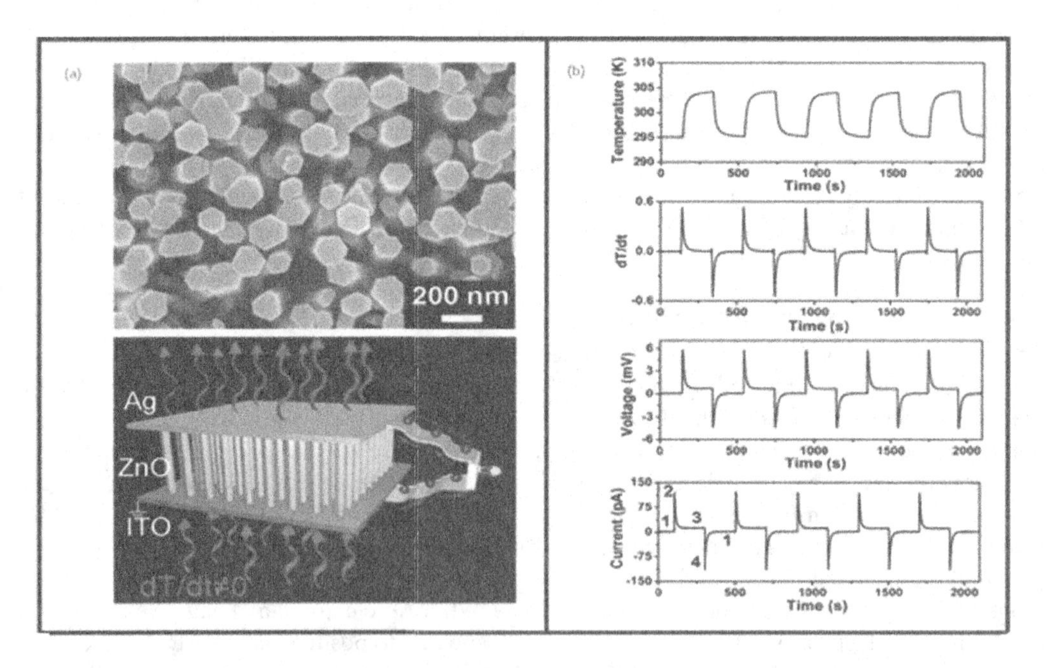

FIGURE 2.9 The working mechanism of secondary pyroelectric effect-based PYNG: (a) scanning electron microscopy (SEM) image of ZnO nanowires and the schematic structure of the nanogenerator and (b) the corresponding output current and voltage as the variation of temperature. (Reprinted with permission from Yang, Guo, et al. (2012).)

2.5 SUMMARY

Harvesting different forms of energy from the environment using nanogenerators is one of the most appropriate alternatives to reach the increasing demand of energy. This chapter reviews the basic principles and the working mechanisms of PENG based on piezoelectric effect, TENG based on the combined effect of triboelectrification and electrostatic induction and PYNG based on pyroelectric effect. All the principles behind these classes of nanogenerators are well known from the past. But their applications in energy harvesting are being explored since the last decades. The rapid development of nanogenerators will be beneficial to evolve a self-powered system and will help to cope with the energy crisis. They have a large scope in the field of biomedical sensing, transportation, rehabilitation devices, etc. and can be certainly used as a replacement of batteries in upcoming future.

REFERENCES

Ahmed, Abdelsalam, Steven L. Zhang, Islam Hassan, Zia Saadatnia, Yunlong Zi, Jean Zu, and Zhong L. Wang. 2017. "A Washable, Stretchable, and Self-Powered Human-Machine Interfacing Triboelectric Nanogenerator for Wireless Communications and Soft Robotics Pressure Sensor Arrays." *Extreme Mechanics Letters* 13: 25–35.

Briscoe, Joe, and Steve Dunn. 2014. "Piezoelectricity and ferroelectricity." *Nanostructured Piezoelectric Energy Harvesters,* Springer Breifs in materials: 3–17.

Chandrasekaran, Sundaram, Chris Bowen, James Roscow, Yan Zhang, Dinh K. Dang, Eui J. Kim, R.D.K. Misra, Libo Deng, Jin S. Chung, and Seung H. Hur. 2019. 2019. "Micro-Scale to Nano-Scale Generators for Energy Harvesting: Self Powered Piezoelectric, Triboelectric and Hybrid Devices." *Physics Reports* 792 (January): 1–33.

Chen, Haotian, Yu Song, Hang Guo, Liming Miao, Xuexian Chen, Zongming Su, and Haixia Zhang. 2018. "Hybrid Porous Micro Structured Finger Skin Inspired Self-Powered Electronic Skin System for Pressure Sensing and Sliding Detection." *Nano Energy* 51 (July): 496–503.

Chen, Jun, Guang Zhu, Weiqing Yang, Qingshen Jing, Peng Bai, Ya Yang, Te C. Hou, and Zhong L. Wang. 2013. "Harmonic-Resonator-Based Triboelectric Nanogenerator as a Sustainable Power Source and a Self-Powered Active Vibration Sensor." *Advanced Materials* 25 (42): 6094–9.

Choi, Min Y., Dukhyun Choi, Mi J. Jin, Insoo Kim, Sang H. Kim, Jae Y. Choi, Sang Y. Lee, Jong M. Kim, and Sang W. Kim. 2009. "Mechanically Powered Transparent Flexible Charge-Generating Nanodevices with Piezoelectric ZnO Nanorods." *Advanced Materials* 21 (21): 2185–9.

Damjanovic, D., M. Budimir, M. Davis, and N. Setter. 2006. "Piezoelectric Anisotropy: Enhanced Piezoelectric Response along Nonpolar Directions in Perovskite Crystals." *Journal of Materials Science* 41 (1): 65–76.

Ehre, David, Etay Lavert, Meir Lahav, and Igor Lubomirsky. 2010. "Water Freezes Differently on Positively and Negatively Charged Surfaces of Pyroelectric Materials." *Engineering* 327 (February): 672–5.

Fan, Feng R., Long Lin, Guang Zhu, Wenzhuo Wu, Rui Zhang, and Zhong L. Wang. 2012. "Transparent Triboelectric Nanogenerators and Self-Powered Pressure Sensors Based on Micropatterned Plastic Films." *Nano Letters* 12 (6): 3109–14.

Fan, Feng R., Zhong Q. Tian, and Zhong L. Wang. 2012. "Flexible Triboelectric Generator." *Nano Energy* 1 (2): 328–34.

Gao, Yifan, and Zhong L. Wang. 2007. "Electrostatic Potential in a Bent Piezoelectric Nanowire. The Fundamental Theory of Nanogenerator and Nanopiezotronics." *Nano Letters* 7 (8): 2499–505.

Grutzmacher, D., M. Mikulics, and H. Hardtdegen. 2016. "Low-Energy Consumption Nano-Opto-Electronics Based on III-Nitride-LED Mesoscopic Structures." *European Solid-State Device Research Conference 2016-October*: 327–9.

Hou, Te C., Ya Yang, Hulin Zhang, Jun Chen, Lih J. Chen, and Zhong L. Wang. 2013. "Triboelectric Nanogenerator Built inside Shoe Insole for Harvesting Walking Energy." *Nano Energy* 2 (5): 856–62.

Lang, Sidney B. 2005. "Pyroelectricity: From Ancient Curiosity to Modern Imaging Tool." *Physics Today* 58 (8): 31.

Leng, Qiang, Lin Chen, Hengyu Guo, Jianlin Liu, Guanlin Liu, Chenguo Hu, and Yi Xi. 2014. "Harvesting Heat Energy from Hot/Cold Water with a Pyroelectric Generator." *Journal of Materials Chemistry A* 2 (30): 11940–7.

Li, Xinda, Yi Chen, Amit Kumar, Ahmed Mahmoud, John A. Nychka, and Hyun J. Chung. 2015. "Sponge-Templated Macroporous Graphene Network for Piezoelectric ZnO Nanogenerator." *ACS Applied Materials and Interfaces* 7 (37): 20753–60.

Lin, Zong H., Gang Cheng, Sangmin Lee, Ken C. Pradel, and Zhong L. Wang. 2014. "Harvesting Water Drop Energy by a Sequential Contact-Electrification and Electrostatic-Induction Process." *Advanced Materials* 26 (27): 4690–6.

Manbachi, Amir, and Richard S.C. Cobbold. 2011. "Development and Application of Piezoelectric Materials for Ultrasound Generation and Detection." *Ultrasound* 19 (4): 187–96.

Menéndez-Manjón, Ana, Kirsten Moldenhauer, Philipp Wagener, and Stephan Barcikowski. 2011. "Nano-Energy Research Trends: Bibliometrical Analysis of Nanotechnology Research in the Energy Sector." *Journal of Nanoparticle Research* 13 (9): 3911–22.

Nalwa, Hari Singh. 2001. "Handbook of Advanced Electronic and Photonic Materials and Devices." In: *Handbook of Advanced Electronic and Photonic Materials and Devices*, Vol. 1: 321–50. Academic Press.

Wang, Zhong Lin; Lin, Long; Chen, Jun; Niu, Simiao; Zi, Yunlong. 2016. "Triboelectrification" *Triboelectric Nanogenerators. Green Energy and Technology, Springer, Cham:* 1–19.

Nie, Jinhui, Zewei Ren, Jiajia Shao, Chaoran Deng, Liang Xu, Xiangyu Chen, Meicheng Li, and Zhong L. Wang. 2018. "Self-Powered Microfluidic Transport System Based on Triboelectric Nanogenerator and Electrowetting Technique." *ACS Nano* 12 (2): 1491–9.

Qin, Yong, Xudong Wang, and Zhong L Wang. 2008. "Microfibre-Nanowire Hybrid Structure for Energy Scavenging." *Nature* 451 (7180): 809–13.

Roji, Ani M.M., G. Jiji, and Ajith B.T. Raj. 2017. "A Retrospect on the Role of Piezoelectric Nanogenerators in the Development of the Green World." *RSC Advances* 7 (53): 33642–70.

Shi, Hu, Zhaoying Liu, and XuesongMei. 2020. "Overview of Human Walking Induced Energy Harvesting Technologies and Its Possibility for Walking Robotics." *Energies* 13: 86.

Tat, Trinny, Alberto Libanori, Christian Au, Andy Yau, and Jun Chen. 2021. "Advances in Triboelectric Nanogenerators for Biomedical Sensing." *Biosensors and Bioelectronics* 171 (August): 112714.

Thomson, William. 1878. "II. On the Thermoelastic, Thermomagnetic, and Pyroelectric Properties of Matter." *The London, Edinburgh, and Dublin Philosophical Magazine and Journal of Science* 5 (28): 4–27.

Wang, Zhong L. 2008. "Towards Self-Powered Nanosystems: From Nanogenerators to Nanopiezotronics." *Advanced Functional Materials* 18 (22): 3553–67.

Wang, Xudong, Jiaming Liang, Yuxiang Xiao, Yichuan Wu, Yang Deng, Xiaohao Wang, and Min Zhang. 2018. "A Flexible Slip Sensor Using Triboelectric Nanogenerator Approach." *Journal of Physics: Conference Series* 986 (1): 012009.

Wang, Sihong, Long Lin, and Zhong L. Wang. 2012. "Nanoscale Triboelectric-Effect-Enabled Energy Conversion for Sustainably Powering Portable Electronics." *Nano Letters* 12 (12): 6339–46.

Wang, Sihong, Long Lin, Yannan Xie, Qingshen Jing, Simiao Niu, and Zhong Li Wang. 2013. "Sliding-Triboelectric Nanogenerators Based on in-Plane Charge-Separation Mechanism." *Nano Letters* 13 (5): 2226–33.

Wang, Sihong, Simiao Niu, Jin Yang, Long Lin, and Zhong L Wang. 2014. "Quantitative Measurements of Vibration Amplitude Using a Contact-Mode Freestanding Triboelectric Nanogenerator." *ACS Nano* 8 (12): 12004–13.

Wang, Zhong L., and Jinhui Song. 2006. "Piezoelectric Nanogenerators Based on Zinc Oxide Nanowire Arrays." *Science* 312 (5771): 242–6.

Wang, Yang, Ya Yang, and Zhong L. Wang. 2017. "Triboelectric Nanogenerators as Flexible Power Sources." *npj Flexible Electronics* 1 (1): 1–9.

Yang, Ya, Wenxi Guo, Ken C. Pradel, Guang Zhu, Yusheng Zhou, Yan Zhang, Youfan Hu, Long Lin, and Zhong L. Wang. 2012. "Pyroelectric Nanogenerators for Harvesting Thermoelectric Energy." *Nano Letters* 12 (6): 2833–8.

Yang, Ya, Jong H. Jung, Byung K. Yun, Fang Zhang, Ken C. Pradel, Wenxi Guo, and Zhong L. Wang. 2012. "Flexible Pyroelectric Nanogenerators Using a Composite Structure of Lead-Free KNbO3 Nanowires." *Advanced Materials* 24 (39): 5357–62.

Yang, Rusen, Yong Qin, Liming Dai, and Zhong L. Wang. 2009. "Power Generation with Laterally Packaged Piezoelectric Fine Wires." *Nature Nanotechnology* 4 (1): 34–9.

Yang, Ya, Hulin Zhang, Jun Chen, Qingshen Jing, Yu S. Zhou, Xiaonan Wen, and Zhong L. Wang. 2013. "Single-Electrode-Based Sliding Triboelectric Nanogenerator for Self-Powered Displacement Vector Sensor System." *ACS Nano* 7 (8): 7342–51.

Yang, Ya, Yu S. Zhou, Hulin Zhang, Ying Liu, Sangmin Lee, and Zhong L. Wang. 2013. "A Single-Electrode Based Triboelectric Nanogenerator as Self-Powered Tracking System." *Advanced Materials* 25 (45): 6594–601.

Ye, Chian P., Takashi Tamagawa, and D.L. Polla. 1991. "Experimental Studies on Primary and Secondary Pyroelectric Effects in $Pb(Zr_xTi_{1-x})O_3$, $PbTiO_3$, and ZnO Thin Films." *Journal of Applied Physics* 70 (10): 5538–43.

Yuan, Zuqing, Guozhen Shen, Caofeng Pan, and Zhong L. Wang. 2020. "Flexible Sliding Sensor for Simultaneous Monitoring Deformation and Displacement on a Robotic Hand/Arm." *Nano Energy* 73 (April): 104764.

Zhao, Gengrui, Yawen Zhang, Nan Shi, Zhirong Liu, Xiaodi Zhang, Mengqi Wu, Caofeng Pan, Hongliang Liu, Linlin Li, and Zhong L. Wang. 2019. "Transparent and Stretchable Triboelectric Nanogenerator for Self-Powered Tactile Sensing." *Nano Energy* 59 (January): 302–10.

Zhu, Guang, Jun Chen, Ying Liu, Peng Bai, Yu S Zhou, Qingshen Jing, Caofeng Pan, and Zhong L. Wang. 2013. "Linear-Grating Triboelectric Generator Based on Sliding Electrification." *Nano Letters* 13 (5): 2282–9.

Zhu, Guang, Caofeng Pan, Wenxi Guo, Chih Y. Chen, Yusheng Zhou, Ruomeng Yu, and Zhong L. Wang. 2012. "Triboelectric-Generator-Driven Pulse Electrodeposition for Micropatterning." *Nano Letters* 12 (9): 4960–5.

Zhu, Guang, Rusen Yang, Sihong Wang, and Zhong L. Wang. 2010. "Flexible High-Output Nanogenerator Based on Lateral ZnO Nanowire Array." *Nano Letters* 10 (8): 3151–5.

Zhu, Guang, Aurelia C. Wang, Ying Liu, Yusheng Zhou, and Zhong L. Wang. 2012. "Functional Electrical Stimulation by Nanogenerator with 58 v Output Voltage." *Nano Letters* 12 (6): 3086–90.

3 Nanogenerators Theories and Applications

Shah Imtiaz, Md Rabiul Islam, Syqa Banoo, Nimra Shakeel, Mohd Imran Ahamed, Muhanna K. Al-Muhanna, and Naushad Anwar

CONTENTS

DOI: 10.1201/9781003187615-3

3.1 INTRODUCTION

Microelectronic devices play an enormously important role in our everyday lives with the exponential growth of the economy and culture. These devices usually can be powered by Li-ion batteries or supercapacitors that are regularly charged by external power sources due to their small capacity [1–3]. It would also cost a considerable amount of workforce, financial capital and time, especially in remote areas. Researchers are exploring new approaches to scavenge renewable fuels from the atmosphere for energy supplies [4–8].

On this basis, piezoelectric (PZE) and triboelectric (TBE) nanogenerators (NGs), which can transform small mechanical energy to electric energy in the atmosphere such as wind energy [4, 5], wave energy [6], droplets and other mechanical energies, were invented in 2006 and 2012, respectively, in the atmospheric environment [7–10]. In our environments, they are clean or wasted energies. NGs not only efficiently scavenge the above-mentioned mechanical energy but also have many benefits, such as simple, small, light, cost-effective, no auxiliary and practical ones. It can be used with microelectronic instruments and wireless sensors.

Currently, scientists are pursuing and developing self-powering electronic devices [11, 12]. Especially in the field of self-charge textiles for wearable electronics, significant progress has been made [13]. It is therefore necessary to study self-charging energy storage systems that combine energy harvesting and storage units into one unit for the supply of renewable energy to some small electronic devices. This study reflects on the development of self-charge energy storage systems based on NGs in recent years, presenting and discussing the manufacturing technologies of nanomaterials, device architectures, operating standards, self-charging efficiency and the possible applications of self-charging stockings. In addition, some viewpoints and issues that need to be addressed are identified and will pave the way for functional implementations.

There are three common approaches to an NG: PZE, TBE and pyroelectric (PE) NGs.

3.1.1 Piezoelectricity Nanogenerators and Effects of Materials on Their Performance

Those functional materials that can be used to transform mechanical energy into electrical energy are PZE materials. These materials are lead (Pb)-containing and lead-free, which were used in the formation of these NGs. Pb-containing materials mainly include lead zirconate titanate ($Pb_xZr_{1-x}TiO_3$, abbreviated as PZT) and different doped models with great PZE characteristics. However, Pb-based products will be significantly reduced to apply in the future because of the adverse impact of Pb-elements on humanity and the environment. High efficiency based on environmentally sustainable materials therefore needs to be explored very urgently. Recently, several types have been explored, like simple structuring, well piezoelectricity, easy to synthesize and cost-saving production and availability for the production of mass and their application [14–18], in order to gradually replace PZE Pb-based materials, including barium titanate ($BaTiO_3$), zinc oxide (ZnO), polyvinylidene fluoride (PVDF) and other Pb-containing materials.

The primary aim of PZE NGs is to improve system stability and voltage/current performance. In view of this, we have summarized many popular methods to improve the efficiency of this NG. Using various PZE materials, chemical doping of these materials, adjusting the microscopic morphology of these materials, choosing suitable substrates and designing composite thin-membrane materials are few examples. This chapter examines the various factors that influence the efficiency of this NG in order to make research recommendations for the construction of recent PZE NGs based on their high performance.

3.1.1.1 Effect of Piezoelectric Material Matrices

Since various PZE materials exhibit diverse PZE properties, such as their own defects, PZE NGs performance is severely restricted. While ZnO PZE materials are simple to make, their PZE characteristics are low, and although $BaTiO_3$-based PZE materials have high piezoelectricity, the prepared

PZE NGs have very low flexibility. Overall, the output efficiency of $BaTiO_3$-based PZE generators is reasonable, and they hold great promise for future applications. Other techniques, such as closely regulating the micromorphologies of PZE materials, can be used to make PZE NGs better in addition to designing PZE materials of high performance.

3.1.1.2 Effect of Material Micromorphology

Various micromorphologies of the same material may be evaluated by regulating the reactions temperature and other parameters during the material synthesis method, such as nano-ribbons, nanowires, nano-fibers, nanotubes, nano-rods and nano-pillars. The PZE characteristics of the same materials with different micromorphologies change greatly and have an effect on the performance of the fabricated PZE NGs. The various micromorphologies of PZE materials like ZnO and $BaTiO_3$, as well as a comparison of their output performance with respective PZE NGs, have been discussed.

3.1.1.3 Effect of Chemical Doping

An Eu^{3+}-doped P (VDF-HFP) matrix will protect the PVDF phase from external polarization conditions and nucleate the phase with high-voltage electroactivity [19]. Eu^{3+} can also cause the electroactive phase of P to self-polarize (VDF-HFP). Furthermore, doping the materials with Eu^{3+} will result in powerful light emission properties, making the materials ideal for versatile optoelectronics. PZE NGs' display efficiency and flexibility can also be greatly increased. When P (VDF-TrFE) was doped with graphene, a bilayer membrane material was developed. Internal stress was produced in P (VDF-TrFE) as a result of the electrostatic effect in the graphene film, which increased Young's modulus at the interface between the graphene membrane and the P (VDF-TrFE) material, increasing the production efficiency of PZE NGs [20, 21]. Doping PZE materials with chemical elements and materials will increase the efficiency of PZE NGs, which is also a significant potential growth path. Such techniques can help to constantly increase the efficiency of PZE generators and help them to be used in everyday life.

3.1.1.4 Effect of Device Substrate

Polyethylene terephthalate (PET) has outstanding versatility and may be used as a versatile base to promote the deformation of engineered machinery under poor mechanical external forces, allowing for the effective transfer of mechanical to electrical energy. Its isolation properties can efficiently reduce the electricity loss during charging flow, thereby enhancing PZE NGs' output efficiency. Polydimethylsiloxane (PDMS) provides strong insulation properties that can minimize electrical energy loss in the transmission method and increase the performance of PZE NGs. It may also avoid mechanical damage as a versatile substrate as it can protect PZE material against emissions. PET and PDMS are also the best materials for making PZE NGs versatile and high-performance.

3.1.1.5 Composite Development Thin Film Materials

To date, four techniques have been explored for improving the performance of PZE NGs: the synthesis of various PZE materials, different micromorphologies of the material, chemical doping and suitable substrates. However, the dual demands for PZE NGs cannot be satisfied: flexibility and high efficiency. Inorganic ceramic materials are the most common PZE materials and lack their mechanical stability and flexibility. In addition, the PZE properties are weak. Researchers have also been suggested that composite film materials should be used with polymers based on organic materials and inorganic particles to increase the stability and efficiency of PZE NGs. Therefore, a further trend in PZE NGs has been the design and manufacturing of composed thin-membrane materials with stability and excellent PZE features. In order to improve the performance of PZE NGs, the thickness of the composite thin film material needs to be increased, the nanoparticles weight ratio of the polymer matrix must be optimized, the electric dipole should be guided into an appropriate electric poling method and surface morphology and interactions between nanoparticles and polymers must also be altered. The challenges facing composite thin film materials in the future however include developing high response thin-membrane PZEs, refined surface roughness control and exploration of PZE systems in nanoscale and microscale [22].

3.1.2 Triboelectric Nanogenerators

The continuing growth of mobile devices, portable electronic devices and everyday health tracking/management has given wide-ranging focus to further monitoring of the microprocessing stage. The decrease in feature size is aimed not only at improving performance but also at taking into account the power usage of devices. If we can replace external dependency with new energy supply mechanisms in the current bottleneck era, this will greatly propel industry growth. Because of the lack of energy efficiencies mainly due to the reduced energy supply or process of conversion, the majority of energy harvesting methods on the market was not enough until Zhong Lin Wang has invented a TBE NG [5]. The TBENG has three main modes of operation: vertical mode of contact isolation, flat mode of sliding and one electromechanical mode.

3.1.2.1 Vertical Contact-Separation Mode

A regular potential difference variation may be caused by cycled separation, and recontact of the inverse TBE charges on both the inner surfaces of the sheets can be described as the working mechanism of the TBENG. The inner surfaces of the two sheets are in contact with each other as a mechanical turbulence is employed to the unit to bend, and the charging switch starts, leaving with the positive loads surface and the other with the negative load surface. This is just called the TBE effect. The two surfaces with opposite charges are immediately separated as the deformation is released such that the opposite TBE charges create an electric field between them and thereby create a potential difference between top and bottom electrodes. The electrons are guided by the external load to flow from one electrode to another in order to monitor this possible difference. The electricity produced will continue until the capacity of both the electrodes is upgraded. So, the potential difference between the TBE charges continues to decrease to zero as both the plates are pushed against each other again, so that the charges transferred are recirculated into the external charges to produce another current pulsation in the opposite direction. The alternating current (AC) signal is continuously produced while this periodic mechanical deflection lasts.

3.1.2.2 Lateral Sliding Mode

Two basic mechanisms of friction occur: regular and sliding lateral. Here, we demonstrated a TENG modeled on the basis of an in-plan sliding laterally between the two surfaces [23]. The report is not available. A continuous variation is made possible between the contact area of two surfaces (vertical mode of slidind and flat mode of sliding) by an intense triboelectrification facilated by sliding friction, which creates a lateral separation of charges centers to develop in the reduction of applied vots between the layers to drive the flow of electrons through the ecterior load.

3.1.2.3 Single-Electrode Mode

As a more viable and practical configuration for some applications including fingertip-powered TBENG, a one electrode TBENG is implemented [24, 25]. The developed TBE charges with opposite polarities have been fully balanced/screened, leading to no external electron flow. When there is a relative detachment of PDMS from the skin, these TBE charges cannot be accounted for. This electrostatic induction procedure may produce a current/outgoing signal if it is significantly corresponding to the size of the PDMS membrane to the distance between the touching skin and the base of the PDMS.

3.1.3 Pyroelectric Nanogenerators

The advancement of green and sustainable nano-energy sources has become one of the most relevant research areas thanks to the huge rise in the energy demand of modern societies [26]. There are some common effects on physics that can be used to manufacture NGs to collect energy from the atmosphere. Mechanical energy can be harvested using both piezoelectric and triboelectric effects [27]. The PE effect is based on the alteration of spontaneous polarization in some anisotropic solids because of the differences in the temperature [28] and can be used in the recovery of thermal energy from temperature shifts. Although the Seebeck effect can be used for the recovery of thermal energy by using two-end temperature differences in the unit for the diffusion of carriers [29] in a climate where the temperature is uniform but time-dependent, the PE effect must be chosen [30].

PZE NGs are currently used in the production of mechanical energy from the atmosphere for driving electrical equipment [31]. Though various designs of PENGs have been published, they still have very low output voltage and current (voltage below 0.1 V and current below 1 mA) [30], which are not sufficient to power commercial electronics. To solve this dilemma, it is urgently necessary to improve the efficiency of the PENGs. Here we showed PENG, a lead zirconate titanate (PZT) film in which the output can be up to 22 V and 171 $mA{\cdot}cm^{-2}$ of the open circuit tension and the short circuit currents density. A single output pulse of PENG will constantly pull an LCD for longer than 60 s while the temperature changes by 45 K. Such PENGs can charge a Li-ion battery under various frequencies. Recently PENGs, such as ZnO [32], PZT [33], $KNbO_3$ [34] and PVDF [35], have evolved for thermal energy scavenging with a gradient in time and have produced PENGs. PENGs have recently been developed. In certain instances, power output signals investigate to significant thresholds.

3.1.3.1 Pyroelectric Materials, ZnO

Pyroelectric ZnO nanowires are used to collect small-scale mechanical energy efficiently. PE in nanowires formed by a mechanical straining to drive electron flow in the external charge is the basis of the PE nanogenerating system. Due to the anisotropic physical properties of ZnO, pyroelectric ZnO nanowire arrays to transform heat energy into electricity can be produced by either strain or temperature. By combining the pyroelectrical and semi-conducting properties in ZnO, the time-dependent temperature shift will lead to polarization of the electric field and load separation along the ZnO nanowire. There is strong stability in the processed NG, with a characteristic heat flow conversion coefficient of approximately 0.050–0.08 $Vm^2\ W^{-1}$. The research reveals that PE nanowires can be used to control X-ray generation and space science. The concept of the PENG is the use of ZnO nanowires anisotropic polarization produced by a time-dependent fluctuation in the temperature for driving electricity.

3.1.3.2 Pyroelectric Materials, $KNbO_3$

After forming a composite of $KNbO_3$ nanowires with PDMS polymer, versatile NGs were easily fabricated. There is a discussion of the dynamics of PENGs. The NGs were also used to produce energy from solar lighting and to create hybrid NGs with solar cells. To give our NGs stability, we created a 3:7 volume ratio composite of $KNbO_3$ nanowires and PDMS polymer. For the first time, PENGs were created using a lead-free $KNbO_3$ nanowire/PDMS-polymer composite. Because of the shift of ferroelectric domains tuned by electric fields, the output voltage/current of the fabricated NGs can be modulated in a controlled manner. Under heating and cooling conditions, the voltage/current outputs of the NGs change in the opposite direction and can be improved by increasing the temperature change. Energy harvesting from sunlight-mediated heat and hybrid NGs with a solar cell is demonstrated using the NGs, suggesting that the PENGs could have several applications in self-powered nano-devices and nano-systems.

3.1.3.3 Pyroelectric Materials PZT

Despite their incredibly strong PZE and PE coefficients, lightweight ferroelectric perovskite film-based NGs have received little attention (NGs). The effective formation of a versatile $Pb(Zr_{0.52}Ti_{0.48})O_3$ (PZT) membrane is present, and it is used as a PZE NG. At room temperature, an extremely flexible Ni-Cr-alloyed metal foil material with a conductive $LaNiO_3$ bottom electrode enables the development of a flexible PZT membrane with high PZE (140 pC N^{-1}) and PE (50 nC·cm^{-2} K^{-1}) coefficients. The versatile PZT-based NG effectively modifies thermal fluctuation and mechanical vibration from a range of sources, including the human body and the environment. Moreover, due to its high Curie temperature and good resistance toward water, it produces electric current even at elevated temperatures of 373 K, pH = 13 and relative humidity of 70%. Illustrated is the production of incredibly large current flow at the exhaust pipe of vehicles, where hot gases like CO and CO_2 are quickly released to the atmosphere, as evidence of power generation in harsh environments. In order to cleave mechanical vibration and thermal fluctuation energy even under intense conditions, this work extends the implementation of versatile PZT membrane-based NGs.

3.1.3.4 Pyroelectric Materials PVDF

In the present analysis, we record a water vapor self-sustaining PNG with omnipresent low-speed airflows (typical 1–2 m·s^{-1} speeds). The water absorbs and releases great quantities of heat during evaporation and steam condensation on a surface as an ideal heat transfer agent. The large-range, flexible thin film PNG with good PE efficiency can be used as commercially available ferroelectric polymers like PVDF, copolymer difluorides and trifluoroethylenes P(VDF-TrFe). The PNG temperature decreases as a low-speed airflow moves by, primarily as a consequence of the fast evaporation of small water droplets. In this way, the temperature oscillation process can be self-supporting without using any mechanical or electronic alternator with just one low-temperature thermal source and omnipresent low-speed air (or a wind) flows in the environment. The electric power collected is used to provide low power, such as a digital watch, unbroken power. The electrical power gained can also be stored in a condenser for immediate use of high power, including the blowing of many LEDs, once per minute. Hot water vapor is widely accessible in manufacturing and our everyday lives, typically liberated to the atmosphere so we do not have an efficient usage process. The water vapor-driven self-supporting PE generator will have a new method to recover the water vapor energy efficiently.

3.2 THEORIES THE BASIS FOR NANOGENERATORS

NGs are considered to convert mechanical energy into electrical energy. The drive force responsible for this effective conversion can be attributed to the displacement current (I_D). It is also well known that Maxwell has worked with I_D and provided us with the different equations. Therefore, the theories of NGs can be well derived from these Maxwell relations. The use of these Maxwell relations can be understood in three steps to understand the basic working of NGs.

To overcome limitations in the continuity equation of electric charges, Maxwell was the first to introduce the term I_D in 1861 [36]. The term I_D or current density (J_D) is denoted as $\frac{\partial D}{\partial t}$.

where D is electric displacement vector and t is time.

The electric displacement vector (D) is given by

$$D = \varepsilon_0 E + P \tag{3.1}$$

And the value of P for isotropic medium (dielectric) is given as

$$P = (\varepsilon - \varepsilon_0) \tag{3.2}$$

Therefore, it can be concluded from earlier equations that

$$D = \varepsilon E \tag{3.3}$$

Thus, the displacement current density (J_D) can be presented as

$$\mathrm{J}_D = \frac{\partial D}{\partial t} = \varepsilon \frac{\partial E}{\partial t} \tag{3.4}$$

Recently, Wang *et al.* and his team have expanded the Maxwell equations to evaluate the power output obtained from NGs, by adding the addition term (P_s) to the electric displacement vector (D) [37, 38]. The term P_s is the polarization because of electrostatic charges over the surface, produced by mechanical triggering. The term P_s should not be confused with P that denotes polarization induced by the electric field. Therefore, after the modifications are done by Wang and his team the electric displacement vector (D) can be written as

$$D = \varepsilon_0 E + P + P_s \quad (3.5)$$

Similarly, the displacement current density (J_D) can be written in modified form as

$$J_D = \frac{\partial D}{\partial t} = \varepsilon \frac{\partial E}{\partial t} + \frac{\partial Ps}{\partial t} \quad (3.6)$$

The same group of workers then wrote the Maxwell equations in a modified form that can be found elsewhere. The term P_s was added to these Equations (3.7)–(3.9). These modified forms of equations act as cornerstones and have been used to derive output characteristics such as current and voltage for NGs. The modified forms of those equations are

$$\varepsilon \nabla \cdot E = \rho - \nabla \cdot P_s \quad (3.7)$$

$$\nabla \cdot B = 0 \quad (3.8)$$

$$\nabla \times E = -\frac{\partial B}{\partial t} \quad (3.9)$$

$$\nabla \cdot H = J + \varepsilon \frac{\partial E}{\partial t} + \frac{\partial Ps}{\partial t} \quad (3.10)$$

3.2.1 The Theory of Polarization (P_s)

According to this theory, polarization (P_s) created because of electrostatic charge over the surface is expressed by the following equation

$$\nabla \times P_s = -\sigma_s(r,t)\, \delta\left[f(r,t)\right] \quad (3.11)$$

where σ_s (r, t) is the charge density function on the surface of media.

$\delta\left[f(r,t)\right]$ is used for the media shape.

On solving the scalar electric potential from the electrostatic charge over the surface, we get the following relation:

$$\Phi_s(r, t) = \frac{1}{4\pi} \int \frac{\sigma_s(r',t')}{|r-r'|} ds' \quad (3.12)$$

where Φ_s (r, t) is scalar electrostatic potential.

Using all these relations, Wang and his group have derived an expression for calculating P_s as shown below [39]:

$$P_s = -\nabla \Phi_s\,(r,t)$$

$$= \frac{1}{4\pi} \sigma_s(r',t') \frac{\sigma_s(r-r')}{|r-r'|^3} ds' + \frac{1}{4\pi c} \int \frac{\partial \sigma_s(r'', t)}{\partial t'} \frac{(r-r')}{|r-r'|^2} ds' \quad (3.13)$$

This expression is a general expression for surface polarization density, and the value calculated from this equation is used in Maxwell relations 7 and 10 to calculate I_D.

3.2.2 Current Transport Equations

It has been observed that in NGs, capacitive current shows dominance in the external circuit while I_D shows dominance in the internal circuit. The expression by I_D in NGs can be obtained from the surface integral of current density (J_D) as shown below:

$$ID = \int J_D . ds = \int \frac{\partial D}{\partial t} ds$$

$$ID = \frac{\partial}{\partial t} \int \nabla . D\, dr = \frac{\partial}{\partial t} \int \rho\, dr$$

$$I_D = \frac{\partial Q}{\partial t} \tag{3.14}$$

where Q is the total amount of free charge on an electrode of a generator.

Using this equation, Wang *et al.* have shown that, generally current transport behavior for any NG is given by [39]

$$\Phi_{AB} = \int_A^B E \cdot dL = \frac{\partial Q}{\partial t} R \tag{3.15}$$

where Φ_{AB} is the potential drop seen from electrodes A to B and dL is the integral over-path between points A and B.

The equation of current transport for various types of NGs, i.e., PZE and TBE NGs have been published by the same group of researchers and can be understood by Figure 3.1.

3.2.3 Projections in Maxwell Equations (Nanogenerators)

It is well known that the birth to electromagnetic theory was given by the first term of Maxwell equation, i.e., $\varepsilon \frac{\partial E}{\partial t}$. This first term of the theory has developed the world by providing us with

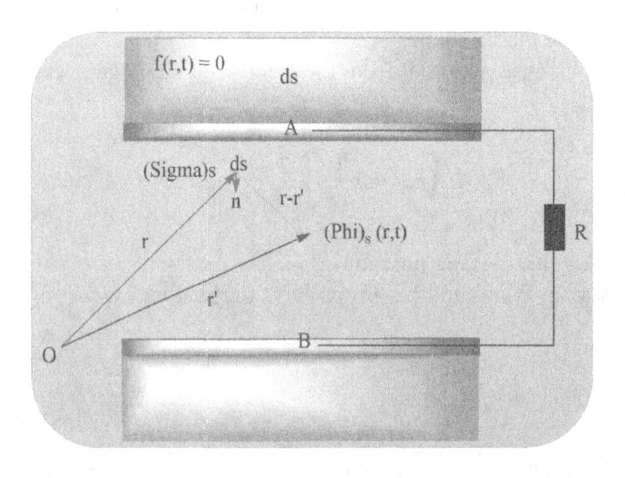

FIGURE 3.1 Schematic representation of nanogenerator with an external load.

antenna, radio, telegram, TV, radar, microwave, space communication and wireless communications. Furthermore, this has provided us with useful inventions such as laser and high-tech photonics.

The second-term $\frac{\partial P_s}{\partial t}$ in this Maxwell relation was inserted by Wang and his team that lead to set the foundation of NGs [38]. The addition of this term to Maxwell relations provided us with an application to generate energy from NGs. Therefore, we can say that NGs are another application of Maxwell relation that resulted into the production of more energy and nanosensors.

3.3 WORKING MECHANISM

3.3.1 Piezoelectric Nanogenerator

The working mechanism of these types of NGs is by two ways.

3.3.1.1 When Force Is Exerted Perpendicular to the Axis of Nanowire

In this case, the nanowire is grown vertically with its tip moving in a lateral manner. It is observed that applying force by the tip of the nanowire over PZE structure leads to deformation of the structure throughout. This deformation is inside the PZE structure results into the generation of the electric field inside the nanostructure. The positive potential is seen on the side where positive strain in structure is produced and vice versa. This generation of potential difference in PZE crystalline structure is because of relative displacement anions and cations with respect to each other. In summary, the nanowire tip with distributing electrical potential on their surface and other end (bottom) of nanowire will neutralize it to ground.

The electrical contact in NG also plays a crucial role to pump charges into the tip surface. Therefore, Schottky contact should be induced between the counter electrode and nanowire tip instead of ohmic contact, because it neutralizes the electric field generated at the nanowire tip. It has been observed that to form effective Schottky contact, metal used to form counter electrode should have a large work function than its electron affinity.

3.3.1.2 When Force Is Exerted Parallel to the Axis of Nanowire

In this case, the nanowire that is vertically grown is stacked between the Schottky contact on the top and ohmic contact on the bottom. On application of external force on the tip of the nanowire, the mono-axial strain is produced. Due to this generated strain, the nanowire tip shows negative potential leading into an increase in fermi level at the tip of the nanowire. As a result of this, the electrons start flowing from the tip of the nanowire to its bottom, through the external circuit generating positive potential at the tip of the nanowire.

The Schottky contact will act as a barricade for the electrons that flow through the interface and maintain the potential at the tip of the nanowire. The PZE effect in NG diminishes as soon as externally applied force is removed, and electrons start moving back to the top from the bottom so that positive potential can be neutralized at the tip of the nanowire.

3.3.2 Pyroelectric Nanogenerator

The working principle of these types of NGs is also believed to work in two ways.

3.3.2.1 Primary Pyroelectric Nanogenerator

The primary PE effect describes the production of charge without any strain. These types of effects dominate the response in PZT, BTO and many other ferromagnetic substances [40]. The working principle of these NGs is based on random electric dipole wobbling, induced thermally around its equilibrium axis. The magnitude of this random electric dipole wobbling increases with an increase in temperature [41]. The electric dipoles oscillate in a random manner due to thermal fluctuations under room temperature. The average strength of spontaneous polarization is constant in electric dipoles under constant temperature; therefore, no output can be seen from PENG at a constant temperature.

Therefore, to get output from NGs, the applied temperature needs to be fluctuated, so that electric dipoles can oscillate in a large degree of spread around aligning axes. Hence, average spontaneous polarization decreases because of spreading oscillation angles. Because of this spread in oscillation angles, the induced charge on the electrodes is reduced which results into the flow of electrons. In the similar fashion if the system is cool, the electric dipoles will oscillate with a smaller degree of spread angles because of low thermal activity, and as a result, this spontaneous polarization will be increased. Therefore, the amount of induced charges will be increased, resulting in the flow of electrons in opposite direction.

3.3.2.2 Secondary Pyroelectric Nanogenerator

In this type of effect, the thermal expansion induces strain as a result of which charge is produced. It has been observed that in this type of effects, the PE response dominates in CdS, ZnO and wurtzite (TiO_2). This type of thermal deformation induces a potential difference (PZE type) across the NG as a result of which electrons flow in the external circuit. Thus, the output of the NG in this secondary effect is because of the PZE coefficient as well as the thermal deformation of these materials. The output current (I) in these PENGs is given as

$$I = pA\frac{\partial T}{\partial t} \tag{3.16}$$

where p, A and $\frac{dT}{dt}$ are PE coefficient, effective area and the rate of change of temperature inside the NG.

3.3.3 Triboelectric Nanogenerator

The basic working of these TBENGs can be explained by continuous variation in the difference in potential. This variation in potential difference is produced as a result of the cycled separation of charges and contact between opposite TBE charges on the inner surfaces of the two sheets. On applying the external pressure to TBE NG, the two opposite plates of NG come in close contact with each other resulting into the transfer of charges from one plate to the other. This charge transfer leaves one plate with positive charges and the other with negative charges. This effect of charge separation is known as the TBE effect. When the external pressure is removed, the opposite charges separate automatically, resulting in generation of an electric field that further produces potential difference across the two electrodes of TBE NGs. Therefore, to neutralize the potential difference between two electrodes, the electrons start flowing from one electrode to the other through the external load. This process restarts again on applying external pressure and so on. When the mechanical deformation lasts on the removal of pressure, the AC current signals are produced continuously [42, 43]. One important thing should always be remembered that one of the plates producing turboelectric charges should be an insulator. There is also a little change in general mechanism as the different modes of TBE NGs are studied.

3.4 APPLICATIONS

3.4.1 Piezoelectric Nanogenerators

These types of NGs show large-scale applications where the periodic kinetic energy can be generated. These periodic kinetic energy sources can be on large scale such as oceans waves and wind waves. These periodic kinetic sources can also be on small scale such as heartbeat, muscle moment, breathing by lungs etc.

One such application was found in nano- or micro-devices where a small amount of continuous energy is required either as a supplementary or independent source. Wang and his coworkers in 2010 introduced a self-powered pH device and a UV sensor having 20–40 mV output voltage. The energy produced by these devices is too small that these cannot support microdevices independently on their own. Therefore, these are used as supplementary sources of energy for the batteries of the nano- or micro-devices. These NGs provide better results when combined with other energy harvesting devices like biochemical energy sources or solar cells. It is expected that in future, these types of NGs can be used to contribute energy in many devices independently.

Another application of these NGs is smart wearables or outfits made of PZE fibers. These PZE fibers convert kinetic energy obtained from the human body to electric energy. This energy can be further used in a lot of electronic devices such as devices used to monitor health systems in the human body. The NGs can be employed in shoes to get kinetic energy requirements that can be converted into electric energy. Another similar power generating device was generated by Wang *et al.* in which human artificial skin was developed to generate power. They have made a system in which an AC voltage of 100 mV is thought to be produced. In this device, they have used flexible SWG attached to running hamster.

One of the paradigms shift in the development of NGs is the production of transparent and flexible devices. These devices can be used to develop self-powered tactile sensors and also found their use in energy efficient screen touch devices. Kim and Choi *et al.* have synthesized a PZE nanostructure with flexible properties and transparent behavior. They have enhanced the cost-effectiveness as well as transparency of this PZE device by substituting indium tin oxide material with a layer of graphene. Furthermore, the NGs can be used with implants in the human body to get information about their functioning. These are used with implants because these can be synthesized on implants and are biocompatible. One of the best examples for such device is NG-based on ZnO nanowire. This implantable device on combination with NG receives ultrasonic vibrations from outside the body of humans. These vibrations are then converted into electrical energy by the PZE nanostructure.

Lu *et al.* have synthesized ZnO nanowires doped with phosphorus. These were grown on the silicon substrate and were first to be used for energy conversion [44]. Chung *et al.* have synthesized a PZE NG (thin film) that was flexible and all solutions processed. This was synthesized using a screen printing method and a reactive zinc-hydroxo condensation. The highly elastic thin film in this NG generates piezoelectricity by muscle stretching and mechanical rolling of the PZE unit. These can be used as a good energy harvester in electronics as well as human wearables [45]. Recently, Saikh *et al.* have synthesized Zr-based PZE NGs, *i.e.*, self-polarized ZrO_2/Poly(vinylidene fluoride-*co*-hexafluoropropylene) nanocomposite-based PZE NG. They were found to show high capacitor charging performance and have seen to charge 2.2-μF capacitor under 3.5 V, in a time interval of 17 s. It was found that single electrode-based TENG when fabricated by this composite film also shows a remarkable output voltage of ~7 V under finger touch [46]. These were used to harvest energy from different human movements.

Jung *et al.* have developed the Pb-free nanowires $NaNbO_3$ for the high-output PZE NGs. The device was found to show an output current of 72 nA and output voltage of 3.2 V under 0.23% compressive strain [47]. Wang *et al.* have developed a bionic single-electrode electronic skin unit that works on the basis of PZE NG. These have been found critical for endowing intelligent robots with tactile and temperature sense on electronic skin [48]. The synthesis of PZE NGs with wearable fabric based on hybrid fiber has also been studied. This was used in energy harvesting applications by attaching it on elbow pad. The output energy harvested was found sufficient to power an LCD [49]. Gosh and Mandal have been prepared bio-piezoelectric NGs with the fish scale (a transparent biowaste). These NGs can be used to scavenge energy from different types of motions from body moments, sound vibrations, body movements and wind flow etc. It was observed that it produces an output current of 1.5 μA and voltage of 4 V under the compressive stress of 0.17 MPa [50].

3.4.2 Pyroelectric Nanogenerators

As suggested in the name, it can be seen that these types of devices are used to detect temperature-related properties. In other words, PENGs show the applications in devices where time-dependent fluctuations are present. For example, the PENG can be used to sense temperature (active sensor) without using any battery. In one of the studies, Wang and his coworkers have devised a self-powered device i.e., PENG, which was used to detect the change in temperature with a response time of 0.9 s and reset time of 3 s [51]. Generally, it is observed that PENGs provide low current output but high voltage; therefore, besides its use as a temperature sensor, it can also be used as a high potential source.

Yang *et al.* have developed PENGs by fabricating lead-free $KNbO_3$ with nanowire-PDMS a polymer composite. The voltage and current output for these NGs was controlled by the electric field and increased with an increase in the rate of change in temperature. These NGs have found their use in harvesting energy from sunlight and further used in many self-powered nanodevices [52]. One study reveals that the development of PENGs based on PZT film. Using this NG, a Li-ion battery was charged at different frequencies that were further used in green color light-emitting diode. These NGs also showed a lot of applications in wireless sensors [53].

Recently, it has been seen that in COVID-19 pandemics, these NGs are integrated into masks and worked as good sensors to detect the temperature of the human body. Other best applications of these NGs are that these can be used to convert waste heat into electrical energy [54]. In one of the recent studies, Li *et al.* have synthesized vibration free PENG that can be used in the regulation of local hot charge density. These NGs are also used as photocatalysts and also acted as surface enhanced Raman scattering substrates for oxidation reactions [55].

Same work has been performed by Zhao *et al.* in the current century to highlight the use of these NGs in self-powered drug delivery systems, which is one of the best uses [56]. In the same studies, they have developed a PENG $Pb(Mg_{1/3}Nb_{2/3})O_3$-$PbTiO_3$ single crystals, which are based on Mn ion substitution. This PENG acts as a good Infra-Red sensor [56]. Recently, Zhang *et al.* were synthesized nanostructures $BiFeO_3$/Au/ZnO that showed coupled PE-photovoltaic effect. These nanostructures also showed improved broadband response as well photoelectric properties [57].

3.4.3 Turboelectric Nanogenerators

These types of NGs convert physical agitation to the electric signal in the internal circuit (by triboelectrification) and in the outer circuit (by electrostatic induction process). This is of significant importance because it has demonstrated two major ways of harvesting energy; first, it harvests energy from external mechanical sources, and second, it acts as a self-powered active sensor and didn't require any external power source to get energy.

One of the best applications of these NGs is to harvest energy from vibrational motions from walking, automobiles, voices, wind aircraft, trains etc. These types of vibrational motions are easily accessible anywhere and at any time. A large number of researchers have demonstrated many aspects of harvesting these mechanical energies. Based on the contact separation mode, a classical approach named the Cantilever-based technique was used for harvesting energy from vibrational motions. This work is based on a design in which the contact surface of the cantilever is formed between the top and bottom surfaces during vibrations [58]. In another study, Yang *et al.* have harvested energy from a backpack in which an NG was designed by integrated rhombic gridding. This rhombic gridding has shown a great improvement in total current output when structurally multiplied cells were connected to each other in parallel circuits [59]. Similarly, Chen *et al.* have fabricated the NG based on harmonic resonance in which four supporting springs were used to induce contact separation between two TBE materials. This system was used to harvest the energy from vibrational motions of automobile engines, a desk and a sofa [60]. Recently, a 3D TBE NG is designed using hybridization mode in which both plane sliding mode and vertical contact separation modes have been integrated. This device can harvest vibrational energy over a wide bandwidth in multiple directions produced randomly. This type of hybrid NGs are expected to harvest vibrational energies even at low frequencies thus can have many applications in infrastructure monitoring, in the internet of things and in charging portable electronic devices.

These TBE NGs have also used in harvesting energy from human body motions. This electrical energy is generated for charging portable electronic devices and many other biomedical applications [61]. It was found that under general conditions, TBENGs can produce an output voltage of 17 V. The TBE NGs showed its important applications in self-powered active strain in themselves and as force sensors. Fan *et al.* have developed a TBENG that can be used to measure the subtle pressure even in real life. It was so sensitive that it even showed a response to a piece of feather having 20-mg weight and 0.4-Pa contact pressure [62]. Majid *et al.* have developed an active pressure sensor by integrating NG with sponge-shaped polymer embedded on to the wire. When pressure is applied on to it, the charge separation occurs between the active wire and composite soft polymer. This charge separation occurs because of air gets filled into the gap produced between the nanowire and polymer. This passive wire as an electrode allows the sensor to work in single-electrode mode [63]. To know the self-powered pressure applied over the large surface area, an array of a large number of TBENGs is used [64].

Majid *et al.* were proposed their work during the study of the application of TBENGs integrated a small belt pulley system to the self-powered TBENG (encoder). This integrated system was used to convert friction obtained from the belt pulley system to harvest energy and then store it to the capacitor. This stored charge was used to power different circuits including LCD and a microcontroller [65]. One of the important applications of these TBE NGs is their use in chemical sensors. In one of the studies, the 3-mercaptopropionic acid (3-MPA) was integrated with gold nanoparticles which were used to modify an NG. This modified form of the NG was used to detect mercury ions because of different TBE polarities. It is further believed that these NGs can be used in future to detect the presence of other metal ions, DNA, proteins etc. [66].

3.5 CONCLUSION

The chapter briefly describes the various types of NGs. Furthermore, we have discussed the basic theory for their functioning with its working mechanism and applications in various fields. We have also summarized the working mechanisms for all the types of NGs such as PZE, PE and TBENGs. We have also discussed about the various possible applications of these NGs in numerous fields such as biomedical, optoelectronics etc. These NGs are found to convert waste energy from the environment to electrical energy. The sources of energy used are wind waves, body motions, vibrational motions from engines, heat from the body, vibrations from automobiles etc. The harvested energy can be used to self-power the micro- and nano-devices, smart wearable systems, transparent and flexible devices, implantable telemetric receiver devices, force sensors, motion sensors and chemical sensors. They are successfully utilized in human wearable devices, touch screens and chemical sensors. In one of the studies, TBENGs have been used to sense mercury ions. These are too sensitive that these can convert a small pressure applied by a feather of bird to electrical energy. It is also believed by many researchers, it can be used to sense biomolecules like DNA, proteins etc. In summary, the NGs are a new path for scientists to provide energy from waste resources.

REFERENCES

1. Tarascon, J.M. and Armand, M. 2001. Issues and Challenges Facing Rechargeable Lithium Batteries, *Nature*, 414:359–367.
2. Goodenough, J.B. and Park, K. 2013. The Li-ion Rechargeable Battery: A Perspective. *J. Am. Chem. Soc.* 135:1167–1176.

3. Zhu, Y., Murali, S., Stoller, M.D., Ganesh, K.J., Cai W. et al. 2011. Carbon-Based Supercapacitors Produced by Activation of Grapheme. *Science* 332:1537–1541.
4. Wang, Z.L. and Song, J.H. 2006. Piezoelectric Nanogenerators Based on Zinc Oxide Nanowire Arrays. *Science* 312:242–246.
5. Fan, F.R., Tian, Z.Q. and Wang, Z.L. 2012. Flexible Triboelectric Generator. *Nano Energy* 1:328–334.
6. Zhao, K., Wang, Z.L. and Yang, Y. 2016. Self-Powered Wireless Smart Sensor Node Enabled by an Ultrastable, Highly Efficient, and Superhydrophobic-surface-based Triboelectric Nanogenerator. *ACS Nano.* 10:9044–9052.
7. Diaz-Gonzalez, F., Sumper, and A. Gomis-Bellmunt, O. 2012. A Review of Energy Storage Technologies for Wind Power Applications. *Renew. Sustain. Energy Rev.* 16:2154–2171.
8. Wang, Z. L., Jiang, T., and Xu, L. 2017. Toward the Blue Energy Dream by Triboelectric Nanogenerator Networks. *Nano Energy* 39:9–23.
9. Wang, Y. and Yang, Y. 2019. Superhydrophobic Surfaces-Based Redoxinduced Electricity from Water Droplets for Self-Powered Wearable Electronics. *Nano Energy* 56:547–554.
10. Quan, T., Wang, X., Wang, Z.L. and Yang, Y. 2015. Hybridized Electromagnetic–Triboelectric Nanogenerator for a Self-Powered Electronic Watch. *ACS Nano* 9:12301–12310.
11. He, H., Fu, Y., Zhao, T., Gao, X., Xing, L., Zhang, Y. and Xue, X. 2017. All-Solid-State Flexible Self-Charging Power Cell Basing on Piezo-Electrolyte for Harvesting/Storing Body-Motion Energy and Powering Wearable Electronics. *Nano Energy* 39:590–600.
12. Zhang, Y., Zhang, Y., Xue, X., Cui, C., He, B., Nie, Y., Deng, P. and Wang, Z.L. 2014. PVDF–PZT Nanocomposite Film Based Self-Charging Power Cell. *Nanotechnology* 25:105401.
13. Pu, X., Hu, W., and Wang, Z.L. 2018. Toward Wearable Self-Charging Power Systems: The Integration of Energy-Harvesting and Storage Devices. *Small* 14:1702817.
14. Xu, S., Wei, Y., Liu, J., Yang, R. and Wang, Z. 2008. Integrated Multilayer Nanogenerator Fabricated Using Paired Nanotip-to-Nanowire Brushes. *Nano Lett.* 8:4027–4032.
15. Ma, J., Hu, J., Li, Z. and Nan, C. 2011. Recent Progress in Multiferroic Magnetoelectric Composites: From Bulk to Thin Films. *Adv. Mater.* 23:1062–1087.
16. Hu, J., Li, Z., Wang, J., Ma, J., Lin, Y.H. and Nan, C.W. 2010. A Simple Bilayered Magnetoelectric Random Access Memory Cell Based on Electric-Field Controllable Domain Structure. *J. Appl. Phys.* 108:043909.
17. Ma, J., Shi, Z. and Nan, C. 2007. Magnetoelectric Properties of Composites of Single Pb(Zr, Ti) O_3 Rods and Terfenol-D/Epoxy with a Single-Period of 1-3-Type Structure. *Adv. Mater.* 19 (2007) 2571–2573.
18. Fujii, I., Nakashima, K., Kumada, N. and Wada, S. 2012. Structural, Dielectric, and Piezoelectric Properties of Batio-3-Bi($Ni_{1/2}Ti_{1/2}$)O_3 Ceramics. *J. Ceram. Soc. Jpn.* 120:30–34.
19. Adhikary, P., Garain, S., Ram, S. and Mandal, D. 2016. Flexible Hybrid Eu^{3+} Doped P (VDFHFP) Nanocomposite Film Possess Hypersensitive Electronic Transitions and Piezoelectric Throughput. *J. Polym. Sci., B.* 54:2335–2345.
20. Bhavanasi, V., Kumar, V., Parida, K., Wang, J. and Lee, P.S. 2015. Enhanced Piezoelectric Energy Harvesting Performance of Flexible P(VDF-TrFE) Bilayer Films with Graphene Oxide. *ACS Appl. Mater. Interfaces* 8:521–529.
21. Bae, S.H., Kahya, O., Sharma, B.K., Kwon, J., Cho, H.J., Ozyilmaz, B. and Ahn, J.H. 2013. Graphene-P (VDF-TrFE) Multilayer Film for Flexible Applications. *ACS Nano* 7:3130–3138.
22. Khan, A., Abbas, Z., Kim, H.S. and Oh, I.K. 2016. Piezoelectric Thin Films: An Integrated Review of Transducers and Energy Harvesting. *Smart Mater. Struct.* 25:053002.
23. Wang, S., Lin, L., Xie, Y., Jing, Q., Niu, S. and Wang, Z.L. 2013. Sliding-Triboelectric Nanogenerators Based on In-Plane Charge-Separation Mechanism. *Nano Lett.*, 2226–2233.
24. Yang, Y., Zhou, Y.S., Zhang, H., Liu, Y., Lee, S. and Wang, Z.L. 2013. A Single-Electrode Based Triboelectric Nanogenerator as Self-Powered Tracking System *Adv. Mat.* 25:6594–6601.
25. Yang, Y., Zhang, H., Chen, J., Jing, Q., Zhou, Y.S., Wen, X. and Wang, Z.L. 2013. Single-Electrode-Based Sliding Triboelectric Nanogenerator for Self-Powered Displacement Vector Sensor System. *ACS Nano.* **7**:7342–7351.
26. Hu, L., Choi, J.W., Yang, Y., Jeong, S., Mantia, F.L., Cui, L. and Cui, Y. 2009. Highly Conductive Paper for Energy-Storage Devices. *PNAS* 22:21490–21494.
27. Qi, Y., Jafferis, N.T., Lyons Jr., K., Lee, C.M., Ahmad, H. and McAlpine, M.C. 2010. Piezoelectric Ribbons Printed onto Rubber for Flexible Energy Conversion. *Nano Lett.* 10:524–528.
28. Lang, S.B. 2005. Pyroelectricity: From Ancient Curiosity to Modern Imaging Tool. *Phys. Today.* 58:31–36.

29. Yang, Y., Pradel, K.C., Jing, Q., Wu, J.M., Zhang, F., Zhou, Y., Zhang, Y. and Wang, Z.L. 2012. Thermoelectric Nanogenerators Based on Single Sb-Doped ZnO Micro/Nanobelts. *ACS Nano.* 6:6984–6989.
30. Yang, Y., Guo, W., Pradel, K.C., Zhu, G., Zhou, Y., Zhang, Y., Hu, Y., Lin., L. and Wang, Z.L. 2012. Pyroelectric Nanogenerators for Harvesting Thermoelectric Energy. *Nano Lett.* 12:2833–2838.
31. Lee, M., Bae, J., Lee, J., Lee, C., Hong, S. and Wang, Z.L. 2011. Self-powered Environmental Sensor System Driven by Nanogenerators. *Energy Environ. Sci.* 4:3359–3363.
32. Wang, Z.L. and Song, J. 2006. Piezoelectric Nanogenerators Based on Zinc Oxide Nanowire Arrays. *H. Sci.*, 14:242–246.
33. Ko, Y.J., Kim, D.Y., Won, S.S., Ahn, C.W., Kim, W., Kingon, A.I., Kim, S., Ko, J. and Jung, J.H. 2016. Flexible $Pb(Zr_{0.52}Ti_{0.48})O_3$ Films for a Hybrid Piezoelectric-Pyroelectric Nanogenerator under Harsh Environments. *ACS Appl. Mater. Interfaces* 8(10):6504–6511.
34. Yang, Y., Jung, J.H., Yun, B.K., Zhang, F., Pradel, K.C., Guo, W. and Wang, Z.L. 2012. Flexible Pyroelectric Nanogenerators Using a Composite Structure of Lead-Free $KNbO_3$ Nanowires. *Adv. Mater.* 9:5357–5362.
35. Li, X., Lu, S., Chen, X., Gu, H., Qian, X. and Zhang, Q.M. 2013. Pyroelectric and Electrocaloric Materials. *J. Mater. Chem. C* 1:23–37.
36. Maxwell, J.C. 1861. Philosophical Magazine and Journal of Science. London: Edinburg and Dubline, 4th series. pp. 161.
37. Wang, Z.L. Jiang, Tao, X.L. 2017. Toward the Blue Energy Dream by Triboelectric Nanogenerator Networks. *Nano Energy* 39:9–23. doi:10.1016/j.nanoen.2017.06.035.
38. Wang, Z.L. 2017. On Maxwell's Displacement Current for Energy and Sensors: The Origin of Nanogenerators. *Mat. Today* 20(2):74–82. doi:10.1016/j.mattod.2016.12.001.
39. Wang, Z.L, 2019. On the First Principle Theory Of Nanogenerators from Maxwell's Equations. *Nano Energy* 68:104272. doi:10.1016/j.nanoen.2019.104272.
40. Ye, C.P., Tamagawa, T. and Polla, D.L. 1991. Experimental Studies on Primary and Secondary Pyroelectric Effects in $Pb(ZrO_xTi_{1-x})O_3$, $PbTiO_3$, and ZnO thin films. *J. Appl. Phys.* 70(10):5538. doi:10.1063/1.350212.
41. Yang, Y., Jung, J.H., Yun, B.K., Zhang, F., Pradel, K.C., Guo, W., and Wang, Z.L. (2012). Flexible Pyroelectric Nanogenerators using a Composite Structure of Lead-Free $KNbO_3$ Nanowires. *Adv. Mater.* 24(39): 5357–5362. doi:10.1002/adma.201201414.
42. Zhu, G., Pan, C., Guo, W., Chen, C.Y., Zhou, Y., Yu, R. and Wang, Z.L. 2012. Triboelectric-Generator-Driven Pulse Electrodeposition for Micropatterning. *Nano Lett.* 12(9):4960–4965. doi:10.1021/nl302560k.
43. Wang, S., Lin, L. and Wang, Z.L. 2012. Nanoscale Triboelectric-Effect-Enabled Energy Conversion for Sustainably Powering Portable Electronics. *Nano Lett.* 12(12):6339–6346. doi:10.1021/nl303573d.
44. Lu, M.P., Song, J., Lu, M.Y., Chen, M.T., Gao, Y., Chen, L.J. and Wang, Z.L. 2009. Piezoelectric Nanogenerator Using p-Type ZnO Nanowire Arrays. *Nano Lett.* 9(3):1223–1227. doi:10.1021/nl900115y.
45. Chung, S.Y., Kim, S., Lee, J.H., Kim, K., Kim, S.W., Kang, C.Y., Yoon, S.K. and Kim, Y.S. 2012. All-Solution-Processed Flexible Thin Film Piezoelectric Nanogenerator. *Adv. Mater.* 24(45): 6022–6027. doi:10.1002/adma.201202708.
46. Saikh, M.M., Hoque, N.A., Biswas, P., Rahman, W., Das, N., Das, S. and Thakur, P. 2021. Self-Polarized ZrO_2/Poly(Vinylidenefluoride-*co*-Hexafluoropropylene) Nanocomposite-Based Piezoelectric Nanogenerator and Single-Electrode Triboelectric Nanogenerator for Sustainable Energy Harvesting from Human Movements. *Nat. Sci.* doi: 10.1002/pssa.202000695.
47. Jung, J.H., Lee, M., Hong, J.I., Ding, Y., Chen, C.Y., Chou, L.J. and Wang, Z.L. 2011. Lead-Free $NaNbO_3$ Nanowires for a High Output Piezoelectric Nanogenerator. *ACS Nano.* 5(12):10041–10046. doi:10.1021/nn2039033.
48. Wang, X., Song, W.Z., You, M.H., Zhang, J., Yu, M., Fan, Z., Ramakrishna S. and Long, Y.Z. 2018. Bionic Single-Electrode Electronic Skin Unit Based on Piezoelectric Nanogenerator. *ACS Nano.* doi:10.1021/acsnano.8b04244.
49. Zhang, M., Gao, T., Wang, J., Liao, J., Qiu, Y., Yang, Q., Xau H., Shi, Z., Zhao, Y., Xiong, Z. and Chen, L. 2015. A Hybrid Fibers Based Wearable Fabric Piezoelectric Nanogenerator for Energy Harvesting Application. *Nano Energy* 13:298–305. doi:10.1016/j.nanoen.2015.02.034.
50. Ghosh, S.K. and Mandal, D. 2016. High-Performance Bio-Piezoelectric Nanogenerator Made with Fish Scale. *Appl. Phys. Lett.* 109(10):103701. doi:10.1063/1.4961623.
51. Yang, Y., Zhou, Y., Wu, J.M. and Wang, Z.L. 2012. Single Micro/Nanowire Pyroelectric Nanogenerators as Self-Powered Temperature Sensors. *ACS Nano.* 6(9):8456–8461. doi:10.1021/nn303414u.

52. Yang, Y., Jung, J.H., Yun, B.K., Zhang, F., Pradel, K.C., Guo, W. and Wang, Z.L. 2012. Flexible Pyroelectric Nanogenerators using a Composite Structure of Lead-Free $KNbO_3$ Nanowires. *Adv. Mat.* 24(39):5357–5362. doi:10.1002/adma.201201414.
53. Yang, Y., Wang, S., Zhang, Y., and Wang, Z.L. 2012. Pyroelectric Nanogenerators for Driving Wireless Sensors. *Nano Lett.*, 12(12):6408–6413. doi:10.1021/nl303755m.
54. Korkmaz, S. and Kariper, A. 2021. Pyroelectric Nanogenerators (PyNGs) in Converting Thermal Energy into Electrical Energy: Fundamentals and Current Status. *Nano Energy*, 84:105888. doi:10.1016/j.nanoen.2021.105888.
55. Li, C., Xu, S., Yu, J., Li, Z., Li, W., Wang, J. and Zhang, C. 2020. Local Hot Charge Density Regulation: Vibration-Free Pyroelectric Nanogenerator for Effectively Enhancing Catalysis and In-Situ Surface Enhanced Raman Scattering Monitoring. *Nano Energy.* 105585. doi:10.1016/j.nanoen.2020.105585.
56. Zhao, C., Cui, X., Wu, Y. and Li, Z. 2020. Recent Progress of Nanogenerators Acting as Self-Powered Drug Delivery Devices. *Adv. Sustain. Syst.* 5(4):2000268. doi: 10.1002/adsu.202000268.
57. Zhang, Y., Su, H., Li, H., Xie, Z., Zhang, Y., Zhou, Y., Yang, L., Lu, H. Yuan, G. and Zheng, H. 2021. Enhanced Photovoltaic-Pyroelectric Coupled Effect of $BiFeO_3$/Au/ZnO Heterostructures, *Nano Energy.* 85:105968. doi: 10.1016/j.nanoen.2021.105968.
58. Yang, W., Chen, J., Zhu, G., Wen, X., Bai, P., Su, Y., Lin, Y. and Wang, Z. 2013. Harvesting vibration energy by a triple-cantilever based triboelectric nanogenerator. *Nano Res.* 6(12):880–886. doi:10.1007/s12274-013-0364-0.
59. Yang, W., Chen, J., Zhu, G., Yang, J., Bai, P., Su, Y., Jing, Q., Cao, X., Wang, Z.L. 2013. Harvesting Energy from the Natural Vibration of Human Walking. *ACS Nano.* 7(12):11317–11324. doi:10.1021/nn405175z.
60. Chen, J., Zhu, G., Yang, W., Jing, Q., Bai, P., Yang, Y., Hou, T.C. and Wang, Z.L. 2013. Harmonic-Resonator-Based Triboelectric Nanogenerator as a Sustainable Power Source and a Self-Powered Active Vibration Sensor. *Adv. Mat.* 25(42):6094–6099. doi:10.1002/adma.201302397.
61. de Medeiros, S., Chanci, M., Moreno, D., Goswami, C., Martinez, D. and Ramses, V. 2019. Waterproof, Breathable, and Antibacterial Self-Powered e-Textiles Based on Omniphobic Triboelectric Nanogenerators. *Adv. Funct. Mat.* 29(42):1904350. doi:10.1002/adfm.201904350.
62. Fan, F.R., Lin, L., Zhu, G., Wu, W., Zhang, R. and Wang, Z.L. 2012. Transparent Triboelectric Nanogenerators and Self-Powered Pressure Sensors Based on Micropatterned Plastic Films. *Nano Lett.* 12(6): 3109–3114. doi:10.1021/nl300988z.
63. Majid, T., Virgilio, M. Ali, S., Barbara, M. and Lucia, B. 2014. A Novel Soft Metal-Polymer Composite for Multidirectional Pressure Energy Harvesting. *Adv. Energy Mat.* 4(12): 1400024. doi:10.1002/aenm.201400024.
64. Lin, L., Xie, Y., Wang, S., Wu, W., Niu, S., Wen, X. and Wang, Z.L. 2013. Triboelectric Active Sensor Array for Self-Powered Static and Dynamic Pressure Detection and Tactile Imaging. *ACS Nano* 7(9): 8266–8274. doi:10.1021/nn4037514.
65. Majid, T., Ali, S., Alessio, M., Barbara, M., Lucia, B. and Virgilio, M. 2015. Triboelectric Smart Machine Elements and Self-powered Encoder. *Nano Energy* 13:92–102. doi:10.1016/j.nanoen.2015.02.011.
66. Lin, Z.H., Zhu, G., Zhou, Y.S., Yang, Y., Bai, P., Chen, J. and Wang, Z.L. 2013. A Self-Powered Triboelectric Nanosensor for Mercury Ion Detection. *Angew. Chem.* 125(19):5169–5173. doi:10.1002/ange.201300437.

4 2D Materials for Nanogenerators

Bhargavi Koneru, Jhilmil Swapnalin, Prasun Banerjee, Manikanta P. Narayanaswamy, Dinesh Rangappa, and Srinivasan Natarajan

CONTENTS

4.1 INTRODUCTION

The depletion of fossil fuels due to their excessive use and environmental impact leads to searching for a sustainable source of energy [1]. The primitive sources like the breeze of air, constantly flowing river and motion of large crowds are always neglected as sources of mechanical energy [2]. These sources can replace the conventional sources of energies with the progress of more innovative technology. The miniaturization of the circuit can be associated with the movement of large crowds to generate a sustainable source of energy [3]. The battery technology indeed has the potential to match our needs, but it is to remember that they have some environmental impact. On the other hand, if we don't take any initiative for conversion of biomechanical energy, we are also wasting a large proportion of readily available energy sources daily. The technologies are available for the conversion of energies presented in Figure 4.1 [4].

The construction of nanogenerators (NGs) with the use of nanoscience made it possible to convert a reasonable amount of mechanical energies into electricity easily [5]. Here, in general, triboelectricity and piezoelectricity mechanisms have been used to convert the energies shown in Figure 4.2 [4]. The first report of such energy-harvesting technologies achieved by placing the nanowires of zinc oxides in between the electrodes [6]. The application of mechanical pressure converts a good amount of energy into electricity by using the piezoelectric mechanism. Way back in the year 2012, the efficiency of such energy production improved further by the combined effect of induction of electrostatic charges and triboelectricity [7]. The polymer-based materials were placed in these cases between the electrodes. The scope for further enhancement of the performances still open as the efficiencies of the NG largely depends on the materials used [8]. Here the surface designs such as nanopatterns, micropatterns etc. enhance the contact area between the materials and electrodes [9–13]. Hence, the output efficiency of the NG largely depends on both of these factors, i.e., material selection and the surface morphologies. Initially, metals were given preferences as the electrode

DOI: 10.1201/9781003187615-4

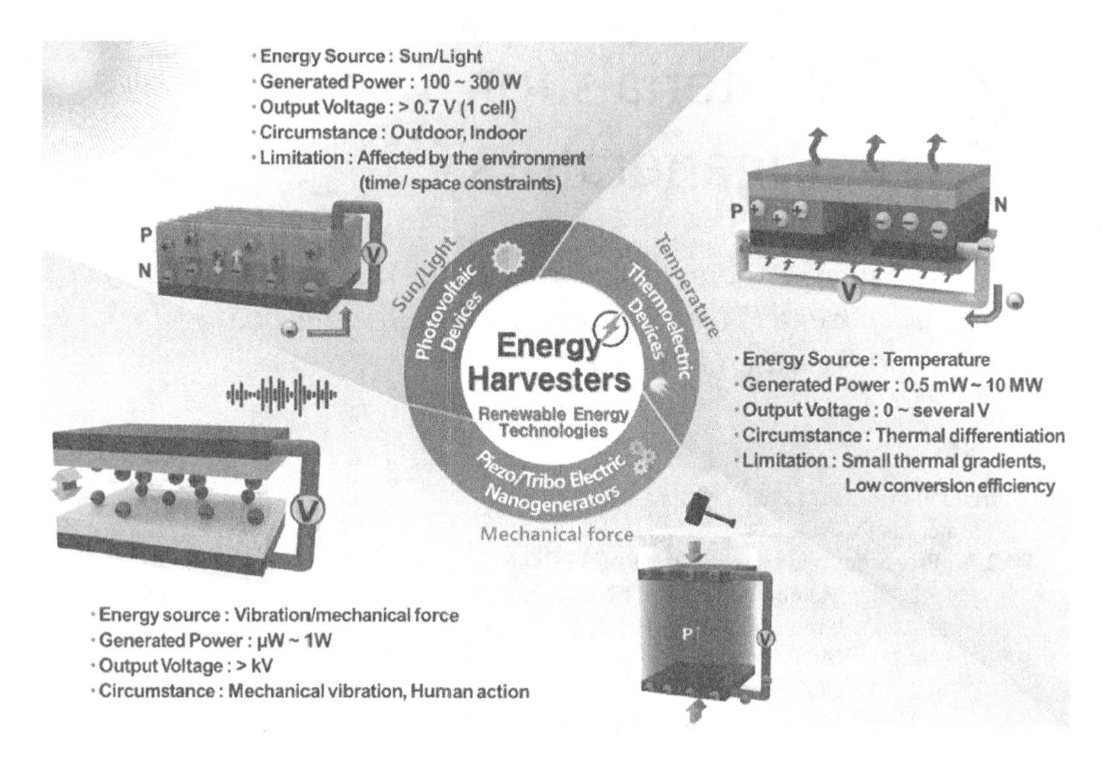

FIGURE 4.1 Classification of energy conversion in accordance with the source of energies [4].

materials and the polymers were the choice for the NG [14–17]. This is because metals had low resistivity and the polymers were highly flexible. But metals are always prone to corrosion, whereas inside the polymer matrix, charge carriers recombine very quickly [18]. Hence, these led to the search for new materials to replace them.

In this sense, the 2D materials offer high output performances for the NG shown in Figure 4.3 [4]. 2D materials such as MXene, graphene etc. offer superior properties such as flexibilities, conductivity, surface area etc. [19–21]. The 2D materials can be used as the NG materials, electrodes and both the combinations as well [22–23]. In this chapter, we highlighted the progress of application of 2D materials for the NG. We have presented different 2D materials for this purpose, such as MXene, graphene, MOF (metal-organic frameworks), BP (black phosphorus) and COF (covalent organic frameworks). The main emphasis has been given to the performances, mechanism, applications and present-day challenges when these materials have been applied in the NG design.

4.2 2D MATERIALS FOR NG

4.2.1 Properties and Classifications of the 2D NG

The available 2D materials for the NG applications can broadly be classified into MXene, graphene, BP, COF and MOF. These materials can be used as the NG materials, electrodes and both the combinations as well. Their unique structure and suitable properties make them ideal for the appropriate role in the energy-harvesting technologies for the NG.

4.2.1.1 MXene Materials for NG

The MXene is the newest members in the 2D layered structure materials [24]. It already demonstrated superior properties for energy storage applications. They are presently a family of odd 30 transition metal nitrides/carbides [25]. But the sheer interest for their broad application already

	Piezoelectric Nanogenerator	Triboelectric Nanogenerator
Device Structure	 Force V	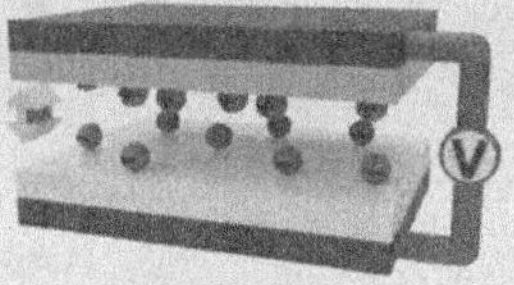 V
Energy Source	Bending/ Vibration	Pressure/ Vibration/ Rotation/ Sliding
Materials	Materials with piezoelectric properties ❖ Insulator: Quartz, PZT[a], BTO[b], PVDF[c], hexagonal boron nitride (h-BN) etc. ❖ Semiconductor: ZnO, GaN, AlN, Transition Metal Dichalcogenides (TMDS) etc.	❖ Positive Materials: nylon, human skin,etc. ❖ Negative Materials: PTFE[d], PFA[e], PDMS[f], etc.
Characteristics	❖ Nature friendly ❖ Efficiency: 25 ~50 % ❖ Output: 0.005 mW/cm^2 ~ 5 mW/cm^2 ❖ Applications: Body implantation devices, Wireless sensor node,Wearable devices	❖ Nature friendly ❖ Efficiency: 30 ~ 60 % ❖ Output: ~ $1W/cm^2$ ❖ Applications: Self-powered devices, Wearable devices

FIGURE 4.2 Result of comparison between triboelectric and piezoelectric NG [4].

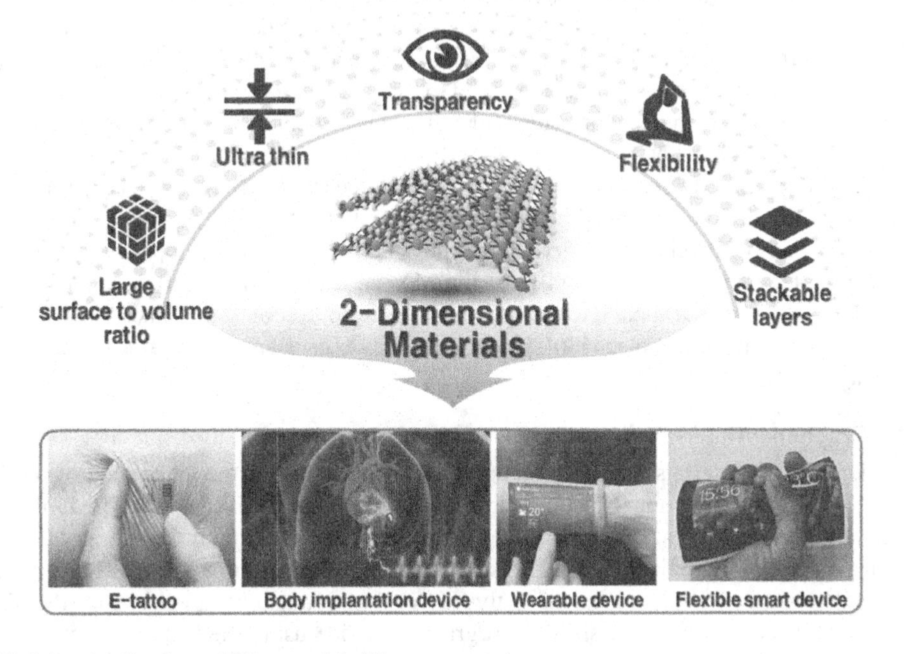

FIGURE 4.3 Applications of 2D materials [4].

FIGURE 4.4 SEM images of the $Ti_3C_2T_x$ 2D MXene material.

speculated a few million combinations of the MXene materials by using the first principle study [26]. They can be presented by the formula $M_{n+1}X_nT_x$ where transition metal ions M can be Sc, Ti, Mo, Cr, Hf, Zr, Ta, Nb etc., and nitrogen/carbon presents X and functional groups such as —F, —OH, —O etc. and offers T in the formula [27]. The initial step of the synthesis of the $Ti_3C_2T_x$ 2D material starts with the removal of the Al layer from the titanium aluminum carbide MXene family materials. This can be achieved by the exposure of the material in the acidic environment for several days. The synthesized layered $Ti_3C_2T_x$ 2D MXene material has been presented in Figure 4.4. Several washes and after that manual shaking for 5 minutes delaminated the 2D flakes. The effective method to spray coat the $Ti_3C_2T_x$ MXene-delaminated 2D flakes was to use the substrates such as PET-ITO and glass substrate. Here the piranha solution can be used to add functional group -OH into the substrate. This makes the substrate highly hydrophilic. The subsequent spraying of the $Ti_3C_2T_x$ MXene delaminated 2D flakes using an airbrush nozzle deposits thin film on the substrate. The thin film can be placed between the substrate for the fabrication of the NG. Four spacers, Kapton tape and Cu connecting wires can be used to transfer electricity from the NG. A motorized impactor can apply the mechanical pressure and a digital oscillator can record the generated signal from the $Ti_3C_2T_x$ 2D MXene NG.

When the $Ti_3C_2T_x$ 2D MXene NG is subject to a 15-N force at 2-Hz frequency, the recorded value of the V_{oc} open-circuit voltage ranged between −180 V and 500 V [28]. The reason for converting the mechanical energy into electrical energy is the presence of functional groups [29]. Here the long etching process for removing aluminum from the MXene leads to the creation of the functional groups. When the pressure is applied to the NG, the top electrode comes in contact with the MXene; hence, transfer of charge takes place [30]. Due to that, we observe an output value of the V_{oc} open-circuit voltage.

4.2.1.2 Graphene Materials for NG

Another 2D class of materials that has grabbed a lot of attention for the fabrication of the NG is graphene. Here also the functional groups help in the conversion of the energies. The fundamental difference between the MXene and graphene principle of transformation is that in the MXene, the -F functional group gas more importance. Because the removal of aluminum from MXene is generally done with the hydrofluoric acid leaching, this leaves the -F functional groups inside the MXene, whereas graphene is usually synthesized from graphene oxide using the Hammer method. These leave oxygen functional groups inside the graphene layers.

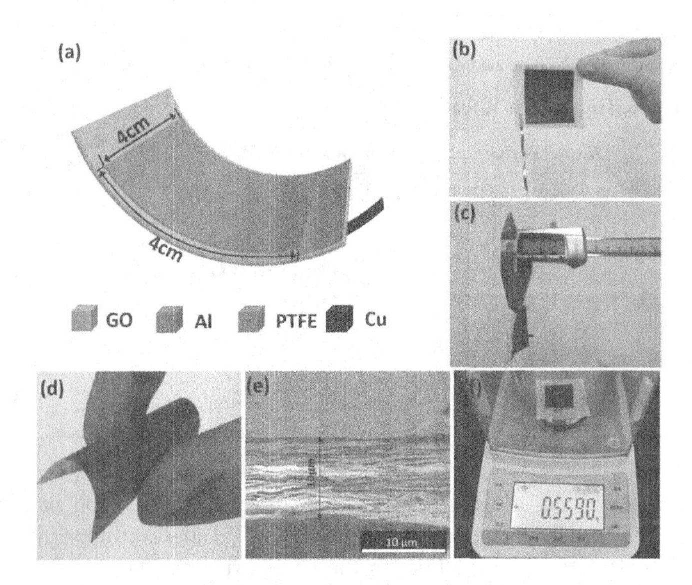

FIGURE 4.5 (a) Schematic, (b) photography, (c) thickness, (d–e) cross section and (f) weight of the graphene NG [31].

The fabrication of the single-layer graphene NG starts with selecting suitable flexible substrate such as PTFE (polytetrafluoroethylene). The main advantages of using the PTFE here are their lightweight, small thickness, impact ability and good working cycles. The grounding electrode for the NG used as aluminum foils with a Cu-connecting wire shown in Figure 4.5 [31]. Here the V_{oc} ranges between –1100 V and 500 V at 5-Hz frequencies when an impact of 37.5 kPa maintained in the graphene NG.

4.2.1.3 MOF Materials for NG

The output performances, flexibility, sustainability and functionality of the NG can further be improved using MOF as the material. Here, the organic ligands act as the linker for the metal nodes inside the structure of the MOF. That led to enhanced porosity and increased surface area for the NG applications. Different MOF such as KUAST-8, $Cu_3(BTC)_2$ and ZIF-8 can act as the MOF material for the NG. The substrate-like aluminum foil, acrylic and PET-ITO also can be used for these purposes for the fabrication of the NG. The thin film of the MOF can also be integrated with polymers like PDMS (polydimethylsiloxane) for the flexibility of the NG. The role of the different matrices such as PDMS, PTFE, silicon rubber, Polyurethane (PU), Polyvinylidene fluoride (PVDF), and Polyacrylonitrile (PAN) switch and without the addition of the MOF on the output performances is represented in Table 4.1 [40].

Although we can add several other nanoparticles with the polymer matrix, the combination of the PDMS matrix with the KUAST-8 MOF can enhance the power density of the NG by a factor of 11 presented in Table 4.2 [40]. The observed 52 microwatt/cm^2 power density with 11-fold jump might be due to the higher capacitance, surface roughness and increased level of functional groups inside the MOF structure. The output performance of such MOF NG can convert from mechanical to electrical energy. The functionalization and new design of the organic ligands inside the MOF structure have the future potential for applying the field of the NG.

4.2.1.4 COF Materials for NG

The effect of the toxicity of the heavy metal ions present in the MOF can further be reduced using the COF as an NG material. Here light elements like H, C, B and N are interconnected via powerful covalent bonds inside the COF. Moreover, the porous structure, low density and high surface area can help to enhance higher contact area for the charge transfer process during the energy conversion process.

TABLE 4.1
Power Densities of MOF NG [40]

Matrix	Output Performance Power Density (mW/cm²) Without F-MOF	With F-MOF (Mass Ratio)	Increasing Ratios (Power Density)
PDMS	4.7	52 (0.5 wt.%)	×11
Si-rubber	9.6	78.8 (0.5 wt.%)	×8.1
PTFE	5.8	20.8 (0.5 wt.%)	×3.6
PVDF	26.3	121 (0.2 wt.%)	×4.6
PAN	21	137 (0.2 wt.%)	×6.5
PU	21.4	140 (0.2 wt.%)	×6.5

The DBBA (dibromobenzene dicarbaldehyde), DP (diamino phenylphenanthridine) or DB (dimidium bromide) can be used as the linker for the COF. TPB (tris-aminophenyl benzene) or TFP (triformylphloroglucinol) can act as the knot in the COF structure. The synthesis process can be carried out using a solvothermal technique with acetic acid as the medium. The reaction process for three days at a temperature of 120°C typically gives around 80% of COF yield. The fabrication step for the COF NG includes the growth of the thin film on a substrate such as Cu tape. Here the Cu tape acts as an electrode. The second electrode consists of the layer of the Cu tape, Kapton tape and PVDF polymer layer. Here the PVDF polymer layer side faces the COF thin film. Subject to applying the periodic mechanical pressure, the electronegative PVDF layer induces positive ions on the topmost Cu layer through Kapton tape. At the same time, the electropositive COF layer causes a negative charge on the bottom layer of the Cu tape. This process builds up a voltage when the mechanical force applied to it and converts it into electrical energy.

The output performances of such COF NG are presented in Figure 4.6 [41]. The output I_{sc} and V_{oc} values determined at an applied frequency of mechanical pressure when different linkers are used in COF. The production V_{oc} and I_{sc}'s measured values were 365 V and 23.5 μA when 5-Hz frequency applied with DP as a linker. In contrast, the output performances change V_{oc} and I_{sc} with 815 V and 47.9 μA when 5-Hz frequency used with DB as the linker. Hence, there is a possibility of

TABLE 4.2
Output Performances from Layered Nanomaterials to MOF NG [40]

Fillers	Matrix	Increasing Ratios: Voltage	Current	Power Density	Output Performance: Volt (V)	Current Density (mA/m²)	Power Density (W/m²)	References
Fe_3O_4 nanoparticles	PVDF	×~1	×~1	×~1	138	5	–	[32]
Graphite particles	PDMS	–	×7	×2.6	286	45	3.7	[33]
GO nanosheets	PVDF	–	×1.47	×5.13	340	48.7	–	[34]
$SrTiO_3$ nanoparticles	PDMS	×1.8	×2.4	–	338	90.6	6.47	[35]
$BaTiO_3$ nanoparticles	PVDF	×3.5	–	–	1130	–	–	[36]
$ZnSnO_3$ nanoparticles	PDMS	×5.5	×5.2	–	330	50	–	[37]
ZnO nanoflakes	PDMS	×2.35	×2.7	–	470	600	282	[38]
Phosphorene nanosheets	Cellulose nanofibrils	×5	×9	×5.3	5.2	18	0.01	[39]
KUAST-8 particles	PVDF	×1.9	×2	×4.6	550	4.3	1.21	[40]
KUAST-8 particles	PDMS	×3.0	×2.9	×11	530	5.3	0.52	[40]

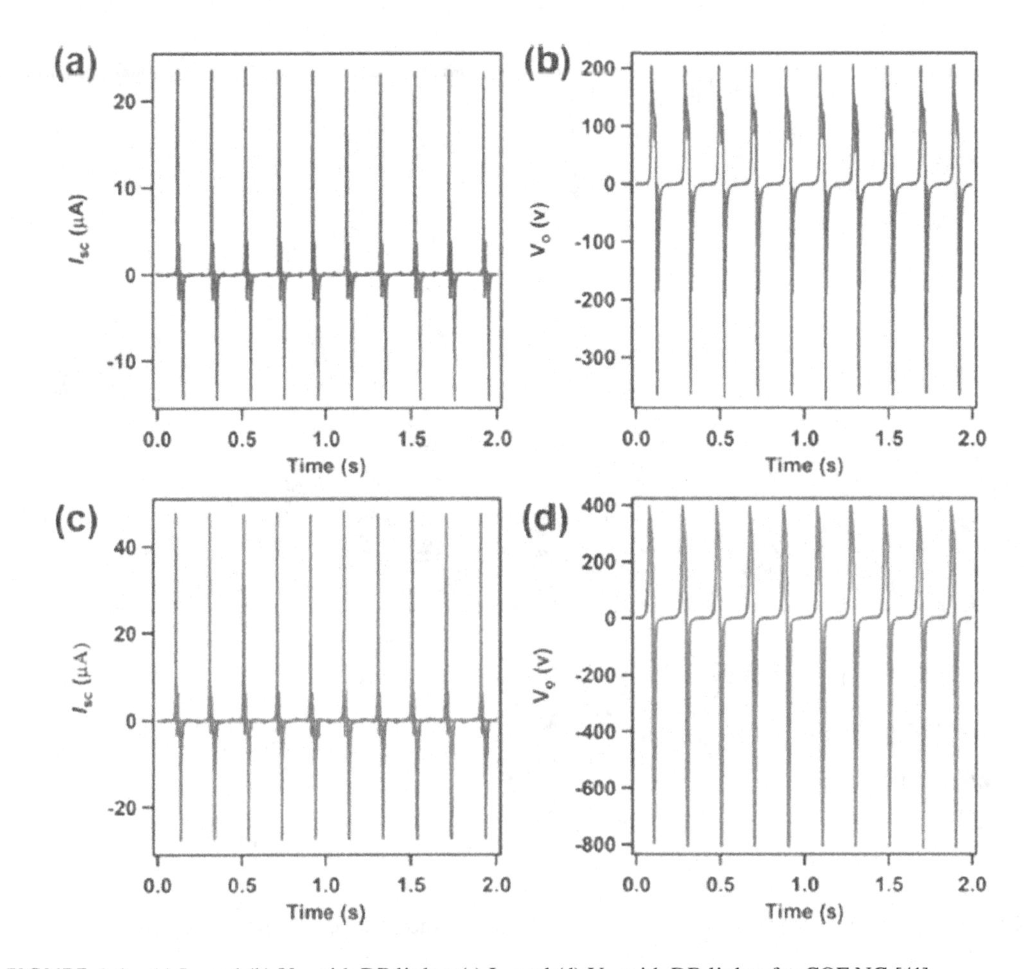

FIGURE 4.6 (a) I_{sc} and (b) V_{oc} with DP linker (c) I_{sc} and (d) V_{oc} with DB linker for COF NG [41].

nearly two times enhancement of the performances of the COF NG if we change the linker from DP to DB. The charge density of such COF NG recorded with a value of 73.5 $\mu C/m^2$ was even higher than that of the MOF-based NG [41]. This indicates the superior performance of the COF NG to any other 2D NG materials.

4.2.1.5 BP and Phosphorene Materials for NG

BP can be treated as a unique semiconducting class of 2D materials that possess some unique properties such as higher mobility of the carriers, tunable energy gap and anisotropic in-plane transport properties. On the other hand, monolayer of the BP known as phosphorene can show higher mobility of the carriers up to 286 $cm^2/(V\ s)$. But the main drawback of the application of the multilayer or monolayer of the BP in NG is that it is highly prone to oxidation even at ambient condition. Hence, a range of encapsulation techniques has been presented to eliminate the oxidation effect of the BP.

Encapsulation process of the BP layer presented in Figure 4.7 [42]. Here dip-coating method is utilized to fabricate BP layer on the PET (polyethylene terephthalate). The protection of the BP layer achieved with the outermost layer of the cellulose-derived hydrophobic nanoparticles to construct the HCOENPs/BP/PET (HBP) fabric. Furthermore, the NG can be fabricated with this HBP fabric with a waterproof layer of material and a Fabric electrode. These help to repel the water and increase the durability of the NG. Here the electropositive human skin layer interacts with the electronegative HBP fabric on the application of biomechanical energy. The electron-trapping

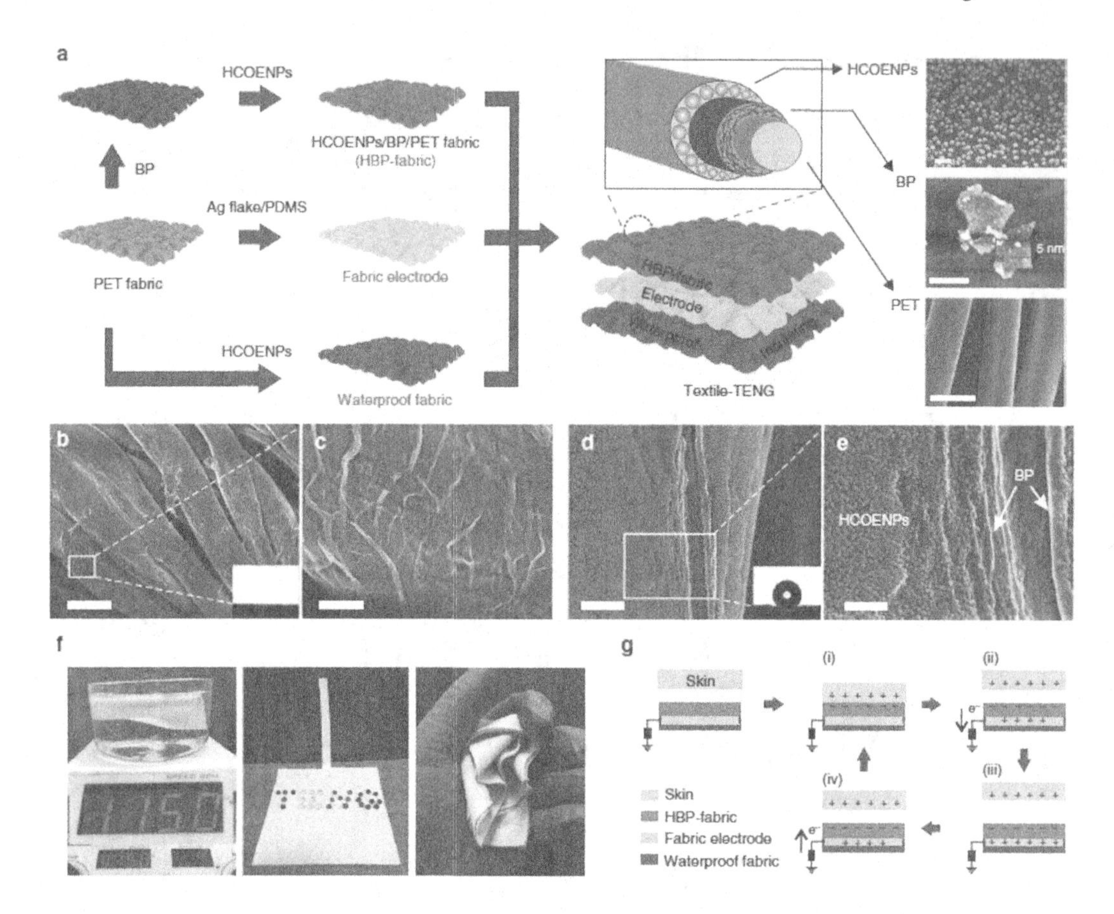

FIGURE 4.7 (a–g) Mechanism, working and configuration of the BP NG. Used under a Creative Commons Attribution 4.0 [42].

mechanism, in this case, converts into electrical energy. The observed V_{oc} and I_{sc} were 880 V and 1.1 μA/cm^2. The output power of the BP NG can be capable enough to handle an external load as well. Hence, the BP NG qualified to run wearable devices by converting the biomechanical energy to supply the proper electrical form.

4.3 CONCLUSIONS

Overuse of fossil fuels and their environmental impact always led to search for new renewable sources of energy. It is the long tendency to neglect biomechanical energy as the source of power. But miniaturization of the devices and search for flexible electronic devices led to the search beyond the battery power. Hence, the more innovative way to harvest energy primarily from biomechanical sources is to use NG. The two possible ways of energy generation through NG are piezoelectricity and triboelectricity. The higher power density, flexibility properties and requirement of less mechanical energy help to popularize the triboelectricity mechanism. The selection of appropriate material can enhance the output power by several factors. Hence, a range of 2D materials show promising efficiency as NG.

MXene is one of the best 2D class materials that show higher output efficiency in NG applications. The Al layer can be removed from the titanium aluminum carbide by etching process to generate functional groups inside the $Ti_3C_2T_x$ MXene material. Such a layer of MXene placed between the two electrodes for the formation of NG shows promising output performances. Subject to a 15-N force at 2-Hz frequency, the recorded value of the V_{oc} open-circuit voltage ranges between −180 V and 500 V. For graphene as the NG material, the V_{oc} ranges between −1100 V and 500 V at 5-Hz frequencies when an impact of 37.5 kPa maintained. Higher capacitance, surface roughness and increased functional groups inside the MOF structure help to show 52 μW/cm^2 power density value. The observed power output of the COF and BP are 815 V, 47.9 μA and 880 V, 1.1 μA, respectively. This indicates the superior performance of the 2D materials for the NG applications.

ACKNOWLEDGMENTS

One of us, P. Banerjee, received SERB, India, grant no, TAR/2021/000032.

REFERENCES

1. Rostami, Raheleh, Seyed M. Khoshnava, Hasanuddin Lamit, Dalia Streimikiene, and Abbas Mardani. “An overview of Afghanistan’s trends toward renewable and sustainable energies.” *Renewable and Sustainable Energy Reviews* 76 (2017): 1440–1464.
2. Wilkie, D. R. “Man as a source of mechanical power.” *Ergonomics* 3, no. 1 (1960): 1–8.
3. Chen, Shu W., Xia Cao, Ning Wang, Long Ma, Hui R. Zhu, Magnus Willander, Yang Jie, and Zhong L. Wang. “An ultrathin flexible single-electrode triboelectric-nanogenerator for mechanical energy harvesting and instantaneous force sensing.” *Advanced Energy Materials* 7, no. 1 (2017): 1601255.
4. Han, Sang A., Jaewoo Lee, Jianjian Lin, Sang-Woo Kim, and Jung H. Kim. “Piezo/triboelectric nanogenerators based on 2-dimensional layered structure materials.” *Nano Energy* 57 (2019): 680–691.
5. Niu, Simiao, and Zhong L. Wang. “Theoretical systems of triboelectric nanogenerators.” *Nano Energy* 14 (2015): 161–192.
6. Wang, Zhong L., and Jinhui Song. “Piezoelectric nanogenerators based on zinc oxide nanowire arrays.” *Science* 312, no. 5771 (2006): 242–246.
7. Fan, Feng-Ru, Zhong-Qun Tian, and Zhong L. Wang. “Flexible triboelectric generator.” *Nano Energy* 1, no. 2 (2012): 328–334.
8. Askari, Hassan, Amir Khajepour, Mir B. Khamesee, Zia Saadatnia, and Zhong L. Wang. “Piezoelectric and triboelectric nanogenerators: Trends and impacts.” *Nano Today* 22 (2018): 10–13.
9. Seung, Wanchul, Manoj K. Gupta, Keun Y. Lee, Kyung-Sik Shin, Ju-Hyuck Lee, Tae Y. Kim, Sanghyun Kim, Jianjian Lin, Jung H. Kim, and Sang-Woo Kim. “Nanopatterned textile-based wearable triboelectric nanogenerator.” *ACS Nano* 9, no. 4 (2015): 3501–3509.
10. Kim, Daewon, Seung-Bae Jeon, Ju Y. Kim, Myeong-Lok Seol, Sang O. Kim, and Yang-Kyu Choi. “High-performance nanopattern triboelectric generator by block copolymer lithography.” *Nano Energy* 12 (2015): 331–338.
11. Wang, Hee S., Chang K. Jeong, Min-Ho Seo, Daniel J. Joe, Jae H. Han, Jun-Bo Yoon, and Keon J. Lee. “Performance-enhanced triboelectric nanogenerator enabled by wafer-scale nanogrates of multistep pattern downscaling.” *Nano Energy* 35 (2017): 415–423.
12. Mule, Anki R., Bhaskar Dudem, and Jae S. Yu. “High-performance and cost-effective triboelectric nanogenerators by sandpaper-assisted micropatterned polytetrafluoroethylene.” *Energy* 165 (2018): 677–684.
13. Lee, Ju-Hyuck, Hong-Joon Yoon, Tae Y. Kim, Manoj K. Gupta, Jeong H. Lee, Wanchul Seung, Hanjun Ryu, and Sang-Woo Kim. “Micropatterned P (VDF-TrFE) film-based piezoelectric nanogenerators for highly sensitive self-powered pressure sensors.” *Advanced Functional Materials* 25, no. 21 (2015): 3203–3209.
14. Wang, Jie, Zhen Wen, Yunlong Zi, Pengfei Zhou, Jun Lin, Hengyu Guo, Youlong Xu, and Zhong L. Wang. “All-plastic-materials based self-charging power system composed of triboelectric nanogenerators and supercapacitors.” *Advanced Functional Materials* 26, no. 7 (2016): 1070–1076.
15. Wang, Zhong L., Long Lin, Jun Chen, Simiao Niu, and Yunlong Zi. *Triboelectric nanogenerators.* Berlin, Germany: Springer International Publishing, 2016.
16. Chang, Chieh, Van H. Tran, Junbo Wang, Yiin-Kuen Fuh, and Liwei Lin. “Direct-write piezoelectric polymeric nanogenerator with high energy conversion efficiency.” *Nano Letters* 10, no. 2 (2010): 726–731.
17. Wang, Zhong L., Guang Zhu, Ya Yang, Sihong Wang, and Caofeng Pan. “Progress in nanogenerators for portable electronics.” *Materials Today* 15, no. 12 (2012): 532–543.

18. Wen, Rongmei, Junmeng Guo, Aifang Yu, Junyi Zhai, and Zhong L. Wang. "Humidity-resistive triboelectric nanogenerator fabricated using metal organic framework composite." *Advanced Functional Materials* 29, no. 20 (2019): 1807655.
19. Fan, Feng R., and Wenzhuo Wu. "Emerging devices based on two-dimensional monolayer materials for energy harvesting." *Research* 2019 (2019): 7367828.
20. Lu, Yanfu, and Susan B. Sinnott. "Density functional theory study of epitaxially strained monolayer transition metal chalcogenides for piezoelectricity generation." *ACS Applied Nano Materials* 3, no. 1 (2019): 384–390.
21. Kocabaş, Tuğbey, Deniz Çakır, and Cem Sevik. "First-principles discovery of stable two-dimensional materials with high-level piezoelectric response." *Journal of Physics: Condensed Matter* 33, no. 11 (2021): 115705.
22. Kim, Kyeong N., Jinsung Chun, Jin W. Kim, Keun Y. Lee, Jang-Ung Park, Sang-Woo Kim, Zhong L. Wang, and Jeong M. Baik. "Highly stretchable 2D fabrics for wearable triboelectric nanogenerator under harsh environments." *ACS Nano* 9, no. 6 (2015): 6394–6400.
23. Dong, Yongchang, Sai Sunil K. Mallineni, Kathleen Maleski, Herbert Behlow, Vadym N. Mochalin, Apparao M. Rao, Yury Gogotsi, and Ramakrishna Podila. "Metallic MXenes: A new family of materials for flexible triboelectric nanogenerators." *Nano Energy* 44 (2018): 103–110.
24. Cheng, Lei, Xin Li, Huaiwu Zhang, and Quanjun Xiang. "Two-dimensional transition metal MXene-based photocatalysts for solar fuel generation." *The Journal of Physical Chemistry Letters* 10, no. 12 (2019): 3488–3494.
25. Anasori, Babak, and Yury Gogotsi. "Introduction to 2D transition metal carbides and nitrides (MXenes)." In *2D Metal carbides and nitrides (MXenes)*, pp. 3–12. Cham: Springer, 2019.
26. Tan, Teck L., Hong M. Jin, Michael B. Sullivan, Babak Anasori, and Yury Gogotsi. "High-throughput survey of ordering configurations in MXene alloys across compositions and temperatures." *ACS Nano* 11, no. 5 (2017): 4407–4418.
27. Jiang, Yanan, Tao Sun, Xi Xie, Wei Jiang, Jia Li, Bingbing Tian, and Chenliang Su. "Oxygen-functionalized ultrathin $Ti_3C_2T_x$ MXene for enhanced electrocatalytic hydrogen evolution." *ChemSusChem* 12, no. 7 (2019): 1368–1373.
28. Dong, Yongchang, Sai Sunil K. Mallineni, Kathleen Maleski, Herbert Behlow, Vadym N. Mochalin, Apparao M. Rao, Yury Gogotsi, and Ramakrishna Podila. "Metallic MXenes: A new family of materials for flexible triboelectric nanogenerators." *Nano Energy* 44 (2018): 103–110.
29. Hope, Michael A., Alexander C. Forse, Kent J. Griffith, Maria R. Lukatskaya, Michael Ghidiu, Yury Gogotsi, and Clare P. Grey. "NMR reveals the surface functionalisation of Ti_3C_2 MXene." *Physical Chemistry Chemical Physics* 18, no. 7 (2016): 5099–5102.
30. Wang, Hsiu-Wen, Michael Naguib, Katharine Page, David J. Wesolowski, and Yury Gogotsi. "Resolving the structure of $Ti_3C_2T_x$ MXenes through multilevel structural modeling of the atomic pair distribution function." *Chemistry of Materials* 28, no. 1 (2016): 349–359.
31. Guo, Huijuan, Tao Li, Xiaotao Cao, Jin Xiong, Yang Jie, Magnus Willander, Xia Cao, Ning Wang, and Zhong L. Wang. "Self-sterilized flexible single-electrode triboelectric nanogenerator for energy harvesting and dynamic force sensing." *ACS Nano* 11, no. 1 (2017): 856–864.
32. Im, Ji-Su, and Il-Kyu Park. "Mechanically robust magnetic Fe_3O_4 nanoparticle/polyvinylidene fluoride composite nanofiber and its application in a triboelectric nanogenerator." *ACS Applied Materials & Interfaces* 10, no. 30 (2018): 25660–25665.
33. He, Xianming, Hengyu Guo, Xule Yue, Jun Gao, Yi Xi, and Chenguo Hu. "Improving energy conversion efficiency for triboelectric nanogenerator with capacitor structure by maximizing surface charge density." *Nanoscale* 7, no. 5 (2015): 1896–1903.
34. Huang, Tao, Mingxia Lu, Hao Yu, Qinghong Zhang, Hongzhi Wang, and Meifang Zhu. "Enhanced power output of a triboelectric nanogenerator composed of electrospun nanofiber mats doped with graphene oxide." *Scientific Reports* 5, no. 1 (2015): 1–8.
35. Chen, Jie, Hengyu Guo, Xianming He, Guanlin Liu, Yi Xi, Haofei Shi, and Chenguo Hu. "Enhancing performance of triboelectric nanogenerator by filling high dielectric nanoparticles into sponge PDMS film." *ACS Applied Materials & Interfaces* 8, no. 1 (2016): 736–744.
36. Seung, Wanchul, Hong-Joon Yoon, Tae Y. Kim, Hanjun Ryu, Jihye Kim, Ju-Hyuck Lee, Jeong H. Lee, et al. "Boosting power-generating performance of triboelectric nanogenerators via artificial control of ferroelectric polarization and dielectric properties." *Advanced Energy Materials* 7, no. 2 (2017): 1600988.

37. Paria, Sarbaranjan, Suman K. Si, Sumanta K. Karan, Amit K. Das, Anirban Maitra, Ranadip Bera, Lopamudra Halder, Aswini Bera, Anurima De, and Bhanu B. Khatua. "A strategy to develop highly efficient TENGs through the dielectric constant, internal resistance optimization, and surface modification." *Journal of Materials Chemistry A* 7, no. 8 (2019): 3979–3991.
38. He, Wen, Yongteng Qian, Byeok S. Lee, Fangfang Zhang, Aamir Rasheed, Jae-Eun Jung, and Dae J. Kang. "Ultrahigh output piezoelectric and triboelectric hybrid nanogenerators based on ZnO nanoflakes/polydimethylsiloxane composite films." *ACS Applied Materials & Interfaces* 10, no. 51 (2018): 44415–44420.
39. Cui, Peng, Kaushik Parida, Meng-Fang Lin, Jiaqing Xiong, Guofa Cai, and Pooi S. Lee. "Transparent, flexible cellulose nanofibril–phosphorene hybrid paper as triboelectric nanogenerator." *Advanced Materials Interfaces* 4, no. 22 (2017): 1700651.
40. Guo, Yinben, Yule Cao, Zixi Chen, Ran Li, Wei Gong, Weifeng Yang, Qinghong Zhang, and Hongzhi Wang. "Fluorinated metal-organic framework as bifunctional filler toward highly improving output performance of triboelectric nanogenerators." *Nano Energy* 70 (2020): 104517.
41. Zhai, Lipeng, Wutao Wei, Baiwei Ma, Wanyu Ye, Jing Wang, Weihua Chen, Xiubei Yang, et al. "Cationic covalent organic frameworks for fabricating an efficient triboelectric nanogenerator." *ACS Materials Letters* 2, no. 12 (2020): 1691–1697.
42. Xiong, Jiaqing, Peng Cui, Xiaoliang Chen, Jiangxin Wang, Kaushik Parida, Meng-Fang Lin, and Pooi S. Lee. "Skin-touch-actuated textile-based triboelectric nanogenerator with black phosphorus for durable biomechanical energy harvesting." *Nature Communications* 9, no. 1 (2018): 1–9.

5 Carbon Materials for Nanogenerators

Mehmet Bugdayci and Levent Oncel

CONTENTS

5.1 INTRODUCTION

Interest in smart electronics such as smart wearable electronic devices, implantable devices, wireless sensor networks, and the like has increased lately. The studies on small-scale renewable and sustainable energy resources continue. Currently, batteries are used as the primary alternative to meet this energy need. However, batteries turn into waste at the end of their life, which creates environmental problems (Wang 2014; Zi et al. 2015).

Nanogenerators are used as an alternative to batteries in small-scale and sustainable energy production. Nanogenerators generate sustainable energy based on electromagnetic, pyroelectric, piezoelectric and triboelectric mechanisms by harvesting energy in the working environment. Triboelectric nanogenerators (TENGs) provide energy conversion with high efficiency. They are seen to have a significant potential among nanogenerators with their high power generation and environmentally friendly structure (Aghaie-Khafri and Daemi 2012; Andrei, Bethke, and Rademann 2016; Briscoe and Dunn 2015; Fan, Tian, and Wang 2012; He et al. 2018; Kim et al. 2017; Kuo 2005; Li et al. 2017; Roy et al. 2000; Wang 2014; Wang and Song 2006; Wang et al. 2018; Xu et al. 2018; Xue et al. 2017; Zhao et al. 2018; Zi et al. 2015).

Piezoelectric materials were discovered nearly a century ago and have played an important role in the development of nanogenerators. However, nanostructured piezoelectric materials have started to be developed in recent years. In this process, the effects of nanoscale size on piezoelectric properties were investigated, and new formulations, applications and ferroelectric materials were developed (Briscoe and Dunn 2015; Roy et al. 2000).

When a certain crystalline material generates electricity because of the electrical polarization occurs as a result of temperature change, it is called pyroelectric material. This effect occurs as a result of heating or cooling the crystal structure. Consequently, polarization takes place in the crystal structure. Pyroelectricity is like thermoelectric power conversion, so an inverse electrocaloric

DOI: 10.1201/9781003187615-5

state occurs according to the Curie principle. The pyroelectric effect mechanism is expressed as the minimum energy and structure of a polar crystal being equal under constant temperature and pressure. In the process, the polar crystal reacts electrically with the equilibrium change that occurs as a result of the temperature change (Wu, Zhu, and Zhang 2015; Zhuo et al. 2017, 2018).

Another method for energy generation using nanomaterials is to generate energy by using the mobility that occurs on the surface where liquid and solid contact. It has been observed that electricity generation is possible when water interacts with nanostructured surfaces (Bahaj 2011; Chen, Xu, and Liu 2014; Fan, Tang, and Wang 2016; Hu et al. 2015; Prudell et al. 2010; Striolo, Michaelides, and Joly 2016; Wang 2017).

Carbon black, carbon nanotube, fullerene and graphene are carbon-based materials used in nanogenerator applications. These materials are used effectively in both energy production and energy transfer. The properties of carbon-based materials used in nanogenerator production are presented in the second section.

5.2 CARBON MATERIALS

Carbon is the basis of all organic compounds. Carbon is included in group 4A in the periodic table, its melting temperature is 3500°C, and its atomic diameter is 0.077 nm. There are around 10,000 compounds formed as a result of the various structural arrangements of carbon. Some of these compounds are shown in Figure 5.1.

Carbon's atomic number is 12 and it has 6 electrons. Carbon has many electron configurations. Accordingly, it has the ability to form compounds in various bond structures. Possible electron configurations of carbon are given in Figure 5.2.

5.2.1 Carbon Black

Carbon black is a black powder that is including elemental carbon. The properties of carbon black can be adjusted by controlling the burning process of petroleum and its derivatives. The usage area of carbon black ranges from ink to high-tech devices such as nanogenerators (Alcántara et al. 2001; Bernal et al. 2011; Borah et al. 2008; Brunauer, Emmett, and Teller 1938; Chen, Xu, and Liu 2014).

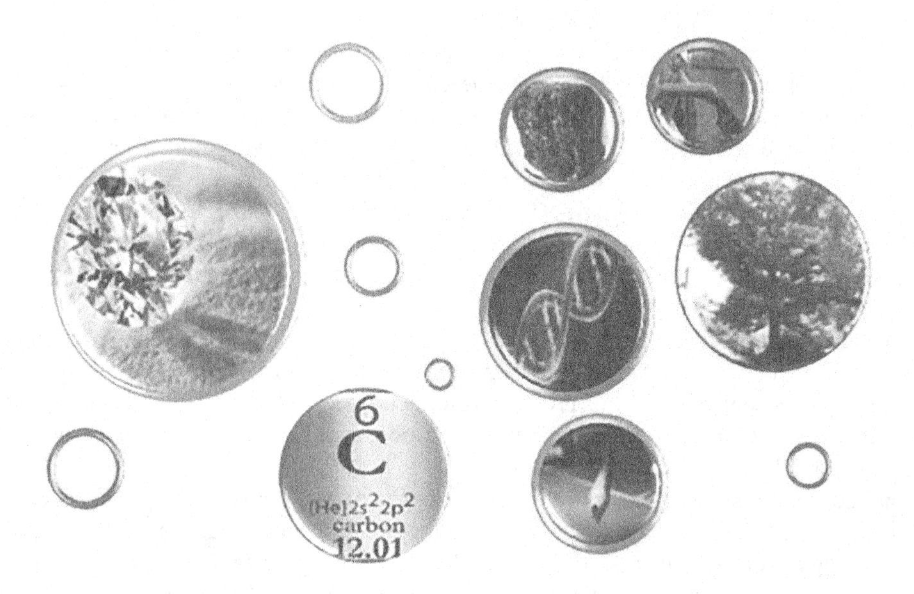

FIGURE 5.1 Examples of carbon-based materials.

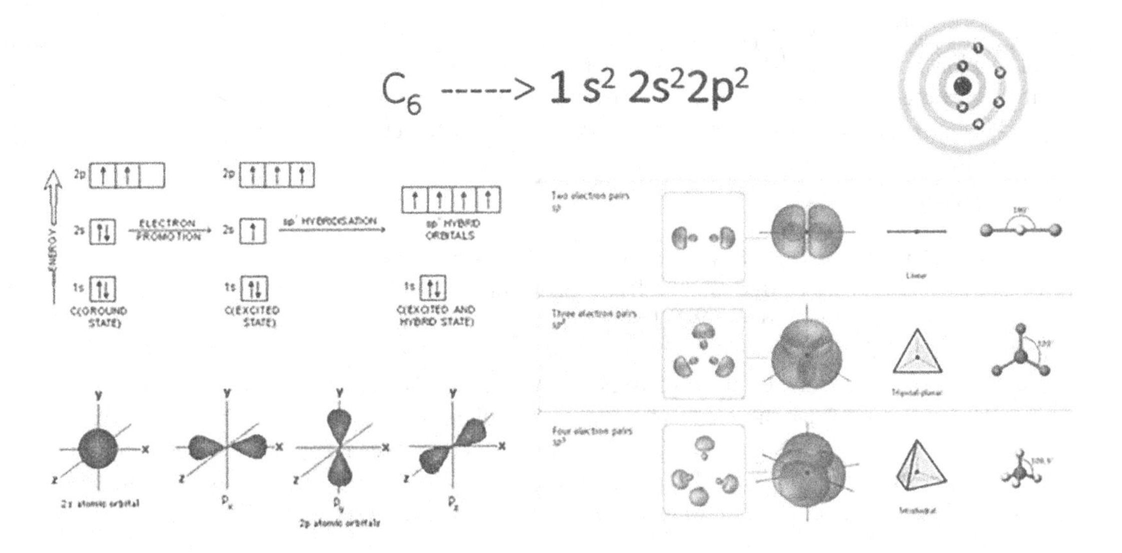

FIGURE 5.2 Various electron configurations of carbon compounds.

The soot obtained as a result of gassing in the casting process was used for writing in ancient Egypt and China. This material exhibits similar properties to carbon black. Carbon black consists of hydroxyl and carboxyl groups. Grain size and surface tension are important properties of carbon black, and these properties affect usage area, blackness and dispersibility of this material (Dong et al. 2017; Dons et al. 2012; Fan and Zhang 2008).

Carbon black is produced by thermal decomposition method or partial combustion method using hydrocarbons such as petroleum or natural gas as raw materials. Another parameter that determines the properties of carbon black other than microstructure is the production method. Production methods of the carbon black are Furnace Black Process, Channel Process, Acetylene Black Process and Lampblack Process. The most common of these processes is the Furnace Black Process, and the quality of the carbon black obtained in this process is higher than the carbon black obtained in other processes (Moulin et al. 2017; Samaržija-Jovanović et al. 2009; Tang et al. 2015; Wang et al. 2011).

5.2.1.1 Furnace Black Process

In this process, carbon black is obtained by the desorption of gases formed at high temperatures as a result of the partial combustion of coal oil. The desired particle size can be achieved by temperature control. The process is efficient and sustainable. Due to its advantages, this process is the most used method in carbon black synthesis (Rumpf, Taylor, and Toombs 2010).

5.2.1.2 Channel Process

In this process, the formation of carbon black occurs as a result of the contact of the product obtained by the combustion of natural gas with the channel steel (H-shaped steel). The biggest problems of this method are low efficiency and the environmental problems it creates (Nester et al. 2011).

5.2.1.3 Acetylene Black Process

In this process, carbon is synthesized by the thermal decomposition of acetylene gas. Obtained carbon has a high degree of crystallization, thus synthesized carbon by this method is commonly used in electrical applications (Nester et al. 2011).

5.2.1.4 Lampblack Process

The Lampblack Process is the method in which carbon black is obtained from the soot formed as a result of burning pine trees with oils. This method was used to obtain ink in ancient times. Its use is limited because it is not convenient for mass production (Nester et al. 2011).

5.2.2 Carbon Nanotube

Carbon nanotubes are produced by bending planar sheets of carbon atoms into cylindrical shapes. In 1991, Sumio Iijima discovered multiwall carbon nanotubes (MWNTs), and in 1993, the discovery of single-wall carbon nanotubes (SWNTs) was reported (Jin, Suenaga, and Iijima 2008; Qin et al. 2000; Yudasaka et al. 1997).

Carbon nanotubes are divided in two classes as SWNTs and MWNTs. SWNTs are obtained by rolling a single layer of graphene sheet to form cylinders with diameters ranging from 0.7 to 2 nm. NWNTs are produced by nesting SWNTs inside each other (the distance between these cylinders is 0.34 nm) to form a tube structure with a diameter between 1.4 and 150 nm (Futaba et al. 2006; Yamada et al. 2006). The types of carbon nanotubes according to the bending vector are presented in Figure 5.3.

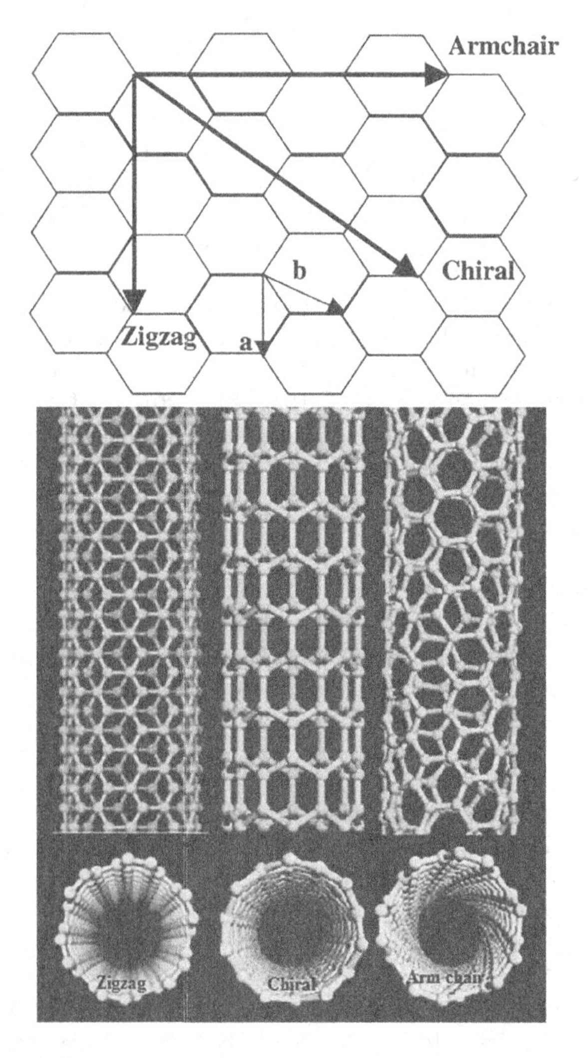

FIGURE 5.3 Carbon nanotube types.

Morphology has an important effect on the properties of CNTs. For example, the electrical conductivity of CNTs depends on their helix structure (Futaba et al. 2005; Jin, Suenaga, and Iijima 2008).

While the current capacity of copper cable is 1 MAmps/cm^2, this value is 1 GAmps/cm^2, for carbon nanotube. Accordingly, carbon nanotube's current capacity is 1000 times higher than copper. Its thermal conductivity is 3320 W/(m K), and this value is same for pure diamond. The tensile strength of carbon nanotube can be up to 63 GPA, and tensile strength of common high-strength steels is around 2 GPA. The density of carbon nanotube (1.33 g/cm^3) is much lower than aluminum, which has a density of 2.7 g/mm^3 and is in the light materials group. Carbon nanotube has become a suitable material for use in many applications, as they are stronger than steel and lighter than aluminum (Firstov, Van Humbeeck, and Koval 2004; Futaba et al. 2006; Izard et al. 2008; Yamada et al. 2008).

5.2.2.1 Carbon Nanotube Production Methods

Carbon nanotube production methods are Arc Discharge Method, Laser Ablation Synthesis, Thermal Synthesis Methods (Chemical Vapor Deposition, High-Pressure Carbon Monoxide Synthesis, Flame Synthesis) and Plasma-Enhanced Chemical Vapor Deposition. Details of these methods are shared in the following section.

5.2.2.1.1 Arc Discharge Method

In this method, carbon nanotubes are produced from carbon gas. This vapor is produced by an arc discharge between two graphite electrodes. This operation is carried out in a noble gas atmosphere. Direct current in the range of 50–100 A and potential difference in the range of 20–30 V are used in the operation. In the arc discharge method, the graphite electrodes are brought close to each other until a plasma is formed, and in general, this distance where the plasma is formed is less than 1 mm.

In the arc discharge method, a potential difference of 20–30 V is applied between two graphite rods in an environment containing noble gas (such as He) under a constant pressure of 0.25–0.65 atm. Graphite electrodes are generally moved close to each other until a plasma is formed at a distance of ~ 1 mm or less. In the plasma environment, C atoms pass from anode graphite to the gas phase and grow on the cathode as MWNT. Also, a graphite anode containing a metal catalyst is used in the synthesis of SWNTs. Inert gas enables CNTs to form on the cathode (Ando and Zhao 2006; Keidar and Waas 2004). Figure 5.4 shows this method's schematic representation.

5.2.2.1.2 Laser Ablation Synthesis

The laser ablation method was initially developed by Kroto et al. for the synthesis of fullerene. The main principle of this method is to vaporize material from a solid surface. In this method, the pulsed laser beam is sent on the metal catalyst-graphite composite material in a high-temperature reactor and evaporation occurs. The laser beam scans the target surface to achieve homogeneous evaporation and reduce carbon deposits in the form of soot. The steam generated is carried from the high-temperature zone to the water-cooled copper collector by argon gas and carbon nanotubes accumulate on the collector surface (Chrzanowska et al. 2015; Morales and Lieber 1998; Semaltianos 2010). Figure 5.5 illustrates the schematic view of the laser ablation synthesis.

5.2.2.1.3 Chemical Vapor Deposition Method

In the chemical vapor deposition (CVD) method, solid phase is deposited on the substrate surface by pyrolysis of gas phase precursors. CVD is also a widely used technique for thin film growth on various substrates. With this method, mass production of carbon film, carbon fiber, carbon-carbon composite and MWNTs can be carried out. The production of SWNTs with the CVD technique has been possible recently. Being able to control the C structure and morphology is the advantage of CVD over other methods (Cai et al. 2018).

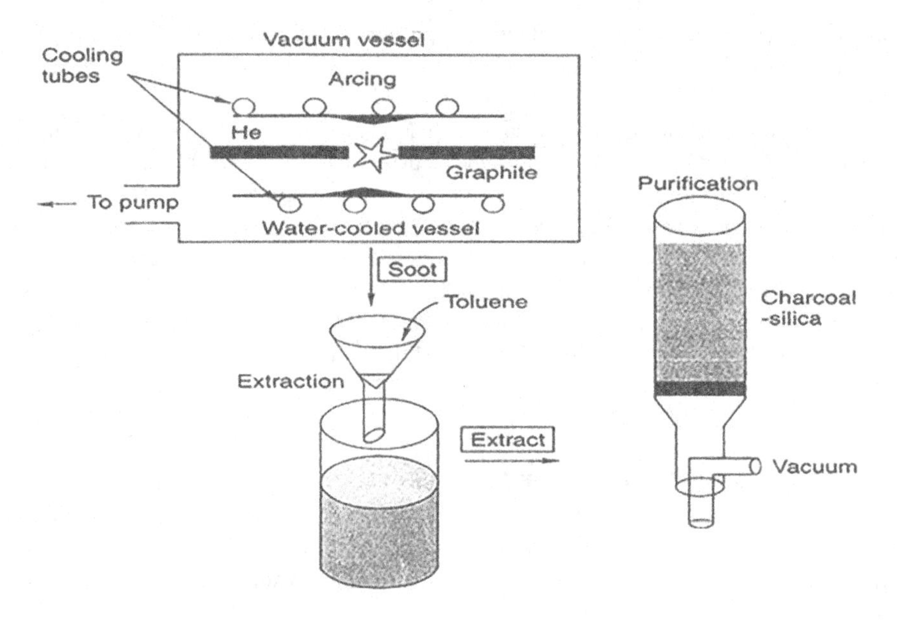

FIGURE 5.4 Schematic view of arc discharge method.

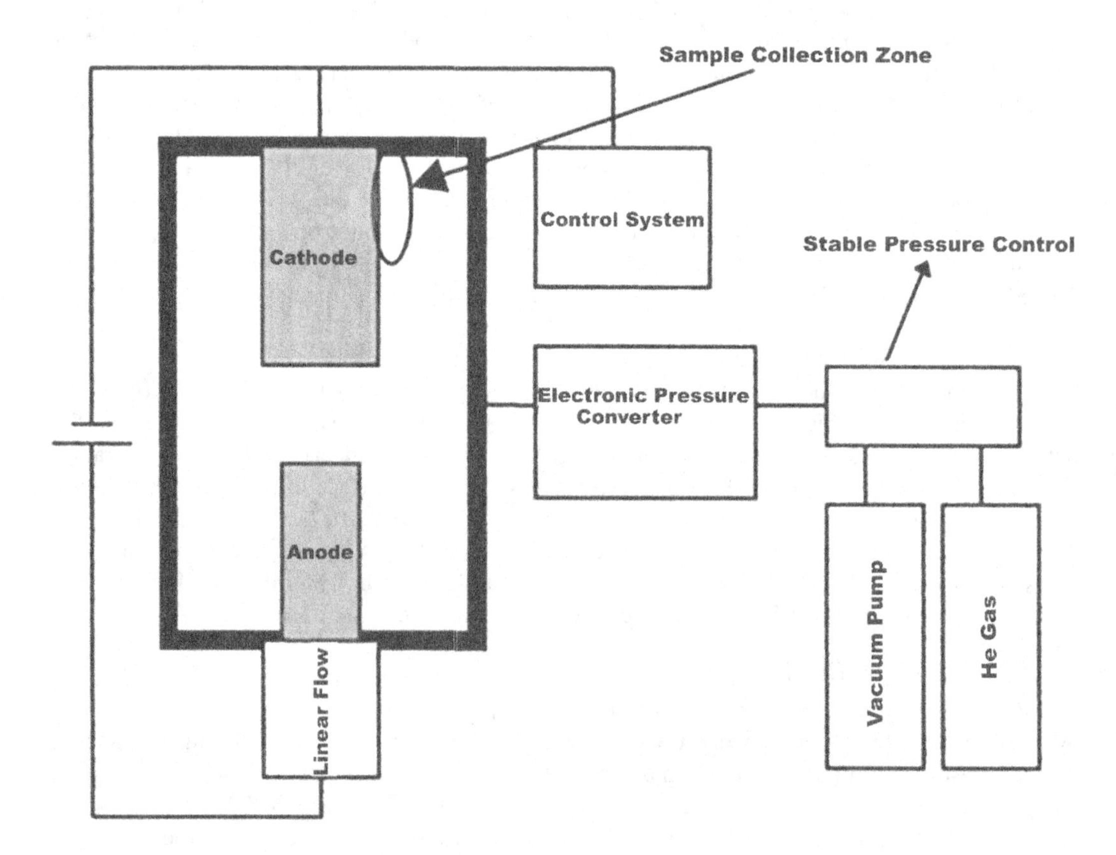

FIGURE 5.5 Schematic view of laser ablation synthesis.

In this method, solid carbon is deposited on the substrate. This deposition is a result of homogeneous pyrolysis reactions together with surface reactions (heterogeneous deposition reactions). Surface reactions are the reaction of carbon-containing gases with metal catalyst particles on the substrate surface at high temperatures. Metal particles (transition elements such as Fe, Co and Ni) are crucial for the growth of carbon nanofilaments. Diluted hydrocarbons (CH_4, C_2H_2, C_2H_4) and CO gas are generally used as carbon sources. In the CVD method, carbon nanotubes are generally synthesized in the temperature range of 500 and 1000°C. SWNTs are produced between 627 and 1027°C with mixtures containing methane (CH_4) gas (Quartiermeister et al. 2020; Mwafy 2020; Reina et al. 2009; Wang et al. 2020).

5.2.3 Fullerene

In 1984, R. E. Smalley and his colleagues noticed that after the graphite crystal is melted and evaporated with a laser beam, released carbon atoms are mixed with a stream of helium gas and form clusters of different sizes. These clusters contain about 20–130 carbon atoms (Liu et al. 1998). H.W. Kroto, R.E. Smalley and R. F. Curl managed to isolate the carbon sphere formed and took the first step toward gaining full knowledge of the structure of carbon nanospheres, and with this study, they received the Nobel Award in 1996 (Chai et al. 1991; Smalley 1997).

The distance between the graphite electrodes placed in the vacuum chamber is adjusted, so that they form an arc with each other. In the process that takes place under helium or argon atmosphere, soot formation occurs. The mixture containing fullerene, which is separated from the soot with the help of soxhlet extraction (toluene or benzene), is subjected to chromatography and the production process is completed. The schematic view of the fullerene production process is shown in Figure 5.6.

Fullerenes are soluble in various solvents. The most common ones are toluene and carbon disulfide. When dissolved in solvents, pure C_{60} fullerene gives purple color and C70 gives red-brown color. Fullerenes are the only allotropes of carbon that are soluble in various solvents at

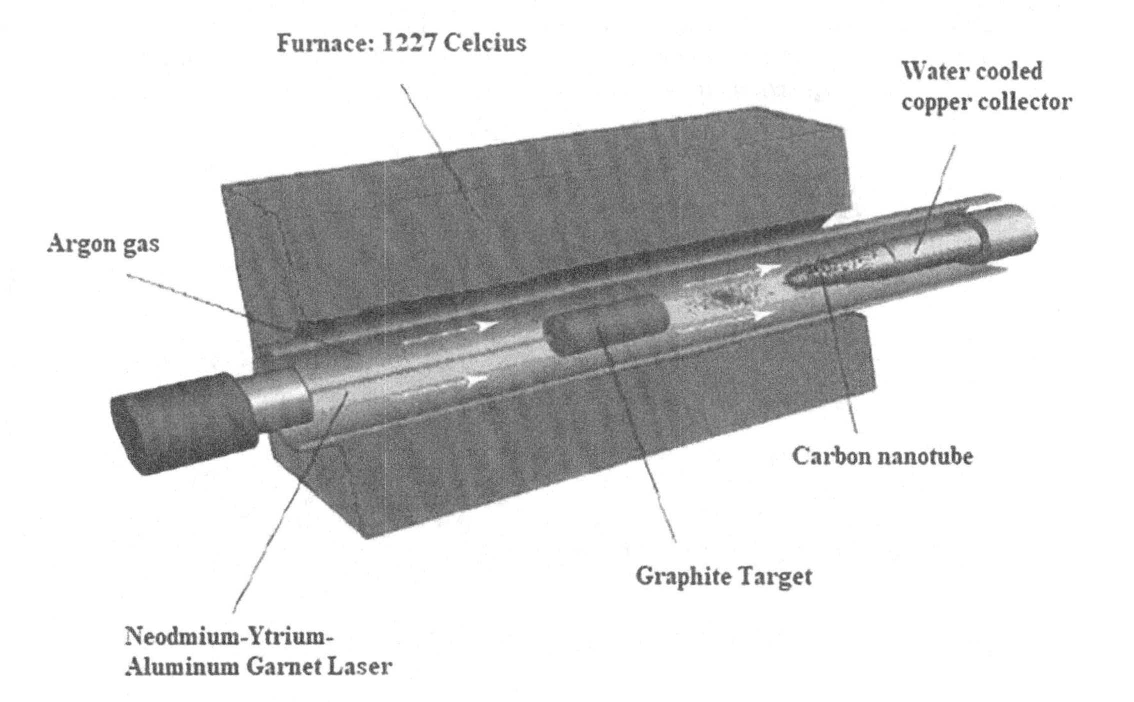

FIGURE 5.6 Schematic view of fullerene production process.

room temperature. Fullerenes are used as lubricants and nanogenerator materials due to their symmetrical shape, while they are used as catalysts due to their large surface area (Balasubramanian 2020a).

Nanoballs protect materials from excessive and harmful light when used as coating material. Polymers-containing fullerene structures show photoconductivity. So, carbon nanoballs are used in photodiodes, transistors and solar cells. Fullerenes are good protectors against oxidation. Also, it has been observed in experiments that a substance created from water-soluble fullerene derivatives limits the activities of the HIV.

Energy storage is one of today's important issues. Energy sources are not infinitely available and the end products of most are not environment friendly. Hydrogen is a good and clean alternative for energy problem, and studies for storing hydrogen is going on. Fullerene is one of the alternatives for storing hydrogen by the physisorption method. Fullerene is also a promising material for high-energy battery production (Balasubramanian 2020b).

5.2.4 Graphene

Until 1980, it was thought that carbon has only three basic forms (diamond, graphite, amorphous carbon). In 2004, it was proved that graphene, which was synthesized experimentally, consists of high-quality two-dimensional crystals. In 2005, Novoselov and his team succeeded synthesizing single-layer graphene and proved that it has different electronic and physical properties than expected (Blake et al. 2007; Ferrari et al. 2006; Geim and Novoselov 2009; Novoselov 2011).

Graphene has a wavy structure, and its use is advantageous due to the small size of the carbon atom. It has a regular crystal structure. Graphene is not very reactive chemically. It has high transparency; a single layer absorbs only 2.3% of white light. Graphene has electronic quality as good as a carbon nanotube and it is smaller in volume. While silicon's mobility is 1360 cm^2/Vs, graphene's mobility is 250,000 cm^2/Vs. Graphene has a high temperature resistance, the properties of graphene do not change between −75 and 200°C. It has high thermal conductivity (5000 W/(m K)) and high tensile strength (1100 GPa) (Bai et al. 2010; Chang and Liao 2016; Lui et al. 2009; Novoselov et al. 2012). Graphene synthesis techniques are presented in Figure 5.7.

The application areas of graphene are nanogenerators, OLED technology, armors, light aircrafts, photovoltaic cells, superconductor batteries and filtration applications (Choi et al. 2010; Olabi et al. 2021; Zhang, Zhang, and Zhou 2013).

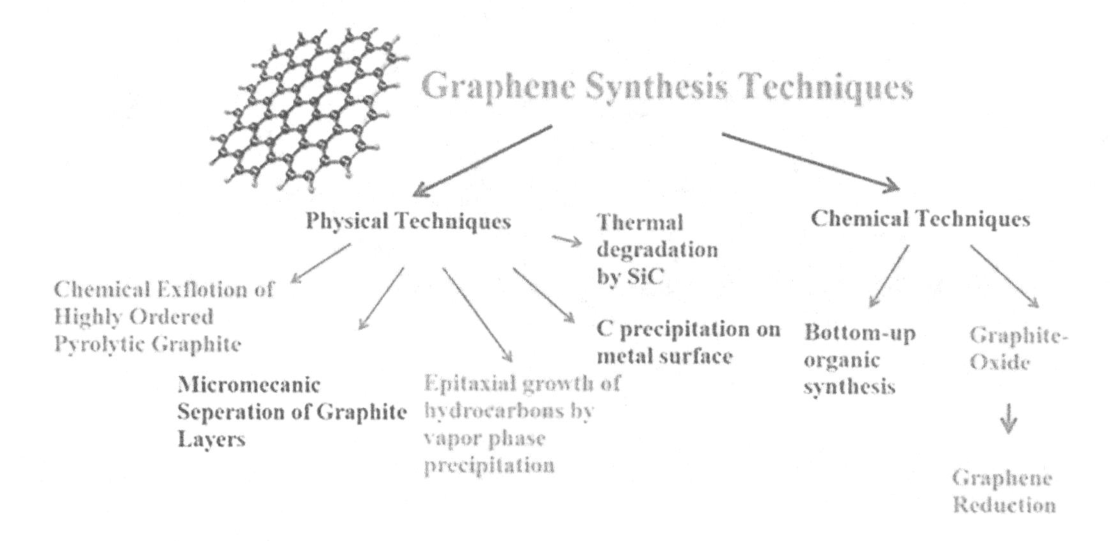

FIGURE 5.7 Graphene synthesis techniques.

5.3 THE USE OF CARBON MATERIALS IN NANOGENERATOR APPLICATIONS

Carbon-based materials have suitable properties for use in nanogenerator production. In recent years, it is seen that interest in the use of carbon-based materials in this field has increased. Nanogenerator applications of carbon-based materials and studies conducted in this field will be shared in this section.

As mentioned in the previous section, carbon black is one of the carbon-based materials used in nanogenerator production. In a study conducted by Xiao et al. (2018), a TENG using carbon black and silicone rubber composite electrode was designed. This study aimed to generate electric energy by converting the energy of the water waves. As stated in this study, the C/PTFE-C (which consists of silicone rubber/carbon black electrode and PTFE thin film) material pair has 75.2% higher charge density, 60.4% higher output current density and 103.9% higher voltage than the Cu/PTFE-Cu pair (which consists of a copper electrode and PTFE thin film). It is also stated that the segmentation on this composite electrode can further increase the output power by 23.5% (Xiao et al. 2018). In a study carried out by Jayababu and Kim (2021), carbon black was employed in a TENG application. Zinc oxide nanorods and conductive carbon black nanocomposite was used together for energy conversion and storage in this study. Initially, nanorods were produced with a facile precipitation method and then carbon black was added for adjusting the work function. The study showed that the addition of conductive carbon black (20 wt.%) improved the open-circuit voltage (V_{oc}) and short circuit current (I_{sc}) of pure zinc oxide nanorods (Jayababu and Kim 2021). Nanocarbon black made from kerosene soot was used as a conductive medium with a bottom electrode screen printed using a silver paste to create a piezoresistive pressure sensor in a study carried out by Hu et al. (2020). This study aimed to design and fabricate a high-performance wearable pressure sensor based on nanocarbon black. Very high sensitivity (31.63 kPa^{-1} within the range of 0–2 kPa), rapid response (15 ms) and a broad workable pressure sensor range (0–15 kPa) values were obtained. Results showed that mechanical signals like wrist movement and faint pulses can be detected by this technology (Hu et al. 2020).

There are various studies about the use of carbon nanotubes in nanogenerator applications. In a study published by Wang et al. (2016), aligned CNTs were embedded on the PDMS surface. Carbon nanotubes were used as the layer which donates electrons. The layer increases the electric production for the output and shows distinguishable stretchability. A high-performance TENG was achieved as a result of this study. The nanogenerator gave 150-V output voltage, 60-mA current density and 4.62 W/m^2 (at an external load of 30 MΩ). When compared with the TENG using directly doped PDMS/MWNTs, the output voltage was improved by 250% and the current density was improved by 300% (Wang et al. 2016). Matsunaga et al. (2020) indicated that power generated by employing a stretchable triboelectric generator is not enough. They developed an idea to use a transparent CNT thin-film-based stretchable triboelectric generator to obtain high output power (8 W/m^2). They produced a highly transparent stretchable TENG employing a CNT film as a transparent electrode for wearable applications. They expressed that carbon nanotube film provides great scalability, optimum transparency (95%) and durability against tensile strains up to 80% (Matsunaga et al. 2020). You et al. (2018) published a research about a self-powered flexible hybrid piezoelectric – pyroelectric NG. This nanogenerator was based on the piezoelectric and pyroelectric effects of an electrospun non-woven PVDF nanofiber membrane. The flexibility of the nanogenerator was improved by employing an electrospun TPU nanofiber membrane as substrate, conductive PEDOT:PSS-PVP nanofiber membrane and CNT layer as the electrodes. It was detected that at impact frequencies of 1.2 and 3 Hz, this nanogenerator could light a white LED. The thermal gradient-induced current output of the nanogenerator was determined, and stable pyroelectric current outputs under different mechanical stresses were shown. Also, a hybrid current (piezoelectric and pyroelectric) was obtained by impacting (and releasing) and heating (and cooling) the hybrid nanogenerator simultaneously (You et al. 2018). Choi et al. (2017) worked on a flexible and ultralight thermoelectric generator based on CNT yarn. The as-prepared CNT yarn had an electrical conductivity of 3147 S/cm. Carbon nanotube yarn was doped into n and p-types using PEI and $FeCl_3$, respectively. This flexible thermoelectric nanogenerator consist of 60 pairs of n- and p-doped carbon nanotube yarns has the maximum power density of 10.85 and 697 μW/g at temperature differences of 5 and 40 K, respectively. Obtained values are the best values among the TENGs made of flexible materials (Choi et al. 2017).

In a study conducted by Parajuli et al. (2020), fullerene was used in a TENG. Their target was to coat zero-dimensional C_{60} fullerene on the triboelectric material surface and enhance the output performance in this way. They built a fullerene-based TENG by using a C_{60}-fullerene-coated PET sheet and Al/Pyrex electrodes. Results of the experimental studies presented that maximum peak power was increased to 3.8 times and capacitor charging with the same experimental parameters showed a sevenfold increase in average output power for the fullerene-based TENG. This nanogenerator was capable of powering a digital watch continuously (Parajuli et al. 2020). Lee et al. (2021) reported about a new dielectric, a C_{60}-containing block polyimide (PI–b–C_{60}) in a study they published in 2021. C_{60} was added into a polymer backbone (6FDA–APS). In this way, high-output power non-contact mode thermoelectric nanogenerators can be built. The casting process was used to build a film based on PI–b–C_{60}, and a TENG, which has a silver nanowire embedded PDMS as a positively charged layer, was built. When it was compared with PFA film-based TENG, it was seen that this nanogenerator produced 4.3 times higher output power. Also, it was determined that 85% of the PI–b–C_{60}-based TENG's maximum charge density was kept after 5 hours. Also, the charge decay rate was three times slower when compared to the PFA-based TENG (Lee et al. 2021).

In 2019, Chen et al. (2019) manufactured a stretchable TENG using stretchable crumpled graphene. They prepared various TENGs with different degrees of crumpling to show the effect of crumple nanostructure on triboelectricity. It was seen in the study that crumple degree influences the output performance. When crumple degree was 300%, the output voltage was 83 V, the output current was 25.78 μA and the power density was 0.25 mW/cm^2. These values are superior when compared to a planer graphene-based TENG. This study shows that this system is capable of harvesting energy under various complex deformations, which can occur in practical wearable situations (Chen et al. 2019). Chandrashekar et al. published a study about a flexible and transparent TENG by employing CVD-grown graphene as one of the friction layers (Shankaregowda et al. 2016). CVD-grown graphene increased the conductivity, the optical transmittance and output efficiency of the TENG. A plasma-treated layer of PDMS was used as another layer. The maximum output voltage of 650 V and a current of 12 μA are achieved at 4.3-Hz frequency. This power is enough for lighting up 50 commercial blue LEDs (Shankaregowda et al. 2016). Bhavanasi et al. (2016) reported about the performance of the bilayer films of poled PVDF-TrFE and graphene oxide. Obtained results showed that bilayer film's energy harvesting performance is better when compared to poled PVDF-TrFE films alone. While the voltage output of bilayer film is 4V, poled PVDF-TrFE's voltage output is 1.9 V. Also, the power outputs of bilayer film and PVDF-TrFE are 4.41 and 1.77 $\mu W/cm^2$, respectively. The reasons for the superior energy values obtained using graphene oxide film are indicated as the electrostatic contribution of the graphene oxide, the improved elastic modulus of the bilayer films, residual tensile stress, and the space charge at the interface between the PVDF-TrFE and graphene oxide films (Bhavanasi et al. 2016). Bharti et al. (2020) published a study about a transparent and flexible piezoelectric nanogenerator based on lead-free Zn_2SiO_4 nanorods and graphene. A mixture of PDMS and Zn_2SiO_4 (70:30) was prepared and then spin-coated on the flexible ITO-coated PET substrate. Graphene was initially CVD-grown on copper foil, and PMMA was used to transfer graphene to the PET substrate. Piezoelectric charge coefficient of the piezoelectric nanogenerator was 117 pM/V. Also, output circuit voltage of 5.5 V and current density of 0.50 $\mu A/cm^2$ values were obtained under low pressure (0.15 kgf) (Bharti et al. 2020). Zabek et al. (2017) produced a pyroelectric nanogenerator based on free-standing PVDF that is taking advantage of the high thermal radiation absorbance of screen-printed graphene ink electrode. Interconnected graphene nanoplatelets provide high thermal radiation absorbance and electrical conductivity. When compared to aluminum electrodes, the graphene electrodes perform 4 and 7 times better under open and closed-circuit conditions. Also, graphene ink electrodes are 25 times better when rectified in pyroelectric NG mode (Zabek et al. 2017). Xie et al. (2017) manufactured a thermoelectric NG based on MoS_2 and graphene nanocomposite. This was a flexible nanogenerator and graphene addition to MoS_2 was expected to improve the thermoelectric performance. The nanogenerator included indium tin oxide layer as the bottom conductive electrode, silver paste as the top conductive electrode and MoS_2/graphene nanocomposite as the layer between. When the temperature difference is less than −35 K, MoS_2/graphene thermoelectric NG produced an output voltage of −0.73 mV. This value is 2 times the value obtained with the pure MoS_2 thermoelectric nanogenerator and 8 times the value obtained with the pure graphene thermoelectric nanogenerator (Xie et al. 2017).

Blue energy has become an important issue lately. The source of blue energy is the movement of natural water such as raindrops, ocean tides and river flow. Harvesting blue energy is carried out by using triboelectric and piezoelectric effects. Due to their properties, carbon nanomaterials are considered to be the ideal materials for blue energy harvesting. Harvesting blue energy is carried out by droplet-based, flow-induced and phase change nanogenerators. CVD-grown monolayer graphene, reduced graphene oxide prepared by Hummers method, monolayer graphene and graphene oxide produced by electrophoresis, graphene and carbon black/PTFE sheet prepared by wet chemistry are used in droplet-based generator studies. CVD-grown materials (graphene sheet, monolayer graphene, graphene film, graphene sheet, monolayer graphene, graphene grid and graphene foam), MWNT yarn, single-wall nanotube bundles prepared by electric arc method, aligned MWNT fiber, graphene hydrogel membrane prepared by vacuum filtration, graphene oxide prepared by Hummers method are used in flow-induced generator studies. Gradient graphene oxide film prepared by vacuum filtration, graphene oxide nanoribbon produced by freezing and self-assembly, gradient graphene oxide frameworks prepared by freeze-drying strategy, porous carbon film and carbon black film produced by ethanol flame method, robust carbon black film produced by printing method are used in phase-change generator studies (Liu et al. 2018).

REFERENCES

Aghaie-Khafri, M., and N. Daemi. 2012. "Characterization of Vanadium Carbide Coating Deposited by Borax Salt Bath Process." *Advances in Materials Research* 1 (3): 233–43. https://doi.org/10.12989/amr.2012.1.3.233.

Alcántara, Ricardo, Juan M. Jiménez-Mateos, Pedro Lavela, and José L. Tirado. 2001. "Carbon Black: A Promising Electrode Material for Sodium-Ion Batteries." *Electrochemistry Communications* 3 (11): 639–42. https://doi.org/10.1016/S1388-2481(01)00244-2.

Ando, Yoshinori, and Zhao, Xinluo. 2006. "Synthesis of Carbon Nanotubes by Arc-Discharge Method." *New Diamond and Frontier Carbon Technology* 16 (3): 123–37.

Andrei, Virgil, Kevin Bethke, and Klaus Rademann. 2016. "Thermoelectricity in the Context of Renewable Energy Sources: Joining Forces Instead of Competing." *Energy & Environmental Science* 9 (5): 1528–32. https://doi.org/10.1039/C6EE00247A.

Bahaj, AbuBakr S. 2011. "Generating Electricity from the Oceans." *Renewable and Sustainable Energy Reviews* 15 (7): 3399–3416. https://doi.org/10.1016/j.rser.2011.04.032.

Bai, Jingwei, Xing Zhong, Shan Jiang, Yu Huang, and Xiangfeng Duan. 2010. "Graphene Nanomesh." *Nature Nanotechnology* 5 (3): 190–4. https://doi.org/10.1038/nnano.2010.8.

Balasubramanian, Krishnan. 2020a. "Topological Peripheral Shapes and Distance-Based Characterization of Fullerenes C20-C720: Existence of Isoperipheral Fullerenes." Polycyclic Aromatic Compounds (Published Online): 1 -19. https://doi.org/10.1080/10406638.2020.1802303.

Balasubramanian, Krishnan. 2020b. "Combinatorics of Supergiant Fullerenes: Enumeration of Polysubstituted Isomers, Chirality, Nuclear Magnetic Resonance, Electron Spin Resonance Patterns, and Vibrational Modes from C 70 to C 150000." *The Journal of Physical Chemistry A* 124 (49): 10359–83. https://doi.org/10.1021/acs.jpca.0c08914.

Bernal, Miguel, Ivan Nenadic, Matthew W. Urban, and James F. Greenleaf. 2011. "Material Property Estimation for Tubes and Arteries Using Ultrasound Radiation Force and Analysis of Propagating Modes." *The Journal of the Acoustical Society of America* 129 (3): 1344–54. https://doi.org/10.1121/1.3533735.

Bharti, Dhiraj Kumar, Manoj Kumar Gupta, Rajeev Kumar, Natarajan Sathish, and Avanish Kumar Srivastava. 2020. "Non-Centrosymmetric Zinc Silicate-Graphene Based Transparent Flexible Piezoelectric Nanogenerator." *Nano Energy* 73 (July): 104821. https://doi.org/10.1016/j.nanoen.2020.104821.

Bhavanasi, Venkateswarlu, Vipin Kumar, Kaushik Parida, Jiangxin Wang, and Pooi See Lee. 2016. "Enhanced Piezoelectric Energy Harvesting Performance of Flexible PVDF-TrFE Bilayer Films with Graphene Oxide." *ACS Applied Materials & Interfaces* 8 (1): 521–9. https://doi.org/10.1021/acsami.5b09502.

Blake, P., E. W. Hill, A. H. Castro Neto, K. S. Novoselov, D. Jiang, R. Yang, T. J. Booth, and A. K. Geim. 2007. "Making Graphene Visible." *Applied Physics Letters* 91 (6): 063124. https://doi.org/10.1063/1.2768624.

Borah, Dipu, Shigeo Satokawa, Shigeru Kato, and Toshinori Kojima. 2008. "Characterization of Chemically Modified Carbon Black for Sorption Application." *Applied Surface Science* 254 (10): 3049–56. https://doi.org/10.1016/j.apsusc.2007.10.053.

Briscoe, Joe, and Steve Dunn. 2015. "Piezoelectric Nanogenerators – A Review of Nanostructured Piezoelectric Energy Harvesters." *Nano Energy* 14 (May): 15–29. https://doi.org/10.1016/j.nanoen.2014.11.059.

Brunauer, Stephen, P. H. Emmett, and Edward Teller. 1938. "Adsorption of Gases in Multimolecular Layers." *Journal of the American Chemical Society* 60 (2): 309–19. https://doi.org/10.1021/ja01269a023.

Cai, Zhengyang, Bilu Liu, Xiaolong Zou, and Hui-Ming Cheng. 2018. "Chemical Vapor Deposition Growth and Applications of Two-Dimensional Materials and Their Heterostructures." *Chemical Reviews* 118 (13): 6091–133. https://doi.org/10.1021/acs.chemrev.7b00536.

Chai, Yan, Ting Guo, Changming Jin, Robert E. Haufler, L. P. Felipe Chibante, Jan Fure, Lihong Wang, J. Michael Alford, and Richard E. Smalley. 1991. "Fullerenes with Metals Inside." *The Journal of Physical Chemistry* 95 (20): 7564–8. https://doi.org/10.1021/j100173a002.

Chang, Chia-Wei, and Ying-Chih Liao. 2016. "Accelerated Sedimentation Velocity Assessment for Nanowires Stabilized in a Non-Newtonian Fluid." *Langmuir* 32 (51): 13620–6. https://doi.org/10.1021/acs.langmuir.6b03602.

Chen, Xi, Baoxing Xu, and Ling Liu. 2014. "Nanoscale Fluid Mechanics and Energy Conversion." *Applied Mechanics Reviews* 66 (5). https://doi.org/10.1115/1.4026913.

Chen, Huamin, Yun Xu, Jiushuang Zhang, Weitong Wu, and Guofeng Song. 2019. "Enhanced Stretchable Graphene-Based Triboelectric Nanogenerator via Control of Surface Nanostructure." *Nano Energy* 58 (April): 304–11. https://doi.org/10.1016/j.nanoen.2019.01.029.

Choi, Jaeyoo, Yeonsu Jung, Seung Jae Yang, Jun Young Oh, Jinwoo Oh, Kiyoung Jo, Jeong Gon Son, Seung Eon Moon, Chong Rae Park, and Heesuk Kim. 2017. "Flexible and Robust Thermoelectric Generators Based on All-Carbon Nanotube Yarn without Metal Electrodes." *ACS Nano* 11 (8): 7608–14. https://doi.org/10.1021/acsnano.7b01771.

Choi, Wonbong, Indranil Lahiri, Raghunandan Seelaboyina, and Yong Soo Kang. 2010. "Synthesis of Graphene and Its Applications: A Review." *Critical Reviews in Solid State and Materials Sciences* 35 (1): 52–71. https://doi.org/10.1080/10408430903505036.

Chrzanowska, Justyna, Jacek Hoffman, Artur Małolepszy, Marta Mazurkiewicz, Tomasz A. Kowalewski, Zygmunt Szymanski, and Leszek Stobinski. 2015. "Synthesis of Carbon Nanotubes by the Laser Ablation Method: Effect of Laser Wavelength." *Physica Status Solidi (B)* 252 (8): 1860–7. https://doi.org/10.1002/pssb.201451614.

Dong, Pengwei, Thawatchai Maneerung, Wei Cheng Ng, Xu Zhen, Yanjun Dai, Yen Wah Tong, Yen-Peng Ting, Shin Nuo Koh, Chi-Hwa Wang, and Koon Gee Neoh. 2017. "Chemically Treated Carbon Black Waste and Its Potential Applications." *Journal of Hazardous Materials* 321 (January): 62–72. https://doi.org/10.1016/j.jhazmat.2016.08.065.

Dons, Evi, Luc Int Panis, Martine Van Poppel, Jan Theunis, and Geert Wets. 2012. "Personal Exposure to Black Carbon in Transport Microenvironments." *Atmospheric Environment* 55 (August): 392–8. https://doi.org/10.1016/j.atmosenv.2012.03.020.

Fan, Feng Ru, Wei Tang, and Zhong Lin Wang. 2016. "Flexible Nanogenerators for Energy Harvesting and Self-Powered Electronics." *Advanced Materials* 28 (22): 4283–305. https://doi.org/10.1002/adma.201504299.

Fan, Feng Ru, Zhong Q. Tian, and Zhong L. Wang. 2012. "Flexible Triboelectric Generator." *Nano Energy* 1 (2): 328–34. https://doi.org/10.1016/j.nanoen.2012.01.004.

Fan, Xiaodan, and Xiangkai Zhang. 2008. "Adsorption Properties of Activated Carbon from Sewage Sludge to Alkaline-Black." *Materials Letters* 62 (10–11): 1704–6. https://doi.org/10.1016/j.matlet.2007.09.085.

Ferrari, A. C., J. C. Meyer, V. Scardaci, C. Casiraghi, M. Lazzeri, F. Mauri, S. Piscanec, et al. 2006. "Raman Spectrum of Graphene and Graphene Layers." *Physical Review Letters* 97 (18): 187401. https://doi.org/10.1103/PhysRevLett.97.187401.

Firstov, G. S., J. Van Humbeeck, and Y. N. Koval. 2004. "High-Temperature Shape Memory Alloys Some Recent Developments." *Materials Science and Engineering A* 378 (1–2 SPEC. ISS.): 210. https://doi.org/10.1016/j.msea.2003.10.324.

Futaba, Don N., Kenji Hata, Takeo Yamada, Kohei Mizuno, Motoo Yumura, and Sumio Iijima. 2005. "Kinetics of Water-Assisted Single-Walled Carbon Nanotube Synthesis Revealed by a Time-Evolution Analysis." *Physical Review Letters* 95 (5): 056104. https://doi.org/10.1103/PhysRevLett.95.056104.

Futaba, Don N., Kenji Hata, Takeo Yamada, Tatsuki Hiraoka, Yuhei Hayamizu, Yozo Kakudate, Osamu Tanaike, Hiroaki Hatori, Motoo Yumura, and Sumio Iijima. 2006. "Shape-Engineerable and Highly Densely Packed Single-Walled Carbon Nanotubes and Their Application as Super-Capacitor Electrodes." *Nature Materials* 5 (12): 987–94. https://doi.org/10.1038/nmat1782.

Geim, A. K., and K. S. Novoselov. 2009. "The Rise of Graphene." In *Nanoscience and Technology*, 11–19. Co-Published with Macmillan Publishers Ltd, UK. https://doi.org/10.1142/9789814287005_0002.

He, Jian, Tao Wen, Shuo Qian, Zengxing Zhang, Zhumei Tian, Jie Zhu, Jiliang Mu, et al. 2018. "Triboelectric-Piezoelectric-Electromagnetic Hybrid Nanogenerator for High-Efficient Vibration Energy Harvesting and Self-Powered Wireless Monitoring System." *Nano Energy* 43 (January): 326–39. https://doi.org/10.1016/j.nanoen.2017.11.039.

Hu, Chuangang, Long Song, Zhipan Zhang, Nan Chen, Zhihai Feng, and Liangti Qu. 2015. "Tailored Graphene Systems for Unconventional Applications in Energy Conversion and Storage Devices." *Energy and Environmental Science* 8 (1): 31–54. https://doi.org/10.1039/c4ee02594f.

Hu, Junsong, Junsheng Yu, Ying Li, Xiaoqing Liao, Xingwu Yan, and Lu Li. 2020. "Nano Carbon Black-Based High Performance Wearable Pressure Sensors." *Nanomaterials* 10 (4): 664. https://doi.org/10.3390/nano10040664.

Izard, N., S. Kazaoui, K. Hata, T. Okazaki, T. Saito, S. Iijima, and N. Minami. 2008. "Semiconductor-Enriched Single Wall Carbon Nanotube Networks Applied to Field Effect Transistors." *Applied Physics Letters* 92 (24): 243112. https://doi.org/10.1063/1.2939560.

Jayababu, Nagabandi, and Daewon Kim. 2021. "ZnO Nanorods@conductive Carbon Black Nanocomposite Based Flexible Integrated System for Energy Conversion and Storage through Triboelectric Nanogenerator and Supercapacitor." *Nano Energy* 82 (April): 105726. https://doi.org/10.1016/j.nanoen.2020.105726.

Jin, Chuanhong, Kazu Suenaga, and Sumio Iijima. 2008. "Plumbing Carbon Nanotubes." *Nature Nanotechnology* 3 (1): 17–21. https://doi.org/10.1038/nnano.2007.406.

Keidar, M, and A M Waas. 2004. "On the Conditions of Carbon Nanotube Growth in the Arc Discharge." *Nanotechnology* 15 (11): 1571–5. https://doi.org/10.1088/0957-4484/15/11/034.

Kim, Jihye, Jeong Hwan Lee, Hanjun Ryu, Ju-Hyuck Lee, Usman Khan, Han Kim, Sung Soo Kwak, and Sang-Woo Kim. 2017. "High-Performance Piezoelectric, Pyroelectric, and Triboelectric Nanogenerators Based on P(VDF-TrFE) with Controlled Crystallinity and Dipole Alignment." *Advanced Functional Materials* 27 (22): 1700702. https://doi.org/10.1002/adfm.201700702.

Kuo, Arthur D. 2005. "Harvesting Energy by Improving the Economy of Human Walking." *Science* 309 (5741): 1686–7. https://doi.org/10.1126/science.1118058.

Lee, Jae Won, Sungwoo Jung, Jinhyeong Jo, Gi Hyeon Han, Dong-Min Lee, Jiyeon Oh, Hee Jae Hwang, et al. 2021. "Sustainable Highly Charged C 60-Functionalized Polyimide in a Non-Contact Mode Triboelectric Nanogenerator." *Energy & Environmental Science* 14 (2): 1004–15. https://doi.org/10.1039/D0EE03057K.

Li, Xuemei, Chun Shen, Qin Wang, Chi Man Luk, Baowen Li, Jun Yin, Shu Ping Lau, and Wanlin Guo. 2017. "Hydroelectric Generator from Transparent Flexible Zinc Oxide Nanofilms." *Nano Energy* 32 (February): 125–9. https://doi.org/10.1016/j.nanoen.2016.11.050.

Liu, Guohua, Ting Chen, Jinliang Xu, and Kaiying Wang. 2018. "Blue Energy Harvesting on Nanostructured Carbon Materials." *Journal of Materials Chemistry A* 6 (38): 18357–77. https://doi.org/10.1039/C8TA07125J.

Liu, Jie, Andrew G. Rinzler, Hongjie Dai, Jason H. Hafner, R. Kelley Bradley, Peter J. Boul, Adrian Lu, et al. 1998. "Fullerene Pipes." *Science* 280 (5367): 1253–6. https://doi.org/10.1126/science.280.5367.1253.

Lui, Chun Hung, Li Liu, Kin Fai Mak, George W. Flynn, and Tony F. Heinz. 2009. "Ultraflat Graphene." *Nature* 462 (7271): 339–41. https://doi.org/10.1038/nature08569.

Matsunaga, Masahiro, Jun Hirotani, Shigeru Kishimoto, and Yutaka Ohno. 2020. "High-Output, Transparent, Stretchable Triboelectric Nanogenerator Based on Carbon Nanotube Thin Film toward Wearable Energy Harvesters." *Nano Energy* 67 (January): 104297. https://doi.org/10.1016/j.nanoen.2019.104297.

Morales, Alfredo M., and Charles M. Lieber. 1998. "A Laser Ablation Method for the Synthesis of Crystalline Semiconductor Nanowires." *Science* 279 (5348): 208–11. https://doi.org/10.1126/science.279.5348.208.

Moulin, L., S. Da Silva, A. Bounaceur, M. Herblot, and Y. Soudais. 2017. "Assessment of Recovered Carbon Black Obtained by Waste Tires Steam Water Thermolysis: An Industrial Application." *Waste and Biomass Valorization* 8 (8): 2757–70. https://doi.org/10.1007/s12649-016-9822-8.

Mwafy, Eman A. 2020. "Eco-Friendly Approach for the Synthesis of MWCNTs from Waste Tires via Chemical Vapor Deposition." *Environmental Nanotechnology, Monitoring & Management* 14 (December): 100342. https://doi.org/10.1016/j.enmm.2020.100342.

Nester, Serguei; Rumpf, Frederick H.; Kutsovsky, Yakov E.; Natalie, Charles A. 2011. Methods for carbon black production using preheated feedstock and apparatus for same. *WO 2011*/103015 A2, issued 2011. https://patentimages.storage.googleapis.com/07/37/65/968339415d24f1/WO2011103015A2.pdf.

Novoselov, K. S. 2011. "Nobel Lecture: Graphene: Materials in the Flatland." *Reviews of Modern Physics* 83 (3): 837–49. https://doi.org/10.1103/RevModPhys.83.837.

Novoselov, K. S., V. I. Fal′ko, L. Colombo, P. R. Gellert, M. G. Schwab, and K. Kim. 2012. "A Roadmap for Graphene." *Nature* 490 (7419): 192–200. https://doi.org/10.1038/nature11458.

Olabi, A.G., Mohammad Ali Abdelkareem, Tabbi Wilberforce, and Enas Taha Sayed. 2021. "Application of Graphene in Energy Storage Device – A Review." *Renewable and Sustainable Energy Reviews* 135 (January): 110026. https://doi.org/10.1016/j.rser.2020.110026.

Parajuli, Prakash, Bipin Sharma, Herbert Behlow, and Apparao M. Rao. 2020. "Fullerene-Enhanced Triboelectric Nanogenerators." *Advanced Materials Technologies* 5 (8): 2000295. https://doi.org/10.1002/admt.202000295.

Prudell, Joseph, Martin Stoddard, Ean Amon, Ted K. A. Brekken, and Annette von Jouanne. 2010. "A Permanent-Magnet Tubular Linear Generator for Ocean Wave Energy Conversion." *IEEE Transactions on Industry Applications* 46 (6): 2392–400. https://doi.org/10.1109/TIA.2010.2073433.

Qin, Lu-Chang, Xinluo Zhao, Kaori Hirahara, Yoshiyuki Miyamoto, Yoshinori Ando, and Sumio Iijima. 2000. "The Smallest Carbon Nanotube." *Nature* 408 (6808): 50. https://doi.org/10.1038/35040699.

Quartiermeister, Marcelo V., Danielle C. C. Magalhães, Guilherme S. Vacchi, Diogo P. Braga, Rodrigo Silva, Andréa M. Kliauga, Vitor L. Sordi, and Carlos A. D. Rovere. 2020. "On the Pitting Corrosion Behavior of Ultrafine-grained Aluminum Processed by ECAP: A Statistical Analysis." *Materials and Corrosion* 71 (8): 1244–56. https://doi.org/10.1002/maco.201911293.

Reina, Alfonso, Xiaoting Jia, John Ho, Daniel Nezich, Hyungbin Son, Vladimir Bulovic, Mildred S Dresselhaus, and Jing Kong. 2009. "Large Area, Few-Layer Graphene Films on Arbitrary Substrates by Chemical Vapor Deposition." *Nano Letters* 9 (1): 30–5. https://doi.org/10.1021/nl801827v.

Roy, S. S., H. Gleeson, C. P. Shaw, R. W. Whatmore, Z. Huang, Q. Zhang, and S. Dunn. 2000. "Growth and Characterisation of Lead Zirconate Titanate (30/70) on Indium Tin Oxide Coated Glass for Oxide Ferroelectric-Liquid Crystal Display Application." *Integrated Ferroelectrics* 29 (3–4): 189–213. https://doi.org/10.1080/10584580008222239.

Rumpf, Frederick H.; Taylor, Roscoe W.; Toombs, Alvin E. 2010. Process for Production of Carbon Black. US 7.655,209 B2, issued 2010. https://patentimages.storage.googleapis.com/44/a1/9e/eec951ae5cb309/US7655209.pdf.

Samaržija-Jovanović, Suzana, Vojislav Jovanović, Gordana Marković, and Milena Marinović-Cincović. 2009. "The Effect of Different Types of Carbon Blacks on the Rheological and Thermal Properties of Acrylonitrile Butadiene Rubber." *Journal of Thermal Analysis and Calorimetry* 98 (1): 275–83. https://doi.org/10.1007/s10973-009-0131-3.

Semaltianos, N. G. 2010. "Nanoparticles by Laser Ablation." *Critical Reviews in Solid State and Materials Sciences* 35 (2): 105–24. https://doi.org/10.1080/10408431003788233.

Shankaregowda, Smitha Ankanahalli, Chandrashekar Banankere Nanjegowda, Xiao-Liang Cheng, Ma-Yue Shi, Zhong-Fan Liu, and Hai-Xia Zhang. 2016. "A Flexible and Transparent Graphene-Based Triboelectric Nanogenerator." *IEEE Transactions on Nanotechnology* 15 (3): 435–41. https://doi.org/10.1109/TNANO.2016.2540958.

Smalley, Richard E. 1997. "Discovering the Fullerenes." *Reviews of Modern Physics* 69 (3): 723–30. https://doi.org/10.1103/RevModPhys.69.723.

Striolo, Alberto, Angelos Michaelides, and Laurent Joly. 2016. "The Carbon-Water Interface: Modeling Challenges and Opportunities for the Water-Energy Nexus." *Annual Review of Chemical and Biomolecular Engineering* 7 (1): 533–56. https://doi.org/10.1146/annurev-chembioeng-080615-034455.

Tang, Qunwei, Jialong Duan, Yanyan Duan, Benlin He, and Liangmin Yu. 2015. "Recent Advances in Alloy Counter Electrodes for Dye-Sensitized Solar Cells. A Critical Review." *Electrochimica Acta* 178 (October): 886–99. https://doi.org/10.1016/j.electacta.2015.08.072.

Wang, Huan, Mayue Shi, Kai Zhu, Zongming Su, Xiaoliang Cheng, Yu Song, Xuexian Chen, Zhiqiang Liao, Min Zhang, and Haixia Zhang. 2016. "High Performance Triboelectric Nanogenerators with Aligned Carbon Nanotubes." *Nanoscale* 8 (43): 18489–94. https://doi.org/10.1039/C6NR06319E.

Wang, Lili, Xiaofeng Wang, Bo Zou, Xiaoyu Ma, Yuning Qu, Chunguang Rong, Ying Li, Ying Su, and Zichen Wang. 2011. "Preparation of Carbon Black from Rice Husk by Hydrolysis, Carbonization and Pyrolysis." *Bioresource Technology* 102 (17): 8220–4. https://doi.org/10.1016/j.biortech.2011.05.079.

Wang, Peihong, Ruiyuan Liu, Wenbo Ding, Peng Zhang, Lun Pan, Guozhang Dai, Haiyang Zou, Kai Dong, Cheng Xu, and Zhong Lin Wang. 2018. "Complementary Electromagnetic-Triboelectric Active Sensor for Detecting Multiple Mechanical Triggering." *Advanced Functional Materials* 28 (11): 1705808. https://doi.org/10.1002/adfm.201705808.

Wang, Qichen, Yongpeng Lei, Yuchao Wang, Yi Liu, Chengye Song, Jian Zeng, Yaohao Song, Xidong Duan, Dingsheng Wang, and Yadong Li. 2020. "Atomic-Scale Engineering of Chemical-Vapor-Deposition-Grown 2D Transition Metal Dichalcogenides for Electrocatalysis." *Energy & Environmental Science* 13 (6): 1593–616. https://doi.org/10.1039/D0EE00450B.

Wang, Zhong Lin. 2014. "Triboelectric Nanogenerators as New Energy Technology and Self-Powered Sensors – Principles, Problems and Perspectives." *Faraday Discussions* 176: 447–58. https://doi.org/10.1039/c4fd00159a.

Wang, Zhong Lin. 2017. "Catch Wave Power in Floating Nets." *Nature* 542 (7640): 159–60. https://doi.org/10.1038/542159a.

Wang, Zhong L., and Jinhui Song. 2006. "Piezoelectric Nanogenerators Based on Zinc Oxide Nanowire Arrays" *Science* 312 (5771): 242–6.

Wu, Hong-Hui, Jiaming Zhu, and Tong-Yi Zhang. 2015. "Pseudo-First-Order Phase Transition for Ultrahigh Positive/Negative Electrocaloric Effects in Perovskite Ferroelectrics." *Nano Energy* 16 (September): 419–27. https://doi.org/10.1016/j.nanoen.2015.06.030.

Xiao, Tian Xiao, Tao Jiang, Jian Xiong Zhu, Xi Liang, Liang Xu, Jia Jia Shao, Chun Lei Zhang, Jie Wang, and Zhong Lin Wang. 2018. "Silicone-Based Triboelectric Nanogenerator for Water Wave Energy Harvesting." *ACS Applied Materials & Interfaces* 10 (4): 3616–23. https://doi.org/10.1021/acsami.7b17239.

Xie, Yannan, Ting-Mao Chou, Weifeng Yang, Minghui He, Yingru Zhao, Ning Li, and Zong-Hong Lin. 2017. "Flexible Thermoelectric Nanogenerator Based on the MoS_2/Graphene Nanocomposite and Its Application for a Self-Powered Temperature Sensor." *Semiconductor Science and Technology* 32 (4): 044003. https://doi.org/10.1088/1361-6641/aa62f2.

Xu, Liang, Tao Jiang, Pei Lin, Jia Jia Shao, Chuan He, Wei Zhong, Xiang Yu Chen, and Zhong Lin Wang. 2018. "Coupled Triboelectric Nanogenerator Networks for Efficient Water Wave Energy Harvesting." *ACS Nano* 12 (2): 1849–58. https://doi.org/10.1021/acsnano.7b08674.

Xue, Hao, Quan Yang, Dingyi Wang, Weijian Luo, Wenqian Wang, Mushun Lin, Dingli Liang, and Qiming Luo. 2017. "A Wearable Pyroelectric Nanogenerator and Self-Powered Breathing Sensor." *Nano Energy* 38 (August): 147–54. https://doi.org/10.1016/j.nanoen.2017.05.056.

Yamada, Takeo, Alan Maigne, Masako Yudasaka, Kouhei Mizuno, Don N. Futaba, Motoo Yumura, Sumio Iijima, and Kenji Hata. 2008. "Revealing the Secret of Water-Assisted Carbon Nanotube Synthesis by Microscopic Observation of the Interaction of Water on the Catalysts." *Nano Letters* 8 (12): 4288–92. https://doi.org/10.1021/nl801981m.

Yamada, Takeo, Tatsunori Namai, Kenji Hata, Don N. Futaba, Kohei Mizuno, Jing Fan, Masako Yudasaka, Motoo Yumura, and Sumio Iijima. 2006. "Size-Selective Growth of Double-Walled Carbon Nanotube Forests from Engineered Iron Catalysts." *Nature Nanotechnology* 1 (2): 131–6. https://doi.org/10.1038/nnano.2006.95.

You, Ming-Hao, Xiao-Xiong Wang, Xu Yan, Jun Zhang, Wei-Zhi Song, Miao Yu, Zhi-Yong Fan, Seeram Ramakrishna, and Yun-Ze Long. 2018. "A Self-Powered Flexible Hybrid Piezoelectric–Pyroelectric Nanogenerator Based on Non-Woven Nanofiber Membranes." *Journal of Materials Chemistry A* 6 (8): 3500–9. https://doi.org/10.1039/C7TA10175A.

Yudasaka, Masako, Toshiki Komatsu, Toshinari Ichihashi, and Sumio Iijima. 1997. "Single-Wall Carbon Nanotube Formation by Laser Ablation Using Double-Targets of Carbon and Metal." *Chemical Physics Letters* 278 (1–3): 102–6. https://doi.org/10.1016/S0009-2614(97)00952-4.

Zabek, Daniel, Kris Seunarine, Chris Spacie, and Chris Bowen. 2017. "Graphene Ink Laminate Structures on Poly(Vinylidene Difluoride) (PVDF) for Pyroelectric Thermal Energy Harvesting and Waste Heat Recovery." *ACS Applied Materials & Interfaces* 9 (10): 9161–7. https://doi.org/10.1021/acsami.6b16477.

Zhang, Yi, Luyao Zhang, and Chongwu Zhou. 2013. "Review of Chemical Vapor Deposition of Graphene and Related Applications." *Accounts of Chemical Research* 46 (10): 2329–39. https://doi.org/10.1021/ar300203n.

Zhao, Pengfei, Navneet Soin, Kovur Prashanthi, Jinkai Chen, Shurong Dong, Erping Zhou, Zhigang Zhu, et al. 2018. "Emulsion Electrospinning of Polytetrafluoroethylene (PTFE) Nanofibrous Membranes for High-Performance Triboelectric Nanogenerators." *ACS Applied Materials & Interfaces* 10 (6): 5880–91. https://doi.org/10.1021/acsami.7b18442.

Zhuo, Fangping, Qiang Li, Jinghan Gao, Qingfeng Yan, Yiling Zhang, Xiaoqing Xi, and Xiangcheng Chu. 2017. "Phase Transformations, Anisotropic Pyroelectric Energy Harvesting and Electrocaloric Properties of (Pb, La)(Zr, Sn, Ti)O_3 Single Crystals." *Physical Chemistry Chemical Physics* 19 (21): 13534–46. https://doi.org/10.1039/C7CP01762F.

Zhuo, Fangping, Qiang Li, Huimin Qiao, Qingfeng Yan, Yiling Zhang, Xiaoqing Xi, Xiangcheng Chu, Xifa Long, and Wenwu Cao. 2018. "Field-Induced Phase Transitions and Enhanced Double Negative Electrocaloric Effects in (Pb, La)(Zr, Sn, Ti)O 3 Antiferroelectric Single Crystal." *Applied Physics Letters* 112 (13): 133901. https://doi.org/10.1063/1.5018790.

Zi, Yunlong, Simiao Niu, Jie Wang, Zhen Wen, Wei Tang, and Zhong Lin Wang. 2015. "Standards and Figure-of-Merits for Quantifying the Performance of Triboelectric Nanogenerators." *Nature Communications* 6 (1): 8376. https://doi.org/10.1038/ncomms9376.

6 Piezoelectric Nanogenerators

Bhavya Padha, Sonali Verma, and Sandeep Arya

CONTENTS

6.1 INTRODUCTION

Electrical energy is produced at power stations by electromechanical generators via chemical combustion, nuclear fission, or via the kinetic energy of flowing water. With the proliferation of wireless microelectromechanical and nanoelectromechanical devices, the demand for clean and efficient power generation has been increased for self-powering such systems using ambient energy sources like thermal gradient, solar, mechanical vibration, and bio-fluid. Piezoelectricity is a highly flexible phenomenon used for powering small-scale electronic devices directly by converting the mechanical energy available in the surroundings. The piezoelectric technique for power generation through mechanical energy harvesting has garnered considerable attention due to its self-power/wireless charging capability and controllability of the output power [1–6]. Power generation through ambient energy harvesting has a number of applications, such as sensor network devices that monitor and collect useful data about their surroundings. A large number of such tiny sensors should be placed in various locations, like in an office or even within a living organism, for monitoring specific variables. Depending on the location of these networks, powering them sometimes becomes difficult.

DOI: 10.1201/9781003187615-6

In such situations, piezoelectric nanogenerators (PENGs) exhibit the potential of powering these networks. Recent advancements in PENGs open up numerous possibilities for power generation via ambient energy harvesting in practical applications [7–12]. The use of PENGs to generate electricity by using the surrounding vibrations is one approach that gains tremendous popularity in power generation. The PENG's active materials exhibit crystalline structures for effectively converting mechanical strain energy to electricity. Such active materials exhibit the property of absorbing even a very small amount of mechanical energy from their surroundings and convert it into electrical signals capable of powering some devices [13].

Furthermore, the development of a wireless nanosystem, i.e. the integration of nano-devices, functional modules, and a power source, is crucial in nanoscience and nanotechnology [14–16]. Despite their compactness and less power consumption, these wireless systems need their power sources. There are two approaches for implanting such wireless nanosystems. The first is to use a battery. Although a battery exhibits large capacitance, it has a finite life, and also for a compact system, it is important to reduce the size of the battery, which results in reduced battery life. As a result, there is a need to develop long-lasting, compact, and probably lightweight batteries that is very challenging. Furthermore, in biomedical applications, the toxicity of materials used in batteries must be taken into consideration. The alternative is to produce electricity via harvesting ambient energy [17], and for this, using PENG is a viable option. Moreover, nanomaterials exhibit the property of energy conversion, such as PENGs, thermoelectric cells, etc. [18, 19]. On the one hand, the mechanism of energy harvesting is application-dependent. Since solar energy is not always available, it is not the best option for mobile as well as for personal electronics. On the other hand, mechanical energy, which includes vibrations, airflow, human activity, etc., is available almost every time as well as everywhere. This chapter addresses various active materials as PENGs as well as their applications, fabrication techniques, and challenges.

6.2 HISTORY OF NANOGENERATORS

In general, NGs are used to generate electricity from waste mechanical energy present in the surrounding environment. Earlier in the 17th century, some efficient methods for generating electrical energy via friction were introduced. Around 1663, a primitive friction machine was invented [20]. Later, several researchers worked on establishing the accuracy of this machine. The electromagnetic generator was invented in 1831 and continues to be the most commonly employed generator at thermal power stations. The popular Van de Graaff generator that generates extremely high-voltage direct current was invented in 1929. Modern Van de Graaff generators are capable of producing up to 25 megavolts. Later, using the concept of nanotechnology, various technologies for piezoelectric, triboelectric, and pyroelectric electricity generations were introduced. The large surface area along with tunable physical as well as chemical characteristics of nanostructures (NSs) offers more efficient technologies for capturing, converting, and storing various types of energy [21].

Several techniques for extracting ambient energy have been developed to date, including the piezoelectric, triboelectric, and pyroelectric effects, as well as electromagnetic induction, for converting mechanical stress to electrical energy. PENG is a rapidly developing energy harvesting technology for converting mechanical energy, including human motion (i.e. breathing, running, vibration, etc.), into electricity. Wang *et al.* reported the first NG in 2006, via employing piezoelectric zinc oxide (ZnO) nanowires (NWs) to achieve a power conversion efficiency of 17–30 per cent [19].

6.3 WORKING PRINCIPLE

PENGs operate on the piezoelectric principle, which refers to the production of electricity when mechanical stress is applied. In PENG, an external strain is applied to two electrodes having balanced Fermi levels, creating a piezo-driven potential difference among internal and external Fermi levels at the contacts [22–27]. To compensate for this difference, charge carriers pass via an external load, resulting in balanced electrostatic level. There are two types of PENGs [28], one in which each NS experiences force perpendicular to its growing axis [29], resulting in the production of an electric field. Figure 6.1(a) illustrates the operation of PENG on applying a force perpendicular to its axis. When an atomic force microscopy probe is used for applying a force perpendicular to direction of the NS's axis, one part of it is stretched (positive strain), whereas the other is compressed (negative strain). The probe initially made contact with the stretched surface and the bias voltage at this interface is negative. Hence, a Schottky diode with reversed bias is produced with small currents. When the probe's negative potential makes contact with the compressed side, a positive biased voltage with a sharp peak output current is generated at the interface. The current produced at the bottom of the NS eventually balances the electric field produced at its tip. Conduction occurs only when the top electrode makes contact with the negative potential. This is true for n-type semiconductor NSs, whereas in p-type, the effect is reversed because the hole is mobile. Another case is when external force is parallel to the growing axis of the NS (Figure 6.1(b)). On applying force to the tip of a laterally synthesized NWs sandwiched between Schottky and ohmic contacts, a uniaxial compressive force is produced. Due to piezoelectric effect, the tip has a negative potential, as a result of which Fermi level is increased. Positive potential is produced at the tip as electrons flow from the tip to the bottom through the external circuit. The Schottky contact prevents electrons from flowing through the NWs and thus directs them to the external circuit. On removing the applied force, the piezoelectric effect vanishes instantly as well as a positive potential at the tip is neutralized due to electron migration from the bottom to top, resulting in an opposite-direction voltage peak. Zhu *et al.* substituted a PMMA layer for the Schottky contact in their work to establish a possible barrier

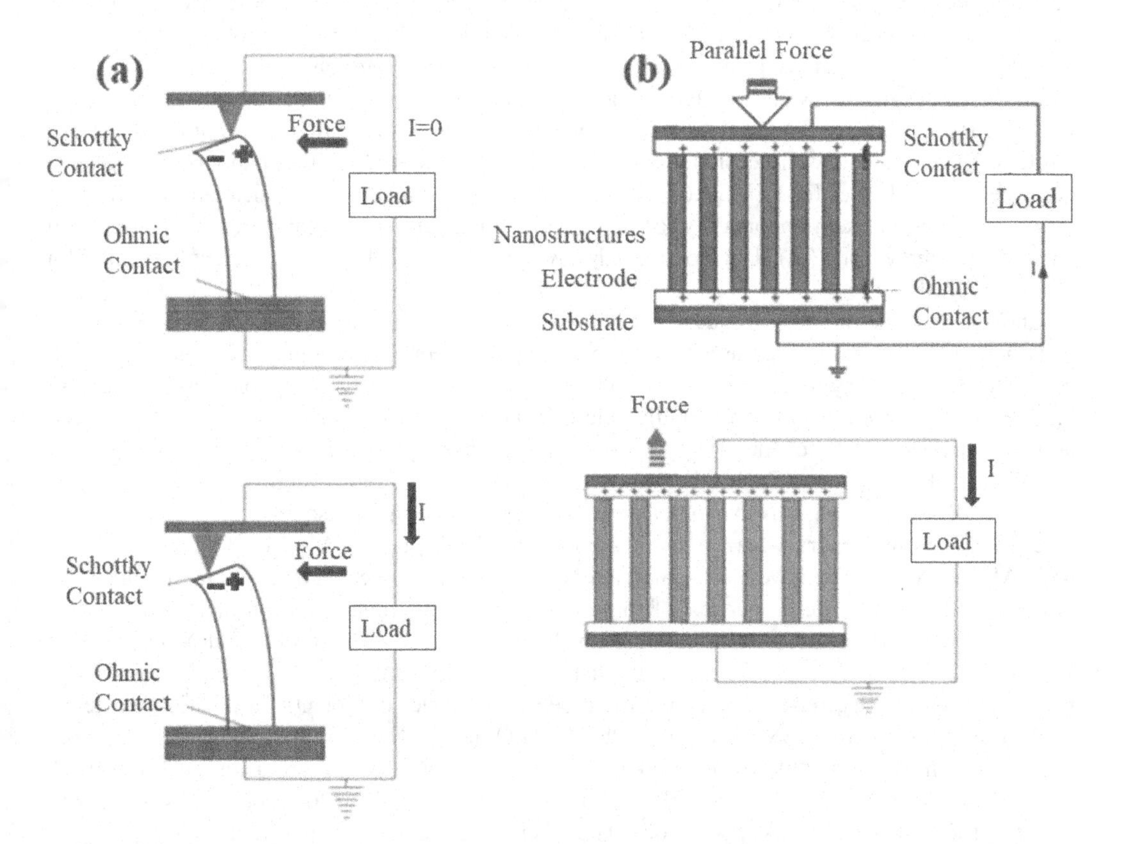

FIGURE 6.1 (a) NW with force applied perpendicular to the growth. (b) NW with force applied parallel to the growth [24]. (Creative Commons Attribution (CC BY) license.)

for charge accumulation [30]. On applying force to the NG, a piezopotential field is created along the NWs. Inductive charges get accumulated on the top and bottom of the electrodes as a result of electrostatic stress. This is analogous to a capacitive configuration. On removing the applied force, the piezopotential vanishes as well as electrons return to the external circuit [31].

6.4 POTENTIAL/ACTIVE MATERIALS FOR PENGs

Different types of materials are explored and utilized as PENGs for energy harvesting. Some of them are as follows:

- Inorganic materials
- Organic materials
- Metal halide perovskites

6.4.1 Inorganic Piezoelectric Materials

Due to their high piezoelectricity, inorganic nanomaterials like $BaTiO_3$ [32], MoS_2 [33, 34], ZnO [35], as well as their composites [36–39] have been used as PENGs. Among all, ZnO has a relatively high piezoelectric constant d_{33} in direction [0 0 0 1]. ZnO is a widely used material for fabricating sensing devices and transistors due to its excellent piezoelectric properties and direct bandgap semiconductor properties [40]. ZnO NSs demonstrated exceptional resilience, withstanding strains of up to 5–7 per cent without visible plastic deformation [41, 42]. ZnO is an excellent PENG candidate due to its flexibility, biosafety, ease of processing, as well as special piezotronic effect. ZnO NSs can be synthesized hydrothermally [43] or via vapour-phase method [44], or a solid-vapour growth technique [45]. Hydrothermal growth in conjunction with photolithography offers vertically aligned ZnO NW array patterns. Numerous techniques, including doping [46–48] and composite structures [49, 50] for improving the properties of ZnO nanomaterials, have been studied. Additionally, due to the following factors, one-dimensional (1D) NSs like NWs, nanorods (NRs), nanotubes (NTs), etc. are excellent candidates for PENGs: (i) they can withstand significant mechanical strains, (ii) their large surface provides an opportunity for surface functionalization for enhancing the physical and chemical properties, and (iii) 1D nanomaterials have a higher piezoelectric coefficient than thin film or bulk equivalents.

ZnO has become the most frequently used material for 1D NSs because it can be easily grown in large amounts at low temperatures using chemical or vapour-solid approaches on a variety of substrates (crystalline, amorphous, rigid, soft, or flexible) with desired morphologies. ZnO is also biocompatible as well as environmentally safe [51]. Due to a lack of inversion symmetry as well as wide bandgap, wurtzite ZnO 1D NSs have been extensively investigated for piezotronics and PENGs [52]. In addition to ZnO, other wurtzite II–VI and III–V piezoelectric semiconductor 1D NSs such as GaN [53–58], InN [59–62], CdS [63], and CdSe [64] have also been investigated in this area. Furthermore, ternary wurtzite semiconductor materials such as $Mg_xZn_{1-x}O$ [65], $In_xGa_{1-x}N$ [66], $Al_xGa_{1-x}N$ [67], etc. have also gained attention, as the piezoelectric coefficient (among other physical and piezoelectric properties) of these materials can be tuned through their composition. Additionally, the effect of axial orientation on the efficiency of piezotronics and NGs has been investigated [68–73]. Non-wurzite materials such as Te [74, 75], $ZnSnO_3$ [76, 77], and $NaNbO_3$ [78] have also been investigated in this area to increase piezoelectric coefficients for the high efficiency of PENGs. Additionally, PENGs based on Pb (Zr, Ti)O_3 (PZT) NWs [8] and nanofibres [79] demonstrated high output voltages of 0.7 and 1.63 V, respectively, but PZT produces toxic lead [80]. Another piezoelectric ceramic material, $BaTiO_3$, has been synthesized for energy harvesting [81], but its brittle nature hinders its applications. Thus, rather than looking for new materials with higher piezoelectric coefficients, it is considered feasible to enhance the piezoelectric properties of the most popular wurtzite piezoelectric semiconductors through microstructure modification.

6.4.2 Organic Piezoelectric Materials

Usually, NGs that convert mechanical energy to electricity are made of metal oxides or lead-based perovskites. However, since these inorganic materials are not biocompatible, a race to develop natural biocompatible piezoelectric materials for energy storage, electronic sensing, and nerve and muscle stimulation is underway. Under applied mechanical stress, reorientation of molecular dipole primarily results in polarization in organic piezo-materials [82, 83]. Such piezo-materials dominate the entire market of electromechanical products, including sensing devices [84–86], energy harvesters [87–89], and storage [90, 91]. Nowadays, medically implantable devices have garnered significant interest [92, 93], indicating that piezoelectric materials are finding new applications. Organic piezoelectric biomaterials exhibit many advantages in comparison to inorganic materials, including a high degree of biocompatibility, outstanding stability, eco-friendly as well as high processability. Since the detection of polarization in asymmetric biological tissue [94], numerous researchers not only worked on understanding the fundamental concept of piezoelectricity in such materials but also tried to improve their physiochemical properties via molecular designing, nanostructuring, and doping [95]. Though organic piezo-materials have lower piezoelectric characteristics than inorganic materials, recent studies indicate that when properly processed, biocompatible piezo-materials that interfaced with the biological systems will act as functional materials in medically implantable and mountable applications. Since organic piezoelectric materials apply to a wide variety of devices, from nano- to millimetre-scaled devices, device fabrication has some challenges due to local damage as well as non-local elasticity [96, 97]. The following section discusses some organic piezoelectric materials.

6.4.2.1 Piezoelectric Proteins

Since collagen, a major structural protein in the extracellular matrix of tissues, is primarily responsible for the piezoelectricity in bones but the mechanism is not clear even till now. Numerous theories describe the origin of the piezoelectricity in collagen fibril, such as non-centrosymmetric structure, presence of polar bondings, reorientation of C=O–NH bonding as well as the polarization of H-bonds in collagen are available in literature [98–102]. According to a recent analysis, the piezoelectricity in collagen is caused by reorientation and magnitude shift of permanent dipoles of every charged and polar residue towards collagen fibril's long axis, as illustrated in Figure 6.2(a). The reported shear piezoelectric constants for collagens impregnated by bone and tendon range from $d_{14} = 0.2$ to 2.0 pC/N [103]. Recently, the M13 bacteriophage has emerged as a useful material for a variety of applications, including energy harvesting [104–106], chemical sensing [107–109], as well as in tissue regeneration [110, 111]. It is a filamentous bacterial virus with definite dimensions, composed of single-stranded DNA wrapped around 2700 copies of major proteins as well as covered at the top or bottom ends by five copies of minor proteins. Every major protein contains a dipole moment in the direction of the DNA axis, resulting in the phage remaining permanently polarized in both the axial and radial directions (Figure 6.2(b)). The piezoelectric constant in the radial direction is $d_{33} = 7.8$ pm/V [104], whereas in the axial direction, it is approximately three times greater [105]. Lee *et al.* recently increased the value for axial direction up to 26.4 pm/V via sticking M13 unidirectionally [106].

6.4.2.2 Piezoelectric Peptides

Glycine (G) is a zwitterionic amino acid that serves as an excellent model for studying polymorphic crystallization [112–114]. At ambient conditions, crystalline glycines exhibit three different structures, α, β, and γ, among which β- and γ-structures exhibit shear piezoelectric effect owing to their acentric configuration [115–118]. As with inorganic materials, piezoelectricity is produced when an ion is displaced inside the crystal, and this displacement results in the formation of the dipole in the local material as well as net polarization in the bulk material, as illustrated in Figure 6.2(c). β-structure exhibits good shear piezoelectric constant, $d_{16} = 190$ pm/V [119], that is analogous to barium titanate's ($BaTiO_3$) piezoelectric constant [120]. Diphenylalanine (FF) consists of

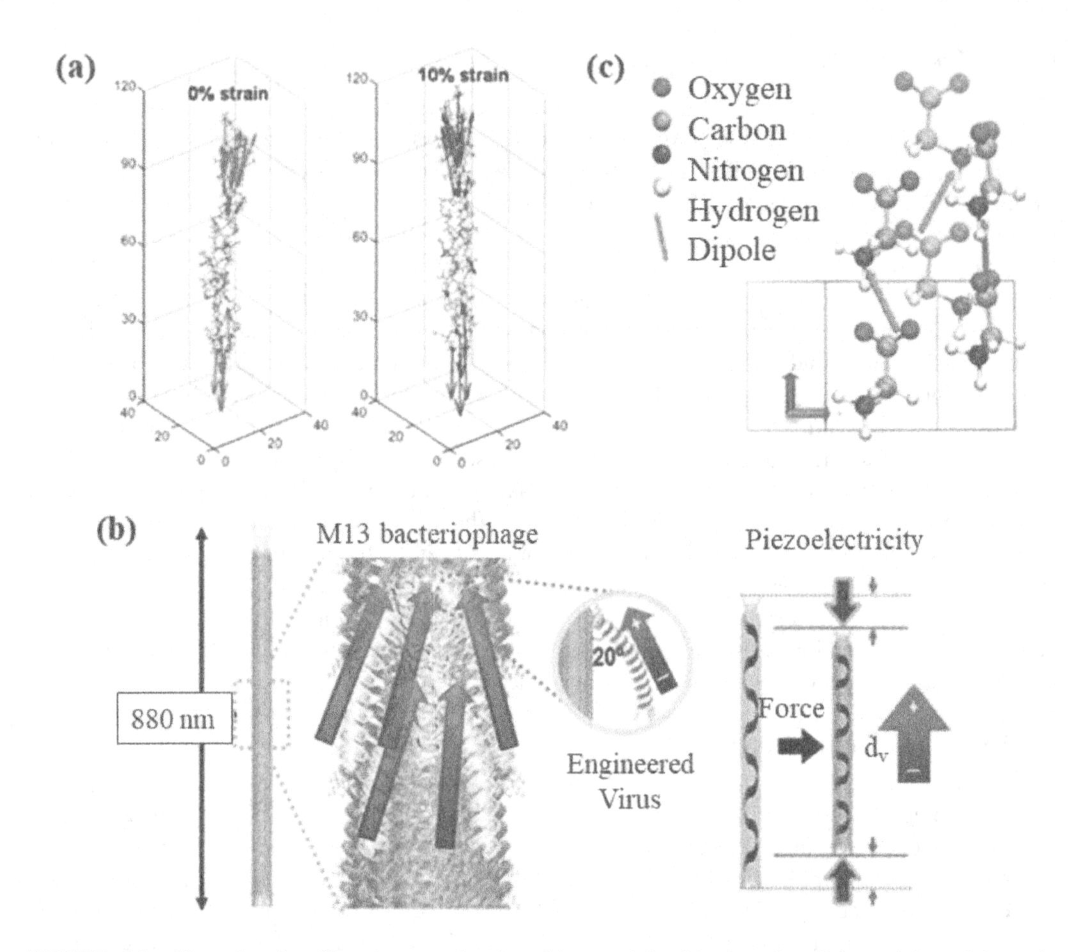

FIGURE 6.2 Piezoelectric effect in organic piezo-biomaterials: (a) piezoelectricity originated by collagen molecules (reproduced from [98] with permission. Copyright 2016, American Chemical Society), (b) piezoelectric M13 bacteriophage (reproduced from [105] with permission. Copyright 2015, Royal Society of Chemistry), and (c) unit cell of β-glycine crystal [118] (Creative Commons Attribution (CC BY) license).

two phenylalanine (F) amino acids and can self-assemble into semi-crystalline peptide NTs and microrods as well as exhibits several desirable properties such as morphological and functional diversity, good biocompatibility as well as a high Young's modulus [121]. Diphenylalanines with NSs are extensively investigated piezo-materials with non-centrosymmetric hexagonal space group (P61) [122]. Such crystalline material demonstrates a variety of physical properties such as piezoelectric, pyroelectric, ferroelectric effects, as well as enantiomorphism [123]. The shear piezoelectric constant of peptide NTs has been increased to $d_{15} = 60$ pm/V [124]. For enhancing scalability and homogeny of the semi-crystalline film, unidirectionally polarized as well as aligned diphenylalanine NTs films were synthesized via meniscus-driven self-assembly method [125], resulting in $d_{15} = 45$ pm/V that is equivalent to highly crystalline materials. Additionally, Nguyen *et al.* focused on achieving piezoelectric effect (up to $d_{33} = 17.9$ pm/V) in nano-scaled peptides through vertical alignment of every microrod [126, 127].

6.4.2.3 Other Piezoelectric Biopolymers

Poly(L-lactic acid) (PLLA) is a polymorphic polymer with tremendous biodegradability as well as biocompatibility. External stimulus, like electrospinning, induces unidirectional orientation of the dipoles in the stretched path [128], resulting in the β-crystalline structure and a shear piezoelectric effect of ~12 pC/N [129, 130]. Interestingly, β-crystalline PLLA, in addition to having reasonable piezoelectricity, does not need polling because of its helical structure, thus expanding its utilization in biocompatible devices [131].

Additionally, poly(vinylidene fluoride) (PVDF) and poly(vinylidene fluoride-trifluoroethylene) (P(VDF-TrFE)) is the most studied piezoelectric polymers and have significant potential for electronic skins due to their structural versatility, biocompatibility, chemical stability, and piezoelectricity [132, 133]. Four different phases, i.e. α, β, γ, and δ, exist in PVDF, in which β-phase shows strong piezoelectricity and α-phase is non-piezoelectric [134]. The α-phase can be obtained easily via solution casting/solvent evaporation and transformed into β-phase via drawing or traditional stretching [135, 136]. In β-phase, orthorhombic structure with parallel fluorine and hydrogen branches contributes to net dipole moment as well as to piezoelectric effect, while in the hexagonal structure of α-phase, the net dipole moment is cancelled out via dipole's anti-parallel arrangement. Electrospinning technology can be used to fabricate 1D PVDF NSs in the β-phase or mixed-phase state. Voltage, solvent, electrospinning temperature, flow rate, and other process parameters can all be used to regulate polymorphism and morphology [137]. For example, during the electrospinning process, high pressure, a high solvent evaporation rate, a high chain stretching ratio, and a high rotation speed all contributed to the development of β-phase PVDF [138, 139]. Electrospinning in combination with poling allowed the fabrication of uniformly polarized β-phase PVDF NWs [140].

A method for preparing free-standing P(VDF-TrFE) nanofibres for pressure/force sensors was published [141]. They used a fast-rotating collector to fabricate versatile and large-area films from aligned P(VDF-TrFE) fibres. The P(VDF-TrFE) textile was utilized to detect pressures as low as 0.1 Pa due to its good piezoelectric response and wide contact area. Additionally, electrospinning was used to fabricate PVDF composites, such as PVDF/$BaTiO_3$ nanocomposite fibres [142]. The electrospun PVDF fibres/wires had higher piezoelectricity than PVDF films, which made them ideal for the application of highly sensitive electronic skins [143].

6.4.3 Metal Halide Perovskites

Due to high electronic as well as optoelectronic applications, halide perovskites garnered considerable interest over the last few years. Their fascinating ferroic properties, particularly ferroelectric and piezoelectric properties, open up new doors for mechanical energy harvesting. Hybrid perovskite PENGs were investigated for mechanical energy harvesting by using their superior ferroic characteristics. Kim *et al.* reported sandwiching of 500-nm thick polycrystalline $MAPbI_3$ films between two PET substrates using PDMS as a binder [144]. PDMS served as a barrier for charge carriers during system usage, thereby reducing current leakage. PENGs demonstrated high output voltage and current density of 2.7 V and 140 nA/cm^2, correspondingly, via poling with strong electric field (80 kV/cm). Using such a device as an example, its working mechanism applies to the majority of PENGs, as illustrated in Figure 6.3(a) [144]. After getting poled in the strong electric field, the domains inside $MAPbI_3$ polycrystalline film with random orientation align parallel to the direction of the electric field. On applying force, a piezoelectric potential is produced inside perovskite films, which pushes electrons through the external circuit from the bottom electrode to the top electrode, generating positive voltage/current signal. After that, negative charge gets accumulated at the top electrode and brings piezopotential to equilibrium. As soon as force is removed, the piezopotential of the $MAPbI_3$ film vanishes, and the stored electrons migrate towards the bottom electrode, producing negative voltage. In comparison to non-poled PENGs, aligned polarization domains result in high output efficiency. The fusion of energy harvesters that incorporate photovoltaic, pyroelectric, as well as piezoelectric effects, has sparked considerable interest in multisource energy harvesting [145–147]. Ferroelectrics are the subsets of pyroelectric and piezoelectric materials. Owing to the promising features of ferroelectric $MAPbI_3$ such as its ambipolar existence, ferroelectricity, and ultrahigh Seebeck coefficient, its use as a fusion material gains attention. $MAPbI_3$-based fusion energy harvesters have been developed by making use of interdigitated electrodes with buffer layers of n-type ZnO along with p-type Cu_2O [148]. These harvesters, which are based on a

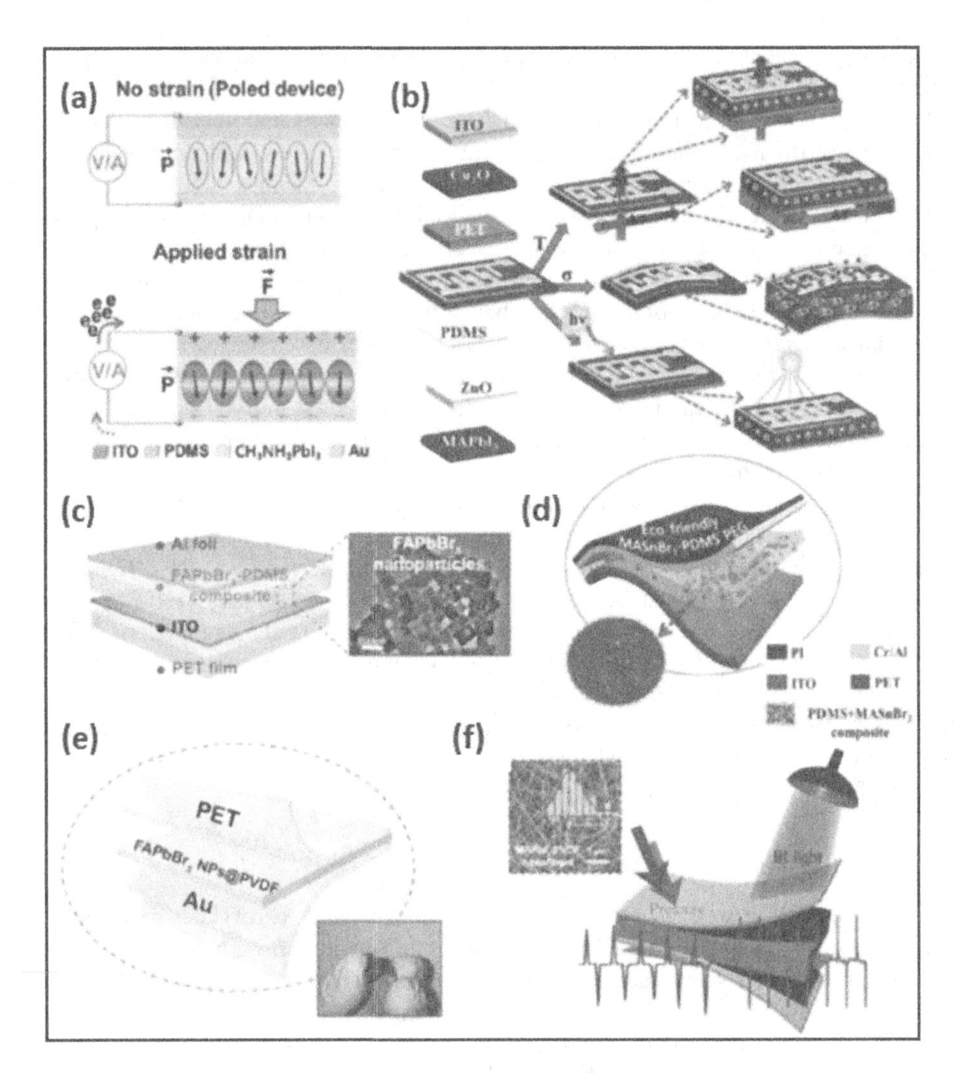

FIGURE 6.3 (a) Power production processes of $MAPbI_3$ thin-film PENGs under poled condition, (b) architecture scheme and operating procedure of fusion energy harvesters based on $MAPbI_3$ polycrystalline films, (c) illustration of $FAPbBr_3$-polydimethylsiloxane (PDMS) nanocomposite PENGs, (d) illustration of $MASnBr_3$-PDMS PENGs, (e) illustration of the $FAPbBr_3$-polyvinylidene difluoride (PVDF) PENGs, (f) illustration of $MAPbI_3$-PVDF composite NFs for piezoelectric-pyroelectric PENGs [162]. (Creative Commons Attribution (CC BY) license.)

single active layer of $MAPbI_3$, are capable of converting thermal, mechanical, as well as solar energy, into electric power (Figure 6.3(b)), as demonstrated by their impressive results. Furthermore, when subjected to temperature difference, the harvester may act as a thermoelectric generator, offering superior thermoelectric production along in-plane temperature gradient. It also acts as a piezoelectric material with the piezoelectric output of 1.47 V and 0.56 μA when subjected to periodic strain. In terms of photovoltaic efficiency, electrical poling is needed to create p-i-n junction in the interior of the cell. The harvester attains an open-circuit voltage of 0.77 V as well as a current density of 0.022 mA/cm^2 under illumination by light. Owing to its greater chemical stability in comparison to other hybrids, the inorganic halide perovskite $CsPbX_3$ has drawn considerable interest as a high-performance perovskite material. The piezoelectric polarization effect has been demonstrated in $CsPbBr_3$ NWs [149]. Recent research has demonstrated the use of $CsPbBr_3$ polycrystalline films as a piezoelectric material in PENGs for electro-mechanical energy harvesting as well as for current sensing devices [150]. On the one hand, the structure of $CsPbBr_3$ films will be optimized for achieving a high piezoelectric coefficient of 40.3 pm/V at subsequent poling field. Due to high piezoelectric effect, the bonding angles of Pb-Br-Pb and Br-Pb-Pb were extended along b-axis in ac-plane of perovskite configuration. Outstanding output results with a voltage of 16.4 V as well as current of 604 nA were achieved in the optimized $CsPbBr_3$-based PENG along with good durability of up to 3000 bending cycles with only 7 per cent deprivation. On the other hand, a serious issue arises in their practical applications due to their brittleness as well as rigidity. The incorporation of piezoelectric perovskite ceramics in a soft polymer matrix to form piezo-nanocomposite was suggested as a solution. $FAPbBr_3$ nanoparticles were prepared and dispersed in PDMS matrix to form a nanocomposite of $FAPbBr_3$ and PDMS [151]. PFM was used to confirm the presence of ferroelectricity in $FAPbBr_3$ nanoparticles, and a relatively high piezoelectric coefficient of 25 pm/V was measured. The PENGs have developed via spin-coating the composite solution directly onto an indium tin oxide (ITO)-coated PET substrate as the bottom electrode and encapsulating it with an aluminium sheet as a top electrode (Figure 6.3(c)). Under periodic vertical compression and release operations, the piezoelectric nanocomposite generators generated a maximum piezoelectric output voltage and current density of 8.5 V and 3.8 A/cm^2, respectively. The mechanism by which cubic $FAPbBr_3$ nanoparticles exhibit ferroelectricity-like behaviour is fascinating. The polarization in the $FAPbBr_3$ perovskite was suggested to be caused by the orientational polarization of FA^+ cations. According to theoretical simulations of Stroppa *et al.*, FA^+ cations within cubic $FASnI_3$ perovskites can arrange perpendicularly in a "weak" but a more stable state with a polarization of 0.88 C/cm^2, ensuring the ferroelectricity of FA^+ cation-based perovskites [152]. The PENG based on $MAPbI_3$-PDMS was optimized on the basis of degree of $MAPbI_3$ crystal defect that is analysed via positron annihilation spectroscopy [153]. Moreover, the lattice imperfections play a critical role in determining the ionic polarization concerning the piezoelectric effect of $MAPbI_3$. Environment-friendly PENGs with an enhanced piezoelectric output voltage of 18.8 V and current density of 13.76 A/cm^2 were also fabricated using lead-free $MASnBr_3$ powders (Figure 6.3(d)) [154]. Under ambient conditions, the low stability of Sn-based perovskites was enhanced via enwrapping in PDMS that resulted in 120-day system stability. Furthermore, owing to its ferroic activity, dielectric properties, as well as ease of solution processing, PVDF polymer exhibits the potential in this field of energy harvesting [155]. As a result, the $FAPbBr_3$-PVDF hybrid with well-dispersed $FAPbBr_3$ nanoparticles incorporated the advantages of $FAPbBr_3$ nanocrystals as well as the piezoelectric properties of PVDF. PENGs were constructed by sandwiching $FAPbBr_3$-PVDF between two Au/Cr-coated PET films (Figure 6.3(e)) [156]. Due to PVDF's high Young's modulus, competent stress was transferred efficiently from rigid PVDF to dispersed $FAPbBr_3$ nanomaterial. Maximum output voltage and current density measured from $FAPbBr_3$-PVDF-based PENG were 30 V and 6.2 A/cm^2, respectively. Additionally, $MAPbI_3$ nanoparticles were also embedded in PVDF [157], and the obtained output voltage and current density of $MAPbI_3$-PVDF-based PENG were 45.6 V and 4.7 A/cm^2, respectively. Self-assembled macro-porous configuration has been found in $FAPbBr_2I$-PVDF composites, which can dramatically increase the bulk strain of piezoelectric composite films, resulting in a fivefold increase in piezopotential [158]. Sultana et al. reported a piezoelectric-pyroelectric NG (PyENGs) that shows output power densities of approximately 0.8 and 0.2 mW/m^2 as a result of mechanical vibrations and temperature fluctuations, respectively, due to excellent electrochemical characteristics of $MAPbI_3$-PVDF composite nanofibres (Figure 6.3(f)) [159]. Additionally, all-inorganic $CsPbBr_3$ perovskite rods embedded in PVDF nanofibres cause nucleation of β-phase in PVDF at a rate of over 90 per cent that makes them highly suitable for PENGs [160]. Furthermore, Pandey *et al.* used first-principles calculations to determine the microscopic fundamental of the piezoelectric effect in lead-free $FASnI_3$ material [161]. The high piezoelectric response of $FASnI_3$ has been attributed to its soft elastic behaviour as well as due to soft polar optic phonons. The piezoelectric coefficient of $FASnI_3$-PVDF composite was about 73 pm/V, indicating that this NG could generate an output voltage of 23 V with a power density of ~35.05 mW/cm^2. On the basis of the earlier discussion, it was concluded that perovskite materials open new doors for facile as well as economical fabrication of highly efficient PENGs.

6.5 FABRICATION TECHNIQUES: PENGs

NGs have evolved in connection to material growth techniques. To fabricate the core NS, vapor-solid-solid systems, vapor-liquid-solid systems, pulse-laser deposition, and various other chemical approaches are employed [163]. Numerous materials, including ZnO, gallium nitride (GaN), indium nitride (InN), and cadmium sulphide (CdSe), were developed and found suitable to act as NGs [2, 3, 64, 164–167]. Although great progress has been made in materials and technologies, minimal growth is seen in structural design. The vertically grown array of NWs, which began with a prototype NG, continues to be the most commonly used core NS [2, 3, 64, 163–168]. One apparent explanation for employing vertically standing structures is that their efficiency metrics are quite impressive. A more plausible explanation is that the methods for fabrication of these structures are limited by bottom-up synthesis methods. As the NG advances more rapidly, structural modelling becomes an extremely critical topic. Cha *et al.* designed an NG using a piezoelectric nanopore array structure in 2011 [169]. About 45 per cent increment in production piezopotential was observed by extensive experimental tests. In particular, the preparation techniques involve hot-pressing, spin-coating, and electrospinning [170]. The following sections address various techniques for fabricating PENGs.

6.5.1 The Sol-Gel Process

The two most powerful strategies for optimizing the output efficiency of flexible PENGs include the usage of crystalline ferroelectric substances having a high piezoelectric coefficient, like barium titanate (BTO/$BaTiO_3$) and lead zirconium titanate $PbZr_xT_{1-x}O_3$ (PZT), to deploy NSs capable of efficiently absorbing applied physical energy [171, 172]. While these strategies help in achieving high output voltages, they have difficulty in achieving uniform output efficiency due to the vaguely distributed nanoparticles exhibiting piezoelectricity in the composite NSs. Hence, a basic, repeatable, and customizable fabrication method is required for the production of low-dimensional NSs over flexible substrates. A suitable strategy for fabricating fairly homogeneous piezoelectric nanocrystals over a plastic substrate includes the sol-gel method, which enables the formation of various piezoelectricity-based metal-oxide films through the simplified methodology of spin-coating [173, 174].

Nonetheless, it remains hard to fulfil both the production of homogeneous NSs and enhance the crystalline character of the sol-gel film deposited over plastic substrates. Though sintering is needed to improve the crystalline nature of the deposited film [175] required by PENGs to generate high piezoelectric output strength, only a specific temperature is apt for plastic substrates. Earlier, inorganic sol-gel precursor systems were commonly used in nanoimprint lithography to create nanopatterns [176–178].

- A report on the significantly improved performance of iron (Fe)-doped $BaTiO_3$ nanopillar array PENGs is presented here. These nanopillars were developed in a uniform pattern over the sol-gel-deposited polyethylene terephthalate (PET) substrate with the help of a polydimethylsiloxane (PDMS) stamp, enabling customizable and improved piezoelectric production efficiency. Additionally, exposure of the ultraviolet (UV) rays over the patterned sol-gel layer aided in the microstructure improvement of the sol-gel $BaTiO_3$ layer. As a result, the current density and output voltage of the PENG encompass the values 1.2 A/cm^2 and 10 V, respectively. The pitch length between the pillars was around 400 nm (area: 1 cm^2).
- **Synthesis of $BaTiO_3$ sol-gel solution**: The Fe-doped $BaTiO_3$ sol-gel was produced with barium hydroxide octahydrate ($Ba(OH)_2{\cdot}8H_2O$, 95 per cent), titanium(IV) butoxide ($Ti(OCH_2CH_2CH_2CH_3)_4$, 97 per cent), and iron(III) nitrate nonahydrate ($FeNO_3{\cdot}9H_2O$, 98.5 per cent). The molar ratio of Ba:Ti:Fe was 1:0.98:0.02. Barium hydroxide was applied

to a 1:1 volume ratio mixture of ethanol and acetic acid, and the combined solution was calcinated to 60°C while stirring until the barium hydroxide was dissolved fully. After bringing the solution to room temperature, titanium butoxide was gradually added followed by iron nitrate. The ageing process was repeated until the desired viscosity was achieved.

- **BTO nanopillar patterns processing:** To pattern barium titanate hexagonal nanopillar array, ITO (130 nm thick)-coated PET was first prepared and cleaned with IPA, acetone, and deionized (DI) water, respectively. Synthesized barium titanate sol-gel was then poured on the pre-cleaned ITO surface, followed by covering with a PDMS mould to pattern nanopillars. Lastly, the PDMS stamp was removed from the ITO/PET substrate, leaving hexagonal BTO pillar patterns on the substrate.
- **Exposure of UV light:** After the high-voltage poling procedure, the nanoimprinted BTO sol-gel on the ITO/PET films was subjected to ultraviolet (UV) light (I~ 365 nm) for 2.5 minutes. The UV light source utilized had an intensity of about 80 mW/cm^2.
- **Fe-doped $BaTiO_3$ sol-gel PENG fabrication procedure:** Polydimethylsiloxane (PDMS) was spin-coated on both bare ITO/PET and BTO sol-gel-coated ITO/PET for about half a minute at 4k rpm. Following that, each substrate was kept for nearly 5 minutes on a hot plate set to 80°C. Following that, two pre-cured substrates were coupled with care to prevent the creation of air bubbles in the PDMS. Following that, the unit was treated effectively for about 10 hours at 95°C. Finally, high-voltage poling was conducted for 10 hours under an applied field of 1 MV/cm and the substrate temperature of about 135°C. A schematic of the same is shown in Figure 6.4.

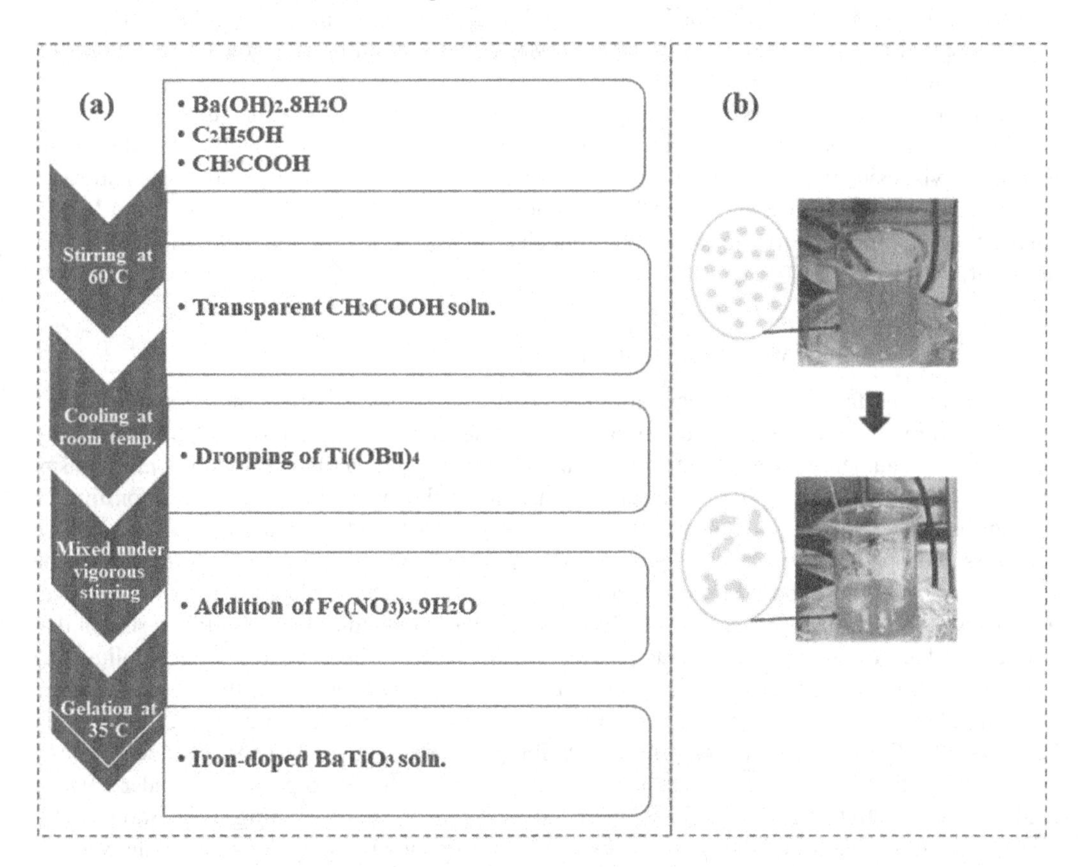

FIGURE 6.4 (a) The synthesis of Fe-doped $BaTiO_3$ sol-gel solution is depicted in the flowchart; (b) picture of sol-gel solution as-synthesized. (Reproduced from [179] with permission. Copyright 2017, American Chemical Society.)

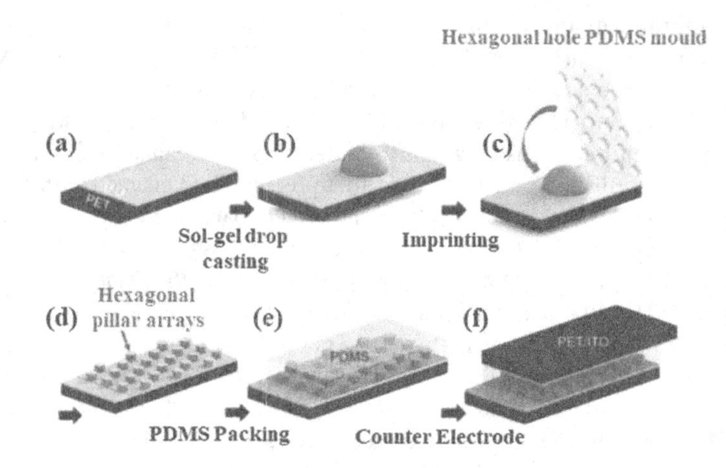

FIGURE 6.5 Schematic diagram for the device fabrication procedure. (Reproduced from [179] with permission. Copyright 2017, American Chemical Society.)

To begin, the ITO/PET substrate was cleaned with IPA, acetone, DI water, and N_2 blow-dried, and the prepared substrate was spin-coated with the synthesized Fe-doped barium titanate sol-gel solution (Figure 6.5a and b). Following that, a relatively populated and homogeneous Fe-doped BTO nanopillar array was developed using a PDMS mould containing hexagonal hole patterns on the sol-gel-coated ITO/PET (Figure 6.5c and d).

We used the soft nanoimprint technique to create sol-gel NSs on flexible substrates because the coated sol-gel layer could be conveniently shaped into the form of the stamp due to its low initial viscosity, mechanical strength, and electrical stability. Following that, the patterned arrays were treated under a variety of UV and poling treatments. Finally, a spin-coated PDMS solution was applied to both the top and bottom electrodes, mixing the top and bottom electrode layers of PENGs [179].

6.5.2 The Methods Based on Electrospinning

It's a widely used method for fabricating fibre-based membranes by stretching polymer droplets or melting directly with the aid of electrical forces. It has received widespread attention in recent years due to its numerous benefits, including simplified machinery, a tunable spinning mechanism, reduced cost, and a diverse material selection. It has in turn become a primary method for directly and consistently producing nanofibres (NFs) [180–183]. It is based upon a mechanism in which compressed drops of polymer or molten materials transform into a "Taylor cone" structure at the spinning nozzle when exposed to a high applied field. Consequently, the constant high electrical force causes droplet that is charged to resolve its surface tension and escape, forming a stream that is spread and narrowed consistently along the motion. Meanwhile, the solvent vaporizes, solidifying the stream and resulting micro/nanosized fibres that are gradually deposited at the receiver yielding a thin film having a spontaneous and non-woven cloth structure [184–186]. Baii *et al.* illustrated in 2011 that PVDF films, post-electrospinning, exhibit great piezopotential in the absence of poling and could be utilized as a film. This is because electrospinning has been performed under a strong field, and there is stretching of PVDF and polarized during the phase of spinning [187]. Additionally, Yee *et al.* demonstrated that the crystallinity of PVDF films made using electrospinning was primarily composed of polar crystallites, which corroborated the preceding findings [138]. Still, there are a few difficulties encountered along with the electrospinning of these films. As a matter of fact, distorted composition of fibres inside the PVDF films has a partially offset effect, thereby reducing the fibre films' overall piezoelectric efficiency. This section discusses four popular electrospinning techniques in depth and summarizes some guidelines for interpreting this fabrication technique.

To begin, Deng *et al.* synthesized the piezoelectric sensors with ZnO/PVDF NFs based on cowpea structure directly by electrospinning the diffused ZnO/PVDF solution (as shown in Figure 6.3(a)) [188]. To ensure a smooth and efficient electrospinning operation, nanoscale ZnO was chosen in addition to increase the crystalline nature of the β-phase. The architecture of distorted nanofibres embedded in ZnO enhanced piezoelectric efficiencies, like sensitivity and response time, allowing controlled movements (Figure 6.6(b and c)). Moreover, the process of deposition of electrode is of widespread importance since it sprays versatile MXene (Ti_3C_2) electrodes on both sides of the microstructure, thus avoiding short circuits triggered by electron beam deposition. Even though the system possesses advantageous characteristics, the distorted configuration of PVDF NFs impairs the device's voltage performance. Different investigations have tried to

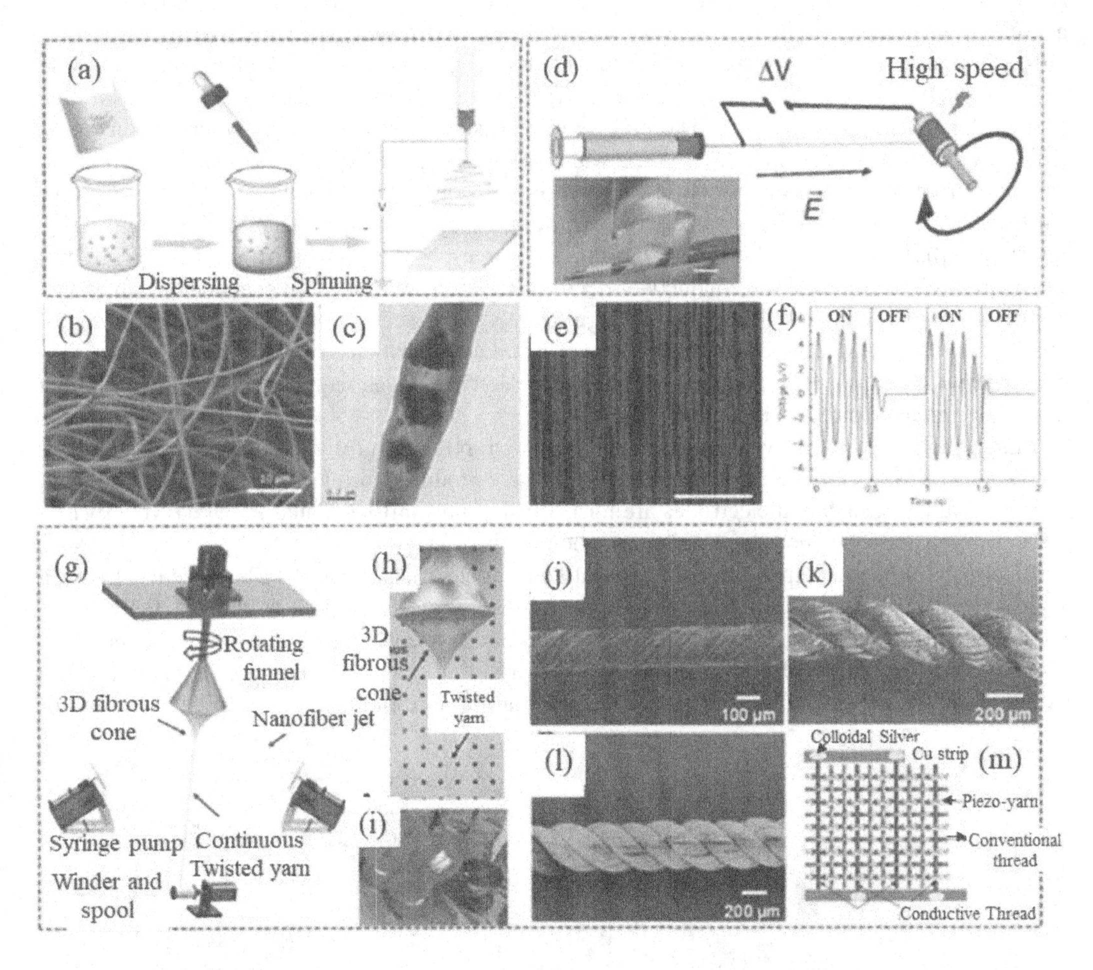

FIGURE 6.6 (a) Electrospun random ZnO/PVDF flexible film fabrication process; (b) NF's SEM image; (c) TEM micrograph of highly aligned P(VDF-TrFE) NFs; (d) schematic representation of electrospun highly aligned P(VDF-TrFE) film; (e) SEM micrograph of highly aligned P(VDF-TrFE) fibres; (f) output voltage measured for this device upon exposure to sound of about 70 dB in magnitude [141] (Creative Commons Attribution (CC BY) license); (g) apparatus employed for producing constantly twisted yarns; (h) 3D fibre-based cone and twisted yarn about the cone; (i) twisted yarns procured on reels. SEM micrographs of (j) one-ply yarn; (k) two-ply yarn; and (l) three-ply yarn of P(VDF-TrFE); and (m) piezoelectricity-based cloth with indicated conductive threads. (Reproduced from [191] with permission. Copyright 2017, American Chemical Society.)

minimize this adverse effect by constructing a closely aligned arrangement of PVDF nanofibres [189]. For instance, Persano *et al.* used a collector with a high linear speed (greater than 16 m/s) in order to fabricate a wide P(VDF-TrFE) sheet with oriented NF structures (Figure 6.6(d and e)) [141]. Owing to the stretching produced by the high rotational speeds, the polymer chains assume a highly preferential configuration, improving the piezoelectric effect in the direction of orientation. Figure 6.6(f) demonstrates that the sensor could detect the minute air vibrations induced through sound having magnitude of 70 dB, highly sensitive. In general, the PVDF film post-electrospinning can be used to create either disorganized or aligned structures depending on the application. In recent times, attempts have been made to create a more adaptable one-dimensional PVDF piezoelectric fibre for wearable applications. Gao *et al.* created a novel mechanism for electrospinning in 2018 for specifically wrapping a sheet of PVDF NFs post-electrospinning on an Ag-deposited filament of nylon serving as the inner electrode [190]. The benefit of the core-sheath arrangement is that the PVDF film is placed in the middle of inner core and outer electrode, maximizing energy harvesting in practice by maximizing the use of piezoelectric substances. Similarly, Yang *et al.* introduced another structure and fabrication method for one-dimensional electrospun P(VDF-TrFE) yarn [191]. In contrast to the core-sheath arrangement, the electrode's inner core was omitted during planning (Figure 6.6(g–i)). The procedure consisted of two stages. To begin, the network, which was electrospun, accumulated over a wide end of a rotating funnel, and the wound yarn was extracted from the three-dimensional cone of fibre and wound about the winder's reel. Adaptable yarns of P(VDF-TrFE) having a variety of twist numbers are available, including one-ply (seen in Figure 6.6(j)), two-ply (seen in Figure 6.6(k)), and three-ply yarns (seen in Figure 6.6(l)). Once woven into cloth, the pair of conductive electrodes were constructed in the manner shown in Figure 6.6(m). Even though the strength is significantly increased after twisting, the processing of these left-right electrodes eventually introduces certain complications, including inefficient energy conservation. As a result, fabricating fibrous piezoelectric devices requires special attention that must be given to electrode placement [192].

Maity *et al.* [193] also developed a self-driven PVDF NF-based fully organic piezoelectric sensor, as illustrated in Figure 6.7, wherein PVDF NFs were produced via electrospinning method work as the active layer, such that electrodes are formed using polyaniline-coated PVDF (PANI-PVDF) NF mats prepared via vapor phase polymerization method.

Electrospinning is a better option for fabricating piezoelectric polymer NFs because it achieves a high aspect ratio by stretching the polymer fibres as they are ejected as jets. As a result, the β-phase induction and in situ arrangement of CF_2/CH_2 dipoles occur concurrently, obviating the need for postdeposition or electrical poling steps. It is worth noting that the polar-phase component and extent of dipole alignment are the critical factors that measure the final piezoelectric response of

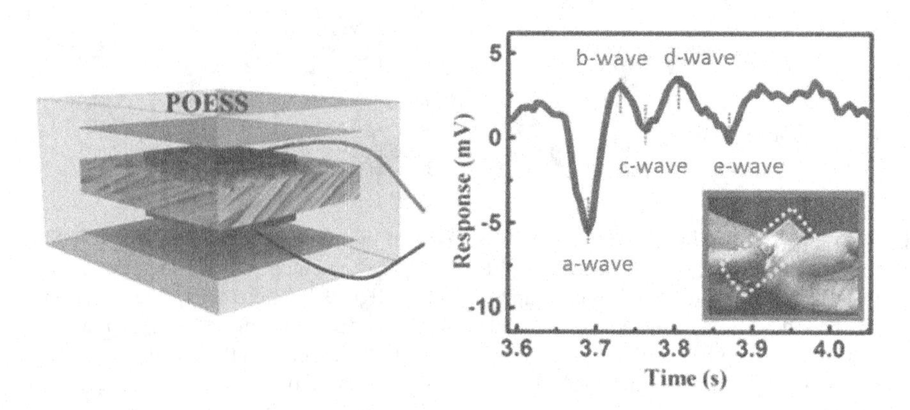

FIGURE 6.7 Piezoelectricity-based organic e-skin sensor design structure. (Reproduced from [193] with permission. Copyright 2020, American Chemical Society.)

PVDF. As a result, electrospun PVDF NFs are incredibly impressive for use in NGs due to their low weight, air permeability, flexibility, and wearability. Nevertheless, shortcomings remain in practice as a result of substandard results and a short lifespan. Due to the fact that piezoelectric devices operate in response to external stimuli like bending, pressure, twisting, and compression, free-standing high-density arrays of piezoelectric PVDF nanofibres having a wide area and higher mechanical durability are needed to fabricate a supersensitive piezoelectricity-based sensor. PVDF NFs produced through the conventional electrospinning process sag quickly in the surrounding atmosphere, reducing their efficacy. Due to the sensor's low output power at low frequencies, it is only useful in a restricted variety of relevant, especially wearable purposes. To improve their response, a novel technique is used in which additional mechanical stretch is applied during the electrospinning, leading to an improved arrangement of the dipoles. The aligned arrays of these NFs exhibit a highly dense and preferred alignment, leading to a better piezoelectric and mechanically robust material. In comparison, once the widely used non-rotating plate collector is employed in the electrospinning method, arbitrary orientation NFs are collected. By and large, these randomly oriented NFs are incapable of forming wide-area mats having an extremely high extent of dipole orientation and homogeneity [193].

6.5.3 Spin-Coating Techniques

Spin-coating was utilized to prepare fine films for a century and has evolved into the chosen mode for processing the majority of polymer usable films. The underlying idea of spin-coating has always been adding a drop of polymer solution amid the smooth surface and then rapidly spinning it. Huge centrifugal force leads to the majority of the solution being wasted, leaving thin layer of substance on substrate's surface for subsequent functional film creation. It should be noticed that volatile solvent is often used for vaporization. The features of the synthesized film could be enhanced by adjusting rotation speed, time, droplet size, and solution composition. Presently, despite some drawbacks like lack of pattern, this strategy has numerous advantages like lesser cost, simple procedure, ease of operation, and low emission, providing a great opportunity as far as the field of the spin-coating system is concerned [194]. The critical procedures include spinning and heating, affecting directly the final thickness and phase of piezoelectric material. When heating methods like hot plate and oven are used, the vaporization and crystallization are very distinct. On the one hand, for instance, when PVDF is isotopically heated in an oven, from surface to interior, it results in a more accelerated crystallization. Even then, in the inner field, vaporization is slowed. In other words, this process enhances the β-phase content while decreasing the electroactive characteristics through the trapping of solvent. And on the other hand, the hot-plate approach will generate a uniform effect of heat starting at the bottom ending at the top, thus avoiding solvent accumulation and maximizing the benefits of rapid crystallization and vaporization from lower to upper end. PVDF's clean and smooth surface has much more desirability for subsequent electrode deposition and poling processes. As a result, Cardoso *et al.* explored the morphology of PVDF surface as a result of thermal annealing [195]. For more spin-coating, a regulated 20 per cent PVDF/DMF solution was used, and the time, rotation speed, and acceleration were set to 1/2 a minute, 1k rpm, and 7.5k/rpms, respectively. At last, these were dried up and heated at various temperatures (i.e. 20, 30, 50, 70, and 80°C). As visible in the SEM pictures, treatment at low temperature resulted in an increased number of pores, attributing to DMF's low vaporization rate. The transmittance spectra of these films showed that the pores were created without heating, whereas the smooth substrate was created with an annealing technique based on higher temperature. Additionally, some particular amount of heating is needed to meet demand of a transparent film. Also, the width of films could be modified after adjusting the solution's angular velocity and ratio of mass. The more the velocity, the thinner the sample for each particular ratio of mass of DMF/PVDF solution, and it must be observed that as the velocity exceeds 4k rpm, the thinning effect of acceleration becomes insignificant. In the meantime, increasing the velocity of rotation benefits the β-phase crystallization, mostly due to increased shear effect on PVDF at a high rotation velocity. Likewise, the mixture containing less PVDF would be thin and have more β-phase material in the same scenario because of its low viscosity. Shaik *et al.* [196] achieved 87 per cent β-phase alongside a 20 per cent by weight PVDF solution spin-coated on the surface at 9k rpm and annealed 100°C, demonstrating that the temperature at which film is annealed affects the amount of β-phase component in PVDF films after spin-coating, especially at 95°C and slower rotating speeds. Whenever the velocity is greater than 8k rpm, it appears as if no additional β-phase material is obtained by increasing the annealing temperature. Consequently, the inference is that the speed of rotation significantly affected more towards β-phase compared to high-speed temperature of annealing. The effect of layers on the crystallinity of β-phase present inside PVDF film (post-spin-coating) used to assemble this system received some publicity. For example, Cardoso *et al.* saw a diminished effect of crystallization in PVDF film [197]. One-layer film often has the maximum amount of β-phase, while three-layer film has the least, at all annealing temperatures. It's because upon spin-coating first layer, the surface resulted in an extremely drastic stretch and shear on film of PVDF, primarily necessary for the creation of β-phase crystallization. As the remaining PVDF layers were accumulated on the previously deposited surface, the polymer solution contact area and surface harshness increased, mitigating the shear impact at top films, causing negligible variability in electroactive phase [192].

6.5.4 Methods Based on Hot-Pressing

It is a process involving heating and pressing a polymer or its composite till its melting point is reached to create a piezocomposite. Piezoelectric behaviours of the composite are not only dependent on the polymer matrix's characteristics but also on the temperature and pressure applied during the moulding. The piezoelectric material is uniformly wrapped as the temperature and pressure rise, thus reducing the internal porosity and enhancing interfacial adhesive force. Consequently, the material's crystal structure was compressed and the dielectric loss was decreased, leading to enhanced polarity and improved piezoelectric efficiency. At very high temperature, PVDF (organic) shrinks, resulting in more intrinsic faults. Consequently, the applied field direction within the specimen gets compromised, affecting polarization and resulting in a reduced piezoelectric efficiency to a certain extent. In particular, the hot-pressing method produces piezoelectric PVDF with a dense population and enhanced piezoelectric efficiency. However, this method is limited in certain applications. For example, the production of piezoelectric coatings is limited to the coating. Meng *et al.* simplified the process of "pressing and folding (P&F)" in 2019 for constructing PVDF films with a huge quantity of β-phase material [198]. The processing technique is specified for a period using hot-pressing procedure at 165°C and a force of 300 kN. Crystallization of β-phase happened alongside cooling method at high pressure (Figure 6.8(a)). The SEM micrographs of as-synthesized PVDF are depicted in Figure 6.8(b). As illustrated in Figure 6.8(c), expanding the P&F cycle count enhances β-phase. Whenever the P&F cycle count is increased to seven, a remarkable 98 per cent β-phase content image is found. In contrast to the generally recorded values for the β-phase material in extended PVDF layers, as shown by the region, shaded in green, in Figure 6.8(c), it gets significantly increased, attaining its peak value.

Fu *et al.* created novel sandwich structure in 2020 after conducting research, hot-pressing three piled up layers composed of one PVDF/FTN($FeTiNbO_6$) composite layer and pure PVDF top and bottom layers at 10 MPa and 200°C (seen in Figure 6.8(d)) [199]. The sandwich structure's interfacial polarization can be improved and electroactive phase material in PVDF is shown in Figure 6.8(e and f). Ultimately, it showed exceptional utility for energy accumulation produced by human activity. The aforementioned scientists placed a premium on designing the macrostructure of a PVDF-based flexible system through the hot-pressing process. However, the configuration at the molecular level is critical in determining the efficiency of PVDF. Thus, Tian *et al.* modelled PVDF/PZT piezocomposite with a rich lamellar crystal structure rearranging

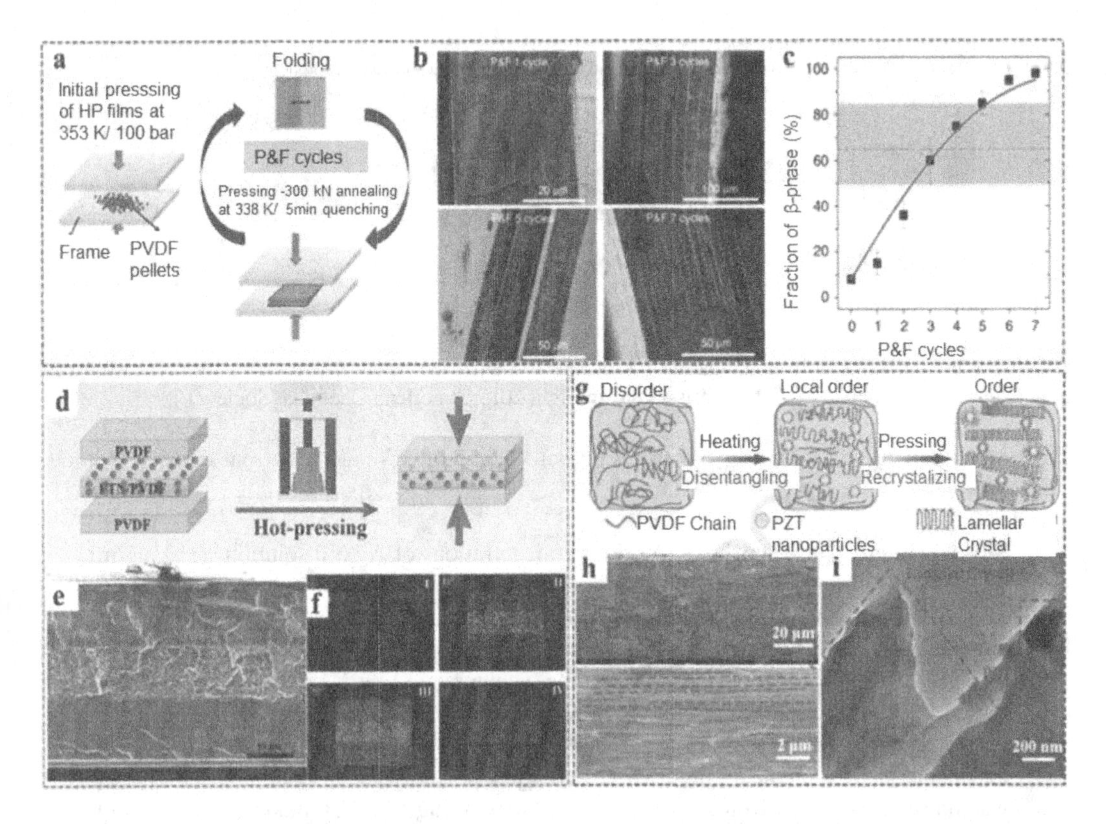

FIGURE 6.8 (a) Illustration of the P&F process; (b) SEM images of cross sections of P&F specimens after folding at 300 kN and 165°C, after varying folding cycle counts; (c) β-phase material after varying numbers of cycles of P&F [198]; (d) graphical representation of the hot-pressing in fabricating a composite of P-FTNx-P having sandwich structure; (e) SEM image of the cross section of composite of P-FTN-(15 per cent) P after hot-pressing; (f) EDS mapping of P-FTN(15 per cent)-P [199]; (g) lamellar β-phase crystal preparation; (h) SEM images of the piezocomposite; (i) TEM micrographs of specimen [200].

the polymer chains [200]. The process is illustrated in Figure 6.8(g), wherein the PVDF/PZT piezocomposite, produced via non-solvent-induced separation of phase process, was first heated for matching the transition temperature of a glass of PVDF, disentangling the tangled molecule chain, turning it into disentangled and local order chain. In the meantime, the inorganic PZT particles implanted will act as a nucleus for the crystallization of PVDF polymer chains. This PVDF/PZT sensors show a sensitivity of about 6.38 mV/N, because of lamellar structure and PZT's synergistic effect (Figure 6.8(h and i)), thus paving the way for a new path in artificial intelligence (AI) [192].

6.5.5 Flow Coating for High-Performance Flexible PENGs

In this article, we describe a straightforward and inexpensive flow-coating process for fabricating transparent and flexible metal electrodes employing silver nanoparticles (AgNPs). The procedure is straightforward and consists of two steps: patterning and sintering the horizontal AgNPs lines, accompanied by patterning and sintering the longitudinal ones. The size of grid is adjusted by adjusting the AgNP solution concentration, and the gap between grid is adjusted by altering the distance travelled by the translation stage between occasional breaks. At 550 nm, the optimized silver grid electrode had an optical transmittance of 86 per cent and sheet resistance of 174 Ω/cm^2. Silver grid electrodes formed as a result of this procedure were successfully used to fabricate a flexible PENG.

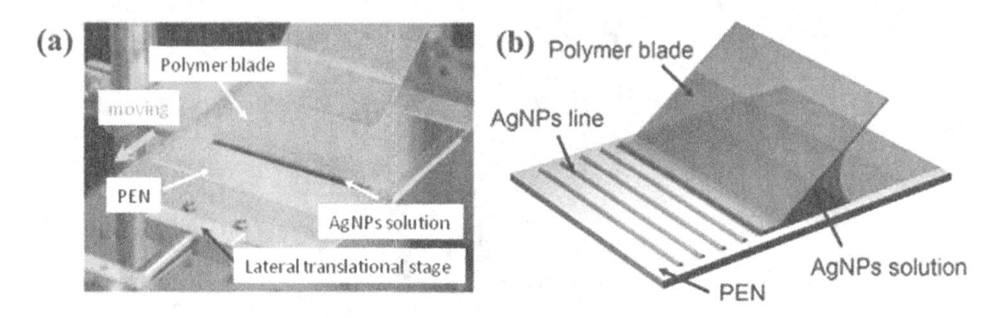

FIGURE 6.9 (a) Home-built flow-coating arrangement picture; (b) AgNP line creation procedure illustration. (Reproduced from [201] with permission. Copyright 2015, American Chemical Society.)

This system performed admirably, with an output voltage of 5 V and a current density of about 0.5 A/cm^2.

- **Silver grid electrode synthesis:** Various concentrations of Ag NP solutions (110 mg/mL) were made by dissolving Ag NPs in toluene. The particles were approximately 10 nm in size. For 24 hours, the solution was stirred to create a uniform suspension of silver nanoparticles. The substrate made of polyethylene naphthalate (PEN) was washed successively using 2-propanol, DI, and acetone, accompanied by N_2 gas drying. As seen in Figure 6.9, the set-up for flow-coating procedure comprised an angled polymer blade connected to a vertical translation stage and a linear translation stage connected to a piezo nanopositioner. To create a hinge, a 7500-cm thick polyethylene terephthalate (PET) blade was marked 0.012 cm from the tip. After rigidly fixing the PET blade at an angle of a 40 degree with respect to the vertical translation point, attached to the substrate. The 6 m^3 of silver nanoparticle solutions were infused between the substrate and PET blade and were trapped underneath the blade through capillary force. The linear translation stage was pushed at a constant velocity of 15×10^{-3} cm/s, having an occasional halt of 1 s and a programmed travel distance of 50250 m. On the PEN substrate, the silver nanoparticle line patterns were grown and then thermally sintered for about 1 hour at 150°C in a nitrogen atmosphere. After rotating the stage 90 degrees, another set of Ag NP lines was modelled, accompanied by thermal sintering of the Ag NPs.
- **Flexible PENG fabrication:** The P(VDF-TrFE) copolymer (20 per cent by weight) was mixed with dimethylformamide (DMF) for nearly a day. Silver metal grid/single-layer graphene (SLG) electrodes were spin-coated with a 0.055 mm P(VDF-TrFE) layer. SLG was produced using chemical vapor deposition (CVD) and was transferred onto silver electrodes using a PMMA layer. The spin-coated P(VDF-TrFE) layer was annealed at 140°C for 180 minutes to obtain the crystalline phase. The 0.1-μm thick uppermost silver electrodes were vaporized [201].

In brief, the manufacturing of wide, versatile, and transparent metal grid electrodes using an AgNPs-based two-step flow-coating process was depicted. The flow-coating technique was straightforward, inexpensive, and won't require the use of lithography to fabricate a metallic grid. The size and spacing of the metal grid lines were specifically regulated by modifying the concentration of AgNP and the scheduled movement path. The resultant silver grid electrodes had an optical transmittance of 86 per cent at 550 nm and sheet resistance of about 174 Ω/cm^2. A versatile piezoelectric NG with a patterned metallic grid system was built using the patterned metallic grid. The flow-coating technique listed here embraces existing standard trichloroethylene (TCE) fabrication methods that may be critical in the preparation of wide-area clear transparent electrodes for application in future electronic devices [201].

6.5.6 Liquid Metal-Based Technique

Together with their inherent versatility, the expected high piezoelectricity of monolayers of group IV monochalcogenides makes them attractive specimens for forming versatile NGs. Tin sulphide (SnS) is a possible candidate for such NGs within this community owing to its unique improved characteristics. Obtaining wide area and a SnS monolayer with high degree of crystallinity was difficult in the past because of the existence of interlayer interactions caused by lone-pair of electrons on S. This chapter describes the production of the single crystal of SnS employing liquid metal-based method.

Despite predictions of high piezoelectricity and thus the potential to reach optimal conversion efficiencies for NGs composed of monolayers of group IV monochalcogenides, their production has been restricted to small-scale surface coverage because of issues, including an exchange between lateral size and thickness and a constrained sliding effect among layers. Traditional two-dimensional exfoliation methods have been unable to achieve wafer-scale measurements, limiting the usage of certain substances. Mechanical exfoliation techniques produce only two-dimensional batches of very limited lateral dimensions at substrates. CVD and physical vapor deposition (PVD) synthesis methods are better for wafer-scale manufacturing. Likewise, liquid-phase exfoliation processes give stacked films of nanosheets, making these unsuitable for energy conservation, dependant on stress acting on 2D sheets. To add to the difficulty, it has been proposed that the production of SnS, through traditional exfoliation is restricted by the intense interlayer correlations between the lone-pair of electrons of S.

These methods are based on the self-limiting deposition of wide surface area SnS on molten tin (Sn). The process occurs according to the Cabrera-Mott framework [202], illustrating the formation of extremely thin interfacial layers on elemental metals having finite thicknesses of about one or a few monolayers. Using a liquid substrate, a relatively high tension and nearly an ideal structure is achieved in liquid metals. The formed interfacial layers could then be easily delaminated from the interface of liquid metal by applying minimum pressures on the interface. These interfacial layers have poor polar character, or more precisely, a diminished van der Waals force applied on substrate. By employing a direct surface reaction, it is possible to fabricate intact, outstanding monolayer Sn compounds with no intervening impurity in the immediate surroundings, which are then used to produce PENG with outstanding results.

The fabrication of super-thin nanosheets presents large lateral dimensions (Figure 6.10).

- **Device fabrication and analysis:** By developing piezoelectric NGs, we investigated the monolayer SnS's piezoelectric characteristics. The SnS layer was delaminated on the appropriate versatile substrate. The systems were then subjected to mechanical stresses in a variety of modes. Since an out-of-plane coefficient is not anticipated because of the

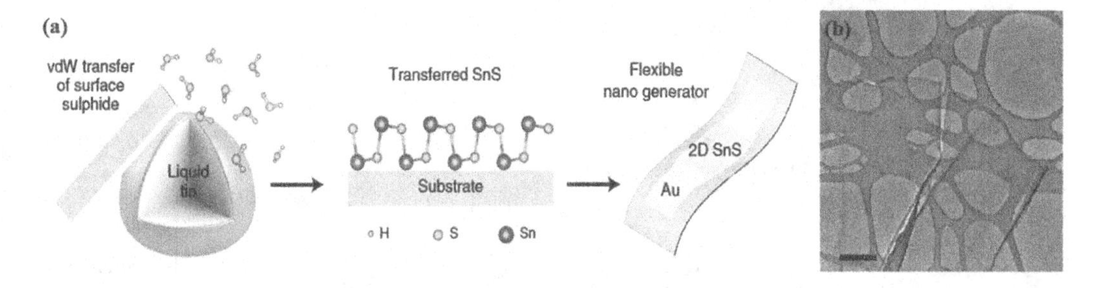

FIGURE 6.10 Fabrication and material characterizations of the two-dimensional SnS illustration: (a) illustration of the fabrication, monolayer of SnS and its usage on an NG transducer; (b) TEM micrograph of monolayered SnS nanosheet [204]. (Creative Commons Attribution (CC BY) license.)

structure of the material, the in-plane coefficients were measured using the lateral piezoresponse force microscopy (LPFM) mode. When LPFM microscopy is used, the stimulus is delivered above the plane (and thus out of plane), while the reaction is evaluated in the plane.

Kelvin probe force microscopy (KPFM) calculations were recorded to account for any electrostatic charge inputs to the output signal [203]. Figure 6.11(a) illustrates the topology of the as-synthesized SnS films having 1-nm width, supporting the films' monolayer origin. When the lateral amplitude and phase are measured, Figure 6.11(b and c) demonstrates a strong distinction between the SiO_2/Si substrate and the SnS film. The lateral piezoresponse is plotted between the driving voltage of SnS and the regularly poled lithium niobate (PPLN) specimen in Figure 6.11(d). Both samples exhibit a unique gradual incline in piezoresponse upon increasing voltage. This indicates

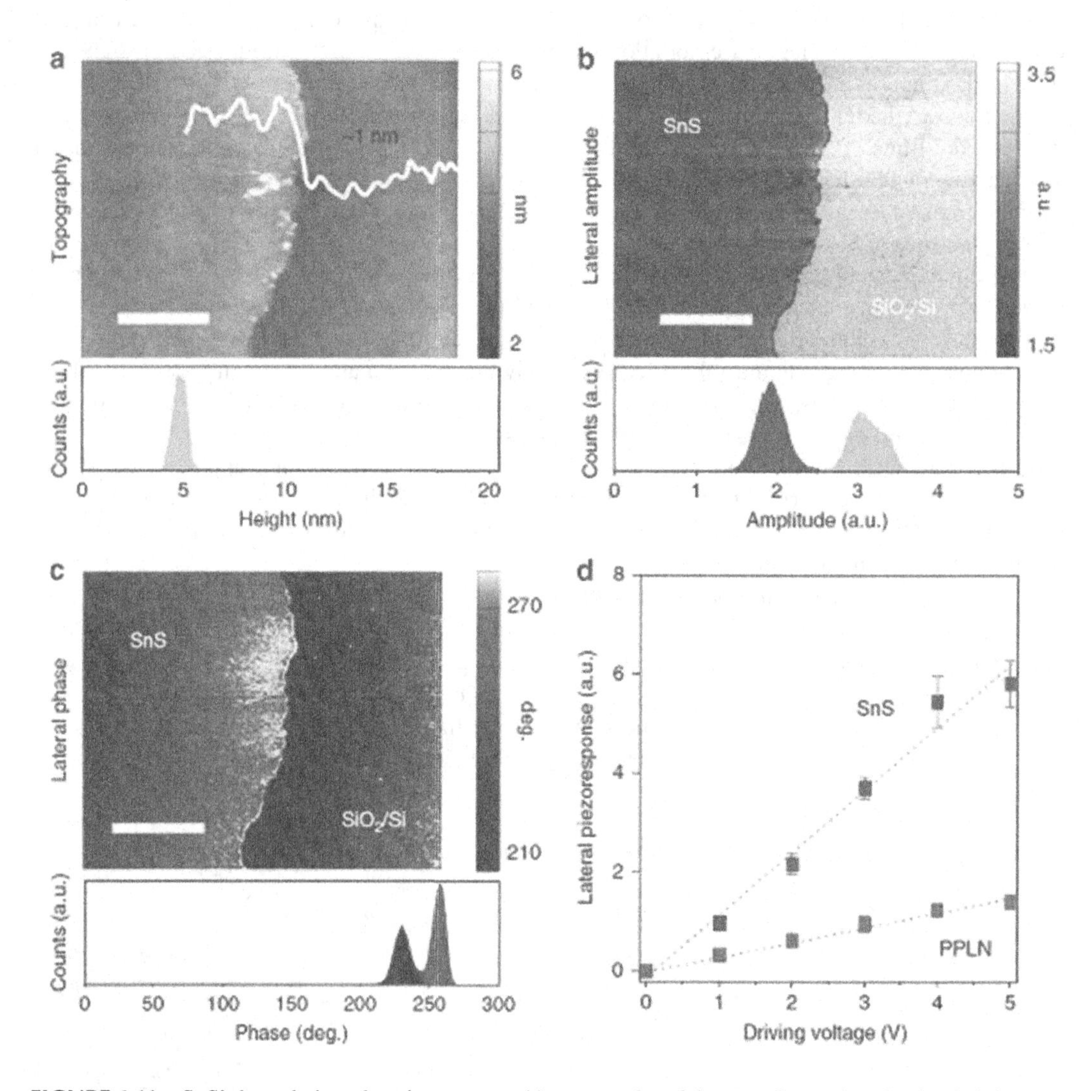

FIGURE 6.11 SnS's lateral piezoelectric response: (a) topography of the monolayer plane having height profile demonstrating sheet thickness of about ~1 nm. It is verified by the histogram; (b) lateral amplitude of the SnS monolayer along with an inset of the histogram; (c) lateral phase of the SnS monolayer with inset of the histogram; (d) lateral piezoresponse of monolayer SnS and PPLN versus the driving voltage [204]. (Creative Commons Attribution (CC BY) license.)

that the source is piezoelectric, and piezoelectric coefficient is given by the slope of plot. The monolayer SnS has a piezoelectric coefficient of 25.99 pm/V.

Two different modes of analysis have been employed to evaluate the systems for (i) d_{31} mode and (ii) d_{11} mode. The first one (for d_{31} mode) involved pounding the device's substrate in a managed way with a maximum load amplitude of 3.99 N and a frequency region of 1–10 Hz. Its aim was to determine the feasibility of encapsulating such an NG in low-frequency systems wherein force is in direction normal to the substrate. To validate the d_{31} mode, devices with two and multiple electrodes were built. Figure 6.12(a) shows an optical image of a dual-electrode system, and when subjected to

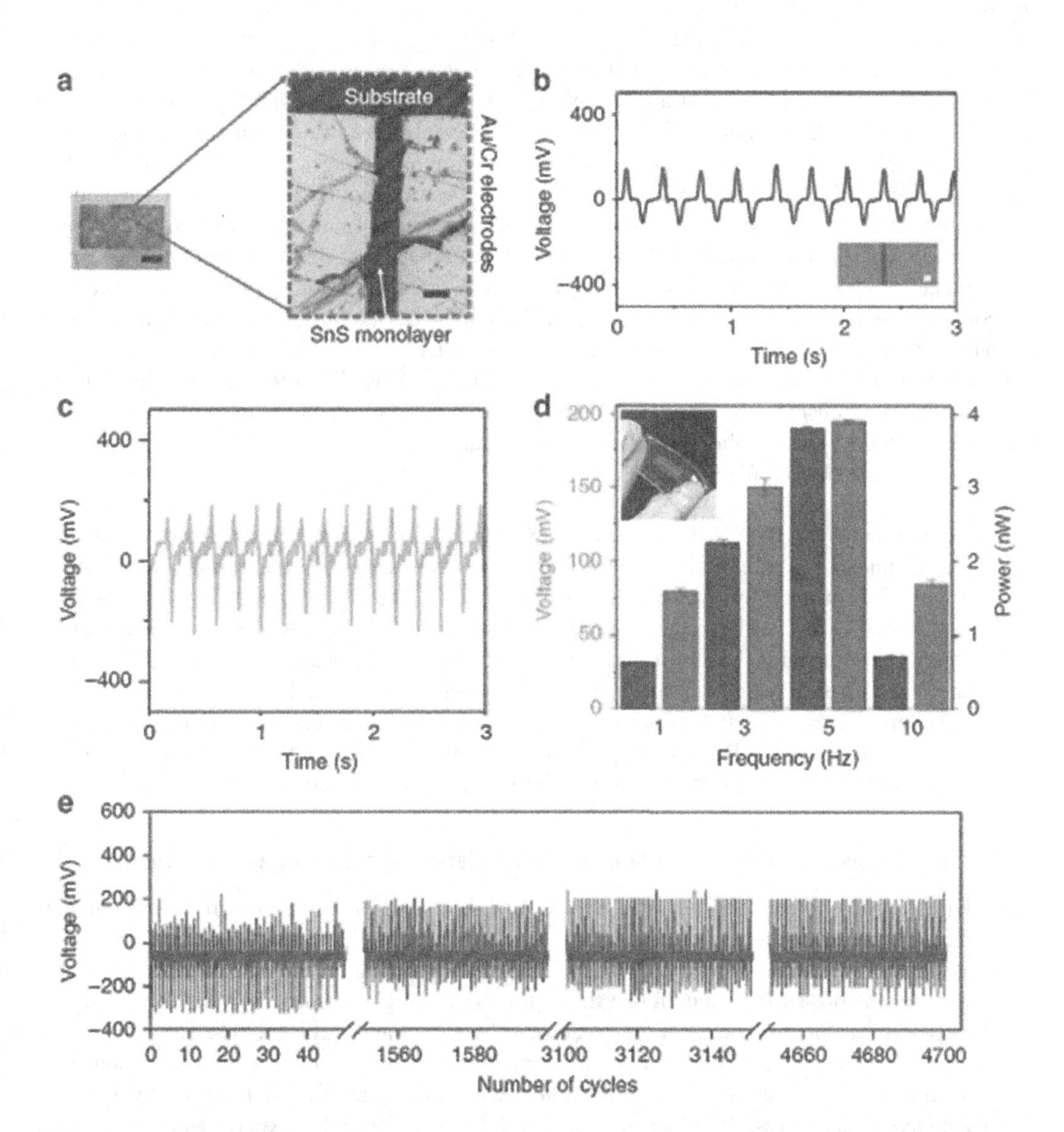

FIGURE 6.12 The outputs of systems having monolayer SnS produced along them using the pounding mode to excite d31: (a) dual-electrode system's optical image; (b) dual-electrode system's output voltage's response. Inset shows an optical image of the system; (c) multiple electrode system's output voltage's responses at 5 Hz; (d) plot demonstrating the variation of average output voltage and power at different tapping frequencies and inset is the photo of the mica flexible NG encapsulating PDMS (multiple electrode systems); (e) stability investigation of output voltage with stimulus applied at 5 Hz for nearly 4.7k cycles (multiple electrode systems) [204]. (Creative Commons Attribution (CC BY) license.)

a pounding force of about 3 Hz, this system produced voltages greater than 150-mV peak value. The voltage increases as the force is exerted and decreases as the field is released, which is a characteristic of piezoelectric devices. Figure 6.12(b) illustrates the output of this dual-electrode unit, that is 160-mV maximum value over a 10-MΩ load resistance at 3 Hz. Figure 6.12(c) illustrates the output of the multi-electrode unit, that is 190-mV maximum value over a 10-MΩ load resistance at 5 Hz. The performance of the unit was characterized at various frequencies Hz (Figure 6.12(d)).

As such, the system demonstrated a high output voltage, making it ideal for realistic energy conservation applications. The intended use of such systems is to conserve energy from naturally occurring excitation in the atmosphere, instead of providing a constant or regular driving force. As a result, we evaluated the NG at a variety of frequencies. This indicates that this NG could be used in situations requiring low-frequency stimulation, like biomedical sensors or airport stairways. The frequency of about 5 Hz produced an average maximum output of 190 mV. Apart from this, the possible explanation for damping of signal is the flexible substrate's failure to react at the excitation rate of applied stimulus force, resulting in a reduced output voltage. It could be easily fixed by selecting a more versatile substrate. Figure 6.12(e) illustrates the output for a continuously applied force applied at 5 Hz for around 4.7k cycles. This consistent performance demonstrates the device's mechanical reliability and its suitability for use in real-world applications requiring it to operate for extended periods and in hazardous conditions. To demonstrate the feasibility of the established flexible SnS monolayers in utilizing energy for wearable NG, the device was analysed in d_{11} mode using a robust configuration of the piezoelectric plane on the device's surface. On an approximated strain of 0.7 per cent, an average maximum output voltage of 0.15 V was observed. Additionally, existing data were measured using a 1-GΩ passive resistance that produced output current of 0.16 nA. This voltage produced is considerably higher compared to any pre-published PENG performance from monolayer-based NGs like WSe_2 or MoS_2.

- **Fabrication of the PENG:** The layers of SnS were delaminated over either 50 mm × 20 mm smooth fluorphlogopite mica sheets or PDMS. By decreasing the height of the mica substrate to 0.02 mm, the systems retained their versatility. The following move was to deposit Cr-Au (10 or 100 nm) electrodes. Devices with two and several electrodes were fabricated. Figure 6.10 illustrates the dual-electrode system. The electrodes are 2 mm wide and separated by a 40-μm distance. The deposition hard mask was aligned to lie above the optically defined nanosheets. Ag paste was used to attach the electrical wires to the electrode contact pads. Finally, a PDMS encapsulation coating was added to the device's top surface to protect the material's surface and for reinforcing the mica layers [204].

6.5.7 Nanoforest Structure Fabricated Using Metal-Assisted Chemical Etching

This chapter describes the fabrication of an NG based on a nanoforest structure through metal-assisted chemical etching (mac-etch). A core NS composed of densely packed 10-nm thick NWs and nanovoids leads to an incredibly broad interfacial region. Following that, a piezoelectric material is dispersed over the pre-existing nanoforest through PVD. Owing to the deposition methods' inherent characteristics, a broader range of piezoelectric substances can be availed than using bottom-up growth. This is important to note that photolithography or plasma etching wasn't used to construct the NG. Additionally, an additional step is irrelevant to extract the pre-existing prototype as the heavily doped Si nanoforest template functions as a bottom electrode. Experimental procedures start with the manufacturing of a 4-inch p-type Si wafer (Figure 6.13(a)). The wafer has a resistivity of 5 mΩ cm, which is sufficient for charge conduction. The mac-etch method is critical for creating the structure of nanoforests [205–207]. To prepare Si substrate for mac-etching, a layer of gold with 6-nm thickness was coated through a thermal evaporator. A thin gold layer with this structure can't form a continuous film-type structure. Other than that, the layer has an nm-scale island-like structure consisting of naturally formed gold clusters and cavities between the clusters

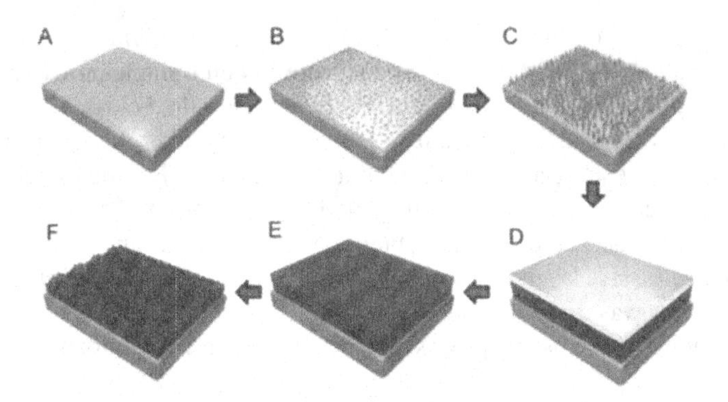

FIGURE 6.13 Fabrication process of the nanoforest-based PENG scheme: (a) heavily doped Si substrate at the beginning; (b) deposition of gold thin film (~6 nm) on the Si substrate. The gold film forms an island-like structure naturally; (c) silicon substrate being mac-etched; (d) nanoforest substrate (thickness of around 100 nm) deposited with $BaTiO_3$; (e) device after the PMMA coating procedure; (f) resultant nanoforest-based PENG after the coating of the top electrode, comprising Au and Cr having thicknesses of about 10 and 200 nm, respectively. (Reproduced from [208] with permission. Copyright 2013, Elsevier.)

(Figure 6.13(b)). Gold's island-type patterns will serve as a basis for the nanoforest's creation. The width of the deposited gold films determines the structural transformation from a film-type to an island-type pattern. As a result, precise control of gold film width is important. The Au-coated Si wafer was immersed in a wet etchant containing H_2O, H_2O_2, and HF. With the help of Au as a catalyst, H_2O_2 partially oxidized the silicon, and the oxidized portion was etched away by HF.

Due to the inability of H_2O_2 to oxidize Si to SiO_2, only the gold deposited portion of the silicon is etched away. Thus, holes between gold clusters were converted to vertically standing NWs, while Si underneath gold clusters was converted to gaps. As a result, they took on the shape of a nanoforest (Figure 6.13(c)). The NWs and nanovoids have dimensions in the 10-nm scale, which is difficult to accomplish using traditional photolithography owing to resolution constraints. Radio frequency (RF) sputtering was then used to mount a BTO film having a thickness of about 100 nm onto the mac-etched Si substrate (Figure 6.13(d)) [7]. A total of 200 W was used as the source power, 3.0 mTorr was used as the deposition pressure, and half-hour was used for the deposition period. The mac-etched pattern serves as a blueprint for the formation of a BTO NS. The BTO film conformally covers the protruding nanoforest. Controlling the thickness of the $BaTiO_3$ film is critical in determining the piezoelectricity of the proposed PENG. Whenever the $BaTiO_3$ layer is too small, the deposited $BaTiO_3$ clusters become unconnected. As a result, it is impossible to make a continuous film. Alternatively, when $BaTiO_3$ is very thick, it significantly reduces the void density inside the NS, weakening the strain confinement effect. Following the deposition process, the $BaTiO_3$-deposited wafer was annealed for 20 minutes at 700°C in O_2. This annealing process transforms the $BaTiO_3$ phase from amorphous to polycrystalline, which is essential for increasing piezoelectricity. The supporting material provides a thorough understanding of the phase shift of the $BaTiO_3$ layer using X-ray diffraction (XRD) and Raman spectroscopy. Before creating the top electrode, the $BaTiO_3$-deposited nanoforest substrate was covered with a PMMA coating (Figure 6.13(e)). It was coated using spin-coating at 4k rpm for 45 s, followed by a half-hour of annealing at 170°C. Four different advantages are given by this PMMA layer. To begin, the film acts as an insulator, preventing a detrimental piezoelectric screening effect caused due to free electrons. Typically, a Schottky barrier is formed at the phase boundary; moreover, the addition of insulation was found to be simpler and more efficient at blocking free-electron flows. Second, the film acts as a barrier, preventing electrical short circuits. A tiny portion of manufacturing flaws will result in an electrical short between the top and bottom electrodes, significantly degrading the output potential. The PMMA layer may be inserted between both electrodes to avoid any close contact. Third, the film acts as a diffuser, dispersing any applied force omnidirectionally. Due to the vertical alignment of the $BaTiO_3$ NSs on the nanoforest substrate, the applied pressure is concentrated on a single protruding area. Inserting a PMMA film will reallocate the applied force axially and radially. Fourth, the layer acts as a plagiarizer, simplifying corresponding procedures. If the top electrode substrate is directly deposited onto the rugged nanoforest substrate, it can't be attached indefinitely. Additionally, electrical open will occur in the worst-case scenario. Following the PMMA coating, the top electrode's bi-metals were deposited through a thermal evaporator (Figure 6.13(f)). Chromium with a thickness of 10 nm was coated first as an adhesive layer, followed by gold having a thickness of 200 nm. Following that, the square-shaped wafer was diced and the top and bottom electrodes were attached using Al wire and Ag paste. Following that, a poling process was carried out for 10 hours with an electric field of 0.1 MV/cm.

As a result, an NG was designed and evaluated with a nanoforest structure. The mac-etch procedure was used to create a Si nanoforest that served as the basis for the NS. The nanoforest-based NG possessed both fabrication and performance virtues. In terms of fabrication, the proposed method allows for a larger spectrum of piezoelectric materials to be used since the bottom-up growth mechanism is not used. In terms of efficiency, densely packed nanovoids within the $BaTiO_3$-coated nanoforest enabled the strain confinement effect, resulting in an increased piezopotential. The engineered NG showed a 4.2-fold increment in output power density as opposed to the bulk-type NG. The nanoforest's height effect was compared using a simulator (COMSOL) and experimental data. Both findings indicated a growth in the number of tiny NWs and saturation in the number of long NWs. Not only will the analyses of the nanoforest generator proposed in this work be directly applicable to energy conservation applications based on silicon chips, but they will also motivate more structural evolution of a functional and effective PENG [208].

6.5.8 Polymer Nanowire Fabrication via Template-Assisted Growth Mechanism

Following solidification from the melt or vaporization, the polymer NSs described by the template's symmetry is created. Numerous templates, including anodic aluminium oxide (AAO), polyimide (PI), and polycarbonate (PC), could be readily formed and are also viable. The method used to remove the template is highly dependent on the quality of the template and the properties of the polymer NWs. In the case of NG production, the template removal may be pointless because the template may help to coordinate the NWs, providing additional mechanical benefits, including sturdiness and stability, as well as protection from environmental deterioration.

In Figure 6.14, a variety of approaches for the template-assisted growth of NWs are identified, with the ultimate goal of fabricating NGs. Figure 6.14(a) illustrates the infiltration of polymer solution [209, 210] or melted polymer [211] into the template, which wets upon filling polymer nanopores. In Figure 6.14(b), template is suspended over a warmed solution of polymer, whereby capillary force draws polymer along the pores [212, 213]. The crystallization throughout infiltration is regulated by exposing gas at the top of the template to stabilize the temperature profile [214] (Figure 6.14(c)). Figure 6.14(d and e) also shows pieces of evidence of nanocomposite of ceramic and polymer NG. In these cases, piezoelectricity-based ZnO NWs were produced within the nanopores of the template through hydrothermal treatment [215] (Figure 6.14(d)) or electrodeposition [216] (Figure 6.14(e)).

PVDF and its copolymers have similar piezoelectric behaviour as certain ceramic substances. Due to their sturdiness and excellent chemical stability, PVDF NWs are especially well suited for energy conservation. The issues involved with fabricating PVDF NW-based piezoelectric NGs are primarily due to two factors: (i) developing scalable manufacturing procedures and (ii) poling the polymer to achieve the necessary efficiency. Whiter *et al.* defined a template-assisted method for growing self-polymerized P(VDF-TrFE) NW arrays and using them in NGs [218]. For initiating the growth methods, an AAO template was formulated using a thin layer (0.015 μm) of Ag-coated on an end of the substrate. The solution was then lowered and kept at 60°C for several hours over silver-coated AAO template to ensure complete infiltration. The mould-grown P(VDF-TrFE) NWs

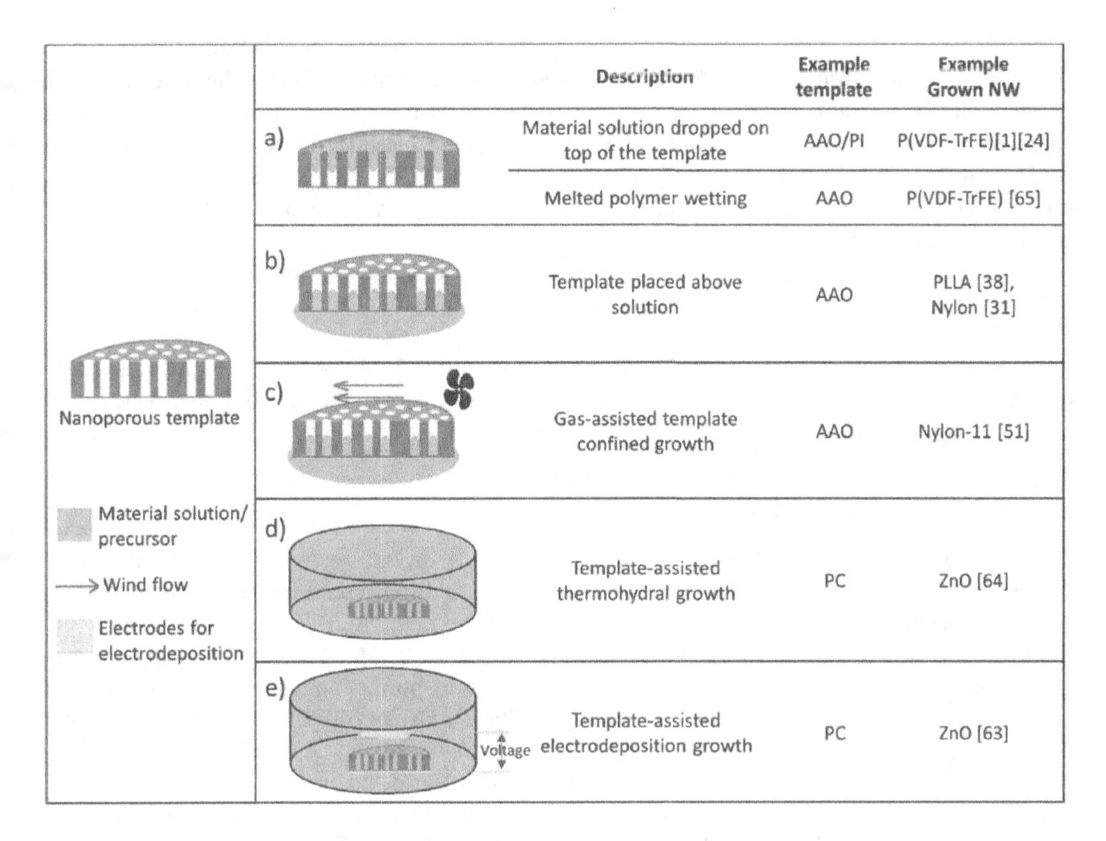

FIGURE 6.14 Template-assisted NW growth techniques especially for energy conservation are enlisted in this figure. (a–e) Various growth procedures describing processes, types of template, and materials grown [217]. (Creative Commons Attribution (CC BY) license.)

were then integrated onto NG by sputtering Pt electrodes on both ends of the NW-filled template and encapsulating them in PDMS. Peak output voltage and current were observed to be 3 V and 5.5 nA, respectively.

Bhavanasi *et al.* also published a template-wetting method for the synthesis of P(VDF-TrFE) NTs, where the NTs have shown predominant crystallinity in the β-phase, polarized along the NT's long axis [211]. The synthesis began with the drop-casting of thin film of P(VDF-TrFE), followed by 135°C annealing. Figure 6.13(d) illustrates XRD plots relating embedded P(VDF-TrFE) NTs to films. Peaks at 19.8 degrees correspond to (2 0 0) or (1 1 0) planes were observed in both NTs and films indicating the piezoelectric β-phase, while the peak corresponding to the c-axis orientation at (2 0 1) or (1 0 1) was visible in the film but negligible in the NT specimen. The authors created an NG in this work by releasing the NTs from the template. For accomplishing this, the template was immersed in a sodium hydroxide (NaOH), but the released NTs remained attached to the residual P(VDF-TrFE) layer at bottom. An energy harvester constructed using a poled P(VDF-TrFE) layer provided a peak output voltage and power density of 800 mV and 0.6 mW/m^2, respectively [217].

6.5.9 Other Techniques

Apart from the conventional processing procedure listed previously, there are numerous other manufacturing techniques for a variety of applications. Numerous methodologies have been introduced in the last few years to acquire β-phase PVDF, including wet spinning [219], quenching of melt [220], expanding [221, 222], and solvent-casting [223, 224]. Even so, β-phase PVDF is most frequently obtained by casting-solvent or melting [70], whereas mechanical methods like expanding or hot-pressing may dramatically improve the β-phase material. Within these, the porous spongy PVDF system is a unique area in which it has been proven that when the efficiency of electromechanical coupling and porous structure of P(VDF-TrFE) are optimized, the output performance could be increased in comparison to P(VDF-TrFE) thin film [225–227]. For instance, Dong *et al.* established a cantilever with pores and exhibiting piezoelectricity for non-solvent-induced phase separation methodology [228].

Additionally, Mao *et al.* [229] used the conventional solvent-casting particulate leaching method to create an integrated NG by acid etching the nanoparticles of ZnO, which were prepared using ZnO/PVDF piezocomposite.

In summary, since the features of the flexible piezoelectric PVDF system are highly dependent on processing techniques, the most suitable methods must be chosen for each building structure in order to accomplish the experiment's or practical application's objective [192].

6.6 APPLICATIONS OF PENGs

PENGs provide a long-term power source for a range of diverse technologies, including self-powered micro/nanosensors, wearable/flexible electronics, self-powered electronics, and biomedical applications.

Hu *et al.* [230] incorporated a ZnO NW-based PENG on inner side of a tire; rotational deformation of the tire results in a 1.5 V and 25 nA output with a peak power content of 70 μW/cm^3. Due to the increased dielectric constant and piezoelectric voltage constant of PZT NFs, they are suitable for NG.

In 2014, ZnO NW-based biomedical sensors were developed, paving the way for future biosensing system creation [231]. PENGs efficiently harness energy from the interior of the body movements and control health tracking systems and medical implants such as neural stimulators, cardioverter-defibrillators, and pacemakers. To leverage the energy produced by finger typing, flexible and transparent PENGs based on ZnO NWs, ITO, and PZT films have been created [1, 7, 232].

Numerous mixed energy collectors incorporating PENG with various energy harvesting technologies such as triboelectric NGs (TENGs) and pyroelectric NGs (PyENGs) have been designed for standardized power generation and increased power conversion efficiency. PENGs were also used to increase the performance of solar photovoltaic cells. By combining a-Si nanopillar solar cell with a PVDF NG, a tandem NG was built.

Zhu *et al.* developed piezo-phototronic photovoltaic system dependent on a Si nano heterostructure. The solar cell's performance was increased from 8.97 to 9.51 per cent [233].

Mandal and Maity developed a PENG using a multilayer structure of PVDF [234]. This performance indicates that it could be used in self-powered flexible electronics. A hybrid organic-inorganic PENG formed of zinc sulphide (ZnS) NRs and electrospun PVDF. It serves as a power source for self-powered sensors. There are many protection device sensors dependent on PENGs, including transport tracking, wireless sensors, and biosensors [24].

The primary step to use an energy harvester is to transform it from arbitrarily changing, dual polarity voltage peaks that represent variations in the input force. The easiest and the most popular way in NGs is using a bridge rectifier for transforming both polarity output voltages to single polarity, attached to a capacitor, charged using rectified output. As the capacitor reaches the appropriate voltage, a switch is turned on, joining it to an external system or, in some cases, a battery conserves the energy.

Typically, such circuits were employed to examine the efficacy for NG applications by flashing an LED/s after an adequate charge was already deposited. It must be remembered that as long as the capacitor stores enough voltage to resolve the LED's switching voltage, LED/s could be enabled to flash momentarily with small overall power. As a result, this is helpful for showing that the system can produce a certain power, but not enough for displaying that accessible levels of power can be produced. Certain appliances, viz. liquid crystal displays (LCDs), were driven by an NG, in certain cases providing power for several seconds post some minutes of charging [43]. This type of illustration of a system being powered for an extended period of time offers a much more identified success of an energy harvester's capability to supply powerful skill, particularly when the length of the input and output duty cycles could be correlated.

Using an NG to power an electrophoretic [235] monitor shows that prudent system selection can maximize the use of produced power since these screens need only power to turn on and thus allow efficient usage of the narrow output pulses from an energy harvester. As mentioned previously [230], one such implementation was shown for a ZnO NRs-based NG using a dual-sided versatile system with PMMA-Au electrodes. The system was flexed continuously, producing a maximum voltage of 10 V that was collected by a capacitor through a bridge rectifier. It was linked to a photodetector and a radio transmitter, which detects and transmits an optical signal. The transmitter needed comparatively little energy, functioning after three device bending cycles, while the photodetector required approximately 1k device bending cycles for functioning. So, either the power output of NGs or the power needs of sensors require improvement. Another self-powered device made use of an Hg sensor, in which the resistance of field-effect transistor (FET) changed in response to the presence of Hg in a water solution, thus illuminating an LED when adequate power was collected [168].

Along with enhancing the performance of an NG, applications of NGs for self-powered sensors were suggested. In this case, the performance of the NG varies in response to an external stimulus, and this variance reflects the stimuli's power. Even so, it must be remembered that when the input force is varied, the performance of the NG will also differ, and hence this change must be carefully monitored to avoid being mistaken with the variation caused by the stimulus.

This is not a problem when the NG is used as a self-powered pressure, movement, or vibration sensor. This was illustrated by depositing a steel spring with ZnO NRs and using spring's bending variation to control a self-powered balance. Owing to the device's thin, flexible nature, it shifted with the eye as it moved under the eyelid. Circumstances wherein the device's output was varied by external stimuli could provide valuable analysis of the impact of screening on the output, which typically changes the output. It was demonstrated that the performance decreased under UV irradiation [10], thereby increasing the internal carrier concentration [236].

- **PVDF-based flexible PENGs applications:** The proliferation of micro-electro-mechanical systems (MEMS), versatile PVDF-based systems, became a viable micro-sized battery substitute. The advancement in microtechnology in recent times leads to numerous inefficient devices posing difficulties. For instance, batteries were primarily used as power sources in conventional small medical equipment due to their stability and large energy content. They are, indeed, unable to fulfil the standards of highly complicated applications due to the battery's regular replacement. As a result, it is extremely desirable for developing a self-powered PENG that is both reliable and safe. We discuss the use of versatile energy harvesters on PVDF based in different contexts: implantable NGs, flexible NGs, and wearable NGs.
- **PENGs that can be worn:** Wearables, including a heart rate monitor and smart band, that assist in the recording of human medical information are inextricably linked to virtual interfacing in order to cater to the requirements of specific user's segments. Due to its capability to link to a network, it can attain multifunctional and sequential data. Until now, these wearables have been operated primarily by two types of conventional chemical batteries. One is the conventional button battery, which is not rechargeable and must be changed on a frequent basis, triggering not only discomfort to consumers and moreover data loss during the transmit mode. The other choice is rechargeable batteries, requiring an external power cord as well as a charging table and providing only just a few weeks of standby time. As a result, it's indeed vital to develop self-powered wearables, and a series of studies have been undertaken on energy storage devices for smart technology. The approach

for structural designing a wearable energy harvesting system would be to gather energy produced by motion while not interfering with the individual body's normal movement or posing a strain. Given the considerable potential for extracting energy from movement, numerous researches on piezoelectric flexible PVDF wearables were put forward. Wang group pioneered the idea of "human-motion-driven energy harvesting" in 2012, developing strategy involving versatile PVDF/ZnO PENG for transforming low-frequency mechanical energy into the electric current at varying deformation degrees [88]. Unusually, these fibre-based PENGs could be connected seamlessly to elbow and simultaneous current and voltage pulses were observed when fibre was folded and extended at a low frequency (<1 Hz), suggesting the viability of supplying portable electronic gadgets. Both electrospinning techniques are designed to avoid the formation of rigid structures, which reduce flexibility while increasing mechanical stability and energy harvesting. The as-synthesized PVDF-based non-woven NG produces electricity upon hitting a hand, while moving. Additionally, increasing the frequency or force of human activity on the system will increase its electrical performance directly. Karan *et al.* developed the PVDF/AlO-rGO NG by employing a steel cloth as the electrode [237].

- **NGs that can be implanted:** Numerous diseased organs need treatment with micro-medical devices in the medical sector. However, energy supply for these devices is a critical issue that must be addressed and therefore using energy transfer in vivo for achieving supply of direct energy to embedded devices that's what we're all worried about. Unlike Pb-based piezoceramics, which can be toxic for living beings, Pb-free biocompatible PVDF is an excellent option for energy harvesters. Clearly, the heart and internal joints are suitable candidates for providing continuous vibration energy in humans, which has tremendous potential for implantable energy harvesters. When it comes to implantable devices, long-term bioavailability and biosecurity are still the primary concerns. Typically, several studies have concentrated on manufacturing materials and processes. For instance, Li *et al.* used PDMS and PDMS/Parylene-C to fully pack the PVDF system prior to implantation [238]. More significantly, regardless of the packaging method used, the unit demonstrated no damping in performance after a 24-week implant placement cycle in rat, demonstrating the device's high strength. It outlines a comprehensive and practical strategy for surgically implanted device packaging. Wang group performed another implant placement analysis on a porous PVDF NG packed in PDMS [238, 239]. The continuous fabrication of this PVDF-based technology provides a pathway for potential applications in this field.
- **Industrially used flexible PENGs:** Choice of batteries for power requirements entails immense human and financial resources for battery replacement. However, rain- and wind-induced vibrations are common in remote areas; by using them and converting them to electrical energy, sensor's self-power feature can be accomplished, which has considerable application potential. Considering these abundant sources of electricity, various gadgets might be directly powered by the piezoelectric energy. To be more precise, Jung *et al.* demonstrated bi-morph PVDF-based NG to be used in roadways [95]. A model mobile load system (MMLS3) was developed in this study to track a moving vehicle on road (Figure 6.15(a)). The system was constructed using PI substrate (functional layers), PVDF thin film, and an Ni-based conductive fabric tapes (electrodes) to increase the energy production (Figure 6.15(b)). As evident from Figure 6.15(c), the analysis showed that the peak output power is 200 mW at 40 kHz, with an associated power density of approximately 8.9 W/m^2. Tang *et al.*, for example, designed a wind energy harvester to power a milliwatt-level sensor in the wild [240]. As illustrated in Figure 6.15(d), the PVDF film is powered using stick impact generated by winds in various directions. The wind energy was then transformed into electrical energy. A temperature sensor was set up during the practical test (shown in Figure 6.15(e)), demonstrating the viability of this wind conversion device for further use. Likewise, Sultana *et al.* electrospun a PVDF-ZnS hybrid PENG for harvesting

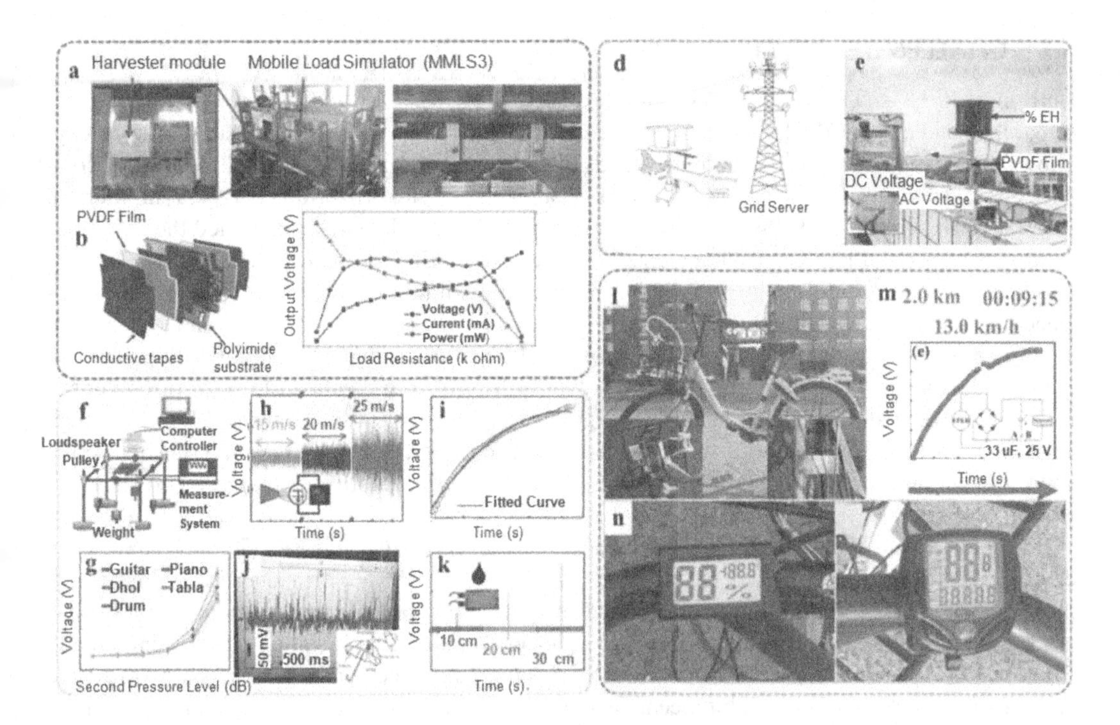

FIGURE 6.15 (a) Model Mobile Load Simulator (MMLS3) system, along with the harvester module located beneath; (b) picture and scheme of the bi-morph structure of energy harvester. The system comprises Ni-based conductive cloth strips (~15 μm), a thin PVDF layer (~110 μm), and substrate of PI (~300 μm); (c) Maximum output power, current, and voltage due to varying load resistors; (d) proposed wind energy harvester for energy supply to wireless sensors in the grid; (e) apparatus for performing an outdoor experiment. The energy collected from PVDF films is used to drive a temperature sensor. Input voltage (1.5–3.3 VDC) was provided by a voltage regulator needed for the sensor; (f) depiction of PENG's acoustic pressure detection analysis. The distance between sensor and speaker is about 5 cm; (g) sound's pressure level effect on output voltages obtained from PENG under different vibration; (h) V_{oc} from PENG by air blown with different velocity. LED lighting under the pressure of the wind is depicted in the inset; (i) a 2.2-μF capacitor's response due to time, driven by PENG; (j) observed V_{oc} of PENG while dropping water. Inset is the depiction of the uses of the PENG to harvest energy from water, using an umbrella; (k) response of individual droplets of water falling from varying heights; (l) rotational energy conservation from riding a bicycle; (m) voltage-time charging plot of 33-μF capacitor. Inset is a circuit diagram of charging the capacitor and lighting the sensors; (n) ambient temperature detecting sensors to record humidity, and bicycle riding speed, driven by the 33-μF charged capacitor [192].

various forms of mechanical energy that are usually overlooked [128]. Figure 6.15(f and g) illustrates scheme for noticing noise in sound and the related voltage output induced by different instruments. From Figure 6.15(h), it is evident that the output voltage increases with the increasing speed of wind acting on the system, owing to the increased deformation of the PVDF-ZnS film caused by the stronger wind. When attached to the wind-driven system, a blue LED can be turned on (Figure 6.15(i)). Surprisingly, wind energy conversion efficiency has been estimated to be as high as 58 per cent. Additionally, this system can be attached to an umbrella to track raindrop energy harvesting caused by water drop tension (Figure 6.15(j)), and the output is improved as the gap between individual drops increases (Figure 6.15(k)), suggesting effective transformation of environment's mechanical energy to electrical one [192].

Apart from the applications discussed earlier, there are numerous other applications of PENGs. These recent applications of PENGs focus on bio-synthesized materials [95].

6.7 CHALLENGES FACED BY PENGs

Until now, PENGs, TENGs, PyENGs, and thermoelectric NGs (TEMNGs) have been thoroughly researched and displayed in practice for their ability to transform waste mechanical/thermal energy into usable electricity. PENG technology stands out due to its high reliability, stable operation, compact system size, and diverse application areas as shown in Figure 6.16. Wang *et al.* [19] announced the first PENG-based ZnO NSs (labelled as inorganic PENG) and then demonstrated the numerous ways to increase the PENG's instantaneous power density. PENG was implemented using flexible PVDF and its copolymers due to their high flexibility, ease of fabrication, low electrical output efficiency, and ability to withstand high mechanical forces. In recent times, our group and a number of several other research teams developed a composite PENG technology that successfully addressed the problems associated with inorganic and polymer PENGs. The creation of mixed or composite piezoelectric devices is a critical factor in designing an effective, flexible composite PENG system. These composite PENGs are well suited for use as self-powered power sources and sensors for the measurement of a variety of physical parameters such as physical, optical, biological, and chemical stimuli [241–243].

Numerous PENGs based on ZnO, PZT, $BaTiO_3$, PVDF, and zinc stannate ($ZnSnO_3$) have been developed. Although PZT-based NGs have a high piezoelectric performance due to their high piezoelectric coefficient, they are toxic and dangerous to living organisms. Additionally, these NGs are fragile and difficult to install on intricate surfaces. As a result, versatile and environmentally friendly NGs are needed, with PVDF-based NGs being a possible candidate. Not only are PVDF NGs versatile, but they also exhibit mechanical stability, chemical inertness, and thermal stability. Nonetheless, PVDF-based NGs have disadvantages such as a low piezoelectric coefficient, time-dependent amplitudes, and a wide frequency range. PVDF-based NGs have been manufactured using a variety of techniques, including solution casting, injection moulding, compression moulding, and spin-coating. These procedures, although, are time-consuming and may require post-processing (like uniaxial stretching and/or high-voltage poling) to improve the electroactive phases in PVDF. It is a well-known fact that the output voltage of PVDF increases linearly with the sum of β-phase. It is critical to remember, however, that the piezoelectric efficiency of PVDF is also dependant on the preferred orientation of the CF_2 units. On the one hand, force spinning is analogous to uniaxial stretching in that the β-phase of PVDF is formed without preferential orientation of PVDF chains; consequently, PVDF fibres demonstrated extremely low voltage production. On the other hand, electrospinning has the unique benefit of simultaneously providing uniaxial stretching and in

FIGURE 6.16 Challenges and drawbacks of PENG.

situ poling, which results in the creation of PVDF's β-phase. As a result, electrospun PVDF nanofibres (PNFs) do not need post-processing [244].

However, traditional piezoelectric materials such as PZT and PVDF are prohibitively costly, non-renewable, non-biodegradable, and lack biocompatibility due to the cytotoxic nature of Pb-based materials [245].

Additionally, various piezoelectric composites are being investigated. Nevertheless, these piezoelectric materials exhibit certain disadvantages due to their toxic nature, lack of eco-friendliness, incompatibility with biological systems, inability to biodegrade, rigidity, and lower piezoelectric properties. Additionally, the development of piezoelectric materials requires a lengthy and dangerous chemical process for their synthesis and fabrication, which is one of their limitations. These disadvantages of these materials precluded their use in biomedical applications [246].

6.8 SUMMARY AND OUTLOOK

6.8.1 Summary

This chapter presents an in-depth analysis of PENGs. This chapter discusses the operation of PENGs, accompanied by a brief overview of the different materials that could be used to fabricate them, including organic, inorganic, and MOF. Currently, piezoelectric materials used in PENG research and development include ZnO, PZT, $BaTiO_3$, and PVDF [246]. Additionally, a thorough overview of fabrication techniques (electrospinning, spin-coating, flow coating, liquid metal-based technique, and template-assisted growth) has been provided. Additionally, some difficulties encountered by these NGs are illustrated. Among them are the piezoelectric NSs' typical growth mechanism, brittleness, lower force limits, failure instability, leakage current issues, and low piezoelectric coefficient.

PENGs serve as a renewable power source for a variety of intelligent applications. The schematic for energy harnessing is shown in Figure 6.17.

PENGs efficiently harness energy from inner body movements and control health tracking systems [24]. The devices described in this document are being manufactured on nearly every substrate imaginable. To meet the actual power requirements, devices must maximize the variation rate of different strains introduced, increasing the polarization produced with help of functional layers. This improves the device's coupling to the atmosphere and overall device performance [236].

The performance of PENGs advanced in many aspects due to developments in the field of piezoelectric materials, fabrication methods, and modelling. As a result, increasing current production is now a significant problem. Producing materials exhibiting exceptional piezoelectric potential, doping them with suitable dopants, and fabricating composite thin films all increase the production efficiency of PENGs. By chemically doping, the output power of PENGs can be increased by several orders of magnitude (mW). By using the double piezoelectric effect characteristics of inorganic

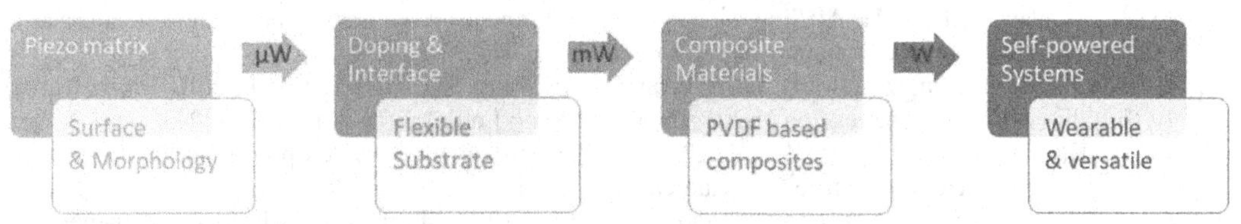

FIGURE 6.17 Steps for improving the PENG output performance (μW to mW to W).

materials and polymers, the performance strength of PENGs can be significantly increased (μW). Taking into account the aforementioned considerations, the output power of the PENG could be significantly increased. Naturally, the production efficiency of the PENG would be significantly improved, owing to the synergy of all influencing parameters. PENG will emerge as fairly high-power in the near future, providing electrical power through self-charged devices [246].

This field is expected to have a more promising future with advancements in existing fabrication techniques and improvements in the properties of materials used to produce PENGs. Concentrating on Pb-free piezoelectric materials with higher output power is critical for the production of flexible NGs capable of a wide range of technology applications.

6.8.2 Outlook

The growing demand for smart electronics has made research into combining energy conversion and storage devices as a hot topic. As the energy conversion unit, the multifunctional PENG must have a high output power and longevity. Additionally, the devices have significant criteria for usability and comfort. Due to the fact that these wearable and implantable devices must be inserted into clothes or footwear and others must be connected to the human body, PENGs must be extremely flexible. The combination of high-efficiency versatile PENGs and flexible platforms will likely result in a shift in their application models in the near future. The following aspects should be considered when developing versatile high-performance PENGs for future applications: (i) the production of highly efficient smart devices for gathering and utilizing mechanical energy; (ii) the enhancement of the performance of flexible PENGs. Along with their output strength, other performance metrics such as flexibility, stability, and durability must meet the demands of practical application; (iii) improving existing integrated technology of PENGs [246].

By using biodegradable, environmentally safe, and non-toxic materials (particularly those that are Pb-free), PENGs can be used in medical applications as well as for patient care and diagnosis.

Additionally, fabricating materials more efficiently allows for an increase in the performance of PENGs, which increases their application potential.

Flexible PENGs are a strong future trend for powering smart devices and microelectronic devices. The more adaptable PENGs are, the broader their application area would be. They are about to enter the age of big data in the physical network.

REFERENCES

1. Yang, R., Qin, Y., Li, C., Zhu, G. and Wang, Z.L., 2009. Converting biomechanical energy into electricity by a muscle-movement-driven nanogenerator. Nano Letters, 9(3), pp.1201–1205.
2. Wang, X., Song, J., Liu, J. and Wang, Z.L., 2007. Direct-current nanogenerator driven by ultrasonic waves. Science, 316(5821), pp.102–105.
3. Cha, S.N., Seo, J.S., Kim, S.M., Kim, H.J., Park, Y.J., Kim, S.W. and Kim, J.M., 2010. Sound-driven piezoelectric nanowire-based nanogenerators. Advanced Materials, 22(42), pp.4726–4730.
4. Gao, P.X., Song, J., Liu, J. and Wang, Z.L., 2007. Nanowire piezoelectric nanogenerators on plastic substrates as flexible power sources for nanodevices. Advanced Materials, 19(1), pp.67–72.
5. Wang, X., Liu, J., Song, J. and Wang, Z.L., 2007. Integrated nanogenerators in biofluid. Nano Letters, 7(8), pp.2475–2479.
6. Li, Z. and Wang, Z.L., 2011. Air/liquid-pressure and heartbeat-driven flexible fibernanogenerators as a micro/nano-power source or diagnostic sensor. Advanced Materials, 23(1), pp.84–89.
7. Park, K.I., Xu, S., Liu, Y., Hwang, G.T., Kang, S.J.L., Wang, Z.L. and Lee, K.J., 2010. Piezoelectric $BaTiO_3$ thin film nanogenerator on plastic substrates. Nano Letters, 10(12), pp.4939–4943.
8. Xu, S., Hansen, B.J. and Wang, Z.L., 2010. Piezoelectric-nanowire-enabled power source for driving wireless microelectronics. Nature Communications, 1(1), pp.1–5.
9. Li, Z., Zhu, G., Yang, R., Wang, A.C. and Wang, Z.L., 2010. Muscle-driven in vivo nanogenerator. Advanced Materials, 22(23), pp.2534–2537.
10. Xu, S., Qin, Y., Xu, C., Wei, Y., Yang, R. and Wang, Z.L., 2010. Self-powered nanowire devices. Nature Nanotechnology, 5(5), pp.366–373.

11. Hu, Y., Zhang, Y., Xu, C., Zhu, G. and Wang, Z.L., 2010. High-output nanogenerator by rational unipolar assembly of conical nanowires and its application for driving a small liquid crystal display. Nano Letters, 10(12), pp.5025–5031.
12. Zhu, G., Yang, R., Wang, S. and Wang, Z.L., 2010. Flexible high-output nanogenerator based on lateral ZnO nanowire array. Nano Letters, 10(8), pp.3151–3155.
13. Wang, Z.L., 2007. The new field of nanopiezotronics. Materials Today, 10(5), pp.20–28.
14. Patolsky, F., Timko, B.P., Zheng, G. and Lieber, C.M., 2007. Nanowire-based nanoelectronic devices in the life sciences. MRS Bulletin, 32(2), pp.142–149.
15. Patolsky, F., Timko, B.P., Yu, G., Fang, Y., Greytak, A.B., Zheng, G. and Lieber, C.M., 2006. Detection, stimulation, and inhibition of neuronal signals with high-density nanowire transistor arrays. Science, 313(5790), pp.1100–1104.
16. Pauzauskie, P.J. and Yang, P., 2006. Nanowire photonics. Materials Today, 9(10), pp.36–45.
17. Paradiso, J.A. and Starner, T., 2005. Energy scavenging for mobile and wireless electronics. IEEE Pervasive Computing, 4(1), pp.18–27.
18. Hochbaum, A.I. and Yang, P., 2010. Semiconductor nanowires for energy conversion. Chemical Reviews, 110(1), pp.527–546.
19. Wang, Z.L. and Song, J., 2006. Piezoelectric nanogenerators based on zinc oxide nanowire arrays. Science, 312(5771), pp.242–246.
20. Schiffer, M.B. Draw the Lightning Down: Benjamin Franklin and Electrical Technology in the Age of Enlightenment; University of California Press: Berkeley, CA, 2006; Volume 408.
21. Kagan, C.R., Fernandez, L.E., Gogotsi, Y., Hammond, P.T., Hersam, M.C., Nel, A.E., Penner, R.M., Willson, C.G. and Weiss, P.S., 2016. Nano day: celebrating the next decade of nanoscience and nanotechnology. ACS Nano, 10, pp. 9093–9103.
22. Askari, H., Khajepour, A., Khamesee, M.B., Saadatnia, Z. and Wang, Z.L., 2018. Piezoelectric and triboelectric nanogenerators: Trends and impacts. Nano Today, 22, pp.10–13.
23. Zi, Y. and Wang, Z.L., 2017. Nanogenerators: an emerging technology towards nanoenergy. APL Materials, 5(7), p.074103.
24. Sripadmanabhan Indira, S., Aravind Vaithilingam, C., Oruganti, K.S.P., Mohd, F. and Rahman, S., 2019. Nanogenerators as a sustainable power source: state of art, applications, and challenges. Nanomaterials, 9(5), p.773.
25. Nechibvute, A., Chawanda, A. and Luhanga, P., 2012. Piezoelectric energy harvesting devices: an alternative energy source for wireless sensors. Smart Materials Research, 2012, pp.1–13.
26. Wang, X., 2012. Piezoelectric nanogenerators—harvesting ambient mechanical energy at the nanometer scale. Nano Energy, 1(1), pp.13–24.
27. Gao, Y. and Wang, Z.L., 2007. Electrostatic potential in a bent piezoelectric nanowire. The fundamental theory of nanogenerator and nanopiezotronics. Nano Letters, 7(8), pp.2499–2505.
28. Jiji, G., 2017. A retrospect on the role of piezoelectric nanogenerators in the development of the green world. RSC Advances, 7(53), pp.33642–33670.
29. Zhang, C., Yan, Y., Zhao, Y.S. and Yao, J., 2013. Synthesis and applications of organic nanorods, nanowires and nanotubes. Annual Reports Section "C" (Physical Chemistry), 109, pp.211–239.
30. Zhu, G., Wang, A.C., Liu, Y., Zhou, Y. and Wang, Z.L., 2012. Functional electrical stimulation by nanogenerator with 58 V output voltage. Nano Letters, 12(6), pp.3086–3090.
31. Wang, Z.L., Zhu, G., Yang, Y., Wang, S. and Pan, C., 2012. Progress in nanogenerators for portable electronics. Materials Today, 15(12), pp.532–543.
32. Shin, S.H., Kim, Y.H., Lee, M.H., Jung, J.Y. and Nah, J., 2014. Hemispherically aggregated $BaTiO_3$ nanoparticle composite thin film for high-performance flexible piezoelectric nanogenerator. ACS Nano, 8(3), pp.2766–2773.
33. Wu, W., Wang, L., Li, Y., Zhang, F., Lin, L., Niu, S., Chenet, D., Zhang, X., Hao, Y., Heinz, T.F. and Hone, J., 2014. Piezoelectricity of single-atomic-layer MoS_2 for energy conversion and piezotronics. Nature, 514(7523), pp.470–474.
34. Han, S.A., Kim, T.H., Kim, S.K., Lee, K.H., Park, H.J., Lee, J.H. and Kim, S.W., 2018. Point-defect-passivated MoS_2 nanosheet-based high performance piezoelectric nanogenerator. Advanced Materials, 30(21), p.1800342.
35. Niu, S., Hu, Y., Wen, X., Zhou, Y., Zhang, F., Lin, L., Wang, S. and Wang, Z.L., 2013. Enhanced performance of flexible ZnO nanowire based room-temperature oxygen sensors by Piezotronic effect. Advanced Materials, 25(27), pp.3701–3706.
36. Dong, C., Fu, Y., Zang, W., He, H., Xing, L. and Xue, X., 2017. Self-powering/self-cleaning electronic-skin basing on PVDF/TiO_2 nanofibers for actively detecting body motion and degrading organic pollutants. Applied Surface Science, 416, pp.424–431.

37. Maity, K., Mahanty, B., Sinha, T.K., Garain, S., Biswas, A., Ghosh, S.K., Manna, S., Ray, S.K. and Mandal, D., 2017. Two-dimensional piezoelectric MoS_2-modulated nanogenerator and nanosensor made of poly (vinlydine Fluoride) nanofiber webs for self-powered electronics and robotics. Energy Technology, 5(2), pp.234–243.
38. Lu, X., Qu, H. and Skorobogatiy, M., 2017. Piezoelectric micro-and nanostructured fibers fabricated from thermoplastic nanocomposites using a fiber drawing technique: comparative study and potential applications. ACS Nano, 11(2), pp.2103–2114.
39. Wang, Z.L., 2010. Piezopotential gated nanowire devices: piezotronics and piezo-phototronics. Nano Today, 5(6), pp.540–552.
40. Pradel, K.C., Wu, W., Ding, Y. and Wang, Z.L., 2014. Solution-derived ZnO homojunction nanowire films on wearable substrates for energy conversion and self-powered gesture recognition. Nano Letters, 14(12), pp.6897–6905.
41. Chen, C.Q. and Zhu, J., 2007. Bending strength and flexibility of ZnO nanowires. Applied Physics Letters, 90(4), p.043105.
42. Li, P., Liao, Q., Yang, S., Bai, X., Huang, Y., Yan, X., Zhang, Z., Liu, S., Lin, P., Kang, Z. and Zhang, Y., 2014. In situ transmission electron microscopy investigation on fatigue behavior of single ZnO wires under high-cycle strain. Nano Letters, 14(2), pp.480–485.
43. Kim, H., Kim, S.M., Son, H., Kim, H., Park, B., Ku, J., Sohn, J.I., Im, K., Jang, J.E., Park, J.J. and Kim, O., 2012. Enhancement of piezoelectricity via electrostatic effects on a textile platform. Energy & Environmental Science, 5(10), pp.8932–8936.
44. Qiu, Y. and Yang, S., 2007. ZnO nanotetrapods: controlled vapor-phase synthesis and application for humidity sensing. Advanced Functional Materials, 17(8), pp.1345–1352.
45. Gao, P.X., Ding, Y., Mai, W., Hughes, W.L., Lao, C. and Wang, Z.L., 2005. Conversion of zinc oxide nanobelts into superlattice-structured nanohelices. Science, 309(5741), pp.1700–1704.
46. Zhang, Y., Liu, C., Liu, J., Xiong, J., Liu, J., Zhang, K., Liu, Y., Peng, M., Yu, A., Zhang, A. and Zhang, Y., 2016. Lattice strain induced remarkable enhancement in piezoelectric performance of ZnO-based flexible nanogenerators. ACS Applied Materials & Interfaces, 8(2), pp.1381–1387.
47. Wang, X.B., Song, C., Li, D.M., Geng, K.W., Zeng, F. and Pan, F., 2006. The influence of different doping elements on microstructure, piezoelectric coefficient and resistivity of sputtered ZnO film. Applied Surface Science, 253(3), pp.1639–1643.
48. Chen, Y.Q., Zheng, X.J. and Feng, X., 2009. The fabrication of vanadium-doped ZnO piezoelectric nanofiber by electrospinning. Nanotechnology, 21(5), p.055708.
49. Wang, Z.L., 2007. Novel nanostructures of ZnO for nanoscale photonics, optoelectronics, piezoelectricity, and sensing. Applied Physics A, 88(1), pp.7–15.
50. Laurenti, M., Canavese, G., Sacco, A., Fontana, M., Bejtka, K., Castellino, M., Pirri, C.F. and Cauda, V., 2015. Nanobranched ZnO structure: p-type doping induces piezoelectric voltage generation and ferroelectric–photovoltaic effect. Advanced Materials, 27(28), pp.4218–4223.
51. Morkoç, H. and Özgür, Ü., 2008. Zinc Oxide: Fundamentals, Materials and Device Technology. John Wiley & Sons, Richmond VA.
52. Xi, Y., Song, J., Xu, S., Yang, R., Gao, Z., Hu, C. and Wang, Z.L., 2009. Growth of ZnO nanotube arrays and nanotube based piezoelectric nanogenerators. Journal of Materials Chemistry, 19(48), pp.9260–9264.
53. Yu, R., Wu, W., Ding, Y. and Wang, Z.L., 2013. GaN nanobelt-based strain-gated piezotronic logic devices and computation. ACS Nano, 7(7), pp.6403–6409.
54. Zhou, Y.S., Hinchet, R., Yang, Y., Ardila, G., Songmuang, R., Zhang, F., Zhang, Y., Han, W., Pradel, K., Montès, L. and Mouis, M., 2013. Nano-Newton transverse force sensor using a vertical GaN nanowire based on the piezotronic effect. Advanced Materials, 25(6), pp.883–888.
55. Wang, C.H., Liao, W.S., Lin, Z.H., Ku, N.J., Li, Y.C., Chen, Y.C., Wang, Z.L. and Liu, C.P., 2014. Optimization of the output efficiency of GaN nanowire piezoelectric nanogenerators by tuning the free carrier concentration. Advanced Energy Materials, 4(16), p.1400392.
56. Chen, C.Y., Zhu, G., Hu, Y., Yu, J.W., Song, J., Cheng, K.Y., Peng, L.H., Chou, L.J. and Wang, Z.L., 2012. Gallium nitride nanowire based nanogenerators and light-emitting diodes. ACS Nano, 6(6), pp.5687–5692.
57. Wang, C.H., Liao, W.S., Ku, N.J., Li, Y.C., Chen, Y.C., Tu, L.W. and Liu, C.P., 2014. Effects of free carriers on piezoelectric nanogenerators and piezotronic devices made of GaN nanowire arrays. Small, 10(22), pp.4718–4725.
58. Chen, J., Oh, S.K., Zou, H., Shervin, S., Wang, W., Pouladi, S., Zi, Y., Wang, Z.L. and Ryou, J.H., 2018. High-output lead-free flexible piezoelectric generator using single-crystalline GaN thin film. ACS Applied Materials & Interfaces, 10(15), pp.12839–12846.

59. Ku, N.J., Wang, C.H., Huang, J.H., Fang, H.C., Huang, P.C. and Liu, C.P., 2013. Energy harvesting from the obliquely aligned InN nanowire array with a surface electron-accumulation layer. Advanced Materials, 25(6), pp.861–866.
60. Ku, N.J., Huang, J.H., Wang, C.H., Fang, H.C. and Liu, C.P., 2012. Crystal face-dependent nanopiezotronics of an obliquely aligned InN nanorod array. Nano Letters, 12(2), pp.562–568.
61. Ku, N.J., Liu, G., Wang, C.H., Gupta, K., Liao, W.S., Ban, D. and Liu, C.P., 2017. Optimal geometrical design of inertial vibration DC piezoelectric nanogenerators based on obliquely aligned InN nanowire arrays. Nanoscale, 9(37), pp.14039–14046.
62. Liu, G., Zhao, S., Henderson, R.D., Leonenko, Z., Abdel-Rahman, E., Mi, Z. and Ban, D., 2016. Nanogenerators based on vertically aligned InN nanowires. Nanoscale, 8(4), pp.2097–2106.
63. Lin, Y.F., Song, J., Ding, Y., Lu, S.Y. and Wang, Z.L., 2008. Piezoelectric nanogenerator using CdS nanowires. Applied Physics Letters, 92(2), p.022105.
64. Zhou, Y.S., Wang, K., Han, W., Rai, S.C., Zhang, Y., Ding, Y., Pan, C., Zhang, F., Zhou, W. and Wang, Z.L., 2012. Vertically aligned CdSe nanowire arrays for energy harvesting and piezotronic devices. ACS Nano, 6(7), pp.6478–6482.
65. Chen, Y.Y., Wang, C.H., Chen, G.S., Li, Y.C. and Liu, C.P., 2015. Self-powered n-$Mg_xZn_{1-x}O$/p-Si photodetector improved by alloying-enhanced piezopotential through piezo-phototronic effect. Nano Energy, 11, pp.533–539.
66. Tangi, M., Min, J.W., Priante, D., Subedi, R.C., Anjum, D.H., Prabaswara, A., Alfaraj, N., Liang, J.W., Shakfa, M.K., Ng, T.K. and Ooi, B.S., 2018. Observation of piezotronic and piezo-phototronic effects in n-InGaN nanowires/Ti grown by molecular beam epitaxy. Nano Energy, 54, pp.264–271.
67. Wang, X., Yu, R., Jiang, C., Hu, W., Wu, W., Ding, Y., Peng, W., Li, S. and Wang, Z.L., 2016. Piezotronic effect modulated heterojunction electron gas in AlGaN/AlN/GaN heterostructure microwire. Advanced Materials, 28(33), pp.7234–7242.
68. Wang, X., Peng, W., Pan, C. and Wang, Z.L., 2017. Piezotronics and piezo-phototronics based on a-axis nano/microwires: fundamentals and applications. Semiconductor Science and Technology, 32(4), p.043005.
69. Wang, X., Yu, R., Peng, W., Wu, W., Li, S. and Wang, Z.L., 2015. Temperature dependence of the piezotronic and piezophototronic effects in a-axis GaN nanobelts. Advanced Materials, 27(48), pp.8067–8074.
70. Yu, R., Wang, X., Peng, W., Wu, W., Ding, Y., Li, S. and Wang, Z.L., 2015. Piezotronic effect in strain-gated transistor of a-axis GaN nanobelt. ACS Nano, 9(10), pp.9822–9829.
71. Zhao, Z., Pu, X., Han, C., Du, C., Li, L., Jiang, C., Hu, W. and Wang, Z.L., 2015. Piezotronic effect in polarity-controlled GaN nanowires. ACS Nano, 9(8), pp.8578–8583.
72. Tsai, C.Y., Gupta, K., Wang, C.H. and Liu, C.P., 2017. Ultrahigh UV responsivity of single nonpolar a-axial GaN nanowire with asymmetric piezopotential via piezo-phototronic effect: dependence of carrier screening effect on strain. Nano Energy, 34, pp.367–374.
73. Wang, X., Peng, W., Yu, R., Zou, H., Dai, Y., Zi, Y., Wu, C., Li, S. and Wang, Z.L., 2017. Simultaneously enhancing light emission and suppressing efficiency droop in GaN microwire-based ultraviolet light-emitting diode by the piezo-phototronic effect. Nano Letters, 17(6), pp.3718–3724.
74. Kou, J., Zhang, Y., Liu, Y., Zhang, K., Liu, W. and Zhai, J., 2017. Nano-force sensor based on a single tellurium microwire. Semiconductor Science and Technology, 32(7), p.074001.
75. Gao, S., Wang, Y., Wang, R. and Wu, W., 2017. Piezotronic effect in 1D van der Waals solid of elemental tellurium nanobelt for smart adaptive electronics. Semiconductor Science and Technology, 32(10), p.104004.
76. Wu, J.M., Xu, C., Zhang, Y. and Wang, Z.L., 2012. Lead-free nanogenerator made from single $ZnSnO_3$ microbelt. ACS Nano, 6(5), pp.4335–4340.
77. Wu, J.M., Chen, C.Y., Zhang, Y., Chen, K.H., Yang, Y., Hu, Y., He, J.H. and Wang, Z.L., 2012. Ultrahigh sensitive piezotronic strain sensors based on a $ZnSnO_3$ nanowire/microwire. ACS Nano, 6(5), pp.4369–4374.
78. Jung, J.H., Lee, M., Hong, J.I., Ding, Y., Chen, C.Y., Chou, L.J. and Wang, Z.L., 2011. Lead-free $NaNbO_3$ nanowires for a high output piezoelectric nanogenerator. ACS Nano 5, pp.10041–10046.
79. Chen, X., Xu, S., Yao, N. and Shi, Y., 2010. 1.6 V nanogenerator for mechanical energy harvesting using PZT nanofibers. Nano Letters, 10(6), pp.2133–2137.
80. Cross, E., 2004. Lead-free at last. Nature, 432(7013), pp.24–25.
81. Wang, Z., Hu, J., Suryavanshi, A.P., Yum, K. and Yu, M.F., 2007. Voltage generation from individual $BaTiO_3$ nanowires under periodic tensile mechanical load. Nano Letters, 7(10), pp.2966–2969.
82. Chorsi, M.T., Curry, E.J., Chorsi, H.T., Das, R., Baroody, J., Purohit, P.K., Ilies, H. and Nguyen, T.D., 2019. Piezoelectric biomaterials for sensors and actuators. Advanced Materials, 31(1), p.1802084.

83. Jacob, J., More, N., Kalia, K. and Kapusetti, G., 2018. Piezoelectric smart biomaterials for bone and cartilage tissue engineering. Inflammation and Regeneration, 38(1), pp.1–11.
84. Sirohi, J. and Chopra, I., 2000. Fundamental understanding of piezoelectric strain sensors. Journal of Intelligent Material Systems and Structures, 11(4), pp.246–257.
85. Cannata, D., Benetti, M., Verona, E., Varriale, A., Staiano, M., D'Auria, S. and Di Pietrantonio, F., 2012, October. Odorant detection via solidly mounted resonator biosensor. In 2012 IEEE International Ultrasonics Symposium (pp. 1537–1540). IEEE, Dresden, Germany.
86. Chen, D., Wang, J. and Xu, Y., 2013. Highly sensitive lateral field excited piezoelectric film acoustic enzyme biosensor. IEEE Sensors Journal, 13(6), pp.2217–2222.
87. Dagdeviren, C., Yang, B.D., Su, Y., Tran, P.L., Joe, P., Anderson, E., Xia, J., Doraiswamy, V., Dehdashti, B., Feng, X. and Lu, B., 2014. Conformal piezoelectric energy harvesting and storage from motions of the heart, lung, and diaphragm. Proceedings of the National Academy of Sciences, 111(5), pp.1927–1932.
88. Hwang, G.T., Park, H., Lee, J.H., Oh, S., Park, K.I., Byun, M., Park, H., Ahn, G., Jeong, C.K., No, K. and Kwon, H., 2014. Self-powered cardiac pacemaker enabled by flexible single crystalline PMN-PT piezoelectric energy harvester. Advanced Materials, 26(28), pp.4880–4887.
89. Zhu, G., Wang, A.C., Liu, Y., Zhou, Y. and Wang, Z.L., 2012. Functional electrical stimulation by nanogenerator with 58 V output voltage. Nano Letters, 12(6), pp.3086–3090.
90. He, Y.B., Li, G.R., Wang, Z.L., Su, C.Y. and Tong, Y.X., 2011. Single-crystal ZnO nanorod/amorphous and nanoporous metal oxide shell composites: controllable electrochemical synthesis and enhanced supercapacitor performances. Energy & Environmental Science, 4(4), pp.1288–1292.
91. Ryu, J., Kim, S.W., Kang, K. and Park, C.B., 2010. Synthesis of diphenylalanine/cobalt oxide hybrid nanowires and their application to energy storage. ACS Nano, 4(1), pp.159–164.
92. Kim, D.H., Lu, N., Ma, R., Kim, Y.S., Kim, R.H., Wang, S., Wu, J., Won, S.M., Tao, H., Islam, A. and Yu, K.J., 2011. Epidermal electronics. Science, 333(6044), pp.838–843.
93. Choi, C., Lee, Y., Cho, K.W., Koo, J.H. and Kim, D.H., 2018. Wearable and implantable soft bioelectronics using two-dimensional materials. Accounts of Chemical Research, 52(1), pp.73–81.
94. Martin, A.J.P., 1941. Tribo-electricity in wool and hair. Proceedings of the Physical Society (1926-1948), 53(2), p.186.
95. Yuan, H., Lei, T., Qin, Y., He, J.H. and Yang, R., 2019. Design and application of piezoelectric biomaterials. Journal of Physics D, 52(19), p.194002.
96. de Sciarra, F.M., 2009. A nonlocal model with strain-based damage. International Journal of Solids and Structures, 46(22–23), pp.4107–4122.
97. Barretta, R., Fabbrocino, F., Luciano, R. and de Sciarra, F.M., 2018. Closed-form solutions in stress-driven two-phase integral elasticity for bending of functionally graded nano-beams. Physica E: Low-Dimensional Systems and Nanostructures, 97, pp.13–30.
98. Zhou, Z., Qian, D. and Minary-Jolandan, M., 2016. Molecular mechanism of polarization and piezoelectric effect in super-twisted collagen. ACS Biomaterials Science & Engineering, 2(6), pp.929–936.
99. Bystrov, V.S., Bdikin, I.K., Heredia, A., Pullar, R.C., Mishina, E.D., Sigov, A.S. and Kholkin, A.L., 2012. Piezoelectricity and ferroelectricity in biomaterials: from proteins to self-assembled peptide nanotubes. In Piezoelectric Nanomaterials for Biomedical Applications (pp. 187–211). Springer, Berlin, Heidelberg.
100. Wojnar, R., 2012. Piezoelectric phenomena in biological tissues. In Piezoelectric Nanomaterials for Biomedical Applications (pp. 173–185). Springer, Berlin, Heidelberg.
101. Lemanov, V.V., Popov, S.N. and Pankova, G.A., 2002. Piezoelectric properties of crystals of some protein aminoacids and their related compounds. Physics of the Solid State, 44(10), pp.1929–1935.
102. Namiki, K., Hayakawa, R. and Wada, Y., 1980. Molecular theory of piezoelectricity of α-helical polypeptide. Journal of Polymer Science, 18(5), pp.993–1004.
103. Fukada, E., 2000. History and recent progress in piezoelectric polymers. IEEE Transactions on Ultrasonics, Ferroelectrics, and Frequency Control, 47(6), pp.1277–1290.
104. Lee, B.Y., Zhang, J., Zueger, C., Chung, W.J., Yoo, S.Y., Wang, E., Meyer, J., Ramesh, R. and Lee, S.W., 2012. Virus-based piezoelectric energy generation. Nature Nanotechnology, 7(6), pp.351–356.
105. Shin, D.M., Han, H.J., Kim, W.G., Kim, E., Kim, C., Hong, S.W., Kim, H.K., Oh, J.W. and Hwang, Y.H., 2015. Bioinspired piezoelectric nanogenerators based on vertically aligned phage nanopillars. Energy & Environmental Science, 8(11), pp.3198–3203.
106. Lee, J.H., Lee, J.H., Xiao, J., Desai, M.S., Zhang, X. and Lee, S.W., 2019. Vertical self-assembly of polarized phage nanostructure for energy harvesting. Nano Letters, 19(4), pp.2661–2667.
107. Moon, J.S., Kim, W.G., Shin, D.M., Lee, S.Y., Kim, C., Lee, Y., Han, J., Kim, K., Yoo, S.Y. and Oh, J.W., 2017. Bioinspired M-13 bacteriophage-based photonic nose for differential cell recognition. Chemical Science, 8(2), pp.921–927.

108. Moon, J.S., Lee, Y., Shin, D.M., Kim, C., Kim, W.G., Park, M., Han, J., Song, H., Kim, K. and Oh, J.W., 2016. Identification of endocrine disrupting chemicals using a virus-based colorimetric sensor. Chemistry–An Asian Journal, 11(21), pp.3097–3101.
109. Kim, W.G., Zueger, C., Kim, C., Wong, W., Devaraj, V., Yoo, H.W., Hwang, S., Oh, J.W. and Lee, S.W., 2019. Experimental and numerical evaluation of a genetically engineered M13 bacteriophage with high sensitivity and selectivity for 2,4,6-trinitrotoluene. Organic & Biomolecular Chemistry, 17(23), pp.5666–5670.
110. Shin, Y.C., Kim, C., Song, S.J., Jun, S., Kim, C.S., Hong, S.W., Hyon, S.H., Han, D.W. and Oh, J.W., 2018. Ternary aligned nanofibers of RGD peptide-displaying M13 bacteriophage/PLGA/graphene oxide for facilitated myogenesis. Nanotheranostics, 2(2), p.144.
111. Raja, I.S., Kim, C., Song, S.J., Shin, Y.C., Kang, M.S., Hyon, S.H., Oh, J.W. and Han, D.W., 2019. Virus-incorporated biomimetic nanocomposites for tissue regeneration. Nanomaterials, 9(7), p.1014.
112. Chew, J.W., Black, S.N., Chow, P.S., Tan, R.B. and Carpenter, K.J., 2007. Stable polymorphs: difficult to make and difficult to predict. CrystEngComm, 9(2), pp.128–130.
113. Poornachary, S.K., Chow, P.S. and Tan, R.B., 2008. Influence of solution speciation of impurities on polymorphic nucleation in glycine. Crystal Growth and Design, 8(1), pp.179–185.
114. Dowling, R., Davey, R.J., Curtis, R.A., Han, G., Poornachary, S.K., Chow, P.S. and Tan, R.B., 2010. Acceleration of crystal growth rates: an unexpected effect of tailor-made additives. Chemical Communications, 46(32), pp.5924–5926.
115. Iitaka, Y., 1954. A new form of glycine. Proceedings of the Japan Academy, 30(2), pp.109–112.
116. Kumar, R.A., Vizhi, R.E., Vijayan, N. and Babu, D.R., 2011. Structural, dielectric and piezoelectric properties of nonlinear optical γ-glycine single crystals. Physica B, 406(13), pp.2594–2600.
117. Iitaka, Y., 1958. The crystal structure of γ-glycine. Acta Crystallographica, 11(3), pp.225–226.
118. Guerin, S., Tofail, S.A. and Thompson, D., 2019. Organic piezoelectric materials: milestones and potential. NPG Asia Materials, 11(1), pp.1–5.
119. Guerin, S., Stapleton, A., Chovan, D., Mouras, R., Gleeson, M., McKeown, C., Noor, M.R., Silien, C., Rhen, F.M., Kholkin, A.L. and Liu, N., 2018. Control of piezoelectricity in amino acids by supramolecular packing. Nature Materials, 17(2), pp.180–186.
120. Newnham, R.E., 2005. Properties of Materials: Anisotropy, Symmetry, Structure. Oxford University Press on Demand, NY.
121. Yan, X., Zhu, P., Fei, J. and Li, J., 2010. Self-assembly of peptide-inorganic hybrid spheres for adaptive encapsulation of guests. Advanced Materials, 22(11), pp.1283–1287.
122. Görbitz, C.H., 2001. Nanotube formation by hydrophobic dipeptides. Chemistry–A European Journal, 7(23), pp.5153–5159.
123. Amdursky, N., Beker, P. and Rosenman, G., 2013. Physics of peptide nanostructures and their nanotechnology applications. Peptide Materials: From Nanostuctures to Applications, 1st ed.; Aleman, C., Bianco, A., Venanzi, M., Eds, Wiley, pp.1–37.
124. Kholkin, A., Amdursky, N., Bdikin, I., Gazit, E. and Rosenman, G., 2010. Strong piezoelectricity in bioinspired peptide nanotubes. ACS Nano, 4(2), pp.610–614.
125. Lee, J.H., Heo, K., Schulz-Schönhagen, K., Lee, J.H., Desai, M.S., Jin, H.E. and Lee, S.W., 2018. Diphenylalanine peptide nanotube energy harvesters. ACS Nano, 12(8), pp.8138–8144.
126. Nguyen, V., Jenkins, K. and Yang, R., 2015. Epitaxial growth of vertically aligned piezoelectric diphenylalanine peptide microrods with uniform polarization. Nano Energy, 17, pp.323–329.
127. Nguyen, V., Zhu, R., Jenkins, K. and Yang, R., 2016. Self-assembly of diphenylalanine peptide with controlled polarization for power generation. Nature Communications, 7(1), pp.1–6.
128. Sultana, A., Ghosh, S.K., Sencadas, V., Zheng, T., Higgins, M.J., Middya, T.R. and Mandal, D., 2017. Human skin interactive self-powered wearable piezoelectric bio-e-skin by electrospun poly-l-lactic acid nanofibers for non-invasive physiological signal monitoring. Journal of Materials Chemistry B, 5(35), pp.7352–7359.
129. Tajitsu, Y., 2008. Piezoelectricity of chiral polymeric fiber and its application in biomedical engineering. IEEE Transactions on Ultrasonics, Ferroelectrics, and Frequency Control, 55(5), pp.1000–1008.
130. Fukada, E., 1998. New piezoelectric polymers. Japanese Journal of Applied Physics, 37(5S), p.2775.
131. Yoshida, T., Imoto, K., Tahara, K., Naka, K., Uehara, Y., Kataoka, S., Date, M., Fukada, E. and Tajitsu, Y., 2010. Piezoelectricity of poly (L-lactic acid) composite film with stereocomplex of poly (L-lactide) and poly(D-lactide). Japanese Journal of Applied Physics, 49(9S), p.09MC11.
132. Tao, M., Xue, L., Liu, F. and Jiang, L., 2014. An intelligent superwetting PVDF membrane showing switchable transport performance for oil/water separation. Advanced Materials, 26(18), pp.2943–2948.

133. Zhang, W., Shi, Z., Zhang, F., Liu, X., Jin, J. and Jiang, L., 2013. Superhydrophobic and superoleophilic PVDF membranes for effective separation of water-in-oil emulsions with high flux. Advanced Materials, 25(14), pp.2071–2076.
134. Nunes, J.S., Wu, A., Gomes, J., Sencadas, V., Vilarinho, P.M. and Lanceros-Méndez, S., 2009. Relationship between the microstructure and the microscopic piezoelectric response of the α-and β-phases of poly(vinylidene fluoride). Applied Physics A, 95(3), pp. 875–880.
135. Gomes, J., Nunes, J.S., Sencadas, V. and Lanceros-Méndez, S., 2010. Influence of the β-phase content and degree of crystallinity on the piezo-and ferroelectric properties of poly(vinylidene fluoride). Smart Materials and Structures, 19(6), p.065010.
136. Martins, P.M., Ribeiro, S., Ribeiro, C., Sencadas, V., Gomes, A.C., Gama, F.M. and Lanceros-Méndez, S., 2013. Effect of poling state and morphology of piezoelectric poly(vinylidene fluoride) membranes for skeletal muscle tissue engineering. RSC Advances, 3(39), pp.17938–17944.
137. Zheng, J., He, A., Li, J. and Han, C.C., 2007. Polymorphism control of poly(vinylidene fluoride) through electrospinning. Macromolecular Rapid Communications, 28(22), pp.2159–2162.
138. Yee, W.A., Kotaki, M., Liu, Y. and Lu, X., 2007. Morphology, polymorphism behavior and molecular orientation of electrospun poly(vinylidene fluoride) fibers. Polymer, 48(2), pp.512–521.
139. Ribeiro, C., Sencadas, V., Ribelles, J.L.G. and Lanceros-Méndez, S., 2010. Influence of processing conditions on polymorphism and nanofiber morphology of electroactive poly(vinylidene fluoride) electrospun membranes. Soft Materials, 8(3), pp.274–287.
140. Hansen, B.J., Liu, Y., Yang, R. and Wang, Z.L., 2010. Hybrid nanogenerator for concurrently harvesting biomechanical and biochemical energy. ACS Nano, 4(7), pp.3647–3652.
141. Persano, L., Dagdeviren, C., Su, Y., Zhang, Y., Girardo, S., Pisignano, D., Huang, Y. and Rogers, J.A., 2013. High performance piezoelectric devices based on aligned arrays of nanofibers of poly(vinylidenefluoride-co-trifluoroethylene). Nature Communications, 4(1), pp.1–10.
142. Guo, W., Tan, C., Shi, K., Li, J., Wang, X.X., Sun, B., Huang, X., Long, Y.Z. and Jiang, P., 2018. Wireless piezoelectric devices based on electrospun PVDF/$BaTiO_3$ NW nanocomposite fibers for human motion monitoring. Nanoscale, 10(37), pp.17751–17760.
143. Chen, X., Song, Y., Su, Z., Chen, H., Cheng, X., Zhang, J., Han, M. and Zhang, H., 2017. Flexible fiber-based hybrid nanogenerator for biomechanical energy harvesting and physiological monitoring. Nano Energy, 38, pp.43–50.
144. Kim, Y.J., Dang, T.V., Choi, H.J., Park, B.J., Eom, J.H., Song, H.A., Seol, D., Kim, Y., Shin, S.H., Nah, J. and Yoon, S.G., 2016. Piezoelectric properties of CH_3 NH_3 PbI_3 perovskite thin films and their applications in piezoelectric generators. Journal of Materials Chemistry A, 4(3), pp.756–763.
145. Ma, N. and Yang, Y., 2017. Enhanced self-powered UV photoresponse of ferroelectric $BaTiO_3$ materials by pyroelectric effect. Nano Energy, 40, pp.352–359.
146. Qian, W., Yang, W., Zhang, Y., Bowen, C.R. and Yang, Y., 2020. Piezoelectric materials for controlling electro-chemical processes. Nano-Micro Lettter, 12(1), pp.1–39.
147. Ma, N. and Yang, Y., 2018. Boosted photocurrent in ferroelectric $BaTiO_3$ materials via two dimensional planar-structured contact configurations. Nano Energy, 50, pp.417–424.
148. Jella, V., Ippili, S., Eom, J.H., Kim, Y.J., Kim, H.J. and Yoon, S.G., 2018. A novel approach to ambient energy (thermoelectric, piezoelectric and solar-TPS) harvesting: realization of a single structured TPS-fusion energy device using MAPbI3. Nano Energy, 52, pp.11–21.
149. Yang, Z., Lu, J., ZhuGe, M., Cheng, Y., Hu, J., Li, F., Qiao, S., Zhang, Y., Hu, G., Yang, Q. and Peng, D., 2019. Controllable growth of aligned monocrystalline $CsPbBr_3$ microwire arrays for piezoelectric-induced dynamic modulation of single-mode lasing. Advanced Materials, 31(18), p.1900647.
150. Kim, D.B., Park, K.H. and Cho, Y.S., 2020. Origin of high piezoelectricity of inorganic halide perovskite thin films and their electromechanical energy-harvesting and physiological current-sensing characteristics. Energy & Environmental Science, 13(7), pp.2077–2086.
151. Ding, R., Liu, H., Zhang, X., Xiao, J., Kishor, R., Sun, H., Zhu, B., Chen, G., Gao, F., Feng, X. and Chen, J., 2016. Flexible piezoelectric nanocomposite generators based on formamidinium lead halide perovskite nanoparticles. Advanced Functional Materials, 26(42), pp.7708–7716.
152. Stroppa, A., Di Sante, D., Barone, P., Bokdam, M., Kresse, G., Franchini, C., Whangbo, M.H. and Picozzi, S., 2014. Tunable ferroelectric polarization and its interplay with spin–orbit coupling in tin iodide perovskites. Nature Communications, 5(1), pp.1–8.
153. Dhar, J., Sil, S., Hoque, N.A., Dey, A., Das, S., Ray, P.P. and Sanyal, D., 2018. Lattice-defect-induced piezo response in methylammonium-lead-iodide perovskite based nanogenerator. ChemistrySelect, 3(19), pp.5304–5312.

154. Ippili, S., Jella, V., Kim, J., Hong, S. and Yoon, S.G., 2020. Unveiling predominant air-stable organotin bromide perovskite toward mechanical energy harvesting. ACS Applied Materials & Interfaces, 12(14), pp.16469–16480.
155. Ueberschlag, P., 2001. PVDF piezoelectric polymer. Sensor review.
156. Ding, R., Zhang, X., Chen, G., Wang, H., Kishor, R., Xiao, J., Gao, F., Zeng, K., Chen, X., Sun, X.W. and Zheng, Y., 2017. High-performance piezoelectric nanogenerators composed of formamidinium lead halide perovskite nanoparticles and poly(vinylidene fluoride). Nano Energy, 37, pp.126–135.
157. Jella, V., Ippili, S., Eom, J.H., Choi, J. and Yoon, S.G., 2018. Enhanced output performance of a flexible piezoelectric energy harvester based on stable MAPbI3-PVDF composite films. Nano Energy, 53, pp.46–56.
158. Khan, A.A., Rana, M.M., Huang, G., Mei, N., Saritas, R., Wen, B., Zhang, S., Voss, P., Rahman, E.A., Leonenko, Z. and Islam, S., 2020. Maximizing piezoelectricity by self-assembled highly porous perovskite–polymer composite films to enable the internet of things. Journal of Materials Chemistry A, 8(27), pp.13619–13629.
159. Sultana, A., Ghosh, S.K., Alam, M.M., Sadhukhan, P., Roy, K., Xie, M., Bowen, C.R., Sarkar, S., Das, S., Middya, T.R. and Mandal, D., 2019. Methylammonium lead iodide incorporated poly(vinylidene fluoride) nanofibers for flexible piezoelectric–pyroelectric nanogenerator. ACS Applied Materials & Interfaces, 11(30), pp.27279–27287.
160. Mondal, S., Paul, T., Maiti, S., Das, B.K. and Chattopadhyay, K.K., 2020. Human motion interactive mechanical energy harvester based on all inorganic perovskite-PVDF. Nano Energy, 74, p.104870.
161. Pandey, R., Gangadhar, S.B., Grover, S., Singh, S.K., Kadam, A., Ogale, S., Waghmare, U.V., Rao, V.R. and Kabra, D., 2019. Microscopic origin of piezoelectricity in lead-free halide perovskite: application in nanogenerator design. ACS Energy Letters, 4(5), pp.1004–1011.
162. Ding, R., Wong, M.C. and Hao, J., 2020. Recent advances in hybrid perovskite nanogenerators. EcoMat, 2(4), p.e12057.
163. Wang, Z.L., 2011. Nanogenerators for self-powered devices and systems.
164. Zhu, G., Wang, A.C., Liu, Y., Zhou, Y. and Wang, Z.L., 2012. Functional electrical stimulation by nanogenerator with 58 V output voltage. Nano Letters, 12(6), pp.3086–3090.
165. Lin, L., Hu, Y., Xu, C., Zhang, Y., Zhang, R., Wen, X. and Wang, Z.L., 2013. Transparent flexible nanogenerator as self-powered sensor for transportation monitoring. Nano Energy, 2(1), pp.75–81.
166. Huang, C.T., Song, J., Lee, W.F., Ding, Y., Gao, Z., Hao, Y., Chen, L.J. and Wang, Z.L., 2010. GaN nanowire arrays for high-output nanogenerators. Journal of the American Chemical Society, 132(13), pp.4766–4771.
167. Huang, C.T., Song, J., Tsai, C.M., Lee, W.F., Lien, D.H., Gao, Z., Hao, Y., Chen, L.J. and Wang, Z.L., 2010. Single-InN-nanowire nanogenerator with upto 1 V output voltage. Advanced Materials, 22(36), pp.4008–4013.
168. Lee, M., Bae, J., Lee, J., Lee, C.S., Hong, S. and Wang, Z.L., 2011. Self-powered environmental sensor system driven by nanogenerators. Energy & Environmental Science, 4(9), pp.3359–3363.
169. Cha, S., Kim, S.M., Kim, H., Ku, J., Sohn, J.I., Park, Y.J., Song, B.G., Jung, M.H., Lee, E.K., Choi, B.L. and Park, J.J., 2011. Porous PVDF as effective sonic wave driven nanogenerators. Nano Letters, 11(12), pp.5142–5147.
170. Ruan, L., Yao, X., Chang, Y., Zhou, L., Qin, G. and Zhang, X., 2018. Properties and applications of the β phase poly(vinylidene fluoride). Polymers, 10(3), p.228.
171. Kim, S.M., Sohn, J.I., Kim, H.J., Ku, J., Park, Y.J., Cha, S.N. and Kim, J.M., 2012. Radially dependent effective piezoelectric coefficient and enhanced piezoelectric potential due to geometrical stress confinement in ZnO nanowires/nanotubes. Applied Physics Letters, 101(1), p.013104.
172. Seol, M.L., Im, H., Moon, D.I., Woo, J.H., Kim, D., Choi, S.J. and Choi, Y.K., 2013. Design strategy for a piezoelectric nanogenerator with a well-ordered nanoshell array. ACS Nano, 7(12), pp.10773–10779.
173. Attia, S.M., 2002. Review on sol–gel derived coatings: process, techniques and optical applications. Journal of Materials Science & Technology, 18(3), pp.211–218.
174. Lee, W., Kahya, O., Toh, C.T., Özyilmaz, B. and Ahn, J.H., 2013. Flexible graphene–PZT ferroelectric nonvolatile memory. Nanotechnology, 24(47), p.475202.
175. Peroz, C., Chauveau, V., Barthel, E. and Søndergård, E., 2009. Nanoimprint lithography on silica sol–gels: a simple route to sequential patterning. Advanced Materials, 21(5), pp.555–558.
176. Barbé, J., Thomson, A.F., Wang, E.C., McIntosh, K. and Catchpole, K., 2012. Nanoimprinted TiO_2 sol–gel passivating diffraction gratings for solar cell applications. Progress in Photovoltaics: Research and Applications, 20(2), pp.143–148.

177. Khan, S.U., Göbel, O.F., Blank, D.H. and ten Elshof, J.E., 2009. Patterning lead zirconate titanate nanostructures at sub-200-nm resolution by soft confocal imprint lithography and nanotransfermolding. ACS Applied Materials & Interfaces, 1(10), pp.2250–2255.
178. Bursill, L.A. and Brooks, K.G., 1994. Crystallization of sol-gel derived lead-zirconate-titanate thin films in argon and oxygen atmospheres. Journal of Applied Physics, 75(9), pp.4501–4509.
179. Shin, S.H., Choi, S.Y., Lee, M.H. and Nah, J., 2017. High-performance piezoelectric nanogenerators via imprinted sol–gel $BaTiO_3$ nanopillar array. ACS Applied Materials & Interfaces, 9(47), pp.41099–41103.
180. Park, S.H., Lee, H.B., Yeon, S.M., Park, J. and Lee, N.K., 2016. Flexible and stretchable piezoelectric sensor with thickness-tunable configuration of electrospun nanofiber mat and elastomeric substrates. ACS Applied Materials & Interfaces, 8(37), pp.24773–24781.
181. Lang, C., Fang, J., Shao, H., Ding, X. and Lin, T., 2016. High-sensitivity acoustic sensors from nanofibre webs. Nature Communications, 7(1), pp.1–7.
182. Li, T., Feng, Z.Q., Yan, K., Yuan, T., Wei, W., Yuan, X., Wang, C., Wang, T., Dong, W. and Zheng, J., 2018. Pure OPM nanofibers with high piezoelectricity designed for energy harvesting in vitro and in vivo. Journal of Materials Chemistry B, 6(33), pp.5343–5352.
183. Lin, M.F., Xiong, J., Wang, J., Parida, K. and Lee, P.S., 2018. Core-shell nanofiber mats for tactile pressure sensor and nanogenerator applications. Nano Energy, 44, pp.248–255.
184. Pan, X., Wang, Z., Cao, Z., Zhang, S., He, Y., Zhang, Y., Chen, K., Hu, Y. and Gu, H., 2016. A self-powered vibration sensor based on electrospun poly(vinylidene fluoride) nanofibres with enhanced piezoelectric response. Smart Materials and Structures, 25(10), p.105010.
185. Siddiqui, S., Lee, H.B., Kim, D.I., Duy, L.T., Hanif, A. and Lee, N.E., 2018. An omnidirectionally stretchable piezoelectric nanogenerator based on hybrid nanofibers and carbon electrodes for multimodal straining and human kinematics energy harvesting. Advanced Energy Materials, 8(2), p.1701520.
186. You, M.H., Wang, X.X., Yan, X., Zhang, J., Song, W.Z., Yu, M., Fan, Z.Y., Ramakrishna, S. and Long, Y.Z., 2018. A self-powered flexible hybrid piezoelectric–pyroelectric nanogenerator based on non-woven nanofiber membranes. Journal of Materials Chemistry A, 6(8), pp.3500–3509.
187. Baji, A., Mai, Y.W., Li, Q. and Liu, Y., 2011. Electrospinning induced ferroelectricity in poly(vinylidene fluoride) fibers. Nanoscale, 3(8), pp.3068–3071.
188. Deng, W., Yang, T., Jin, L., Yan, C., Huang, H., Chu, X., Wang, Z., Xiong, D., Tian, G., Gao, Y. and Zhang, H., 2019. Cowpea-structured PVDF/ZnO nanofibers based flexible self-powered piezoelectric bending motion sensor towards remote control of gestures. Nano Energy, 55, pp.516–525.
189. Sharma, T., Naik, S., Langevine, J., Gill, B. and Zhang, J.X., 2014. Aligned PVDF-TrFE nanofibers with high-density PVDF nanofibers and PVDF core–shell structures for endovascular pressure sensing. IEEE Transactions on Biomedical Engineering, 62(1), pp.188–195.
190. Gao, H., Minh, P.T., Wang, H., Minko, S., Locklin, J., Nguyen, T. and Sharma, S., 2018. High-performance flexible yarn for wearable piezoelectric nanogenerators. Smart Materials and Structures, 27(9), p.095018.
191. Yang, E., Xu, Z., Chur, L.K., Behroozfar, A., Baniasadi, M., Moreno, S., Huang, J., Gilligan, J. and Minary-Jolandan, M., 2017. Nanofibrous smart fabrics from twisted yarns of electrospun piezopolymer. ACS Applied Materials & Interfaces, 9(28), pp.24220–24229.
192. Lu, L., Ding, W., Liu, J. and Yang, B., 2020. Flexible PVDF based piezoelectric nanogenerators. Nano Energy, 78, p.105251.
193. Maity, K., Garain, S., Henkel, K., Schmeißer, D. and Mandal, D., 2020. Self-powered human-health monitoring through aligned PVDF nanofibers interfaced skin-interactive piezoelectric sensor. ACS Applied Polymer Materials, 2(2), pp.862–878.
194. Ribeiro, C., Costa, C.M., Correia, D.M., Nunes-Pereira, J., Oliveira, J., Martins, P., Goncalves, R., Cardoso, V.F. and Lanceros-Mendez, S., 2018. Electroactive poly(vinylidene fluoride)-based structures for advanced applications. Nature Protocols, 13(4), p.681.
195. Cardoso, V.F., Minas, G., Costa, C.M., Tavares, C.J. and Lanceros-Mendez, S., 2011. Micro and nanofilms of poly(vinylidene fluoride) with controlled thickness, morphology and electroactive crystalline phase for sensor and actuator applications. Smart Materials and Structures, 20(8), p.087002.
196. Shaik, H., Rachith, S.N., Rudresh, K.J., Sheik, A.S., Raman, K.T., Kondaiah, P. and Rao, G.M., 2017. Towards β-phase formation probability in spin coated PVDF thin films. Journal of Polymer Research, 24(3), p.35.
197. Cardoso, V.F., Minas, G. and Lanceros-Méndez, S., 2013. Multilayer spin-coating deposition of poly(vinylidene fluoride) films for controlling thickness and piezoelectric response. Sensors and Actuators A, 192, pp.76–80.

198. Meng, N., Ren, X., Santagiuliana, G., Ventura, L., Zhang, H., Wu, J., Yan, H., Reece, M.J. and Bilotti, E., 2019. Ultrahigh β-phase content poly(vinylidene fluoride) with relaxor-like ferroelectricity for high energy density capacitors. Nature Communications, 10(1), pp.1–9.
199. Fu, J., Hou, Y., Zheng, M. and Zhu, M., 2020. Flexible piezoelectric energy harvester with extremely high power generation capability by sandwich structure design strategy. ACS Applied Materials & Interfaces, 12(8), pp.9766–9774.
200. Tian, G., Deng, W., Gao, Y., Xiong, D., Yan, C., He, X., Yang, T., Jin, L., Chu, X., Zhang, H. and Yan, W., 2019. Rich lamellar crystal baklava-structured PZT/PVDF piezoelectric sensor toward individual table tennis training. Nano Energy, 59, pp.574–581.
201. Park, J.H., Lee, D.Y., Seung, W., Sun, Q., Kim, S.W. and Cho, J.H., 2015. Metallic grid electrode fabricated via flow coating for high-performance flexible piezoelectric nanogenerators. The Journal of Physical Chemistry C, 119(14), pp.7802–7808.
202. Cabrera, N.F.M.N. and Mott, N.F., 1949. Theory of the oxidation of metals. Reports on Progress in Physics, 12(1), pp.163–184.
203. Kim, S., Seol, D., Lu, X., Alexe, M. and Kim, Y., 2017. Electrostatic-free piezoresponse force microscopy. Scientific Reports, 7(1), pp.1–8.
204. Khan, H., Mahmood, N., Zavabeti, A., Elbourne, A., Rahman, M.A., Zhang, B.Y., Krishnamurthi, V., Atkin, P., Ghasemian, M.B., Yang, J. and Zheng, G., 2020. Liquid metal-based synthesis of high performance monolayer SnS piezoelectric nanogenerators. Nature Communications, 11(1), pp.1–8.
205. Li, X. and Bohn, P.W., 2000. Metal-assisted chemical etching in HF/H_2O_2 produces porous silicon. Applied Physics Letters, 77(16), pp.2572–2574.
206. Huang, Z., Geyer, N., Werner, P., De Boor, J. and Gösele, U., 2011. Metal-assisted chemical etching of silicon: a review: in memory of Prof. Ulrich Gösele. Advanced Materials, 23(2), pp.285–308.
207. Seol, M.L., Ahn, J.H., Choi, J.M., Choi, S.J. and Choi, Y.K., 2012. Self-aligned nanoforest in silicon nanowire for sensitive conductance modulation. Nano Letters, 12(11), pp.5603–5608.
208. Seol, M.L., Choi, J.M., Kim, J.Y., Ahn, J.H., Moon, D.I. and Choi, Y.K., 2013. Piezoelectric nanogenerator with a nanoforest structure. Nano Energy, 2(6), pp.1142–1148.
209. Crossley, S., Whiter, R.A. and Kar-Narayan, S., 2014. Polymer-based nanopiezoelectric generators for energy harvesting applications. Materials Science and Technology, 30(13), pp.1613–1624.
210. Whiter, R.A., Boughey, C., Smith, M. and Kar-Narayan, S., 2018. Mechanical energy harvesting performance of ferroelectric polymer nanowires grown via template-wetting. Energy Technology, 6(5), pp.928–934.
211. Bhavanasi, V., Kusuma, D.Y. and Lee, P.S., 2014. Polarization orientation, piezoelectricity, and energy harvesting performance of ferroelectric PVDF-TrFE nanotubes synthesized by nanoconfinement. Advanced Energy Materials, 4(16), p.1400723.
212. Datta, A., Choi, Y.S., Chalmers, E., Ou, C. and Kar-Narayan, S., 2017. Piezoelectric nylon-11 nanowire arrays grown by template wetting for vibrational energy harvesting applications. Advanced Functional Materials, 27(2), p.1604262.
213. Smith, M., Calahorra, Y., Jing, Q. and Kar-Narayan, S., 2017. Direct observation of shear piezoelectricity in poly-L-lactic acid nanowires. APL Materials, 5(7), p.074105.
214. Choi, Y.S., Jing, Q., Datta, A., Boughey, C. and Kar-Narayan, S., 2017. A triboelectric generator based on self-poled Nylon-11 nanowires fabricated by gas-flow assisted template wetting. Energy & Environmental Science, 10(10), pp.2180–2189.
215. Ou, C., Sanchez-Jimenez, P.E., Datta, A., Boughey, F.L., Whiter, R.A., Sahonta, S.L. and Kar-Narayan, S., 2016. Template-assisted hydrothermal growth of aligned zinc oxide nanowires for piezoelectric energy harvesting applications. ACS Applied Materials & Interfaces, 8(22), pp.13678–13683.
216. Boughey, F.L., Davies, T., Datta, A., Whiter, R.A., Sahonta, S.L. and Kar-Narayan, S., 2016. Vertically aligned zinc oxide nanowires electrodeposited within porous polycarbonate templates for vibrational energy harvesting. Nanotechnology, 27(28), p.28LT02.
217. Jing, Q. and Kar-Narayan, S., 2018. Nanostructured polymer-based piezoelectric and triboelectric materials and devices for energy harvesting applications. Journal of Physics D, 51(30), p.303001.
218. Whiter, R.A., Narayan, V. and Kar-Narayan, S., 2014. A scalable nanogenerator based on self-poled piezoelectric polymer nanowires with high energy conversion efficiency. Advanced Energy Materials, 4(18), p.1400519.
219. Tascan, M. and Nohut, S., 2014. Effects of process parameters on the properties of wet-spun solid PVDF fibers. Textile Research Journal, 84(20), pp.2214–2225.
220. Lee, H.Y. and Choi, B., 2013. A multilayer PVDF composite cantilever in the Helmholtz resonator for energy harvesting from sound pressure. Smart Materials and Structures, 22(11), p.115025.

221. Sharma, M., Madras, G. and Bose, S., 2014. Process induced electroactive β-polymorph in PVDF: effect on dielectric and ferroelectric properties. Physical Chemistry Chemical Physics, 16(28), pp.14792–14799.
222. Wu, L., Huang, G., Hu, N., Fu, S., Qiu, J., Wang, Z., Ying, J., Chen, Z., Li, W. and Tang, S., 2014. Improvement of the piezoelectric properties of PVDF-HFP using AgNWs. RSC Advances, 4(68), pp.35896–35903.
223. Cai, X., Lei, T., Sun, D. and Lin, L., 2017. A critical analysis of the α, β and γ phases in poly (vinylidene fluoride) using FTIR. RSC Advances, 7(25), pp.15382–15389.
224. Singh, P., Borkar, H., Singh, B.P., Singh, V.N. and Kumar, A., 2014. Ferroelectric polymer-ceramic composite thick films for energy storage applications. AIP Advances, 4(8), p.087117.
225. Constantino, C.J.L., Job, A.E., Simoes, R.D., Giacometti, J.A., Zucolotto, V., Oliveira Jr, O.N., Gozzi, G. and Chinaglia, D.L., 2005. Phase transition in poly(vinylidene fluoride) investigated with micro-Raman spectroscopy. Applied Spectroscopy, 59(3), pp.275–279.
226. Chen, D., Chen, K., Brown, K., Hang, A. and Zhang, J.X., 2017. Liquid-phase tuning of porous PVDF-TrFE film on flexible substrate for energy harvesting. Applied Physics Letters, 110(15), p.153902.
227. Hu, N., Chen, D., Wang, D., Huang, S., Trase, I., Grover, H.M., Yu, X., Zhang, J.X. and Chen, Z., 2018. Stretchable kirigami polyvinylidene difluoride thin films for energy harvesting: design, analysis, and performance. Physical Review Applied, 9(2), p.021002.
228. Dong, L., Han, X., Xu, Z., Closson, A.B., Liu, Y., Wen, C., Liu, X., Escobar, G.P., Oglesby, M., Feldman, M. and Chen, Z., 2019. Flexible porous piezoelectric cantilever on a pacemaker lead for compact energy harvesting. Advanced Materials Technologies, 4(1), p.1800148.
229. Mao, Y., Zhao, P., McConohy, G., Yang, H., Tong, Y. and Wang, X., 2014. Sponge-like piezoelectric polymer films for scalable and integratable nanogenerators and self-powered electronic systems. Advanced Energy Materials, 4(7), p.1301624.
230. Hu, Y., Xu, C., Zhang, Y., Lin, L., Snyder, R.L. and Wang, Z.L., 2011. A nanogenerator for energy harvesting from a rotating tire and its application as a self-powered pressure/speed sensor. Advanced Materials, 23(35), pp.4068–4071.
231. Zhao, Y., Deng, P., Nie, Y., Wang, P., Zhang, Y., Xing, L. and Xue, X., 2014. Biomolecule-adsorption-dependent piezoelectric output of ZnO nanowire nanogenerator and its application as self-powered active biosensor. Biosensors and Bioelectronics, 57, pp.269–275.
232. Wang, Z.L., 2008. Towards self-powered nanosystems: from nanogenerators to nanopiezotronics. Advanced Functional Materials, 18(22), pp.3553–3567.
233. Zhu, L., Wang, L., Pan, C., Chen, L., Xue, F., Chen, B., Yang, L., Su, L. and Wang, Z.L., 2017. Enhancing the efficiency of silicon-based solar cells by the piezo-phototronic effect. ACS Nano, 11(2), pp.1894–1900.
234. Maity, K. and Mandal, D., 2018. All-organic high-performance piezoelectric nanogenerator with multilayer assembled electrospun nanofiber mats for self-powered multifunctional sensors. ACS Applied Materials & Interfaces, 10(21), pp.18257–18269.
235. Sohn, J.I., Cha, S.N., Song, B.G., Lee, S., Kim, S.M., Ku, J., Kim, H.J., Park, Y.J., Choi, B.L., Wang, Z.L. and Kim, J.M., 2013. Engineering of efficiency limiting free carriers and an interfacial energy barrier for an enhancing piezoelectric generation. Energy & Environmental Science, 6(1), pp.97–104.
236. Briscoe, J. and Dunn, S., 2015. Piezoelectric nanogenerators – a review of nanostructured piezoelectric energy harvesters. Nano Energy, 14, pp.15–29.
237. Karan, S.K., Bera, R., Paria, S., Das, A.K., Maiti, S., Maitra, A. and Khatua, B.B., 2016. An approach to design highly durable piezoelectric nanogenerator based on self-poled PVDF/AlO-rGO flexible nanocomposite with high power density and energy conversion efficiency. Advanced Energy Materials, 6(20), p.1601016.
238. Li, J., Kang, L., Yu, Y., Long, Y., Jeffery, J.J., Cai, W. and Wang, X., 2018. Study of long-term biocompatibility and bio-safety of implantable nanogenerators. Nano Energy, 51, pp.728–735.
239. Yu, Y., Sun, H., Orbay, H., Chen, F., England, C.G., Cai, W. and Wang, X., 2016. Biocompatibility and in vivo operation of implantable mesoporous PVDF-based nanogenerators. Nano Energy, 27, pp.275–281.
240. Tang, M., Guan, Q., Wu, X., Zeng, X., Zhang, Z. and Yuan, Y., 2019. A high-efficiency multidirectional wind energy harvester based on impact effect for self-powered wireless sensors in the grid. Smart Materials and Structures, 28(11), p.115022.
241. Alluri, N.R., Chanderashkear, A. and Kim, S.J., 2018. Hybrid Structures for Piezoelectric Nanogenerators: Fabrication Methods, Energy Generation, and Self-Powered Applications, Energy Harvesting IntechOpen, United Kingdom. DOI: 10.5772/intechopen.74770

242. Khalifa, M., Mahendran, A. and Anandhan, S., 2019. Durable, efficient, and flexible piezoelectric nanogenerator from electrospunPANi/HNT/PVDF blend nanocomposite. Polymer Composites, 40(4), pp.1663–1675.
243. Akmal, M.H.M. and Ahmad, F.B., 2020. Bionanomaterial thin film for piezoelectric applications. In Advances in Nanotechnology and Its Applications (pp. 63–82). Springer, Singapore.
244. Bairagi, S., Ghosh, S. and Ali, S.W., 2020. A fully sustainable, self-poled, bio-waste based piezoelectric nanogenerator: Electricity generation from pomelo fruit membrane. Scientific Reports, 10(1), pp.1–13.
245. Hu, D., Yao, M., Fan, Y., Ma, C., Fan, M. and Liu, M., 2019. Strategies to achieve high performance piezoelectric nanogenerators. Nano Energy, 55, pp.288–304.
246. Wang, A., Liu, Z., Hu, M., Wang, C., Zhang, X., Shi, B., Fan, Y., Cui, Y., Li, Z. and Ren, K., 2018. Piezoelectric nanofibrous scaffolds as in vivo energy harvesters for modifying fibroblast alignment and proliferation in wound healing. Nano Energy, 43, pp.63–71.

7 Triboelectric Nanogenerators

Ritvik B. Panicker, Ashish Kapoor,
Kannan Deepa, and Prabhakar Sivaraman

CONTENTS

7.1 INTRODUCTION

Triboelectric nanogenerators (TENG) were first invented by Wang and his co-workers. The main goal was to convert irregular low-frequency energy, which was randomly distributed into electric power. TENG is a great opportunity to harvest energy from a myriad of sources such as mechanical vibrations, water droplets, and even ultrasound. The driving force for TENG is Maxwell's displacement current (Wang, 2017; Wang, 2020). TENG works on the combination of contact electrification and electrostatic induction. It is easier to fabricate and also cost effective. Through proper structure and design, the TENG can be tailor-made for a large number of applications. There is a requirement for a cost-effective solution to power health-care monitor sensors, wearable electronics, sensors in Global Positioning System (GPS), and microelectromechanical systems (MEMS). TENG also has a high stability and high efficiency. Hence, extensive research is being done in the field of TENG. There are several modes of operation of TENG and their applications range from healthcare to generating energy from the movement of water or wind.

7.2 CONSTRUCTION AND MODES OF OPERATION

In the fabrication of TENG, accurate handling of the different layers and the electrodes is of utmost importance. When the surface is very rough, the electrical output is higher. The high electrical

DOI: 10.1201/9781003187615-7

output is due to the increase in contact points on the rough surfaces of the TENG devices. TENG has triboelectric layers, electrodes, and a spacer. The stretchable component is important when considering a flexible TENG. It is to be noted that all materials show triboelectrification; however, the polarity dictates the ability of a material in the transfer of electrons. One of the first triboelectric series on static charges was published in 1757 by John Carl Wilcke. A material near the lowermost part of the series when contacted with a material toward the upper portion of the series receives a more negative charge. More charge is transferred when the two materials are farther from each other (Bera, 2016).

TENG works on the principle of a combination of triboelectrification and electrostatic induction effects. There is a break in the electrostatic status due to the movements in the electrostatic charge caused by the displacement in any of the layers, eventually leading to a potential difference. Contact electrification and triboelectrification have differences among them. Contact electrification occurs by contact, while triboelectrification occurs due to friction and the rubbing of two materials (Ahmed et al., 2020). Different modes of TENG have been recognized based on the concept of triboelectrification (Vivekananthan et al., 2020).

7.2.1 Contact-Separation Mode

The contact and separation of layers lead to triboelectrification. In contact-separation mode, the layers can be of two different dielectric layers or one dielectric layer and a metallic layer. The advantage of this mode is that it has a simple design and is easy to fabricate. Its main application is in low-power electronics.

7.2.2 Linear Sliding Mode

In this mode, the charges are generated by relative sliding between the layers of TENG in a to-and-fro manner. It is modeled similar to the contact sliding mode. The demerit of this arrangement is that it requires a sliding action to generate electricity. The advantage on the other hand is that it produces more charge density due to a high contact area.

7.2.3 Single-Electrode Mode

This arrangement is the simplest mode of TENG but the output performance is very less. The voltage output is less. In this mode, contact electrification occurs followed by electrostatic induction. Here, there is a layer of poly dimethyl siloxane (PDMS) and an electrode that gets into contact with a surface and then contact electrification occurs, and later, when the surface is removed, the redundant charges then move to an external load. Single-electrode mode finds widespread usage in skin-inspired TENG due to its simple nature.

7.2.4 Freestanding Mode

Here, a single electrode moves freely in the center of two triboelectric layers. The electrodes are static, while the triboelectric layer in the absence of an electrode moves over it. The major merit of this mode is that the output efficiency is very high.

7.3 ADVANTAGES OF TENG

Various sources of energy can be used in TENG, right from human motion (walking, clapping, pressing, etc.) to the motion of water, wind, engines, etc. (Wang, 2013). TENG is also very cost effective compared to many other energy sources. TENG is majorly employed in energy harvesting applications and also as self-powered sensors since one can get data regarding the mechanical motion that is applied. Besides having great strides in harvesting various kinds of mechanical energy, much emphasis is laid on electrification efficiency as well. Selecting two triboelectric materials that have a large difference in their affinities leads to better electrification performance. Parameters such as contact surface area, resistance, surface roughness, and the electron affinity are taken into consideration while choosing the material.

Small-sized electronics that operate under low power can employ energy harvested from the environment. TENG can be used to generate electricity from movements such as touching, impact, vibration, and sliding. Flexible TENG is cheap, cost effective, and easy to implement in comparison to the already existing solutions in the market. All the components should be flexible in a stretchable TENG. In a flexible TENG, the stretchable conductor is the current collector for the TENG. The conductor in a flexible TENG must have certain parameters such as high conductivity, high stretchability, very low hysteresis, and good mechanical strength (Hinchet et al., 2015).

Since TENG saves a lot of space, they can be used to power microelectronics. TENG has less parts compared to an electromagnetic generator (EMG) and is also an eco-friendly alternative. Hence, TENG is preferable to EMGs as a power source. Transient electronics disappear over a fixed period of time in controlled manner. External stimulus initiates the dissolution. A TENG based on polyvinyl chloride, sodium alginate, lithium, and aluminum as the current collectors were able to dissolve in a step-by-step manner. This technology has many applications especially in the biomedical field where power sources are needed for devices inside the human body and have to be disposed of after a certain time period. To design an environment-friendly nanogenerator, Yang and his co-workers developed a leaf-based TENG (Jie et al., 2018). This leaf-based TENG has a simple structure and makes use of green raw materials. This natural TENG works in the single-electrode mode; the peak voltage is 230 V. This leaf-based TENG can also work in the freestanding mode. Here, a metal electrode is connected to a leaf for electrical connectivity. The contact layer is the polymer sheet, which has a significant difference in electron affinity relative to the leaf surface. This is a simple and cost-effective TENG that can harvest mechanical energy from the environment.

7.4 ENERGY HARVESTING APPLICATIONS

7.4.1 Medicine

7.4.1.1 Pacemakers

They use electrical impulses that are used to regulate the heartbeat of humans by stimulating the heart muscles. A pacemaker is used so as to control abnormal functioning of the heart, which could result in dizziness, syncope, or eventually heart failure. The main issue that the present pacemaker technology is dealing with is that the battery lasts only up to 7–10 years and the patient has to get a surgery after that and this could be risky especially after a certain age (Mallela et al., 2004). The solution is to get an energy supply that is self-powered to do away with the risk of surgery. Li et al. (2010) demonstrated the application of a single-wire generator that harvests energy from inside a live animal, a Sprague Dawley rat (Li et al., 2010). In 2014, an implantable triboelectric nanogenerator (iTENG) was fabricated by Zheng et al. This iTENG harvests mechanical potential within a small living animal (Zheng et al., 2014). Here, the iTENG was implanted under the skin of the left chest of a live rat for the first time. The expansion and contraction of the lungs of the rat resulted in the triboelectric layers in periodic contact and separation, which helped in generating the required energy. Ouyang et al. (2019) had reported an implantable TENG that was based on a symbiotic pacemaker. The implantable TENG was placed in an adult Yorkshire pig and here the rhythmic beating of the heart caused periodic motion between the triboelectric layers. Issues such as lack of high stability, the correct amount of power, and size are what are keeping this technology from clinical applications (Feng et al., 2018).

7.4.1.2 Nerve Stimulation

The nervous system is responsible for coordinating body functions by conveying and accepting signals. In many injuries of the nerves, there occurs severe damage in the relay of neural signals to the muscles. Hence, less voltage stimulation would aid the patients having injuries to their nervous systems. Lee et al. (2017) used TENG having a zigzag configuration for stimulation of the nerves using neural interfaces. They used sling electrodes to neural interface. It was noticed that the stimulation of sciatic nerve using electrical input by TENG induced changes in muscle action. Zheng et al. (2016) had reported a biodegradable and implantable TENG for repairing neurons. Here, the alignment of the cells is one of the vital parameters for repairing the neurons.

Obesity is a very significant disease that affects the world (Singh et al., 2013). Vagus nerves are responsible for the transfer of signals to and from the brain and other parts of the body. The stimulation of the vagus nerves can help reduce weight. Yao et al. (2018) reported a stimulation device eliminating the need for a battery. The vagus nerves were stimulated using a flexible TENG positioned on the abdomen. The biological process that is responsible for the memory and learning of an individual is termed synaptic plasticity. The synaptic activity alters the strength of the synaptic signals. Guan et al. (2020) demonstrated an e-skin based on wireless stimulation of the nerves and powered by TENG. It comprises a photosensitive flexible perovskite unit. This kind of skin TENG harnesses the biomechanical form of energy from bodily movements.

7.4.1.3 Wound Healing and Repairing

Electrical stimulation is used to heal wounds, which can lead to tissue healing by appropriate regulation of the cells. TENG is a cost-effective and easier way to generate electric pulses. Jiang et al. (2018) has reported a bioabsorbable TENG that was fabricated using natural material. The electrical stimulation was implemented to repair cardiomyocytes and myocardial tissue. Li et al. (2018) had reported the fabrication of an iTENG that works on the principle of photothermally tunable biodegradation for repairing tissues. An electrical bandage based on a wearable TENG had two parts: the TENG followed by the dressing (Zargham et al., 2012).

7.4.1.4 Drug Delivery

Implantable drug delivery is a very promising solution to treat diseases on site and can help in treating cancer, diabetes, etc. It is site-specific (Li et al., 2008). One of the demerits of this technology is that it requires a power source that is mostly lithium. Lithium has a limited lifetime and requires its prompt removal later on. This issue can be solved using a TENG that will self-power the system. The TENG will transfer the mechanical energy to electrical energy and can power the drug delivery system. Song et al. (2017) have reported the functioning of a rotary TENG that is based on a drug delivery system, which is self-powered. The TENG acts as the energy source for an electrochemical microfluidic pump. A microneedle integrated into TENG was proposed for drug delivery (Bok et al., 2018). Wu et al. (2020) made a wearable TENG self-powered drug delivery system that did not have an energy storage unit. Liu et al. (2019) have demonstrated a TENG based on drug delivery to individual cells. It consisted of the following parts: (i) TENG, (ii) bridge rectifier, (iii) nano-needle electrode, and (iv) aluminum electrode. This was used for in vitro studies.

7.4.1.5 Pulse Sensor

It is necessary to keep track of various physiological signals in real time so as to make a diagnosis. Pulse sensor approach is used to keep track of the cardiovascular system since the pulse rate is an indicator of the functioning of the heart. Having a pulse sensor that is self-powered can be a great solution to energy consumption and environmental pollution, as using lithium batteries to power such devices will eventually lead to the battery's life span ending. Ouyang et al. (2017) developed a TENG-centered pulse sensor with ultrasensitivity. The triboelectric layers used nanostructured Kapton and Cu films. It was then encapsulated using PDMS.

7.4.1.6 Gene Delivery

Electro-transfection is a common gene delivery method, which uses electroporation. Electroporation is a method that employs the use of electric field to make the cell membrane more permeable. The permeable cells can take genes, drugs, and proteins. Yang et al. (2019) reported the application of TENG-powered nanowire electrode array for electro-transfection into a variety of mammalian cells. The nanowires were made on a copper mesh. The TENG had PTFE/Cu as a negative TENG layer and aluminum that served as the electrode and frictional layer.

7.4.1.7 Microbial Disinfection

Sterilization using high-intensity electric field can be used to disinfect food. It is a very short treatment, which generates no unwanted by-products. Microbial disinfection in combination with nanogenerators can be used to bring about decontamination in a self-powered environment. Jiang et al. (2015a,b) reported a water treatment system that works on the electrochemical principle for sterilization removal of algae and is self-powered. The TENG in this case is driven by water waves. This TENG was fabricated using triboelectric layers composed of Poly tetrafluoroethylene (PTFE) and Indium tin oxide (ITO). Tian et al. (2017) supplied electric pulse from the wave-based TENG to the nanowires of ZnO present along with silver nanoparticles to develop a water sterilization system that is self-powered. Ding et al. (2019) have developed a rotating TENG which can be used to power a cost-effective, point-of-usage water sterilization system. The triboelectric pump had three parts that were disk TENG, a coaxial electrode copper ionization cell for sterilization purpose, and the mechanical configuration. This technology was very cost effective and self-powered for point-of-usage water purification.

7.4.2 Energy from Water Waves

We are in desperate need of renewable energy sources that are having less carbon emissions and are cost effective for a more sustainable development of humankind. Traditional EMGs are bulkier and cannot be used to generate power from the movement of water. It is seen that by using a hybrid of TENG and EMG, we can generate power. Shao et al. (2018) reported a hybrid generator based on contact-separation technique. Tribo-charges are generated on the solid surface when water is in contact with it. A current flow is generated between the layers. This is the main principle behind generating power from water using TENG. Contact electrification taking place between the surface of the solid/water and air/gas causes the water molecules to be charged. Water TENG devices consist of two major devices, which are the electrodes and the hydrophobic coatings (Jiang et al., 2019). When it comes to water TENG, aluminum, ITO, and copper are the typical electrodes (Helseth and Guo, 2015). Water TENG devices that work in sliding mode or single-electrode mode are more preferable for energy harvesting applications. Another way to harness hydropower is using raindrops. The TENG is presented above the umbrella or in solar cells. They could be used to harness energy from the sun, wind, and raindrops (Zheng et al., 2014; Jeon et al., 2015; Kwon et al., 2016). A flowing stream is a solution to harness hydropower. Sensors can directly consume the energy harvested by water TENG devices and can be used to record flow speed, gas flow, or wave height (Park et al., 2018).

Monitoring ocean waves is necessary for marine engineering and is important that we keep track of it as it helps in avoiding disasters. By using water TENG, these waves can be monitored. A self-powered sensing system for monitoring waves has been proposed (Zhang et al., 2019). The power output differs depending on the working conditions. In the current stages of development, the output power is quite low. It is established that tribo-charges on the hydrophobic layer are crucial for water TENG. Hence, it is noticed that upon increasing the ion concentration, the charge count on the hydrophobic layer is considerably lowered.

7.4.3 Wearable TENG

With the development of Internet of Things (IOT), technology has developed into a more portable nature. Wearable TENG devices are those that can be worn directly on the different parts of the body or can be integrated into the accessories worn by the user. There is a major power supply issue. By harnessing the energy of human motion such as hand gestures, touch, walking, and running, these wearable devices can generate their own power. A stack of integrated multilayered TENG was proposed to harvest energy from vibrations. It has an output of 303 volt and a peak power intensity of 104.6 Wm^{-2} (Zhang et al., 2014). Wang et al. suggested a stretchable TENG tactile sensor for tracking motion trajectory. It could detect finger operation while playing Pac-Man. The human feet are a very active part of the body with continuous movement. Much research regarding embedding TENG to shoes has been done to harness the energy of walking. The electrical energy is stored in a charging device (which is wearable) for portable electronics by harvesting typical walking energy. By integrating this with a power management system, one can make self-powered system (Zhu et al., 2013).

Textile-based TENG is a suitable alternative for harvesting energy from human motion. The working principle is the same as that of a normal TENG. It uses contact-separation mode. They can be used to harvest energy by using it as a fabric that can then collect energy from the human motion. T-TENG based on single-electrode mode is also a suitable option to harvest energy (Huang et al., 2021). T-TENG has three crucial parts that are the main parts of the fabric, triboelectric materials, and the electrodes. The fabric should be stretchable, flexible, and lightweight. In 2019, Ye et al. (2020) adopted a layered structure format to make a silk fiber, polytetrafluoroethylene fiber, and a stainless steel fiber into a TENG yarn, which had a core-shell structure. Ma et al. (2020) used electrospinning to generate a single-electrode triboelectric yarn, which has a spiral core-shell structure. Its inner layer used silver nanowires and the outer shell used polyvinylidene fluoride and polyacrylonitrile nanofibers. Fabric-based TENG is very suitable for collecting energy from human motion. Dong et al. (2020a) used silver yarn wrapped with nylon and PTFE threads for knitting and obtaining a fabric-based TENG. A 3D weaving technology can also be applied to produce fabric-based TENG, which will give a higher output of TENG. T-TENG can also be used to monitor physical functions (Dong et al., 2020b). Jao et al. (2018) made a TENG sensor based on chitosan. It can be used to detect humidity, sweat, and gait. Wearable technology is a fast-growing industry with a much need for a power source and hence TENG is a suitable solution to replace the existing ones. To be able to power portable electronics with the power of human motion is a very sustainable and environment-friendly solution. From flexible TENG to textile-based TENG, the implications in wearable technology are numerous and hence a lot of research work is being carried out in this field.

7.4.4 Self-Powered Electrochemical Systems

Using different modes of TENG, electrochemical systems can be created. A water splitting system that was self-powered was reported by Tang et al. (2015), which can produce hydrogen from 30% KOH. In this, a water-driven TENG was implemented into the system, and when the disk TENG rotated, hydrogen was made in a tube. Therefore, a TENG-enabled system is suitable for the generation of hydrogen. Considering the water shortage, desalination is a solution to get clean water. A new concept of combining TENG with an electrodialysis cell has been developed. In this disk TENG, there are two parts that are the stator and a pair of rotators (Jiang et al., 2015a,b).

Pollution is one of the major issues we face right now. Air and water pollution have affected cities drastically. Electrochemical treatment is one of the solutions to control air and water pollution. But it usually requires external power that makes the entire process expensive. As TENG research progressed, electrochemical systems that are self-powered have been reported. Chen et al. (2015) developed an air cleaning system that is self-powered by a rotating TENG. This was meant to remove sulfur dioxide and dust from air. Similarly, for water pollution, a unique system that had a self-powered phenol degradation system by using β-cyclodextrin to enhance triboelectrification has been reported.

Corrosion is a major threat to metals and costs globally billions of dollars in infrastructure maintenance. Guo and co-workers utilized a disk TENG to make a self-powered cathodic protection for stainless steel. This design is durable, cost effective, and uses pulsed current to offer protection. It is a solution to the corrosion problem faced in the ocean (Guo et al., 2014).

7.5 CHALLENGES AND FUTURE PERSPECTIVES

There is a great need for suitable renewable energy sources. From powering microelectronics to harvesting energy from water and wind, nanogenerators show promise as a sustainable option. In the present, the use of lithium as a power source in electronics is proving to be harmful to the environment as it has a limited life span. That being said nanogenerators still have a long way to go before they can be a major power source alternative to lithium. The lack of understanding of the charging mechanisms has slowed down the development of nanogenerator devices. In the case of biomedical applications, there is a lack of sufficient understanding to go for clinical applications. Small animals have a different energy and cell regulation as compared to humans and hence there will be major differences when it comes to designing devices meant for humans. At the same point when it comes to water TENG, the output is still low. With a great stride in IOT, there will be a requirement to power microelectronics. One should remember that TENG has been only around for 10 years, and as time progresses, there will be development.

REFERENCES

Ahmed, A., Hassan, I., Pourrahimi, A.M., Helal, A.S., El-Kady, M.F., Khassaf, H. and Kaner, R.B., 2020. Toward high-performance triboelectric nanogenerators by engineering interfaces at the nanoscale: looking into the future research roadmap. *Advanced Materials Technologies*, *5*(11), p.2000520.

Bera, B., 2016. Literature review on triboelectric nanogenerator. *Imperial Journal of Interdisciplinary Research (IJIR)*, *2*(10), pp.1263–1271.

Bok, M., Lee, Y., Park, D., Shin, S., Zhao, Z.J., Hwang, B., Hwang, S.H., Jeon, S.H., Jung, J.Y., Park, S.H. and Nah, J., 2018. Microneedles integrated with a triboelectric nanogenerator: an electrically active drug delivery system. *Nanoscale*, *10*(28), pp.13502–13510.

Chen, S., Gao, C., Tang, W., Zhu, H., Han, Y., Jiang, Q., Li, T., Cao, X. and Wang, Z., 2015. Self-powered cleaning of air pollution by wind driven triboelectric nanogenerator. *Nano Energy*, *14*, pp.217–225.

Ding, W., Zhou, J., Cheng, J., Wang, Z., Guo, H., Wu, C., Xu, S., Wu, Z., Xie, X. and Wang, Z.L., 2019. TriboPump: a low-cost, hand-powered water disinfection system. *Advanced Energy Materials*, *9*(27), p.1901320.

Dong, K., Peng, X., An, J., Wang, A.C., Luo, J., Sun, B., Wang, J. and Wang, Z.L., 2020b. Shape adaptable and highly resilient 3D braided triboelectric nanogenerators as e-textiles for power and sensing. *Nature Communications*, *11*(1), pp.1–11.

Dong, S., Xu, F., Sheng, Y., Guo, Z., Pu, X. and Liu, Y., 2020a. Seamlessly knitted stretchable comfortable textile triboelectric nanogenerators for E-textile power sources. *Nano Energy*, *78*, p.105327.

Feng, H., Zhao, C., Tan, P., Liu, R., Chen, X. and Li, Z., 2018. Nanogenerator for biomedical applications. *Advanced Healthcare Materials*, *7*(10), p.1701298.

Guan, H., Lv, D., Zhong, T., Dai, Y., Xing, L., Xue, X., Zhang, Y. and Zhan, Y., 2020. Self-powered, wireless-control, neural-stimulating electronic skin for in vivo characterization of synaptic plasticity. *Nano Energy*, *67*, p.104182.

Guo, W., Li, X., Chen, M., Xu, L., Dong, L., Cao, X., Tang, W., Zhu, J., Lin, C., Pan, C. and Wang, Z.L., 2014. Electrochemical cathodic protection powered by triboelectric nanogenerator. *Advanced Functional Materials*, *24*(42), pp.6691–6699.

Helseth, L.E. and Guo, X.D., 2015. Contact electrification and energy harvesting using periodically contacted and squeezed water droplets. *Langmuir*, *31*(10), pp.3269–3276.

Hinchet, R., Seung, W. and Kim, S.W., 2015. Recent progress on flexible triboelectric nanogenerators for self-powered electronics. *ChemSusChem*, *8*(14), pp.2327–2344.

Huang, P., Wen, D.L., Qiu, Y., Yang, M.H., Tu, C., Zhong, H.S. and Zhang, X.S., 2021. Textile-based triboelectric nanogenerators for wearable self-powered microsystems. *Micromachines*, *12*(2), p.158.

Jao, Y.T., Yang, P.K., Chiu, C.M., Lin, Y.J., Chen, S.W., Choi, D. and Lin, Z.H., 2018. A textile-based triboelectric nanogenerator with humidity-resistant output characteristic and its applications in self-powered healthcare sensors. *Nano Energy*, *50*, pp.513–520.

Jeon, S.B., Kim, D., Yoon, G.W., Yoon, J.B., & Choi, Y.K. (2015). Self-cleaning hybrid energy harvester to generate power from raindrop and sunlight. *Nano Energy*, *12*, 636–645.

Jiang, Q., Han, Y., Tang, W., Zhu, H., Gao, C., Chen, S., Willander, M., Cao, X. and Wang, Z.L., 2015a. Self-powered seawater desalination and electrolysis using flowing kinetic energy. *Nano Energy*, *15*, pp.266–274.

Jiang, Q., Jie, Y., Han, Y., Gao, C., Zhu, H., Willander, M., Zhang, X. and Cao, X., 2015b. Self-powered electrochemical water treatment system for sterilization and algae removal using water wave energy. *Nano Energy*, *18*, pp.81–88.

Jiang, W., Li, H., Liu, Z., Li, Z., Tian, J., Shi, B., Zou, Y., Ouyang, H., Zhao, C., Zhao, L. and Sun, R., 2018. Fully bioabsorbable natural-materials-based triboelectric nanogenerators. *Advanced Materials*, *30*(32), p.1801895.

Jiang, D., Xu, M., Dong, M., Guo, F., Liu, X., Chen, G. and Wang, Z.L., 2019. Water-solid triboelectric nanogenerators: an alternative means for harvesting hydropower. *Renewable and Sustainable Energy Reviews*, *115*, p.109366.

Jie, Y., Jia, X., Zou, J., Chen, Y., Wang, N., Wang, Z.L. and Cao, X., 2018. Natural leaf made triboelectric nanogenerator for harvesting environmental mechanical energy. *Advanced Energy Materials*, *8*(12), p.1703133.

Kwon, S.H., Kim, W.K., Park, J., Yang, Y., Yoo, B., Han, C.J. and Kim, Y.S., 2016. Fabric active transducer stimulated by water motion for self-powered wearable device. *ACS Applied Materials and Interfaces*, *8*(37), pp.24579–24584.

Lee, S., Wang, H., Shi, Q., Dhakar, L., Wang, J., Thakor, N.V., Yen, S.C. and Lee, C., 2017. Development of battery-free neural interface and modulated control of tibialis anterior muscle via common peroneal nerve based on triboelectric nanogenerators (TENGs). *Nano Energy*, *33*, pp.1–11.

Li, Z., Feng, H., Zheng, Q., Li, H., Zhao, C., Ouyang, H., Noreen, S., Yu, M., Su, F., Liu, R. and Li, L., 2018. Photothermally tunable biodegradation of implantable triboelectric nanogenerators for tissue repairing. *Nano Energy*, *54*, pp.390–399.

Li, P.Y., Shih, J., Lo, R., Saati, S., Agrawal, R., Humayun, M.S., Tai, Y.C. and Meng, E., 2008. An electrochemical intraocular drug delivery device. *Sensors and Actuators A: Physical*, *143*(1), pp.41–48.

Li, Z., Zhu, G., Yang, R., Wang, A.C. and Wang, Z.L., 2010. Muscle-driven in vivo nanogenerator. *Advanced Materials*, *22*(23), pp.2534–2537.

Liu, Z., Nie, J., Miao, B., Li, J., Cui, Y., Wang, S., Zhang, X., Zhao, G., Deng, Y., Wu, Y. and Li, Z., 2019. Self-powered intracellular drug delivery by a biomechanical energy-driven triboelectric nanogenerator. *Advanced Materials*, *31*(12), p.1807795.

Ma, L., Zhou, M., Wu, R., Patil, A., Gong, H., Zhu, S., Wang, T., Zhang, Y., Shen, S., Dong, K. and Yang, L., 2020. Continuous and scalable manufacture of hybridized nano-micro triboelectric yarns for energy harvesting and signal sensing. *ACS Nano*, *14*(4), pp.4716–4726.

Mallela, V.S., Ilankumaran, V. and Rao, N.S., 2004. Trends in cardiac pacemaker batteries. *Indian Pacing and Electrophysiology Journal*, *4*(4), p.201.

Ouyang, H., Liu, Z., Li, N., Shi, B., Zou, Y., Xie, F., Ma, Y., Li, Z., Li, H., Zheng, Q. and Qu, X., 2019. Symbiotic cardiac pacemaker. *Nature Communications*, *10*(1), pp.1–10.

Ouyang, H., Tian, J., Sun, G., Zou, Y., Liu, Z., Li, H., Zhao, L., Shi, B., Fan, Y., Fan, Y. and Wang, Z.L., 2017. Self-powered pulse sensor for antidiastole of cardiovascular disease. *Advanced Materials*, *29*(40), p.1703456.

Park, H.Y., Kim, H.K., Hwang, Y.H. and Shin, D.M., 2018. Water-through triboelectric nanogenerator based on Ti-mesh for harvesting liquid flow. *Journal of the Korean Physical Society*, *72*(4), pp.499–503.

Shao, H., Cheng, P., Chen, R., Xie, L., Sun, N., Shen, Q., Chen, X., Zhu, Q., Zhang, Y., Liu, Y. and Wen, Z., 2018. Triboelectric–electromagnetic hybrid generator for harvesting blue energy. *Nano-Micro Letters*, *10*(3), pp.1–9.

Singh, G.M., Danaei, G., Farzadfar, F., Stevens, G.A., Woodward, M., Wormser, D., Kaptoge, S., Whitlock, G., Qiao, Q., Lewington, S. and Di Angelantonio, E., 2013. The age-specific quantitative effects of metabolic risk factors on cardiovascular diseases and diabetes: a pooled analysis. *PloS One*, *8*(7), p.e65174.

Song, P., Kuang, S., Panwar, N., Yang, G., Tng, D.J.H., Tjin, S.C., Ng, W.J., Majid, M.B.A., Zhu, G., Yong, K.T. and Wang, Z.L., 2017. A self-powered implantable drug-delivery system using biokinetic energy. *Advanced Materials*, *29*(11), p.1605668.

Tang, W., Han, Y., Han, C.B., Gao, C.Z., Cao, X. and Wang, Z.L., 2015. Self-powered water splitting using flowing kinetic energy. *Advanced Materials*, *27*(2), pp.272–276.

Tian, J., Feng, H., Yan, L., Yu, M., Ouyang, H., Li, H., Jiang, W., Jin, Y., Zhu, G., Li, Z. and Wang, Z.L., 2017. A self-powered sterilization system with both instant and sustainable anti-bacterial ability. *Nano Energy*, *36*, pp.241–249.

Vivekananthan, V., Chandrasekhar, A., Alluri, N.R., Purusothaman, Y., Khandelwal, G. and Kim, S.J., 2020. Triboelectric nanogenerators: design, fabrication, energy harvesting, and portable-wearable applications. In *Nanogenerators*. IntechOpen.

Wang, Z.L., 2013. Triboelectric nanogenerators as new energy technology for self-powered systems and as active mechanical and chemical sensors. *ACS Nano*, *7*(11), pp.9533–9557.

Wang, Z.L., 2017. On Maxwell's displacement current for energy and sensors: the origin of nanogenerators. *Materials Today*, *20*(2), pp.74–82.

Wang, Z.L., 2020. Triboelectric nanogenerator (TENG)—sparking an energy and sensor revolution. *Advanced Energy Materials*, *10*(17), p.2000137.

Wu, C., Jiang, P., Li, W., Guo, H., Wang, J., Chen, J., Prausnitz, M.R. and Wang, Z.L., 2020. Self-powered iontophoretic transdermal drug delivery system driven and regulated by biomechanical motions. *Advanced Functional Materials*, *30*(3), p.1907378.

Yang, C., Yang, G., Ouyang, Q., Kuang, S., Song, P., Xu, G., Poenar, D.P., Zhu, G., Yong, K.T. and Wang, Z.L., 2019. Nanowire-array-based gene electro-transfection system driven by human-motion operated triboelectric nanogenerator. *Nano Energy*, *64*, p.103901.

Yao, G., Kang, L., Li, J., Long, Y., Wei, H., Ferreira, C.A., Jeffery, J.J., Lin, Y., Cai, W. and Wang, X., 2018. Effective weight control via an implanted self-powered vagus nerve stimulation device. *Nature Communications*, *9*(1), pp.1–10.

Ye, C., Dong, S., Ren, J. and Ling, S., 2020. Ultrastable and high-performance silk energy harvesting textiles. *Nano-Micro Letters*, *12*(1), pp.1–15.

Zargham, S., Bazgir, S., Tavakoli, A., Rashidi, A.S. and Damerchely, R., 2012. The effect of flow rate on morphology and deposition area of electrospun nylon 6 nanofiber. *Journal of Engineered Fibers and Fabrics*, *7*(4), p.155892501200700414.

Zhang, H., Yang, Y., Su, Y., Chen, J., Adams, K., Lee, S., Hu, C. and Wang, Z.L., 2014. Triboelectric nanogenerator for harvesting vibration energy in full space and as self-powered acceleration sensor. *Advanced Functional Materials*, *24*(10), pp.1401–1407.

Zhang, X., Yu, M., Ma, Z., Ouyang, H., Zou, Y., Zhang, S.L., Niu, H., Pan, X., Xu, M., Li, Z. and Wang, Z.L., 2019. Self-powered distributed water level sensors based on liquid–solid triboelectric nanogenerators for ship draft detecting. *Advanced Functional Materials*, *29*(41), p.1900327.

Zheng, L., Lin, Z.H., Cheng, G., Wu, W., Wen, X., Lee, S. and Wang, Z.L., 2014. Silicon-based hybrid cell for harvesting solar energy and raindrop electrostatic energy. *Nano Energy*, *9*, pp.291–300.

Zheng, Q., Shi, B., Fan, F., Wang, X., Yan, L., Yuan, W., Wang, S., Liu, H., Li, Z. and Wang, Z.L., 2014. In vivo powering of pacemaker by breathing-driven implanted triboelectric nanogenerator. *Advanced Materials*, *26*(33), pp.5851–5856.

Zheng, Q., Zou, Y., Zhang, Y., Liu, Z., Shi, B., Wang, X., Jin, Y., Ouyang, H., Li, Z. and Wang, Z.L., 2016. Biodegradable triboelectric nanogenerator as a life-time designed implantable power source. *Science Advances*, *2*(3), p.e1501478.

Zhu, G., Bai, P., Chen, J. and Wang, Z.L., 2013. Power-generating shoe insole based on triboelectric nanogenerators for self-powered consumer electronics. *Nano Energy*, *2*(5), pp.688–692.

8 Hybrid Nanogenerators

Jocelyne Estrella-Nuñez, Francisco García-Salinas, Moises Bustamante-Torres, Jorge Cárdenas-Gamboa, and Emilio Bucio Carrillo

CONTENTS

DOI: 10.1201/9781003187615-8

8.1 INTRODUCTION

8.1.1 Environmental and Technological Issues

For the past few centuries, fossil fuels were the principal source of affordable energy which enabled industrialization around the world. Nowadays, one of the greatest challenges of society is related to the increasing energy demand produced by human development. There is an accelerated demand for energy sources and infrastructures that are being stretched to new limits. To face the huge growth, it is indispensable to harness new resources and the improvement of the existing ones. Major areas of research are based on parameters as mega-scale energy conversion, efficiency, storage, energy harvesting, and the most important the use of renewable and green energy, which are capable to reduce greenhouse gas emissions (Hansen et al., 2010). These issues present a field of opportunities for innovations to face energy challenges.

Nanotechnology has developed a relevant role in this problem. This novel technology allows the creation of nanosystems characterized by their intelligence, multifunctionality, reduced size, higher sensibility, and low power consumption (Pan et al., 2011). The principal technological tendency is miniaturization and portability. The evolution of computers is the most common example (Wang et al., 2012). It is notable that the continuous advancements in circuitry efficiency and miniaturization have been achieved much lower consumption in microelectronic devices. However, the power sources like batteries and supercapacitors for sensors are still life-limited, and the generated waste brings a great environmental impact. Now, rechargeable batteries with high energy storage have been studied and applied in portable and wearable electronics. As electronic devices became smaller and use less power, batteries could grow smaller, which allows today a great explosion of wireless and mobile applications (Paradiso, 2005). However, one of the most important developments is related to the search for green and renewable energy. A huge effort has been devoted to self-powered systems that obtain energy from the environment through different conversion mechanisms. Energy harvesters studied are based on different working principles such as electromagnetism, piezoelectricity, triboelectricity, photovoltaic, thermoelectricity, among others (Han, 2013). These green harvesting systems take advantage of different sources such as solar, wind, waves, heat, and vibrations. Ideal energy harvesting systems require to be highly flexible, stretchable, transformable, and mechanically durable to be used in a wide range of applications (Lee, 2013).

The development of new technologies is given by the relationship between energy and materials. Materials are capable to produce energy or allow the transference of energy into useful forms. One example is photovoltaic silicon that converts solar energy into electrical power. The different kinds of energy, in turns, have made possible a broad range of materials known today (Arunachalam, 2008). With the pass of the years, the diversity of materials has been increased and the technologies had changed their main energy source. There has been an evolution in the materials used, from coal, iron, uranium, silicon, until superalloys. The scientific study of potential materials is based on thermodynamic, catalytic, electrical, electronic, and mechanical properties. The choice of materials used for energy production has been associated with the availability, accessibility, and their properties. There is a technological trend, in which increasingly smaller devices have been created with novel properties, which are associated with the use of nanomaterials.

8.1.2 Nanogenerators

Worldwide energy needs require the developing technologies associated with the area of nanoenergy to the sustainable maintenance of micro and nanosystems. In the last years, nanogenerators (NGs) have been studied as a self-powering technology, which use different sources (Wang, 2012). In 2006, the first NG based on ZnO nanowire was studied, which presents a coupling between the piezoelectric and semiconducting properties. Moreover, this material is biocompatible for different applications, and ZnO presents the most diverse configuration of nanostructures (nanobelts, nanowires, nanosprings, nanorings, nanohelices, and nanobows) (Wang, 2006). From that moment, nanowires became relevant. The use of nanowires has the advantage that can be triggered by tiny physical motions, and the excitation frequency can be a few Hz to multiple MHz. To get a high efficiency and to avoid energy lost caused by friction, system with noncontact approach has been developed.

8.1.3 Hybrid Nanogenerators (HNGs)

Hybrid NGs (HNGs) are the combination of two or more NGs to increase the stability and the efficacy to harvest energy. HNGs create sustainable energy taking advantage of the coexistence of different harvesting systems. Hybridization occurs among different sources such as mechanical, solar, and thermal energy. The use of HNGs helps to generate energy in a continuous form due to the harnessing of different energy sources in a single system. For example, in the daytime, photovoltaic systems generate energy through sunlight source. However, in the nigh or during the rain, the principal systems used could be mechanical or thermal (Ryu, 2019).

There is a real challenge with the integrated fabrication method through different energy harvest mechanisms are combined. In the first place, it is indispensable to understand the principles and working mechanism of each power generator mechanism separately (Li, 2014). Chen et al. developed a theoretical analysis that has relevant importance to describe and understand the relationship among materials parameters such as voltage, transfer charges, current, and average output power (Chen et al., 2016).

8.1.4 Sources

Human activities or phenomena around us became an alternative for energy generation. Mechanical vibrations, human motions, sunlight, waves, biomass, airflow, among others have been explored by the academy and the industry as sustainable route for energy supplies.

Ocean keeps a huge source of energy. Wave, tidal, current, salinity gradient, and thermal energy are some types of blue energy that can be collected from the ocean. Sea could provide enough energy for global consumption without environmental consequences (Tollefson, 2014; Yang et al., 2019). Wen et al. studied a hybrid material based on spiral interdigital-electrode triboelectric NG (S-TENG) in combination with a wraparound electromagnetic generator (W-EMG). The use of both results in an NG with a working capacity in a broad frequency range (Wen et al., 2016). However, this kind of energy has not been exploited due to difficult engineering, high cost, irregular, and low efficiency.

Solar energy has been the most common energy source because is the most abundant. However, just 0.04% of the energy used by humans is obtained by solar energy systems (Mohtasham, 2015). The use of solar energy in photovoltaic cells has been studied for many purposes. The technologies used are considered environmentally friendly, easy to fabricate or maintain, but nanostructured solar cells need deeper research due to low efficiency and poor stability obtained. Additionally, harvesting this type of energy could be affected by the weather on cloudy days or during the night (Tian et al., 2007).

The harvesting energy from friction and electrostatic induction is an alternative low-frequency energy source. This kind of energy occurs when two surfaces with opposite polarity are in contact

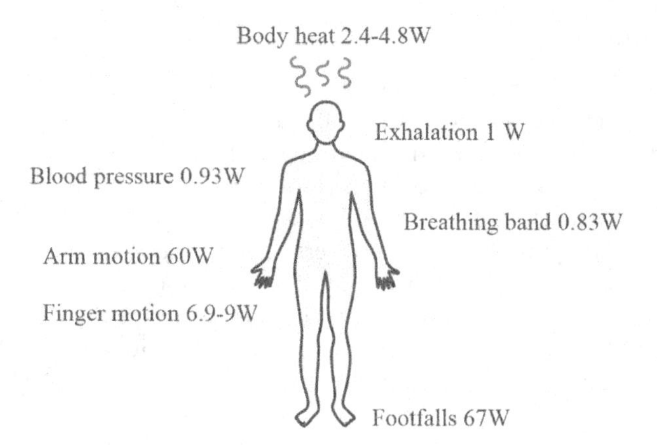

FIGURE 8.1 Energy resources from body activities.

periodically and produce a potential difference that generates an electrical output (Zheng et al., 2014). There are some advantages related to this energy like its high-power density, lightweight, low cost, high efficiency, and simple production (Yang et al., 2019).

Biomechanical has been studied as an energy source by its continuous power which can be applied in implantable and wearable devices (Qi & McAlpine, 2010). As Figure 8.1 shows, different body activities mean a significant energy source. Even 1–5% of body power could be harvested to be used in some corporal devices (Yang et al., 2016). Yan et al. synthesize a linear-to-rotary HNG (LRH-NG) that efficiently harvests biomechanical corporal energy of low frequency (Yang et al., 2020).

Moreover, there are not conventional energy sources like sound, friction, noise, among others, which exist around humans and have a significant potential for novel electronic devices (Beeby et al., 2008).

8.2 PROPERTIES

Crystal materials have a general classification based on 32 categories, 11 of which have centrosymmetric structures. Of the remaining 21 categories, 20 are piezoelectric, and 10 of the 20 piezoelectrics are pyroelectric (Bikramjit, 2014).

8.2.1 PIEZOELECTRICITY

The process in which an electric charge is generated by applying mechanical stress to a solid (crystal, ceramics, biological, proteins, among others) is known as the piezoelectric effect (Skoog, Crouch, & Holler, 2017). To understand this process, it is shown graphically in Figure 8.2, in the

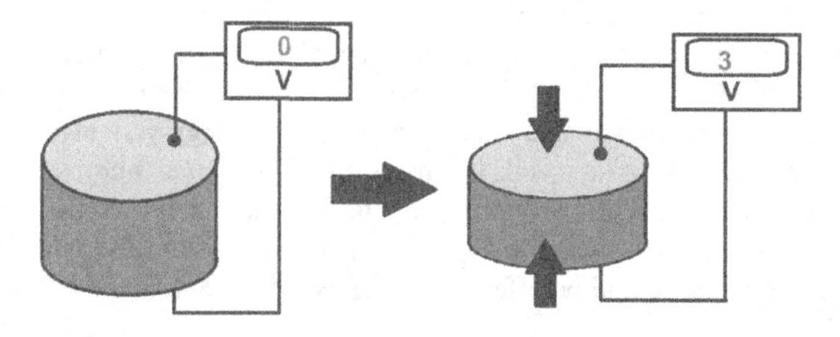

FIGURE 8.2 Piezoelectric effect applied into a material. In the left side without a mechanical stress, and in the right side with the mechanical stress.

left of the figure a solid with a voltage of zero (which is expected); however, if we apply mechanical stress (along the Y axis), the solid generates an electrical response to this effect. For a piezoelectric material, if the mechanical stress produces an electrical polarization and it is proportional to the magnitude and sing of the strain, it is called the direct piezoelectric effect; on the other side, the inverse effect takes place when the opposite polarizing electric field produces mechanical stress in the material, and it is called the converse dielectric effect (Alfredo, 2000).

Pierre Curie and Jacques Curie showed the first principles of piezoelectricity in single crystals, mainly in quartz in 1880 (Curie & Curie, 1880). Their experiments consisted of the measurement of charged surface of prepared crystals (quartz, tourmaline, topaz, among others) that were being subjected to mechanical stress. Curie brothers showed direct piezoelectricity (electricity from mechanical stress) successfully in their experiments; however, they did not predict the converse piezoelectric effect (stress from an applied electric field). The last property was shown mathematically by Lippmann in 1881 in their works based on thermodynamic principles (Badiali & Goodisman, 1975). After that, Curie's brothers shown experimentally the "converse piezoelectric effect" based on the reversibility of electro/elastic mechanical deformation in piezoelectric materials. For the next decades, the piezoelectric effect was a focus of attention. From 80s and on, piezoelectric research has taken place in several laboratories testing novel piezoelectric materials looking for new properties focused their research on composites of ceramic, polymers, glass-ceramic, and NGs.

The piezoelectric effect can exist on natural or synthetic crystals. Once the external field is removed, materials such as quartz will not exhibit remnant polarization. However, other crystalline materials, such as barium titanate (BT), lead zirconate titanate (PZT), potassium-sodium niobate (KNN), among others, show polarization even in the absence of an electric or magnetic field. In addition, the polarization can be reversed by the reverse external applied field that changes the direction of the spin. The change behavior of spin is usually represented by a hysteresis curve, which is associated to ferroelectric and ferromagnetic materials. The piezoelectricity in the crystal is due to the generation of asymmetric unit cells and electric dipoles (Alfredo, 2000). The effect is practically linear and the polarization changes directly with the applied stress and is dependent on the direction, so that stress generates an electric field and hence a voltage of opposite polarity (direct piezoelectricity effect). It is well replicated in the case that the crystal is exposed to an electric field, the crystal will experience an elastic strain; hence, the system will increase or decrease in size according to the field polarity.

The use of nanostructured piezoelectric is relatively a developed area currently, where the goal to understand the effects on ferromagnetic and piezoelectric effect at nanoscale size looking for new applications. Despite the huge range of nanostructures related to piezoelectricity, most of the works have been focused on zinc oxide. Furthermore, materials investigated for energy harvest based on PZT and $BaTiO_3$ (Bowen, Kim, Weaver, & Dunn, 2014).

An increasing number of nanostructures piezoelectric have been developed by using ZnO nanorods. It is reported a previous configuration that was used with a platinum-coated silicon surface as is observed in Figure 8.3 (Wang, Song, Liu, & Wang, 2007). Recently, devices based on lead zirconate titanate and barium titanate are being explored. One example is PZT, which is commonly developed in macro and micro scale with high electromechanical coupling coefficients (Alfredo, 2000). The use of PZT fibers in several configurations and diameters to form an energy harvester was studied. The fibers were synthetized mixing PZT with poly(vinyl pyrrolidone) as is reported in Chen, Xu, Yao, and Shi (2010).

Like PZT, BTO is also developed in nanoscale for energy applications. An important example was demonstrated through the hydrothermal grown of BTO nanorods (Koka, Zhou, & Sodano, 2014). Due to its high electrochemical coupling coefficient, BTO devices generate 16 times the power of ZnO devices and may be used as a lead-free piezoelectric substitute for ZnO. There is a growing interest in the development of piezoelectric materials related to the use of NGs. Although at the beginning all the efforts were placed on ZnO nanorods in the substrate of all kinds, many

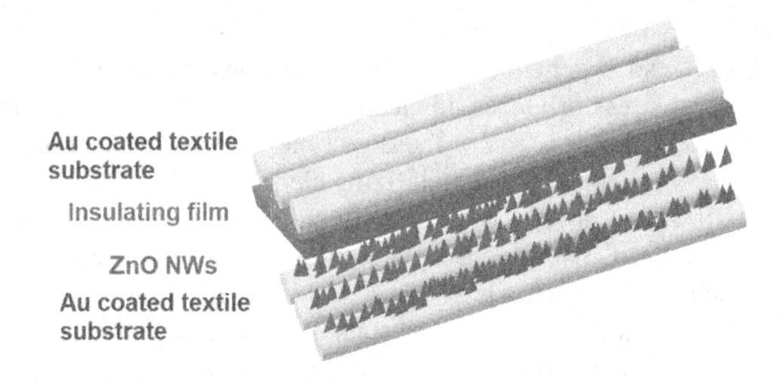

FIGURE 8.3 Schematic of ZnO nanorod grow on an Au-coated polyester.

new applications have been shown to explore other ranges of materials as are the cases of PZT, BTO, among others. Their applicability in energy harvesting has shown over the years a significant motivation to be immersed in this topic. In order to satisfy the needs for real power delivery, the system needed to be maximized in order to increase the polarization and improve the coupling of the device to the environment.

8.2.2 Pyroelectricity

Pyroelectricity is a property related to materials that experiment a spontaneous polarization by a change in the temperature (Cha & Ju, 2013). In 1969, the strong piezo- and pyroelectricity of uniaxially drawn polyvinylidene fluoride (PVDF or PVF2) was discovered after poling in a suitable electric field. It is also known that other plastics, such as nylon or poly vinyl chloride (PVC) display a piezoelectric effect but with less intensity than the PVDF (Alfredo, 2000). Pyroelectric behavior is generated by applying temperature (either heating or cooling) to these materials, and certain electric fields are generated. The representative mechanism is shown in Figure 8.4. The change in polarization is accompanied by the movement of the compensation charge, which induces current in the external circuit.

Figure 8.5 shows the crystal structure of barium titanate. In the left side of the picture, we observe the crystal of the structure above the Curie temperature (temperature needed to lose the permanent magnetic properties) around 120°C, and it is well observed that the crystal structure is a cubic system with Ba^{2+} ions in the corner, Ti^{4+} ions in the center, and O^{2-} ions in the face of each side. However, on the right side, the same perovskite is shown in the case below the Curie temperature, and the structure observed is slightly deformed, where Ba^{2+} and Ti^{4+} ions are displaced concerning to O^{2-} ions which will cause a dipole moment in the crystal.

In the past decade, extensive efforts are being used to study high-performance pyroelectric materials. Single crystals are being widely studied to their excellent detectivity and stability by controlling the orientation as domain structures. Some crystals that have provided good results are pyroelectric figures of merit (FOMs); however, most of them are limited by their low Curie

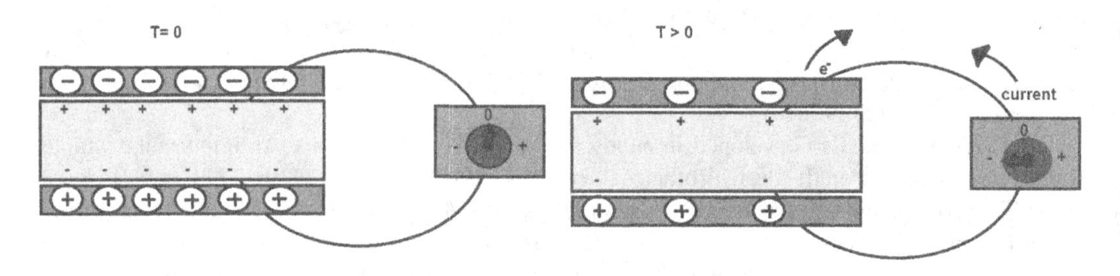

FIGURE 8.4 Mechanism of piezoelectricity.

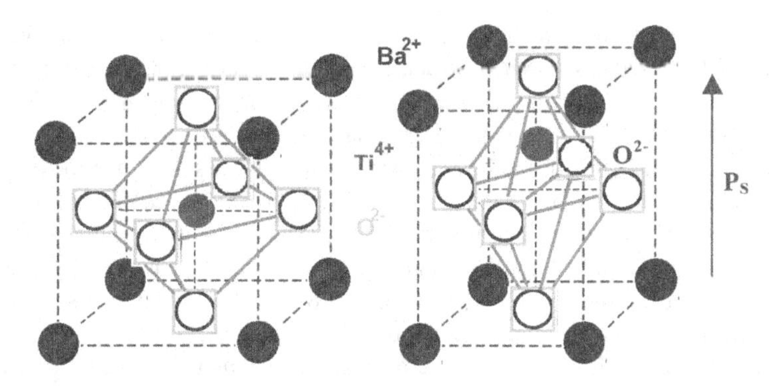

FIGURE 8.5 The crystal structure of perovskite titanate. In the left, the crystal is above Curie temperature and in the right below the Curie temperature.

temperature (49°C) and hygroscopic property. One of the most representative crystals studied over the years is triglycine sulfate (TGS) used as a pyroelectric detector (PD) from 1960. It is obtained at decreasing temperatures or by solution evaporation (He et al., 2020). Another example related to pyroelectric crystal applications is research focused on lithium tantalate ($LiTaO_3$) and lithium niobate ($LiNbO_3$). Both were developed using the Czochralski method and resulted in distorted oxygen octahedral structures (Hossain & Rashid, 1991). Other promising pyroelectric FOMs were reported by Sun et al. for Mn-doped in a material composed of $Na_{0.5}Bi_{0.5}TiO_3$ and $BaTiO_3$. Top Seeded Solution Growth (TSSG) method showed their associated orientations <0 0 1>, <1 1 0>, and <1 1 1> (Sun et al., 2014). In their results, they show that the samples in <1 1 1> orientation develop the highest pyroelectric coefficient (588 μC/(m^2 K)) and the lowest dielectric constant (2 7 9), becoming a promising FOMs for further applications. Single-crystal materials and FOMs are promising pyroelectric materials but are limited by the expensive growth techniques and the difficulty to obtain big-size products; further investigations are being applied to solve this problem.

Pyroelectric ceramics present different advantages such as reliable electric performance, low cost, easy to produce, and potential mechanical properties. The most important pyroelectric ceramics are based on perovskite structure (ABX_3) as is observed in Figure 8.6. In this structure, the ferroelectric behavior results from the permanent electric dipole generated by the displacements of the B ion. Moreover, the properties could be modified by the variation of the A or B sites (He et al., 2020).

Several perovskite pyroelectric materials are reported in the literature, some of the most important examples are $BaTiO_3$ ceramics, which have been applied to transducers, capacitors, among other devices. The $BaTiO_3$ ceramic showed a pyroelectric coefficient of 200 μC/(m^2 K) at room temperature (Bowen et al., 2014). Enhancement related to $BaTiO_3$ perovskite was reported by ion substitution in A for Ca^{2+}, Sr^{2+}, B^{4+}, and Sn^{4+} (He et al., 2020). The composition of $Ba(Zr_{0.3}Ti_{0.8})O_3$–$(Ba_{0.7}Ca_{0.3})TiO_3$ (BCZT–BCT) was reported by Liu et al., which showed a tricritical triple point of the cubic, rhombohedral, and tetragonal phases and a piezoelectric coefficient (d33~620 pCN^{-1}) (Liu & Ren, 2009).

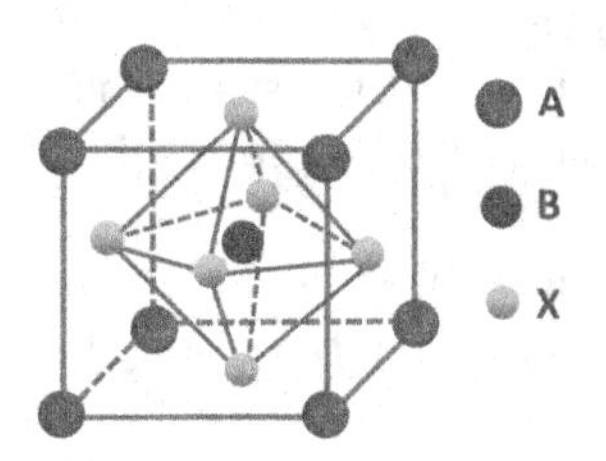

FIGURE 8.6 Crystal structure for a perovskite material.

Inorganic films have been used for pyroelectric properties due to that these provide smaller heat capacity and lower fabrication costs. When pyroelectric materials are implemented in thin films, there are some determinant factors such as pyroelectric structure, substrate, technique, thickness, among others (He et al., 2020). Of the several materials that have been studied in the field of inorganic films, the pyroelectric material $Ba_{0.64}Sr_{0.36}TiO_3$ thin film grown through the sol-gel method is one of the most representative materials developed by Zhang et al. showing a pyroelectric coefficient of 1860 μC/(m^2 K) at room temperature. Further research in this thin film reached a pyroelectric coefficient around of 15 000 μC/(m^2 K) at room temperature. The high pyroelectric coefficient obtained was attributed to free charges accumulated on the electrode (Jin et al., 2018). Further research has to be immersed to achieve higher temperatures with the same degree of pyroelectric coefficient.

Recently, polymers were discovered to have properties such as piezoelectricity, pyroelectricity, ferromagnetic, and conductive properties, among others, which polymers were supposed to be used as an insulator. The polymer in comparison to inorganic pyroelectric materials shows low dielectric permittivity, low acoustic impedance; furthermore, these are flexible, light, and easily fabricated. Because of these properties, they have been implemented in different industrial fields (He et al., 2020).

Pyroelectric materials are a wide field to be studied, in all the examples given in this section, each one has its one advantages and disadvantages. However, their applicability is continuously being researched due to their promising results experimentally. The factor of cost, temperature, pyroelectric constant, dielectric constant, and size manufacturing is important in order to develop a high-efficient pyroelectric material.

8.2.3 Triboelectricity

Triboelectricity is an electrification effect produced by the interaction of two different materials that come into contact, by friction, floating, or sliding (dielectric or paramagnetic). In each material, charges of opposite polarity are generated (Diaz & Navarro, 2004). This phenomenon occurs in nature such as lightning, electrostatic charge in dust explosion, the behavior of sand atoms, volcano, among others. Similarly, to the nature of technology, this phenomenon is applied to electrophotography, electrostatic coating with powders, electrets, and electrostatic filtration and separations (Gishiwa, 2016).

As is observed in Figure 8.7, if the material is at the bottom of the series (positive end), it will, on the one hand, gain a positive charge once in contact, and on the other hand, the top of the series will gain a negative charge. The triboelectric series allows us to identify which materials could present triboelectric effect once in contact; for example, glass that is located at the bottom of the series will gain positive charge in case if it were in contact with Teflon or polytetrafluoroethylene (PTFE) which therefore to be in the top of the list will charge negatively.

In recent years, research focused on energy harvesting is interested in the use of triboelectric materials. TENGs use electrical field for energy collection and their use in different applications such as chemical sensing, charging portable devices, sustainable power sourcing, light sensing, magnetic sensing, and other self-powered systems based on coupling between triboelectrification and electrostatic induction (Gishiwa, 2016).

The working principle of TENG can be realized in two ways. The first method is conjugation through contact electrification, which provides a static polarized charge. The second method is conjugation by electrostatic induction, which transforms the energy from mechanical to electrical (Niu et al., 2013).

Due to the rapid development of technology and high energy consumption, renewable energy technology is becoming the focus of attention. Harvesting mechanical energy is an effective suggestion for this purpose because they have low-cost, high efficiency, easy-to-maintain, clean, and sustainable power sources that can be used in wearable, portable, and wireless electronics. An example of TENGs was developed in 2015 by Zhang et al. An electromagnetic-triboelectric HNG

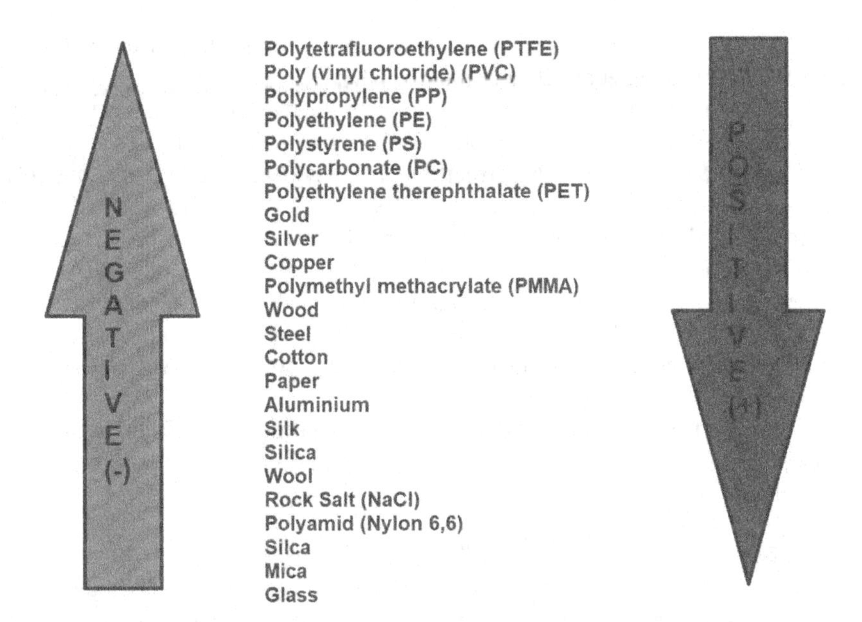

FIGURE 8.7 Triboelectric series.

for scavenging biochemical energy to wearable electronics by human walking was studied (Ru et al., 2015). The energy recollected directly light up tens of light-emitting diodes to sustainably power a smart pedometer which measure walking steps, distance, and energy consumption. A TENG with dimensions 5 cm × 5 cm × 2.5 cm delivered a peak of the power of 4.9 mW under the resistance of 6 MΩ. The hybridized NG exhibited good stability and much better performance.

Another interesting example of TENGs research is in the study of the importance of charge accumulation in the enhancing of output performance developed by Chen et al. (2015). In the study, a TENG composed of polydimethylsiloxane (PDMS) films and nanoparticles of high permittivity was developed, producing a porous structure that is associated with the dielectric behavior of the material. This work highlights the effectiveness of modification of the material and the surface in TENGs as a proper way for further research (Chen et al., 2015).

8.3 DEVICE FABRICATION

8.3.1 Nanogenerator Structure

To know how an HNG can be structured, it is necessary to remember how the NGs are individually structured. We will begin briefly remembering the structure for TENGs, piezoelectric NGs (PENGs), pyroelectric NGs (PyENG), and electromagnetic NGs (EMNGs), and finally, we are going to delve into HNGs.

8.3.1.1 Triboelectric Nanogenerators (TENG)

There are the four basic structures in the development of TENG: contact-separation, sliding, single-electrode, and freestanding triboelectric layer (Wang, 2020; Wu et al., 2019; Zhou et al., 2020; Zhu et al., 2015).

Contact-separation mode is that in which two electrodes are needed to collect the charge and dielectric material in contact with the electrodes. In the active material, the change of electrical potential occurs due to physical contact by an external force that induces an electrostatic polarized charge. The contact mode has been used in many studies with different applications, such as pressure sensors based on micropatterned plastic films (Fan et al., 2012), wearable NG

(Seung et al., 2015), self-powered nano-sensor for mercury ion detection (Lin et al., 2013), self-powered glucose sensor (Zhang et al., 2013), among others.

TENG has been used as energy storage, and Liu et al. use a freestanding triboelectric layer mode with a rotatory disc to produce high output voltage (750 V) and 15 μA; this device shows the efficiency that is possible to achieve with this structure, also they demonstrate that is possible reach a DC-output voltage without any rectifier (Liu et al., 2019).

TENG-based lateral-sliding mode was developed in different studies.

Xia et al. obtained a TENG with an output voltage in an open circuit of 1000 V and a short-circuit current of 42 μA. The potential application includes a self-powered speed sensor and force sensor (Xia et al., 2018). Wang et al. in their research obtained the output voltage in an open circuit of 1300 V, and the energy generated is capable of light around 160 LEDs, with a potential application as a self-powered sensor (Wang et al., 2013).

Another device based on the lateral-sliding mode is the proposed by Zhang et al.; the TENG that they developed was used in a wearable wireless respiration sensor to monitor respiratory rates (Zhang et al., 2019).

8.3.1.1.1 Piezoelectric Nanogenerators (PNGs)

PNGs use a sandwich configuration, the piezoelectric material between the electrodes (top and bottom electrodes). PNGs use an external force applied on a piezoelectric material to convert mechanical energy into electricity. We can find an example of the PNGs in the investigation carried out by Lu M.-P. et al. (2009). They demonstrated the energy conversion using phosphorus-doped ZnO nanowires deposited over silicon substrate. Through atomic force microscopy (AFM), a positive output voltage generated by the NG for the P-ZnO and a negative output for n-type ZnO nanowires were obtained. The PENGs are an alternative to be considered in the design to enhance the harvest of energy in a device (Lu et al., 2009).

8.3.1.2 Pyroelectric Nanogenerators

PyNGs take advantage of the temperature gradient in the material between two points where contacts are used to collect charge carriers. It has been demonstrated that PyNGs can be used to provide energy to sensors and potential to wireless sensors (Yang, Wang, et al., 2012). Another example is the structure presented by Yang et al. who used Indium tin oxide (ITO) deposited over glass substrate as bottom contact, ZnO nanowires were grown up over ITO, and Ag was used as top contact (Yang, Guo, et al., 2012). Xue et al. developed an NG and self-powered breathing sensor by using an N95 respirator with a maximal power of 8.31 μW, and the PyNG was made by a metal-coated PVDF film (Xue et al., 2017).

8.3.1.3 Electromagnetic Generators (EMNGs)

The EMNGs consist basically in the movement of a magnet through a coil considering Faraday's law of electromagnetic induction, the movement of the magnet induces a current in the coil that can be used in different applications (Zhang et al., 2014). A good example is the EMNG proposed by Saha et al.; a free permanent magnet is placed in a tube between two fixed magnets with reverse polarization and a coil around the tube. The NG is designed to harvest energy from the body motion, and the device can generate 300 μW to 2.5 mW (Saha et al., 2008).

8.3.2 Hybrid Nanogenerators

After briefly recalling the basic structures of NGs, it is time to talk about HNGs. As we already mentioned, HNGs are those composed of two or more NGs, so we can say that the energy collection is carried out taking into account two or more mechanisms; therefore, we must consider that the design of the NG must have the possibility of taking advantage of the energy of the environment to convert it into electrical energy.

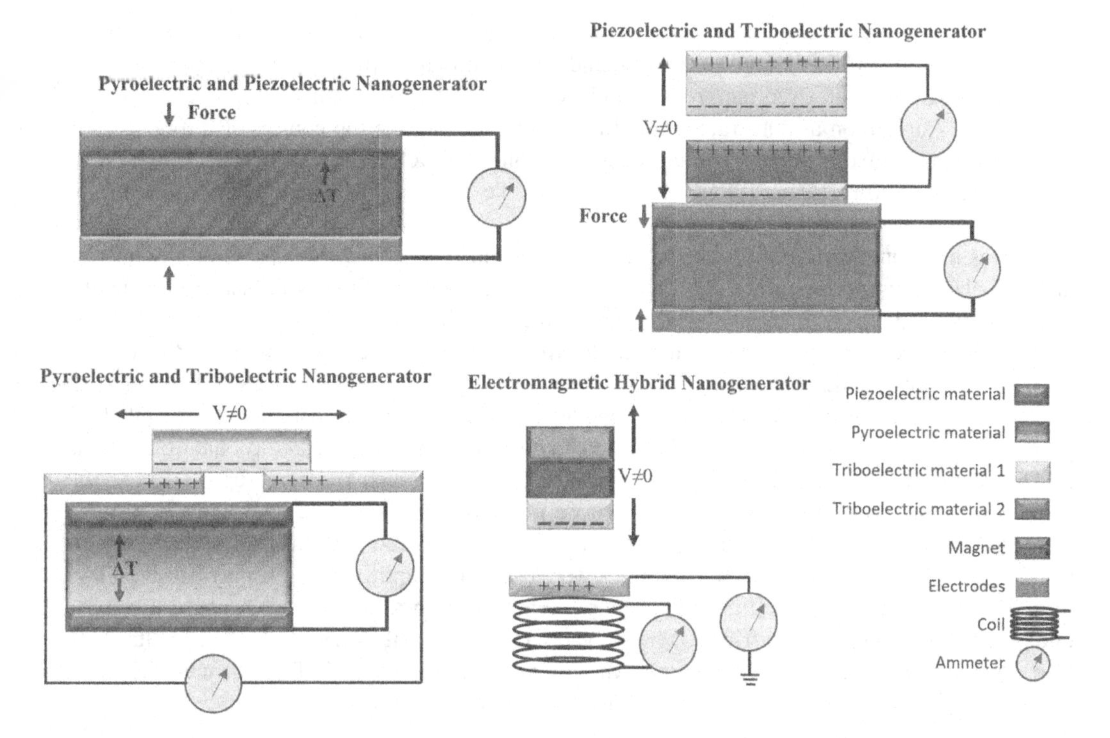

FIGURE 8.8 Configuration for the HNGs with different principles for energy harvesting. Pyroelectric and piezoelectric nanogenerators (PPNGs), piezoelectric and triboelectric nanogenerators (PTENGs), pyroelectric and triboelectric nanogenerators (PyTENGs), and electromagnetic hybrid nanogenerators (EHNGs).

In Figure 8.8, there are examples of the different possible configurations for HNGs. Pyroelectric and piezoelectric NGs (PPNGs) for the harvesting energy with the same material, based on two principles (pyroelectric and piezoelectric effect). The example of piezoelectric and TENG (PTENG) presented in the figure has a contact-separation structure for the TENG and a sandwich structure for the piezoelectric NG. The pyroelectric and TENG (PyTENG) has a freestanding triboelectric layer for the TENG and a sandwich structure for the pyroelectric NG. Electromagnetic HNG (EHNG) has a single-electrode structure for the TENG, and a magnet coupled to the triboelectric material to use the movement for inducing a current in the fixed coil. We can analyze the three-last HNGs mentioned as two coupled individual NGs, considering the principles and configurations previously discussed. All the HNGs are structured to enhance the energy harvesting for the application in different devices.

8.3.2.1 PPNGs (Pyroelectric and Piezoelectric Nanogenerators)

As we mentioned before, in the design of an HNG, it is necessary to consider the mechanisms involved in the harvest of energy, so, in this part, we describe the structure that we can find in the development of PPNGs.

HNGs use the basic functional principles of the single NGs to get a higher output electrical efficiency for different applications like self-powered biomedical applications and wireless sensors. PPNGs are structured considering the pyroelectric and the piezoelectric properties and its corresponding structures, such as Lee et al. in 2014, that developed a PPNG using a composite of PDMS and carbon-nanotubes as a flexible bottom electrode, poly(vinylidene fluoride-*co*-trifluoroethylene) as a piezo- and pyroelectric material and graphene electrode. Using these materials and configurations makes it possible to get a PPNG highly stretchable and flexible (Lee et al., 2013). Ko et al. developed a PPNG using $Pb(Zr_{0.52}Ti_{0.48})O_3$ (PZT) and a highly flexible Ni-Cr foil with a conductive

$LaNiO_3$ as substrate/bottom contact and Pt as top contact. Considering the properties of the NG, it can be used in different applications, including the in-car muffler, radiator, and engine (Ko et al., 2016). Chen et al. used nanorods of poly(vinylidene fluoride-*co*-trifluoroethylene) [P(VDF-TrFE)] over Au/Kapton as substrate and bottom electrode, and the top contact electrode used was poly(3,4-ethylenedioxythiophene):poly(styrenesulfonate) (PEDOT:PSS); in their conclusions, they remark the possibility to use PPNG for biocompatible and flexible energy harvesting for self-powered systems (Chen et al., 2016).

As was mentioned before, ferromagnetic and ferroelectric materials are capable of present piezo- and pyroelectric properties. ZnO, PZT, and PVF2 can be useful to PPNGs fabrication (Yang et al., 2012). In 1920, the first material studied with both properties was the Rochelle salt, when Joseph Valasek discovered ferroelectricity (Lang, 2005). Ko et al. developed NGs based on PZT perovskite film, which worked in harsh environments (extreme humidity, pH, and temperatures) (Ko et al., 2016).

Piezoelectric materials such as poly(vinylidene fluoride-*co*-trifluoroethylene) [P(VDF-TrFE)] polymer, zinc oxide (ZnO) potassium niobate ($KNbO_3$) (Yang, 2012), lead zirconate titanate, and PVDF were previously used like piezoelectric and pyroelectric NGs. You et al. reported a nanofiber membrane based on PVF2 that showed improved flexibility by the use of TPU and PEDOT. This material can be used to obtain energy from human movement and hot and cold air currents (You et al., 2018). Mechanical and thermal energies were used for a liquid crystal display based on PVF2 film. This was considered the first fully integrated flexible hybrid energy cell because it consisted of a pyroelectric and piezoelectric NG, ensembled with a solar cell (Yang et al., 2012). Sultana et al. also developed a PVF2 HNG but with methylammonium lead iodide (MAPI). The material synthesized showed high sensitivity and fast response (Sultana et al., 2019).

8.3.2.2 PTENGs (Piezoelectric and Triboelectric Nanogenerators)

It is important to consider the triboelectric and piezoelectric effects in the design of these HNGs. Jung et al. use the contact-separation structure for the TENG and a sandwich structure for the PENG. The PENG responds by the compressive and tensile stress generated by an applied force, and by using the movement the TENG that generates energy, in this description, it is important to mention that the PENG is a "mobile" part and the PENG is designed with a curvature to use the force applied. PVDF is used for PENG and Teflon for the TENG. The output voltage generated is around 370 V, and it is possible to turn on 600 LEDs (Jung et al., 2015). Zhao et al. use the contact-separation mode for the TENG part, a sandwich structure for the PENG, and a rotation mechanism to push on the HNG. The TENG consists of polyethylene terephthalate substrate with Al electrode and Teflon with Au electrode with one-end fixed connecting the two parts of the TENG. The output voltage generated is around 210 V and an output current of 395 μA; with this PTENG, it is possible to light 50 LEDs (Zhao et al., 2019). Wang et al. developed a PTENG composed of $Ba_{0.838}Ca_{0.162}$ $Ti_{0.9072}Zr_{0.092}$ O_3(BCZTO)/polydimethylsiloxane (PDMS) as PENG component and $Ba(Ti_{0.8}Zr_{0.2})O_3$ (BZTO)/PDMS as TENG component. The PENG is structured in a sandwich configuration with Cu electrodes and the TENG is based on contact-separation mode, all the hybrid generators consist in an arrangement of three-electrode configuration, and it is possible to get an output voltage of 390 V (Wang, Zhang, et al., 2020).

Previously, hybrid generators with piezoelectric and triboelectric units working in contact mode were studied for more advantages. These systems were useful for technologies covering a large area (Song, 2016). It was believed that hybridization affects the piezoelectric output of material by enhancing or reducing the piezoelectric potential. An r-shaped HNG that consists of piezoelectric NG (PVDF film, Al electrodes, polyethylene terephthalate (PET) film) and TENG (PDMS, PET/ITO electrode, Al electrode) (Han, 2013) was reported. In this material, the friction phenomena are used as electric charge source to enhance the piezoelectricity in the system (Kim, 2012). Positive triboelectric charges accumulated on the bottom aluminum electrode flow to the other aluminum electrode, resulting in an additional potential difference (Han, 2013). Piezoelectric and triboelectric nanostructures have been integrated into carbon fibers to convert the mechanical energy into

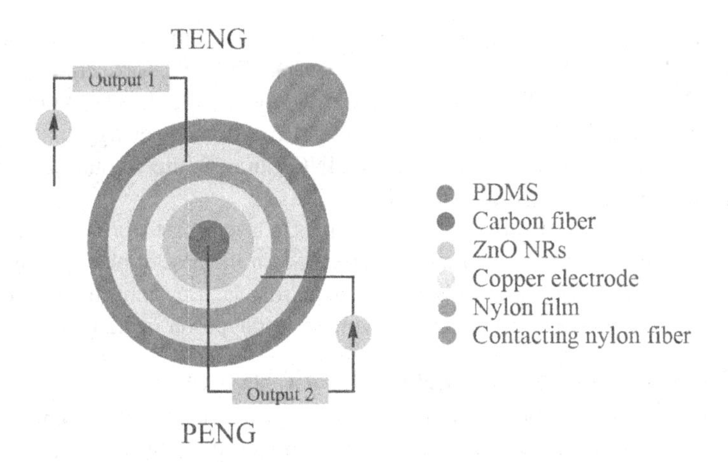

FIGURE 8.9 Example of a fiber-based hybrid nanogenerator based on piezoelectric and triboelectric nanostructures.

electricity; the system is shown in Figure 8.9. The power density obtained through the fiber-based HNG (FBHNG) can be enhanced by increasing the area and number of fibers (Li, 2014). Yang et al. reported a PTENG with a sandwich structure of Al-PPCF-carbon. The results obtained showed that the triboelectric potential is 12.7 times higher when it is part of the hybrid structure studied (Yang & Daoud, 2016).

8.3.2.3 PyTENGs (Pyroelectric and Triboelectric Nanogenerators)

Jiang et al. developed a PTENG to recover energy from low-grade waste fluids, and this device uses hot drops sliding down to the bottom. The device is inclined to promote the movement of the drops along the surface. The device can be described as two NGs united by isolating interphase, and the pyroelectric NG (PyNG) is made from a film of p-PVF2 between silver electrodes; meanwhile, the TENG is over the PyNG and is made with a freestanding triboelectric-layer structure, and a hydrophobic layer is deposited over two lateral silver electrodes. The peak power density is 2.6 μW/cm^2 from a PyTENG with a size of 3 × 20 cm, and the device can light up 28 LEDs (Jiang et al., 2020). Shing et al. use poly(vinylidene fluoride-*co*-trifluoroethylene) (P(VDF-TrFE)) as PyENG and TENG in different configurations. Contact-separation structure is for the TENG, and sandwich electrodes structure is for the PyENG.

8.3.2.4 Electromagnetic Hybrid Nanogenerators

The electromagnetic and triboelectric effects are used in an HNG with a power generation of 0.25 mW and an open-circuit voltage (peak to peak) of 600 V; the novel structure consists of an acrylic tube surrounded with a coil as a fixed part and mobile part made with a magnet connected with two acrylic sheets by a spring, and the S-TENG was put at the bottom and top of the acrylic tube; with this configuration, the HNG uses the vibration of the floating part to generate electrical energy (output voltage/current was measured for the bottom S-TENG) (Wu et al., 2015).

One application of EM-HNGs is in smart wearable smart electronics, as Liu et al. report, the EM-HNG can be embedded inside a shoe. The basic structure consists of a lateral-sliding mode for the TENG, and this NG is placed up and down of a round box; for the EMG, a permanent magnet was placed between the TENGs, and two coils were placed around the box to harness the energy generated by the electromagnetic induction (Liu et al., 2018).

An HNG reported by Wang et al. could be divided in a TENG based on a vertical contact-separation mode in the top and the bottom of the device; in middle acting as dielectric, there is Kapton covered by copper electrodes at both sides and over the electrodes fluorinated ethylene propylene (FEP); meanwhile, the EMG consists of a magnet in both sides of the FEP and a coil over the top/bottom electrodes, and the movement by the vibrations induces the electricity generation (Wang & Yang, 2017).

Zhang et al. in 2016 used an HNG in form of a rotating disc, using the freestanding triboelectric-layer mode as a principle to TENG with static electrodes to collect the energy; in the EMG magnets in the rotating disc and static coils where used, the HNG can generate an output power of 17.5 mW at a rotating rate of 1000 rpm (Zhang et al., 2016).

Quan et al. used a suspended magnet from four springs, a layer made from Kapton, aluminum, and nanostructured PDMS is hanging on the magnet, separated, in the base is placed a layer of aluminum and Kapton, and below there is a coil. The electromagnetic triboelectric nanogenerator (EMTENG) with this structure uses the vibration and generates a maximum output power of 1.2 mW (Quan et al., 2015).

8.3.2.5 ESHNGs (Electromagnetic Shielding Hybrid Nanogenerators)

The development of technology and the greater acquisition of devices and equipment have increased electromagnetic pollution, so we have an alternative to protect ourselves. NGs and HNGs not only serve to generate energy to supply batteries or devices, but also it is possible to take advantage of the material properties. Some NGs or HNGs could act as electromagnetic shielding like some investigations report the efficiency to protect.

The PENG developed by Kar et al. is capable to reach 99% of shielding in the X-band (Kar et al., 2018). Also, Im Ji-Su and Park Il-Kyu design TENG-based nanofibers, Fe_3O_4 nanoparticles, and polyvinylidene fluoride (PVDF) remark the capacity of the device as shielding efficiency (ES) (Im & Park, 2018).

HNGs are used as a protector of electromagnetic energy produced by the devices and scavenging energy. Zhang et al. developed an ES-HNG using a TENG and a PPENG. In this study, they get protection over 99.99% of electromagnetic waves in the range of 0 and 1.5 GHz, the device was used in a computer keyboard, the output voltage generated is around 3 V in 200 s in constant typing. The main application is focused to protect pregnant women from possible damages due to the increase in the use of technology (Zhang et al., 2018).

8.3.2.6 Electromagnetic-Triboelectric Nanogenerator

Quan et al. worked with an HNG of simultaneous energy from EMG and a TENG. The device performed exhibited a higher output performance than the NGs in an individual way. The results could be improved by multilayer structures, especially for TENGs (Quang et al., 2015). Liu et al. use the vibration energy from a magnet to activate the functions of TENGs and EMGs. The energy obtained was useful to small devices like a GPS (Liu et al., 2018).

8.3.2.7 Hybrid Nanogenerators Based on Three Types of Energy Harvesting

8.3.2.7.1 Triboelectric-Pyroelectric-Piezoelectric Hybrid Nanogenerator

Zi et al. integrated a TENG and PPENG in an HNG. The integration of PPENG enhanced the harvesting efficiency up to 26.2% (Zi et al., 2015).

8.3.2.7.2 Triboelectric-Electromagnetic-Piezoelectric Hybrid Nanogenerator

It was reported an HNG to improve the TENG performance. The HNG consisted on TENG, EMG, and PENG, which was applied in a shoe sole. The system created increased 20% of the charge than TENG alone (Rodrigues et al., 2019).

8.3.3 Process

In the development of NGs, different areas of material engineering are involved. We can start describing some chemical and physical methods such as thermal vapor deposition, hydrothermal method, sputtering, inductively coupled plasma (ICP) reactive ion etching, anodization, among others.

In Figure 8.10, there is an example of different possible steps involved in the development of an NG. The example shows the deposition of a triboelectric polymer over a patterned substrate using

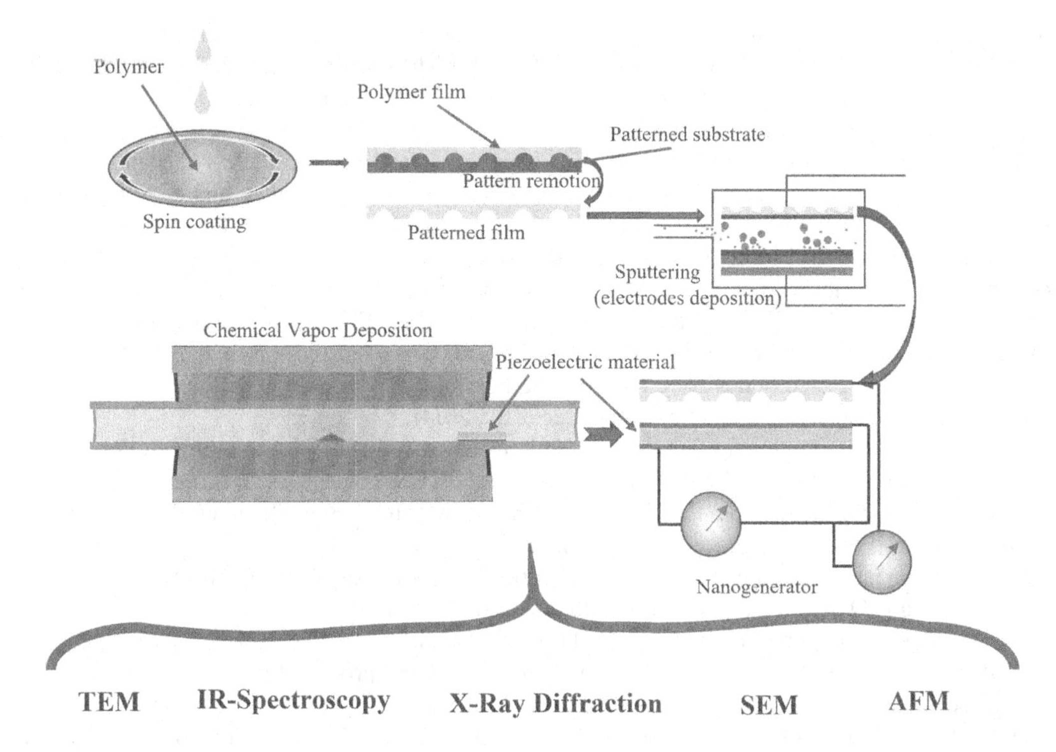

FIGURE 8.10 Process for development on an HNG and characterization techniques to improve the energy harvesting.

the spin coating process, the remotion of the substrate, and the electrode deposition by sputtering. Also, in the example, there is the deposition of a piezoelectric material using chemical vapor deposition, and the substrate can work as the bottom electrode, and the top electrode can be deposited by different processes such as ion-beam or sputtering. In the bottom of Figure 8.10, there are some examples of possible morphological, structural, or compositional characterization.

Chen et al. built a PPNG, a P(VDF-TrFE) film was deposited over Au-coated Kapton substrate by spin-coating, and then the solvent was evaporated. The Au-coated Kapton was made by sputtering. Commercial anodic aluminum oxide (AAO) with nanochannels was employed to form the nanowires, pressing the AAO against the P(VDF-TrFE). The sample was kept at 170°C for 1 h to form the nanowires, after that, annealing was carried out to improve the crystallinity. The AAO was removed using a mixture of $CuCl_2$ and HCl, followed by NaOH. The nanowires resulting were coated with a layer of poly(methyl methacrylate) (PMMA) using the spin-coating. Finally, a conducting polymer was used as the top contact. In the fabrication of this device, it is possible to find different techniques used to improve efficiency (Chen et al., 2016).

Thermal vapor deposition is used to deposit some materials like phosphorus-doped ZnO to be used as an NG, using a silicon substrate. A solution of zinc acetate dihydrate 20 mM in ethanol is dropped on the silicon wafer until it reaches a thickness between 100 and 200 nm after annealing at 300°C in the air for 30 min, this film acts as a seed layer. A mixture of ZnO and graphite powder was used as a precursor, and zinc phosphate (Zn_3P_2) was added as a dopant. The precursor was placed on an alumina boat in the high-temperature zone of the furnace. A mixture of Ar and O_2 was used as a carrier gas, and the growth temperature was kept at 600°C for 60 min. The ZnO nanorods act like a PENG with a Si/ZnO bottom contact (Lu et al., 2009).

The hydrothermal method also has been used to grow ZnO and dip-coating to deposit PDMS over a flexible substrate of Ag-coated textile, which was immersed for 10 min in a solution of zinc acetate 0.03 M to make a seed layer, then dried at 100°C on a hot plate (this process was repeated three times). To grow the nanorods, the Ag-coated textile with seed layer was treated with a zinc nitrate hexahydrate and deionized water at 95°C for 3 h. The Ag-coated textile with nanorods was covered with PDMS using dip-coating process, followed by thermal treatment at 80°C for 2 h (Seung et al., 2015).

Silicon wafers are used usually for getting patterns in materials like PDMS-carbon nanotube (CNT). The patterns are fabricated in the silicon wafer by photolithography and etching process (Fan et al., 2012; Lee et al., 2014). These silicon patterns can be used in the design of some NGs micropatterned, such as Lee et al. demonstrated in the PPNG, PDMS-CNT was deposited by spin-coating on the Si-patterned, the layer of PDMS-CNT was peeled off and then used as substrate and electrode. P(VDF-TrFE) was spin-coated over the PDMS-CNT micropatterned and annealed. Finally, a multilayer graphene growth by CVD was transferred to the top of the structure. The PPNG built was completely flexible and offers the possibility to be used in different applications, such as self-powered sensors, among others (Lee et al., 2014).

Dip-coating and spin-coating are two methods to cover substrates homogeneously, similar to Jiang et al. in 2020, who used dip-coating to fabricate the TENG on a glass substrate with Ag electrodes on each side. Meanwhile, the PyENG was made by spin-coating (Jiang et al., 2020). Spin-coating was used by Fan et al. to deposit a mixture of PDMS and cross-linker (Sylgard 184, Tow Corning) over a patterned Si wafer at 500 rpm for 60s. After incubation, the PDMS was peeled and placed over an ITO-coated polyester film with uncured PDMS (Fan et al., 2012).

For the fabrication of a rotating-disk-based EMTENG, a film of PTFE was cleaned and covered with Au using a mask, then, using ICP reactive ion etching was produced nanowires structures on the surface of PTFE. The nanopore structures in the aluminum electrodes were made by anodization. For the fabrication of the HNG, the acrylic sheets were proceed using laser cutting to make the tailored arrangement to put the coils, and the aluminum electrodes embed in the acrylic. The NG was tough to be used in a self-powered active wireless traffic volume sensor, considering the wind generated by the movement of vehicles through a tunnel (Zhang et al., 2016).

Zhu et al. built a PTENG using a D-arched configuration to separate and improve the energy harvesting. The PTENG was composed of two layers, and the upper layer is made of PVDF film with electrodes of Al electrodes deposited by magnetron sputtering and a PET film as support. The lower layer consists of a silicon rubber micropatterned, Al electrode deposited by magnetron sputtering and PET film. The silicon rubber micropatterned was made by casting a solution onto the top of a Cu patterned. After curing the rubber at 70°C for 2 h, it was peeled off from the pattern. The NG was designed to harvest mechanical energy, the NG responds to external vibration, and it is possible to use the PTENG as a self-powered monitoring system (Zhu et al., 2017). Reactive sputtering is one of the techniques used to deposit thin films, Kim et al. use radio-frequency sputtering to deposit ZnO films with an oxygen atmosphere, the excess of oxygen improves the output performance of the PENG (Kim et al., 2014).

8.4 CHARACTERIZATION

An appropriate characterization is needed for understanding the work mechanisms behind electrical energy harvesting and improve the NG. Some measurements are basics in the development of an NG, such as scanning electron microscopy (SEM), electrical output parameters, and some could be considered specific according to the materials used and the possible applications; in this category, processes that can be mentioned are Fourier transform infrared spectroscopy (FTIR), UV-Vis spectroscopy, X-ray diffraction (XRD), photoluminescence, among others.

8.4.1 Structural Analysis

XRD is one of the preferred techniques to determine and characterize the structure in crystalline and noncrystalline materials. XRD is based on the interference of monochromatic radiation in the

X-ray zone, the constructive interference of the diffracted rays takes place when conditions satisfy the Bragg's law: $n\lambda = 2d\sin\theta$, where n is an integer, λ is the wavelength of the X-rays incident, d is the interplanar spacing, and θ is the diffraction angle (Bunaciu et al., 2015).

Numerous studies use XRD such as basic characterization. In this sense, Kar et al. remark the importance of the content of β phase of the PVF2 in the improvement of the efficiency of the piezoelectric NG (Kar et al., 2018). Khan et al. to demonstrate a single phase of ZnO nanorods growth on Ag-coated textile cotton. The analysis evidence the main orientation to the c-axis of the ZnO nanorods and the Ag of the substrate (Khan et al., 2012).

Also, Chen et al. use XRD to determine the phase of interest in the PPNG, the growth of the nanorods highly orientated to the (1 1 0) direction due to the confinement used in the synthesis, the preferred orientation, and an electrical poling are necessary to get a good piezoelectric and pyroelectric response (Chen et al., 2016).

Transmission electron microscope (TEM) also is used to determine the structure, and one advantage is that is the possibility to measure the interplanar distances. Lu et al. used high resolution-TEM to characterize ZnO nanorods, the image shows the characteristic ordering of atoms in the lattice of the wurtzite structure and an orientation to the (0 0 0 1) plane, also with the analysis they determine the spacing between the planes. Part of the characterization is made with the selective area electron diffraction (SAED) technique inside the TEM, the analysis shows the single-crystal structure and the d-spacing (Lu et al., 2009).

8.4.2 Morphological Characterization

The main technique used to characterize any material is SEM. The SEM is based on electrons accelerated due to the potential and the interaction of that electrons with the samples. Remembering that the interaction of the electron with samples can originally auger electrons, X-ray continuous, characteristic X-rays, backscattered electrons, and secondary electrons. The secondary electrons are some of the most important because they provide information about the surface. Another product of the interaction electrons-sample is the backscattered electrons, which give composition and topographic information (Zhou et al., 2006).

Fan et al. demonstrated that micropatterned film has a bigger efficiency electrical output than flat films, and it was supported by morphological characterization with SEM images (Fan et al., 2012). Wang et al. built a TENG with a porous interface of carboxymethylcellulose (CMC), the porous are evidenced in the morphological characterization made through SEM. The porous CMC benefits the absorption of poly(butyl acrylate and butyl methacrylate) (Wang, Pan, et al., 2020).

Morphological characterization determined by AFM helps to describe the topography, roughness, and other material parameters. Singh and Khare use AFM to compare the morphology between a structure of PVDF with ZnO nanorods, the roughness increases with the addition of the ZnO nanorods, the increase of roughness increases the effective area where the charge can reside and this could lead to an increase in the charge density, and this is evidenced in the power output of the PTENG which increases around 2.5 compared with the bare HNG (Singh & Khare, 2018). Also, Quan et al. use the AFM to confirm the size of the nanostructured AAO template and the PDMS nanostructures obtained. Analyzing the information proportionated by AFM, the authors sized the height of the holes and the nanostructures. The EMTENG generates an output voltage of 9.9 V with a vibration frequency of 14 Hz (Quan et al., 2015).

8.4.3 FTIR Characterization

IR spectroscopy is used to determine structural composition due to the absorption of the energy by the bonds between the atoms, and the absorption of the radiation results in transition between quantized vibrational energy states (Gaffney et al., 2002).

In the NGs or HNGs development in the IR spectroscopy, some studies remark the content of the electroactive-phase in the NG, crystal phases of a polymer in an HNG or confirm the existence of certain phase determined by other characterization technique (Chen et al., 2016; Hansen et al., 2010; Kar et al., 2018).

An important point to mention is the tracing of the reaction in the polymerization when the material is synthesized, and this is important to get the desires properties in the NG, as Wang et al. demonstrates in the development of the TENG based on contact-separation mode (Wang, Pan, et al., 2020).

8.4.4 Photoluminescence Measurements

In the photoluminescence coexist, the absorption of energy, radiative and non-radiative transitions. When a material is excited with enough energy to promote an electron to a higher level of energy, there are two possible results, the emission of a photon with the same energy or a non-radiative transition followed by the emission of a photon of lower energy than the incident photon, the last process is known as luminescence. If the luminescence process is induced by a photon, then, the process is called photoluminescence (Gilliland, 1997).

Alam et al. demonstrate the importance of the photoluminescence (PL) measurement, and in their case, they found a relation between the TiO_2 content and their insertion on the PVF2 nanocomposite (PNC) films (Alam et al., 2017). In other materials, like semiconductors, the PL is important to relate defects to the material, transitions, passivation of vacancies, among others (Han et al., 2018; Kim et al., 2014). The passivation of vacancies is important in PNGs because it is related to the electron concentration, when the passivation increase the free-electron decrease, such as Han et al. demonstrated the output voltage increase over two times and currently more than three times in the PNG of MoS_2 (Han et al., 2018).

8.4.5 UV-Vis Absorbance Measurements

UV-Vis spectroscopy gives us information about absorption or transmittance of radiation in the ultraviolet and visible region, due to the interaction or non-interaction of energy in the UV-Vis region and the material of interest. This is important when the application of the NG is directed to self-powered electronic devices such as a touchscreen, or other personal devices even flexible electronics (Fan et al., 2012; Kang et al., 2017).

8.4.6 Electrical Measurements

The electrical measurements could be considered basic, but there are some considerations to keep in mind, the electrical output expected and the design to use that electrical energy and the way to measurement. Sometimes, the experimental design requires the use of a rectifier bridge to get a rectified DC output voltage.

It is common to measure the electrical output parameters of NGs with a voltmeter, ammeter (or picometer), high-speed voltage-current measurement unit (potentiostat/galvanostat/ELS analyzer), electrometers, oscilloscopes, or even it is possible to use AFM to determine these properties.

AFM has been used to measure piezoelectric response in PNGs, the determination done by Lu et al. was applying a constant force of 5 nN in contact mode, the output voltage was measured with a load of 500 MΩ, and the conductive bottom was grounded. This method generates a 3D plot of the output voltage (Lu et al., 2009). Piezoresponse force microscopy (PFM) is another way to configure AFM equipment, and the PFM gives information about piezoelectricity, Han et al. used a lock-in amplifier to detect the piezo response signal, a picometer for low-noise current measurement, and an electrometer for measuring the output voltage of the PENG (Han et al., 2018).

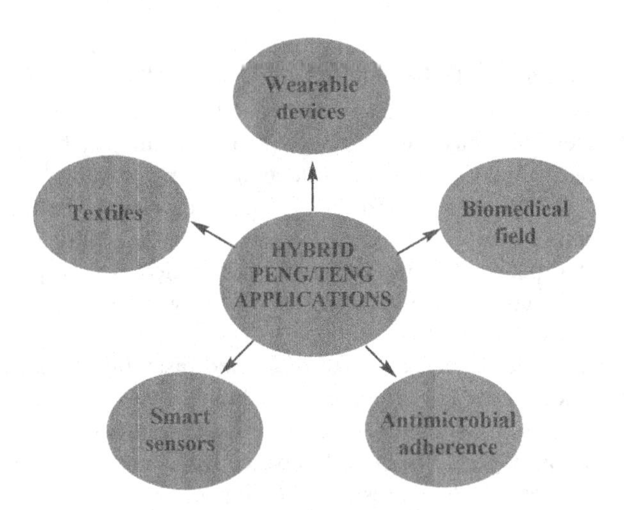

FIGURE 8.11 Applications of hybrid nanogenerators.

8.5 APPLICATIONS

Self-powered hybrid systems are sustainable materials, working without any external power supply. Due to their versatile properties, they can be applied in various fields, being a promising material for the future. Figure 8.11 illustrates some fields where HNGs are usually applied.

8.5.1 Wearable Devices

NGs are smart and wearable devices used as power sources. Fabric-based smart wearable devices, combined with internet technology and modern micro-smart manufacturing technology in the textile and apparel field, put forward the requirements for a smaller size, longer service life, more flexible, and easy-to-integrate power system (Zhao & Zhu, 2020). The vital characteristics are the low costs, clean, and environmentally friendly.

Smart homes are a clear example of wearable devices because they are externally powered. They can be equipped with sensors that can respond to changes in the environment without a power supply to control them (Zhao & Zhu, 2020).

8.5.2 Biomedical Applications

NGs are important because they can analyze abnormal vital signs symptoms in real time, even if the patients are not close to any hospital. This new proposal device will enhance the health-care industry. Commonly, TENGs are also used in biomedical fields such as drug delivery, cell modulation, circulatory system, antimicrobial, and implantable devices.

Parandeh et al. (2020) reported a hybrid material composed of TENG based on fibrous layers of silk fibroin and a polycaprolactone/graphene oxide. The surface modification through graphene oxide nanosheets improved the output of the TENG (Parandeh et al., 2020). This eco-friendly and cost-effective technique is usually applied as a power source for biomedical applications.

8.5.2.1 Circulatory Systems

The circulatory system consists of blood, the heart, and lymphatic vessels (Carlson, 2019). During an inefficient function from any of these circulatory parts, surgery is an optimal pathway to solve that concern. Electronic implants are commonly used to treat adverse effects in the circulatory system. However, these electronic devices must be replaced after a certain time, because of the battery lifetime, through surgery. Hence, NGs are a suitable solution, because they can detect and subsequently harvest the mechanical vibrations of heartbeats, cardiovascular fluids, or sound waves.

Self-powered devices can prolong the pacemaker's battery replacement, avoiding unnecessary surgery. The implant pacemaker is used to restore the heart's natural rhythm. A unipolar or bipolar electrode is implanted on the myocardium to create an action potential to control certain arrhythmias (Collins and Dias). TENGs are used as a solution because it shows a higher output. However, it needs to be well encapsulated to prevent leakage (Li et al., 2020). NGs have not only been employed to supply energy for cardiac pacemakers but also have been applied as a self-powered cardiac sensor implanted in the heart to directly report its beating conditions, without the requirement of any other energy supply (Feng et al., 2018).

Besides, hypertension is the leading cause of death and disability worldwide and is the most important global risk factor for cardiovascular risk (Habib, 2018). Research has shown that exposure to chronic life stress is associated with elevated blood pressures and an increased incidence of hypertension (Larkin & Cavanagh, 2016). A self-powered NG could be implanted and resulting in an early diagnosis, monitoring the blood pressure at real time.

Wearable TENG achieves a good signal detection coverage like heartbeat, pulse signal, limb flexion and extension, and even can be used to realize human sensory functions. An HNG with triboelectric, piezoelectric, and pyroelectric structures was obtained. It was studied in conjunction with skin-like soft objects to monitor different human physiological signals. The created HNG shows a potential use in medical purposes, as the generation of illness diagnostics at low cost and with high efficiency (Sun et al., 2018).

8.5.2.2 Skin

Biocompatible NGs are potential materials to provide power energy to biomedical or skin-contact electronic devices (Wang et al., 2021). Tactile sensors are highly attractive. For instance, Li et al. (2017) used TENGs for the fabrication of a smart tactile e-skin that was able to detect contact, hardness, and location, simultaneously (Li et al., 2017).

A sensor characterized by self-powered, flexible, tactile, and antibacterial properties was developed. Its working principle is associated with the triboelectric, piezoelectric, and pyroelectric components and with the multi-effect coupling mechanisms. Furthermore, the sensor performs excellent antibacterial property by doping Ag nanowires into the contact layer (Ma et al., 2019).

Gong et al. (2020) reported a hybrid generator composed of TENG (meso-poly lactic acid) and PENG (double-layered poly-L-lactic acid). The output power generated correspond to 0.3 mW. Moreover, the hybrid generator has a potential use as e-skin for wearable technological applications (Gong et al., 2020). This technique exhibited high biocompatibility as well as biodegradability. Cao et al. reported a hybrid sensor based on triboelectric fibers to monitor the respiratory rate through a mask, giving information about health conditions (Cao et al., 2017).

8.5.2.3 Cell Differentiation

Cell differentiation is known as a process in which cells become specialized (Ng et al., 2019). NG provides electrical stimulation for cells. TENGs boost electrical stimulation, which is a method to enhance cell differentiations during cell proliferation. Cell proliferation is the increase in cell numbers resulting from cell division, which is a complex, tightly controlled, and a well-defined process (Yang et al., 2014).

Self-powered systems have been employed to treat muscle function loss. TENG is used for rehabilitative and therapeutic purposes through a flexible multiple-channel intramuscular electrode, boosting the electrical muscle stimulation (Wang et al., 2019). According to Long et al., NGs are applied for skin wound healing through electrical stimulations, which have been known to be effective for this field, by converting the mechanical displacement into electricity. The results showed electric field-facilitated fibroblast migration, proliferation, and transdifferentiations (Long et al., 2018).

Guo et al. (2016) performed a hybrid system of TENG, PEDOT, and reduced graphene oxide (rGO) hybrid microfiber as a scaffold. Mesenchymal stem cells cultured on this highly conductive rGO-PEDOT hybrid microfiber possess enhanced proliferation ability and good neural differentiation tendency. Besides, through the electric pulses generated by TENGs by mechanical movements the system is improved (Guo et al., 2016).

8.5.2.4 Drug Delivery System (DDS)

Drug delivery system (DDS) is a formulation or a device that intervenes in substance delivery. It promotes a therapeutic substance to selectively reach its site of action (Rojo, Sousa-Herves, & Mascaraque, 2017). NGs are known as a promising technology in DDS field. For instance, TENG act as a self-powered source for electrochemical microfluidic during DDS. TENG converts the mechanical energy into electricity that is used in medicine as implantable systems.

Hybrid materials can enhance the results of DDS. A synergetic combination of DDS with membrane, microneedle, electrochemistry, microfluidics, and electroporation techniques is highly recommended (Zhao et al., 2021). For instance, Ouyang et al. (2019) reported a miniaturized TENG and a homebuilt power management circuit that were designed to trigger the electric-responsive drug carrier for controlled drug release and activate the iontophoresis treatment for enhanced drug delivery efficiency. The results showed a 3 $\mu g/cm^2$ of transdermal drug release during 1.5 min (Ouyang et al., 2019).

Another example of hybrid system consisted of PTFE and triboelectric layers, which were combined to harvest energy and generate high output voltage (1200 V). The electricity produced was employed in a hydrogel drug patch attached to the skin of the ankle, where the positively charged drug moved from one side to the other (Wu et al., 2019). This new TENG device was employ in transdermal DDS.

The hybrid systems of poly(3-hexythiophene) (P3HT) films and TENGs are highly recommended for DDS. By harvesting the biomechanical energy from clapping motions, the use of TENG provide a stable voltage supply for sustained drug release, with the advantage of adjust the unique switchable wettability of films (Liu et al., 2020).

8.5.3 Textiles

Electrospinning technology has been investigated in these textile applications. It is a method to produce ultrafine fibers by charging and ejecting a polymer melt or solution through a spinneret under a high-voltage electric field and to solidify or coagulate it to form a filament (Zhong, 2016). Textile products of great interest include individual fibers, yarns, fabrics, apparel household textiles, and furnishings and are hierarchical structures (Carr, 2017). Yarn can be incorporated with separate electrodes, and a triboelectric material, that can harvest energy.

Ma et al. (2020) used electrospinning technology to manufacture a HNGs of a single-electrode triboelectric yarn with a spiral core-shell structure, based on PVDF and polyacrylonitrile (PAN) hybrid nanofibers (Ma et al., 2020). Guo et al. (2018) developed a textile-based hybrid TENGs-PENGs. Silk fibroin nanofibers and PVDF nanofibers are electrospun on the conductive fabrics, respectively. To optimize the reaction between two effects and enhance the output performance, the current directions of triboelectricity and piezoelectricity are accordant by finding the reasonable polarization direction. Surface-modified Li-ZnO nanowires are induced into PVDF to increase its piezoelectric response (Guo et al., 2018). Significant results are obtained through this process.

Mokhtari et al. (2020) developed a lightweight NG package based on PVF2/LiCl electrospun nanofibers. The piezoelectric properties of the NG developed were studied to evaluate the influence of the thickness of the as-spun mat felt on the output voltage through vibration and shock tests (Mokhtari et al., 2020). Besides, Ye et al. (2019) reported a hybrid system that harvested textile using different fibers such as natural silk, PTFE, and stainless-steel (SSF), becoming into TENG yarn with a core-shell structure (Ye et al., 2019).

8.5.4 Sensor

The first sensors based on NGs were applied to self-actuated pressure and tactile sensors. A TENG sensor is sensitive at low-pressure, while a PENG sensor works effectively at the high-pressure environment (Li et al., 2020).

PDs have found a huge range of applications in products ranging from fire alarms to intruder detectors, gas analysis and characterization, and in military/paramilitary applications such as thermal imaging (Whatmore & Watton, 2001). One of the most popular applications of IR pyroelectric sensors are the detection and monitoring of gases. IR sensors are of great interest during the characterization of new materials previously synthesized. PDs can convert the energy from the IR to heat. As a result, PDs are more sensitive in a large spectral bandwidth, sensitive in a very wide temperature range, fast response, and cheaper (Lang & Das-Gupta, 2001). These sensors operate by using an IR transmitter to direct IR radiation through a gas sample, and then detect whether a specific IR wavelength is received on the other side. If it is not received, there must be a gas that absorbs that wavelength in the sample. The sensor is tuned to a specific wavelength using an optical IR filter on the sensor, which allows only the desired wavelength to reach the sensing element.

The HNGs can get energy from mechanical energy based on human movements. According to Xu et al. (2016), a hybrid material containing polyvinyl alcohol (PVA)-based hydrogels was synthesized in conjunction with TENGs. These self-powered materials get the energy from human motions including bending, twisting, stretching (Xu et al., 2016). Hansen et al. fabricated a hybrid piezoelectric based on PVDF, PENG, and a flexible biofuel cell. This HNG collects mechanical and biochemical energy to power a single nanowire-based UV sensor as a self-powered electronic device (Hansen et al., 2010). Besides, PVDF was used as a self-powered pedometer (Ishida et al., 2013), where it acted either as a footstep sensor or a gatherer of mechanical energy. Lee et al. reported a hybrid generator developed with TENG and PENG, which is capable of harvesting the vibrational energy from human walking. The proposed structure consisted of two PVDF films, three Al electrodes, and two acrylic supports (Lee et al., 2020). This wearable device can generate enough energy to light up LEDs based on normal walking. Fang et al. (2020) developed a self-powered sensor with piezoelectric and triboelectric elements. The sensor obtained was useful to extract energy from biomechanical motions. The triboelectric sensor was obtained by contact between Cu and PDMS, while the piezoelectric sensor was achieved by depositing PVDF fiber arrays on the substrate (Shu Fang et al., 2020).

He et al. (2018) fabricated an HNG using ZnO nanoflakes (NFs)/PDMS composite films through a facile, cost-effective fabrication method. This innovative design can scavenge ambient mechanical energy from human motions in our daily life (He et al., 2018).

An HNG composed of TENGs-poly(caprolactone), graphene oxide, and cellulose was studied. In this case, higher performance was associated with nanopores structure and the negative strengthened charges on the caprolactone and to the functional groups of graphene oxide (Parandeh, Kharaziham, & Karimzadeh, 2019). Significant performance, low costs, and eco-friendliness were demonstrated through this process.

Wang et al. (2020) reported an HNG composed of one piezoelectric layer of BCZTO/PDMS and a triboelectric layer of BZTO/PDMS. The HNG presented a higher output voltage and current density (Wang et al., 2020). The results provide a novel alternative to self-powered wearable devices.

An HNG based on lead-free $ZNSnO_3$ nanocubes and PDMS-based TENGs. The results indicated a maximum electrical output signal up to 400 V and 28 μA at a current density of 7 $\mu A/cm^2$. The harvested energy is utilized to drive 106 blue LEDs (Yi et al., 2015). This process is highly suitable for large-scale and easy device fabrications.

8.5.5 Antimicrobial Adherence

Bacterial contamination is a huge problem for the environment. Therefore, medical devices with antimicrobial properties are urgent. Biofouling is the development of a colony of microorganisms embedded in a matrix of an organic polymer film formed on heat-exchange surfaces (Panchal & Knudsen, 1998). To understand the phenomenon of biofouling, it is required to know about microbial biofilm development (Rao, 2015).

Polypropylene (PP) is nontoxic and biocompatible with a wide range of applications. Its surface is readily contaminated. However, nanofiller is not suitable for the long-term antimicrobial application. Zhao et al. reported the use of water-wave-driven TENG for biofouling prevention on the insulating surface (Zhao et al., 2016). The electrostatic-induction-reduced adherence of microorganisms prevents biofilm formation (Niu et al., 2013).

Gu et al. (2017) reported a hybrid composite-film-based TENG with antibacterial activity, which employed zeolite material (dielectric permittivity enhanced), silver (Ag^+) (antibacterial property) ions, and PP composite films. The composite films exhibited excellent antibacterial results against *Escherichia coli* and fungus (Gu et al., 2017). This work is an example of new design to develop new self-powered antibacterial devices.

The prevention of infections caused by microbial strains is still a great problem for medicine. Peng et al. developed a new way for mimicking the human sense as electronic skin using hybrid systems. By sandwiching the silver nanowire electrode between polylactic-*co*-glycolic acid and PVA substrate, they showed feasible applications for the electronic skin against *E. coli* and *Staphylococcus aureus* (Peng et al., 2020). The results indicate bacterial growth inhibition. Another hybrid system contains genetically engineering recombinant spider silk proteins and TENG. This technique was improved using water lithography. The patch-free drug was performed, showing potential antimicrobial activity against *E. coli* and *S. aureus*, either in vitro and in vivo (Zhang et al., 2018).

8.6 CONCLUSION

HNGs are eco-friendly materials that give a real alternative to harvesting energy. The reported studies mainly use pyroelectricity, piezoelectricity, triboelectricity, and electromagnetism to create HNGs based on the combination of two or three of these mechanisms. It is relevant to know the application that is going to be given to the HNG to make a correct design and thus take advantage of most of the available energy. In the same way, we can consider the appropriate characterization techniques to understand the operation of NG and improve energy harvesting. These are suitable materials, for a huge range of fields. The biomedical applications are based on the circulatory system, skin, cell proliferation, DDS. Besides, HNGs can also be applied to sensors intelligence, textiles, and even antimicrobial adherence.

REFERENCES

Air Force Aero Propulsion Laboratory. 1975. Polymeric Pyroelectric Sensors for Fire Protections. Air Force Aero Propulsion Laboratory. Retrieved from https://www.electronicsforu.com/electronics-projects/pyroelectric-fire-alarm

Alam, M.M. et al. 2017. "Electroactive β-crystalline phase inclusion and photoluminescence response of a heat-controlled spin-coated PVDF/TiO_2 free-standing nanocomposite film for a nanogenerator and an active nanosensor". Nanotechnology 28 (36): 365401. doi:10.1088/1361-6528/aa7b25.

Alfredo, V. 2000. Novel piezoelectric transducers for high voltage measurements. Barcelona: Universitat Politècnica de Catalunya.

Arunachalam, V.S. & Fleischer, E.L. 2008. "The global energy landscape and materials innovation". MRS Bulletin 33 (4): 264–288. doi:10.1557/mrs2008.61.

Badiali, J.P. & Goodisman, J. 1975. "The Lippmann equation and the ideally polarizable electrode". Chemistry – Faculty Scholarship 73: 223–232.

Beeby, S.P.; Tudor, M.J. & White, N.M. 2006. "Energy harvesting vibration sources for microsystems applications". Measurement Science and Technology 17 (12): R175–R195. doi:10.1088/0957-0233/17/12/r01.

Bikramjit, B. 2014. Nanometer-thick oxide films for pyroelectric energy conversion. Urbana, IL: University of Illinois, 1–2. http://hdl.handle.net/2142/50524.

Bok, M. et al. 2018. "Microneedles integrated with a triboelectric nanogenerator: an electrically active drug delivery system". Nanoscale 10 (28): 13502–13510. doi:10.1039/c8nr02192a.
Bowen, C.R. et al. 2014. "Piezoelectric and ferroelectric materials and structures for energy harvesting applications". Energy & Environmental Science 7 (1): 25–44. doi:10.1039/c3ee42454e.
Bowen, C.R. et al. 2014. "Pyroelectric materials and devices for energy harvesting applications". Energy & Environmental Science 7 (12): 3836–3856. doi:10.1039/c4ee01759e.
Bunaciu, A.A. et al. 2015. "X-ray diffraction: instrumentation and applications". Critical Reviews in Analytical Chemistry 45 (4): 289–299. doi:10.1080/10408347.2014.949616.
Cady, W.G. 1946. Piezoelectricity: an introduction to the theory and applications of electromechanical phenomena in crystals. New York, NY: McGraw-Hill.
Cao, R. et al. 2018. "Self-powered nanofiber-based screen-print triboelectric sensors for respiratory monitoring". Nano Research 11 (7): 3771–3779. doi:10.1007/s12274-017-1951-2.
Carlson, B.M. 2019. "The circulatory system". The Human Body: 271–301. doi:10.1016/b978-0-12-804254-0.00010-7.
Carr, D.J. 2017. "Fibres, yarns and fabrics". Forensic Textile Science: 3–14. doi:10.1016/b978-0-08-101872-9.00001-7.
Cha, G. & Ju, Y.S. 2013. "Pyroelectric energy harvesting using liquid-based switchable thermal interfaces". Sensors and Actuators A 189: 100–107. doi:10.1016/j.sna.2012.09.019.
Chen, J. et al. 2015. "Enhancing performance of triboelectric nanogenerator by filling high dielectric nanoparticles into sponge PDMS film". ACS Applied Materials & Interfaces 8 (1): 736–744. doi:10.1021/acsami.5b09907.
Chen, S. et al. 2016. "Quantifying energy harvested from contact-mode hybrid nanogenerators with cascaded piezoelectric and triboelectric units". Advanced Energy Materials 7 (5): 1601569. doi:10.1002/aenm.201601569.
Chen, X. et al. 2010. "1.6 V nanogenerator for mechanical energy harvesting using PZT nanofibers". Nano Letters 10 (6): 2133–2137. doi:10.1021/nl100812k.
Chen, X. et al. 2016. "A flexible piezoelectric-pyroelectric hybrid nanogenerator based on P (VDF-TrFE) nanowire array". IEEE Transactions on Nanotechnology 15 (2): 295–302. doi:10.1109/TNANO.2016.2522187.
Collins, S.M. & Dias, K.J. 2014. "Cardiac system". Acute care handbook for physical therapists. St. Louis, MI: Elsevier, 15–51. doi:10.1016/b978-1-4557-2896-1.00003-2.
Curie, J. & Curie, P. 1880. "Développement par compression de l'électricité polaire dans les cristaux hémièdres à faces inclinées". Bulletin de la Société minéralogique de France 3 (4): 90–93. doi:10.3406/bulmi.1880.1564.
Diaz, A.F. & Felix-Navarro, R.M. 2004. "A semi-quantitative tribo-electric series for polymeric materials: the influence of chemical structure and properties". Journal of Electrostatics 62 (4): 277–290. doi:10.1016/j.elstat.2004.05.005.
Eichholz, D.E. 1958. "Theophrastus, De Lapidibus – Earle R. Caley and John F. C. Richards: theophrastus, on stones. Introduction, Greek text, English translation and commentary. Pp. vii + 238. Columbus, Ohio: Ohio State University, 1956. Cloth, $6". The Classical Review 8 (1): 38–39. doi:10.1017/s0009840x0016370x.
Fan, F.-R. et al. 2012. "Transparent triboelectric nanogenerators and self-powered pressure sensors based on micropatterned plastic films". Nano Letters 12 (6): 3109–3114. doi:10.1021/nl300988z.
Fan, F.-R.; Tian, Z.-Q. & Lin Wang, Z. 2012. "Flexible triboelectric generator". Nano Energy 1 (2): 328–334. doi:10.1016/j.nanoen.2012.01.004.
Feng, H. et al. 2018. "Nanogenerator for biomedical applications". Advanced Healthcare Materials 7 (10): 1701298. doi:10.1002/adhm.201701298.
Fritsch, R.J. & Krause, I. 2003. "Electrophoresis". Encyclopedia of food sciences and nutrition. San Diego, CA: Elsevier, 2055–2062. doi:10.1016/b0-12-227055-x/01409-7.
Gaffney, J.S. et al. 2002. "Fourier transform infrared (FTIR) spectroscopy". Characterization of Materials: 1–33. doi:10.1002/0471266965.com107.pub2.
Gilliland, G.D. 1997. "Photoluminescence spectroscopy of crystalline semiconductors". Materials Science Engineering 18 (3–6): 99–399. doi:10.1016/S0927-796X(97)80003-4.
Gishiwa, M. 2016. Mechanism of triboelectricity: a novel perspective for studying contact electrification based on metal-polymer and polymer-polymer interactions. Ankara, Turkey: University of Bilkent.
Gong, S. et al. 2020. "Biocompatible poly(lactic acid)-based hybrid piezoelectric and electret nanogenerator for electronic skin applications". Advanced Functional Materials 30 (14): 1908724. doi:10.1002/adfm.201908724.

Gu, G.Q. et al. 2017. "Antibacterial composite film-based triboelectric nanogenerator for harvesting walking energy". ACS Applied Materials & Interfaces 9 (13): 11882–11888. doi:10.1021/acsami.7b00230.
Guo, Y. et al. 2018. "All-fiber hybrid piezoelectric-enhanced triboelectric nanogenerator for wearable gesture monitoring". Nano Energy 48: 152–160. doi:10.1016/j.nanoen.2018.03.033.
Guo, W. et al. 2016. "Self-powered electrical stimulation for enhancing neural differentiation of mesenchymal stem cells on graphene–poly(3,4-ethylenedioxythiophene) hybrid microfibers". ACS Nano 10 (5): 5086–5095. doi:10.1021/acsnano.6b00200.
Habib, G.B. 2018. "Hypertension". Cardiology secrets. Philadelphia, PA: Elsevier, 369–376. doi:10.1016/b978-0-323-47870-0.00041-6.
Han, S.A. et al. 2018. "Point-defect-passivated MoS_2 nanosheet-based high performance piezoelectric nanogenerator". Advanced Materials 30 (21): 1800342. doi:10.1002/adma.201800342.
Han, M. et al. 2013. "r-Shaped hybrid nanogenerator with enhanced piezoelectricity". ACS Nano 7 (10): 8554–8560. doi:10.1021/nn404023v.
Hansen, B.J. et al. 2010. "Hybrid nanogenerator for concurrently harvesting biomechanical and biochemical energy". ACS Nano 4 (7): 3647–3652. doi:10.1021/nn100845b.
He, H. et al. 2020. "Advances in lead-free pyroelectric materials: a comprehensive review". Journal of Materials Chemistry C 8 (5): 1494–1516. doi:10.1039/c9tc05222d.
He, W. et al. 2018. "Ultrahigh output piezoelectric and triboelectric hybrid nanogenerators based on ZnO nanoflakes/polydimethylsiloxane composite films". ACS Applied Materials & Interfaces 10 (51): 44415–44420. doi:10.1021/acsami.8b15410.
Holden, A.J. 2011. "Applications of pyroelectric materials in array-based detectors". IEEE Transactions on Ultrasonics, Ferroelectrics and Frequency Control 58 (9): 1981–1987. doi:10.1109/tuffc.2011.2041.
Hossain, A. & Rashid, M.H. 1991. "Pyroelectric detectors and their applications". IEEE Transactions on Industry Applications 27 (5): 824–829. doi:10.1109/28.90335.
Hu, S. et al. 2020. "Superhydrophobic liquid–solid contact triboelectric nanogenerator as a droplet sensor for biomedical applications". ACS Applied Materials & Interfaces 12 (36): 40021–40030. doi:10.1021/acsami.0c10097.
Im, Ji-Su, & Il-Kyu Park. 2018. "Mechanically robust magnetic Fe3O4 nanoparticle/polyvinylidene fluoride composite nanofiber and its application in a triboelectric nanogenerator". ACS applied materials interfaces 10 (30): 25660-25665. doi:10.1021/acsami.8b07621.
Ishida, K. et al. 2013. "Insole pedometer with piezoelectric energy harvester and 2 V organic circuits". IEEE Journal of Solid-State Circuits 48 (1): 255–264. doi:10.1109/jssc.2012.2221253.
Jaffe, B.; Roth, R.S. & Marzullo, S. 1954. "Piezoelectric properties of lead zirconate-lead titanate solid-solution ceramics". Journal of Applied Physics 25 (6): 809–810. doi:10.1063/1.1721741.
Jin, F. et al. 2018. "Flexocaloric response of epitaxial ferroelectric films". Journal of Applied Physics, 2838–2840. doi:10.1063/1.5009121.
Jung, W. et al. 2015. "High output piezo/triboelectric hybrid generator". Scientific Reports 5 (1): 1–6. doi:10.1038/srep09309.
Kang, J.-H. et al. 2017. "Transparent, flexible piezoelectric nanogenerator based on GaN membrane using electrochemical lift-off". ACS Applied Materials Interfaces 9 (12): 10637–10642. doi:10.1021/acsami.6b15587.
Khan, A. et al. 2012. "Piezoelectric nanogenerator based on zinc oxide nanorods grown on textile cotton fabric". Applied Physics Letters 101 (19): 193506. doi:10.1063/1.4766921.
Khassaf, H. et al. 2018. "Flexocaloric response of epitaxial ferroelectric films". Journal of Applied Physics 123 (2): 024102. doi:10.1063/1.5009121.0.100882.
Kar, E. et al. 2018. "MWCNT@ SiO_2 heterogeneous nanofiller-based polymer composites: a single key to the high-performance piezoelectric nanogenerator and X-band microwave shield". ACS Applied Nano Materials 1 (8): 4005–4018. doi:10.1021/acsanm.8b00770.
Katzir, S. 2012. "Who knew piezoelectricity? Rutherford and Langevin on submarine detection and the invention of sonar". Notes and Records of the Royal Society 66 (2): 157. doi:10.1098/rsnr.2011.0049.
Kim, D. et al. 2014. "Self-compensated insulating ZnO-based piezoelectric nanogenerators". Advanced Functional Materials 24 (44): 6949–6955. doi:10.1002/adfm.201401998.
Kim, H. et al. 2012. "Enhancement of piezoelectricity via electrostatic effects on a textile platform". Energy & Environmental Science 5 (10): 8932. doi:10.1039/c2ee22744d.
Ko, Y.J. et al. 2016. "Flexible Pb $(Zr_{0.52}Ti_{0.48})O_3$ films for a hybrid piezoelectric-pyroelectric nanogenerator under harsh environments". ACS Applied Materials & Interfaces 8 (10): 6504–6511. doi:10.1021/acsami.6b00054.Zi,
Koka, A.; Zhou, Z. & Sodano, H.A. 2014. "Vertically aligned $BaTiO_3$ nanowire arrays for energy harvesting". Energy & Environmental Science 7 (1): 288–296. doi:10.1039/c3ee42540a.

Kumar, M. 2018. "Pyroelectric fire alarm | Detailed circuit diagram available". Electronics for you. https://www.electronicsforu.com/electronics-projects/pyroelectric-fire-alarm.
Lang, S.B. & Das-Gupta, D.K. 2001. "Pyroelectricity". Handbook of advanced electronic and photonic materials and devices. Elsevier, 1–55. doi:10.1016/b978-012513745-4/50036-6.
Li, T. et al. 2017. "From dual-mode triboelectric nanogenerator to smart tactile sensor: a multiplexing design". ACS Nano 11 (4): 3950–3956. doi:10.1021/acsnano.7b00396.
Li, Z. et al. 2018. "A hybrid piezoelectric-triboelectric generator for low-frequency and broad-bandwidth energy harvesting". Energy Conversion and Management 174: 188–197. doi:10.1016/j.enconman.2018.08.018.
Lang, S. 2005. "Pyroelectricity: from ancient curiosity to modern imaging tool". Physics Today, 58 (8), 31–36. doi:10.1063/1.2062916.
Larkin, K.T. & Cavanagh, C. 2016. "Hypertension". Encyclopedia of mental health. Morgantown, WV: Elsevier, 354–360. doi:10.1016/b978-0-12-397045-9.00008-2.
Lee, D.W. et al. 2020. "Polarization-controlled PVDF-based hybrid nanogenerator for an effective vibrational energy harvesting from human foot". Nano Energy 76: 105066. doi:10.1016/j.nanoen.2020.105066.
Lee, J. et al. 2013. "Highly stretchable piezoelectric-pyroelectric hybrid nanogenerator". Advanced Materials 26 (5), 765–769. doi:10.1002/adma.201303570.
Lepak, L.V. 2007. "Localized inflammation". Physical rehabilitation. St. Louis, MI: Elsevier, 117–139. doi:10.1016/b978-072160361-2.50010-7.
Li, P.; Ryu, J. & Hong, S. 2020. "Piezoelectric/triboelectric nanogenerators for biomedical applications". Nanogenerators. doi:10.5772/intechopen.90265.
Li, X. et al. 2014. "3D fiber-based hybrid nanogenerator for energy harvesting and as a self-powered pressure sensor". ACS Nano 8 (10): 10674–10681. doi:10.1021/nn504243j.
Li, Z. et al. 2020. "Nanogenerator-based self-powered sensors for wearable and implantable electronics". Research 2020: 1–25. doi:10.34133/2020/8710686.
Lin, Z.-H. et al. 2013. "A self-powered triboelectric nanosensor for mercury ion detection". Angewandte Chemie 125 (19): 5169–5173. doi:10.1002/ange.201300437.
Liu, D. et al. 2019. "A constant current triboelectric nanogenerator arising from electrostatic breakdown". Science Advances 5 (4): eaav6437. doi:10.1126/sciadv.aav6437.
Liu, G. et al. 2020. "Flexible drug release device powered by triboelectric nanogenerator". Advanced Functional Materials 30 (12): 1909886. doi:10.1002/adfm.201909886.
Liu, L. et al. 2018. "Self-powered versatile shoes based on hybrid nanogenerators". Nano Research 11 (8): 3972–3978. doi:10.1007/s12274-018-1978-z.
Liu, W. & Ren, X. 2009. "Large piezoelectric effect in Pb-free ceramics". Physical Review Letters. doi:10.1103/PhysRevLett.103.257602.
Long, Y. et al. 2018. "Effective wound healing enabled by discrete alternative electric fields from wearable nanogenerators". ACS Nano 12 (12): 12533–12540. doi:10.1021/acsnano.8b07038.
Lu, M.-P. et al. 2009. "Piezoelectric nanogenerator using p-type ZnO nanowire arrays". Nano Letters 9 (3): 1223–1227. doi:10.1021/nl900115y.
Ma, L. et al. 2020. "Continuous and scalable manufacture of hybridized nano-micro triboelectric yarns for energy harvesting and signal sensing". ACS Nano 14 (4): 4716–4726. doi:10.1021/acsnano.0c00524.
Mohtasham, J. 2015. "Review article-renewable energies". Energy Procedia 74: 1289–1297. doi: 10.1016/j.egypro.2015.07.774.
Mokhtari, F. et al. 2020. "Nanofibers-based piezoelectric energy harvester for self-powered wearable technologies". Polymers 12 (11): 2697. doi:10.3390/polym12112697.
Ng, I.C. et al. 2019. "Anatomy and physiology for biomaterials research and development". Encyclopedia of biomedical engineering. Amsterdam: Elsevier, 225–236. doi:10.1016/b978-0-12-801238-3.99876-3.
Niu, S. et al. 2013. "Theoretical study of contact-mode triboelectric nanogenerators as an effective power source". Energy & Environmental Science 6 (12): 3576. doi:10.1039/c3ee42571a.
Niu, S. et al. 2013. "Theory of sliding-mode triboelectric nanogenerators". Advanced Materials 25 (43): 6184–6193. doi:10.1002/adma.201302808.
Olenick, R., Apostol, T. & Goodstein, D. 1986. Beyond the mechanical universe: from electricity to modern physics. Cambridge: Cambridge University Press.
Ouyang, Q. et al. 2019. "Self-powered, on-demand transdermal drug delivery system driven by triboelectric nanogenerator". Nano Energy 62: 610–619. doi:10.1016/j.nanoen.2019.05.056.
Pan, C. et al. 2011. "Fiber-based hybrid nanogenerators for/as self-powered systems in biological liquid". Angewandte Chemie International Edition 50 (47): 11192–11196. doi:10.1002/anie.201104197.

Panchal, C.B. & Knudsen, J.G. 1998. "Mitigation of water fouling: technology status and challenges**work supported by the U.S. Department of Energy, Assistant Secretary of Energy Efficiency and Renewable Energy, and the Office of Industrial Technologies, under Contract W-31-109-Eng-38. Accordingly, the U.S. Government retains a nonexclusive, royalty-free license to publish or reproduce the published form of this contribution, or allow others to do so, for U.S. Government purposes". Advances in Heat Transfer: 431–474. doi:10.1016/s0065-2717(08)70244-2./admi.201600187.

Pang, Y. et al. 2021. "Hybrid energy-harvesting systems based on triboelectric nanogenerators". Matter 4 (1): 116–143. doi:10.1016/j.matt.2020.10.018.

Paradiso, J. & Starner, T. 2005. "Energy scavenging for mobile and wireless electronics". IEEE Pervasive Computing 4 (1): 18–27. doi:10.1109/mprv.2005.9.

Parandeh, S.; Kharaziha, M. & Karimzadeh, F. 2019. "An eco-friendly triboelectric hybrid nanogenerators based on graphene oxide incorporated polycaprolactone fibers and cellulose paper". Nano Energy 59: 412–421. doi:10.1016/j.nanoen.2019.02.058.

Peng, X. et al. 2020. "A breathable, biodegradable, antibacterial, and self-powered electronic skin based on all-nanofiber triboelectric nanogenerators". Science Advances 6 (26): eaba9624. doi:10.1126/sciadv.aba9624.

Qi, Y. & McAlpine, M.C. 2010. "Nanotechnology-enabled flexible and biocompatible energy harvesting". Energy & Environmental Science 3 (9): 1275. doi:10.1039/c0ee00137f.

Quan, T. et al. 2015. "Hybrid electromagnetic–triboelectric nanogenerator for harvesting vibration energy". Nano Research 8 (10): 3272–3280. doi:10.1007/s12274-015-0827-6.

Quan, T.; Wu, Y. & Yang, Y. 2015. "Hybrid electromagnetic–triboelectric nanogenerator for harvesting vibration energy". Nano Research 8 (10): 3272–3280. doi:10.1007/s12274-015-0827-6.

Rao, T.S. 2015. "Biofouling in industrial water systems". Mineral Scales and Deposits: 123–140. doi:10.1016/b978-0-444-63228-9.00006-1.

Rodrigues, C. et al. 2019. "Power-generating footwear based on a triboelectricelectromagnetic-piezoelectric hybrid nanogenerator". Nano Energy 62: 660–666. Elsevier BV. doi:10.1016/j.nanoen.2019.05.063.

Rojo, J.; Sousa-Herves, A. & Mascaraque, A. 2017. "Perspectives of carbohydrates in drug discovery". Comprehensive Medicinal Chemistry III: 577–610. doi:10.1016/b978-0-12-409547-2.12311-x.

Ryu, H.; Yoon, H.-J. & Kim, S.-W. 2019. "Hybrid energy harvesters: toward sustainable energy harvesting". Advanced Materials 31 (34): 1802898. doi:10.1002/adma.201802898.

Saha, C.R. et al. 2008. "Electromagnetic generator for harvesting energy from human motion". Sensors Actuators A 147 (1): 248–253. doi:10.1016/j.sna.2008.03.008.

Seung, W. et al. 2015. "Nanopatterned textile-based wearable triboelectric nanogenerator". ACS Nano 9 (4): 3501–3509. doi:10.1021/nn507221f.

Shi, Q. et al. 2021. "Triboelectric nanogenerators and hybridized systems for enabling next-generation IoT applications". Research 2021: 1–30. doi:10.34133/2021/6849171.

Singh, H.H. & Khare, N. 2018. "Flexible ZnO-PVDF/PTFE based piezo-tribo hybrid nanogenerator". Nano Energy 51:216–222. doi:10.1016/j.nanoen.2018.06.055.

Skoog, D.; Crouch, S. & Holler, J. 2017. Principles of instrumental analysis. Boston, MA: Brooks/Cole.

Sun, J.-G. et al. 2018. "A flexible transparent one-structure tribo-piezo-pyroelectric hybrid energy generator based on bio-inspired silver nanowires network for biomechanical energy harvesting and physiological monitoring". Nano Energy 48: 383–390. doi:10.1016/j.nanoen.2018.03.071.

Sun, R. et al. 2014. "Pyroelectric properties of Mn-doped $94.6Na_{0.5}Bi_{0.5}TiO_3$-$5.4BaTiO_3$ lead-free single crystals". Journal of Applied Physics 115 (7): 074101. doi:10.1063/1.4866327.

Tian, B. et al. 2007. "Coaxial silicon nanowires as solar cells and nanoelectronic power sources". Nature 449 (7164): 885–889. doi:10.1038/nature06181.

Tollefson, J. 2014. "Power from the oceans: blue energy". Nature 508 (7496): 302–304. doi:10.1038/508302a.

Wang, J. et al. 2019. "Self-powered direct muscle stimulation using a triboelectric nanogenerator (TENG) integrated with a flexible multiple-channel intramuscular electrode". ACS Nano 13 (3): 3589–3599. doi:10.1021/acsnano.9b00140.

Wang, M. et al. 2017. "Triboelectric nanogenerator based on 317L stainless steel and ethyl cellulose for biomedical applications". RSC Advances 7 (11): 6772–6779. doi:10.1039/c6ra28252k.

Wang, M. et al. 2020. "High-performance triboelectric nanogenerators based on a mechanoradical mechanism". ACS Sustainable Chemistry Engineering 8 (9): 3865–3871. doi:10.1021/acssuschemeng.9b06986.

Wang, S. et al. 2013. "Sliding-triboelectric nanogenerators based on in-plane charge-separation mechanism". Nano Letters 13 (5): 2226–2233. doi:10.1021/nl400738p.

Wang, W. et al. 2020. "Remarkably enhanced hybrid piezo/triboelectric nanogenerator via rational modulation of piezoelectric and dielectric properties for self-powered electronics". Applied Physics Letters 116 (2): 023901. doi:10.1063/1.5134100.

Wang, X. et al. 2007. "Direct-current nanogenerator driven by ultrasonic waves". Science 316 (5821): 102–105. doi:10.1126/science.1139366.
Wang, X. & Yang, Y. 2017. "Effective energy storage from a hybridized electromagnetic-triboelectric nanogenerator". Nano Energy 32: 36–41. doi:10.1016/j.nanoen.2016.12.006.
Wang, Y.-M. et al. 2021. "Fabrication and application of biocompatible nanogenerators". iScience 24 (4): 102274. doi:10.1016/j.isci.2021.102274.
Wang, Z.L. 2006. "Piezoelectric nanogenerators based on zinc oxide nanowire arrays". Science 312 (5771): 242–246. doi:10.1126/science.1124005.
Wang, Z.L. 2020. "On the first principle theory of nanogenerators from Maxwell's equations". Nano Energy 68: 104272. doi:10.1016/j.nanoen.2019.104272.
Wang, Z.L. et al. 2012. "Progress in nanogenerators for portable electronics". Materials Today 15 (12): 532–543. doi:10.1016/s1369-7021(13)70011-7.
Wen, Z. et al. 2016. "Harvesting broad frequency band blue energy by a triboelectric–electromagnetic hybrid nanogenerator". ACS Nano 10 (7): 6526–6534. doi:10.1021/acsnano.6b03293.
Whatmore, R.W. & Watton, R. 2001. "Pyroelectric materials and devices". Infrared detectors and emitters: materials and devices. Boston, MA: Springer, 99–147. doi:10.1007/978-1-4615-1607-1_5.
Wu, C. et al. 2019. "Self-powered iontophoretic transdermal drug delivery system driven and regulated by biomechanical motions". Advanced Functional Materials 30 (3): 1907378. doi:10.1002/adfm.201907378.
Wu, C. et al. 2019. "Triboelectric nanogenerator: a foundation of the energy for the new era". Advanced Energy Materials 9 (1): 1802906. doi:10.1002/aenm.201802906.
Wu, R. et al. 2015. "Metallic WO_2–carbon mesoporous nanowires as highly efficient electrocatalysts for hydrogen evolution reaction". Journal of the American Chemical Society 137 (22): 6983–6986. doi:10.1021/jacs.5b01330.
Wu, Y. et al. 2015. "Hybrid energy cell for harvesting mechanical energy from one motion using two approaches". Nano Energy 11: 162–170. doi:10.1016/j.nanoen.2014.10.035.
Xia, K. et al. 2018. "Sliding-mode triboelectric nanogenerator based on paper and as a self-powered velocity and force sensor". Applied Materials Today 13: 190–197. doi:10.1016/j.apmt.2018.09.005.
Xu, W. et al. 2016. "Environmentally friendly hydrogel-based triboelectric nanogenerators for versatile energy harvesting and self-powered sensors". Advanced Energy Materials 7 (1): 1601529. doi:10.1002/aenm.201601529.
Xue, H. et al. 2017. "A wearable pyroelectric nanogenerator and self-powered breathing sensor". Nano Energy 38: 147–154. doi:10.1016/j.nanoen.2017.05.056.
Yan, C. et al. 2020. "A linear-to-rotary hybrid nanogenerator for high-performance wearable biomechanical energy harvesting". Nano Energy 67: 104235. doi:10.1016/j.nanoen.2019.104235.
Yang, B. et al. 2016. "A fully verified theoretical analysis of contact-mode triboelectric nanogenerators as a wearable power source". Advanced Energy Materials 6 (16): 1600505. doi:10.1002/aenm.201600505.
Yang, H. et al. 2019. "A nonencapsulative pendulum-like paper–based hybrid nanogenerator for energy harvesting". Advanced Energy Materials 9 (33): 1901149. doi:10.1002/aenm.201901149.
Yang, N.; Ray, S.D. & Krafts, K. 2014. "Cell proliferation". Encyclopedia of toxicology. London: Elsevier, 761–765. doi:10.1016/b978-0-12-386454-3.00274-8.
Yang, X. & Daoud, W.A. 2016. "Triboelectric and piezoelectric effects in a combined tribo-piezoelectric nanogenerator based on an interfacial ZnO nanostructure". Advanced Functional Materials 26 (45): 8194–8201. doi:10.1002/adfm.201602529.
Yang, Y. et al. 2012. "Flexible hybrid energy cell for simultaneously harvesting thermal, mechanical, and solar energies". ACS Nano 7 (1): 785–790. doi:10.1021/nn305247x.
Yang, Y. et al. 2012. "Flexible pyroelectric nanogenerators using a composite structure of lead-free $KNbO_3$ nanowires". Advanced Materials 24 (39): 5357–5362. doi:10.1002/adma.201201414.
Yang, Y. et al. 2012. "Pyroelectric nanogenerators for driving wireless sensors". Nano Letters 12 (12): 6408–6413. doi:10.1021/nl303755m.
Yang, Y. et al. 2012. "Pyroelectric nanogenerators for harvesting thermoelectric energy". Nano Letters 12 (6): 2833–2838. doi:10.1021/nl3003039.
Ye, C. et al. 2019. "Ultrastable and high-performance silk energy harvesting textiles". Nano-Micro Letters 12 (1). doi:10.1007/s40820-019-0348-z.
Yunlong et al. 2015. "Triboelectric-pyroelectric-piezoelectric hybrid cell for high-efficiency energy-harvesting and self-powered sensing". Advanced Materials 27 (14): 2340–2347. doi:10.1002/adma.201500121.

Zhang, B. et al. 2016. "Rotating-disk-based hybridized electromagnetic–triboelectric nanogenerator for sustainably powering wireless traffic volume sensors". ACS Nano 10 (6). 6241–6247. doi:10.1021/acsnano.6b02384.
Zhang, C. et al. 2014. "Theoretical comparison, equivalent transformation, and conjunction operations of electromagnetic induction generator and triboelectric nanogenerator for harvesting mechanical energy". Advanced Materials 26 (22): 3580–3591. doi:10.1002/adma.201400207.
Zhang, H. et al. 2013. "Triboelectric nanogenerator built inside clothes for self-powered glucose biosensors". Nano Energy 2 (5): 1019–1024. doi:10.1016/j.nanoen.2013.03.024.
Zhang, Q. et al. 2018. "Electromagnetic shielding hybrid nanogenerator for health monitoring and protection". Advanced Functional Materials 28 (1): 1703801. doi:10.1002/adfm.201703801.
Zhang, Y. et al. 2018. "'Genetically engineered' biofunctional triboelectric nanogenerators using recombinant spider silk". Advanced Materials 30 (50): 1805722. doi:10.1002/adma.201805722.
Zhang, Z. et al. 2019. "A portable triboelectric nanogenerator for real-time respiration monitoring". Nanoscale Research Letters 14 (1): 1–11. doi:10.1186/s11671-019-3187-4.
Zheng, L. et al. 2014. "Silicon-based hybrid cell for harvesting solar energy and raindrop electrostatic energy". Nano Energy 9: 291–300. doi:10.1016/j.nanoen.2014.07.024.
Zhao, C. et al. 2021. "Recent progress of nanogenerators acting as self-powered drug delivery devices". Advanced Sustainable Systems: 2000268. doi:10.1002/adsu.202000268.
Zhao, C. et al. 2019. "Hybrid piezo/triboelectric nanogenerator for highly efficient and stable rotation energy harvesting". Nano Energy 57: 440–449. doi:10.1016/j.nanoen.2018.12.062.
Zhao, K. & Zhu, Z. 2020. "Application of triboelectric nanogenerator in smart home and clothing". Frontiers in Mechanical Engineering 6. doi:10.3389/fmech.2020.576896.
Zhong, W. 2016. "Nanofibres for medical textiles". Advances in Smart Medical Textiles: 57–70. doi:10.1016/b978-1-78242-379-9.00003-7.
Zhou, L. et al. 2020. "Triboelectric nanogenerators: fundamental physics and potential applications". Friction 8 (3): 481–506. doi:10.1007/s40544-020-0390-3.
Zhou, W.; Apkarian, R.; Lin Wang, Z. & Joy, D. 2006. "Fundamentals of scanning electron microscopy (SEM)". Scanning Microscopy for Nanotechnology: 1–40. doi:10.1007/978-0-387-39620-0_1.
Zhu, G. et al. 2015. "Triboelectric nanogenerators as a new energy technology: from fundamentals, devices, to applications". Nano Energy 14: 126–138. doi:10.1016/j.nanoen.2014.11.050.
Zhu, J. et al. 2017. "The d-arched piezoelectric-triboelectric hybrid nanogenerator as a self-powered vibration sensor". Sensors Actuators A 263: 317–325. doi:10.1016/j.sna.2017.06.012.
Zi, Y. & Wang, Z.L. 2017. "Nanogenerators: an emerging technology towards nanoenergy". APL Materials 5 (7): 074103. doi:10.1063/1.4977208.

9 Pyroelectric Nanogenerators in Energy Technology

Ampattu R. Jayakrishnan, José P. B. da Silva, Sugumaran Sathish, Koppole Kamakshi, and Koppole C. Sekhar

CONTENTS

9.1 INTRODUCTION

International crisis such as global warming, environmental pollution, and e-waste significantly increased the demand of electronic devices powered by renewable and sustainable energy sources [1, 2]. It is known that batteries are key renewable energy sources for powering electronic systems, which suffer from drawbacks like environmental issues, limited life-time, weight, and periodic maintenance or replacement [3]. Therefore, energy harvesting from the most reliable sources such as green energy (solar, wind, biomass, and hydrogen) and mechanical energy (water flow, mechanical vibrations, and human motions) are inevitable. In our daily life, different forms of energies like thermal, mechanical, and photovoltaic are scattered around or wasted. Thus, commercialization of wasted energy into electrical energy has become imperative to build a clean and sustainable world [4, 5]. Moreover, thermal energy exists everywhere in the form of sun light, wind, human body, etc. [6–8]. For instance, our human body is an indispensable source of thermal energy, and we are simply wasting it without knowing the importance. However, harvesting of electrical energy from thermal energy requires a thermal fluctuation or temperature gradient. For instance, the human respiration and locomotion create fluctuations in the thermal energy and can be converted into electrical energy [7, 8]. The process of generating electrical energy using the temperature gradient is known as pyroelectric effect or pyroelectricity [8]. Besides, the pyroelectric materials can generate electrical energy from the temperature fluctuations induced due to small-scale physical changes. These pyroelectric materials that can generate electrical energy using small physical-change-induced thermal fluctuation are known as pyroelectric nanogenerators (PENGs). Converting such a feasible energy source to electrical energy using PENGs under ambient conditions is of great importance for long-term self-powered energy systems. Figure 9.1 illustrates the schematic representation of various applications of PENGs.

DOI: 10.1201/9781003187615-9

FIGURE 9.1 Schematic representation of the conversion of wasted energy into electrical energy for various applications.

Therefore, the present chapter highlights the progress and achievements of PENGs in energy technology. A brief discussion about pyroelectric materials, pyroelectric effect, and the working principle of PENGs is included. A wide range of pyroelectric materials as PENGs is discussed in detail. Besides, their recent applications in the energy harvesting technology are also highlighted. Finally, a comprehensive analysis of the results, challenges, and future perspectives is also remarked.

9.2 PYROELECTRIC EFFECT

Pyroelectric materials are polar materials that have the ability to generate an electric current or potential when they are heated or cooled, and this phenomenon is known as pyroelectric effect or pyroelectricity [9, 10]. The temperature variation induces a net dipole moment change (or bound charges) and consequently, a gross P_s change. The P_s change establishes charges (electrons and ions) on the surface of the pyroelectric material, and these surface charges are responsible for the pyroelectric nature [11].

The generated current (i) and the net charge $\left(Q_{net}\right)$ of the pyroelectric material corresponding to the temperature change have the following expressions [12, 13]:

$$i = \frac{dQ}{dt} = pA\frac{dT}{dt} \text{ (short circuit)} \tag{9.1}$$

$$Q_{net} = pA\Delta T \text{ (open circuit)} \tag{9.2}$$

where p represents the pyroelectric coefficient, A represent surface area of pyroelectric material, $\frac{dT}{dt}$ is the rate of change of temperature when the circuit is short, and ΔT is the temperature change experienced by Q_{net} in an open circuit [12, 13].

The pyroelectric coefficient is a function of electric field (E) and elastic stress (σ) [14]:

$$p^{(\sigma,E)} = \left(\frac{dP_s}{dt}\right)_{(\sigma,E)} \tag{9.3}$$

A pyroelectric material with constant E and σ is free to expand or contract under a thermal change, then the pyroelectric effect is known as primary pyroelectric effect. On the other hand, when the pyroelectric material (with asymmetric crystal structure) under a thermal change is clamped by a constant stress (κ) to hinder the expansion and compression, the material experiences an electric displacement via piezoelectric effect. Consequently, the electrons flow through the external circuit by the piezoelectric response established under thermal deformation and signifies the primary pyroelectric effect. This is known as secondary pyroelectric effect. In such cases, the pyroelectricity is not only a function of σ and E but also depends on the factors such as the κ and crystal deformation [15, 16]. A schematic representation of the primary and secondary pyroelectric effect is shown in Figure 9.2.

Since the pyroelectric materials are also a kind of dielectric materials, the capacitance is expressed as [16]:

$$C = \frac{A\varepsilon_\sigma}{d} \tag{9.4}$$

where ε_σ is the permittivity under constant stress and d is the thickness.

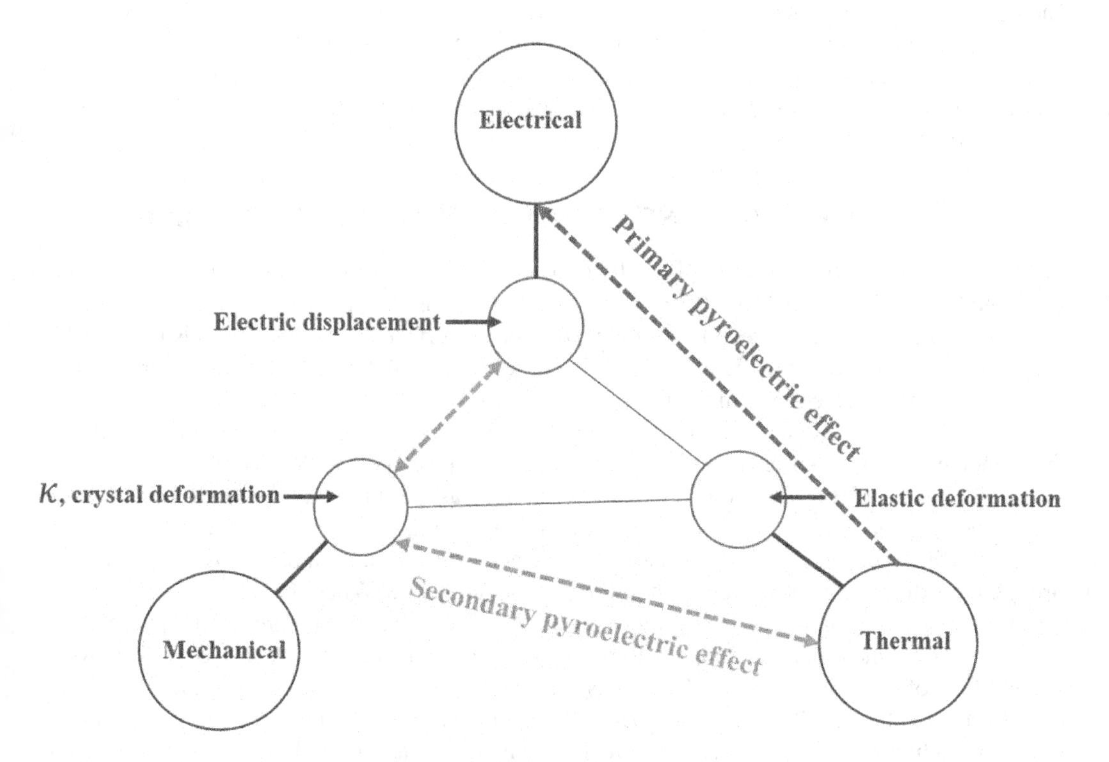

FIGURE 9.2 Schematic representation of primary and secondary pyroelectric effect.

Then, the voltage (V) in an open circuit applied across the electrode is as follows:

$$V = \frac{p}{\varepsilon_\sigma} d\Delta T \tag{9.5}$$

And the corresponding electric field E is

$$E = \frac{V}{d} = \frac{p}{\varepsilon_\sigma} \Delta T \tag{9.6}$$

Therefore, the total energy (U) stored in the pyroelectric material under the temperature change ΔT is given by [5]:

$$U = \frac{1}{2} CV^2 = \frac{1}{2} \frac{p^2}{\varepsilon_\sigma} Ad(\Delta T)^2 \tag{9.7}$$

Furthermore, the real-world applications demand pyroelectric materials with very good figure of merits (FOMs) and can be derived from pyroelectric coefficient as follows [17–20]:

$$F_i = \frac{p}{C_v} \tag{9.8}$$

$$F_v = \frac{p}{C_v \varepsilon_0 \varepsilon_r} \tag{9.9}$$

$$F_D = \frac{p}{C_v \sqrt{\varepsilon_0 \varepsilon_r tan\delta}} \tag{9.10}$$

where F_i is potential current responsivity performance, F_v is voltage responsivity, and F_D is detection capability. Here, C_v corresponds to the volume specific heat and $tan\delta$ the dielectric loss. Therefore, a pyroelectric material to show better pyroelectric sensitivity requires high p, low C_v, low ε_r, and low $tan\delta$. The typical values of pyroelectric coefficient obtained in different pyroelectric materials are summarized in Table 9.1.

9.3 PYROELECTRIC NANOGENERATOR AND ITS WORKING PRINCIPLE

The PENG works on the principle of the thermal change $\left({}^{dT}\!/_{dt}\right)$, induced P_s, and the schematic representation as shown in Figure 9.3 [30]. In ideal conditions, the dipoles align in one direction, and thus, P_s is in the dipole direction. Realistically, such a perfect dipole orientation is not possible owing to the presence of vibrating atoms [30]. The directionless motion of the vibrating or wiggling atoms disturbs the perfection of the dipole orientation. However, both the P_s and the wiggling angle (θ) are constant at a fixed temperature (Figure 9.3(a)). With increasing temperature ($dT/dt > 0$), the dipole moment of the atoms changes due to the thermal change and wiggle around their respective polar axis (Figure 9.3(b)). Consequently, the P_s level decreases at higher temperature and is attributed to the large disturbance ($\theta_1 > \theta$) caused by the wiggling atoms. Thus, there is a reduction in charges (for instance, electrons) on surface of pyroelectric material. As a result, the electrons flow through the external circuit via top electrode to bottom electrode. When the temperature of material brought to its original value, a current flow exists in the reverse direction as the pyroelectric material recovers the dipole alignment. However, when the temperature is decreased ($dT/dt < 0$), a reverse effect of the previous case ($dT/dt > 0$) is expected with a reduced wiggling angle ($\theta_2 < \theta$) and a significantly enhanced P_s level due to reduced temperature change (Figure 9.3(c)). This is attributed to the enhanced surface charges of pyroelectric material and induced the electrons flow from bottom to top electrode. As the temperature retains its initial condition, the pyroelectric material

TABLE 9.1
Materials, Their Corresponding Point Groups, and Pyroelectric Coefficient

Materials	Ferroelectric	Point Group	Pyroelectric Coefficient ($\mu C/m^2$ K)	Reference
$Pb_5Ge_3O_{11}$	Yes	3	95	[15]
$PbZrO_3$	Yes	–	40	[21]
$PbTiO_3$	Yes	–	400	[21]
$PbZr_{0.95}Ti_{0.05}O_3$ (PZT)	Yes	∞m	268	[15]
PZT (Toshiba)	Yes	–	350	[22]
$LiNbO_3$	Yes	$3m$	83	[15]
$LiTaO_3$	Yes	$3m$	176	[15]
$Na_{0.5}La_{0.5}Bi_4Ti_4O_{15}$	Yes	–	74	[23]
$CaBi_4Ti_4O_{15}$	Yes	–	36	[24]
$SrBi_4Ti_4O_{15}$	Yes	–	70	[25]
$BiFeO_3$ (BFO)	Yes	–	30	[26]
$BaTiO_3$ (BTO)	Yes	∞m	200	[15]
$Ba_{0.85}Ca_{0.15}Ti_{0.9}Zr_{0.1}O_3$	Yes	–	584	[27]
ZnO	No	6 mm	9.4	[15]
CdSe	No	6 mm	3.5	[15]
Cds	No	6 mm	4	[15]
Triglycine sulfate TGS	Yes	2	270	[15]
PVDF	Yes	–	27	[28]
P(VDF-TrFE) 50/50	Yes	–	40	[29]

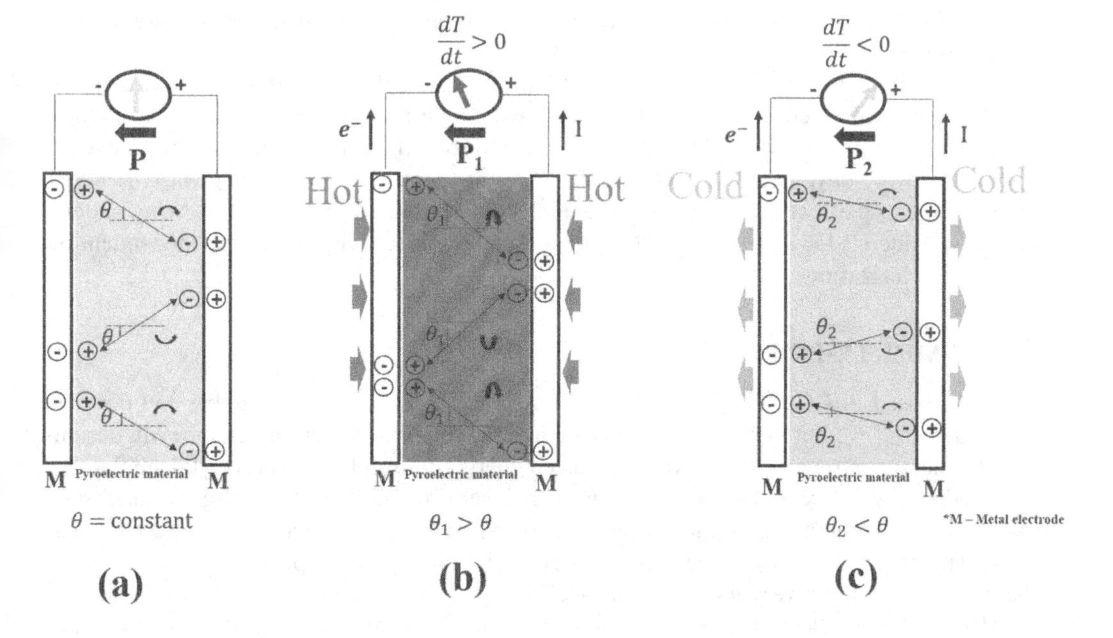

FIGURE 9.3 Working principle of PENG at (a) constant temperature, (b) heating, and (c) cooling.

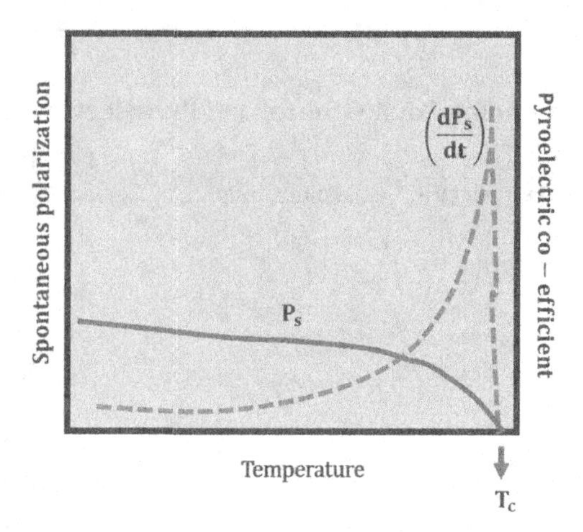

FIGURE 9.4 Spontaneous polarization and pyroelectric coefficient as a function of temperature [31].

regains the dipole alignment through surface charge curtailment. As a consequence, a reverse current flows through the external circuit. In this way, a PENG generates electrical energy and strongly depended on pyroelectric coefficient [30].

From Equation (9.1), the pyroelectric coefficient can be rewritten as follows:

$$p = \frac{i}{A.{}^{dT}\!/_{dt}} \tag{9.11}$$

and the pyroelectric coefficient is one of the important parameters in evaluating the performance of a PENG. Figure 9.4 represents the temperature dependence of P_s and p in a pyroelectric material [31]. It is known that value of P_s decreases slowly with increase in temperature and rapidly diminishes in the vicinity of Curie temperature (T_c). This in turn gives a significant increase in the pyroelectric coefficient. Hence, these materials have a potential of discharging large amount of surface charges and are receiving greater interest in energy harvesting technology at high temperatures [31].

Different sources of energy such as sun, wind, etc. cause fluctuations in heat and can be explored to generate electrical energy. Currently, researchers are concentrated on developing electrical energy from such wasted heat energy. Furthermore, the demand for sustainable energy is increasing day by day. Thus, the method of generating electrical energy using wasted energy through pyroelectric effect can open a new era of electronic devices [2, 3]. The following materials can be considered for developing PENG devices as pyroelectricity is strongly dependent on spontaneous polarization.

9.4 POLYMER-BASED PENGs

The polymer-based PENGs (PPENGs) have unlimited scope in mounting wearable and portable electronic devices [16]. Furthermore, PPENGs are asset to the future electronic industry in designing flexible electronic devices by converting wasted energy to useful one. Polyvinylidene fluoride (PVDF) and poly(vinylidene fluoride-*co*-trifluoroethylene) [P(VDF-TrFE)] are of great interest as PPENGs for flexible device applications like energy harvesters and actuators infrared (IR) sensors [16]. These ferroelectric polymers are very good in maintaining a spontaneous polarization. Furthermore, they exhibit very good pyroelectric and piezoelectric properties. In fact, PVDF and P(VDF-TrFE) have excellent mechanical strength, flexibility, and are biocompatibility, which makes them promising candidates for wearable devices [16].

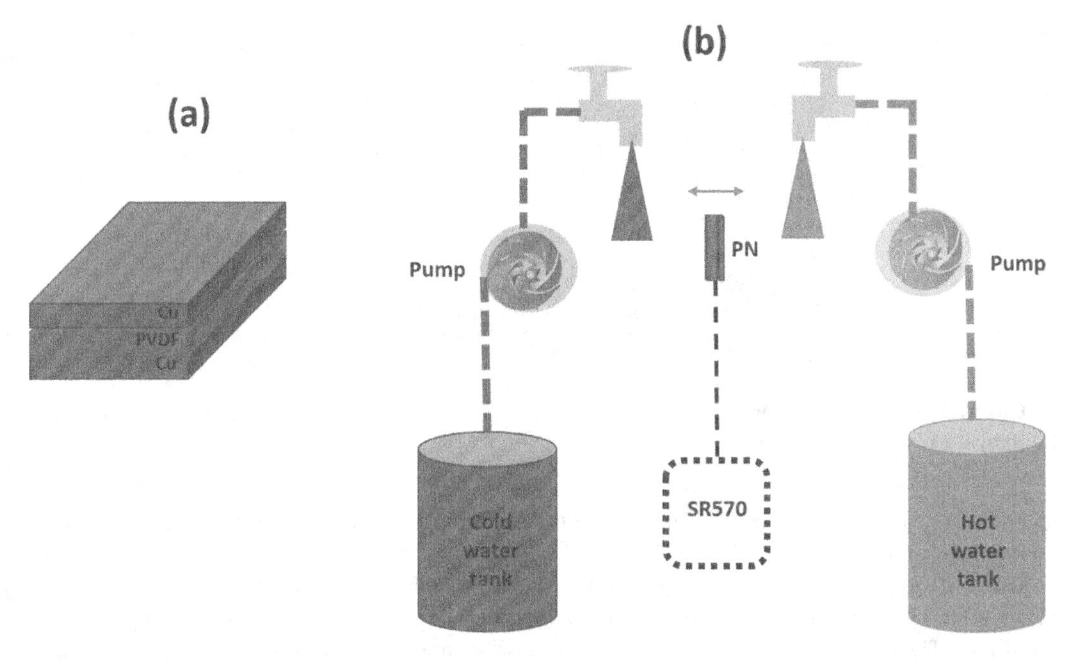

FIGURE 9.5 (a) Schematic diagram and (b) experimental set-up showing the working of PENG under hot- and cold-water flow [30].

In 2014, a PVDF thin-film-based PENG was fabricated by Leng et al., in which the PVDF film (thickness ~110 μm) is sandwiched between two copper (Cu) electrodes as shown in Figure 9.5(a) [30]. In this work, the water is used as the thermal source for generating pyroelectric current. To prevent the Cu/PVDF/Cu device from conductivity and oxidation by the water, the device is covered with polyvinyl chloride (PVC) thin film (30 μm thickness). The PENG is allowed to move right and left via an electric oscillator. This makes the device come in contact with cold and hot water flow. When the contact is made, a temperature change took place and induces a pyroelectric effect and thereby, generating an output current. Since the device is continuously moving under an electric oscillator, an uninterrupted output can be sensed and generated. The working principle is shown schematically in Figure 9.5(b). The device performance is estimated by measuring the temperature gradient developed by the flow of hot water and cold water. For example, the contact between the 40 and 0°C produces a temperature change (ΔT) of 40°C. In this way, the increase in the hot flow created a large temperature change and the PENG produced an increase in output current from 6 to 11 μA. Furthermore, the mechanical deformation due to the expansion and contraction of the PENG enhanced the change in polarization, and thus, the output voltage increased from 100 to 190 V as the temperature of hot water varies from 40 to 80°C. The maximum output power produced by the device is 126 μW. More importantly, the Cu/PVDF/Cu device charged a 100 μF capacitor and lighted 42 light-emitting diodes (LEDs).

Raouadi et al. introduced a PVDF-based energy harvesting PENG by using wind turbulence and vortex generator as the base for generating pyroelectricity [32]. A commercial poled flexible PVDF film (9 μm thickness) with aluminum electrodes deposited on both surfaces through vacuum sputtering were used as the PENG. The schematic representation of the working of the PENG is shown in Figure 9.6. The vortex generator located between the PENG and wind improves the heat transfer between the PENG and wind. As the air is passed through the vortex, a turbulence is formed. This turbulence affects air flow physical properties and causes a temperature change. Such a temperature gradient is proportional to flux of heat and velocity gradient of the wind. Consequently, the pyroelectric effect and, thereby, an uninterrupted energy generation is noticed. The wind-driven PENG

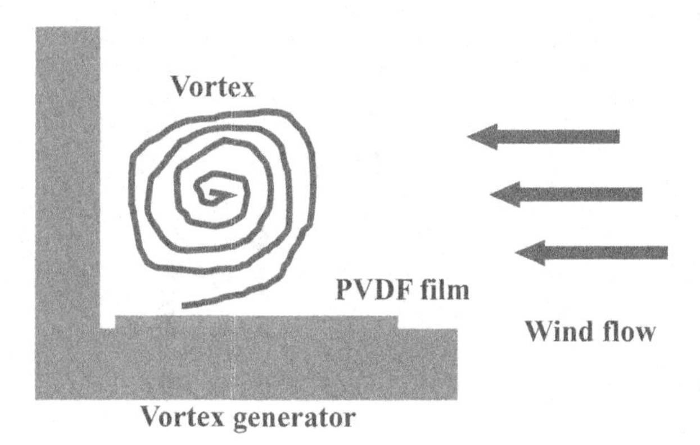

FIGURE 9.6 Vortex generator mechanism in PENG. (Reprinted with permission from [32]. Copyright (2018) by the Elsevier Ltd.)

generated a peak power of 2.8 μW/cm under optimum load resistance. With this power, the PENG operated a sound buzzer as well as a white LED.

Xue et al. prepared a wearable PENG and installed it in an N95 mask to harvest the energy from our breathing [33]. At first, both surface of the PVDF film of size 3.5 cm × 3.5 cm was coated with aluminum. The film is then attached to two copper wires and fixed using silver paste. Afterward, the film is packaged by Kapton tapes. Finally, the film is integrated to an N95 respirator to form a PENG working as a self-powered breathing sensor and temperature sensor from human respiration (Figure 9.7(a and b)). The schematic representation of the working of PENG by human respiration is represented in Figure 9.7(c). The inhaling and exhaling process during human breathing creates a temperature change and generated a peak output current of 2.5 μA and a voltage of 42 V. The maximum output power of 8.31 μW was obtained at a load 50 MΩ. The device successfully charged a 1-V 10-μF capacitor and turned on LEDs and liquid crystal displays (LCDs).

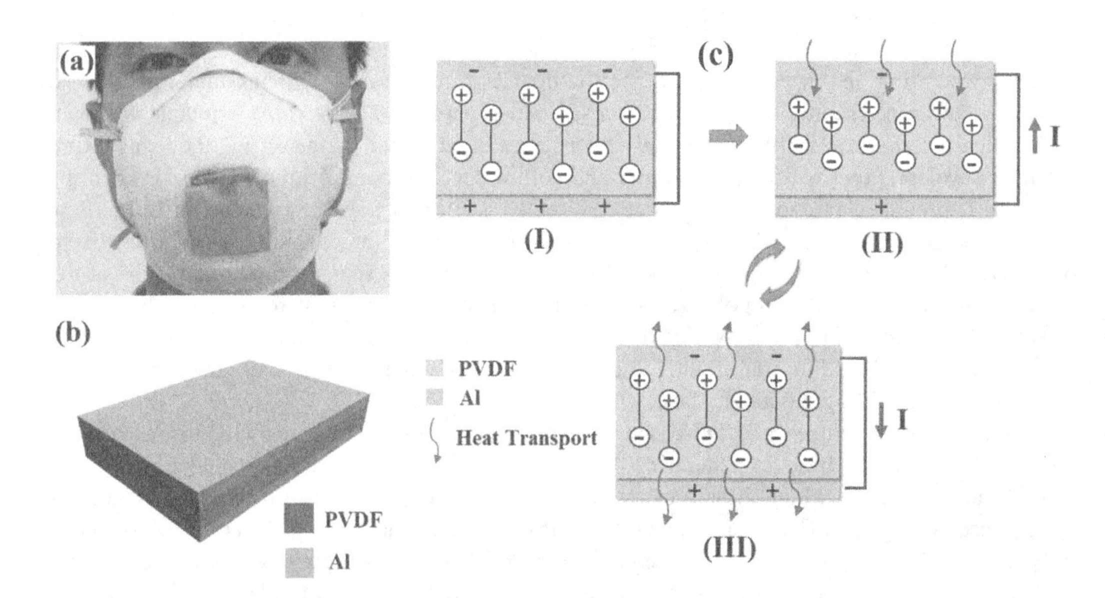

FIGURE 9.7 (a) Wearable PENG, (b) its schematic structure, and (c) working mechanism of PENG: (I) equilibrium state, (II) exhaling, (III) inhaling. (Reprinted with permission from [33]. Copyright (2017) by the Elsevier Ltd.)

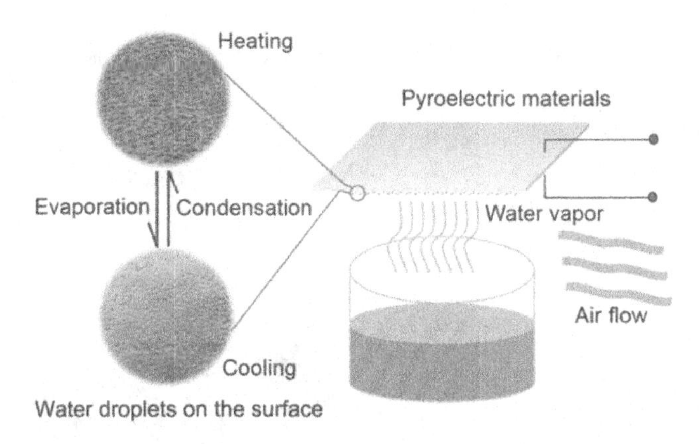

FIGURE 9.8 Schematics of water vapor driven PENG. (Reprinted with permission from [34]. Copyright (2016) by the Elsevier Ltd.)

Gao et al. proposed a combined PVDF and P(VDF-TrFE)-based PPENG operated using the temperature change based on hot water vapor [34]. The ability of water to absorb and release heat by the phenomenon of evaporation and condensation makes them a perfect heat transfer source for the generation of pyroelectricity. Gao et al. prepared a flexible thin film made of PVDF and P(VDF-TrFE), which was then sealed with polyimide (PI) tape. The entire device in maintained in an atmosphere of relative humidity less than 50% and provided with a continuous supply of water vapor from a hot reservoir. A schematic representation of the PENG operated by the water vapor is shown in Figure 9.8. An oscillating airflow of hot water vapor with a temperature of 65°C is permitted to fall on the PENG surface using a small fan with an oscillating flow speed of 1–2 m/s to commence the pyroelectricity. The hot water vapor is quickly cooled to 40°C by the air movement, resulting in a temperature gradient. This develops a pyroelectric current in the PENG through internal polarization change. Thus, the water vapor effectively played the role of heating and cooling for harvesting electricity. Consequently, the PENG produced a peak voltage of 160 V, a peak current of ~5.5 μA, and an optimum output power of ~220 μW. Using this power, the PPENG successfully charged a 2.2 μF capacitor, lighted blue LEDs, and operated a digital watch.

Lee et al. introduced a hybrid stretchable PPENG (S-PPENG) based on P(VDF-TrFE) and polydimethylsiloxane (PDMS) [35]. The P(VDF-TrFE) thin film generates electrical energy using pyroelectricity, while the PDMS induces piezoelectricity through compressive strain as well as the stretchability to the P(VDF-TrFE), when subjected to a high temperature. Thus, a combination of pyroelectric and piezoelectric effect is utilized for the operation of this PENG. A comparison with a normal PPENG (N-PPENG) device is done to understand the remarkable sensitivity of the S-PPENG device even at modest temperature changes. In N-PPENG, the P(VDF-TrFE) is coated onto a flat and rigid Ni/SiO_2/Si substrate. With a temperature change ranging from low ($\Delta T \approx$ 0.64K) to high ($\Delta T \approx$ 18.5K), S-PPENG delivered an output voltage ranging from 8 mV to 2.48 V, while N-PPENG generated an output voltage ranging from 2 mV to 0.54 V. Thus, the S-PPENG shows a clear enhancement of output performance, almost four to five times higher than that of the N-PPENG. Based on micro-patterned P(VDF-TrFE), micro-patterned PDMS, and a stretchable Ag/AgNWs electrode, this work enabled the development of a new type of thermally induced strain coupled highly stretchable and responsive PPENG. The S-PPENG effectively demonstrated the coupling of piezoelectric and pyroelectric effects (hybrid nature) using varied thermal expansion coefficients and micro-patterned designs to dramatically improve power-generating capability. The current density of S-PPENG is found to be 570 nA/cm^2, which suggested a high reactive response to the temperature fluctuations ranging from extremely low to high values. Because of its high flexibility, stretchability, robustness, and mechanical durability, S-PPENG is helpful for prospective device

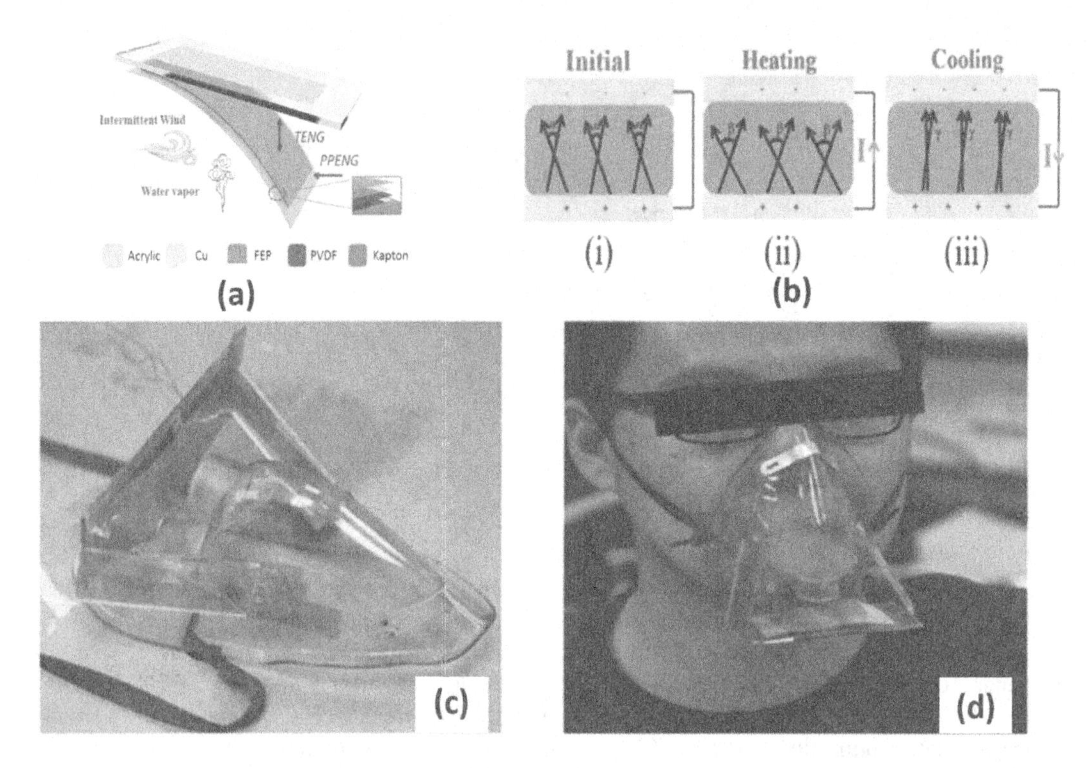

FIGURE 9.9 (a) Schematic representation, (b) mechanism of pyroelectric effect in hybrid PENG, (c) hybrid PENG-integrated face mask, and (d) its demonstration as a wearable device. (Reprinted with permission from [39]. Copyright (2018) by the American Chemical Society Ltd.)

applications such as self-powered wearable devices, and e-skin applications. In another work, Lee et al. proposed a P(VDF-TrFE)-PDMS-carbon nanotube (CNT)-based S-PPENG [36]. A thin layer of graphene is used as the top electrode owing to its excellent flexibility and high thermal conductivity [37, 38]. Furthermore, a silicon master mold is used to fabricate flexible substrate. This device works on the principle of mechanical motion and heat change produced by the human body. The human body provides mechanical deformation and temperature change on the hybrid PENGs. As a result, the hybrid PENG generated an output voltage of 1.4 V, which was increased up to 3.5 times compared to the normal PENG operated under temperature change only. Hence, the S-PPENG demonstrated a stable output performance due to simultaneous effects of pyro and piezo. Therefore, this work can promote the designing of stretchable polymer and graphene-based PENGs for wearable electronic device applications.

Zheng et al. developed a PENG composed of triboelectric, piezoelectric, and pyroelectric effect. Triboelectric effect is one type of generation of electricity. In this case, the materials become electrically charged because of frictional contact with another material. Zheng et al. designed this PENG using a polarized PVDF thin film with Cu electrode on the film (Figure 9.9(a)) [39]. To generate triboelectricity, the PVDF film is covered with fluorinated ethylene propylene (FEP) film. Furthermore, the entire film is covered using a PI tape to avoid moisture contamination. The schematic representation of the wind driven PENG is shown in Figure 9.9(b). When the wind strikes, a frictional contact is formed between the Cu and FEP film. As a result, a triboelectric current is generated in the device due to interaction and dissociation between the electrode and the film. Consequently, a peak output voltage of 350 V and current of 30 μA were generated. Furthermore, the mechanical deformation and the temperature change develop piezoelectricity and pyroelectricity by inducing internal polarization change of PVDF. This enhanced polarization changes due to the hybrid nature (both piezoelectricity and pyroelectricity) enlarged the output to 350 V and

41 μA. The device successfully turned on 31 green LEDs and 24 small white LEDs using human breath. Furthermore, the hybrid nanogenerator was installed to a face mask, so that the energy in the hot air exhaled during human respiration can be effectively harvested using the pyroelectric, piezoelectric, and triboelectric effect for applications in biomedical sensors (Figure 9.9(c and d)). Furthermore, Wang et al. demonstrated tribo-piezo-pyroelectric-based hybrid PENG using PVDF nanowires, PDMS, and indium tin oxide (ITO) electrodes [40]. A flexible nylon contact is made with the PVDF/PDMS composite to induce the triboelectric and piezoelectric current. Under 400 kΩ, a tribo-piezo current is generated with a 15 m/s air flow, and the device produces a 20 A and 0.6 V output. Thus, the polymer-based hybrid PENG energy harvesters have better charging performance than individual PENG and can be effectively implemented for wearable sensor, actuators, and power sources [40].

9.5 CERAMIC-MATERIAL-BASED PENGs (CPENGs)

Ceramic-based pyroelectric materials such as PZT, $KNbO_3$, $BiFeO_3$, $BaTiO_3$, $NaNbO_3$, ZnO, etc. are very good in fabricating PENGs as compared to PPENGs, owing to their large p value and low thermal expansion coefficient [41–48]. A brief discussion on different ceramic-material-based PENGs is given below.

9.5.1 Ferroelectric Ceramic-Material-Based PENGs

In 2012, Yang et al. fabricated lead zirconate titanate (PZT)-based PENG with exceptionally good pyroelectric coefficient that can operate an LCD for up to 60 s [47]. The PZT film has a dimension of 21 mm × 12 mm and a thickness ~175 μm. Nickel layer was deposited as electrodes on the both surfaces of the PZT film. The nickel-coated PZT film was first poles at a high voltage of 4 kV, and then the electrical measurements were performed. The PZT-based PENG achieved a pyroelectric coefficient of 80 nC/cm^2 K. The PENG generated a peak voltage and current of 2.8 V and 42 nA at 0.2 K/s. Furthermore, an output voltage of 2.8 V and a current of 42 nA are achieved at high gradient of temperature. The maximum output power density produced by the device was 0.2 mW/cm^3. Using this power, the device charged a lithium-ion battery from 650 to 810 mV in about 3 h and lighted a green LED. Besides, the device can be operated as a wireless temperature sensor. The circuit comprises a temperature sensor, signal processing unit (SPU), and a flexible antenna as shown in Figure 9.10. A Li-ion battery of voltage 2.8 V was used to operate the sensor. When the sensor is turned on, the received signals from the temperature sensor displayed a temperature of 23.2°C (Figure 9.10(b)). The obtained temperature data is shown in Figure 9.10(c and d). Moreover, the wireless temperature sensor exhibited an accuracy of ±1K with a working distance of ≥50 m. Thus, the PZT-based PENG is a promising candidate for energy harvesting as well as environmental temperature monitoring.

Furthermore, Yang et al. designed a self-powered PENG to detect temperature change for the first time [49]. The device is based on PZT micro/nanowires bonded on a glass substrate, and the microwires were fixed at the ends of glass substrate by silver paste. The entire device is covered using PDMS, and its schematic representation is depicted in Figure 9.11. The PDMS is used as a package for the PZT to prevent the corrosion or contamination from the atmosphere. For the sensor application, the PENG as the temperature sensor is attached to a metallic body. Before conducting the experiment, the room temperature was maintained at 296K. To provide thermal change in the sensor-attached metallic body, a contact is made with a heater (acts as a heat source). When the sensor touches the heat sources, the output voltage linearly rose with increasing rate of temperature change at alternate intervals. This indicated that the device is not only a good temperature sensor but also can detect the rate of change in temperature. It is important to note that a fingertip touch is enough to detect the temperature as well as the temperature changing rate in this sensor. The sensor exhibited a very good response time (0.9 s) and reset time (3 s). At room temperature, the sensor

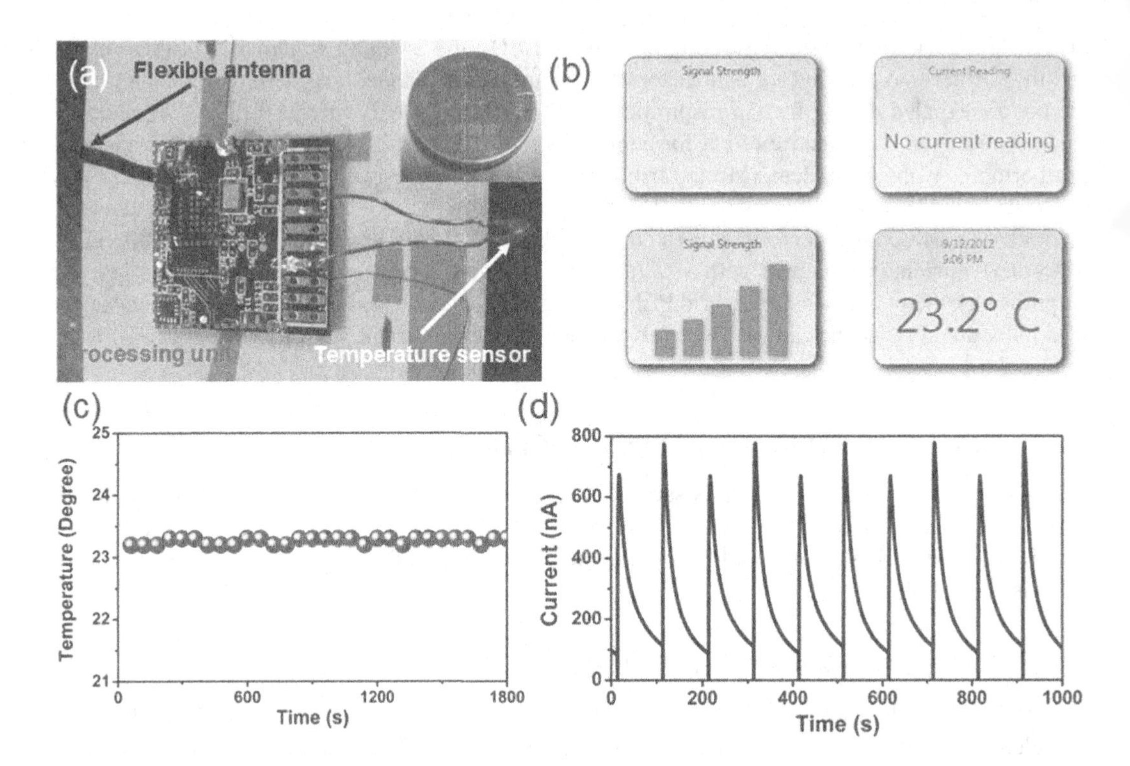

FIGURE 9.10 (a) Wireless temperature sensor with an image of a rechargeable Li-ion battery in the inset, (b) its temperature response in on-off state, (c) temperature as a function of time, (d) output current as a function of time. (Reprinted with permission from [47]. Copyright (2012) by the American Chemical Society Ltd.)

has a minimum detection limit of 0.4K. Furthermore, the maximum temperature limit detected for the finger surfaces was about 303K with a peak temperature change rate approximately 0.68K/s. Indeed, the PENG gave an output voltage of 3 V as heat source temperature reached to 473K (maximum reaching temperature). This voltage was utilized to light an LCD.

However, the brittle nature of ceramic PENG limits them in flexible electronic device applications. In this regard, as shown in Figure 9.12, Ko et al. [50] designed a flexible PZT-film-based hybrid (piezoelectric-pyroelectric) grown on a flexible Ni-Cr metal foil substrate coated with $LaNiO_3$ (LNO) as bottom electrode [50]. The flexible PENG showed a high remanent polarization of ~28 $\mu C/cm^2$, piezoelectric coefficient ~140 pC/N, and a very good pyroelectric coefficient (~50 nC/cm^2 K). This device generated mechanical and temperature-change-induced pyroelectricity

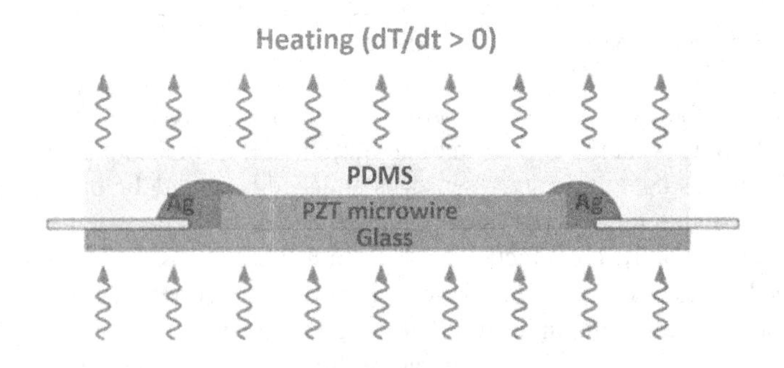

FIGURE 9.11 Schematics of the PZT microwire PENG. (Reprinted with permission from [49]. Copyright (2012) by the American Chemical Society Ltd.)

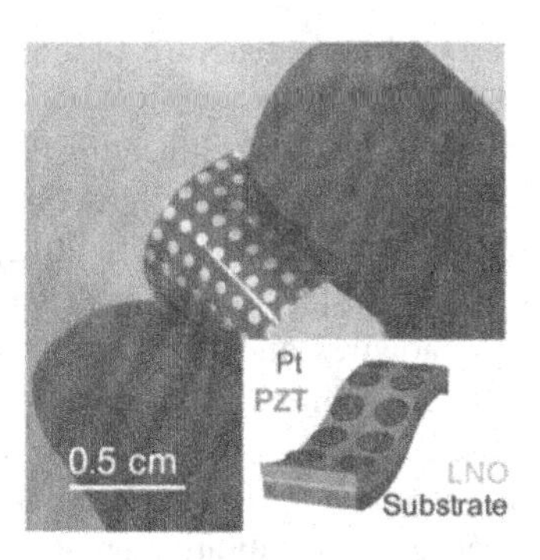

FIGURE 9.12 Image showing the flexibility of PZT thin-film-based PENG and the schematics of its structure is shown as inset. (Reprinted with permission from [50]. Copyright (2016) by the American Chemical Society Ltd.)

from various sources such as human finger contact and cold/hot winds in ambient atmosphere, demonstrating the PENG's energy harvesting capabilities. When the finger touches the PENG, it will bend and unbend. If this process is continued with a thermal fluctuation, a strain-induced mechanical vibration develops, and thereby, a current flows through the external circuit. A short-circuit current density and an open-circuit voltage of 0.34 μA/cm^2 0.34 V were achieved with this device. If the device is subjected to cold ($\Delta T < 0$) and hot winds ($\Delta T > 0$) at a speed of 7 m/s, a current flows through the external circuit due to thermal deformation. It is observed that the output performance of the PENG increased linearly with the velocity of the wind and thermal change rate. In this way, the hybrid PENG attained a peak output voltage of 0.34 V and a current of 0.34 VμA/cm^2. Moreover, flexible PZT-based PENG are appropriate for energy harvesting in hoarse environmental conditions like strong base conditions and high humidity, since these inorganic materials are generally stronger compared to organic materials.

In 2016, Zhang et al. [51] proposed a multifunctional PZT-based PENG that works on the basis of pyroelectric, piezoelectric, and triboelectric effect [51]. The PENG is composed of PZT block (acts as the pyroelectric and piezoelectric active material), flexible polyamide film with FEP (acts as the triboelectric active material), a combination of thin ITO film and Ag nanowires/PDMS film as top electrode, and Ag as the bottom electrode under the PZT. This single nanogenerator can work based on the different effects including pyroelectric, piezoelectric, triboelectric, and photoelectric effects. At room temperature, the electric dipoles are in equilibrium state, and there is no net electron flow due to constant spontaneous polarization. When the thermal change occurs (heating or cooling), spontaneous polarization change takes place resulting in an electron flow in an external circuit. During the cooling process, a negative pyroelectric potential is obtained across the electrodes. At the same time, the ferroelectric PZT perovskite showed a high photoelectric effect. This indicates the device's sensitivity toward visible light and can be utilized for solar cell applications. The triboelectric effect is initiated by the polyamide film through a compressive strain on FEP. When the FEP is in contact with the polyamide film, there will be no electron flow due to electrostatic equilibrium. Simultaneously, a positive piezoelectric potential is experienced between the top and bottom electrodes due to the compressive strain. Consequently, an electron flow takes place from Ag to ITO. Triboelectrification is commenced as the polyamide film is separated or move apart from FEP. As a result, the charges on FEP generate induced charges on ITO. This electric potential drives the electrons from ITO to Ag. Thus, the device utilizes different physical principles for scavenging the thermal, solar, and mechanical energies. The single device hybrid nanogenerator integrated with pyroelectric, photoelectric, piezoelectric, and triboelectric effect produced a peak current of ~5 μA and voltage of ~80 V.

Besides, the multiple effects coupled with one structure-based nanogenerator charged a capacitor of 10 μF up to a potential of 5.1 V in just 90 s.

Even though PZT-based nanomaterials are good choice for PENGs, their toxic nature and high cost limit them in the fabrication of PENGs. Therefore, alternative ferroelectric materials with non-toxic nature and cost-effectivity are important to explore. In 2012, Yang et al. fabricated a Pb-free $KNbO_3$-based flexible PENG for the first time [52]. They formed a composite in the ratio of 7:3 with single-crystalline structure $KNbO_3$ nanowires and PDMS. The top and bottom electrodes were formed of Ag and ITO films, respectively. When the device is subjected to a thermal change, the dipoles in the $KNbO_3$ nanowires rotate as $KNbO_3$ is composed of multidomain. As the temperature increases, the dipoles experience an increased lattice thermal vibration with increased amplitude. This in turn enlarges the internal electric dipole vibration creating a net P_s change and thereby, the movement of surface charges. As a result, a current is generated in the external circuit with a peak voltage of 10 mV and a peak current of 120 pA.

Isakov et al. [53] introduced another Pb-free nanofiber PENG based on 1,4-diazabicyclo[2.2.2]-octane perrhenate (dabcoHReO_4) synthesized using electrospinning method [53]. The precursor aqueous solution was prepared by mixing dissolved dabcoHReO_4 crystals with polyvinyl alcohol (PVA) in 1:1 wt.%. The networks of NH-N hydrogen bond in dabcoHReO_4 crystal contribute to P_s and the ferroelectric crystal shows a P_s ~17 μC/cm^2 [54]. The device exhibited a very good piezoresponse with a value of 20 pm/V. The pyroelectricity is created by the thermal-change-induced spontaneous polarization in dabcoHReO_4 nanofiber array. This develops free charges on the surface layers, and thereby, an electric field is generated by the charge flow. The device achieved a maximum output current of 200 pA with a pyroelectric coefficient of 8.5 μC/m^{2}K. Though these materials have low pyroelectric coefficient, they can be easily incorporated into flexible devices.

Qi et al. [55] developed an ITO/BFO/Ag hybrid PENG as photodetector using the photovoltaic-pyroelectric coupling effect [55]. The peak voltage, peak current, and photovoltaic current generated by the PENG are 0.13 V, 8.8 nA, and 2 nA, respectively, under 450-nm light illumination and 0.2K/s temperature gradient conditions. BTO ceramic are also receiving greater attention in the fabrication of PENGs [56–58]. Ji et al. fabricated BTO ceramic disk-based pyro-photo-triboelectric hybrid nanogenerator through dry press method [56]. It is packed with a PDMS layer to protect the ceramic disk. Furthermore, to generate triboelectricity, an FEP is attached to the PDMS. The nanogenerator exhibited a peak output voltage of 1.5 V and current of 15 nA, respectively. Furthermore, a 0.33 μF capacitor was successfully charged up to potential of 1.1 V in 10 s. Thus, the tetragonal BTO-based hybrid nanogenerator simultaneously scavenges the thermal, mechanical, and light, which can be integrated to operate sensor system. Zhao et al. achieved a strong illumination of 83.2 mW/cm^2 in the Ag/BTO/ITO device [57]. Very recently, Zhang et al. introduced $NaNbO_3$ thin-film-based nanogenerator by combining the pyroelectric and photoelectrochemical (PEC) properties for water splitting [59]. They installed a three-electrode assembly with $NaNbO_3$ film as the working electrode as shown in Figure 9.13(a). The temperature range of 20–50°C is provided between the two terminals to understand the pyro-photoelectric catalysis (PREC) near room temperature. The proposed mechanism behind the PREC water splitting in the presence of both light and thermal change is diagrammatically represented in Figure 9.13(b). Initially, the $NaNbO_3$ is at thermodynamic equilibrium. As the temperature increases, there will be a polarization change and the excess charges released on the surface generate pyroelectric charge carriers to take part in the redox reaction. Simultaneously, the polarization imbalance and the induced pyroelectric potential steer to a band alignment change at $NaNbO_3$/electrolyte. This favors the transport of photogenerated charge carriers. Afterward, the cooling promotes the spontaneous polarization and improves the performance of the PREC device once again. Therefore, this work provides a realization for designing environment friendly PENGs. It also provides an insight for designing new and eco-friendly photoelectrodes for water splitting applications.

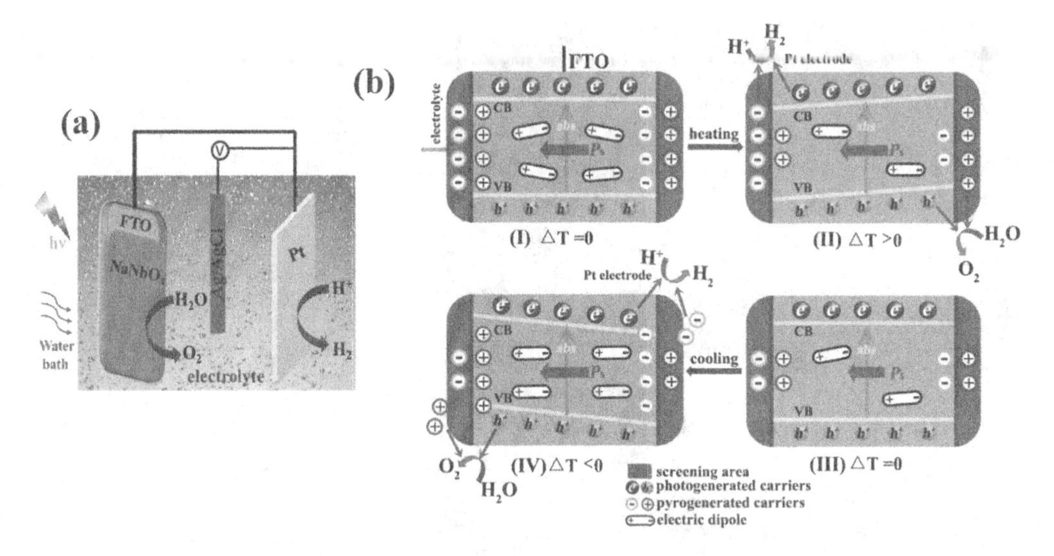

FIGURE 9.13 (a) Three electrode assembly of PEC, (b) proposed mechanism of PREC catalysis process. (Reprinted with permission from [59]. Copyright (2021) by the Elsevier Ltd.)

9.5.2 ZnO-Based PENGs

Semiconductor materials like, ZnO, CdS, CdSe, etc. exhibit the secondary pyroelectric effect and thus induces the pyroelectric current [15]. In such materials, energy harvesting is carried out through the temperature-change-induced piezoelectric potential. Among these materials, wurtzite ZnO is receiving greater attention owing to its nonsymmetric structure [60]. Yang et al. [60] constructed a pyroelectric energy harvester by combining the semiconductor and pyroelectric properties of ZnO nanowires [60]. The ZnO was grown on ITO substrate using solution-based growth technique (Figure 9.14(a)) and the top and bottom electrodes were made of Ag and ITO, respectively. An electron flow is observed in an external circuit, when the Ag/ZnO nanowire/ITO device is subjected to a temperature change (heating and cooling). The ZnO nanowire-based PENG exhibited a pyroelectric coefficient of ~1.2–1.5 nC/cm^{2}K energy conversion characteristic coefficient ~2.5–4.0 × 10^4 V/m. Furthermore, Wang et al. reported ZnO-nanowire-based nanogenerator as self-powered system for applications such as signal processing, active sensing, and energy harvesting [61]. A magnetron sputtering was used to deposit ZnO (n-type) seed layer on p-Si substrate to form a PENG heterostructure. Then, ZnO nanowires were grown using hydrothermal technique. Furthermore, an ITO thin layer is deposited as top electrode and Cu was sputtered on the back of p-Si as bottom electrode as shown in Figure 9.14(b and c). In this way, Ag/p-Si/n-ZnO nanowire/ITO-based PENG is fabricated. The PENG generate current by means of near IR (NIR) triggered pyroelectric current as shown in Figure 9.14(d). Under thermal equilibrium, there will be a built-in electric field E_b at the p-n junction (bias-free condition). The temperature in the ZnO nanowires increases upon the exposure to NIR light illumination and a pyro-potential (E_{py}) in the E_b direction is developed. As a result, the total electric field ($E_{total} = E_{py} + E_b$) in the heterostructure increases followed by the electron migration through external circuit. Furthermore, the conduction current evolved in the external circuit is attributed to the displacement current inside the PENG-heterostructure device. On shutting down the NIR illumination, a reverse pyro-potential ($-E_{py}$) resulting in a decrease in total electric field occurs and thus, a reverse electrons flow takes place through the external circuit. This clearly demonstrates a self-powered NIR photosensor by generating an output current without external bias. Under strong NIR illumination, the device achieved a peak current of 1 mA with an on/off photocurrent ratio of order 10^7.

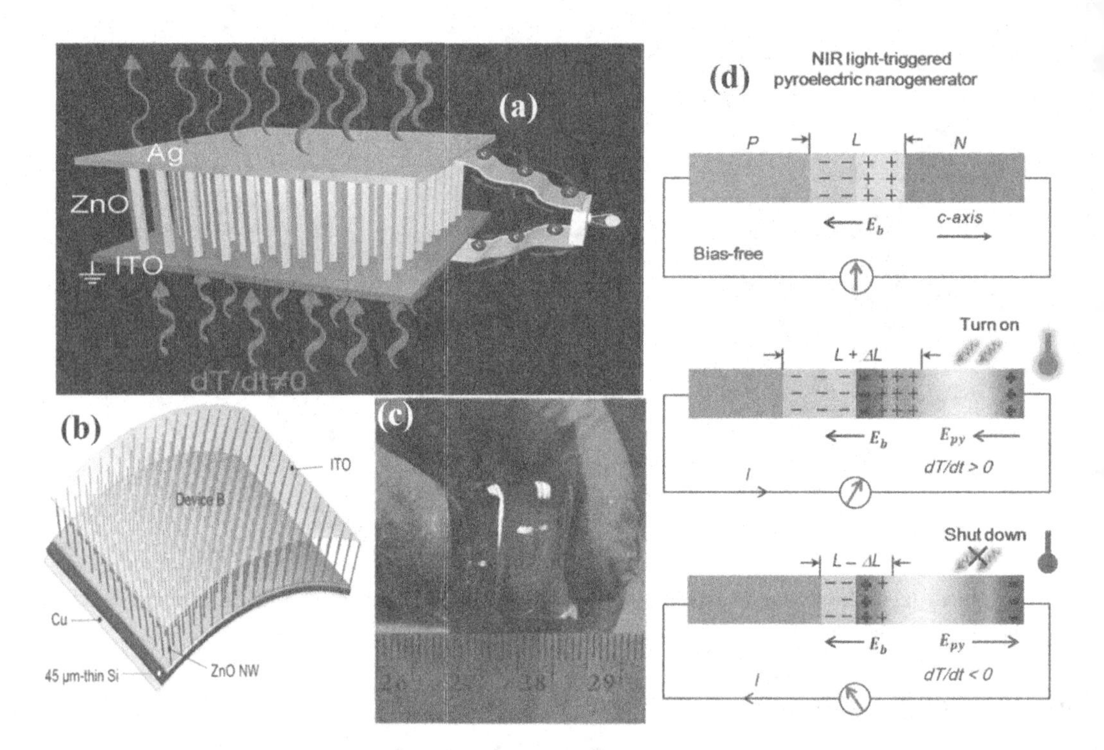

FIGURE 9.14 (a) Schematic diagram of ZnO nanowire-based PENG, (b) schematic and (c) photograph of p-Si/n-ZnO nanowire heterostructure PENG, (d) proposed mechanism. (Reprinted with permission from [61]. Copyright (2017) by the American Chemical Society Ltd.)

9.6 APPLICATIONS

9.6.1 Electrochemical Catalysis

The pyroelectric charge generated by a ferroelectric material can be utilized to control the electrochemical applications such as water splitting for hydrogen generation, photodegradation of pollutants from water, water purification, and water disinfection [62]. From Equations (9.2) and (9.5), it is clear that both charge and electric potential of the pyroelectric material are proportional to the surface area and thickness, respectively. Hence, it is possible to optimize the pyroelectric performance by tuning the geometry of pyroelectric material. For instance, the critical potential required for the splitting of water is in the range 1.23–1.5 V [62]. Thus, the electric charge and the electric potential generated by the pyroelectric material can be used as a power source for various applications such as water splitting, photodegradation, water purification, and water disinfection. Furthermore, a pyroelectric material with low Curie temperature ($T_c < 100°C$) provide maximum pyroelectricity owing to an intrinsically large polarization change with temperature. Consequently, the pyroelectricity can enhance the electrochemical process. Figure 9.15(a) represents a work on water splitting experiment through thermally oscillating the pyroelectric material surface above and below its T_c. Besides, the proposed work enables the purification of air by the oxidation or decomposition of gaseous pollutants [63]. Figure 9.15(b) demonstrates the working of a pyroelectric material as a simple thermal energy harvester that drives an electrochemical reaction. Here, the material has no direct contact with the electrolyte [64]. Figure 9.15(c) illustrates a Pb-free ferroelectric $NaNbO_3$ nanofiber-based PENG for pyro-electrochemical process [65]. The material releases surface charges when subjected to heating. Furthermore, there is no need of poling to create polarization as the single domains of $NaNbO_3$ owing to the direct contact with the electrolyte. This generates the required potential to cycle the material above and below T_c. By utilizing this large polarization changes

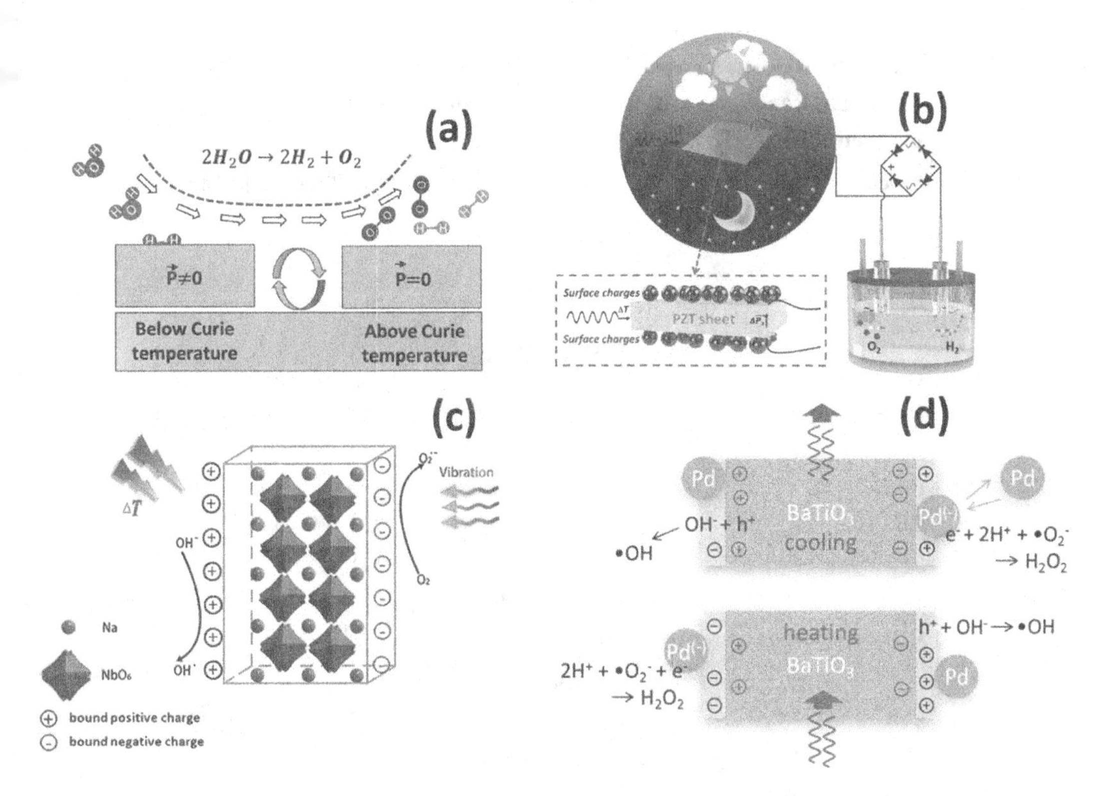

FIGURE 9.15 Mechanism of water splitting in (a) pyroelectricity induced cyclic catalysis (reprinted with permission from [63]); (b) pyroelectric material as an external source (reprinted with permission from [64]. Copyright (2019) by the Elsevier Ltd); (c) pyrocatalysis based on $NaNbO_3$ nanofiber (reprinted with permission from [65]. Copyright (2018) by the Elsevier Ltd); and (d) $BaTiO_3$@Pd nanoparticles induced pyrocatalysis (reprinted with permission from [66]. Copyright (2015) by the American Chemical Society Ltd).

near Curie point, large amount of surface charges can be released during phase transition [65]. Figure 9.15(d) shows the coupling of BTO and a noble metal (Pd) to enhance the catalytic activity for dye degradation [66]. Thus, these works demonstrated that pyroelectric materials can play a crucial role in driving electrochemical reactions using wasted heat energy in the environment.

9.6.2 Reaction Monitoring of Chemical Reactions

It is essential to monitor the reactors and the catalyst properties in order to measure the reactor condition and for enhancing the chemical processes [67, 68]. Reaction monitoring provides the direct information about the catalyst site properties via a change in temperature, composition, and or pressure. Thus, a PENG-based self-powered wireless nanosensor network (WNSN) plays an important role in direct detection of the surface reactions on the catalyst [67, 68]. Zarepour et al. [69] proposed a remote detector based on a PENG fitted with graphene-based nano-antennas [69]. The graphene antenna enables radiation in the range (0.1–10 THz). Figure 9.16 shows a schematic illustration of the WNSN-based reaction monitoring architecture. In order to understand the working mechanism, the authors selected Fischer-Tropsch synthesis (FTS) as a case study. Here, an iron-based fixed-bed FTS reactor is used and the catalyst sites have a dimension of 0.3 μm×0.3 μm with each site having a mass ~1.5 fg. A rod made up of macro-scale wireless remote sinks is also deployed through the catalyst tube's axis. The internet is then connected to this wireless remote sink. FTS involves the process of converting natural gas to liquid hydrocarbons in a chemical reactor [69]. The different chemical reactions and the corresponding chemical species evolved are shown in Table 9.2.

TABLE 9.2
Chemical Reactions and the Corresponding Chemical Species Evolved during FTS

Reaction	Released Energy (KJ/mol)
Adsorption Phase	
$CO \rightarrow C + O$	56.81 ± 0.96
$H_2 \rightarrow H + H$	10 ± 0.5
Water Formation	
$O + H \rightarrow OH$	103.80 ± 0.96
$OH + H \rightarrow H_2O$	86.22 ± 0.62
Chain Initiation	
$C + H \rightarrow CH$	77.66 ± 0.7
$CH + H \rightarrow CH_2$	11.94 ± 0.1
$CH_2 + H \rightarrow CH_3$	61.88 ± 0.5
Chain Growth	
$C_nH_{2n+1} + CH_2 \rightarrow C_mH_{2m+1} (m = n + 1)$	44.79 ± 0.43
Hydrogenation to Paraffin (HTP)	
$C_nH_{2n+1} + H \rightarrow C_nH_{2n+2}$	117.75 ± 0.67
β-Dehydrogenation to Olefin (DTO)	
$C_nH_{2n+1} \rightarrow C_nH_{2n} + H$	96.27 ± 0.5

The rate of temperature change (ΔT) at a time t during a chemical reaction can be directly calculated using the heat formula:

$$\Delta T = \frac{H}{C_p . m}$$

where H is the released amount of heat, C_p is the specific heat capacity of catalyst material (0.45 J/g K for iron), and m is the mass of the catalyst site (1.5 fg). Suppose the time taken for the completion of one reaction is 1 ps, then

$$\frac{dT}{dt} = \frac{\Delta T}{t} = \Delta T \times 10^{12}$$

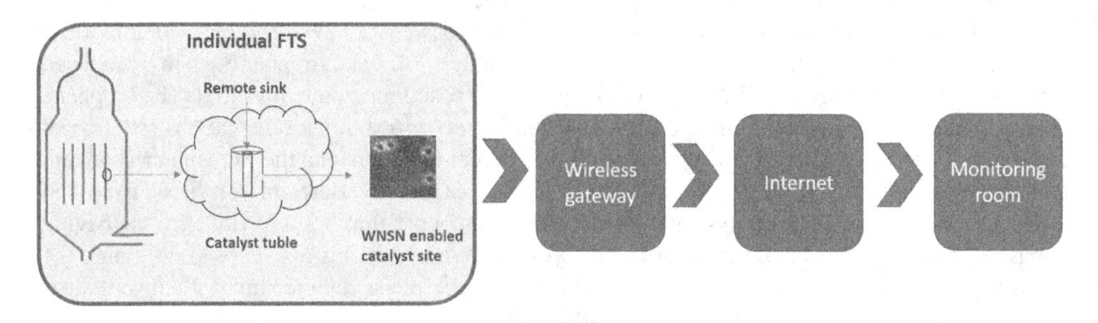

FIGURE 9.16 Flow chart of the monitoring of FTS reactor using WNSN [69].

Hence, a remarkable change in temperature is expected and eventually huge increase in the generated power. Thus, the instantaneous change in temperature is converted into electrical energy by the PENG during a reaction in the catalyst site. A nanoradio uses this harvested power to deliver a radio pulse to a nearby remote sink. Since the harvested energy is proportional to the power amplitude of the pulse, the received energy of different types of reactions can be distinguished in FTS at the remote station [69].

9.6.3 Health-Care Monitoring Devices

In the current scenario, the real time monitoring of health status of human being is inevitable due to the impact of COVID-19 pandemic. However, the main problems associated with the current sensors are their low flexibility, less accuracy, and low sensitivity [70]. This paved the development of wearable nanogenerators using pyroelectric effect and hybrid nanogenerators (using a combined pyroelectricity, piezoelectricity, and triboelectric effect) for health-care monitoring. The hybrid nanogenerators use a single material that responds to all biophysical signals (temperature, mechanical, and pressure changes) in human body. Furthermore, such a combined effect in a single material is beneficial for miniaturization of electronic devices. For instance, a commercially available N95 mask based on Al-coated PVDF film PENG was commercially demonstrated [33]. The breathing (inhaling and exhaling) creates a time-dependent temperature change and generates pyroelectricity-induced huge output responses. Thus, the clinician can easily identify the change in the temperature gradient of the patients; especially during COVID-19 pandemic. Furthermore, a commercially available self-powered temperature sensor was introduced to harvest wasted heat from human body and from the breathing [33, 39]. Furthermore, a transparent pyro-piezo-tribo-based hybrid nanogenerator was introduced to sense the temperature and pressure simultaneously even under stretching condition [36]. This can be used to understand the early intervention of COVID-19 by monitoring the heartbeat, swallowing, and respiration. Overall, pyroelectric and hybrid nanogenerator-based energy harvesters are efficient in monitoring human physiological signals. Figure 9.17 displays a

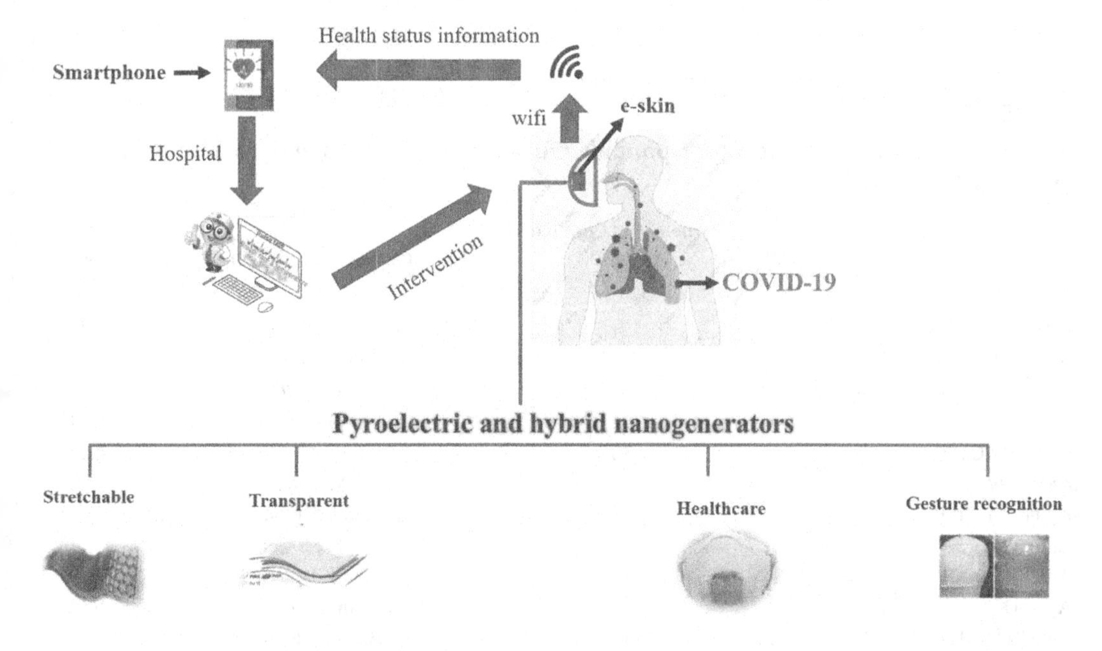

FIGURE 9.17 A summary of pyroelectric and hybrid nanogenerators as health-monitoring devices against COVID-19 pandemic.

schematic representation of the use of pyroelectric and hybrid nanogenerators as health-monitoring devices against COVID-19 pandemic.

9.7 SUMMARY AND FUTURE DIRECTIONS

This chapter discusses the basis of PENGs and their current advancements in the energy harvesting technology. Energy harvesting using pyroelectric effect from wasted energy is relatively simple and can be used to operate electronic devices under different environmental and temperature conditions. The temperature-gradient-induced spontaneous polarization change is the key to harvest pyroelectric electrical energy. Different pyroelectric materials such as polymer-based, ceramic-based, and their combinations can be utilized for fabricating PENGs for various applications. PPENGs are found to be the suitable candidates for wearable applications owing to their exceptionally good flexibility, stretchability, and biocompatibility. On the other hand, ceramic-based PENGs (CPENGs) are cost-effective, possesses a wide range of working temperature, and moderate pyroelectric coefficient compared to the PPENGs. Therefore, CPENGs can be implemented for high-temperature applications.

Currently, the PENGs and hybrid PENGs are receiving greater attention in areas of research such as electrochemical catalysis, reaction monitoring, and health-care monitoring. Table 9.3 shows a summary of some of the best PENG and their output performance. Still, PENGs face challenges that limit them from practical applications.

Like pyroelectric materials, substrate, and electrodes have equal importance in designing a PENG. Therefore, the materials chosen as substrate and electrodes must have the ability to sustain the mechanical property against thermal- and light-induced mutilation. Furthermore, the PPENGs suffer limitations such as low thermal cycle, low light irradiation stability, and low pyroelectric coefficient. To obtain very high FOMs in CPENGs, it is necessary to control the dielectric constant without much affecting the pyroelectric coefficient. Furthermore, the stability of the pyroelectric materials needs to be taken into account during the phase-transition temperature for superior temperature output performances.

TABLE 9.3
Pyroelectric Coefficient and Corresponding Output Performances of Pyroelectric and Hybrid Nanogenerators

Device	Pyroelectric Coefficient nC/cm^2 K	Output Performance	Reference
Cu/PVDF/Cu	2.7	190 V, 11 A	[30]
Al/Ti/PVDF/Ti/Al	3.3	60 V, 30 nA	[71]
Au + Cr/PVDF/Au + Cr	2.72	145 V, 120 nA	[34]
Al/PVDF/Al	2.7	42 V, 2.5 μA	[33]
Cu/PVDF/Cu	2.64 ± 0.15	–, 14 μA	[39]
Cu/PVDF/Cu	3.73	120 V, 60 nA	[72]
Al/PVDF/Al	2.7	5.7 V, 109 nA	[32]
Ag/P(VDF-TrFE)/Al	4.39	13.65 V, 2.69 μA	[73]
Ni/PZT/Ni	80	22 V, 171 nA/cm^2	[47]
Ag/PZT/Ag NW/PDMS-ITO	–	100 V, 480 nA	[51]
Ag/ZnO/ITO	1.2–1.5	18 mV, 0.4 nA	[60]
Ag/$KNbO_3$/ITO	0.8	10 mA, 120 pA	[52]
ITO/BTO/Ag	22.5–25.9	6 V, 80 nA	[56]

Different synthesis approaches such as micro- and or nanowires, nanofibers, and micro-patterned structure were introduced to increase the performance of PENG as energy harvesters using wasted energy. Furthermore, the combination of pyro-piezo-tribo electric effects into a single material not only enhances the energy harvesting but also reduces the device size. High energy harvesting performance can be achieved near morphotropic phase boundary (MPB) through altering the composition to attain a desired phase structure. Furthermore, integration of a low permittivity material at the grain boundaries of pyroelectric material might be a possible way to increase the output performance of the PENG as FOMs require low ε_r. Besides, the amalgamation of PENGs with energy storage devices such as batteries, supercapacitors, etc. might allow the storage of generated pyroelectric electrical energy. Thus, we believe a systematic planning of optimizing the device structure and device packing can overcome the challenges faced by the PENGs and can commercialize a futuristic eco-friendly electronics world.

REFERENCES

1. Zhang, Tongtong, Tao Yang, Mei Zhang, Chris R. Bowen, and Ya Yang. "Recent progress in hybridized nanogenerators for energy scavenging." *iScience* 11 (2020): 101689.
2. Sharma, Manish, Aditya Chauhan, Rahul Vaish, and Vishal Singh Chauhan. "Pyroelectric materials for solar energy harvesting: a comparative study." *Smart Materials and Structures* 24, no. 10 (2015): 105013.
3. Sripadmanabhan Indira, Sridhar, Chockalingam Aravind Vaithilingam, Kameswara Satya Prakash Oruganti, Faizal Mohd, and Saidur Rahman. "Nanogenerators as a sustainable power source: state of art, applications, and challenges." *Nanomaterials* 9, no. 5 (2019): 773.
4. Xu, Chen, Caofeng Pan, Ying Liu, and Zh L. Wang. "Hybrid cells for simultaneously harvesting multi-type energies for self-powered micro/nanosystems." *Nano Energy* 1, no. 2 (2012): 259–272.
5. Khan, Asif Abdullah, Alam Mahmud, and Dayan Ban. "Evolution from single to hybrid nanogenerator: a contemporary review on multimode energy harvesting for self-powered electronics." *IEEE Transactions on Nanotechnology* 18 (2018): 21–36.
6. Guo, Dong, Fei Zeng, and Brahim Dkhil. "Ferroelectric polymer nanostructures: fabrication, structural characteristics and performance under confinement." *Journal of Nanoscience and Nanotechnology* 14, no. 2 (2014): 2086–2100.
7. Cauda, Valentina, Giancarlo Canavese, and Stefano Stassi. "Nanostructured piezoelectric polymers." *Journal of Applied Polymer Science* 132, no. 13 (2015): 41667.
8. Chen, Xiaoliang, Jinyou Shao, Xiangming Li, and Hongmiao Tian. "A flexible piezoelectric-pyroelectric hybrid nanogenerator based on P (VDF-TrFE) nanowire array." *IEEE Transactions on Nanotechnology* 15, no. 2 (2016): 295–302.
9. Whatmore, R. W. "Pyroelectric devices and materials." *Reports on Progress in Physics* 49, no. 12 (1986): 1335.
10. Lang, Sidney B. "Pyroelectricity: from ancient curiosity to modern imaging tool." *Physics Today* 58, no. 8 (2005): 31.
11. Erhart, Jiří. "Experiments to demonstrate piezoelectric and pyroelectric effects." *Physics Education* 48, no. 4 (2013): 438.
12. Mane, Poorna, Jingsi Xie, Kam K. Leang, and Karla Mossi. "Cyclic energy harvesting from pyroelectric materials." *IEEE Transactions on Ultrasonics, Ferroelectrics, and Frequency Control* 58, no. 1 (2011): 10–17.
13. Cuadras Tomas, Angel, Manuel Gasulla Forner, and Vittorio Ferrari. "Thermal energy harvesting through pyroelectricity." *Sensors and Actuators A: Physical* 158, no. 1 (2010): 132–139.
14. Bowen, Chris R., John Taylor, E. LeBoulbar, D. Zabek, A. Chauhan, and R. Vaish. "Pyroelectric materials and devices for energy harvesting applications." *Energy & Environmental Science* 7, no. 12 (2014): 3836–3856.
15. Lang, Sidney B. "Pyroelectricity: from ancient curiosity to modern imaging tool." *Physics Today* 58, no. 8 (2005): 31.
16. Lingam, Devashish, Ankit R. Parikh, Jiacheng Huang, Ankur Jain, and Majid Minary-Jolandan. "Nano/microscale pyroelectric energy harvesting: challenges and opportunities." *International Journal of Smart and Nano Materials* 4, no. 4 (2013): 229–245.

17. Aggarwal, M. D., A. K. Batra, P. Guggilla, M. E. Edwards, B. G. Penn, and J. R. Currie Jr. "Pyroelectric materials for uncooled infrared detectors: processing, properties, and applications." *NASA/TM* (2010): 216373. https://ntrs.nasa.gov/citations/20110008068.
18. Nalwa, Hari Singh, ed. *Handbook of Advanced Electronic and Photonic Materials and Devices, Ten-Volume Set.* Vol. 1. Academic Press, London, UK, 2000.
19. Lee, M. H., R. Guo, and Amar S. Bhalla. "Pyroelectric sensors." *Journal of Electroceramics* 2, no. 4 (1998): 229–242.
20. Muralt, Paul. "Micromachined infrared detectors based on pyroelectric thin films" Rep. Prog. Phys. 64, no. 10 (2001): 1339-1388. https://iopscience.iop.org/article/10.1088/0034-4885/64/10/203/pdf.
21. Takenaka, T., and K. Sakata. "Piezoelectric and pyroelectric properties of calcium-modified and grain-oriented (NaBi) $1/2Bi_4Ti_4O_{15}$ ceramics." *Ferroelectrics* 94, no. 1 (1989): 175–181.
22. Takenaka, Tadashi, and Koichiro Sakata. "Pyroelectric properties of grain-oriented bismuth layer-structured ferroelectric ceramics." *Japanese Journal of Applied Physics* 22, no. S2 (1983): 53.
23. Takenaka, Tadashi, and Koichiro Sakata. "Pyroelectric properties of bismuth layer-structured ferroelectric ceramics." *Ferroelectrics* 118, no. 1 (1991): 123–133.
24. Tang, Yanxue, Zong-Yang Shen, Shujun Zhang, and Thomas R. Shrout. "Improved pyroelectric properties of $CaBi_4Ti_4O_{15}$ ferroelectrics ceramics by Nb/Mn Co-doping for pyrosensors." *Journal of the American Ceramic Society* 99, no. 4 (2016): 1294–1298.
25. Zhao, M. L., C. L. Wang, W. L. Zhong, P. L. Zhang, J. F. Wang, and H. C. Chen. "Dielectric and pyroelectric properties of $SrBi_4Ti_4O_{15}$-based ceramics for high-temperature applications." *Materials Science and Engineering: B* 99, no. 1–3 (2003): 143–146.
26. Yuan, G. L., Siu Wing Or, J. M. Liu, and Z. G. Liu. "Structural transformation and ferroelectromagnetic behavior in single-phase $Bi_{1-x}Nd_xFeO_3$ multiferroic ceramics." *Applied Physics Letters* 89, no. 5 (2006): 052905.
27. Yao, Shanshan, Wei Ren, Hongfen Ji, Xiaoqing Wu, Peng Shi, Dezhen Xue, Xiaobing Ren, and Zuo-Guang Ye. "High pyroelectricity in lead-free $0.5Ba(Zr_{0.2}Ti_{0.8})O_3$–0.5 $(Ba_{0.7}Ca_{0.3})TiO_3$ ceramics." *Journal of Physics D* 45, no. 19 (2012): 195301.
28. McFee, J. H., J. G. Bergman Jr, and G. R. Crane. "Pyroelectric and nonlinear optical properties of poled polyvinylidene fluoride films." *Ferroelectrics* 3, no. 1 (1972): 305–313.
29. Neumann, N., R. Köhler, and G. Hofmann. "Pyroelectric thin film sensors and arrays based on P (VDF/TrFE)." *Integrated Ferroelectrics* 6, no. 1–4 (1995): 213–230.
30. Leng, Qiang, Lin Chen, Hengyu Guo, Jianlin Liu, Guanlin Liu, Chenguo Hu, and Yi Xi. "Harvesting heat energy from hot/cold water with a pyroelectric generator." *Journal of Materials Chemistry A* 2, no. 30 (2014): 11940–11947.
31. Lang, Sidney B., and Dilip K. Das-Gupta. "Pyroelectricity: fundamentals and applications." In *Handbook of Advanced Electronic and Photonic Materials and Devices*, pp. 1–55. Academic Press, 2001.
32. Raouadi, M. H., and O. Touayar. "Harvesting wind energy with pyroelectric nanogenerator PENGG using the vortex generator mechanism." *Sensors and Actuators A* 273 (2018): 42–48.
33. Xue, Hao, Quan Yang, Dingyi Wang, Weijian Luo, Wenqian Wang, Mushun Lin, Dingli Liang, and Qiming Luo. "A wearable pyroelectric nanogenerator and self-powered breathing sensor." *Nano Energy* 38 (2017): 147–154.
34. Gao, Fengxian, Wanwan Li, Xiaoqian Wang, Xiaodong Fang, and Mingming Ma. "A self-sustaining pyroelectric nanogenerator driven by water vapor." *Nano Energy* 22 (2016): 19–26.
35. Lee, Ju-Hyuck, Hanjun Ryu, Tae-Yun Kim, Sung-Soo Kwak, Hong-Joon Yoon, Tae-Ho Kim, Wanchul Seung, and Sang-Woo Kim. "Thermally induced strain-coupled highly stretchable and sensitive pyroelectric nanogenerators." *Advanced Energy Materials* 5, no. 18 (2015): 1500704.
36. Lee, Ju-Hyuck, Keun Young Lee, Manoj Kumar Gupta, Tae Yun Kim, Dae-Yeong Lee, Junho Oh, Changkook Ryu et al. "Highly stretchable piezoelectric-pyroelectric hybrid nanogenerator." *Advanced Materials* 26, no. 5 (2014): 765–769.
37. Ghosh, D. S., I. Calizo, D. Teweldebrhan, Evghenii P. Pokatilov, Denis L. Nika, Alexander A. Balandin, Wenzhong Bao, Feng Miao, and C. Ning Lau. "Extremely high thermal conductivity of graphene: prospects for thermal management applications in nanoelectronic circuits." *Applied Physics Letters* 92, no. 15 (2008): 151911.
38. Balandin, Alexander A., Suchismita Ghosh, Wenzhong Bao, Irene Calizo, Desalegne Teweldebrhan, Feng Miao, and Chun Ning Lau. "Superior thermal conductivity of single-layer graphene." *Nano Letters* 8, no. 3 (2008): 902–907.

39. Zheng, Haiwu, Yunlong Zi, Xu He, Hengyu Guo, Ying-Chih Lai, Jie Wang, Steven L. Zhang, Changsheng Wu, Gang Cheng, and Zhong Lin Wang. "Concurrent harvesting of ambient energy by hybrid nanogenerators for wearable self-powered systems and active remote sensing." *ACS Applied Materials & Interfaces* 10, no. 17 (2018): 14708–14715.
40. Wang, Shuhua, Zhong Lin Wang, and Ya. Yang. "A one-structure-based hybridized nanogenerator for scavenging mechanical and thermal energies by triboelectric–piezoelectric–pyroelectric effects." *Advanced Materials* 28, no. 15 (2016): 2881–2887.
41. Olsen, R. B., and D. D. Brown. "High efficieincy direct conversion of heat to electrical energy-related pyroelectric measurements." *Ferroelectrics* 40, no. 1 (1982): 17–27.
42. Lau, Sien Ting, C. H. Cheng, S. H. Choy, D. M. Lin, Kin Wing Kwok, and Helen L. W. Chan. "Lead-free ceramics for pyroelectric applications." *Journal of Applied Physics* 103, no. 10 (2008): 104105.
43. Zhang, Yaju, Huanxin Su, Hui Li, Zhongshuai Xie, Yuanzheng Zhang, Yan Zhou, Liya Yang, Haowei Lu, Guoliang Yuan, and Haiwu Zheng. "Enhanced photovoltaic-pyroelectric coupled effect of $BiFeO_3$/Au/ZnO heterostructures." *Nano Energy* (2021), no. 85: 105968. https://www.sciencedirect.com/science/article/abs/pii/S2211285521002263
44. Srikanth, Keshavmurthy, Satyanarayan Patel, and Rahul Vaish. "Pyroelectric performance of $BaTi_{1-x}Sn_xO_3$ ceramics." *International Journal of Applied Ceramic Technology* 15, no. 2 (2018): 546–553.
45. You, Huilin, Zheng Wu, Lang Wang, Yanmin Jia, Sheng Li, and Jun Zou. "Highly efficient pyrocatalysis of pyroelectric $NaNbO_3$ shape-controllable nanoparticles for room-temperature dye decomposition." *Chemosphere* 199 (2018): 531–537.
46. Wang, Zhaona, Ruomeng Yu, Caofeng Pan, Zhaoling Li, Jin Yang, Fang Yi, and Zhong Lin Wang. "Light-induced pyroelectric effect as an effective approach for ultrafast ultraviolet nanosensing." *Nature Communications* 6, no. 1 (2015): 1–7.
47. Yang, Ya, Sihong Wang, Yan Zhang, and Zhong Lin Wang. "Pyroelectric nanogenerators for driving wireless sensors." *Nano Letters* 12, no. 12 (2012): 6408–6413.
48. Ye, Chian-Ping, Takashi Tamagawa, and D. L. Polla. "Experimental studies on primary and secondary pyroelectric effects in $Pb(Zr_xTi_{1-x})O_3$, $PbTiO_3$, and ZnO thin films." *Journal of Applied Physics* 70, no. 10 (1991): 5538–5543.
49. Yang, Ya, Yusheng Zhou, Jyh Ming Wu, and Zhong Lin Wang. "Single micro/nanowire pyroelectric nanogenerators as self-powered temperature sensors." *ACS Nano* 6, no. 9 (2012): 8456–8461.
50. Ko, Young Joon, Dong Yeong Kim, Sung Sik Won, Chang Won Ahn, Ill Won Kim, Angus I. Kingon, Seung-Hyun Kim, Jae-Hyeon Ko, and Jong Hoon Jung. "Flexible $Pb(Zr_{0.52}Ti_{0.48})O_3$ films for a hybrid piezoelectric-pyroelectric nanogenerator under harsh environments." *ACS Applied Materials & Interfaces* 8, no. 10 (2016): 6504–6511.
51. Zhang, Kewei, Shuhua Wang, and Ya Yang. "A one-structure-based piezo-tribo-pyro-photoelectric effects coupled nanogenerator for simultaneously scavenging mechanical, thermal, and solar energies." *Advanced Energy Materials* 7, no. 6 (2017): 1601852.
52. Yang, Ya, Jong Hoon Jung, Byung Kil Yun, Fang Zhang, Ken C. Pradel, Wenxi Guo, and Zhong Lin Wang. "Flexible pyroelectric nanogenerators using a composite structure of lead-free $KNbO_3$ nanowires." *Advanced Materials* 24, no. 39 (2012): 5357–5362.
53. Isakov, D., E. de Matos Gomes, B. Almeida, A. L. Kholkin, P. Zelenovskiy, M. Neradovskiy, and V. Ya Shur. "Energy harvesting from nanofibers of hybrid organic ferroelectric $dabcoHReO_4$." *Applied Physics Letters* 104, no. 3 (2014): 032907.
54. Szafrański, Marek, Andrzej Katrusiak, and Garry J. McIntyre. "Ferroelectric order of parallel bistable hydrogen bonds." *Physical Review Letters* 89, no. 21 (2002): 215507.
55. Qi, Jia, Nan Ma, and Ya Yang. "Photovoltaic–pyroelectric coupled effect based nanogenerators for self-powered photodetector system." *Advanced Materials Interfaces* 5, no. 3 (2018): 1701189.
56. Ji, Yun, Kewei Zhang, and Ya Yang. "A one-structure-based multieffects coupled nanogenerator for simultaneously scavenging thermal, solar, and mechanical energies." *Advanced Science* 5, no. 2 (2018): 1700622.
57. Zhao, Kun, Bangsen Ouyang, Chris R. Bowen, Zhong Lin Wang, and Ya Yang. "One-structure-based multi-effects coupled nanogenerators for flexible and self-powered multi-functional coupled sensor systems." *Nano Energy* 71 (2020): 104632.
58. Zhao, Kun, Bangsen Ouyang, Chris R. Bowen, and Ya Yang. "Enhanced photocurrent via ferro-pyro-phototronic effect in ferroelectric $BaTiO_3$ materials for a self-powered flexible photodetector system." *Nano Energy* 77 (2020): 105152.

59. Zhang, Shaoce, Bo Zhang, Dong Chen, Zhengang Guo, Mengnan Ruan, and Zhifeng Liu. "Promising pyro-photo-electric catalysis in $NaNbO_3$ via integrating solar and cold-hot alternation energy in pyroelectric-assisted photoelectrochemical system." *Nano Energy* 79 (2021): 105485.
60. Yang, Ya, Wenxi Guo, Ken C. Pradel, Guang Zhu, Yusheng Zhou, Yan Zhang, Youfan Hu, Long Lin, and Zhong Lin Wang. "Pyroelectric nanogenerators for harvesting thermoelectric energy." *Nano Letters* 12, no. 6 (2012): 2833–2838.
61. Wang, Xingfu, Yejing Dai, Ruiyuan Liu, Xu He, Shuti Li, and Zhong Lin Wang. "Light-triggered pyroelectric nanogenerator based on a PENG-junction for self-powered near-infrared photosensing." *ACS Nano* 11, no. 8 (2017): 8339–8345.
62. Zhang, Yan, Pham Thi Thuy Phuong, Eleanor Roake, Hamideh Khanbareh, Yaqiong Wang, Steve Dunn, and Chris Bowen. "Thermal energy harvesting using pyroelectric-electrochemical coupling in ferroelectric materials." *Joule* 4, no. 2 (2020): 301–309.
63. Kakekhani, A., and S. Ismail-Beigi. "Ferroelectric oxide surface chemistry: water splitting via pyroelectricity." *Journal of Materials Chemistry A* 4, no. 14 (2016): 5235–5246.
64. Zhang, Yan, Santosh Kumar, Frank Marken, Marcin Krasny, Eleanor Roake, Salvador Eslava, Steve Dunn, Enrico Da Como, and Chris R. Bowen. "Pyro-electrolytic water splitting for hydrogen generation." *Nano Energy* 58 (2019): 183–191.
65. You, Huilin, Xinxiu Ma, Zheng Wu, Linfeng Fei, Xiaoqiu Chen, Jie Yang, Yongsheng Liu et al. "Piezoelectrically/pyroelectrically-driven vibration/cold-hot energy harvesting for mechano-/pyro-bi-catalytic dye decomposition of $NaNbO_3$ nanofibers." *Nano Energy* 52 (2018): 351–359.
66. Benke, Annegret, Erik Mehner, Marco Rosenkranz, Evgenia Dmitrieva, Tilmann Leisegang, Hartmut Stöcker, Wolfgang Pompe, and Dirk C. Meyer. "Pyroelectrically driven •OH generation by barium titanate and palladium nanoparticles." *The Journal of Physical Chemistry C* 119, no. 32 (2015): 18278–18286.
67. Zarepour, Eisa, Adesoji A. Adesina, Mahbub Hassan, and Chun Tung Chou. "Nano sensor networks for tailored operation of highly efficient gas-to-liquid fuels catalysts." *Chemeca* 2013 (2013): 91.
68. Zarepour, Eisa, Adesoji A. Adesina, Mahbub Hassan, and Chun Tung Chou. "Innovative approach to improving gas-to-liquid fuel catalysis via nanosensor network modulation." *Industrial & Engineering Chemistry Research* 53, no. 14 (2014): 5728–5736.
69. Zarepour, Eisa, Mahbub Hassan, Chun Tung Chou, and Adesoji A. Adesina. "Remote detection of chemical reactions using nanoscale terahertz communication powered by pyroelectric energy harvesting." In *Proceedings of the Second Annual International Conference on Nanoscale Computing and Communication*, pp. 1–6. 2015.
70. Luo, W. B., Y. C. Yu, Y. Shuai, X. Q. Pan, Q. Q. Wu, C. G. Wu, and W. L. Zhang. "Enhanced pyroelectric properties of lead free KNN/P (VDF-TrFE) composite film by optimizing KNN sintering temperature." *Journal of Materials Science: Materials in Electronics* 27, no. 3 (2016): 2288–2292.
71. Zabek, Daniel, John Taylor, Emmanuel Le Boulbar, and Chris R. Bowen. "Micropatterning of flexible and free standing polyvinylidene difluoride (PVDF) films for enhanced pyroelectric energy transformation." *Advanced Energy Materials* 5, no. 8 (2015): 1401891.
72. Zi, Yunlong, Long Lin, Jie Wang, Sihong Wang, Jun Chen, Xing Fan, Po-Kang Yang, Fang Yi, and Zhong Lin Wang. "Triboelectric–pyroelectric–piezoelectric hybrid cell for high-efficiency energy-harvesting and self-powered sensing." *Advanced Materials* 27, no. 14 (2015): 2340–2347.
73. Kim, Jihye, Jeong Hwan Lee, Hanjun Ryu, Ju-Hyuck Lee, Usman Khan, Han Kim, Sung Soo Kwak, and Sang-Woo Kim. "High-performance piezoelectric, pyroelectric, and triboelectric nanogenerators based on P(VDF-TrFE) with controlled crystallinity and dipole alignment." *Advanced Functional Materials* 27, no. 22 (2017): 1700702.

10 Wearable Nanogenerators

Md Mazbah Uddin, Tanvir Mahady Dip, and Suraj Sharma

CONTENTS

10.1 INTRODUCTION TO WEARABLE NANOGENERATORS

The advancement of portable electronic devices has changed the way people communicate with each other and their surroundings. Self-powered devices capable of harvesting energy from the environment, especially from human biomechanics, have become desirable in the past decade. The harvested energy from the human body can be utilized directly by wearable electronics to perform numerous functional activities (Covaci and Gontean 2020; Martin et al. 2003). Such wearable power technology will play a paramount role in wearable electronics for communication, safety and security, physiological monitoring, health care, the intelligent humanoid robot, and human-interactive

DOI: 10.1201/9781003187615-10

interfaces (Bai et al. 2013; Dong, Peng, and Wang 2020a; Hussain et al. 2015; Lee and Lee 2015; Ma et al. 2018; Yang et al. 2009).

The world is progressing toward the Internet of things (IoTs), where wireless sensor networks, portable electronics, big data, robotics, and artificial intelligence are creating a buzz. IoTs will revolutionize our traditional communication and lifestyles greatly (Chen et al. 2020; Razavi, Iannucci, and Greenhalgh 2020). Therefore, miniaturized and low-power-consuming personal portable electronics are on the rise due to emerging IoTs (Razavi, Iannucci, and Greenhalgh 2020; Siddiqui et al. 2015). But these portable electronics are incompatible with conventional centralized power systems. Traditional chemical batteries are considered to power up the wearable electronics. Unfortunately, most of the batteries require chemical energy to generate electricity (Bairagi and Ali 2019; Elahi et al. 2020; Sultana et al. 2019). Moreover, the battery has a limited lifetime, subsequently requires recharging or replacement, thereby disrupting the continuous functionality of wearable electronics. Besides, chemical batteries are hazardous, structurally inflexible, and bulky (Chen et al. 2017; Kim et al. 2015; Ruiz et al. 2018).

We are living in an era of rapid information exchange, where information comes from data, and these data are measured and transmitted by distributed sensors. Since self-powered integrated sensor networks are desirable for futuristic wireless connectivity, sustainable energy solutions are a prerequisite (Chen et al. 2020). A viable and economical solution is to harness environmental energy to power up on-body wearable electronics. Therefore, numerous environmental energy sources, such as electromagnetic, solar, thermal, mechanical, and microbial/biofuel chemical energies (Figure 10.1), have been explored to fabricate wearable nanogenerators (WNGs) (Hu and Wang 2014). Electromagnetic generators (EMGs) are the main energy source of human civilization that is continuously revolutionizing the advancement of the world. The EMGs are heavy, spatial, require large magnets, and intricated coiling, greatly affecting their suitability as WNGs (Gu et al. 2020; Lee and Roh 2019; Halim et al. 2016). Recent decades have seen an insurgence of other WNGs based on harvesting mechanical (e.g., via piezoelectric [Fan, Tian, and Lin Wang 2012] and triboelectric [Wang and Song 2006] nanogenerators [NGs]), solar (i.e., via solar cells [SCs] [Jinno et al. 2017]), thermal (e.g., via pyroelectric [Leng et al. 2014], and thermoelectric nanogenerators [TEGs] [He et al. 2015]), chemical (i.e., via biofuel cells [Bandodkar et al. 2016]), and hybrid energies (i.e., via hybrid NGs [Jung et al. 2015]). The underlying shortcoming of SCs, pyroelectric nanogenerators (PyNGs), TEGs, and biofuel cells (BFCs) is their dependency on external factors of sunlight,

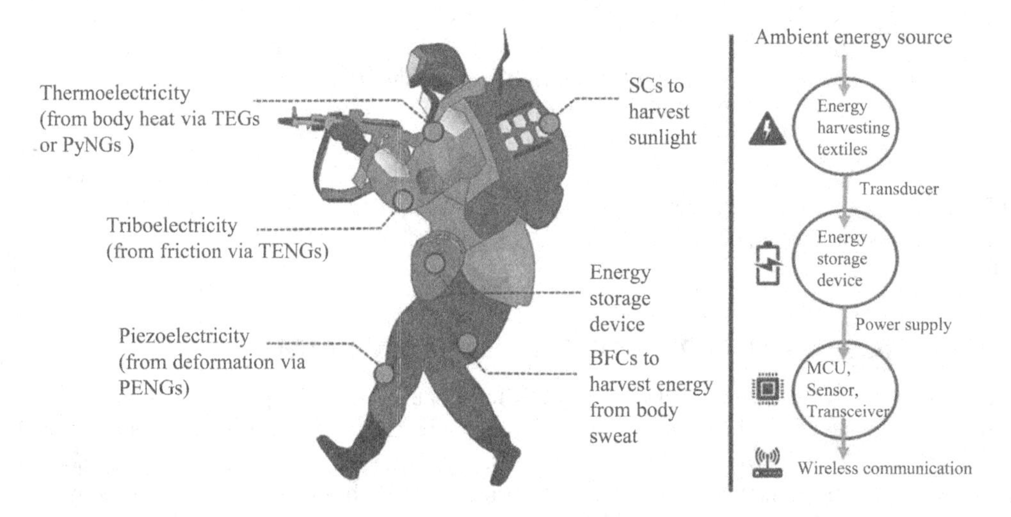

FIGURE 10.1 Different kinds of WNGs attached to a military uniform. MCU (microcontroller unit). (Reused under creative common license from Shuvo, Decaens and Lachapelle (2021). Copyright 2021 The Author(s). Licensee IntechOpen, distributed under the terms of the Creative Commons Attribution 3.0 License.)

FIGURE 10.2 Major locations of generating mechanical energy by the human body. (Based on Gogurla et al. (2019); Guo, Nie, and Wang (2020); Khalid et al. (2019); Zhang et al. (2020).)

temperature, and catalyst, respectively. Despite the shortcomings, the applicability of these devices could be extended to numerous wearable applications.

Mechanical energy is the most pervasive source of energy in the environment, which is independent, unlike other energy sources. The most viable techniques of harvesting environmental mechanical energy are triboelectric (Fan, Tian, and Lin Wang 2012) and piezoelectric (Wang and Song 2006) effects. The simplistic working mechanism has made these two techniques a desirable choice for fabricating mechanical energy harvesting WNGs. The ability of the human body to generate mechanical energy (Figure 10.2) and the capability of triboelectric and piezoelectric nanogenerators (TENGs and PENGs, respectively) to convert such low-frequency energy has led to the inception of small-scale sustainable energy harvesting technologies.

WNGs are mostly fabricated as flexible sheets in planar structures or integrated into traditional textile structures. In this chapter, we will briefly review and discuss the energy sources (Figure 10.3) and energy conversion techniques of WNGs along with their numerous structures, energy output performance, applications, challenges, and limitations.

10.2 MAJOR ENERGY SOURCES FOR WEARABLE NANOGENERATORS

10.2.1 Mechanical Energy

Energy generated by the human body is the most pervasive and one of the easily accessible, renewable, and sustainable energy sources for WNGs. The energy generated by body mechanics may reach up to 67 W during walking (Tao 2019; Wang 2011). Biomechanical energy of breathing, blood flow, elbow and knee bending, shoulder movement, finger typing, and walking can be harvested through WNGs that respond to these kinetic movements. The major techniques of harvesting energy from biomechanics or any kinds of mechanical sources are electromagnetism, triboelectric, and piezoelectric effects.

10.2.2 Solar Energy

Solar energy is also regarded as a pervasive, easily accessible, and renewable energy source (Kannan and Vakeesan 2016). Being non-exhaustible, eco-friendly, and sustainable, solar energy has picked up the credibility to be called the energy of the future (Alanne and Saari 2006; Herzog, Lipman,

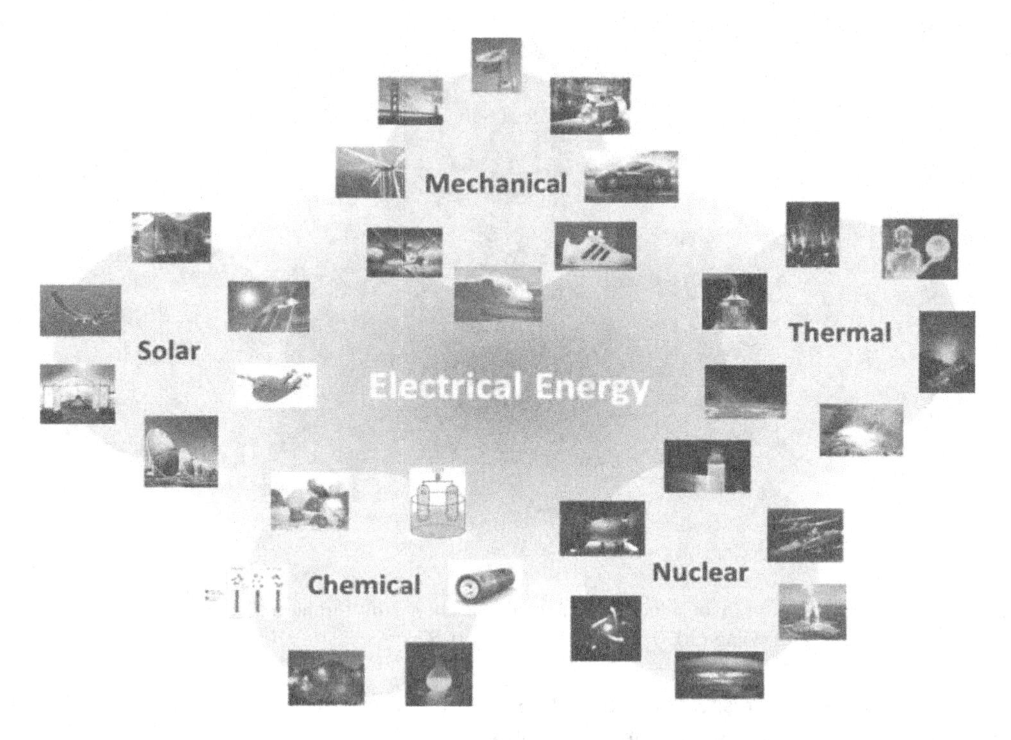

FIGURE 10.3 Various available energy sources in the environment and their representative forms. (Reused with permission from Hu and Wang (2014). Copyright Elsevier 2014.)

and Kammen 2001). As a prime source of energy, the Sun not only can provide enough power for mankind but also takes care of the pollution problem created by conventional electricity-generating methods (Varma et al. 2018). Solar energy can be intercepted in the earth, generally in the form of sunlight, radiation, and heat (Lewis 2007; Löf, Duffie, and Smith 1966). Materials like photosensitive polymers, perovskite, organic dyes, etc. are capable of being used as the harvesting part of wearable solar cells (WSCs) to tap energy from sunlight (Varma et al. 2018). To ensure flexibility and wearability, the fabricated WSCs must be bendable, twistable, stretchable, and lightweight up to a certain limit (Hashemi, Ramakrishna, and Aberle 2020).

10.2.3 Thermal Energy

Recently, thermal energies are receiving much attention due to their high potential to be converted into electrical energy through the thermoelectric (also known as the Seebeck effect) and pyroelectric effects (Ryu and Kim 2019). The human body is a great source of thermal energy since everyday heat is generated in the body due to the continuous metabolism of food, which can be utilized as a renewable source of energy (Leonov and Vullers 2009; Siddique, Mahmud, and Heyst 2017). This renewable source of energy is gaining popularity as they are a clean source and exhibit advantages like no risk of toxicity, high reliability, solid operation state, abundance, maintenance-free performance, etc. (Elsheikh et al. 2014; Li et al. 2017). Wasted heat energy can very easily be converted to electrical energy via both Seebeck and pyroelectric effects and used to power up wearable devices (Korkmaz and Kariper 2021; Xue et al. 2017). According to a recent study, the human body is capable of providing heat energy in a range of 60–180 W based on different day-to-day activities (Nozariasbmarz et al. 2020). Via blood flow, generated heat energy reaches different parts of the body. Under the excited physiological condition, the human body is capable of generating the most amount of heat energy (Proto et al. 2018). However, this energy can be released to the ambient environment through convection, radiation, or evaporation (Gagge and Gonzalez 2011;

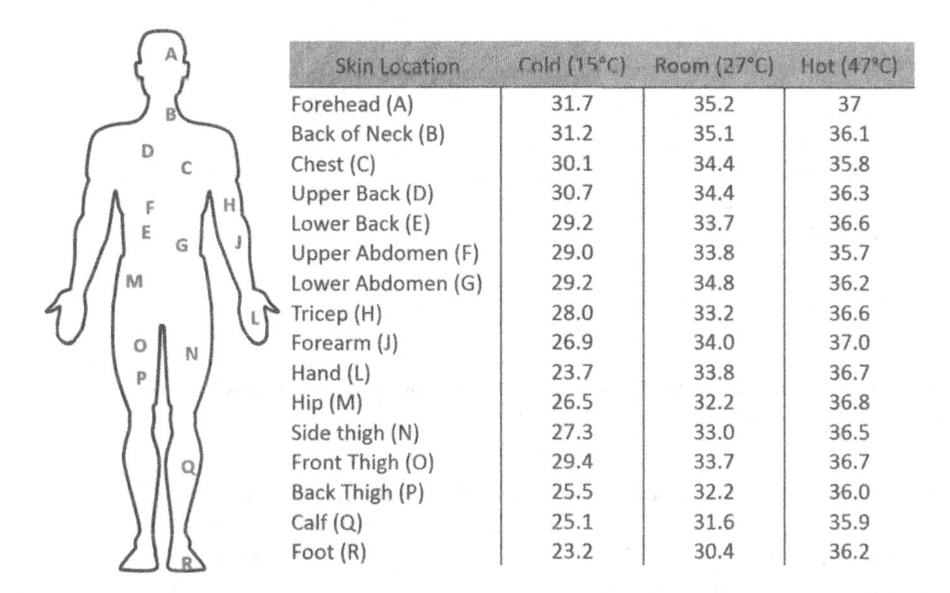

Skin Location	Cold (15°C)	Room (27°C)	Hot (47°C)
Forehead (A)	31.7	35.2	37
Back of Neck (B)	31.2	35.1	36.1
Chest (C)	30.1	34.4	35.8
Upper Back (D)	30.7	34.4	36.3
Lower Back (E)	29.2	33.7	36.6
Upper Abdomen (F)	29.0	33.8	35.7
Lower Abdomen (G)	29.2	34.8	36.2
Tricep (H)	28.0	33.2	36.6
Forearm (J)	26.9	34.0	37.0
Hand (L)	23.7	33.8	36.7
Hip (M)	26.5	32.2	36.8
Side thigh (N)	27.3	33.0	36.5
Front Thigh (O)	29.4	33.7	36.7
Back Thigh (P)	25.5	32.2	36.0
Calf (Q)	25.1	31.6	35.9
Foot (R)	23.2	30.4	36.2

FIGURE 10.4 Reported skin temperature at different parts of the body. (Reused with permission from Suarez et al. (2016). Copyright Royal Society of Chemistry 2016.)

Suarez et al. 2016). Kim et al. (2015) highlighted that the active most way of dissipating thermal energy is by perspiration during forced convection. Figure 10.4 depicts skin temperature at different body parts.

10.2.4 Biochemical Energy

Chemical energy stored in a biological fluid is gaining much attention as an organic renewable source of electrical energy over the last few decades. Their availability in abundance as well as being a pollution-free, environment-friendly, clean energy source have certainly contributed toward the attainment of their popularity (Slaughter and Kulkarni 2015). Biofluids-containing metabolites are the primary source of energy for a BFC to convert into electrical energy (Bandodkar and Wang 2016). The major sources for chemical energy that can be transformed into electrical output are sweat lactate from the perspiring human body (Jia et al. 2014), glucose, interstitial skin fluids (Valdés-Ramírez et al. 2014), tears, multiple oxidoreductase enzymes, serum, blood (Bandodkar et al. 2017; Sim et al. 2018), cerebrospinal fluids (Slaughter and Kulkarni 2015), cellulose fluids, pyruvate, etc. (Bandodkar and Wang 2016).

10.3 MAJOR ENERGY HARVESTING TECHNIQUES OF WEARABLE NANOGENERATORS

10.3.1 Mechanical Energy Harvesting Techniques

10.3.1.1 Electromagnetism

The technique of generating electricity via relative positioning and movement of magnetic and conductive materials can be incorporated into clothing to utilize wasted biomechanical energy (Beeby et al. 2007; Liu et al. 2015). Electromagnetic induction occurs when any conductor moves in a changing magnetic field, thus causing electrons to flow and thereby generating electricity (Williams, Mandal, and Sharma 2011). If magnets and conductive flexible wires are attached to clothing in such a way that causes electromagnetism, the generation of electricity from human biomechanics

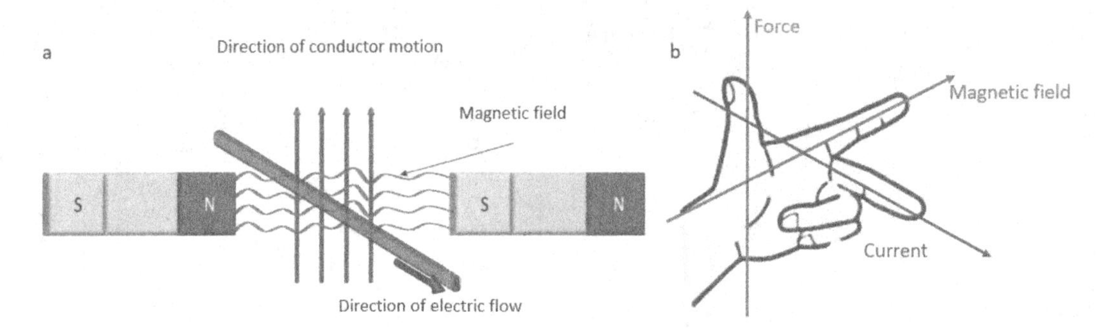

FIGURE 10.5 (a) Schematic of the induction-based electromagnetic current generation. (b) Fleming's right-hand rule of the electromagnetic current generation.

is possible. In such a system, the coil turn number, the magnetic field, the speed of coil movement, magnet arrangement, space between coil and magnet, etc. can impact the output performance. This system follows Fleming's right-hand rule, i.e., when the conductor moves in the area of a magnetic field, electricity is generated (Figure 10.5) (Lee and Roh 2019). Furthermore, to cope with the irregular human body motion, the one-way drive of magnets in wearable devices has proven to increase output performance (Liu et al. 2015).

Another technique that can harness electrical output from intermittent human movement is formed of an electromagnetic energy harvester that contains an AC-DC converter. The low-frequency AC signals received as input can be readily converted into electrical output through the functionality of the converter (Bolt et al. 2017).

10.3.1.2 Triboelectric Effect

The four fundamental working modes (Figure 10.6) of triboelectric nanogenerators (TENGs), namely, the vertical contact-separation mode, the single-electrode mode, the lateral sliding mode, and the free-standing triboelectric layer mode, were first explained by Dr. Wang of Georgia Tech

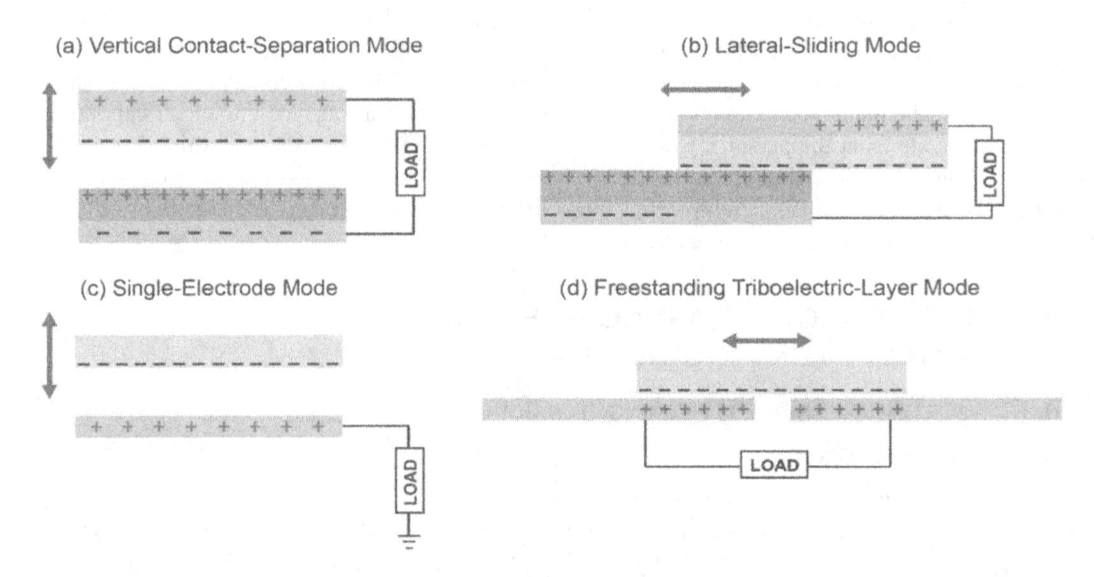

FIGURE 10.6 Four working modes of TENGs. (a) Vertical contact-separation mode. (b) Lateral sliding mode. (c) Single-electrode mode. (d) Free-standing triboelectric layer mode. (Reused with permission from Wu et al. (2019). Copyright John Wiley and Sons 2018.)

(Wang 2017). The underlying principle of triboelectricity was explained as when two distinct materials are physically contacted, electrostatic charges are generated based on the electronegativity or the electron affinity of the materials.

When a slight gap or sliding is facilitated by mechanical force, a potential difference is induced, which causes electrons to transfer across the electrodes through an external electrical circuit. A reverse flow of electrons also occurs across the electrodes when the materials are contacted or slided back again (Chen et al. 2020; Dong, Peng, and Wang 2020b). The contact separation and/or sliding may occur during finger tapping, elbow and knee-bending, swinging arm, and stepping foot. A TENG generates energy in an alternating manner due to cyclic contact-separation or sliding between two distinct materials in triboelectric series (Zou et al. 2019). Furthermore, the active materials are apart in the triboelectric series, the greater electrical output is generated by the TENGs.

10.3.1.3 Piezoelectric Effect

The generation of electric charge under mechanical deformation was first discovered by French physicists Pierre and Jacques in 1880 and was termed as the converse piezoelectric effect. In 1881, Lipmann theoretically discovered the inverse piezoelectric effect, wherein, mechanical deformation occurs as well under applied electric field (Bauer and Bauer 2008; Gupta, Suman, and Yadav 2004; Qin et al. 2016; Ramadan, Sameoto, and Evoy 2014). A piezoelectric material has non-inversion symmetry and dipoles, which endows it with the piezoelectric effect under straining. There exist different types of piezoelectric materials in the form of crystals, ceramics, and polymers, which respond to mechanical deformation and generate electric charges (Gupta, Suman, and Yadav 2004; Papagiannakis et al. 2016). Due to such a facile working mechanism (Figure 10.7) and abundance of cheap materials, piezoelectric materials have been an excellent candidate to fabricate piezoelectric nanogenerators (PENGs). A typical PENG is constituted of a piezoelectric film sandwiched between two metal films. When a mechanical force (i.e., deformation) is absent, the PENG does not exhibit any electrical output in the measurement instrument. But when an external force is applied, opposite charges are found at the film surfaces. This causes an induced electric potential across the electrodes, thereby causing electrons to flow in a direction. A peak voltage/current signal is found until saturation. Afterward, upon withdrawing the mechanical force, the piezoelectric material starts to return to its original charge-neutral position, this causes an opposite peak voltage/current signal as well (Chou et al. 2018; Jin et al. 2018). Such a working mechanism works like an AC generator.

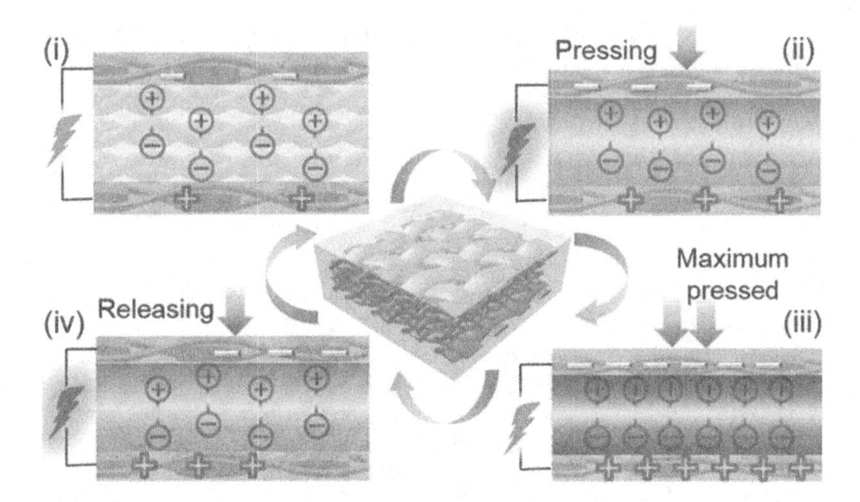

FIGURE 10.7 Working principle of PENGs. (Reused with permission from Dong, Peng, and Wang (2020a). Copyright John Wiley and Sons 2019.)

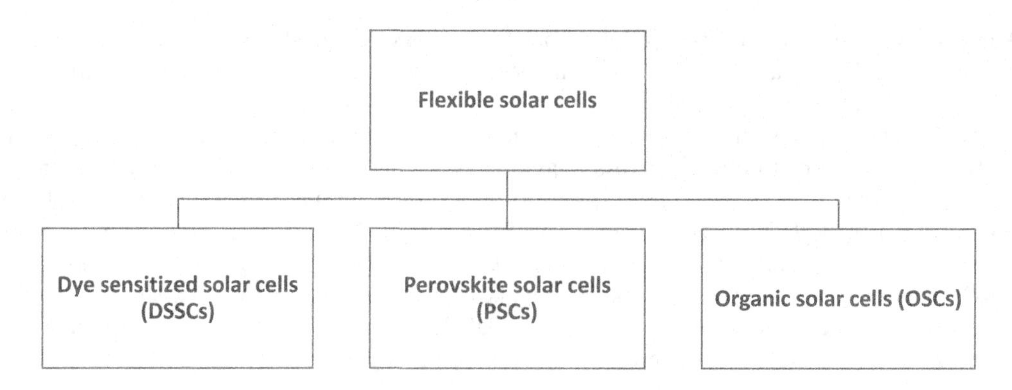

FIGURE 10.8 Major classifications of flexible SCs.

10.3.2 Solar Energy: Photovoltaic Effect

The photovoltaic (PV) effect is the main working principle of SCs, which is mainly based on the photoelectric effect introduced by A.E. Becquerel back in 1839 (Fatet 2005). It takes place upon light irradiation on any semiconducting surface through released electron flow (Satharasinghe, Hughes-Riley, and Dias 2020). Three groups of SCs are mentioned utilizing the PV effect to convert solar energy to electric energy capable of powering up flexible wearable devices (Abate et al. 2018; Hashemi, Ramakrishna, and Aberle 2020; Kumavat, Sonar, and Dalal 2017). The classification of SCs, different constituent elements, and conceptual mechanisms are demonstrated in Figures 10.8–10.10, respectively.

In the case of a DSSC, the electrical energy harnessing principle is based on the photoelectrochemical process, for instance, light-induced redox reactions (Baxter 2012). The typical DSSC operation involves a semiconducting material known as a photoanode that is flexible and mesoporous structured. This material contains electrolytes having redox couple, which is coated with photoactive dyes, generally organic. It also has a counter electrode (Ye et al. 2015). Upon coming into contact with light energy, the generation and migration of electrons in oxidized dye particles help to harness electricity in the system. The dye injects electrons in the conduction band of the semiconductor. After migration to the external circuit, the counter electrode acts to return the electrons to the electrolyte. However, the use of liquid electrolyte creates difficulties in the case of wearable applications (Fu et al. 2018).

The mechanism of PSC is dependent mainly on the functionality of the photoactive perovskite layer. The general structure is a sandwich condition where the perovskite layer resides between a

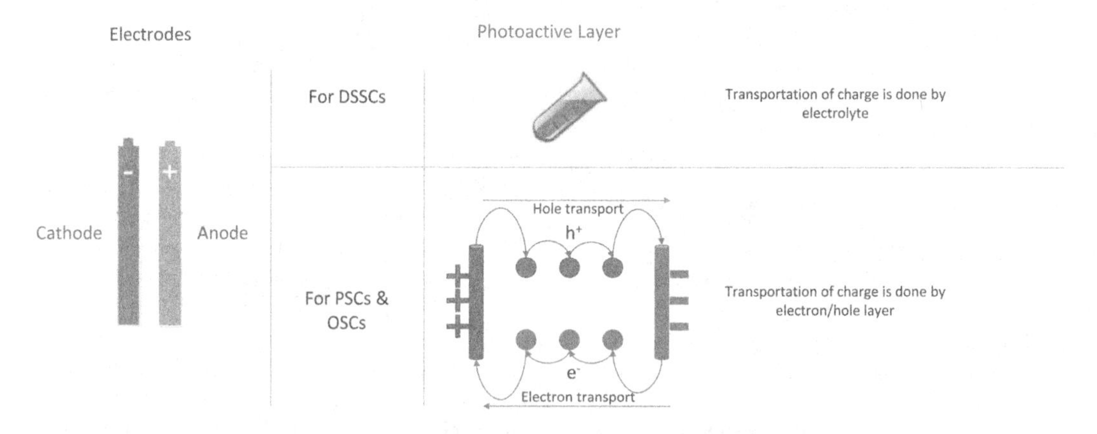

FIGURE 10.9 Constituent elements of DSSCs, PSCs, and OSCs.

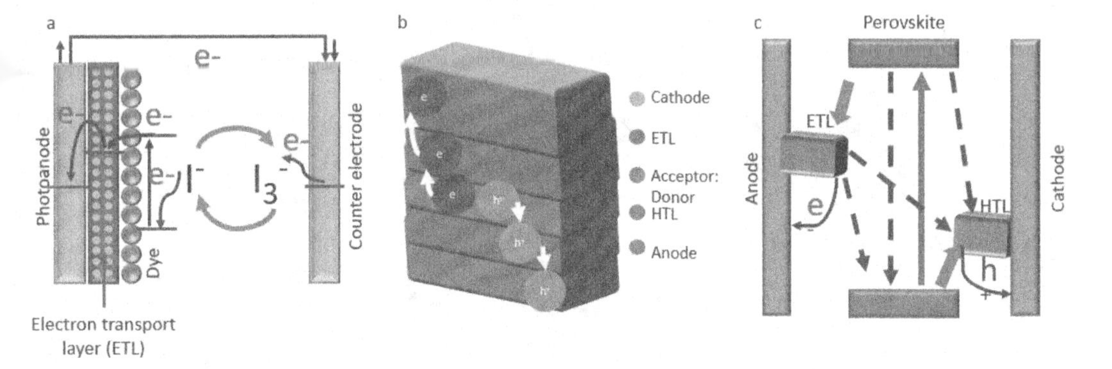

FIGURE 10.10 Conceptual mechanisms of (a) DSSCs, (b) OSCs, and (c) PSCs. (Recreated with permission from Fu et al. (2018). Copyright Elsevier 2018.)

couple of charges transporting layers. In the event of getting excited by energy from sunlight, the absorbed electrons in the valence shell of the photoactive layer go up to the conduction layer. There electrons and holes get transported via activities of electron transport layer (ETL) and hole transport layer (HTL), respectively, before they are collected in anode and cathode electrodes (Fu et al. 2018; Hashemi, Ramakrishna, and Aberle 2020). This basic structure can be in any one of the following sequences: ETL-perovskite layer-HTL or HTL-perovskite layer-ETL. Such a PV device facilitates the wearable application as it can be found in solid-state as well as the temperature for production is quite low (Di Giacomo et al. 2016; Roldán-Carmona et al. 2014; Zhang et al. 2016).

OSC is another form of PV device that is being utilized heavily due to its flexibility. A typical structure of OSC generally comprises a photoactive layer that contains a p-type electron donor and an n-type electron acceptor. Like PSC, this layer is also sandwiched between a couple of charge transport systems. The ETL is responsible for transporting electrons to the cathode and is found between the photoactive layer and cathode. On the other hand, the HTL transports holes to the anode and lies between the photoactive layer and anode (Li et al. 2018). When the layer gets activated by illumination, the electron-hole pair is generated and reaches the donor-acceptor interface, where later, they get separated before subjected to transportation by ETL and HTL. However, such a mechanism delivers a very low power conversion efficiency (PCE) compared to other similar techniques (Fu et al. 2018).

10.3.3 Thermal Energy

10.3.3.1 Thermoelectric Effect

The NGs based on the thermoelectric effect are similar to heat engines that can transform temperature difference into an electrical output which is otherwise known as the 'Seebeck effect' (Indira et al. 2019). This Seebeck effect is the principal mechanism for wearable thermoelectric nanogenerators (WTEGs). This effect was observed for the very first time by Thomas Johann Seebeck in 1821 (Gould et al. 2008; Nolas, Sharp, and Goldsmid 2001; Zhang, Wang, and Yang 2019), and the first TEG was fabricated in 2012 (Yang, Pradel et al. 2012). The basis of such a mechanism was the creation of thermoelectric potential due to the difference of temperature across a couple of thermoelectric materials. The spatial temperature gradient between the human body and the environment is a great prospect in this regard (Deng et al. 2017). The electrical output of such a mechanism has a positive correlation with a temperature gradient (Champier 2017; Patel, Mehta, and Shah 2015; Siddique, Mahmud, and Heyst 2017).

A typical schematic of a TEG is shown in Figure 10.11. The mechanism takes place through the functionality of two semiconductors containing excess electrons and holes. They are called n-type and p-type semiconductors, respectively. They are connected both in series and parallel from

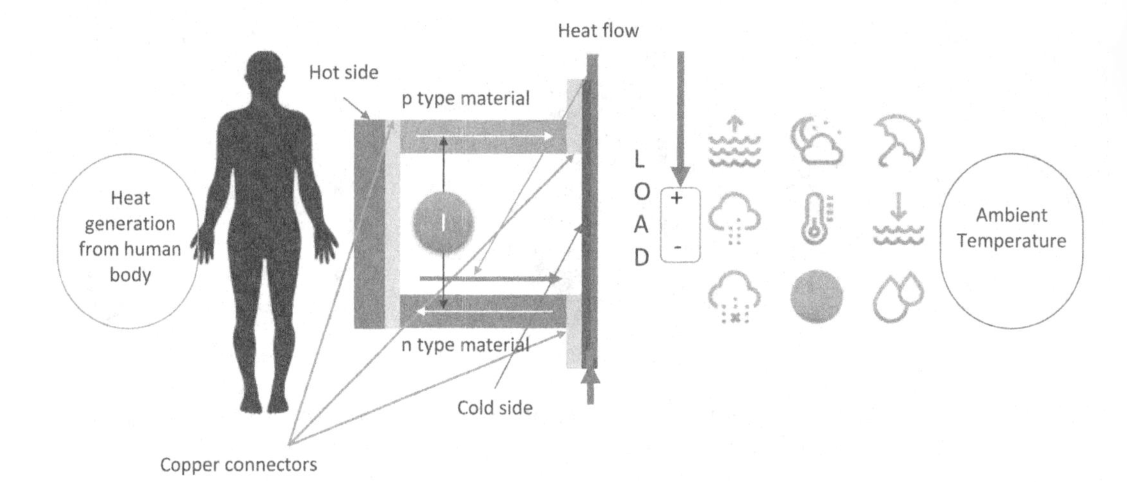

FIGURE 10.11 Schematics of Seebeck mechanism in a WTEG. (Adapted with permission from Siddique, Mahmud, and Heyst (2017). Copyright Elsevier 2017.)

the electrical and thermal points of view, respectively (Nozariasbmarz et al. 2020). The general assembly also contains electrodes, insulators, and some bonding materials along with thermoelectric materials (Kim, Liu, and Ren 2017; Yang et al. 2018). Electricity is generated when the heat is flown from the hot side of the TEG toward the cold side. Due to the movement in the electrons and holes during heat flow in the thermoelectric semiconductors, thermal energy is transformed into electrical output (Zheng et al. 2013). So the crucial factor for a TEG to perform is to create a substantial temperature difference across the device (Kraemer et al. 2011; He et al. 2018). They can play a crucial role in developing a self-powered body area network (Li et al. 2017). The semiconductor's thermal conductivity, physical dimension, fill factor, external spreaders, and lateral heat flow have important contributions in deciding the TEG's performance (Suarez et al. 2016). Moreover, when the TEG is in contact with the human body, the body's internal heat transmission parameters are also important factors (Deng et al. 2017). The flexible and wearable TEGs are finding their applications in waste heat recovery, artificial intelligence, self-powered monitoring and sensing systems, mobile low power consumable electronics, etc. (Zhang, Wang, and Yang 2019).

Another concept of this principle is the spin Seebeck effect based on the thermocouple and magnetic element. In the presence of a magnetic multilayer, the Seebeck coefficient is affected by spin. When the thermocouple has a different Seebeck coefficient, the temperature difference generates voltage through a circuit. Under a temperature gradient when the magnet element is placed, the charge carriers shift from the hot side to the cold side and are gathered there as spin voltage (Shinjo 2013; Zhao et al. 2020). The schematics of the principle are depicted in Figure 10.12.

10.3.3.2 Pyroelectric Effect

The pyroelectric effect is another technique of harvesting thermal energy from the environment (Indira et al. 2019). The very first PyNG was developed back in 2012 (Yang, Guo et al. 2012). The PyNGs function through the pyroelectric effect that is converting thermal energy from different sources (mentioned in Section 10.2.3) into electrical output. The basis of this mechanism is the reorganization of existing dipole orientation due to temperature fluctuations over time (Nozariasbmarz et al. 2020). With varying temperatures over time, a variation in the polarization of material occurs, which causes the surface charge to transform (Li et al. 2020). So by temporarily altering the existing polarization by introducing a change in temperature of pyroelectric material, such an effect can be generated (Korkmaz and Kariper 2021; Lang 2005; Ryu

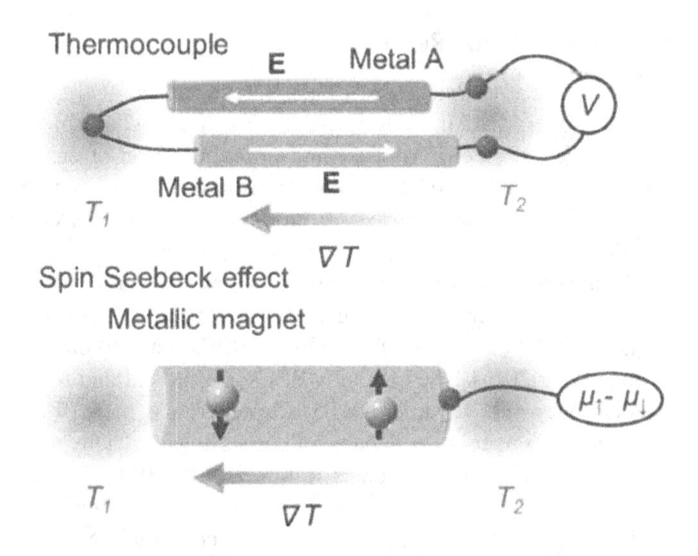

FIGURE 10.12 Schematics of spin Seebeck effect. (Reused under creative common license attribution 4.0 (https://creativecommons.org/licenses/by/4.0/) from Zhao et al. (2020). Copyright John Wiley and Sons 2020.)

and Kim 2019). A typical PyNG construction is illustrated in Figure 10.13. One of the specialties of this device is that it is capable of sensing fluctuations of temperature and so can be utilized as a thermal sensor. The performance of such a device depends significantly on the pyroelectric coefficient of the material used. This coefficient improvement can be brought about by modifying the structure of the material (Ryu and Kim 2019).

TEGs and PyNGs are popular prospects to acquire heat energy from the surrounding and generate electrical power for small self-powered devices, sensors, and monitors via Seebeck and pyroelectric effects, respectively (Yang, Wang et al. 2012; Zhang, Song et al. 2019; Zhang, Wang, and Yang 2019; Zhang, Zhang et al. 2019). Although both TEGs and PyNGs are capable of converting thermal energy into electrical energy, the distinction lies in the fact that the Seebeck effect cannot access temperature fluctuation that is dependent on time, whereas the pyroelectric effect can (Indira et al. 2019; Yang, Guo et al. 2012). In the case of a PyNG, the generation of electricity is observed at the surfaces of the polarized materials due to asymmetric potential, whereas, in the case of TEGs, electricity is generated across the active thermoelectric materials due to temperature gradient (Kim, We, and Cho 2014; He et al. 2018).

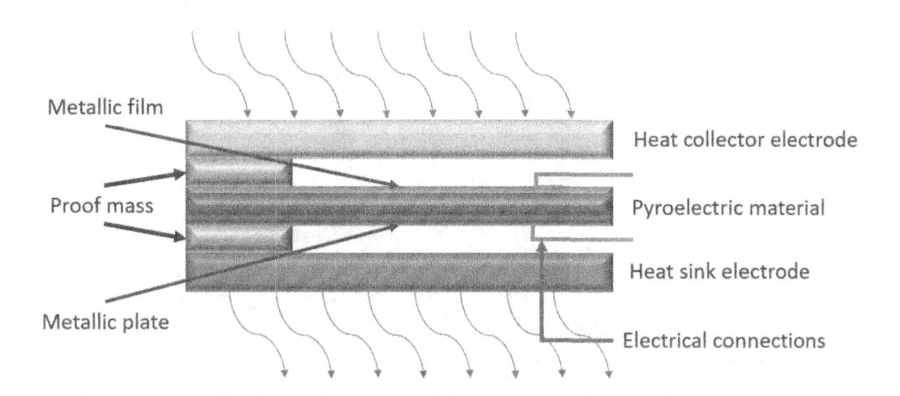

FIGURE 10.13 Schematics of a typical pyroelectric effect in a PyNG. (Recreated with permission from Bowen et al. (2014). Copyright Royal Society of Chemistry 2014.)

10.3.4 Biochemical Energy: Redox Reactions

The naturally occurring bio-fluids are harnessed to generate electrical energy by redox reactions in BFCs. The two principal operations that are impactful for the mechanism of a stable biofluid harvesting unit are the kinetics of electric transfer and immobilization of enzyme under various mechanical deformations and variation of operating environments (Jeerapan, Sempionatto, and Wang 2020). The concept of generating electricity from biofluid was also envisioned for the very first time back in 1970. This idea was based on using blood glucose as a biofluid. However, the very first BFC in the form of the contact lens was first introduced by Falk et al. (2012). This BFC was operated utilizing human tears (Falk et al. 2012). Such biotic energy harvesting devices can be classified into two main categories: (a) microbial fuel cells (MBFCs) and (b) enzymatic biofuel cells (EBFCs) (Bandodkar and Wang 2016). The idea of MBFCs is quite old and the initial exploration dates back to the 1970s (Rabaey and Verstraete 2005). However, it is the EBFCs that are popular with human wearable applications as they are capable of harnessing electricity via biocatalytic reactions of the metabolites present in biofluids (Valdés-Ramírez et al. 2014). Enzymes are capable of carrying out an elevated amount of electrolysis when connected efficiently with the active electrodes (Cosnier, Le Goff, and Holzinger 2014). An EBFC functions in combination with a bio-anode and a bio-cathode which are functionalized with enzymes. So electrodes and enzymes are integral parts of the system. The principal distinctive character of such devices is that they exploit the utilization of biocatalyst instead of normal metallic catalysts (Jeerapan, Sempionatto, and Wang 2020). At the bio-anode part, electro-oxidization of biofluid takes place. At the cathode part, usually, electro-reduction of oxygen takes place, propagated either by oxidase enzymes or metal-based catalysts (Bandodkar and Wang 2016). During oxidation, the electron is released, which travels through a circuit toward the bio-cathode. The flow of electrons during this phenomenon is responsible for generating bio-electricity (Slaughter and Kulkarni 2015). The intensity of the generated electricity has a positive correlation with the concentration of biofuel used. The mechanism is depicted below in Figure 10.14.

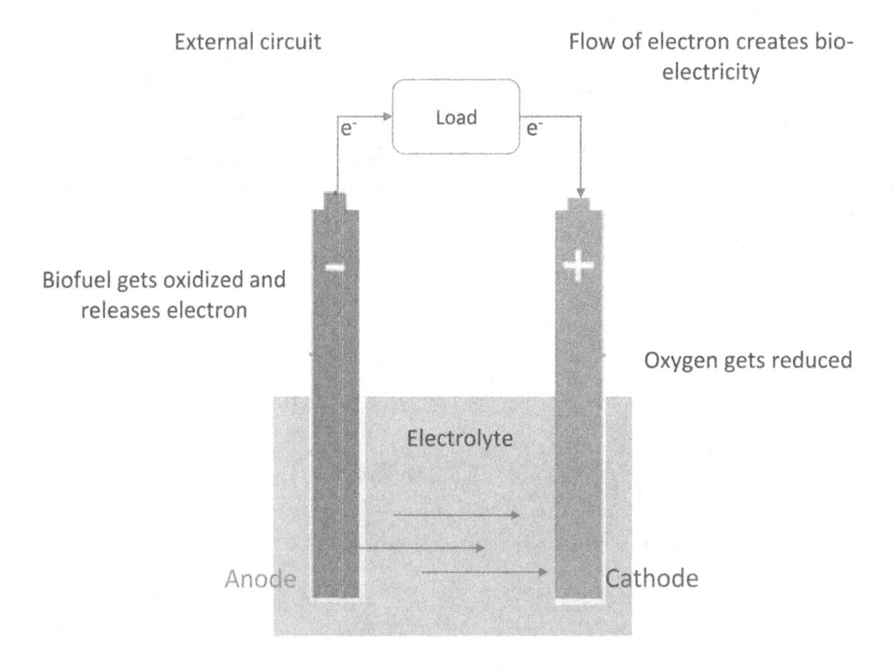

FIGURE 10.14 Mechanism of electricity generation in a biofuel cell, adapted from Slaughter and Kulkarni (2015). (Adapted under creative common license. Copyright 2015 Slaughter and Kulkarni.)

10.4 CLASSIFICATION OF WEARABLE NANOGENERATORS

10.4.1 Textiles-Based Wearable Nanogenerators

Textiles have been a feature of human society that can be traced back to 34,000 BCE. The usefulness of textiles is no longer limited to skin protection, body temperature regulation, esthetics, and decorations. The development of materials science and engineering has provided textiles with smart functionalities that enable these to interact with their surrounding stimuli (i.e., biomechanics, heat, sunlight, bodily secretions, etc.) (Chen et al. 2020). The insurgence of smart wearable electronics and multifaceted artificial intelligent systems, capable of performing multiple functionalities has initiated the field of WNGs. Harvesting energy from the human body, and its surroundings, has been acknowledged as a viable idea to power up these wearable electronics (Dong, Peng, and Wang 2020a). Herein, textiles have resumed their traditional functionalities but additionally providing sufficient electrical energy to run on body electronics. Typical power requirements of wearable electronics and IoTs components are summarized in Figure 10.15.

10.4.1.1 Wearable Nanogenerators Based on Biomechanics of the Human Body

The human body generates enough mechanical energy that can be utilized to run low-power-consuming wearable electronics and IoTs components. The calculated mechanical energies generated by different biomechanics of the human body are summarized in Figure 10.16. Conversion of such biomechanical energy into electrical energy by the smart textiles provides an alternative, green, more economical, and more convenient source of energy to run on body electronics. The most common techniques of harvesting mechanical energy are based on electromagnetic, triboelectric, and piezoelectric effects (Wang 2017).

10.4.1.1.1 Wearable Nanogenerators Based on Electro-Magnetic Effect (WEMGs)

The report of WNGs based on electromagnetic effect is less frequent than triboelectric and piezoelectric effects due to structural complexities. The basic electromagnetic mechanism is based on the relative motion between magnets and conductive coils generating magnetic and electric fields.

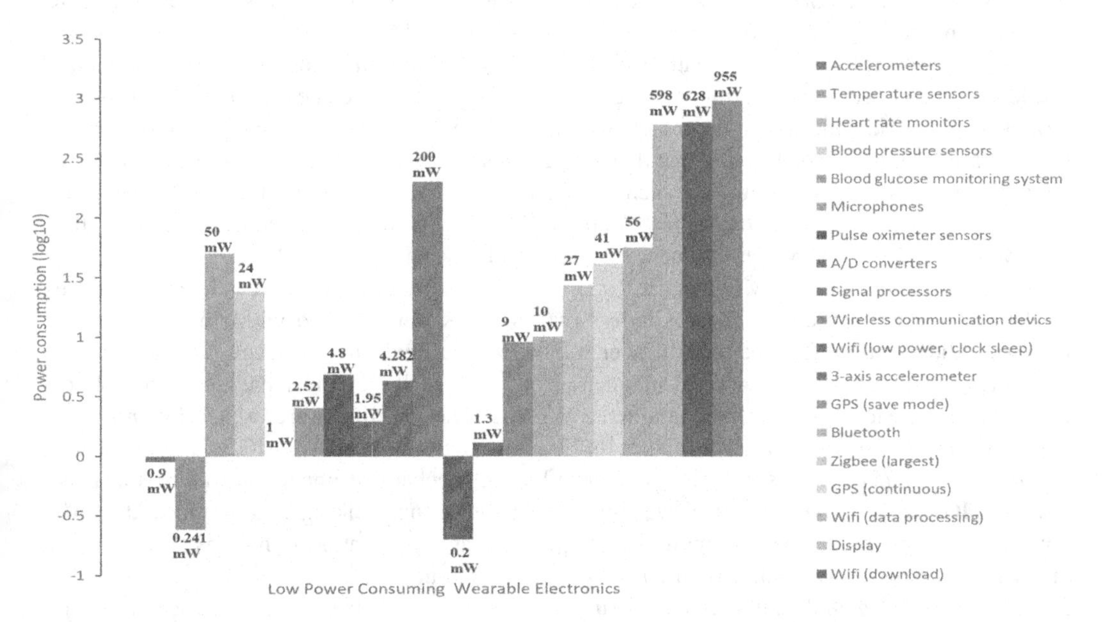

FIGURE 10.15 Reported power consumption of typical wearable electronics and IoTs components. (Based on Gljušćić et al. (2019); Tao (2019).)

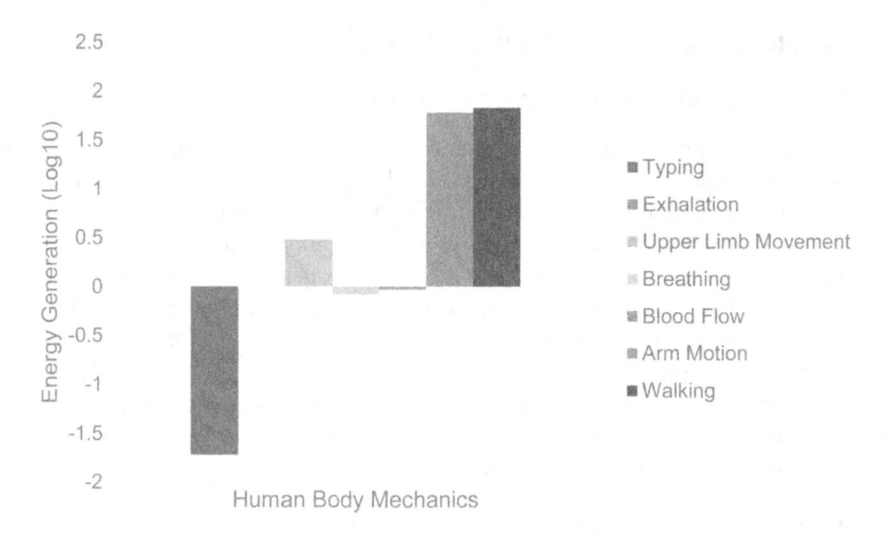

FIGURE 10.16 Reported energy generated by human body mechanics. (Based on Wang (2011) and Tao (2019).)

A wearable electromagnetic generator (WEMG) integrated into an outdoor garment capable of converting swinging motion of arms and legs into electrical energy while walking was reported by Lee et al. Flat magnets and flat conductive yarn-based coils were used to fabricate the WEMG. The magnets were placed on the side of the torso and the coils were placed inside the upper arm. Numerous design parameters such as conductive yarn types, the number of turns of the coil, circuitry, magnet arrangement, speed of the swing, and distance between magnets and coils affect the performance of the WEMG. Tests were performed for slow (i.e., 1.0-Hz frequency) and fast (i.e., 1.4-Hz frequency) walking modes. Design construction including 3-plied conductive thread turned into a coil of 200 turns along with two magnets at 1.4-Hz frequency exhibited about peak voltage of ~15 V and average current of 399.42 μA (Lee and Roh 2019). Another low-frequency WEMG capable of harnessing energy from handshaking was developed by Liu et al. The WEMG consists of a Teflon-based cylindrical pipe with two disk magnets fixed at the ends. A bulk magnet placed inside the pipe between the fixed magnets can travel freely across the ends. Conductive coils were wrapped outside the pipe within the travel distance of the bulk magnet. The magnets at the ends of the pipe are arranged in such a way that they work as a spring to suspend the bulk magnet between the ends. Various process parameters such as pipe height (H_p), coil height (H_c), coil diameter, and coil length were assessed to optimize power output. The WEMG demonstrated a maximum peak to peak voltage of 5.64 V at 5.0-Hz frequency during handshaking (Liu et al. 2015).

A similar cylindrical WEMG (Figure 10.17(a)) was fabricated by Halim et al. with high flux stacked magnets at the ends of the cylinder. Two magnets were stacked with similar poles facing each other, while a ferromagnetic spacer was placed between stacked faces. Conductive coils were then wound around the stacked magnets. Both ends of the cylinder had such an architecture. Therefore, each cylinder end contained a WEMG. The WEMGs were excited by an inertial mass capable to move freely across the cylinder ends. Between each WEMG and cylinder end, spring-mass-damper systems were introduced to handle mechanical impact by the inertial mass. The WEMGs demonstrated energy harvesting from handshaking, walking, and running. At 43-MΩ optimum load resistance, the maximum peak to peak voltage for mentioned motions was 1.15, 1.17, and 1.21 V, respectively. Similarly, the average power output was 203 μW from shaking hands, 32 μW from walking, and 72 μW from slow running (Halim et al. 2016). Gu et al. fabricated a resin-based 3D-printed goblet-shaped WEMG capable of harvesting energy from the wrist, arm, ankle, and knee movements. The WEMG was designed with a neodymium (NdFeB) ball-shaped magnet

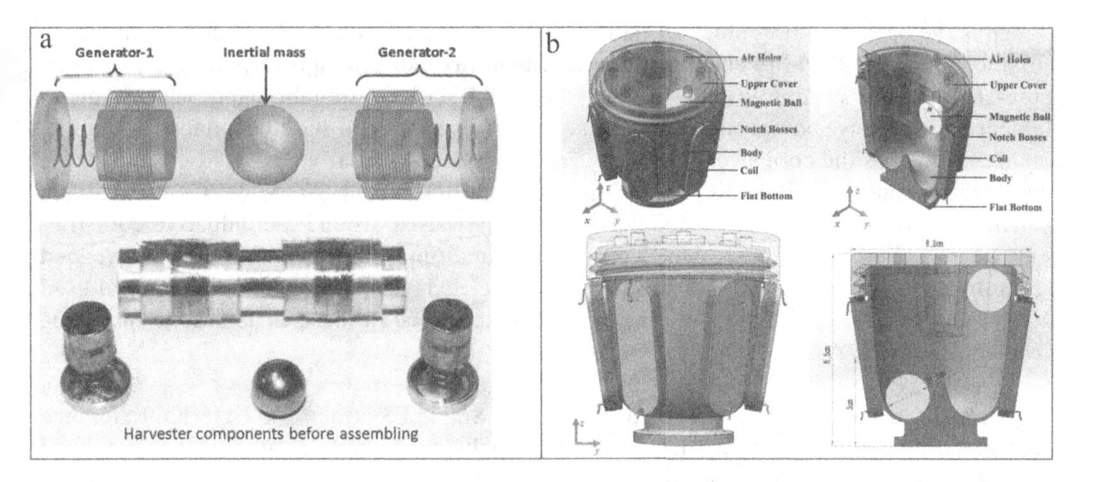

FIGURE 10.17 (a) An AAA-battery-sized WEMG capable of harvesting energy from handshaking, walking, and running. (Reused with permission from Halim et al. (2016). Copyright Elsevier 2016.) (b) A goblet-shaped nonlinear WMEG to harvest planar multidirectional vibrational energy. (Reused with permission from Gu et al. (2020). Copyright Elsevier 2020.)

with Cu-coils wound around the goblet as illustrated in Figure 10.17(b). While attached to the ankle of a man walking at a normal speed of 5 km h^{-1}, the WEMG generated 1.4 mW power at 17.6 Ω load resistance (Gu et al. 2020).

In summary, though reported WEMGs showed the capability of harvesting various mechanical motions of the human body into electrical energy, the design constructions are bulky, and complex due to the presence of inflexible magnets and coils. Such architecture of the WEMG could significantly reduce the wearable comforts of the wearers. This could be the main reason for less development of the WEMGs when different WNGs based on other mechanisms had progressed during the last decade.

10.4.1.1.2 Wearable Nanogenerators Based on Triboelectric Effect (WTENGs)

The mechanism behind TENG originates from Maxwell's displacement current. Since its inception in 2012, a plethora of textiles-based TENGs structures have been developed. Mostly, textiles with dissimilar electron affinities are assembled in various modes of TENGs (illustrated in Figure 10.6) to fabricate wearable triboelectric nanogenerators (WTENGs) (Wang 2017). WTENGs are mostly in the form of coaxial yarn-like structures, yarns interlaced or interloped into woven or knitted traditional fabric-based structures, textiles-based stacked layered structures, and textiles-based in-plane sliding mode structures. WTENGs based on different 3D-printed geometries are also on the rise due to technological development in 3D printing (Chen et al. 2018).

10.4.1.1.2.1 WTENGs Based on Coaxial Yarn (CY-WTENGs) The definition and identification of fiber, filament, yarn, and thread are very explicit in the textile industry. Since researchers of diverse backgrounds are developing such WNGs, the identification of the coaxial yarn-based WTENGs (CY-WTENGs) is not always followed conventionally. Besides, categorizing these as fiber, filament, yarn, or thread is difficult due to atypical constructions of these structures (e.g., triboelectric-layer-coated metal filament, metal spun yarn, neat metal thread, conductive composite filament, metal cladded filament, etc.) (Chen et al. 2020; Dong, Peng, and Wang 2020a). Since these structures are widely considered as 1D dimensional and all such WTENGs are fabricated with a core-shell design. We will address any of the final CY-WTENG structures found in literature as core-shell yarn. Since many of these core-shell structures are also woven into 2D and 3D fabrics, the yarn seems an appropriate identifier.

Yarn is the basic building block of fabric/textiles. It has an excellent aspect ratio of length to diameter. Such physicality endows it greater flexibility and wearability as textiles. Usually, a core-shell coaxial structure is approached when CY-WTENGs are fabricated, wherein metal filament or conductive composite filament (i.e., polymer and metal-based conductive composites) are utilized as the core electrode. The CY-WTENGs can usually be classified into single electrode integrated CY-WTENGs and dual electrode integrated CY-WTENGs. In a single electrode CY-WTENG, a single triboelectric layer is coated around a conductive core filament, whereas, in the latter case, two electrodes with multiple triboelectric layers are arranged in four different arrangements of electrodes, triboelectric layers, and protective sheath if used (Dong, Peng, and Wang 2020a). An air gap is also facilitated in these structures to introduce triboelectrification.

10.4.1.1.2.1.1 CY-WTENGs Based on an Integrated Single Electrode and External Reference Electrode This configuration of CY-WTENG includes a single-core electrode wrapped with a triboelectric layer, wherein the second electrode is not a part of the actual CY-WTENG, rather part of the circuit connected with a load as a reference electrode (Yang, Sun et al. 2018).

Lai et al. reported a CY-WTENG with multi-twisted stainless-steel as the core thread and silicone rubber as the triboelectric layer. The human skin functioned as the second triboelectric layer for the CY-WTENG. When sewn onto an elastic textile in a serpentine path (Figure 10.18(a)), the CY-WTENG was able to harness various kinds of human body motions. A peak open-circuit voltage (V_{oc}) of ~200 V and a peak short circuit current (I_{sc}) of 200 μA were reported from a 6.5 × 4.5 in^2 CY-WTENG sewn textile (Lai et al. 2017). Dong et al. fabricated a similar CY-WTENG with stainless steel/polyester (PET) fiber blended 3-plied thread as the core electrode and silicone rubber as the triboelectric material. In this case, the single CY-WTENG yarn was weft-knitted into a wrist band and a wearable fabric (Figure 10.18(b)). The final fabric-based WTENG demonstrated a maximum instantaneous power density (PD) of ~85 mW m^{-2} at 100-MΩ load resistance. Objects such as hand, foot, glove, and so on can function as the second triboelectric layer (Dong, Wang et al. 2017). Injecting Galinstan liquid metal-based highly stretchable electrode into a silicone rubber tube to fabricate a CY-WTENG was demonstrated by Yang et al. The single CY-WTENG can be further woven into a piece of fabric (Figure 10.18(c)). Herein, human skin functioned as the second triboelectric layer. The output from a bracelet (Figure 10.18(c)) like WTENG was found to be peak V_{oc} of 64 V and peak I_{sc} of 1.5 μA (Yang, Sun et al. 2018).

A similar approach was also undertaken by Park et al. to fabricate a woven WTENG, where human skin was one of the triboelectric layers. Herein, three Au-coated Cu metal-based filament electrode was wrapped spatially onto silicone filament; the assembly was further coated with the silicone rubber. When fabricated into a woven WTENG (Figure 10.18(d)), it provided a PD of ~34.4 μW cm^{-2} at a load of 10^6 Ω when contacted the human skin (Park et al. 2018). Yu et al. demonstrated a CY-WTENG that could easily be woven and knitted into WTENG textiles. They adapted the traditional core-shell yarn spinning technique of the textile industry. This facilitates the utilization of various fibers of natural and synthetic origins. The constructed fabrics were utilized as single electrode mode, vertical-contact separation mode, and free-standing mode WTENGs with different commercially available fabrics (Figure 10.18(e)) (Yu et al. 2017). Liu et al. made common PET yarns conductive by coating them with Ni/Cu. Afterwards, polydimethylsiloxane (PDMS) silicone rubber was coated around the metal cladding yarn to form the CY-WTENG. They also integrated the CY-WTENG into a woven structure (Figure 10.18(f)), which with other common woven fabric can be used as WTENGs (Liu, Cong et al. 2019a). Single CY-WTENGs have lower energy output, but when introduced as fabric-like structures, the output could be greatly enhanced. Table 10.1 summarizes the recently developed CY-WTENGs based on single electrode coaxial yarn configurations.

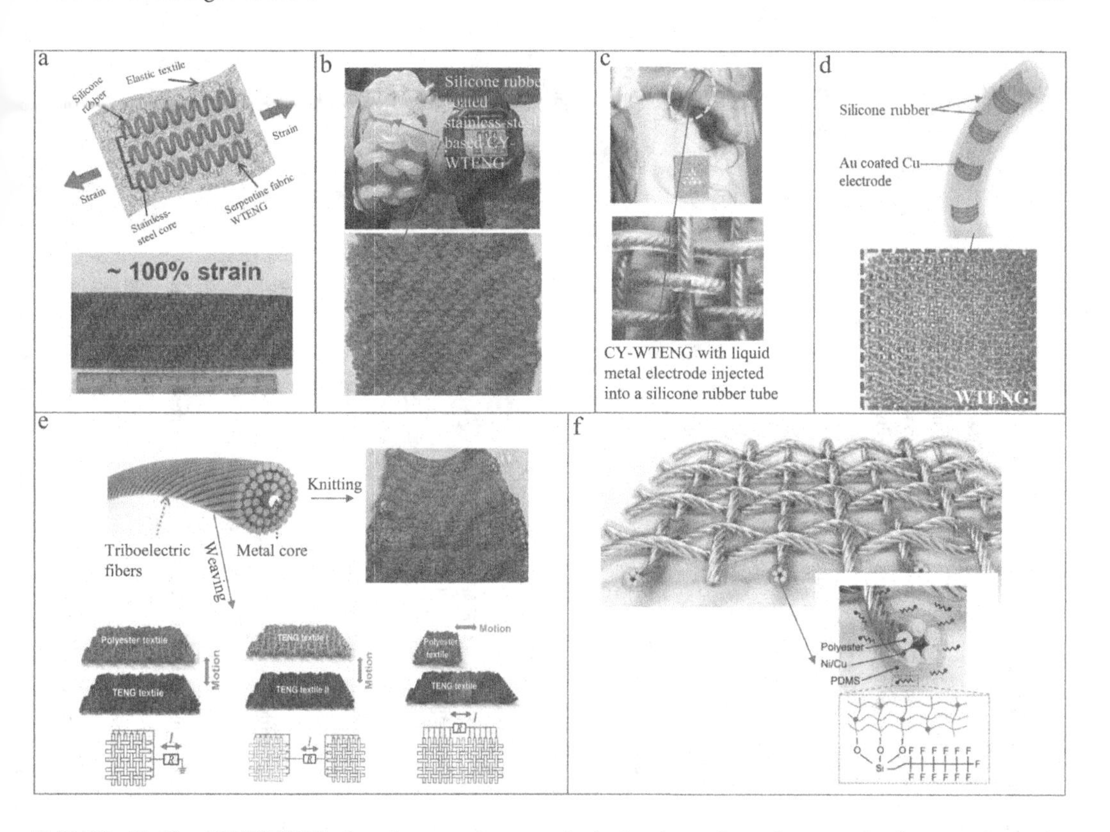

FIGURE 10.18 CY-WTENGs based on an integrated single electrode and external reference electrode. (a) WTENG fabricated by sewing a silicone-rubber-coated multi twisted stainless steel CY-WTENG into serpentine patterns. (Reused with permission from Lai et al. (2017). Copyright John Wiley and Sons 2016.) (b) WTENG fabricated by knitting a silicone-rubber-coated 3-ply stainless steel/PET blended CY-WTENG yarn into a wrist band and knitted fabrics. (Reused with permission from Dong, Wang et al. (2017). Copyright American Chemical Society 2017.) (c) A bracelet/woven fabric WTENG fabricated from Liquid metal injected silicone rubber tube-based CY-WTENG. (Reused with permission from Yang, Sun et al. (2018). Copyright American Chemical Society 2018.) (d) A woven WTENG based on the spatially wrapped Au-coated Cu electrode onto silicone filaments. (Reused under creative common license attribution 4.0 (https://creativecommons.org/licenses/by/4.0/) from Park et al. (2018). Copyright 2018 author(s).) (e) Woven and knitted power cloths fabricated from core-spun yarns. (Reused with permission from Yu et al. (2017). Copyright American Chemical Society 2017.) (f) A woven WTENG of Ni/Cu-coated PET-based core electrode and PDMS shell triboelectric material. (Reused with permission from Liu, Cong et al. (2019a). Copyright John Wiley and Sons 2019.)

10.4.1.1.2.1.2 CY-WTENGs Based on Integrated Dual Electrodes The construction of dual-electrode-based CY-WTENGs usually include four types of configurations, wherein both the electrodes are part of the CY-WTENGs and placed in different position of the coaxial design constructions. The four configurations are illustrated in Figure 10.19.

In a study, tube-like, highly stretchable CY-WTENG (Figure 10.20(a)) was fabricated in two steps. Firstly, silicone rubber and silicone-rubber-based conductive outer electrode were fabricated and turned into a tube structure, respectively. The silicone rubber material functioned as a triboelectric layer. The composite outer electrode was fabricated by mixing silicone rubber with conductive carbon black (CB) and carbon nanotubes (CNTs) materials. Finally, a belt-shaped helical core electrode was then placed at the center of the tube-like structure to fabricate the CY-WTENG. The CY-WTENG can harness energy from bending, twisting, and stretching motions when triboelectric phenomena between the core electrode and silicone rubber incur (Wang et al. 2016). Kim et al. grew ZnO nanowires (NWs) vertically along the surface of an Al filament, which was further coated with

TABLE 10.1
Summary of Recently Developed CY-WTENGs Based on Single Electrode Coaxial Yarns

Active Materials	Electrode Materials	Output	Applied As	Special Attributes	Ref.
Silicone rubber vs human skin	Multi twisted stainless-steel thread	Peak V_{oc} of ~15 V and peak I_{sc} of ~7 μA form a 5 cm single yarn at 3 cm contact separation distance between silicone rubber and skin	Commercial smartwatch, integration with wireless wearable keyboards, and smart beds	Identifying digital gestures, monitoring a human physiological signal	Lai et al. (2017)
Silicone rubber vs acrylic plate	Three-ply twisted stainless steel/PET fiber blended yarn	Peak V_{oc} of ~150 V, peak I_{sc} of ~2.9 μA, and peak instantons PD of ~85 mW m^{-2} (at ~100 MΩ) from a 40 × 40 mm^2 single CY-WTENG integrated knitted fabric at 5-Hz frequency and 11 N force	Powering up a temperature-humidity meter or a calculator by hand tapping and running an LED warning sign	–	Dong, Wang et al. (2017)
Silicone rubber vs human skin	Conductive thread (unidentified)	Peak V_{oc} of ~28 V and peak I_{sc} of ~0.56 μA from a single CY-WTENG of 8 cm length and 1.2 mm diameter at 1 kgf. Besides, a woven structure with 45 mm × 45 mm, generated peak V_{oc} of 170 and peak I_{sc} of 6 μA	Powering up LEDs and electronic watch	–	Park et al. (2017)
Silicone rubber vs animal skin	Liquid galinstan	Peak V_{oc} of ~354.5 V, peak I_{sc} of ~18.6 μA, and average peak PD of ~8.43 mW m^{-2} (at ~1 GΩ) from a woven 3 × 6 cm^2 active area at 3.0-Hz frequency	Powering up LEDs, pedometer, electronic watch, and a mini calculator	It can be worn as a bracelet	Yang, Sun et al. (2018)
Silicone rubber vs human skin	Multiple twisted Au-coated Cu threads	Peak V_{oc} of ~42 V, peak I_{sc} of ~5 μA, and peak PD of ~34.4 μW cm^{-2} (at 1 MΩ) from 45 × 45 mm^2 woven active area at 1 kgf	Powering up LEDs and electronic watch	–	Park et al. (2018)
Silicone rubber vs Nylon	Carbon fiber bundles	Peak V_{oc} of ~42.9 V, peak I_{sc} of ~ 0.51 μA, and power of ~1.12 μW at 2.5-Hz frequency from a 2 mm diameter CY-WTENG	Powering up electronic watch	Single coaxial CY-WTENG with storing ability	Yang, Xie et al. (2018)
Fluoridized PDMS and cotton fabric	Cu/Ni/PET yarn	Peak V_{oc} of ~60 V, peak I_{sc} of ~3 μA, and peak PD of ~127 mW m^{-2} (at ~130 MΩ) at 4 Hz from a CY-WTENG	Powering up electronic watch	–	Liu, Cong et al. (2019b)
Silicone rubber and stainless-steel yarns	Stainless steel yarns	The peak voltage of ~12 V at 100% strain from 30 cm single yarn	Powering up LCDs, electric watch, and monitoring of body movements	Worked in N_2 and H_2O	Gong et al. (2019)

Abbreviations: PDMS (polydimethylsiloxane), Au (gold), Cu (copper), Ni (nickel), (PET) polyester, V_{oc} (open circuit voltage), I_{sc} (short circuit current), PD (power density).

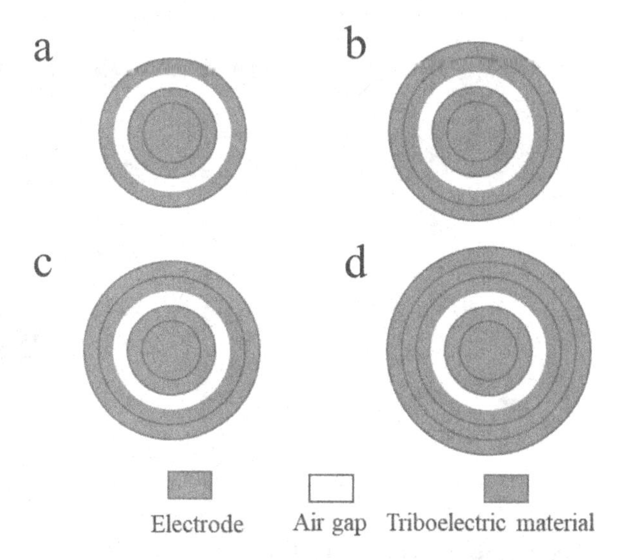

FIGURE 10.19 Different constructions of dual-electrode integrated CY-WTENGs. (Reproduced with permission from Dong, Peng, and Wang (2020a). Copyright John Wiley and Sons 2019.) (a) The conductive electrode at the center is wrapped with single triboelectric material as the core, with the outer electrode as the shell (Kim et al. 2015). (b) The conductive electrode at the center is wrapped with a triboelectric material as the core and a similar configuration as the shell (Wang et al. 2016). (c) Triboelectric/non-conductive material at the center coated with a conductive material as the core and similar configuration as the shell (Cheng, Lu, Chan et al. 2017; He et al. 2017; Zhao et al., n.d.). (d) When a configuration of (c) is further coated with an outermost triboelectric or insulation layer (Dong et al. 2018; Gong et al. 2017b; Yu et al. 2017). *Configurations are more of a generalized structure; therefore, these may not reflect exactly as in the literature.

Au nanoparticles (NPs) to constitute the core electrode. The core electrode was then inserted into a nanostructured PDMS tube, which further was wrapped by an Al film. A 5 cm single CY-WTENG (Figure 10.20(b)) when subjected to 50 N force and 10-Hz frequency, it generated a peak voltage of 40 V and a peak current of 10 μA current. Several CY-WTENGs were woven into a fabric structure and bonded with a piece of waterproof fabric to make it functional even in harsh environmental conditions (Kim et al. 2015). Gong et al. fabricated a CY-WTENG (Figure 10.20(c)) wherein a conductive metal core was wrapped with nylon fibers as the first triboelectric material. A PDMS silicone rubber tube treated with fluoroalkyl silanes (FAS) worked as the second triboelectric layer. Two bamboo fibers were wrapped around the rubber tube in opposite directions to facilitate protection and stretching of the electrode. The outer electrode was made of Ag NWs. The whole assembly was then sprayed with PDMS to further increase the overall robustness of the device and anchoring the outer electrode to the bamboo fibers. An instantaneous peak PD of 1.3 mW m^{-2} at 500-MΩ load resistance was achieved when subjected to 44% pre-stretching with 100% in-test stretching at 0.5-Hz frequency (Gong et al. 2017b).

Zhao et al. used a silicone rubber filament, which was further coated with multi-walled carbon nanotubes (MWCNTs). Polyaniline (PANI) fibers were wrapped around the silicone rubber filament and functioned as the core electrode and first triboelectric layer. An insulated varnished filament wrapped around the entire assembly functioned as the second electrode and second triboelectric layer. The formed CY-WTENG (Figure 10.20(d)) demonstrated sensitivity against gaseous molecules and different body motions (T. Zhao et al., n.d.). Similar CY-WTENG (Figure 10.20(e)) was fabricated by He et al., where a silicone rubber filament was prepared and coated with MWCNTs/Parafilm-based composite conductive ink as a flexible electrode. This assembly was further coated with a thin silicone rubber layer as the triboelectric layer. A stretchable Cu filament participated in triboelectrification and also acted as the second electrode when wrapped around

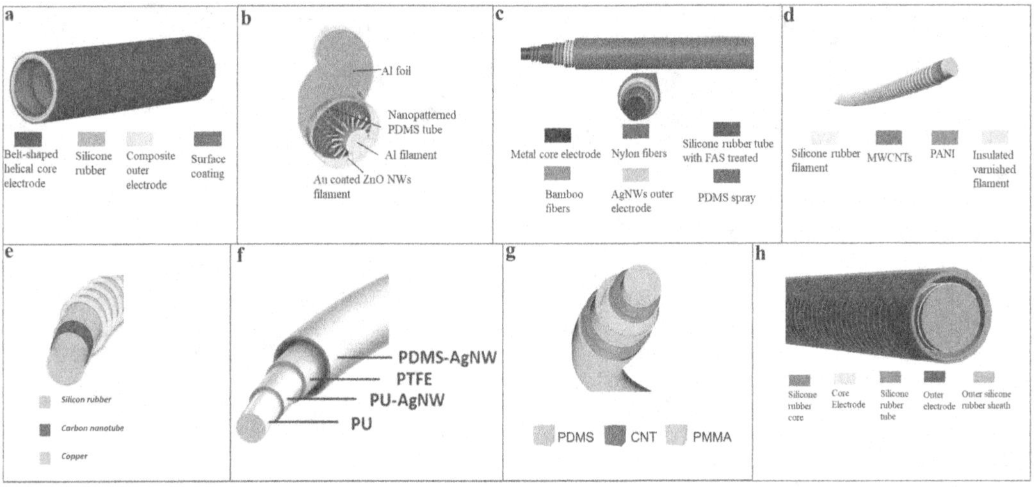

FIGURE 10.20 CY-WTENGs based on dual-electrode coaxial yarns. (a) CY-WTENG based on a belt-shaped core electrode inserted into a silicone rubber tube with an outer silicone-based composite electrode and encapsulation layer. (Reproduced with permission from Wang et al. (2016). Copyright Springer Nature 2016.) (b) CY-WTENG based on Au-coated ZnO NWs grown Al filament-based core electrode with nanopatterned PDMS tube and Al foil-based outer electrode. (Reused with permission from Kim et al. (2015). Copyright American Chemical Society 2015.) (c) CY-WTENG based on the nylon fibers wrapped metal core electrode, and FAS treated silicone rubber tube with bamboo protection layered AgNWs outer electrode and PDMS encapsulation. (Reproduced with permission from Gong et al. (2017b). Copyright Elsevier 2017.) (d) CY-WTENG based on the PANI wrapped MWCNTs-coated silicone rubber core electrode with an insulated varnished filament-based outer electrode. (Reused with permission from T. Zhao et al. (n.d.). Copyright IOP Publishing Ltd 2018.) (e) CY-WTENG based on the MWCNTs-coated silicone rubber core electrode with Cu filament wound thin layer silicone-based outer electrode. (Reused with permission from He et al. (2017). Copyright John Wiley and Sons 2016.) (f) CY-WTENG based on the PTFE coating of PU-AgNWs wrapped PU core electrode and insertion into a PDMS-AgNWs tube-based outer electrode. (Reused with permission from Cheng, Lu, Chan et al. (2017). Copyright Elsevier 2017.) (g) CY-WTENG based on the coating of PMMA microspheres onto a CNTs wrapped PDMS filament with deposition of patterned PDMS layer onto the microspheres and further wrapping with CNTs. (Reused with permission from Yu et al. (2017). Copyright Royal Society of Chemistry 2017.) (h) WTENG based on the 3-ply Ag-coated nylon thread wrapped silicone rubber filament-based core electrode inserted into a 3-ply Ag-coated nylon thread wrapped silicone rubber tube with further silicone rubber encapsulation. (Reproduced with permission from Dong et al. (2018). Copyright John Wiley and Sons 2018.)

the whole assembly. The resultant CY-WTENG produced V_{oc} of ~140 V at frequencies between 0.5 and 5.0 Hz (He et al. 2017). Another CY-WTENG (Figure 10.20(f)) developed by Cheng et al. had an exact similar arrangement of electrodes and triboelectric layers. A polyurethane (PU) filament wrapped with AgNWs film formed the core electrode. It was further coated with polytetrafluoroethylene (PTFE) and finally wrapped by an AgNWs-coated PDMS conductive-tube-based electrode. The single CY-WTENG exhibited a peak PD of 2.25 nW cm^{-2} at 50-MΩ load resistance and 1-Hz frequency with 135° twisting (Cheng, Lu, Chan et al. 2017). Yu et al. developed a dual integrated electrode CY-WTENG (Figure 10.20(g)) with PDMS and polymethyl methacrylate (PMMA) as the triboelectric layers. A PDMS filament was wrapped with CNTs sheet to form the core electrode. On the surface of the core electrode, PMMA microspheres were deposited and further coated with a patterned PDMS layer. Afterwards, another CNTs sheet was assembled to form the outer electrode, and then the whole assembly was coated with PDMS. During fabrication, a sacrificial layer was created between PDMS and PMMA to introduce an air gap between the triboelectric layers. A 5 cm long CY-WTENG under cyclic compression of 25 N force generated peak V_{oc} of 5 V and

I_{sc} of 240 nA (Yu et al. 2017). In more recent work, Dong et al. fabricated a CY-WTENG (Figure 10.20(h)) by winding a 3-plied Ag coated nylon thread as the core electrode around a silicone rubber filament. This assembly was inserted into a commercially available silicone elastomeric tube; before insertion, the tube was wrapped with a similar Ag-coated nylon thread as the second electrode. The final CY-WTENG was fabricated by coating the whole assembly with silicone rubber. A single CY-WTENG under 1–5-Hz frequency showed peak V_{oc} of 19 V and peak I_{sc} of 0.1–0.43 μA. Demonstrations of integrating the CY-WTENG into different woven structures and the capability to operate under all triboelectric modes were also demonstrated (Dong et al. 2018). The dual-electrode CY-WTENGs have higher output even without associating with fabric structures. Table 10.2 summarizes the recently developed CY-WTENGs based on dual electrode coaxial yarn configurations.

TABLE 10.2
Summary of Recently Developed CY-WTENGs Based on Dual-Electrode Coaxial Yarns

Active Materials	Electrode Materials	Output	Applied As	Special Attributes	Ref.
Silicone rubber Vs core electrode	Silicone rubber/ CB/CNTs composite	Peak V_{oc} of ~ 145 V and peak J_{sc} of ~ 16 mA m^{-2} at 10-Hz frequency from 7 mm diameter CY-WTENG	Powering up LED Electronic watch, and fitness tracker	–	Wang et al. (2016)
PDMS vs Au/ZnO/ Al wire	Au/ZnO/Al wire core electrode and Al film outer electrode	Peak V_{oc} of ~ 40 V and peak I_{sc} of ~10 μA from a 5 cm CY-WTENG at 50 N force and 10-Hz frequency. When woven into a piece of fabric at the same parameters, the fabric generated peak V_{oc} ~40 V, peak I_{sc} of ~210 μA, and instantaneous peak power of 4mW at 10-MΩ load resistance	Powering up LED	–	Kim et al. (2015)
Silicone rubber vs nylon	Conductive thread (unidentified) core electrode and Ag NWs/ Bamboo outer electrode	4.16 V (at 68% pre-stretched) and instantaneous peak PD of 1.3 mW m^{-2} at 100% tensile strain with 0.5-Hz frequency and 500-MΩ load resistance	Kinematic sensing	–	Gong et al. (2017a)
PANI vs insulated varnished wire outer electrode	PANI inner electrode and insulated varnished wire outer electrode	Peak I_{sc} of ~23.51 nA at 250 % stretch and 1-Hz frequency	Detecting environmental atmosphere and body motions	Distinguish different gas species (e.g., ethanol, ammonia, acetone, and formaldehyde) in the environment	Zhao et al., n.d.

(Continued)

TABLE 10.2 *(Continued)*
Summary of Recently Developed CY-WTENGs Based on Dual-Electrode Coaxial Yarns

Active Materials	Electrode Materials	Output	Applied As	Special Attributes	Ref.
Silicone rubber vs Cu microwire outer electrode	CNTs/polymer composite core electrode and Cu microwire outer electrode	Peak V_{oc} of ~140 V and peak I_{sc} of ~0.15 μA at 50% strain and 5-Hz frequency	Powering up LCD screen and digital watch/ calculators	Self-powered acceleration sensor	He et al. (2017)
PTFE vs PDMS-AgNWs outer electrode	PU-AgNWs core electrode and PDMS-AgNWs outer electrode	Peak I_{sc} of ~20 nA at 1-Hz frequency and 50% strain. Maximum PD of 2.25 nW cm^{-2} was demonstrated at 50-MΩ load resistance, 1-Hz frequency, and 135° twisting	–	Personal healthcare monitoring	Cheng, Lu, Hoe et al. (2017)
PDMS vs PMMA	Multi-layer aligned CNTs sheets inner and outer electrodes	Peak V_{oc} of ~5 and peak I_{sc} of ~240 nA from a 5 cm CY-WTENG at 25 N force	Powering up LED and LCD	Detecting moving direction	Yu et al. (2017)
Silicone rubber vs 3-ply-twisted Ag-coated nylon yarn	3-ply-twisted Ag-coated nylon yarn (core and outer electrode)	Peak V_{oc} of 19 V and peak I_{sc} of 0.43 μA at 5-Hz frequency, 20N force, and 20 mm contact length. The peak PD of 0.88 W m^{-3} was recorded at 3-Hz frequency, 20 N force, 20 mm contact length, and ~1000 MΩ load resistance	Powering LED and smartwatch	Self-counting skipping rope, a self-powered gesture-recognizing glove, and a real-time golf scoring system	Dong et al. (2018)

Abbreviations: PDMS (polydimethylsiloxane), Au (gold), ZnO (zinc oxide), Al (aluminum), PANI (polyaniline), PTFE (polytetrafluorethylene), Ag (silver), NWs (nanowires), PMMA (polymethyl methacrylate), CB (carbon black), CNTs (carbon nanotubes), Cu (copper), PU (polyurethane), V_{oc} (open circuit voltage), I_{sc} (short circuit current), J_{sc} (current density), PD (power density).

10.4.1.1.2.2 WTENGs Based on Fabric/Textiles Textiles' surfaces are known to be rough; therefore, they could be conducive to excellent contact electrification. By utilizing various fabric forming techniques (e.g., weaving, knitting, braiding, and non-woven techniques mainly), conductive and triboelectric yarns can easily be integrated into different 2D and 3D structures to fabricate fabric/textiles WTENGs.

10.4.1.1.2.2.1 Striped-Fabric-Based 2D WTENGs (2D SF-WTENGs) Since yarn-like structures are the basic textiles unit and are not truly wearable, turning these into the traditional fabric like textiles is appropriate for true wearability. To this end, numerous researchers developed 2D-fabric-based WTENGs by assembling flexible metal and fabric strips.

Ning et al. fabricated a 2D SF-WTENG by utilizing PTFE strips as one of the triboelectric layers. Each PTFE strip was wrapped around a Cu foil strip to get the PTFE wrapped Cu fabric strips.

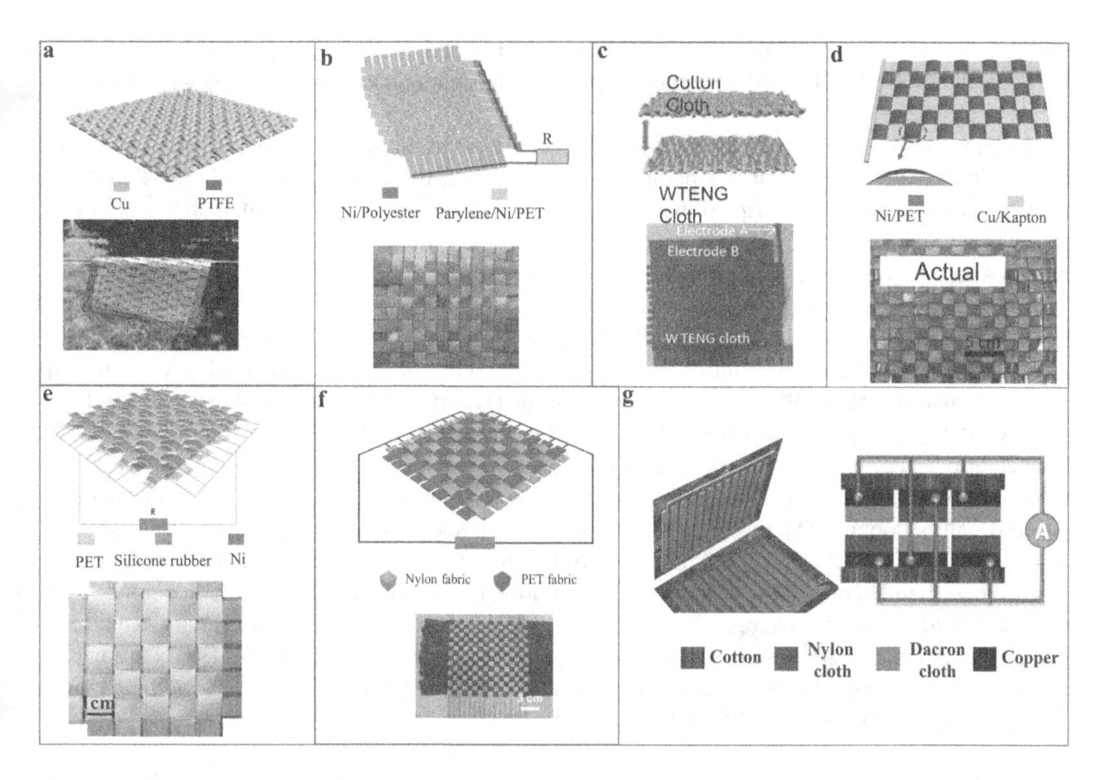

FIGURE 10.21 Striped-fabric-based 2D WTENGs. (a) PTFE fabric strip wrapped Cu-foil-based woven 2D SF-WTENG. (Reused with permission from Ning et al. (2018). Copyright Royal Society of Chemistry 2018.) (b and c) Ni-coated PET, and parylene/Ni-coated PET fabric strips woven 2D SF-WTENG. (Reused with permission from Pu, Li, Song et al. (2015) and Pu, Li, Liu et al. (2016). Copyright John Wiley and Sons 2015.) (d) Ni-coated PET and Kapton-coated Cu fabric strips woven 2D SF-WTENG. (Reused with permission from Zhao et al. (2016). Copyright American Chemical Society 2016.) (e) Ni-coated PET, and silicone rubber/Ni-coated PET fabric strips woven 2D SF-WTENG. (Reused with permission from Tian et al. (2017). Copyright Elsevier 2017). (f) Commodity nylon and PET fabric strips woven 2D SF-WTENG sandwiched between cotton/Ag-based conductive fabric (not visible in the image). (Reused with permission from Zhou et al. (2014). Copyright American Chemical Society 2014.) (g) 2D SF-WTENG-based on the interdigitated cotton substrate electrodes attached with alternatively placed nylon and Dacron fabric strips. (Reused with permission from Cui et al. (2015). Copyright American Chemical Society 2014.)

Several PTFE/Cu strips were woven into a 2D fabric structure (Figure 10.21(a)). When PTFE/Cu strips woven fabric attached to a cotton-made lab coat, triboelectrification occurred between PTFE and cotton yarns. Herein, an Al foil attached inside the lab coat functioned as the reference electrode. While working in the single-electrode mode, the resultant 2D SF-WTENG generated peak V_{oc} of 1050 V and peak I_{sc} of 22 μA while swinging arm during walking or running (Ning et al. 2018). 2D SF-WTENGs were fabricated by Pu et al., wherein PET fabric was coated at first with Ni to get Ni-coated PET fabric. An additional parylene coating was applied to the Ni-coated fabric to prepare the parylene/Ni/PET fabric. The individual fabrics were cut into fabric strips and were woven into a 5 cm × 5 cm 2D-WTENG (Figure 10.21(b)). Two 2D SF-WTENGs working in the contact-separation mode exhibited peak V_{oc} of 50 V, peak I_{sc} of 4 μA, and peak PD of 393.7 mW m^{-2} (Pu, Li, Song et al. 2015). In a further study, the 2D SF-WTENG (Figure 10.21(c)) has also demonstrated power generation in contact-separation mode with available cotton fabrics (Pu, Li, Liu et al. 2016). Similarly, fabric strips of Ni-coated PET fabric and Kapton-coated Cu-film were loosely woven into 2D SF-WTENG (Figure 10.21(d)), avoiding the necessity of another separate 2D SF-WTENG fabric. This loose structure provided an inherent contact-separation mode between the Ni-coated PET and Kapton-coated Cu fabric strips when subject to wind vibration (Zhao et al. 2016).

Similar structures were also developed by Tian et al., but in this case, human skin was the second triboelectric layer in the contact-separation mode. In brief, the design includes Ni/PET and silicone rubber/Ni/PET fabric strips in warp and weft directions (Figure 10.21(e)), respectively (Tian et al. 2017). Zhou et al. fabricated a 2D SF-WTENG (Figure 10.21(f)) in free-standing TENG mode, wherein commodity nylon and PET fabrics were used as the triboelectric layers. Ag and cotton-yarn-based conductive fabric strips were utilized as the electrodes. The strips of conductive fabric were sandwiched between PET and nylon fabrics separately. Such 16 nylon and 16 PET fabric stripes were then woven into a 2D SF-WTENG. When attached inside a lab coat, onto arm and leg joints, the S2DF-WTENG could generate electricity (Zhou et al. 2014). 2D SF-WTENG (Figure 10.21(g)) based on in-plane or lateral sliding mode was fabricated by Cui et al., herein ten strip electrodes (2 cm × 28 cm × 0.05 mm) were attached to cotton cloth substrate in interdigitated fashion by keeping 1 mm gap between the electrodes. The strip electrodes were then alternatively covered with nylon and Dacron (PET) fabric strips. Two such fabrics were attached to a lab coat, wherein one fabric to the waist and the other to the inner forearm. The fabricated 2D SF-WTENG worked based on lateral sliding by the swinging motion of the arm. The peak V_{oc} and peak I_{sc} were recorded as 700 V and 50 μA, respectively (Cui et al. 2015). 2D SF-WTENGs are simple to fabricate and also demonstrate greater output. But the majority of the reported 2D SF-WTENGs are woven manually, hampering bulk production and wide applicability. Table 10.3 summarizes the recently developed striped fabric-based 2D WTENGs.

10.4.1.1.2.2.2 Yarn Interlaced or Interlooped Fabric-Based 2D WTENGs (2D YI-WTENGs) 2D SF-WTENGs fabrication processes are manual; therefore, conventional continuous fabric manufacturing processes were applied to fabricate yarn interlaced (i.e., woven) or interlooped (i.e., knitted) fabric-based WTENGs. Such manufacturing approaches are more suitable for large-scale production.

Zhong et al. fabricated a 2D YI-WTENG utilizing CNTs-based electrodes. Here, regular cotton yarns were coated with CNTs to get CNTs/cotton yarns (CCYs), which were further coated with PTFE to get PTFE/CNTs/cotton yarns (PCCYs). To get the final 2D YI-WTENG, CCYs and PCCYs were 2-plied and woven into a commercial fabric (Figure 10.22(a)). The final 2D YI-WTENG demonstrated an average PD of ~0.1 μW cm^{-2} (Zhong et al. 2014). Zhao et al. demonstrated 2D YI-WTENG (Figure 10.22(b)) wherein, Cu-PET-based 2-plied threads were used in the warp direction, and polyimide (PI)-coated Cu-PET-based 2-plied threads were used in the weft direction. Herein, contact separation between Cu and PI interfaces resulted in triboelectrification. Tapping at 10 cm s^{-1} generated peak V_{oc} of 4.98 V and peak PD of 33.16 mW m^{-2} at 60-MΩ load resistance (Zhao et al. 2016).

2D YI-WTENGs (Figure 10.22(c)), based on three common knitted structures (i.e., plain knit, double knit, and rib-knit), were developed by Kwak et al. by integrating PTFE and Ag yarns as the triboelectric materials. The knitting was performed in such a way that allowed an Ag fabric layer in the middle to get sandwiched between two PTFE fabric layers in a buckled shape. The top and bottom surfaces of the sandwich structure hold Ag-fabric layers. The 2D YI-WTENG generated a peak V_{oc} of 23.5 V and peak I_{sc} of 1.05 μA when stretched to 30% (Kwak et al. 2017). A more simplistic approach was undertaken by Li et al. by adopting cone knitting technology to prepare Ag/polyamide 6 (PA6) and Ag/PTFE core-shell yarns. The 2D YI-WTENG was prepared in plain structure (Figure 10.22(d)) using the core-shell triboelectric yarns in opposite directions. The 2D YI-WTENG demonstrated peak V_{oc} of 50 V, peak I_{sc} of 2.5 μA, and PD of 11.0 mW m^{-2} at 10-MΩ load resistance (Li et al. 2018). In summary, utilization of established textiles forming techniques is more desirable to fabricate versatile 2D YI-WTENGs.

TABLE 10.3
Summary of Recently Developed Striped-Fabric-Based 2D WTENGs

Active Materials	Electrode Materials	Output	Applied As	Special Attributes	Ref.
PTFE vs cotton	Cu working electrode and Al foil reference electrode	Peak V_{oc} of 780 V, peak I_{sc} of 4.9 μA, and peak PD of 0.56 W m^{-2} (at 100-MΩ load resistance) from a 5 × 5 cm^2 at 30-Hz frequency	Powering up LEDs, night running light, and digital watch without any energy storage electronics	–	Ning et al. (2018)
Parylene and Ni vs Skin	Ni/PET	Peak V_{oc} of 50 V and peak I_{sc} of 4.5 μA from a 5 × 5 cm^2 WTENG	Powering up LEDs and smartphone	–	Pu, Li, Song et al. (2015)
Parylene and Ni vs cotton	Ni/PET	Peak V_{oc} of 40 V and peak I_{sc} of 5 μA at 5-Hz frequency. The peak I_{sc} could be increased to 17 μA at 20-Hz frequency from a 10 × 10 cm^2 active area	–	Integrated supercapacitor	Pu, Li et al. (2016)
Ni vs Kapton	Ni and Cu	Peak V_{oc} of 40 V and peak I_{sc} of 30 μA from a 1.5 × 1.5 cm^2 active area at a wind speed of 22 ms^{-1}	Powering up LEDs, wireless temperature, and humidity sensors	Flag-shaped	Zhao et al. (2016)
Ni and PDMS vs skin	Ni/PET	Peak V_{oc} of 500 V and peak I_{sc} of 60 μA from a 5 × 5 cm^2 at 3-Hz frequency and 300 N force. But with a double-stacked WTENG, peak V_{oc} and peak I_{sc} reached 540 V and 140 μA. At a load of 10-MΩ, PD reached 0.892 mW cm^{-2}	Powering up LEDs, competition timer, digital clock, and electronic calculator	–	Tian et al. (2017)
Nylon and PET fabric vs acrylic plate	Ag and cotton-yarn-based fabric	Peak V_{oc} of 95 V and peak I_{sc} of 2.5 μA from a 16 × 16 strips made WTENG at 50 ms^{-2} acceleration and 1.2 ms^{-1} speed	Powering up LEDs	–	Zhou et al. (2014)
Nylon vs Dacron (PET)	Cu film	Peak V_{oc} of 2 kV and peak I_{sc} of 0.2 mA at an average sliding velocity of 1.7 ms^{-1} between two WTENG fabrics	-	Electroluminescent tube-like lamp	Cui et al. (2015)

Abbreviations: PTFE (polytetrafluorethylene), Ni (nickel), PDMS (polydimethylsiloxane), PET (polyethylene terephthalate), Cu (copper), Al (aluminum), Ag (silver), V_{oc} (open circuit voltage), I_{sc} (short circuit current), PD (power density).

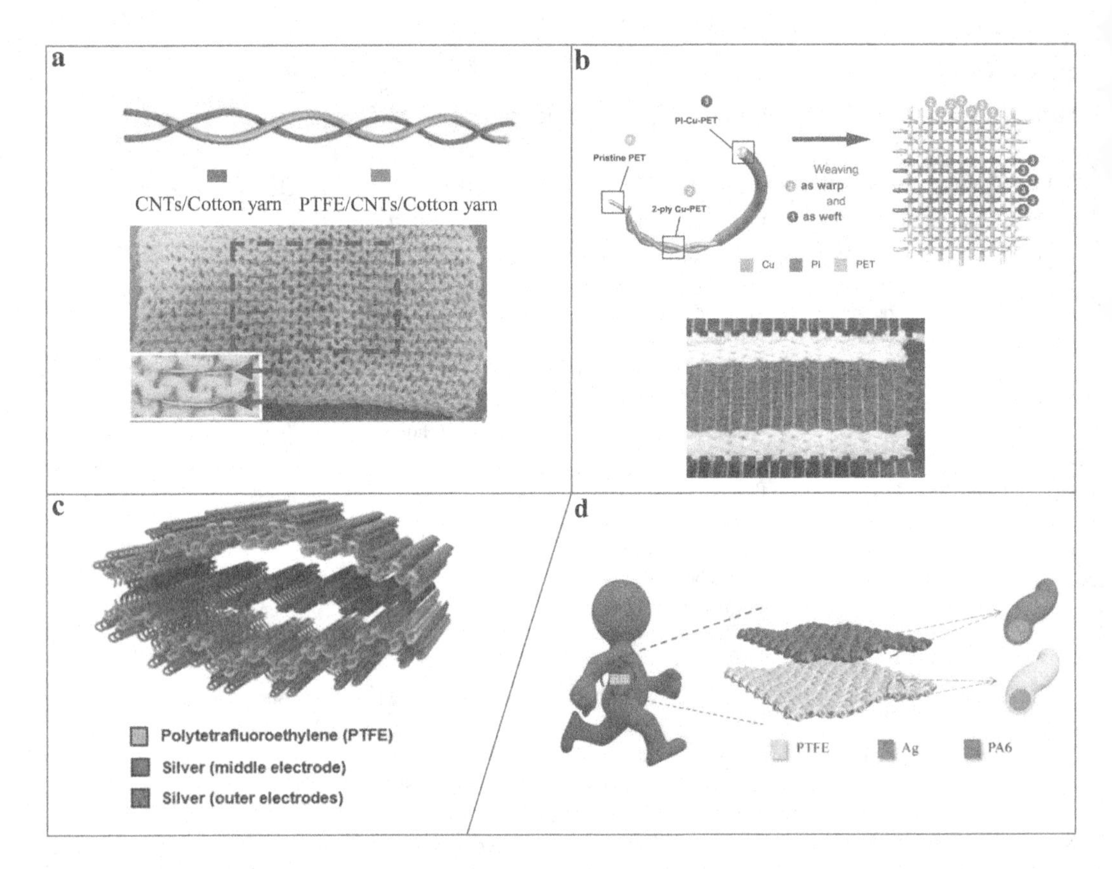

FIGURE 10.22 Yarn interlaced or interlooped fabric-based 2D WTENGs. (a) 2D YI-WTENG based on weaving CNTs/Cotton and PTFE/CNTs cotton yarns made 2-ply thread into a commercial fabric. (Reused with permission from Zhong et al. (2014). Copyright American Chemical Society 2014.) (b) 2D YI-WTENG woven with 2-ply Cu-PET threads and PI-coated 2-ply Cu-PET threads. (Reused with permission from Zhao et al. (2016). Copyright John Wiley and Sons 2016.) (c) Knitted 3-layered buckled-shaped 2D YI-WTENG. (Reused with permission from Kwak et al. (2017). Copyright American Chemical Society 2017.) (d) Ag/nylon and Ag/PTFE core-shell yarns woven 2D YI-WTENG. (Reused with permission from Li et al. (2018). Copyright John Wiley and Sons 2018.)

10.4.1.1.2.2.3 Yarn Integrated or Interlooped Fabric-Based 3D WTENGs (3D YI-WTENGs) Yarn interlaced or interlooped 3D WTENGs are fabricated to enhance the contact between triboelectric layers and thereby increasing charge density, resulting in greater energy output than 2D structures (Gong et al. 2019). Compared to 2D YI-WTENGs, the fabrication of 3D YI-WTENGs are fairly easy and also compatible with the established textiles machinery. The ability to increase triboelectric layers into thickness direction has endowed 3D YI-WTENG to produce more energy than the 2D YI-WTENG.

One of the most popularly adopted techniques for fabricating 3D YI-WTENG involves spacer knitting technology. Usually, a spacer layer separates the top and bottom layers, facilitating effective contact and separation for triboelectrification. For example, Gong et al. developed a 3D YI-WTENG (Figure 10.23(a)) from three sets of commercially available yarns by knitting through computerized programming. Herein, Ag-plated nylon yarns formed the top conductive layer (tends to charge positively in triboelectrification), and polyacrylonitrile yarns (tends to charge negatively in triboelectrification) formed the bottom layer. For the middle layer, a neutral or same as the bottom layer could hold the top and the bottom layers together, which further enhanced the mechanical robustness of the 3D YI-WTENG. In this case, regular cotton yarns were used to form the middle layer.

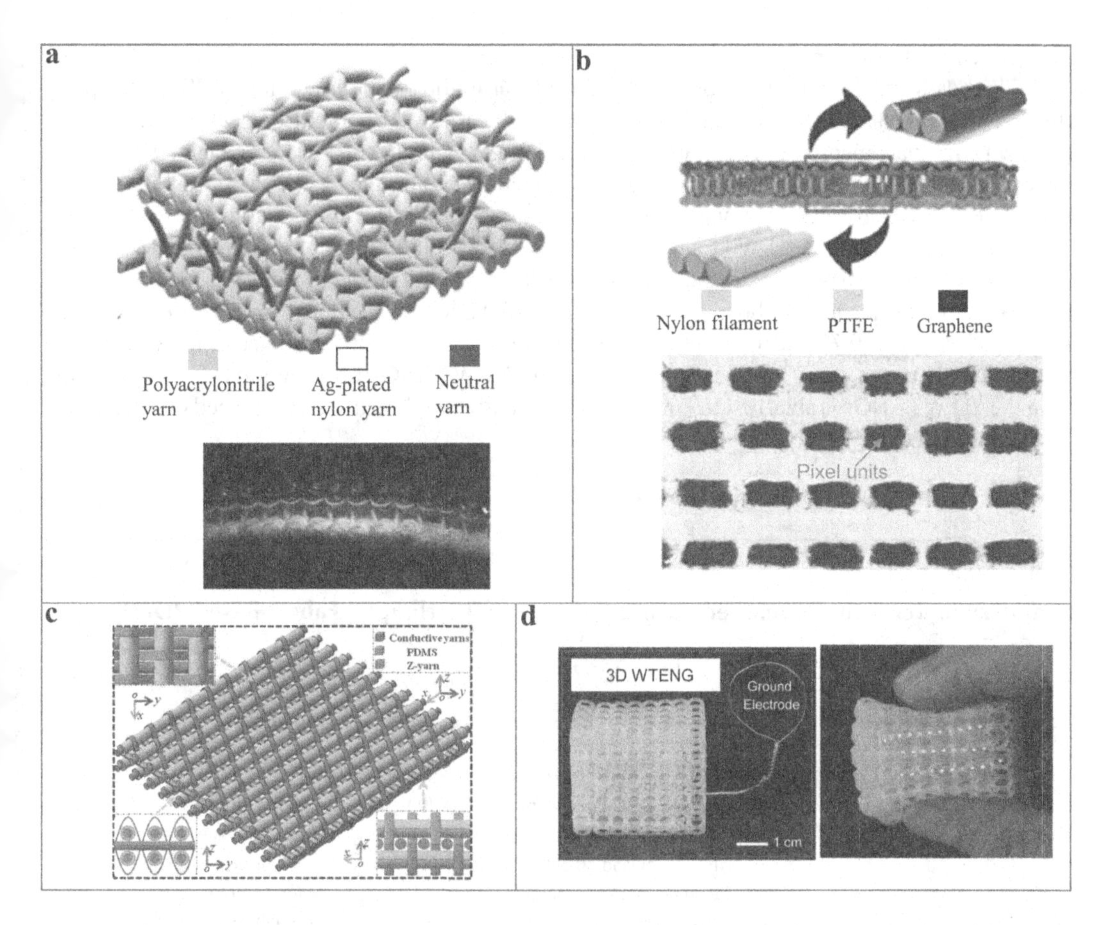

FIGURE 10.23 Yarn integrated or interlooped fabric-based 3D WTENGs. (a) Spacer knitted 3D YI-WTENG based on Ag-plated nylon integrated top and polyacrylonitrile integrated bottom layers with interlacing neutral yarns. (Reused with permission from Gong et al. (2019). Copyright Elsevier 2019.) (b) Spacer knitted 3D YI-WTENG based on nylon yarns integrated top and bottom layers with interlacing PET yarns. (Reused with permission from Zhu et al. (2016). Copyright Elsevier 2016.) (c) Spacer knitted 3D YI-WTENG based on PDMS-coated stainless steel/PET blended conductive yarn (top and bottom layers), uncoated conductive yarn (middle layer), and neutral interlacing yarns. (Reused with permission from Dong, Deng et al. (2017). Copyright John Wiley and Sons 2017.) (d) Resin-based 3D-printed 3D YI-WTENG. (Reused with permission from Chen et al. (2018). Copyright Elsevier 2018.)

While working in the single-electrode mode, the 3D YI-WTENG was subjected to cyclic contact separation with a PDMS film. A 3D YI-WTENG of 56.7 cm^2 effective area produced an excellent maximum PD of 1768.2 mWm^{-2} at 1200 N and 50-MΩ load resistance (Gong et al. 2019). Similar knitting and constructional approach were also approached by Zhu et al.; the 3D YI-WTENG also showed energy generation in the vertical contact-separation mode. Herein, the top and bottom fabric layers were knitted from nylon multifilament and the spacer layer was fabricated from PET monofilament. The upper surface of the top layer was coated with graphene ink as the top electrode, and the bottom layer was coated with PTFE to work as the second triboelectric material. The 3D YI-WTENGs (Figure 10.23(b)) were pixelated to work as small individual units. Such a unit exhibited a peak V_{oc} of 3.3 V at 1-Hz frequency and peak I_{sc} of 0.3 μA at 3-Hz frequency (Zhu et al. 2016).

Dong et al. fabricated a 3D YI-WTENG (Figure 10.23(c)) by weaving three different sets of yarns. The structural design was similar to the above-mentioned knitted 3D YI-WTENG, with two triboelectric fabric layers at the top and the bottom. A spacer fabric layer at the middle linked the top and the bottom fabric layers. When tapped with a 3-Hz frequency, the 3D YI-WTENG's

PD peaked up to 263.36 mW m^{-2} at 132-MΩ load resistance. Stainless steel/PET fiber blended 3-ply threads formed the middle conductive fabric layer in the warp direction. PDMS-coated 3-ply threads were inserted in the weft direction to fabricate the top and bottom triboelectric layers. There exist zero interlacement between warp and weft threads. Therefore, a non-conducive thread was introduced as the interlacing between the top and the bottom fabric layers (Dong, Deng et al. 2017). Cheng et al. demonstrated an unconventional technique of additive manufacturing to fabricate a single-electrode mode 3D YI-WTENG (Figure 10.23(d)) based on UV-cured 3D-printed WTENG. Two distinct resin parts were assembled as triboelectric materials with an ionic hydrogel-based electrode. At a frequency of about 1.3 Hz, the 3D YI-WTENG showed peak V_{oc} of ~60 V and peak PD ~10.98 W m^{-3} at 0.75 TΩ (Chen et al. 2018). In summary, due to the presence of more yarns and interactions between the triboelectric layers, the 3D YI-WTENGs produce greater energy output than the 2D WTENGs. Table 10.4 summarizes the recently developed yarn interlaced or interlooped fabric-based 2D/3D WTENGs.

TABLE 10.4
Summary of Recently Developed Yarn Interlaced or Interlooped Fabric-Based 2D/3D WTENGs

Active Materials	Electrode Materials	Output	Applied As	Special Attributes	Ref.
		Yarn interlaced or interlooped fabric-based 2D WTENGs			
PTFE vs CNTs ink	CNTs ink/cotton	A CY-WTENG of ~9 cm with 2.15% strain at 5-Hz frequency and 80-MΩ load resistance, generated a peak current of 11.22 nA and instantaneous peak power of 11.08 nW (at 100 MΩ). The CY-TENG was integrated into a WOVEN 2D WTENG to power different electronics when attached to a lab coat	Powering up LCDs and LEDs	Body motion Detection and wireless body temperature monitoring system	Zhong et al. (2014)
Cu vs PI	Cu/PET	Peak V_{oc} of ~4.98 V, peak J_{sc} of 15.5 mA m^{-2}, and peak PD of 33.16 mW m^{-2} (at 60-MΩ load resistance) at 10 cm s^{-1} tapping speed	–	Monitoring human respiratory rate and depth	Zhao et al. (2016)
PTFE vs Ag	Ag yarns	Under 30% stretching, a 10 × 10 cm^2 WTENG generated 23.5 V at 40-MΩ load resistance and 1.05 μA at 100-MΩ load resistance	Powering up LEDs	–	Kwak et al. (2017)
PTFE vs PA6	Ag yarns	Peak V_{oc} of 80 V and peak I_{sc} of 2.5 μA at 1.5-Hz frequency	Powering up LEDs, digital watch, and calculator	–	Li et al. (2018)

(Continued)

TABLE 10.4 *(Continued)*
Summary of Recently Developed Yarn Interlaced or Interlooped Fabric-Based 2D/3D WTENGs

Active Materials	Electrode Materials	Output	Applied As	Special Attributes	Ref.
		Yarn integrated or interlooped fabric-based 3D WTENGs			
Polyacrylonitrile vs Ag vs PDMS	Ag/Nylon	Maximum PD of 1768.2 mW m^{-2} at 1400 N force and 50-MΩ load resistance	Powering up LEDs, propeller, and buzzer alarm	–	Gong et al. (2019)
Nylon vs PTFE	Graphene ink/ nylon and graphene/PTFE	Peak V_{oc} of 3.3 V and peak I_{sc} of 0.2 μA from 1 cm^2 active area at 1-Hz frequency. The power can reach up to 16 μW at 0.6-MΩ load resistance	Powering up LEDs	Motion monitoring and sensing system	Zhu et al. (2016)
PDMS vs skin	3-ply stainless-steel/PET fiber blended yarn	Peak PD of 263.36 mW m^{-2} at 3-Hz tapping frequency and 132-MΩ load resistance	Powering up LEDs, watch, and lighting up a warning indicator	Tracking human motion signals	Dong, Deng et al. (2017)
Acrylonitrile butadiene styrene-based resins	Ionic hydrogel PAAm-LiCl	Peak V_{oc} of ~62 V, peak J_{sc} ~26 mA m^{-3}, and peak PD of 10.98 W m^{-3} (at 0.75 TΩ) under ~1.3-Hz frequency from a 3.5 × 3.5 × 3.5 cm^3 WTENG	LEDs flickering and buzzing SOS distress systems, watch and smart LEDs lighting shoes	Temperature sensor	Chen et al. (2018)

Abbreviations: PTFE (polytetrafluoroethylene), CNTs (carbon nanotubes), Cu (copper), PI (polyimide), Ag (silver), PDMS (polydimethylsiloxane), PET (polyethylene terephthalate), PAAm (polyacrylamide), V_{oc} (open circuit voltage), I_{sc} (short circuit current), PD (power density), J_{sc} (short circuit density), PA6 (polyamide 6).

10.4.1.1.2.2.4 Multi-Layer Stacked Fabric-Based 3D WTENGs (3D ML-WTENGs) Multi-layer stacked 3D WTENGs are an alternative to 3D YI-WTENGs and could be a faster bulk production technique. Qiu et al. fabricated a 3D ML-WTENG working in the single-electrode mode (Figure 10.24(a)). Herein, a conductive breathable fabric as the electrode was attached to a PET fabric substrate. Another PET fabric was simultaneously electro-sprayed and electrospun with PTFE NPs and PVDF (polyvinylidene fluoride) nanofibers, respectively, to make negative triboelectric material. Apart from PET, other commercially available fabric such nylon, silk, cotton, and polypropylene (PP) were also used as the substrate and similarly coated with PTFE NPs and PVDF nanofibers as the negative triboelectric materials. The coated PET fabric was then attached to the surface of the fabric electrode to fabricate the final 3D ML-WTENG. When contacted with nylon fabric at 3-Hz frequency, the 3D ML-WTENG generated peak V_{oc} of 112.7 V, peak I_{sc} of 1.98 μA, and peak PD of 80 mW m^{-2} at 50-MΩ load resistance (Qiu et al. 2019).

Single-electrode mode 3D ML-WTENG (Figure 10.24(b)) fabricated by Cao et al. utilized a nylon fabric coated with PU mixed CNTs ink as the electrode and substrate. A silk fabric layer placed on top of the electrode served as one of the triboelectric layers. The human skin was considered as the second triboelectric layer (Cao et al. 2018). To develop a 3D ML-WTENG, Xiong et al. coated a PET fabric with hydrophobic cellulose oleoyl ester (HCOE) encapsulated black phosphorus

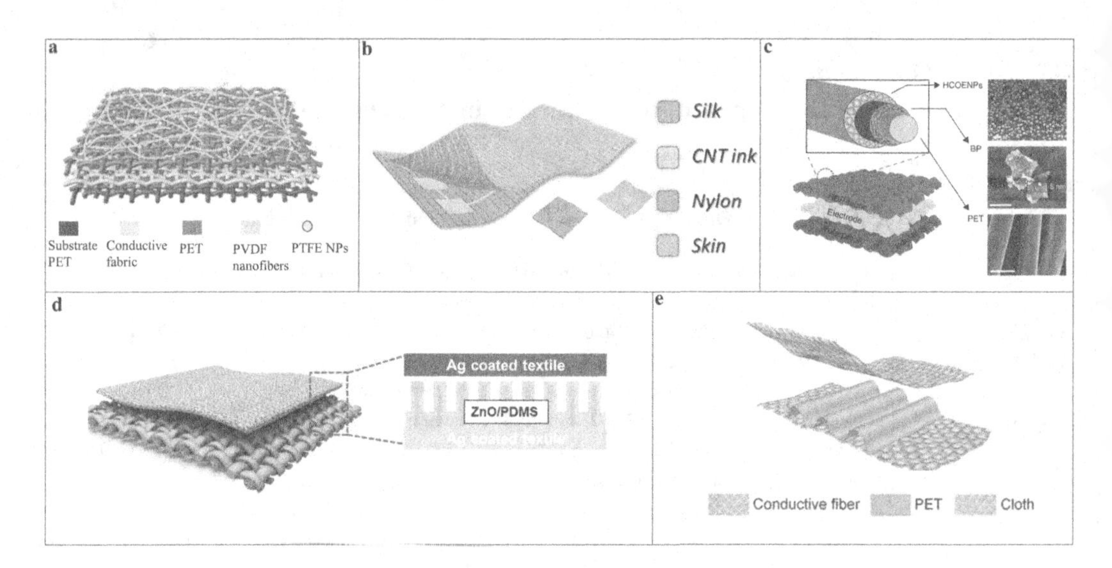

FIGURE 10.24 Multi-layer stacked fabric-based 3D WTENGs. (a) 3D ML-WTENG based on PVDF nanofibers and PTFE NPs deposited PET fabric substrate. (Reused with permission from Qiu et al. (2019). Copyright Elsevier 2019.) (b) 3D ML-WTENG based on PU-CNTs ink-coated nylon fabric-based electrode and silk fabric-based triboelectric layer. (Reused with permission from Cao et al. (2018). Copyright American Chemical Society 2018.) (c) 3D ML-WTENG based on coating PET fabric with HCOE encapsulated BP NPs (HBP) and stacking onto an Ag-flakes/PDMS-coated PET fabric. (Reused with permission from Xiong et al. (2018). Copyright Springer Nature 2018). (d) 3D ML-WTENG based on stacked Ag-coated textile and ZnO/PDMS grown Ag-coated textile. (Reused with permission from Seung et al. (2015). Copyright American Chemical Society 2015.) (e) 3D ML-WTENG based on middle wavelike PET fabric sandwiched between two conductive fabrics. (Reused with permission from Lin et al. (2018). Copyright John Wiley and Sons 2017.)

(BP) NPs. Another PET fabric was coated with Ag flakes/PDMS to fabricate a conductive electrode. The final 3D ML-WTENG (Figure 10.24(c)) was fabricated by encapsulating the conductive fabric electrode with a waterproof fabric and placing the hydrophobic fabric layer on the conductive fabric electrode. The maximum instantaneous PD of 0.52 mW cm^{-2} was demonstrated at 100-MΩ load resistance (Xiong et al. 2018). Seung et al. fabricated a 3D ML-WTENG (Figure 10.24(d)), which worked in a vertical-separation contact mode. PDMS-coated ZnO nanorod arrays vertically grown on an Ag-coated fabric formed one of the electrodes and triboelectric layers. Another Ag-coated fabric served as the second electrode and triboelectric layer. No-spacer layer was used since ZnO nanorod arrays facilitated triboelectrification. The 3D ML-WTENG generated peak V_{oc} of 120 V and peak I_{sc} of 65 μA. When stacked in four layers, the peak V_{oc} and peak I_{sc} reached 170 V and 120 μA (Seung et al. 2015). 3D WTENG developed by Lin et al. used two conductive fabrics as electrodes, which also participated in triboelectrification. A PET fabric with a wavelike structure (Figure 10.24(e)) was sandwiched between the fabric electrodes to function as the other triboelectric layer (Lin et al. 2018). In conclusion, 3D ML-WTENG demonstrated a simple fabrication process by employing conductive fabrics/coatings with/onto commercial fabrics. The energy output is also significantly higher than the 2D WTENGs.

10.4.1.1.2.2.5 Textiles Based in Plane Sliding Mode 3D WTENGs WTENGs developed based on the in-plane sliding mode omit the requirement of spacer or air gap between the triboelectric layers. The major advantage of the in-plane sliding mode lies in its capability to harness energy from the swinging motion of arms while walking or running. In most cases, the on-body cloths function as the second triboelectric layer for in-plane sliding mode.

Jung et al. fabricated a 3D WTENG (Figure 10.25(a)) capable of harnessing mechanical energy of friction between the arm and the torso by employing the sliding mode of TENG. A conductive

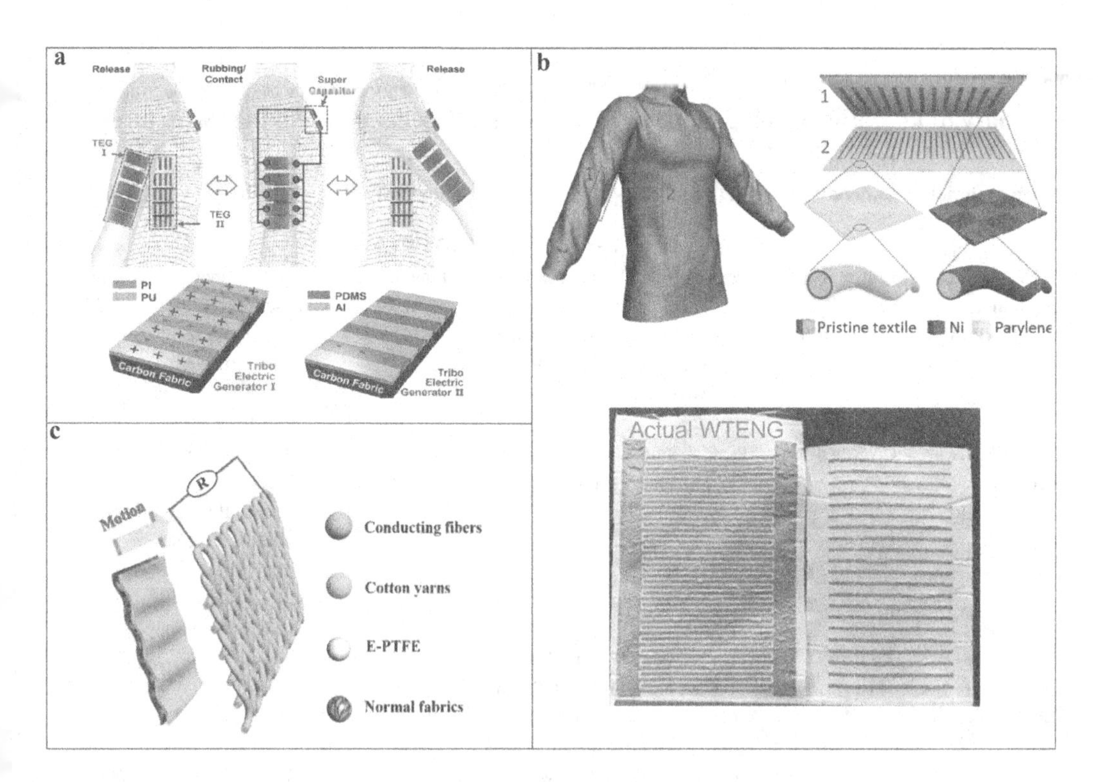

FIGURE 10.25 Textiles based in-plane sliding mode 3D WTENGs. (a) Sliding 3D WTENG based on alternatively patterned carbon fabrics with PI, PU (TEG-I) and PDMS, Al (TEG-II). (Reused with permission from Jung et al. (2014). Copyright John Wiley and Sons 2014.) (b) Sliding 3D WTENG based on striped-patterned Ni/textile and parylene/Ni/textile. (Reused with permission from Pu, Song et al. (2016). Copyright John Wiley and Sons 2016). (c) Freestanding sliding 3D WTENG based on the knitting of Ag-plated nylon and cotton yarns with E-PTFE-coated PET fabric. (Reused with permission from Huang et al. (2019). Copyright Elsevier 2019.)

carbon fabric was alternatively patterned with PU and PI. Similarly, another conductive carbon fabric was alternatively coated with PDMS and Al. Forward and reverse peaks were observed when Al was aligned with PU and PI, respectively, in the sliding mode. In a typical running condition (i.e., 1.5-Hz frequency), the 3D WTENG generated an average PD of ~0.18 μW cm^{-2} (Jung et al. 2014). Similarly, Pu et al. described a 3D WTENG (Figure 10.25(b)) generating energy due to frictional force created between the relative motion of fabrics at the arm and side of the torso. The fabric at the torso was interdigitated with Ni-coating and the fabric at the arm was coated with Ni and parylene consecutively in a parallel grated structure. The observed peak PD was 3.2 W m^{-2} at a sliding speed of 0.75 ms^{-1} (Pu, Song et al. 2016). Huang et al. fabricated a 3D WTENG (Figure 10.25(c)) as freestanding sliding mode TENG, wherein Ag-plated nylon and cotton yarns were weft-knitted in such a way that cotton yarns are positioned in the middle. A PET fabric layer coated with expanded PTFE (E-PTFE) worked as one of the triboelectric layers. When subjected to 5 MPa pressure, the 3D WTENG showed peak V_{oc} and peak I_{sc} of 800 V and 15 μA, respectively. A peak PD of 203 mW m^{-2} at 80-MΩ load resistance was also reported (Huang et al. 2019). Table 10.5 summarizes the recently developed 3D WTENGs with miscellaneous textiles structures.

10.4.1.1.2.2.6 Textiles-Based 3D WTENGs with Nanofibrous or Membranous Structures These types of textiles-based 3D WTENGs are different from conventional yarns and fabrics-based WTENGs. Since nanofibrous or membranous structures are used with conventional textile structures and exhibit features of textiles. These are included under 3D textiles-based WTENG.

TABLE 10.5
Summary of Recently Developed 3D WTENGs with Miscellaneous Textiles Structures

Active Materials	Electrode Materials	Output	Applied As	Special Attributes	Ref.
		Multi-layer stacked fabric-based 3D WTENGs			
PVDF nanofibers/ PTFE NPs onto PET vs skin	Conductive fabric (unidentified)	Peak V_{oc} of 112.7 V, peak I_{sc} of 1.98 μA, and peak PD of 80 mW m^{-2} at 50-MΩ load resistance from a 16 cm^2 active area	Powering up LEDs	Human motion sensor to monitor the amplitude of movements and postures	Qiu et al. (2019)
Cu film (as a substitute for skin) vs silk	CNTs/nylon	Peak V_{oc} of 6 V and peak I_{sc} of 120 nA under 2-Hz frequency	–	Touch/gesture sensor, controlling light bulbs, electric fans, and microwave ovens wirelessly	Cao et al. (2018)
HCOE encapsulated BP NPs vs skin	Ag flake-PDMS/PET	Peak V_{oc} of 1860 V, peak J_{sc} of 1.1 μA cm^{-2}, and peak PD of 0.52 mW cm^{-2} (at 100-MΩ load resistance) during fast contact/separation between HBP fabric and skin (~5N force and ~4-Hz frequency)	Powering up LEDs and digital-watch	–	Xiong et al. (2018)
Ag vs PDMS	Ag-coated textile	Peak V_{oc} of 120 V and peak I_{sc} of 65 μA at 10 kgf. When stacked in four layers, the respective output increased to 170 V and 120 μA at 10 kgf. The power output was also maximized to 1.1 mW at 1-MΩ load resistance	Powering up LEDs and LCDs	Keyless vehicle entry system	Seung et al. (2015)
Ag vs PET	Ag	Pressure sensitivity of 0.77 V Pa^{-1}	–	Sleep behavior monitoring and self-powered warning system for emergency conditions	Lin et al. (2018)
		Textiles based on plane sliding mode 3D WTENGs			
PU and PI vs PDMS and Al	Carbon fabric	Peak V_{oc} of ~6 and peak I_{sc} of ~55 nA were recorded for a 1.5 cm × 9 cm area. Besides at 1.5-Hz frequency, average PD was recorded to ~0.18 μW cm^{-2}	Powering up LEDs	–	Jung et al. (2014)
Ni vs parylene	Ni/PET	Peak V_{oc} of ~100 V, Peak I_{sc} of ~55 μA, and peak PD of 3.2 W m^{-2} (at 8-MΩ load resistance) at 0.75 ms^{-1} sliding speed	Powering up LCDs and LEDs	–	Pu, Song et al. (2016)

(Continued)

TABLE 10.5 *(Continued)*
Summary of Recently Developed 3D WTENGs with Miscellaneous Textiles Structures

Active Materials	Electrode Materials	Output	Applied As	Special Attributes	Ref.
Ag vs E-PTFE	Ag-plated nylon fabric	Peak V_{oc} 800 V and peak I_{sc} 15 μA at 5 MPa. A PD of 203 mW m^{-2} at 80-MΩ load resistance was reported	Powering up LEDs, caution sign, and smartwatch	Motion sensor to monitor human body	Huang et al. (2019)
		Textiles based 3D WTENGs with nanofibrous or membranous structures			
Cellulose vs fluorinated ethylene propylene	Ag nanofibers membrane	Peak V_{oc} of 21.9 V_{oc}, peak I_{sc} of 0.73 μA, and peak PD of 7.68 μW cm^{-2} (at 20-MΩ load resistance) under 5-Hz frequency contact/separation with a 3 mm operating gap	–	Particulate matter ($PM_{2.5}$), antibacterial, and self-powered breathing monitoring	He et al. (2018)
Cellulose acetate vs polyethersulfone	Cu foil	Peak V_{oc} of 115.2 V, peak I_{sc} of 9.21 μA, and the peak PD reached to 0.13 W m^{-2} at 30-MΩ load resistance under 3-Hz frequency from a 3 × 3 cm^2 area	Powering up LEDs, electronic watch, commercial thermometer, and calculator	–	Li et al. (2018)

Abbreviations: PVDF (polyvinylidene fluoride), PTFE (polytetrafluoroethylene), PET (polyethylene terephthalate), Cu (copper), HCOE (hydrophobic cellulose oleoyl ester), BP (black phosphorus), NPs (nanoparticles), HBP (HCOE encapsulated BP NPs), PDMS (polydimethylsiloxane), Ag (silver), PU (polyurethane), PI (polyimide), Al (aluminum), Ni (nickel), CNTs (carbon nanotubes), V_{oc} (open circuit voltage), I_{sc} (short circuit current), PD (power density), J_{sc} (short circuit current density).

A cellulose filter paper was structured into a cellulose microfibers (CMFs) skeleton, which was further covered with cellulose nanofibers (CNFs) on both sides to fabricate a nano-patterned positive triboelectric 2D textile. Further coating with an Ag-nanolayer on the top surface formed the electrode for the 3D WTENG. On the bottom layer of 2D textile, fluorinated ethylene propylene (FEP) film with punched micro-holes utilized as the negative triboelectric fabric. The 3D WTENG simultaneously showed potential for removing particulate matter (PM), inhibiting bacterial growth, and monitoring health (Figure 10.26a). At a 3 mm displacement amplitude between the triboelectric layers and 3-Hz frequency, the 3D WTENG generated peak V_{oc} of 21.9 V, peak I_{sc} of 0.73 μA, and a peak PD of 7.68 μW cm^{-2} at 20-MΩ load resistance (He et al. 2018).

Li et al. developed a multi-nanofibrous layered 3D WTENG (Figure 10.26b) that demonstrated a peak PD of 0.13 W m^{-2} at 3-Hz frequency. Herein, a nanofiber membrane of cellulose acetate (CA) taped with a Cu film formed the top layer of the 3D WTENG as the triboelectric layer and the top electrode, respectively. A polyethersulfone (PES) nanofibrous layer was used as the second triboelectric layer. Additionally, polystyrene/CB (PS + C) layers were attached to the bottom surface of PES as enhanced charge transfer and storage layers. To get the final 3D WTENG, another Cu film as the second electrode adhered to the bottom side of the stacked nanofibrous layers (Li et al. 2018).

10.4.1.1.3 WNGs Based on Piezoelectric Effect (WPENGs)

The presence of non-inversion symmetry in piezoelectric crystalline materials causes displacement of positive and negative charges under mechanical stimulation. This charge displacement can be

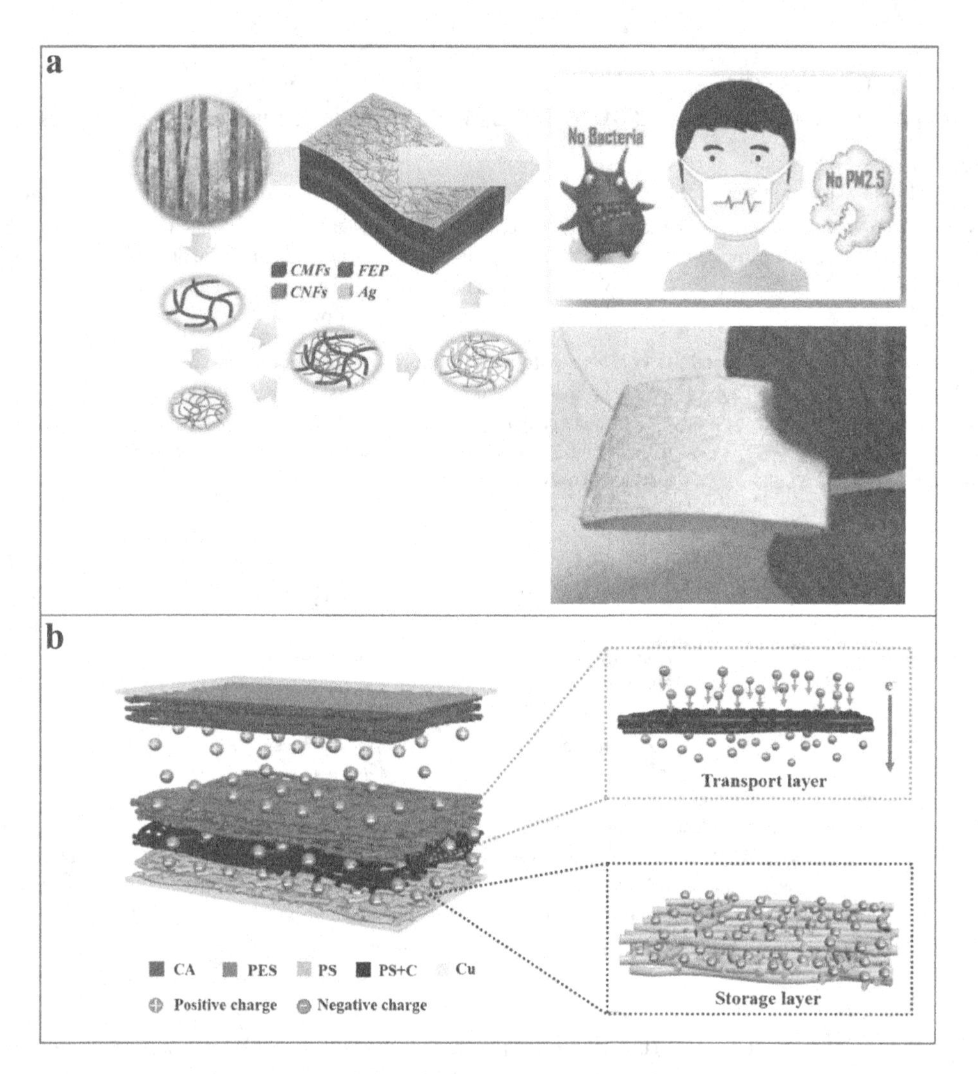

FIGURE 10.26 Textiles-based 3D WTENGs with nanofibrous or membranous structures. (a) 3D WTENG based on Ag-coated nano/microfibrous cellulose paper and FEP film. (Reused with permission from He et al. (2018). Copyright John Wiley and Sons 2018.) (b) 3D WTENG based on CA membrane and PES nanofibrous layer with Cu electrodes and charge transfer/storage layers. (Reused with permission from Li et al. (2018). Copyright Elsevier 2018.)

utilized to harness ambient mechanical energy by using the concept of PENG. The classification and identification aspects of textile-based WPENGs are somewhat similar to those of textile-based WTENGs. Mainly textile-based WPENGs could be classified as coaxial yarn-based WPENGs (CY-WPENGs), yarn integrated fabric/textile-based PENG (YF-WPENGs), and multi-layered stacked WPENGs (ML-WPENGs). Versatility in device structures for textile WPENGs is not as great as textile WTENGs.

10.4.1.1.3.1 Coaxial Yarn-Based WPENGs (CY-WPENGs) The most common structures for CY-WPENGs are coaxial in single, plied, and braided structures utilizing conductive materials/polymer-based composite electrodes, metal-coated polymeric electrodes, and neat metal-based electrodes. The active piezoelectric materials are certain ceramics (e.g., ZnO, $BaTiO_3$, KNN, $KNbO_3$, and $LiNbO_3$) and polymers (e.g., PVDF and polyvinylidene fluoride trifluoroethylene [PVDF-Tr-Fe]).

Lund et al. demonstrated melt co-extrusion of conductive CB and high-density polyethylene (HDPE) matrix-based composite core electrode with PVDF as the sheath. When an array of such coaxial yarns were arranged, as illustrated in Figure 10.27(a), a peak V_{oc} of 20 mV at 0.75 N was observed (Lund et al. 2012). A flexible 2-ply structure-based CY-WPENG was demonstrated by Jun Sim et al., wherein an electrospun PVDF-Tr-Fe nanofibrous sheet was wrapped around Ag-coated multifilament nylon as the core electrode. A similar electrode was further plied with the coated electrode to fabricate the final CY-WPENG. A peak V_{oc} of 0.7 V and peak I_{sc} of 0.18 nA were demonstrated (Sim et al. 2015).

Razavi et al. developed a braided structure-based coaxial CY-WPENG (Figure 10.27(b)). Herein, a PU-enameled Cu filament was wrapped with eight PVDF filaments as the active piezoelectric material. The co-axial structure was further braided with sixteen PU-enameled Cu filaments, which formed the outer electrode. A PD of 2.2 mW cm^{-3} was reported by the braided CY-WPENG (Razavi, Iannucci, and Greenhalgh 2020). Similarly, Mokhtari et al. developed a braided CY-WPENG (Figure 10.27(c)), wherein an Ag-coated nylon filament was braided with twelve PVDF filaments as the shell in a programmed way that the nylon electrode stayed in the center. The resultant structure was again braided with twelve Ag-coated nylon filaments forming the outer electrode. A peak V_{oc} of ~380 mV and a PD of 29.62 μW cm^{-3} at 0.023 MPa were demonstrated (Mokhtari et al. 2019).

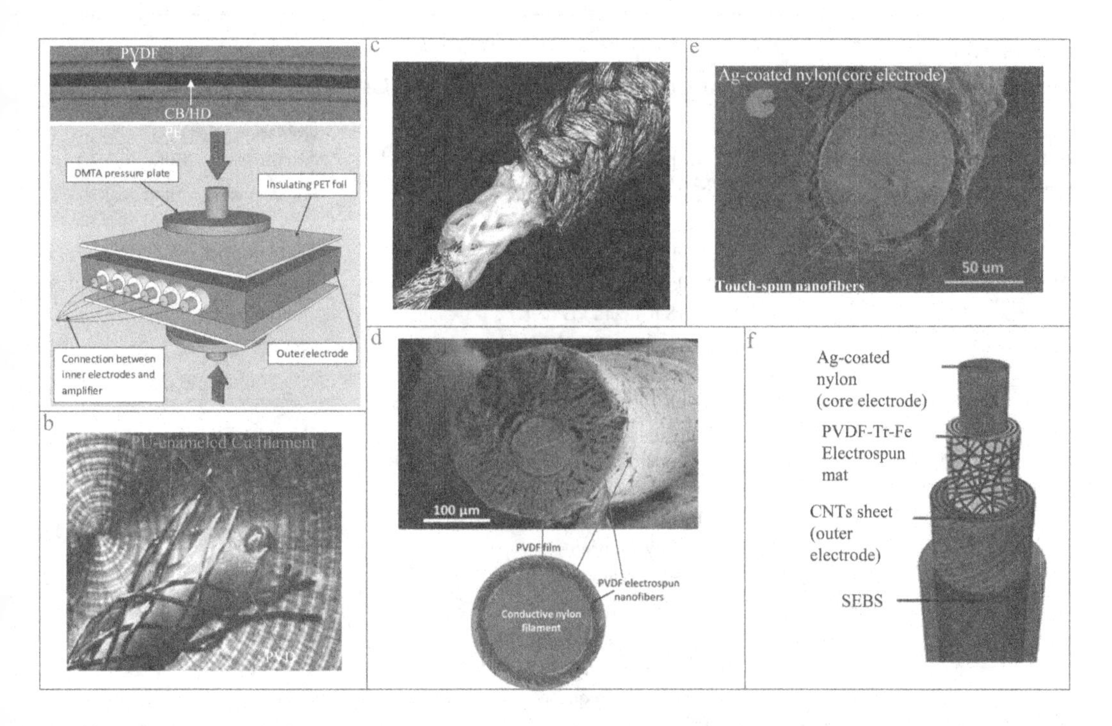

FIGURE 10.27 Coaxial yarn-based WPENGs. (a) CY-WPENG with CB/HDFE core and PVDF sheath. (Reused with permission from Lund et al. (2012). Copyright John Wiley and Sons 2012.) (b) Braided CY-WPENG with PVDF sheath and PU-enameled Cu-based core and outer electrodes. (Reused with permission from Razavi, Iannucci, and Greenhalgh (2020). Copyright John Wiley and Sons 2020). (c) Braided CY-WPENG with PVDF sheath Ag-plated nylon core and outer electrodes. (Reused with permission from Mokhtari et al. (2019). Copyright Royal Society of Chemistry 2019.) (d) CY-WPENG based on electrospun PVDF sheath with Ag-plated nylon core and metal deposited outer electrode. (Reused with permission from (Gao et al. 2018). Copyright IOP Publishing, Ltd 2018.) (e) CY-WPENG based on touch spun PVDF sheath with Ag-plated nylon core and metal deposited outer electrode. (Reused with permission from Gao et al. (2019). Copyright Taylor & Francis 2020.) (f) CY-WPENG based on PVDF-Tr-FE electrospun mat with Ag-coated nylon core and CNTs sheet wrapped outer electrode. (Reused with permission from Sim et al. (2015). Copyright John Wiley and Sons 2015.)

Huipu et al. fabricated CY-WPENGs based on PVDF nanofibers via electrospinning (Figure 10.27(d)) and touch spinning (Figure 10.27(e)) techniques. Wherein an Ag-coated nylon filament served as the core electrode. After wrapping the core electrode with nanofibers, the electron-beam evaporation technique was adopted to coat the outer surface of the PVDF nanofibers with metal particles to form the outer electrode. The reported average power output was 5.54 μW cm^{-3} under 0.02 MPa and 1.85-Hz frequency Gao et al. (2018, 2019). Sometimes, a protection layer is also applied to increase the mechanical robustness of the CY-WPENGs structures. For instance, Sim et al. developed a CY-WPENG (Figure 10.27(f)) with an Ag-coated nylon multifilament yarn core and PVDF-Tr-Fe nanofibers shell. The nanofibers-coated core electrode was further wrapped with a CNTs sheet to develop the final CY-WPENG. To make the device mechanically robust and electrically insulated, the CY-WPENG was coated with an elastomeric styrene-ethylene-butylene- styrene (SEBS) (Sim et al. 2015). The overall performance of CY-WPENGs is low and the comparative energy output of CY-WPENGs is also lower than the CY-WTENGs. Table 10.6 summarizes the recently developed coaxial yarn-based WPENGs.

TABLE 10.6
Summary of Recently Developed Coaxial Yarn-Based WPENGs

Active Materials	Electrode Materials	Output	Applied As	Special Attributes	Ref.
PVDF	CB-HDPE	Peak V_{oc} of 20 mV under compression at 1 N force and 1- or 10-Hz frequency	–	Force sensor	Lund et al. (2012)
PVDF–Tr-FE	Ag-coated nylon	Peak V_{oc} of 0.7 V, peak I_{sc} of 0.18 nA, and PD of 1.22 W m^{-3} (at 100-MΩ load resistance) under impact with 100 g object from 10 mm height The peak V_{oc} of 0.55 V when tension was applied by the same object with 20 mm cyclic drop	–	Human motion sensor	Sim et al. (2015)
PVDF	PU-enameled Cu wire	Under 100 kΩ, the voltage reached 3.8 V with 3-Hz frequency and under 1 kΩ, the voltage reached 4.15 V with 2-Hz frequency	–	–	Razavi, Iannucci, and Greenhalgh (2020)
PVDF	Ag-coated nylon	The peak of V_{oc} of ~380 mV, peak I_{sc} of 2 μA, and a maximum power of 0.16 μW at 0.023 MPa from a 2 cm CY-WPENG	Charging commercial battery	–	Mokhtari et al. (2019)
PVDF	Ag-coated nylon yarn as core and e-beam evaporated Ag-particles-based outer electrode	Average peak V_{oc} of 0.52 V, the peak current of 18.76 nA, and average PD of 5.54 μW cm^{-3} under 0.02 MPa and 1.85-Hz frequency from a 3 cm yarn	Powering up LEDs	–	Gao et al. (2018)

(Continued)

TABLE 10.6 *(Continued)*
Summary of Recently Developed Coaxial Yarn-Based WPENGs

Active Materials	Electrode Materials	Output	Applied As	Special Attributes	Ref.
PVDF	Ag-coated nylon yarn as core and e-beam evaporated Ag-particles-based outer electrode	Average peak V_{oc} of 0.72 V could be achieved at 0.33 MPa from a 3 cm yarn	–	–	Gao et al. (2019)
PVDF–Tr-FE	Multifilament Ag-coated nylon yarn as core electrode and CNTs-based outer electrode	Peak V_{oc} of 2.6 V, peak I_{sc} of 15 nA, and PD 52 μW cm^{-3} (at 100-MΩ load resistance) from a 10 mm long CY-WPENG under 160 kPa pressure and 100 mm s^{-1} impact speed. In longitudinal extension, the voltage could be reached to 1.4 V	–	–	Sim et al. (2015)

Abbreviations: PVDF (polyvinylidene fluoride), PVDF-Tr-FE (polyvinylidene fluoride trifluoroethylene), CB (carbon black), HDPE (high-density polyethylene), Ag (silver), PU (polyurethane), Cu (copper), CNTs (carbon nanotubes), V_{oc} (open circuit voltage), I_{sc} (short circuit current), PD (power density).

10.4.1.1.3.2 WPENGs Based on Fabric/Textiles

10.4.1.1.3.2.1 Yarn Integrated 2D Fabric/Textile-Based WPENGs (2D YF-WPENGs): Single CY-PENGs exhibited relatively lower output; therefore, integration into 2D or 3D textile structures from yarn-based WPENGs has been explored greatly.

The facile way to fabricate a YF-WPENG is through integrating single piezoelectric yarns into fabric-based structures. Both single piezoelectric filament and core-shell structures are integrated into the fabric to develop YF-WPENGs. For instance, Zhou et al. developed a 2D YF-WPENG as pressure-sensitive electronic skin (Figure 10.28(a)), wherein electrospun PVDF nanofibrous single yarns were coated with PEDOT [poly(3,4-ethylene dioxythiophene)] and woven into a double-layered fabric. Under a cyclic pressure of 10 kPa, peak V_{oc} reached 300 mV (Zhou et al. 2017). Since single piezoelectric yarns have low electrical output, using two or more piezoelectric yarns is also very common to fabricate YF-WPENGs. YF-WPENG constructed by weaving two sets of co-axial yarns into a 2D structure was demonstrated by Bai et al. (Figure 10.28(b)). Herein, Kevlar filaments were coated with ZnO NWs as the one set of yarns, and subsequent coating with Pd formed the second set of yarns. The peak V_{oc} and peak I_{sc} were reported as 3 mV and 17 pA, respectively (Bai et al. 2013). The coated structures are significantly prone to surface scraping, which can lead to short-circuiting. Therefore, piezoelectric yarns are also integrated into 2D YF-WPENGs by interweaving with other conductive and/or conventional threads. For instance, Yang et al. developed a 2D YF-WPENG (Figure 10.28(c)), wherein in the warp direction, non-conductive 2-ply PET and conductive 2-ply stainless threads were arranged alternatively. For every conductive thread, there were two non-conductive yarns (1:2 ratio). In the weft direction, 2-ply PVDF-Tr-Fe piezoelectric threads and non-conductive yarns were arranged alternatively in a 1:1 ratio. Herein, Cu-strips were used as the interdigitated electrodes. A peak V_{oc} of ~16.2 mV was demonstrated (Yang et al. 2017).

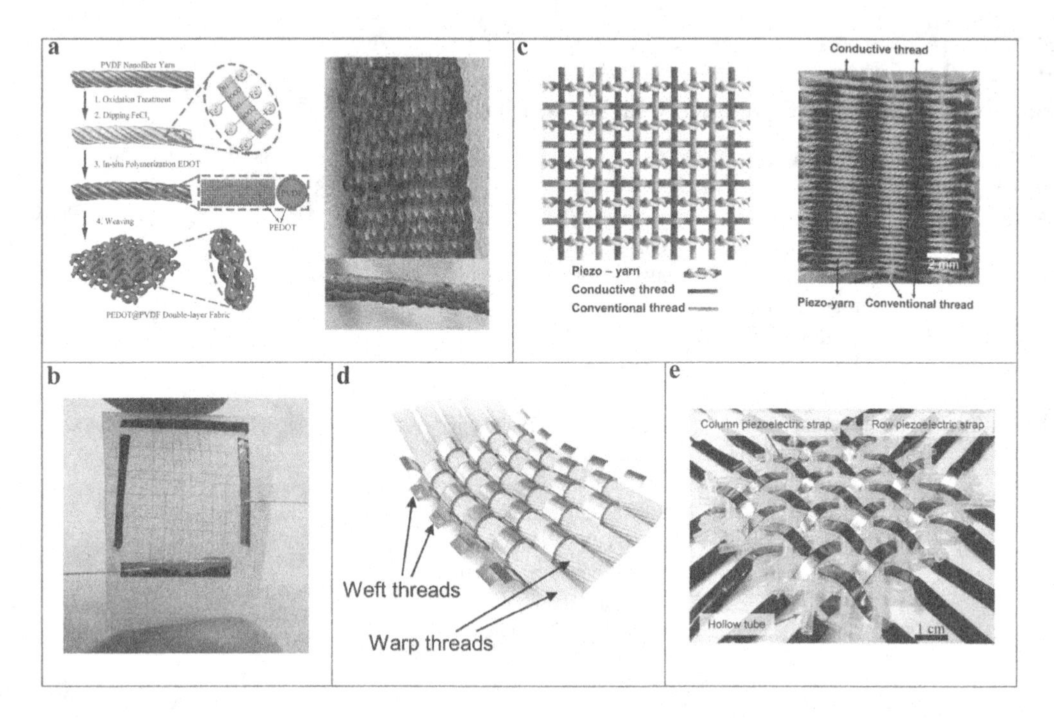

FIGURE 10.28 Yarn integrated 2D fabric/textile-based WPENGs. (a) Woven 2D YF-WPENGs based on PEDOT-coated PVDF electrospun nanofibrous yarns. (Reused with permission from Zhou et al. (2017). Copyright Springer Nature, The Author(s) 2017.) (b) Woven 2D YF-WPENGs based on ZnO NWs-coated Kevlar, and Pd/ZnO NWs-coated Kevlar filaments. (Reused with permission from Bai et al. (2013). Copyright Elsevier 2013.) (c) Woven 2D YF-WPENG based on 2-ply PET and 2-ply stainless steel threads in the warp direction with 2-ply PVDF-Tr-Fe threads and non-conductive yarns in the weft direction. (Reused with permission from Yang et al. (2017). Copyright American Chemical Society 2017.) (d) Woven 2D YF-WPENG-based flat bands of Ni/Cu-alloy-coated PVDF in the weft direction with conventional yarn in the warp direction. (Reused with permission from Song and Yun (2015). Copyright IOP Publishing Ltd 2015.) (e) 2D YF-WPENG of orthogonally woven hollow elastic tube-based mesh with piezoelectric bands in warp and weft directions. (Reused with permission from Ahn, Song, and Yun (2015). Copyright IOP Publishing Ltd 2015.)

2D YF-WPENGs based on flat bands were also developed. Song and Yun fabricated a 2D YF-WPENG (Figure 10.28(d)) using regular circular-shaped yarn in the warp direction and both sides Ni/Cu-alloy-coated PVDF film bands in the weft direction. A maximum peak PD of 125 μW cm^{-2} at 6-Hz frequency and 6.6-MΩ load resistance were demonstrated (Song and Yun 2015). Similarly, piezoelectric flat bands with metal deposition on both sides were integrated into an orthogonally woven hollow elastic tubes-based mesh. The final 2D YF-WPENG (Figure 10.28(e)) exhibited a peak PD of 105 mW m^{-2} at 1-MΩ load resistance (Ahn, Song, and Yun 2015).

10.4.1.1.3.2.2 Yarn Integrated 3D Fabric/Textile-Based WPENGs (3D YF-WPENGs) In efforts to continue improving the power output, 3D YF-WPENGs were also explored. For instance, Dong et al. developed WPENGs and compared the output between 2D (Figure 10.29(a)) and 3D (Figure 10.29(b)) yarn integrated fabric structures. Herein, PVDF multifilaments were woven into a plain 2D fabric, and a 3D diagonally interlock woven fabric. For electrical measurements, the fabrics were sandwiched between two Cu-film-based electrodes. The peak V_{oc} was recorded 2.3 V for a 3D WPENG, which was 16 times higher than a 2D WPENG (Talbourdet et al. 2018).

Soin et al. fabricated a 3D YF-WPENG (Figure 10.29(c)) in double jersey knit fabric construction by employing three sets of filaments. The knitting was conducted in such a way that Ag-coated polyamide-66 (PA66) filaments-based fabric layers at the top and the bottom sandwiched the PVDF melt

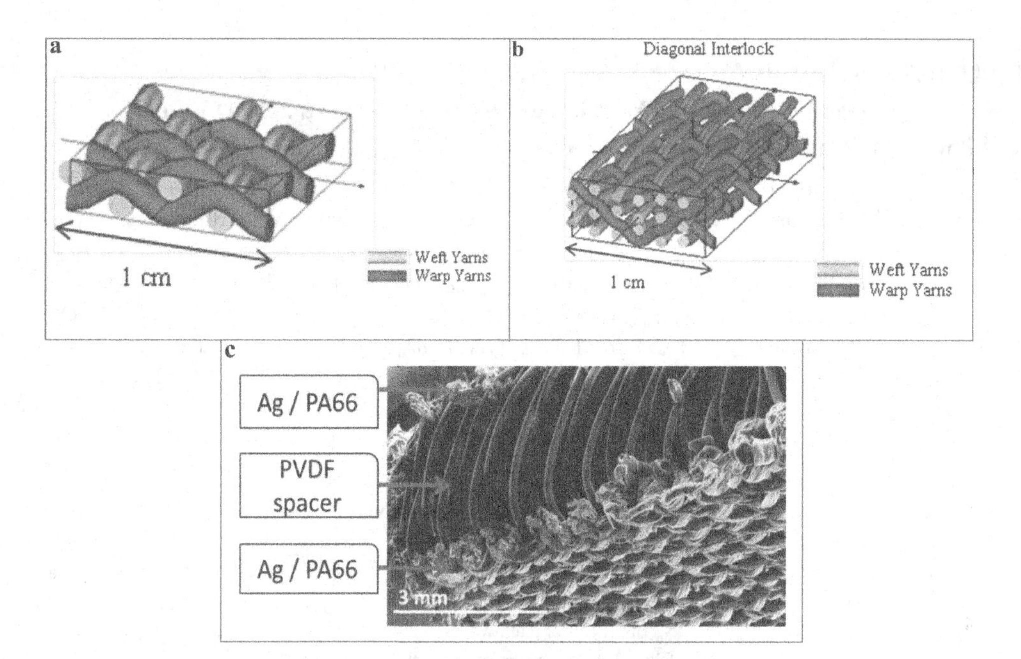

FIGURE 10.29 Yarn integrated 3D fabric/textile-based WPENGs. (a) Diagonally interlock woven fabric-based 3D YF-WPENG of PVDF multifilaments. (Reused with permission from Talbourdet et al. (2018). Copyright IOP Publishing Ltd 2018.) (b) Double jersey knitted spacer fabric-based 3D YF-WPENG of PVDF filaments. (Reused with permission from Soin et al. (2014). Copyright Royal Society of Chemistry 2014.)

filaments at the middle. The third set of yarns (PET) filaments, when used, functioned as spacers to prevent short-circuiting. Without the PET spacer filaments, often short-circuit occurred, making the WPENG unusable. The 3D YF-WPENG managed to generate PD of 5.10 μW cm^{-2} at 0.10 MPa and 470 kΩ load resistance (Soin et al. 2014). Table 10.7 summarizes the recently developed yarn integrated textiles-based 2D/3D WPENGs and multilayer stacked textiles-based WPENGs.

TABLE 10.7
Summary of Recently Developed Yarn Integrated Textiles-Based 2D/3D WPENGs and Multilayer Stacked Textiles-Based WPENGs

Active Materials	Electrode Materials	Output	Applied As	Special Attributes	Ref.
		Yarn integrated 2d fabric/textile-based WPENGs			
PVDF	PEDOT	Peak V_{oc} of 300 mV and peak I_{sc} of 0.16 mA at 10 kPa	Pressure sensor	Monitoring subtle human motion, muscle vibration, and health in daily life	Zhou et al. (2017)
ZnO	Pd and Cu wires	Peak V_{oc} of 3 mV and peak I_{sc} of 17 pA	–	Detect the intensity of UV light	Bai et al. (2013)
PVDF-Tr-FE	Stainless-steel yarn	Peak V_{oc} of ~16.2 mV under a periodic pressure	–	–	Yang et al. (2017)

(Continued)

TABLE 10.7 *(Continued)*
Summary of Recently Developed Yarn Integrated Textiles-Based 2D/3D WPENGs and Multilayer Stacked Textiles-Based WPENGs

Active Materials	Electrode Materials	Output	Applied As	Special Attributes	Ref.
PVDF	Ni/Cu alloy	Maximum PD to 125 μW cm^{-2} at 6.6-MΩ load resistance and 6-Hz frequency	Powering up LEDs	–	Song and Yun (2015)
PVDF	Ni/Cu alloy	Peak V_{oc} of 51 V, peak I_{sc} of 28.5μA, and PD of 105 mW m^{-2} (at 1-MΩ load resistance) at 6 mm stretching with 6-Hz frequency from an active area of 9 × 9 cm^2	–	Tactile sensor array	Ahn, Song, and Yun (2015)
		Yarn integrated 3d fabric/textile-based WPENGs			
PVDF	Cu	Peak V_{oc} of 2.3 V at 100-Hz frequency, 4% strain, and 5 N force for woven interlock fabric and peak V_{oc} of 0.14 V for plain-woven fabric. From 10 cycles of stress, the energy output was reported as 10.5 and 2.0 μJ m^{-2} for interlock and plain-woven fabrics, respectively	–	–	Talbourdet et al. (2018)
PVDF	Ag/PA 66	PD of 1.10–5.10 mW cm^{-2} at 0.02 and 0.10 MPa, respectively	–	–	Soin et al. (2014)
		Multi-layers stacked fabric/textile-based WPENGs			
ZnO	Ag/nylon fabric	Peak V_{oc} of 4 V, peak I_{sc} of 20 nA and peak V_{oc} of 0.8 V, peak I_{sc} of 5 nA were reported from palm clapping and finger bending, respectively	Powering up miniature displays and LEDs	–	Zhang, Chen, and Guo (2019)
ZnO for piezoelectric and PE vs Au for triboelectric	Au/textile	The individual triboelectric and piezoelectric devices exhibited peak to peak V_{oc} of ~ 1 and ~ 1.6 V, respectively. Upon combining the peak-to-peak V_{oc} reached ~ 8 V. The tests were performed under 100 dB and 100-Hz frequency	Organic LEDs (OLED) and LCDs	–	Kim et al. (2012)
ZnO/PVDF	Cotton and Ag blend double jersey knit or Al foil	Average peak V_{oc} of 6.36 V for knitted electrodes and 10.84 V for Al foil electrodes at 1-Hz frequency and 0.10 MPa from a 15 cm^2 active area	–	–	Kim et al. (2018)
PVDF–NaNbO$_3$	Segmented PU and Ag-coated PI multifilament yarns knitted electrode	Peak V_{oc} of 3.4 V and I_{sc} of 4.4 μA at 1-Hz frequency and 0.2 MPa	–	–	Zeng et al. (2013)

Abbreviations: PVDF (polyvinylidene fluoride), PVDF-Tr-FE (polyvinylidene fluoride trifluoroethylene), ZnO (zinc oxide), PEDOT [poly(3,4-ethylene dioxythiophene)], Pd (palladium), Cu (copper), Ni (nickel), Ag (silver), Au (gold), PU (polyurethane), Al (aluminum), PI (polyamide), V_{oc} (open circuit voltage), I_{sc} (short circuit current), PD (power density).

10.4.1.1.3.3 Multi-Layers Stacked Fabric/Textile-Based WPENGs (ML-WPENGs) The fabric-based substrates can also be treated with piezoelectric materials to induce piezoelectric properties. Without using any textile forming technique (e.g., weaving, knitting, braiding, etc.), direct fabric substrate can be utilized to fabricate WPENGs. In this case, stacked layers of multiple fabrics of traditional textiles, non-woven mats, and nano-structure mats are used to facilitate piezoelectric behavior.

Zhang et al. developed a nanorods patterned-textile ML-WPENG (Figure 10.30(a)) by coating the surface of two nylon fabrics with Ag-paste to be used as conductive electrodes. A dense ZnO nanorod array was grown vertically on the surface of a fabric-based electrode. These surface-treated fabrics were stacked together to fabricating the final ML-WPENG. When palm clapped, the ML-WPENG exhibited peak V_{oc} of 4 V and peak I_{sc} of 20 nA. While finger bending, the WPENG generated peak V_{oc} of 0.8 V and peak I_{sc} of 5 nA (Zhang, Chen, and Guo 2019). A similar design construction was also demonstrated by Kim et al. though they used ZnO NWs as the piezoelectric material. Herein, two PET fabrics were Au-coated to be utilized as the conductive fabric-based electrode. A piezoelectric ZnO NWs layer was grown vertically on the surface of one of the electrodes. Afterwards, a self-adhesive charged polyethylene (PE) film was attached to the ZnO NWs layer.

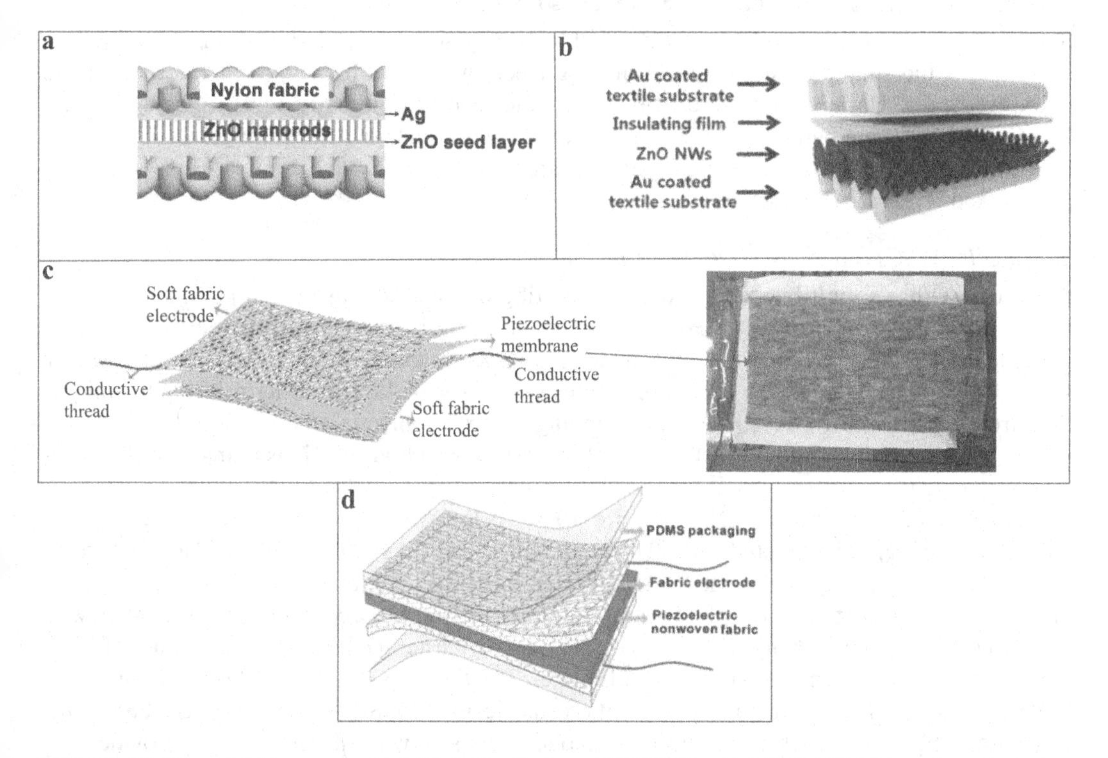

FIGURE 10.30 Multi-layers stacked fabric/textile-based WPENGs. (a) ML-WPENG based on Ag/nylon and ZnO nanorods grown Ag/nylon fabrics. (Reused with permission from Zhang, Chen, and Guo (2019). Copyright Elsevier 2018.) (b) ML-WPENG based on Au/textile and ZnO nanorods grown Au/textile sandwiching a PE charged film. (Reused with permission from Kim et al. (2012). Copyright Royal Society of Chemistry 2012.) (c) ML-WPENG based on ZnO NWs/PVDF mat sandwiched between two conductive fabric electrodes. (Reused under creative commons license attribution 4.0 (https://creativecommons.org/licenses/by/4.0/) from Kim et al. (2018). Copyright MDPI 2018.) (d) ML-WPENG based on $NaNbO_3$-PVDF mat sandwiched between two conductive fabric electrodes. (Reused with permission from Zeng et al. (2013). Copyright Royal Society of Chemistry 2013.)

Similar to the first one, these surface-treated fabrics were stacked to get the final ML-WPENG (Figure 10.30(b)). Since the presence of PE and ZnO NWs also introduced triboelectrification, therefore, the ML-WPENG also served as a hybrid NG as well. A 10 cm^2 hybrid ML-WPENG when subjected to a sound of 100 dB and 100-Hz frequency, generated a peak to peak V_{oc} and a peak to peak I_{sc} of 8 V and 2.5 μA, respectively (Kim et al. 2012).

A rather straightforward approach could be using metal-based fabric directly as the electrodes. Kim et al. developed such an ML-WPENG (Figure 10.30(c)), wherein ZnO NWs were grown on PVDF nanofibrous mat. The mat was further sandwiched between two knitted fabric-based electrodes to fabricate the final ML-WPENG. The ML-WPENG of 15 cm^2 area demonstrated an average peak V_{oc} of 6.36 V at 1-Hz frequency and 0.10 MPa impact pressure (Kim et al. 2018). Zeng et al. showed a similar design with an additional encasing of PDMS for robustness and protection from water and dust (Figure 10.30(d)). Nanofibrous mat of $NaNbO_3$ incorporated PVDF was the active piezoelectric fabric layer. The mat was sandwiched between two knitted fabric-based electrodes of segmented PU and Ag-coated polyamide multifilament. When subjected to 1-Hz frequency and 0.12 MPa pressure, the ML-WPENG exhibited a peak V_{oc} of 3.4V and a peak I_{sc} of 4.4 μA (Zeng et al. 2013). The overall output of 2D, 3D, and multi-layered fabric-based WPENGs are higher than single yarn WPENGs but compared to WTENGs, the output of WPENGs is lower.

10.4.1.2 Textiles-Based Wearable Solar Cells (WSCs)

Solar energy is one of the clean and renewable energy sources around the human body. The PV effect is the simplest mechanism of moving electron/holes to harness solar electricity (Cho et al. 2019). The earlier SCs were developed in wafer and thin-film structures. Recently, OSCs, DSSCs, and PSCs are being developed with flexible, lightweight, and cheap materials (Chen et al. 2020; Zhang et al. 2014). The textile-based SCs are typically fabricated from yarn-like structures and stacked layers-based structures.

10.4.1.2.1 WSCs Built Directly onto Fabric Substrates

A typical SC has a multilayered structure consisting of electrode layers and photoactive material layers. Thus, to construct a textile SC, layer stacking is a simple and effective approach. Technically, it could be fabricated by transferring an as-prepared SC vertically onto the textile substrate or building the SC on textile substrates via a layer-by-layer coating/printing technique. SCs are built onto textile substrates by employing coating or printing techniques. As integrated into textiles, it provides better mechanical robustness against body motions than directly attaching SCs on textiles.

Zhang et al. developed an OSC (Figure 10.31(a)) by using Ti-wire-based textile cathode, which was sequentially treated with TiO_2 nanotubes and NPs. This layer of TiO_2 helped in better charge separation and transfer along with increased polymer load as well as reduced electrical resistance for the OSC. The modified Ti-textile was sequentially dip-coated with poly(3-hexylthiophene):phenyl-C71-butyric acid methyl ester (P3HT:PC_{71}BM) and poly(3,4-ethyl-enedioxythio-phene):poly(styrene sulfonate) (PEDOT:PSS) as the PV layers. Finally, the structure was sandwiched with CNTs sheets to fabricate the OSC. Herein, PEDOT:PSS aided in transferring holes to CNTs sheet-based anode. This sandwiched construction provided the opportunity to harness sunlight from any side of the OSC. A maximum PCE of 1.08% was demonstrated, along with a peak V_{oc} of ~ 0.53 V (Zhang et al. 2014). But the structure is rigid to the point that it can create wearable discomfort.

For that purpose, OSC (Figure 10.31(b)) developed by Kim et al. utilized a PET-based conductive composite coaxial structure working as the anode to fabricate a flexible traditional textile substrate-based SC. The coaxial yarn was fabricated by coating PET sequentially with Ag NWs and reduced graphene oxide (rGO). The first stage involved coating the fabric substrate with PEDOT:PSS and P3HT: PC_{61}BM [1-(3-methoxycarbonyl)-propyl-1-phenyl-(6,6) C_{61}]

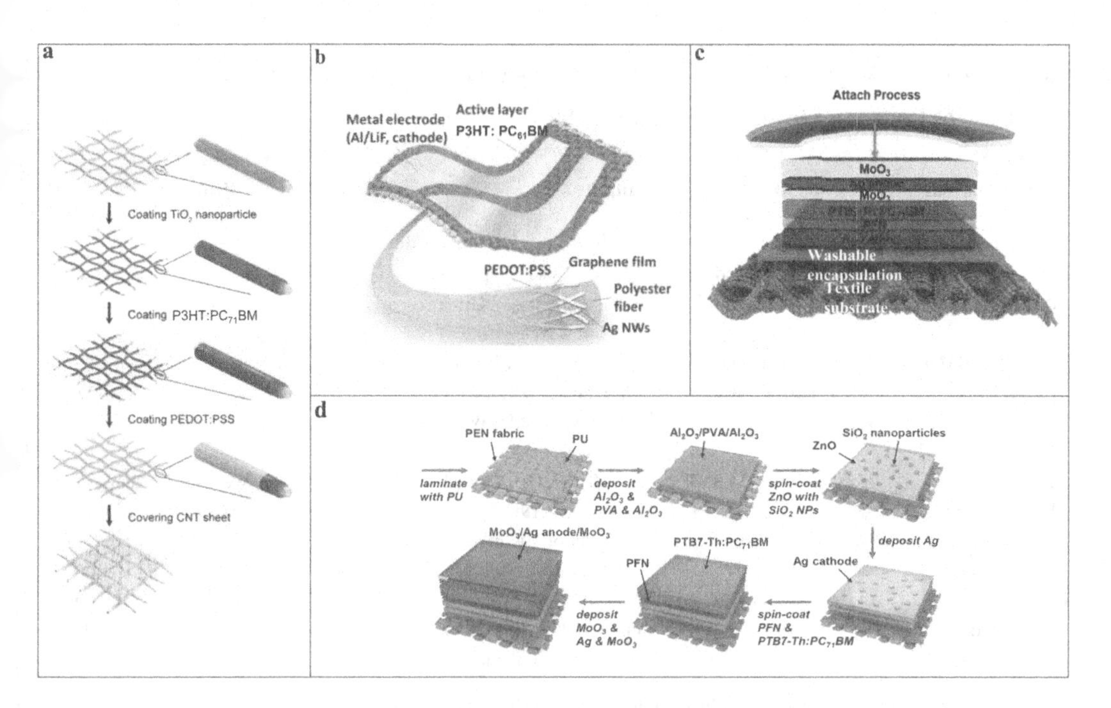

FIGURE 10.31 WSCs built directly onto fabric substrates. (a) SC based on P3HT: PC_{71} BM/PEDOT:PSS with TiO_2/Ti cathode and CNTs sheet anode. (Reused with permission from Zhang et al. (2014). Copyright John Wiley and Sons 2014.) (b) SC based on PEDOT:PSS/graphene/Ag NWs-coated PET yarn-based fabric-based anode with P3HT: PC_{61}CBM coating and Al/LiF-based cathode. (Reused with permission from Wu et al. (2017). Copyright Elsevier 2016.) (c and d) SC based on PTB_7-Th:PC_{71}BM with Ag-based cathode and MnO_3/Ag/MnO_3-based anode. (Reused with permission from Cho et al. (2019) and Jeong et al. (2019), respectively. Copyright American Chemical Society 2019 and © Royal Society of Chemistry 2019, respectively.)

in sequence. Layers of LiF and Al were deposited onto P3HT:PC_{61}BM as the cathode. The device demonstrated a peak V_{oc} of 0.55 V and a PCE of 2.27%. Also, after 400 cycles of curling and folding, the device retained PCEs of 87% and 73%, respectively (Wu et al. 2017). Washability is one of the important attributes of wearable textiles; OSC (Figure 10.31(c)) developed by Jeong et al. has incorporated the washable feature into the SC. Herein, a textile substrate was coated with a nanostratified/SiO_2-polymer composite-based barrier layer to protect against moisture, washing, and oxygen. The fabrication process followed by coating the barrier substrate with Ag as the cathode and poly[(9,9-bis(3′-(*N*,*N*-dimethylamino) propyl)-2,7-fluorene)-alt-2,7-(9,9-dioctylfluorene)] (PFN) as the electron transfer layers. Further coating with a blend of poly[4,8-bis[5-(2-ethylhexyl)thiophen-2-yl]benzo[1,2-b:4,5-b′] dithiophene-2,6-diyl-alt-3-fluoro-2-[(2-ethylhexyl)carbonyl]thieno[3,4-b]thio-phene-4,6-diyl] (PTB7-Th) and (6,6)-phenyl C71 butyric acid methyl ester (PC_{71}BM, Nano-C) as the active layer. A dielectric/metal/dielectric layer such as MoO_3/Ag/MoO_3 served as the anode for the OSC. The device demonstrated a peak V_{oc} of 0.77 V and PCE of 7.19% (Jeong et al. 2019). A similar OSC (Figure 10.31(d)) was also developed by Cho et al. by using a similar electrode, active layers, and barrier layers. The OSC demonstrated a PCE and peak V_{oc} of 0.77 V and 7.3%, respectively (Cho et al. 2019). The top layers of the SCs must be transparent, allowing the light as much as possible to travel to the photoactive layers and vice-versa for the bottom layers. Though these SCs are built onto textiles substrate, such stacked structures may reduce the breathability of the textiles and cause wearable discomfort. Table 10.8 summarizes the recently developed wearable SCs.

TABLE 10.8
Summary of Recently Developed Wearable SCs

Active Materials	Electrode Materials	Output	Applied As	Special Attributes	Ref.
		Solar cells built onto fabric substrates			
P3HT:PC_{71}BM and PEDOT:PSS	TiO_2/Ti cathode and CNTs/TiO_2/Ti anode	V_{oc} of ~0.53 V, J_{sc} of 5.2 mA cm^{-2}, and 1.08% PCE at 550 nm wavelength	–	–	Zhang et al. (2014)
P3HT:PC_{61}BM and PEDOT:PSS	LiF/Al cathode and graphene/Ag NWs/PET anode	V_{oc} of 0.55 V, J_{sc} of 9.31 mA cm^{-2}, and 2.27 % PCE at an intensity of 100 mW cm^{-2} illumination	–	–	Wu et al. (2017)
PFN, PTB7-Th:PC_{71}BM, and MoO_3	Ag cathode and MnO_3/Ag anode	V_{oc} of 0.76 V, J_{sc} of 14.43 mA cm^{-2}, and 7.19% PCE at an intensity of 100 mW cm^{-2} illumination	–	Washable SCs	Jeong et al. (2019)
PFN and PTB7-Th: PC_{71}BM	Ag cathode and MoO_3/Ag	V_{oc} of 0.769 V, J_{sc} of 15.57 mA cm^{-2}, and 8.3 % PCE at 100 mW cm^{-2} illumination	–	–	Cho et al. (2019)
		Solar cells built onto yarn substrates			
N719 dye	TiO_2/Ti working electrode and MWCNTs counter electrode	V_{oc} of 0.71 V, J_{sc} of 16 mA cm^{-2}, and 7.13% PCE at 100 mW cm^{-2} illumination	–	–	Yang et al. (2014)
N719 dye and CuI	ZnO/Mn-plated polymer anode and Cu-coated polymer wires counter electrode	V_{oc} of 4.6 V, J_{sc} of 7.8 mA cm^{-2}, and 1.3% PCE at 100 mW cm^{-2} illumination	Powering digital calculator	Splitting the water	Zhang et al. (2016)
N719 dye and CuI	ZnO/Mn-plated polymer anode and Cu-coated polymer wires counter electrode	With six strings of photoanodes (each with a length of ~2 cm), I_{sc} of 0.28 mA and V_{oc} of 2.6 V were achieved under 100 mW cm^{-2} illumination	–	–	Chai et al. (2016)
PTB7:PC_{71}BM and PEDTOT:PSS	ZnO NCs/Ti wire cathode and Ag-plated nylon yarn	V_{oc} of 0.48 V, J_{sc} of 7.39 mA cm^{-2}, and 1.62 % at an intensity of 100 mW cm^{-2} illumination	Powering digital watch		Liu et al. (2018)
N719 dye and CuI	TiO_2/Ti photoanode and Ag-plated nylon counter electrode	V_{oc} of 0.4 V, J_{sc} of 9.42 mA cm^{-2}, and 1.62 % PCE at an intensity of 100 mW cm^{-2} illumination	–	–	Gao et al. (2019)

Abbreviations: P3HT:PC_{71} BM [poly(3-hexylthiophene):phenyl-C71-butyric acid methyl ester], PEDOT:PSS [poly(3,4-ethylenedioxythio-phene):poly(styrene sulfonate)], PC_{61}CBM [1-(3-methoxycarbonyl)-propyl-1-phenyl-(6,6) C61], PFN [poly[(9,9-bis(3′-(*N*,*N*-dimethyl amino)propyl)-2,7-fluorene)-alt-2,7-(9,9-dioctylfluorene)]], PTB7-Th:PC71BM [poly[4,8- bis[5-(2-ethylhexyl)thiophen-2-yl]benzo[1,2-b:4,5-b′]dithiophene-2,6-diyl-alt-3-fluoro-2-[(2-ethylhexyl)carbonyl]thieno[3,4-b]thio-phene-4,6-diyl] (PTB7-Th), MoO_3 (molybdenum trioxide), CuI [copper(I) iodide], Ti (titanium), TiO_2 (titanium dioxide), CNTs (carbon nanotubes), MWCNTs (multiwalled CNTs), LiF (lithium fluoride), Al (aluminum), Ag (silver), PET (polyethylene terephthalate), ZnO (zinc oxide), Mn (manganese), Cu (copper), NCs (nanocrystals), NWs (nanowires), V_{oc} (open circuit voltage), I_{sc} (short circuit current), PD (power density).

10.4.1.2.2 WSCs Built Directly onto Yarn Substrates

Despite providing comparatively simple fabrication of textile-based SCs by directly attaching to textile substrates or built onto textiles substrates, these structures may impair the breathability of clothing and create wearable discomfort. Besides, the minimized transmission of light through the top electrode and stacked layers may reduce the PCE of such SCs. Imparting PV property to yarn-like structures and adopting conventional textile manufacturing techniques of weaving and knitting could lead to breathable and wearable SCs. Yarn integrated fabric-based SCs can enhance PCE due to the likeliness of the direct impact of sunlight onto the surface of the yarns. Typical design construction involves a core yarn coated with PV materials and further twisted or coated with another electrode.

Yang et al. used a rubber filament substrate and wrapped it with an MWCNTs sheet as the counter electrode to fabricate a DSSC (Figure 10.32(a)). The resulting yarn exhibited flexibility and stretchability through knotting and winding onto a substrate and by stretching to 100%. This assembly was inserted into a Ti-wire-based spring. Before insertion, the spring was treated with TiO_2 nanotubes (as the working electrode) and N719 dye. Afterwards, the current assembly was inserted into a PET tube to introduce an electrolyte. When several of these were loosely woven into a fabric, a PCE of 7.13%, V_{oc} of 3.71 V, and I_{sc} of 4.27 mA were achieved (Yang et al. 2014). Despite such higher output, the device seems bulky and could be discomforting when worn.

To that end, Zhang et al. fabricated textile-based DSSCs (Figure 10.32(b)) of yarns by weaving them into plain, twill, and satin fabric structures. Herein, the warp threads worked as the counter

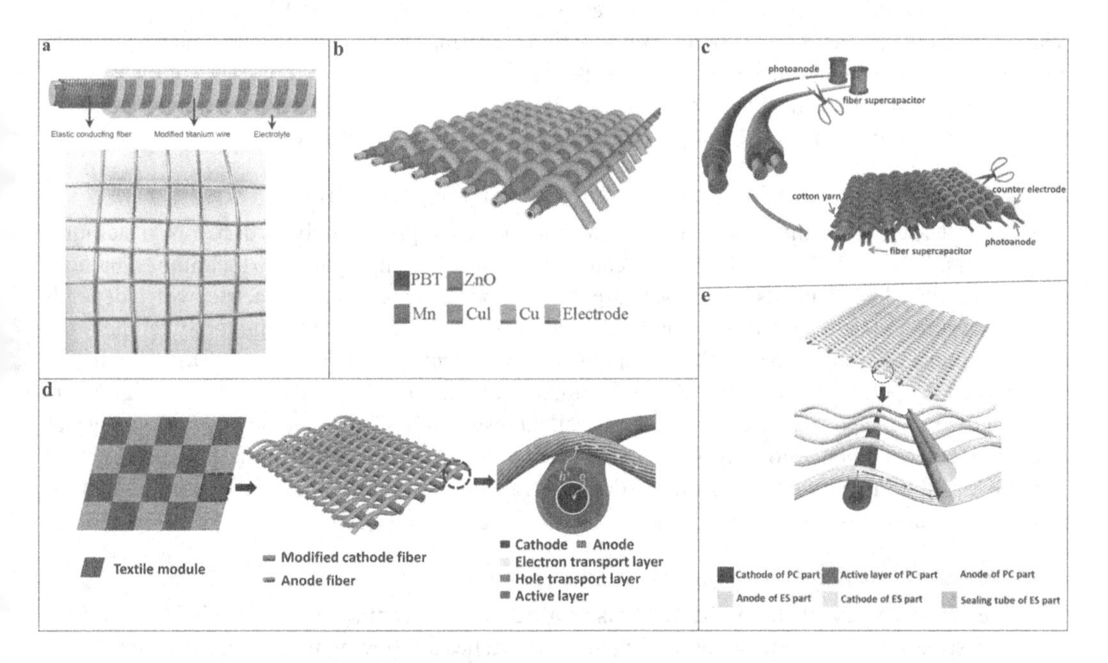

FIGURE 10.32 SCs built onto yarn substrates. (a) SC based on MWCNTs-based counter and TiO_2/Ti-based working electrode with N719 dye. (Reused with permission from Yang et al. (2014). Copyright John Wiley and Sons 2014.) (b) SC based on Cu/PBT counter electrode and ZnO NWs/Mn/PBT-based working electrode with N719 dye. (Reused with permission from Zhang et al. (2016). Copyright John Wiley and Sons 2015.) (c) SC with similar construction as (b) and additionally integrated with a supercapacitor. (Reused with permission from Chai et al. (2016). Copyright American Chemical Society 2016). (d) SC based on Ag-coated nylon yarn cathode and PTB7:PC_{71}BM-coated ZnO/Ti-wire-based anode. (Reused with permission from Liu et al. (2018). Copyright Royal Society of Chemistry 2018.) (e) SC based on Ag-coated nylon yarn cathode and N719 dye-coated TiO_2/Ti wire-based anode. (Reused with permission from Gao et al. (2019). Copyright Royal Society of Chemistry 2019.)

electrode and were fabricated from a Cu-coated polybutylene terephthalate (PBT) filament. The weft threads as the working electrode were formed using Cu-PBT filament but sequentially coated with Mn, ZnO NWs, and N719 dye. A CuI layer was also deposited onto the working electrode to function as a hole transfer layer between the electrodes. The device reported a peak V_{oc} of 4.6 V, a peak J_{sc} of 7.8 mA cm^{-2}, and a PCE of 1.3% (Zhang et al. 2016). The main advantage of these DSSCs is scaled-up production processes by using commercial textile machinery with low-cost all-solid-state materials. The research group also demonstrated similar fabric-based SCs with energy-storing (ES) supercapacitors, as illustrated in (Figure 10.32(c)) (Chai et al. 2016).

A similar interlaced woven structure was adopted by Liu et al., wherein warp yarns and weft yarns functioned as anode and cathode, respectively. The warp yarns were formed by coating ZnO nanocrystals treated Ti-wire sequentially with photoactive (PTB7-Th: $PC_{71}BM$) and hole transfer (PEDOT: PSS) layers. Commercially available Ag-coated nylon yarns were used as the anode and interlaced with the cathode in a weaving loom. The final fabric-based SC was constructed into alternate modules of SCs and conventional fabric as illustrated in (Figure 10.32(d)); in an optimized solar fabric, a V_{oc} of 0.48 V, J_{sc} density of 7.39 mA cm^{-2}, and PCE of 1.62% were reported (Liu et al. 2018). The same group also integrated ES supercapacitors along with power conversion (PC) segments using a similar fabrication approach (Figure 10.32(e)). In this case, the PV yarns (i.e., cathode) were fabricated from TiO_2 nanotubes treated Ti-wire coated with N719 dye and CuI layers.

Similarly, Ag-plated commercially available nylon filaments were used as the anode. An optimized design exhibited a peak V_{oc} of 0.4 V, a peak J_{sc} of 9.42 mA cm^{-2}, and a PCE of 1.92% (Gao et al. 2019). In conclusion, yarn integrated fabric-based SCs have allowed large-scale production with comforting and flexible attributes. But functional coating of these textile substrates is prone to coating defects and could reduce the overall functionality and mechanical robustness of the coating. The energy conversion efficiency of SCs is still lower, which further needs to be improved along with ensuring a more controlled and precise coating of the textile substrate.

10.4.1.3 Textiles-Based Wearable Thermoelectric Nanogenerators (WTEGs)

The heat released from the human body can reach up to approximately 100–525 W (Siddique, Mahmud, and Heyst 2017). The body heat can be harvested by employing two techniques, namely, thermoelectric, which depends on temperature gradient over space, and pyroelectric, which depends on temperature gradient over time (Bhatnagar and Owende 2015). Since human body temperature remains almost constant at 37°C, the mechanism of pyroelectric is ineffective due to insignificant variation of temperature over time. But the human body and its surrounding can have significant temperature variance over space. Therefore, excellent research has been conducted to develop wearable thermoelectric generators. Similar to other wearable textile constructions discussed above, WTEGs could be fabricated either on textiles substrates or yarns integrated fabrics in 2D and 3D textiles constructions (Chen et al. 2020).

10.4.1.3.1 WTEGs Built Directly onto Textiles Substrates

To harvest heat energy of the body, WTEGs require to be in contact with the skin. In that case, textile substrates are an excellent choice. Typical approaches are sewing thermoelectric yarns into a textile substrate or coating a textile substrate with thermoelectric materials.

Kim et al. fabricated a fabric-based WTEG by employing a dispenser printing technique. P-type ($Bi_{0.5}Sb_{1.5}Te_3$) and n-type ($Bi_2Sb_{0.3}Te_{2.7}$) materials were printed alternatively with a gap onto a meshed fabric substrate (Figure 10.33(a)). Herein, electrical connections were made by sewing Ag-based conductive thread on both sides of the fabric. Finally, a PI film was attached to the bottom surface of the WTEG to avoid physical and electrical contact with the wearer. When worn by a wearer at 32°C body temperature and subjected to 5 and 25°C ambient temperatures, the WTEG with 12 pairs of thermocouples generated an output voltage of 11.5 and 2.7 mV, respectively. The PD was reported to be 292.4 nW cm^{-2} for the earlier case (Kim et al. 2014).

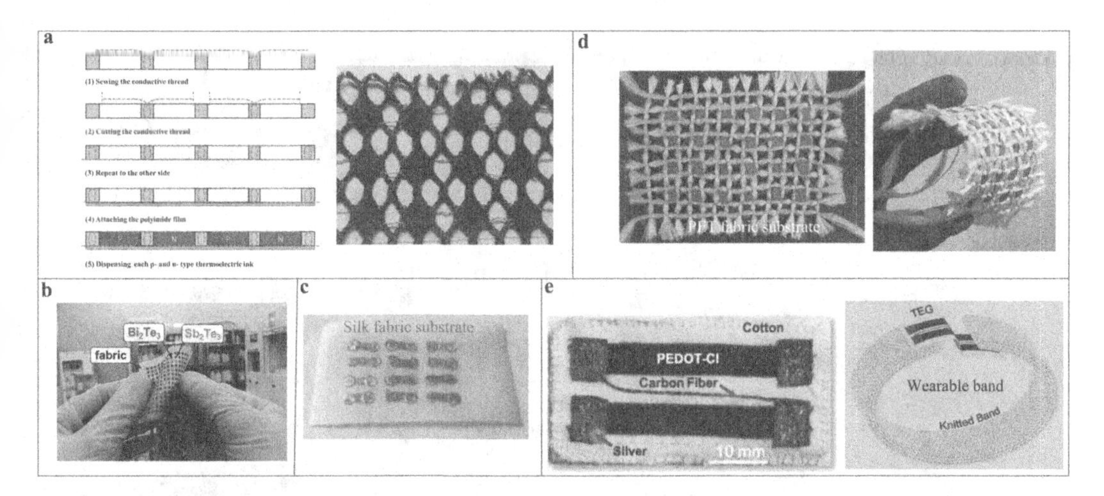

FIGURE 10.33 WTEGs built directly onto textiles substrates: Part 1. (a) Meshed fabric substrate-based WTEG of p-type ($Bi_{0.5}Sb_{1.5}Te_3$) and n-type ($Bi_2Sb_{0.3}Te_{2.7}$) materials. (Reused with permission from Kim et al. (2014). Copyright IOP Publishing Ltd 2014.) (b) Glass fabric substrate-based WTEG of p-type (Sb_2Te_3) and n-type (Bi_2Te_3) materials. (Reused with permission from Kim, We, and Cho (2014). Copyright Royal Society of Chemistry 2014.) (c) Silk fabric substrate-based WTEG of p-type (Sb_2Te_3) and n-type (Bi_2Te_3) materials. (Reused with permission from Lu et al. (2016). Copyright Elsevier 2015.) (d) PET fabric substrate-based WTEG of p-type [$(0.25Bi,0.75Sb)_2(0.95Te,0.05Se)_3$] and n-type [$(0.98Bi,0.02Sb)_2(0.9Te,0.1Se)_3$] materials. (Reused with permission from Siddique et al. (2016). Copyright Elsevier 2016.) (e) Cotton fabric substrate-based WTEG with p-type (CNTs based) and n-type (PEDOT-Cl-coated CNTs) materials. (Reused with permission from Allison and Andrew (2019). Copyright John Wiley and Sons 2019.)

The overall fabrication process involves heating to 100°C for curing and binding thermoelectric materials onto the fabric substrate. Besides, the dispenser printing technique is inappropriate for mass production. The screen printing is more facile and appropriate for mass production. J. Kim et al. reported a WTEG with p-type (Sb_2Te_3) and n-type (Bi_2Te_3) materials screen printed as dots onto a glass fabric substrate alternatively with a gap between the distinct dots (Figure 10.33(b)) (Kim, We, and Cho 2014). Herein Cu-based conductors were used to making the electrical connections. Though the device showed decent flexibility to be wearable, the processing temperature during fabrication reached 600°C. This could be a drawback to be built onto conventional textiles. Nevertheless, the WTEG with 11 thermocouples and at 15°C ambient temperature demonstrated V_{oc} of 2.9 mV and power of 3 μW. Lu et al. developed WTEG on a piece of conventional silk fabric with nanostructure Bi_2Te_3 (n-type) and Sb_2Te_3 (p-type) as thermoelectric materials (Figure 10.33(c)). Ag foils were used to connect the p-type and n-type materials with adhesive Ag-paste. A WTEG with 12 thermocouples was constructed. When subjected to temperature gradients of 5-35 DK, the WTEG exhibited a V_{oc} of ~5–10 mV and power of ~14–15 nW, respectively (Lu et al. 2016). The WTEG seems to be providing more wearable comfort than the other two devices mentioned earlier. But mechanical strength of Ag-foils during the body movement could hinder device functionality. Similar constructions and working mechanisms were also adopted by Siddique et al., as demonstrated in Figure 10.33(d)) (Siddique et al. 2016), respectively. So far, we have been talking about WTEG based on n-type and p-type materials (Figure 10.33(e)), but it is also possible to harness thermal energy by employing only p-type or n-type materials. For instance, Allison et al. fabricated a WTEG by coating a cotton substrate with p-doped PEDOT (PEDOT-Cl). Subsequently, coated strips were electrically connected by carbon fiber. The carbon fiber was attached to the thermoelectric edges with Ag-paste. When integrated with a wearable band, the WTEG generated a V_{oc} of 23mV (Allison and Andrew 2019).

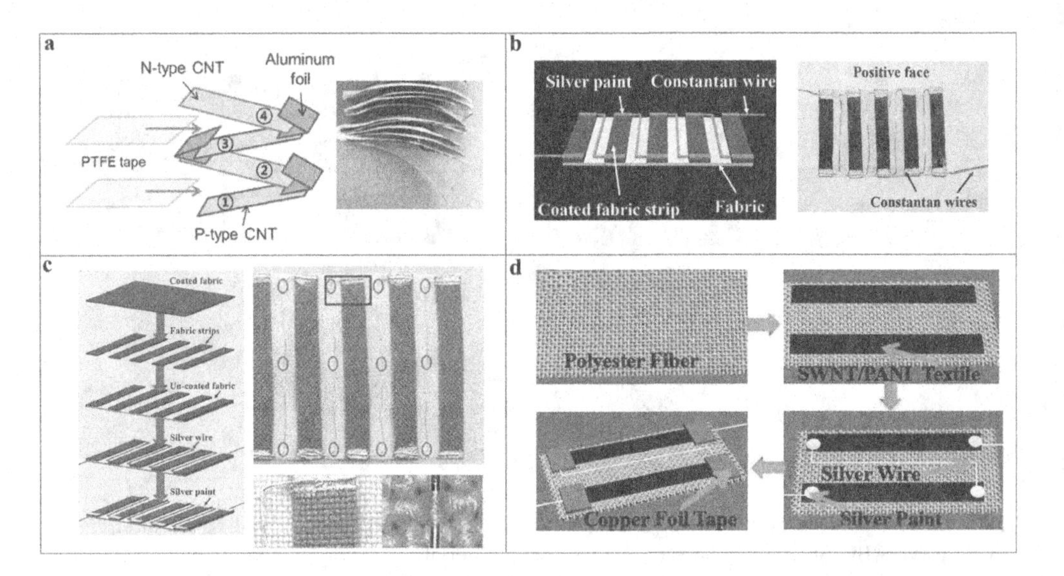

FIGURE 10.34 WTEGs built directly onto textiles substrates: Part 2. (a) WTEG based on stacked p-type (CNTs sheet) and n-type (n-doped CNTs sheet) materials with PTFE as insulation. (Reused with permission from Kim et al. (2014). Copyright American Chemical Society 2014.) (b) WTEG based on p-type (PEDOT:PSS) material-coated PET fabric strips. (Reused with permission from Du et al. (2015). Copyright Springer Nature 2015.) (c) WTEG based on p-type (PEDOT:PSS) material-coated cotton fabric strips and Constantan wire. (Reused with permission from Du et al. (2017). Copyright Royal Society of Chemistry 2017.) (d) WTEG based on p-type (CNTs/PANI) materials-coated PET fabric substrate. CNTs/PANI. (Reused with permission from Li et al. (2016). Copyright Royal Society of Chemistry 2016.)

A stacked WTEG was developed by Kim et al. with CNTs and treated CNTs sheets as the p-type and n-type thermoelectric materials, respectively (Figure 10.34(a)). To get the n-type material, the CNTs sheet was treated with polyethyleneimine, diethylenetriamine, and $NaBH_4$. The individual materials were then connected alternatively in series with Al foil/Ag-adhesive and stacked p-type on top of each other with PTFE insulating film between them (Kim et al. 2014). A WTEG with 72 n-type and 72 p-type as in pairs generated V_{oc} of 465 mV at 49 DK temperature gradient. Du et al. approached a very easy fabrication process to fabricate a WTEG with only a p-type material (Du et al. 2015). Herein, a PET fabric was coated with PEDOT:PSS as the p-type material. Then fabric strips were cut from the coated fabric and attached to another PET fabric substrate with Ag-adhesive (Figure 10.34(b)). The electrical connections were made through Ag wires. Such a device with five thermoelectric strips generated a voltage of 4.3 mV at 75.2 DK. More recently, the group used the same design approach, except the PET fabric substrate was replaced with a cotton fabric substrate, and the electrical connections were made with Constantan wire (Figure 10.34(c)). When subjected to 74.3 DK, the voltage output of the WTEG reached 18.7 mV (Du et al. 2017). Since Constantan wire as an n-type material participated in the thermoelectric effect, the voltage had increased by 4.34 factor. Similarly, WTEG with single-walled CNTs (SCNTs)/PANI as the p-type material was developed (Figure 10.34(d)). When a WTEG with two thermoelectric strips were connected in series, the device produced a V_{oc} of 3.82 mV (Li et al. 2016). Table 10.9 summarizes the recently developed WTEGs built directly onto textiles substrates.

10.4.1.3.2 WTEGs Built Directly onto Yarn Substrate

Despite showing significant thermal insulation and considerable flexibility as wearables, WTEGs-based on textile substrate lacks breathability. Therefore, fabrication of textiles WTEGs by adopting traditional weaving and knitting approach seems more appropriate than textile substrate-based WTEGs. WTEGs based on yarns are generally woven or knitted into 2D and 3D fabric structures.

TABLE 10.9
Summary of Recently Developed WTEGs Built Directly onto Textiles Substrates

Active Materials	Electrode Materials	Output	Applied As	Special Attributes	Ref.
$Bi_{0.5}Sb_{1.5}Te_3$ (p-type) and $Bi_2Se_{0.3}Te_{2.7}$ (n-type)	Ag-based conductive threads	When worn onto a body, 12 thermocouples with 6 × 25 mm^2 generated 2.7 mV, 11.5 mV and 8.1 nW, 146.9 nW power at 25°C and 5°C ambient temperatures, respectively. The PD could be calculated to 292.4 nW cm^{-2}	–	–	Kim et al. (2014)
Bi_2Te_3 (n-type) and Sb_2Te_3 (p-type)	Cu/Ni/Ag paste-based electrodes	A medical bandage-like WTEG with 11 thermocouples generated 2.9 mV and 3 μW at 15°C ambient temperature	–	–	Kim, We, and Cho (2014)
Bi_2Te_3 (n-type) and Sb_2Te_3 (p-type)	Ag foils/Ag paste	A WTEG with 11 thermocouples generated 6.02 mV when worn on the arm. Besides another WTEG with 12 thermocouples at 35 DK, generated ~10 mV and ~15 nW power	–	–	Lu et al. (2016)
$Bi_1.8Te_{3.2}$ (n-type) and Sb_2Te_3 (p-type)	SbTe or Cu	A WTEG built onto a glass fiber textile substrate with 8 thermocouples, generated ~25 mV and ~2304 nW power at 20 DK	–	–	Cao et al. (2016)
n-type $(0.98Bi,0.02Sb)_2$ $(0.9Te,0.1Se)_3$ and p-type $(0.25Bi,0.75Sb)_2$ $(0.95Te,0.05Se)_3$	Ag conductive threads	Maximum 23.9 mV and 3.107 nW power were reported from a prototype WTEG at 22.5 DK	–	–	Siddique et al. (2016)
PEDOT-Cl	Carbon fiber/ Ag paste	WTEG when worn indoors (25°C) or outdoors (2°C) onto palm generated ~10 mV and. During perspiration in outdoors, the voltage reached ~24 mV	–	–	Allison and Andrew (2019)
p-type (oxygen doped CNTs) n-type CNTs (polyethyleneimine-/ diethylenetriamine-coated CNTs)	Al foil/Ag adhesive	A WTEG with 72 thermocouples demonstrated 465 mV at 49 DK and when integrated to power up a glucose sensor, a power of 1.8 μW was generated at 32 DK	–	Electrochromic glucose sensor	Kim et al. (2014)
PEDOT:PSS (p-type)	Ag wires/Ag paste (induced p-type)	~4.3 mV and 12.29 nW power at 75.2 DK with five fabric strips (40 mm × 5 mm each) of thermoelectric materials	–	–	Du et al. (2015)
PEDOT:PSS (p-type)	Constantan wires/Ag paint (n-type as well)	~18.7 mV and 212.6 nW power at 74.3 DK with five fabric strips (35 mm × 5 mm each) of thermoelectric materials	–	–	Du et al. (2017)
SCNTs/PANI (p-type)	Cu/Ag wires/ Ag paste	~3.82 mV and 48 nW power at 75 DK	–	–	Li et al. (2016)

Abbreviations: PEDOT [poly(3,4-ethylenedioxythiophene)], CNTs (carbon nanotubes), SCNTs (single walled CNTs), PEDOT:PSS [poly(3,4-ethylenedioxythiophene):polystyrene sulfonate], Ag (silver), Cu (copper), Ni (nickel), Al (aluminum).

10.4.1.3.2.1 Yarn-Based WTEGs These 1D yarn structures are typically based on coating filament with metals (Yadav, Pipe, and Shtein 2008), organic/inorganic materials, and/or by twisting thermoelectric materials into spun yarns (Liang et al. 2012). For instance, Yadav et al. demonstrated a coaxial WTEG (Figure 10.35(a)) of silica filament coated with alternative Ni and Ag segments. Overlapping junctions (0.5 mm) of Ni-Ag were also introduced in the structure. Then insulated posts were placed at the junctions for heating the Ni-Ag junctions alternatively. At 6.6 DK, a thermocouple junction exhibited a V_{oc} of 19.6 μVK and a TEG with seven thermocouples generated 2 nW power (Yadav, Pipe, and Shtein 2008). Liang et al. coated 10 μm glass fibers with a 300 nm layer of p-type (PbTe) thermoelectric nanocrystals. The fibers were then turned into 1D thermoelectric yarns. A demonstration as a proof of concept for harnessing heat from industrial pipes was conducted (Figure 10.35(b)), though the output was merely 1.7 mV (Liang et al. 2012). Wu and Hu took a different approach by coating conventional cotton and PET yarns with thermoelectric materials. A water-based polyurethane (WPU) matrix with MWCNTs and PEDOT: PSS as fillers (Figure 10.35(c)) was coated onto the yarns. The resultant electrical conductivity of ~13,826 S m^{-1}, a Seebeck coefficient of ~10 mV K^{-1}, and a power factor of ~1.41 mW m^{-1} K^{-2} at room temperature were achieved (Wu and Hu 2016).

10.4.1.3.2.2 Yarn Integrated 2D WTEGs Since yarn WTEGs have low output, approaching fabric-based structure to enhance the power output seems appropriate. Interlacing yarns by weaving or knitting technologies is the appropriate choice for fabricating 2D WTEGs. The yarns are typically interlaced in the in-plane or thickness directional mechanism of TEGs.

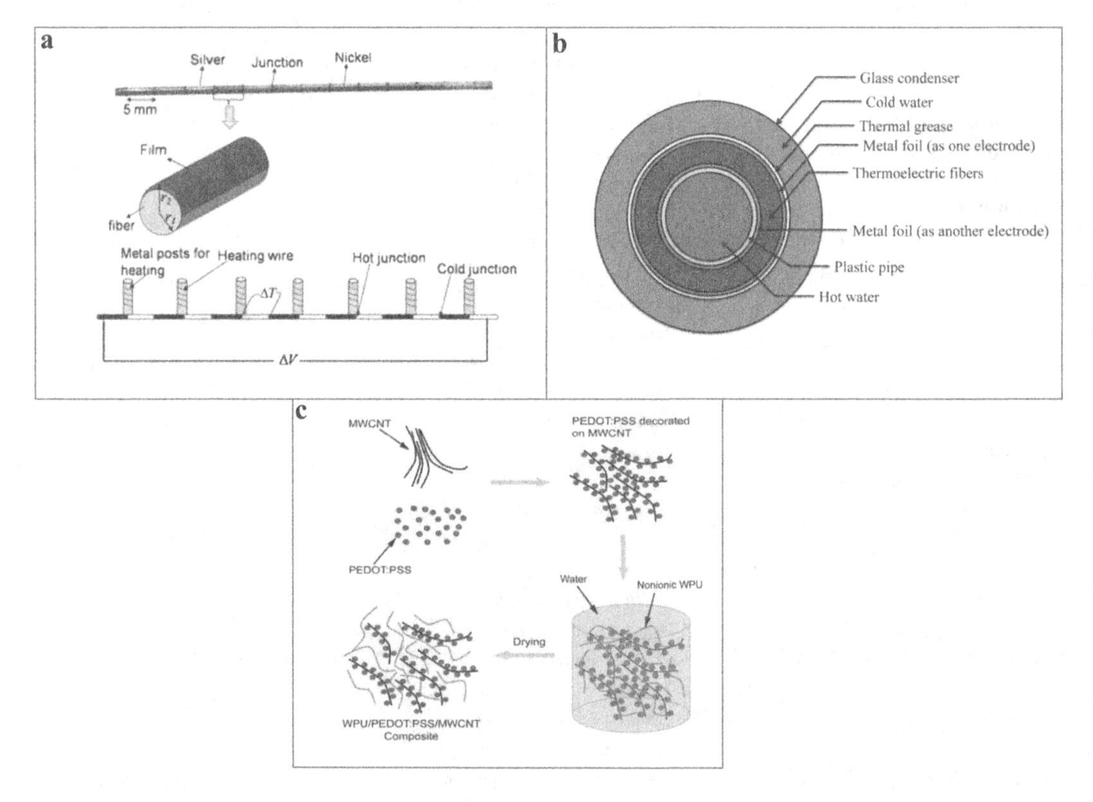

FIGURE 10.35 Yarn-based WTEGs. (a) WTEG based on alternatively coated silica filament with Ag and Ni. (Reused with permission from Yadav, Pipe, and Shtein (2008). Copyright Elsevier 2007.) (b) WTEG based on p-type (PbTe nanocrystals)-coated glass fiber-based yarn. (Reused with permission from Liang et al. (2012). Copyright American Chemical Society 2012.) (c) WTEG based on coating cotton and PET yarns with thermoelectric materials. (Reused with permission from Wu and Hu (2016). Copyright Elsevier 2016.)

Lee et al. developed 2D WTEGs via interlacing/interloping of n-type and p-type yarns using knitting (e.g., zigzag and garter stitching) and weaving (e.g., plain weave) techniques (Figure 10.36(a)). The yarns for weaving were based on twisting an electrospun polyacrylonitrile mat coated on both sides with alternating strips of Bi_2Te_3 (n-type) and Sb_2Te_3 (p-type) with gold junctions between the alternating thermoelectric materials. For knitting, separate yarns as n-type and p-type yarns were fabricated by coating individual polyacrylonitrile nanofiber mats with Bi_2Te_3 and Sb_2Te_3. At 55 DK temperature difference, the plain-woven fabric, zigzag, and garter knitted fabrics exhibited PD of 0.62, 0.11, and 0.09 W m^{-2}, respectively (Lee et al. 2016). In all fabric constructions, spacer yarns were also included either to separate heat sources or to ensure protection against short-circuiting. The majority of the 2D WTEGs are formed by sewing thermoelectric yarns into a 2D fabric substrate. For instance, Qu et al. coated a cotton yarn with P3HT (p-type) and Argentum paste (n-type) (Figure 10.36(b)). The cotton yarn was dipped into the P3HT; afterwards, one side was doped, and the other side was brushed with Argentum paste. The yarn was sewed into a flexible textile substrate, and when tested with 13 junctions under 50 DK, the device generated a power output of 1.15 μW at the matched internal resistance (Qu et al. 2018). Ito et al. demonstrated a similar thermoelectric yarn sewn textile substrate using alternatively coated CNTs/polyethylene glycol (PEG) p-type and n-doped-CNTs/PEG n-type segmented yarn (Figure 10.36(c)). The $[BMIM]PF_6$ was used for n-type doping of CNTs/PEG-coated yarn. At 5 DK temperature difference, the WTEG generated a V_{oc} of 2.3 mV (Ito et al. 2017).

On the other hand, Zhang et al. developed a 2D textile substrate woven WTEG with p-type ($Bi_{0.5}Sb_{1.5}Te_3$) and n-type (Bi_2Se_3) yarns produced by the technique of thermal drawing

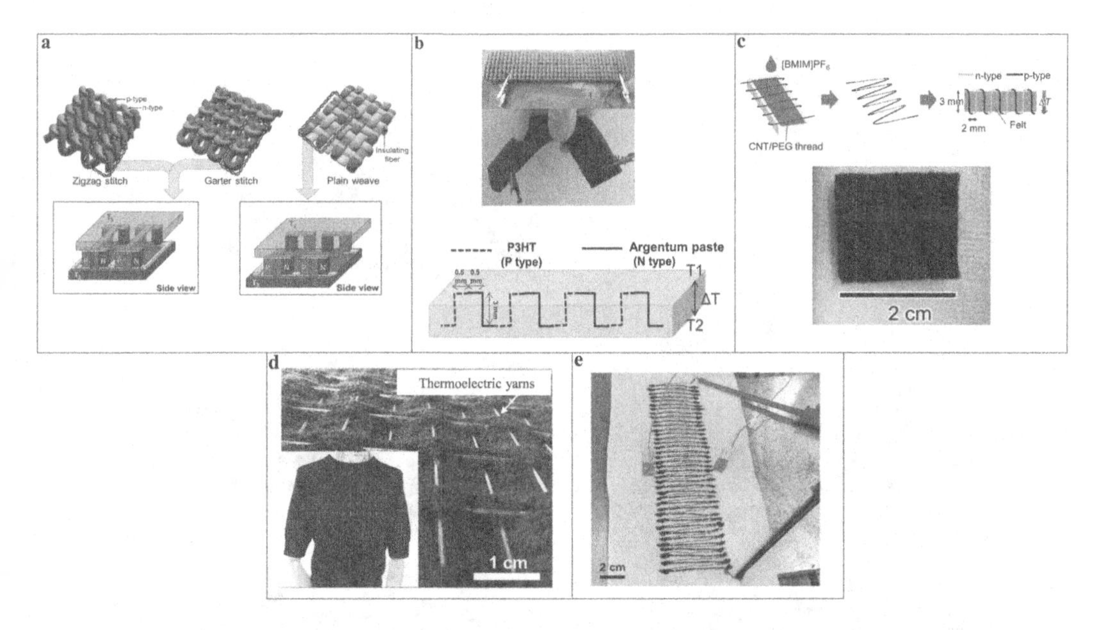

FIGURE 10.36 Yarn integrated 2D WTEGs. (a) 2D WTEGs based on n-type (Bi_2Te_3) and p-type (Sb_2Te_3) yarns integrated knitted and woven fabrics. (Reused with permission from Lee et al. (2016). Copyright John Wiley and Sons 2016.) (b) 2D WTEG based on p-type (P3HT) and n-type (Argentum paste) materials-coated cotton yarn sewn textile. (Reused with permission from Qu et al. (2018). Copyright Elsevier 2018.) (c) 2D WTEG based on p-type (CNTs) and n-type ($[BMIM]PF_6$/CNTs) materials-coated PEG yarn sewn textile. (Reused with permission from Ito et al. (2017). Copyright Royal Society of Chemistry 2017.) (d) 2D WTEG based on the woven fabric of p-type ($Bi_{0.5}Sb_{1.5}Te_3$) and n-type (Bi_2Se_3) yarns. (Reused with permission from Zhang et al. (2017). Copyright Elsevier 2017.) (e) 2D WTEG based on p-type (MWCNTs/PVP)-coated PET and n-type (PEDOT: PSS) dyed silk yarns sewn textile. (Reused with permission from Ryan et al. (2018). Copyright American Chemical Society 2018.)

(Figure 10.36(d)) (Zhang et al. 2017). Similarly, Ryan et al. developed a 2D WTEG by sewing n-type and p-type yarns onto a textile substrate (Figure 10.36(e)). Herein, the n-type yarns were fabricated by coating a PET yarn with MWCNTs and poly-N-vinylpyrrolidone (PVP). To fabric the p-type yarns, silk yarns were dyed with PEDOT: PSS. The output voltage reached 143 mV at 80 DK (Ryan et al. 2018). Table 10.10 summarizes the recently developed WTEGs built directly onto yarn substrates.

10.4.1.3.2.3 Yarn Integrated 3D WTEGs 3D textiles structures are regarded to have a better thermal retention property capable of maintaining a stable temperature gradient between the body and the environment. Wu and Hu developed p-type yarns based on WPU/PEDOT:PSS/MWCNTs yarns and n-type yarns by coating PET yarn substrates with n-doped MWCNTs. The yarns were fabricated based on their early reports of thermoelectric yarns in Section 4.1.3.2.1. The fabrication steps involved embroidering p-type and n-type yarns alternatively into a spacer fabric substrate. Ag-paste connected the adjacent alternative loops of n-type and p-types yarns in the top and the bottom layers in series. To draw the overall power output Cu-wires were attached with Ag-paste to the first p-type and last n-type yarns (Figure 10.37(a)) (Wu and Hu 2017). A 3D WTEG fabric with 10 p-n couplings at 66 ΔK generated a V_{oc} of ~800 μV and a power of ~2.6 nW. So far, 2D and 3D textiles WTEGs have been discussed based on thermoelectric yarn sewn into 2D and 3D fabrics.

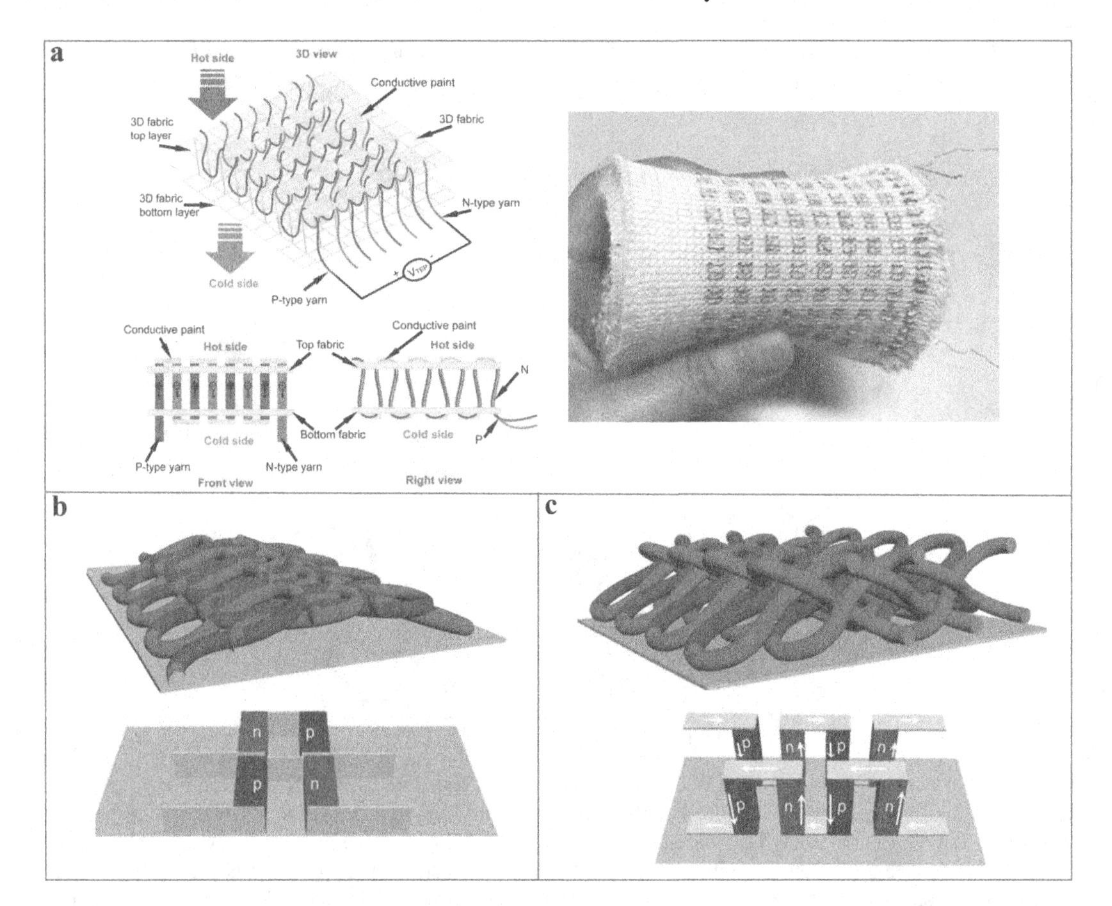

FIGURE 10.37 Yarn integrated 3D WTEGs. (a) 3D WTEG based on sewing p-type and n-type yarns alternatively into a spacer fabric substrate. (Reused with permission from Wu and Hu (2017). Copyright IOP Publishing Ltd 2017.) (b and c) WTEG based on knitting p-type and n-type yarns into 2D and 3D fabrics. (Reused under creative common license attribution 4.0 (https://creativecommons.org/licenses/by/4.0/) from Sun et al. (2020). Copyright Springer Nature 2020.)

TABLE 10.10
Summary of Recently Developed WTEGs Built Directly onto Yarn Substrate

Active Materials	Electrode Materials	Output	Ref.
		Yarn-based WTEGs	
Ag/Ni	Overlapping Ag/Ni junctions	19.6 μVK per thermocouple and 2 nW for 7 thermocouples at 6.6 DK	Yadav, Pipe, and Shtein (2008)
PbTe (p-type)	Metal foil	1.7 mV at 58 DK	Liang et al. (2012)
		Yarn integrated 2D WTEGs	
Bi_2Te_3 (n-type) and Sb_2Te_3 (p-type)	Au or separate Ag-paste electrical contact/Cu wires	15.8, 14.8, 11.9 mV and PD of ~0.62 W m^{-2}, 0.11 W m^{-2}, and 0.09 W m^{-2} at 55 DK for woven, zigzag, and garter stitch, respectively	Lee et al. (2016)
P3HT (p-type) and Argentum (n-type)	Conductive P3HT	~13 mV, ~1.6 mA, and 1.15 μW power with 13 thermocouples at a temperature gradient of 50 DK at matched the resistance	Qu et al. (2018)
([BMIM]PF6) doped-CNTs/PEG (n-type) and CNTs/PEG (p-type)	CNTs/PEG	2.3 mV at <5 DK	Ito et al. (2017)
$Bi_{0.5}Sb_{1.5}Te_3$ (p-type) and Bi_2Se_3 (n-type)	Cu/Ag paste	At 50K, two thermocouples generated 24.2 mV, 59.1 μA, and 0.36 μW power	Zhang et al. (2017)
PVP/MWCNTs) (n-type) and PEDOT:PSS (p-type)	Ag-paste or carbon-based paste	~2.5 mV, ~0.9 μA, and ~0.2 nW power with 4 thermocouples at ~25 DK. With 38 thermocouples, the energy output reached to ~35 mV, ~0.7 μA, and ~2.0 nW power at ~25 DK	Ryan et al. (2018)
		Yarn integrated 3D WTEGs	
NWPU/PEDOT: PSS/MWCNTs (p-type) and NWPU/n-doped MWCT (n-type)	Ag paint	~800 μV and ~2.6 nW power at 66 DK from a WTEG of 10 thermocouples with 6 × 6 × 7 cm^3. When this device was worn as a wrist band, it generated 0.025 mV at 6.4 DK between body and environment.	Wu and Hu (2017)
PEDOT: PSS (p-type) and oleamine doping (n-type)	CNTs	With 15 thermocouples, while attached to the elbow, the device generated ~2.8 mV and when touched by fingertip, 3.5 mV was generated. The PD of the device was recorded 70 mW m^{-2} at 44 DK	Sun et al. (2020)

Abbreviations: Ag (silver), Ni (nickel), P3HT, CNTs (carbon nanotubes), MWCNTs (multi-walled CNTs) PEG (polyethylene glycol), PVP (poly-N-vinylpyrrolidone), PEDOT:PSS [(poly(3,4-ethylenedioxythiophene):polystyrene sulfonate)], NWPU (nonionic waterborne polyurethane), Au (gold), Cu (copper).

Sun et al. developed a textiles substrate-free 3D WTEG by alternatively doping a CNTs fiber twisted yarn. The yarn was turned into a thermoelectric yarn by alternatively doping with PEDOT:PSS (p-type) and oleamine (n-type) materials (Sun et al. 2020). To avoid short circuits, n-doped sections were coated with acrylic fibers. While doping, undoped sections were also included between p-type and n-type segments for making electrical connections. 2D and 3D WTEG fabric developed by warp/weft knitting (Figure 10.37(a and b)) and weaving were also demonstrated. The resultant 3D WTEGs showed greater stretchability at 80% straining, with PD reaching 70 mW m^{-2} at ΔK 44K.

10.4.1.4 Textiles-Based Wearable Biofuel Cells (WBFCs)

The idea of harnessing chemical energy of sweat, tears, and saliva have been regarded as renewable and eco-friendly energy sources. Most biofuels like glucose, lactate, and fructose can be oxidized via biocatalysts in the BFCs (Bandodkar 2017; Xiao et al. 2019). The underlying principle involves the oxidation of biofuels at an anode, such as glucose, to donate electrons that move toward the cathode via an external electrical circuit. The cathode receives the electrons and reduces the oxygen, thereby converting biochemical energy into electrical energy (Cosnier, Le Goff, and Holzinger 2014; Sakai et al. 2009). BFCs could be categorized as microbial MBFCs (Xie et al. 2011, 2012) and EBFCs (Minteer, Liaw, and Cooney 2007; Rasmussen, Abdellaoui, and Minteer 2016). The surface roughness of textiles facilitates the ease of incorporating enzymes to fabricate BFCs. Besides, excellent flexibility, porousness, and ability to conform to the body shape have endowed textiles with large contact areas with the body to absorb bodily fluids and generate electricity (Chen et al. 2020). EBFCs have been recommended for wearable textiles because of higher conversion efficiency, biocompatibility, and miniaturization. Textiles or yarns are regarded as the appropriate substrates to incorporate enzymes for fabricating BFCs into textiles.

10.4.1.4.1 WBFCs Built Directly onto Fabric Substrates

Readymade fabric substrates are a desirable choice for the fast fabrication of textile-based BFCs by using facile coating and printing techniques. Using coating or printing techniques, functional materials (electrodes, fuel, biocatalyst, and enzyme) could easily be integrated into the textile substrate.

A typical sandwich structured BFC was demonstrated by Miyake et al., wherein anode and cathode were prepared from CNTs treated carbon filament (CF) fabrics (Figure 10.38(a)). The active middle layer was a double networked hydrogel, which contained the electrolyte and the fuel (i.e., fructose). After the CNTs treatments, the conductive fabrics were treated with fructose dehydrogenase (FDH) and bilirubin oxidase (BOD) respectively to get the final anode (CNTs/FDH/CNTs/CF) and with an additional CNTs treatment to get the cathode as CNTs/BOD/CNTs/CF. The fabrication technique has endowed the cathode with O_2 diffusivity ability. A single such sandwiched device exhibited a peak V_{oc} of 0.74 V, but when three units were layered, it demonstrated a peak V_{oc} of 2.09 V and peak PD of 2.55 mW cm^{-2} (Miyake et al. 2013). But the device lacked sufficient stretchability to be utilized as wearables. Recently, the group utilized a stretchable pantyhose meshed textile (SPT) of nylon/PU co-filament to fabricate a stretchable wearable BFC (Figure 10.38(b)). Variations were also brought into the anode and the cathode as SCNTs/MWCNTs/FDH/MWCNTs/SCNTs/SPT and PTFE-MWCNTs/BOD/MWCNTs/SCNTs/SPT, respectively. While tested, the conductivity of the base fabric electrode (MWCNTs/SCNTs/SPTs) was reduced to one-third at 50% stretching after 30 cycles and remained stable afterwards (Ogawa, Takai et al. 2015).

Jia et al. developed BFCs that could be printed onto a regular T-shirt (cotton), socks (PET/cotton), and care label (PET). They fabricated BFCs onto textiles-based head and wrist bands and performed in-vitro and on-body energy conversion tests (Figure 10.38(c)). Herein, the anode was functionalized with CNTs, tetrathiafulvalene 7,7,8,8-tetracyanoquinodimethane (biocatalyst), and lactate oxidase (LO_x) as the enzyme. A biocompatible layer of chitosan was also applied to enhance the anode stability and avoiding direct contact with the skin. Nafion®, known as an oxygen reduction catalyst, was used along with Pt-black to develop the cathode. Demonstration of lighting up an LED and powering up a digital watch was conducted via the wearable headband and wrist band,

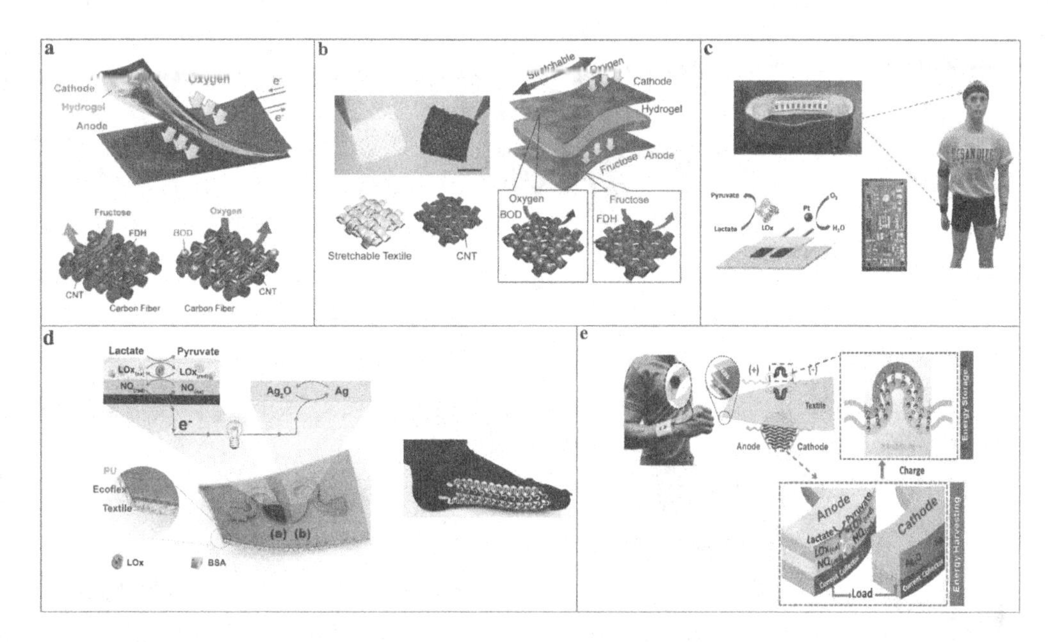

FIGURE 10.38 WBFCs built directly onto fabric substrates. (a) BFC based on FDH and BOD enzymes with CNTs-based electrodes built onto CF fabric substrate. (Reused with permission from Miyake et al. (2013). Copyright Elsevier 2012.) (b) BFC based on FDH and BOD enzymes with CNTs-based electrodes built onto CF fabric substrate pantyhose meshed textile of nylon/PU coaxial yarns. (Reused with permission from Ogawa, Takai et al. (2015). Copyright Elsevier 2015.) (c) BFC based on LO_x enzyme with CNTs and Pt-black-based electrodes built onto head and wrist bands textiles. (Reused with permission from Jia et al. (2014). Copyright Royal Society of Chemistry 2014.) (d) BFC based on LO_X or GO_x enzyme with CNTs composite anode and AgO_2/Ag cathode. (Reused with permission from Jeerapan et al. (2016). Copyright Royal Society of Chemistry 2016.) (e) A similar BFC, as (d), integrated with a stretchable supercapacitor to store energy output. (Reused with permission from Lv et al. (2018). Copyright Royal Society of Chemistry 2018.)

respectively (Jia et al. 2014). The major drawback of these types of BFCs is the dilution of lactate with the increased sweating rate that hampers the convection of lactate and oxygen at the respective electrodes. The performance of these devices was based on the oxygen diffusivity ability of the cathode. To avoid that, Jeerapan et al. demonstrated energy generation from lactate-containing sweat secretion of the human body during cycling, excluding the need for oxygen. Herein, a CNTs composite-based anode was functionalized with LO_x [or glucose oxidase (GO_x)] and 1,4-naphthoquinone (NQ) as the biocatalyst. On the other hand, the cathode was functionalized with AgO_2/Ag to perform in anaerobic conditions without the need for O_2 to receive electrons from the anode. All the materials were deposited onto flexible textiles substrates via screen printing in a serpentine pattern to fabricate the final BFC (Figure 10.38(d)). The device has shown excellent flexibility and maintained its performance after 100 cycles of 100% stretching. The BFCs based on GO_x and LO_x reported peak PDs of 160 and 250 μW cm^{-2} at V_{oc} of 0.44 and 0.46 V, respectively (Jeerapan et al. 2016). The group further developed their BFC by integrating a supercapacitor for energy storage purposes (Figure 10.38(e)). Similarly, lactate of human sweat was the working fuel for the BFC with similar anode and cathode materials (Lv et al. 2018).

10.4.1.4.2 WBFCs Built Directly onto Yarn Substrates

Like other yarn integrated NGs, weaving, knitting, and sewing into the textile substrate are the techniques that have been also explored to fabricate BFCs. Kwon et al. demonstrated a biscrolled yarn-based BFC (Figure 10.39(a)) that could generate a peak V_{oc} of 0.7 V and PD of 2.18 mW cm^{-2} utilizing human body fluid in in-vitro conditions. The yarn electrodes were fabricated by twisting PEDOT-coated MWCNTs sheets-host with enzyme (GO_x), redox mediator-I [poly(N-vinyl

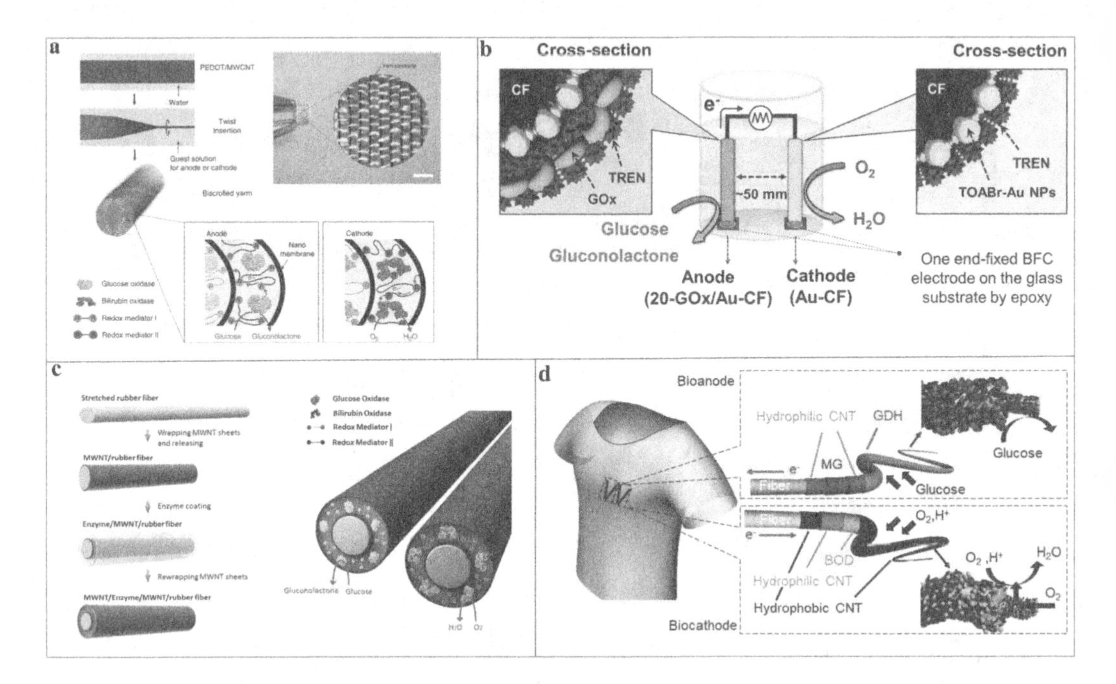

FIGURE 10.39 WBFCs built directly onto yarn substrates. (a) BFC based on biscrolled yarn as the host and thermoelectric materials as the guests. (Reused with permission from Kwon et al. (2014). Copyright Springer Nature 2014.) (b) BFC based on Au NPs-coated CNTs cathode and GO_x-coated Au NPs/CNTs anode. (Reused with permission from Kwon et al. (2019). Copyright Royal Society of Chemistry 2019.) (c) BFC based on the MWCNTs/GOx/MWCNTs/silicone rubber anode and MWCNTs/BOD/MWCNTs/silicone rubber cathode. (Reused with permission from Sim et al. (2018). Copyright American Chemical Society 2018.) (d) Sewable BFC based on GDH and CNTs-based anode and cathode. (Reused with permission from Yin, Jin, and Miyake (2019). Copyright Elsevier 2019).

imidazole)-[Os(4,4′-dimethoxy-2,2′-bipyridine)2 Cl])$^{+/2+}$] (PNVIODMBP)-guests for anode; BOD with redox mediator-II [poly(acrylamide)-poly(N-vinyl imidazole)-[Os(4,4′-dichloro-2,2′-bipyridine)2])$^{+/2+}$] (PAAmPNVIODMBP)-guests for cathode. In both cases, the enzymes (GOx and BOD) were cross-linked by poly(ethylene glycol) diglycidyl ether (PEGDGE). The resultant yarn-based electrodes showed excellent flexibility with mechanical robustness for weaving and knitting suitability. A woven textile was fabricated by placing anode and cathode (50 μm diameter) electrodes in the warp direction and PET yarns (100 μm diameter) in the weft direction. The textile with a total cathode and anode length of 56 mm generated V_{oc} of 0.7 V, but when two such textiles were connected in series, the V_{oc} was peaked at 1.32 V with 229 μW power (Kwon et al. 2014). Kwon et al. developed another BFC with $GO_{x,}$ but this time CNTs yarn was coated with Au NPs for cathode and GOx/Au-NPs for the anode (Figure 10.39(b)). Herein, the cathode also performed as the oxygen reduction mediator due to the presence of pores in Au NPs arrays. The final BFC exhibited a peak PD of ~2.1 mW cm^{-2} (Kwon et al. 2019). The researchers aimed at achieving greater electron transfer between the enzyme and the anode, increasing biocompatibility, and discarding the use of toxic mediators/cross-likers.

Stretchability is important for these wearable BFCs to withstand different kinds of body movement. The active enzyme material and electrode should match in mechanical properties to avoid slow delamination and decreased BFCs performance. To that end, Sim et al. used a commercial rubber filament as the substrate material to fabricate the anode and the cathode for the BFC (Figure 10.39(c)). To develop the anode, silicone rubber was wrapped with a porous MWCNTs sheet and then deposited with a layer of GO_x enzyme, redox mediator-I (PNVIODMBP), and crosslinker (PEGDGE) mix. Finally, an additional MWCNTs sheet was wrapped to fabricate the anode. Similarly, to develop the cathode, a mixture of BOD enzyme, redox mediator-II ((PAAmPNVIODMBP), and crosslinker (PEGDGE) were incorporated between the two MWCNTs sheets, while the inner sheet having wrapped around another rubber filament. The electrodes were able to withstand repetitive 100% stretching and retained their performance up to ~99% in glucose and ~86% in human serum. Such high retention of power densities is due to the presence of outer MWCNTs sheet that traps the enzyme. Without the outer MWCNTs sheet, the PD in glucose was 63% and in human serum was 29%. Overall, human serum reduces the performance of the BFC due to the presence of anionic chemical species. The BFC generated a PD of 39.5 mW cm^{-2} in glucose and 36.6 μW cm^{-2} in human serum (Sim et al. 2018).

A direct apparel sewable yarn-based BFC (Figure 10.39(d)) was demonstrated by Yin et al. Herein, CF was successively treated in the following order to get the anode: CF/acid-treated-CNTs/polyMG redox mediator/acid-treated CNTs/GDH (glucose dehydrogenase) [(CF/A-CNTs/PolyMG/A-CNTs/GDH)]. Similarly, to fabricate the cathode, another CF was treated in the following order: PTFE-CNTs/BOD/A-CNTs/CF. Though demonstration of wearability was confirmed, the main biofuel (i.e., glucose) was poured externally. Herein, four BFCs units were able to power up an LED continuously (Yin, Jin, and Miyake 2019).

BFCs built onto yarn substrate and integrated into fabrics seem like a promising way of harvesting biofuel energy of the human body. To that end, yarns with stable electrodes capable of withstanding current textile manufacturing processes need to be ensured in the future. Table 10.11 summarizes the recently developed textiles-based BFCs.

TABLE 10.11
Summary of Recently Developed Textiles-Based BFCs

Active Materials	Electrode Materials	Output	Applied As	Special Attributes	Ref.
		WBFCs built directly onto fabric substrates			
McIlvaine buffer solution (pH 5.0) containing 500 mM fructose	FDH/CNTs (bioanode) and CNTs/BOD/CNTs (biocathode) CNTs	~0.74 V and 0.95 mW cm^{-2} PD were achieved from a single BFC. When three BFCs were stacked, the V_{oc} and PD reached 2.09 V and 2.55 mW cm^{-2}, respectively	Powering up LED	–	Miyake et al. (2013)
McIlvaine buffer solution (pH 5.0) containing 200 mM fructose	FDH/CNTs (bioanode) and CNTs/BOD/CNTs (biocathode)	~0.74 V, current density of 1.0 mA cm^{-2}, and PD of 0.25 mW cm^{-2} from single BFC	Powering up LED	–	Ogawa, Takai et al. (2015)
Glucose or lactate	GO_x or LO_x as enzymes. For bioanode OH-CNTs, NQ, and GO_x or LO_x-based ink and for biocathode, COOH-CNTs/ silicone/AgO_2/Ag-based ink were used respectively	0.44 V and PD of 160 μW cm^{-2} with glucose, when lactate was used, the voltage and PD increased to 0.46 V and 250 μW cm^{-2}, respectively	Powering up LED	Self-powered lactate sensor	Jeerapan et al. (2016)
Lactate	LO_x as enzyme for bioanode OH-CNTs, NQ, and LO_x-based ink and for biocathode COOH-CNTs/ silicone/ AgO_2/Ag-based ink were used respectively	0.49 V, current density of 1.1 mA cm^{-2}, and PD of 252 μW cm^{-2} were reported	Powering up LED	–	Lv et al. (2018)

(Continued)

TABLE 10.11 ***(Continued)***
Summary of Recently Developed Textiles-Based BFCs

Active Materials	Electrode Materials	Output	Applied As	Special Attributes	Ref.
Lactate	Chitosan/LO_x/TTF. TCNQ/CNTs bioanode and Nafion® Pt-based cathode	0.67 V and PD of 100 $\mu W\ cm^{-2}$ were reported	Powering up LED and digital watch	–	Jia et al. (2014)
		WBFCs built directly onto yarn substrates			
Glucose	GO_x-redox mediator/ PEDOT/MWCNTs and BOD-redox mediator / PEDOT/MWCNTs	A woven textile with a total 56 mm anode and cathode length, generated V_{oc} of 0.7 V and power of 128 μW	–	–	Kwon et al. (2014)
Glucose	GO_x/Au-CFs anode and Au-CNTs	PD of ~1.2 $mW\ cm^{-2}$ and ~0.4 V were reported	–	–	Kwon et al. (2019)
Glucose	MWCNTs/GO_x/ MWCNTs and MWCNTs /BOD/ MWCNTs	V_{oc} of 0.55 V and PD of 42 $\mu W\ cm^{-2}$ were reported	–	–	Sim et al. (2018)
Glucose	GDH/A-CNTs/ PolyMG/A-CNTs and PTFE-CNTs/ BOD/A-CNTs	PD of 48 $\mu W\ cm^{-2}$ at 0.24 V in 0.1 mM human sweat glucose. PD of 216 $\mu W\ cm^{-2}$ at 0.36 V with 200 mM glucose	Powering up LED	–	Yin, Jin, and Miyake (2019)

Abbreviations: FDH (fructose dehydrogenase), CNTs (carbon nanotubes), MWCNTs (multi walled CNTs), BOD (bilirubin oxidase), GO_x (glucose oxidase), LO_x (lactate oxidase), NQ (1,4-naphthoquinone), TTF.TCNQ (tetrathiafulvalene 7,7,8,8-tetracyanoquinodimethane), Ag (silver), Pt (platinum), PEDOT [poly(3,4-ethylenedioxythiophene): polystyrene sulfonate], Au (gold), CF (carbon filament), GDH (glucose dehydrogenase), PTFE (polytetrafluoroethylene), PD (power density), V_{oc} (open circuit voltage).

10.4.1.5 Textiles-Based Wearable Hybrid Nanogenerators (WHNGs)

Boosting energy output by combining multiple energy conversion abilities into a single NG is desirable to sufficiently powering up wearable electronics. For instance, SCs and TEGs are more dependent on ambient conditions than the mechanical energy converting NGs. A combination of SCs/TEGs with TENGs or PENGs can compensate for that weather-dependent power generating aspect.

For instance, A WHNG based on SC and TENG (Figure 10.40(a)) with a supercapacitor was developed by Wen et al. Herein, the SC part was constituted of N719 dye-sensitized TiO_2 nanotube arrays/Ti wire-based working electrode, Pt-coated carbon fiber-based counter electrode, and $I^-/I3^-$-based electrolyte. The electrolyte was contained in a Cu-coated ethylene vinyl acetate (EVA) tubing. A triboelectric mechanism was introduced with the SC and the supercapacitor. The supercapacitor was fabricated by coating two CFs with $RuO_2{\cdot}xH_2O$. Which was further assembled into H_3PO_4/PVA electrolyte and packaged into a PDMS-covered Cu-coated EVA tubing to get the supercapacitor. For the TENG, Cu coatings of SC and supercapacitor functioned as the electrodes (one of the Cu electrodes also served as the triboelectric material). PDMS of the supercapacitor part served as the second triboelectric material. The SC and TENG showed the capability to generate energy simultaneously. The SC part was able to generate 1.8 V, which is low but could be compensated by the

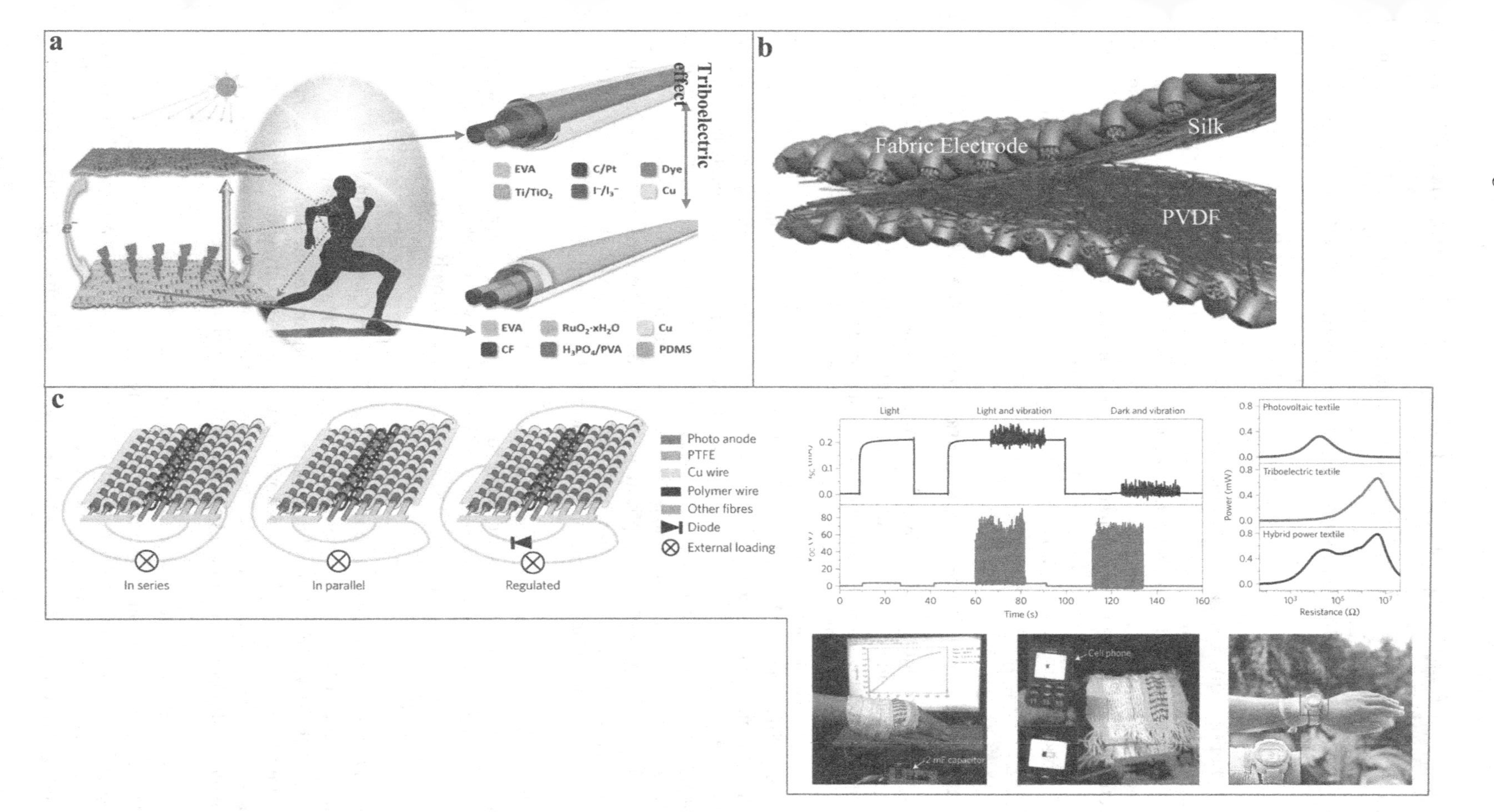

FIGURE 10.40 WNGs-based on hybrid nanogenerators. (a) WHNG based on SC (N719 dye) and TENG (Cu vs PDMS). (Reused under creative common license attribution 4.0 (https://creativecommons.org/licenses/by/4.0/) from Wen et al. (2016). Copyright Science Advances 2016.) (b) WHNG based on TENG (silk vs PVDF) and PENG (PVDF). (Reused with permission from Guo et al. (2018). Copyright Elsevier 2018.) (c) WHNG based on SC (N719 dye) and TENG (PTFE vs Cu). (Reused with permission from Chen et al. (2016). Copyright Springer Nature 2016.)

TENG, which generated ~12.6 V (Wen et al. 2016). Humidity impact the performance of TENG, for instance, in Japan, humidity reaches as high as ~70%, but sunshine is far intense in Summer. In that case, the reduced performance of TENG could be compensated by SC exposure to higher sunshine during summer (Pu, Song et al. 2016). Guo et al. developed a multilayer stacked WHNG based on TENG and PENG (Figure 10.40(b)). Herein, triboelectrification between PVDF and silk fibroin was utilized for the TENG part, and the inherent piezoelectric property of PVDF was utilized for the PENG part. Electrospun nanofibrous mats of silk fibroin and PVDF were collected onto conductive fabric, respectively, and stacked to fabricate the WHNG. The WHNG generated V_{oc} of 500 V, I_{sc} of 12 μA, and PD of 0.31 mW cm^{-2} at 2-Hz frequency and 25.7 N force (Guo et al. 2018). The SC developed by Zhang et al. in Section 4.1.2.2, was further developed to include a TENG by utilizing triboelectrification between PTFE and Cu, as illustrated in Figure 10.40(c) (Chen et al. 2016). The WHNG was able to charge up a 1 mF capacitor to 2 V in a minute. Besides, the HNG was also able to drive an electronic watch continuously, charge a cell phone, and driving water splitting reactions. It seems like WHNG is a promising candidate to harness the energy of multiple sources and sustainably powering on body electronics. The compensating attributes of WHNGs could be the potential WNGs to supply power continuously regardless of the environmental conditions.

10.4.2 Flexible-Sheet Based Wearable Nanogenerators (FS-WNGs)

Principally, the wearable application of FS-WNGs is directly on the body, either outside on the skin (indicated as in vitro application) or inside the physical structure (indicated as in vivo applications). From another perspective, the body implantable WNGs can be exploited in two ways. In the case of human body-affiliated applications, they either can act directly on the human body as physiological signals processors, or they can be utilized indirectly as a long-term power source for other functioning devices (Shawon et al. 2021). Since the last decade, wearable and implantable NGs as a medium of harnessing micro/nanoscale as well as self-powered sensors are being explored (Li et al. 2020). Many of these find application in medical wearable devices that can effectively track physiological systems, perform real-time detection of diseases, provide the power required to drive some of the implanted organs, and even treat some nervous or cardiovascular anomalies (Zhang, Wang, and Yang 2019). The major area of such application includes a self-powered medical information system. Therefore, apart from performing the functionalities of an individual sensor, they can also be utilized as a source of power for driving other commercial sensors and devices (Niu et al. 2015; Zhao et al. 2020).

As a potential and ever-growing field of interest, an uncountable amount of research studies can be found in this area. Some of the most significant and recent studies are referred to in this section. These human body integrated devices can either be worn on the body, which is included in the in vitro area of application, otherwise, they can be implanted inside the body, which is termed as the in vivo application (Wang, He, and Lee 2019).

10.4.2.1 FS-WNGs Attached Directly onto Body (i.e., Skin or Clothing) (In Vitro)

In this section, a review of NGs that can be directly applied to human skin or can be attached to any human wearable stuff via ultrathin polymer or plastic platforms are mentioned. Stretchability and flexibility are imperatives for a NG to be worn on the skin to comply with the body contours (Bandodkar and Wang 2016; Han et al. 2017). When the harvesters conform to the curves of the human body aptly, the discomfort minimizes significantly (Kim et al. 2018).

10.4.2.1.1 In Vitro FS-WNGs Based on Triboelectric Effect (FS-WTENG)

An FS-WTENG can perform the dual functionality of being a self-powered sensor and a wearable power source to drive wearable electronics (Yu et al. 2020). For instance, Wang et al. developed self-powered cardiovascular monitoring systems based on biocompatible gelatin-polyvinyl alcohol (PVA) composite films. The device was highly sensitive and capable of sensing slight skin

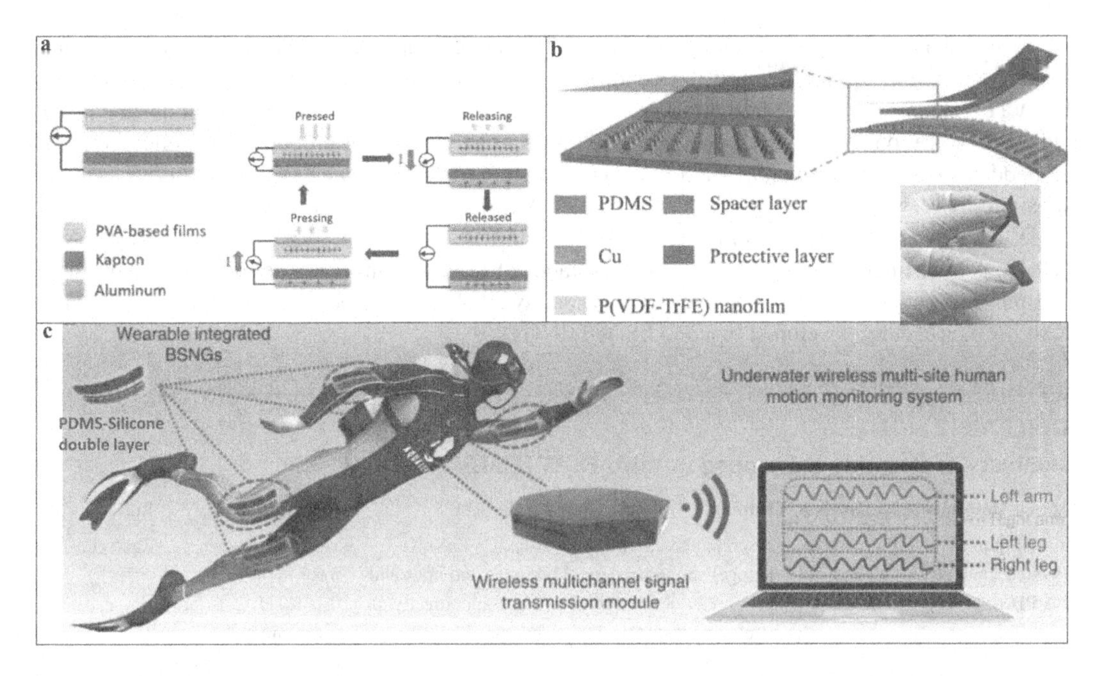

FIGURE 10.41 In vitro FS-WNGs based on triboelectric effect. (a) Schematic diagram and working mechanism of PVA-based TENG, (Reused with permission from Wang et al. (2020). Copyright John Wiley and Sons 2020.) (b) The schematics and photographs of micro-frustum arrays-PDMS based TENG. (Reused with permission from Yu et al. (2020). Copyright Elsevier 2019.) (c) Illustration of an underwater wireless multi-site human motion monitoring system based on BSNG and photographs of integrated wearable BSNG worn on the arthrosis of human. (Reused under the terms of the Creative Commons CC BY license from Zou et al. (2019). Copyright Springer Nature 2019.)

deformations induced by a human pulse. It was appropriate to monitor vital physiological signals continuously for self-powered health diagnostics and therapeutics. Herein, a gelatin-PVA blended film and a Kapton layer acted as the triboelectric surfaces, while two aluminum foils were utilized as the electrodes (Figure 10.41(a)). This device was attached to the human wrist via a band-aid, while in this case a Cu tape was utilized as the counter electrode. When attached to the human wrist, the device was able to sense blood flow and skin deformations, unlike an electrocardiogram (ECG), which is based on a heartbeat. Under 8-Hz mechanical stimulation frequency, the device generated 1.6 V_{oc} and 5.3 nA I_{sc} (Wang et al. 2020). In a similar study, Yu et al. fabricated an ultrasensitive PDMS film-based FS-WTENG that functioned through the contact-separation principle. The elastic-natured micro-frustum arrays-based PDMS and copper electrodes acted as the triboelectric materials. Additionally, a PVDF-TrFE nanofiber layer was electrospun on the PDMS surface to reduce adhesion between triboelectric surfaces (Figure 10.41(b)). Like the device mentioned above, it was capable of sensing subtle changes in human wrist pulse and thus generate output from physiological signals. The sensitivity of this sensor was recorded as 5.67 V/10^5 Pa (Yu et al. 2020). Zou et al. developed a wearable bionic stretchable nanogenerator (BSNG) based on triboelectric effect imitating the ion-concentration gradients-based electricity generation mechanism of eel. The Na^+, K^+ ion channels, and protein-based electrolytes at the cell membrane are the major players to generate electrical output. The mechanism was mimicked using PDMS and silicone-based mechanical channels, which open/close due to mismatch in mechanical stresses. Herein, triboelectrification occurs between PDMS-silicone double layer and deionized (DI) water. The electricity is harvested through two NaCl solution-based electrodes. Such design configuration endows the BSNG to work in both dry and wet conditions. The device demonstrated the capability to harness mechanical energy from human physical movement underwater and was also able to track human location underwater (Figure 10.41(c)). The FS-WTENG generated V_{oc} of 170 V in dry conditions and approximately 10 V in underwater conditions (Zou et al. 2019). An ultrasensitive self-powered FS-WTENG wearable on the human wrist or chest to sense and generate electrical signals driven by pulse or respiration was reported by Liu et al. This pressure sensor was prepared from microspheres embedded PDMS mixture and FEP. In the first step, a Cu electrode was spin-coated onto an FEP substrate. Afterwards, microspheres embedded PDMS mixture was spin coated on to the Cu electrode. Finally, another FEP layer coated with Cu as the other electrode was attached to the PDMS layer to get the final FS-WTENG. The peak voltages observed from such a device upon respiration and pulse are approximately 4 V and 0.3 V respectively (Liu, Zhao et al. 2019). Table 10.12 summarizes the recently developed in Vitro FS-WTENGs.

TABLE 10.12
Summary of Recently Developed in Vitro FS-WTENGs

Active Material	Electrodes	Output	Applied As	Special Attributes	Ref.
Ionic liquid locked ionogel vs PDMS	Ionic liquid locked ionogel	The conductivity of the ionogel varied between 1.7 and 2.4 S m^{-1} upon stretching and releasing; Generated V_{oc} and I_{sc} were found ~0.3 V and ~2.3 nA at 0.1 N impulse force of 1-Hz frequency	Tracking various human activities and adaptable directly on human skin	Can detect impact force in the range of 0.1–1.0 N; Can sense and collect data from various bodily functions, for example, bending, breathing, pulse wave, finger touching, etc.	Zhao et al. (2019)
PVDF-HFP vs nylon 6	rGO	For applied external force the V_{oc} and I_{sc} were recorded as ~80 V and ~1.67 μA for a contact area of 3 × 3 cm^2. Maximum instantaneous power was recorded 45 mW under 1 GΩ load resistance	A highly stretchable self-powered TENG directly attachable to skin fueled by mechanical energy from daily activities	Apart from energy harvesting, such a device is capable of self-powered detection of motion	Qin et al. (2019)
Chitosan vs Kapton	Cu layer	A PD of ~2.1 μW m^{-2} was obtained for pure chitosan film and ~17.5 μW m^{-2} for chitosan nanocomposite with 10% acetic acid. The optimized peak V_{oc} and I_{sc} were 13.5 V and 42 nA, respectively	Can harness energy from various deformation occurring on skin surface efficiently	Laser treated biopolymer film has tunable biodegradation property and offers the potential for surface engineering	Wang et al. (2018)
Nanostructured Kapton super-thin film vs nanostructured Cu film	Ultrathin Cu layer	The test was conducted on the 24 years older adult's radial arteria. The V_{oc} and I_{sc} were ~1.52 V and ~5.4 nA, respectively	Self-powered pulse sensor for real-time cardiovascular signal monitoring applied on a human wrist	By real-time and wireless tracking of physiological signal from human pulse wave velocity, it can provide important information about heart rate variability	Ouyang et al. (2017)

(Continued)

TABLE 10.12 ***(Continued)***
Summary of Recently Developed in Vitro FS-WTENGs

Active Material	Electrodes	Output	Applied As	Special Attributes	Ref.
Al-plastic laminated film	Cu and Al layers	The average V_{oc} and I_{sc} are recorded to be 55 V and 0.9 μA, respectively	Monitoring body movement during sleeping	The nanostructure-patterned active material layer improved the output performance greatly	Song et al. (2016)
Kapton vs PDMS	Serpentine shaped Cu	At 44-MΩ load resistance, PD of ~5 W m^{-2} was observed in compression mode. Recorded average V_{oc} and I_{sc} were approximately 700 V and 75 μA	Self-powered stretchable, shape adaptive, flexible biomedical monitoring TENG device directly applied on human skin	Can operate flexibly in both stretching and compressing mode. It can be attached to monitor motions of muscle, joints as well as Adam's apple	Yang et al. (2015)

Abbreviations: PDMS (polydimethyl siloxane), PVDF [poly(vinylidene fluoride)], (PVDF-HFP [poly(vinylidene fluoride-hexafluoropropylene], rGO (reduced graphene oxide), Al (aluminum), Ag (silver), Cu (cupper), V_{oc} (open circuit voltage), I_{sc} (short circuit current), PD (power density), TENG (triboelectric nanogenerator).

10.4.2.1.2 In Vitro FS-WNGs Based on Piezoelectric Effect (FS-WPENGs)

In a recent study, Hanani et al. designed a Pb-free, non-toxic, and bio flexible FS-WPENG of ceramic-based piezoelectric [$Ba_{0.85}Ca_{0.15}Zr_{0.10}Ti_{0.90}O_3$ (BCZT)] material. Herein, polydopamine (PDA)-coated BCZT NPs were dispersed in a polylactic acid (PLA) polymer matrix. The polar carbonyl group present in PLA aided to self-pole PDA/BCZT NPs. The final FS-WPENG was fabricated by sandwiching a drop-casted PDA/BCZT film (Figure 10.42(a)) between two Cu electrodes. Upon finger tapping gently, the device exhibited V_{oc}, I_{sc}, and maximum PD of 14.4 V, 0.55 μA, and 7.54 mW cm^{-3}, respectively, at 3.5-MΩ load resistance (Hanani et al. 2021). Guo et al. reported an FS-WPENG (Figure 10.42(b))-based pressure sensor fabricated from a nanocomposite of electrospun PVDF and $BaTiO_3$ NWs. The FS-WPENG demonstrated the capability to supervise real-time human motion. With an impact force of 9-N and 3.5-Hz frequency, the peak current output was recorded 105 nA with 3% $BaTiO_3$ NWs. While attached to the human body, the current outputs were 97, 241, and 31 nA for walking, running, and squatting up-down, respectively. The device additionally contained a wireless circuit, data conversion control panel, and a wireless data transmission module (Guo et al. 2018; Shi et al. 2018). Siddiqui et al. reported an omnidirectional stretchable FS-WPENG made of electrospun $BaTiO_3$-PU and PVDF-TrFE nanofibrous mats. Herein graphite electrodes were formed onto two-patterned PDMS substrates. One of the substrates was coated with $BaTiO_3$-PU nanofiber mat, and the other was coated with PVDF-Tr-Fe nanofiber mat. The final FS-WPENG was built by stacking $BaTiO_3$-Pu and PVDF-Tr-Fe containing electrodes/substrates. Such a wearable device (Figure 10.42(c)) demonstrates peak V_{oc} and I_{sc} of 9.3V and 189 nA at a stretching strain of 40%, respectively. Besides, when attached to a kneecap, the device generated a maximum of peak V_{oc} 10.1 V during walking (Siddiqui et al. 2018). Du et al., in their recent study, demonstrated a self-adhesive FS-WPENG-based protective patch (Figure 10.42(d)). The patch contained a polydopamine-polyacrylamide (PDA-PAAm) hydrogel capable of skin wound healing via electrical stimulation provided by a piezoelectric segment. The piezoelectric segment was fabricated by sandwiching an aligned PVDF nanofibrous mat between two Al tape electrodes. The low-frequency pulse voltages generated by the FS-WPENG were responsible for boosting the regeneration of the skin at the wound site. With self-adhesive and protective functionality provided by the hydrogel, the FS-WPENG can adhere to human skin as well as generate an electrical voltage

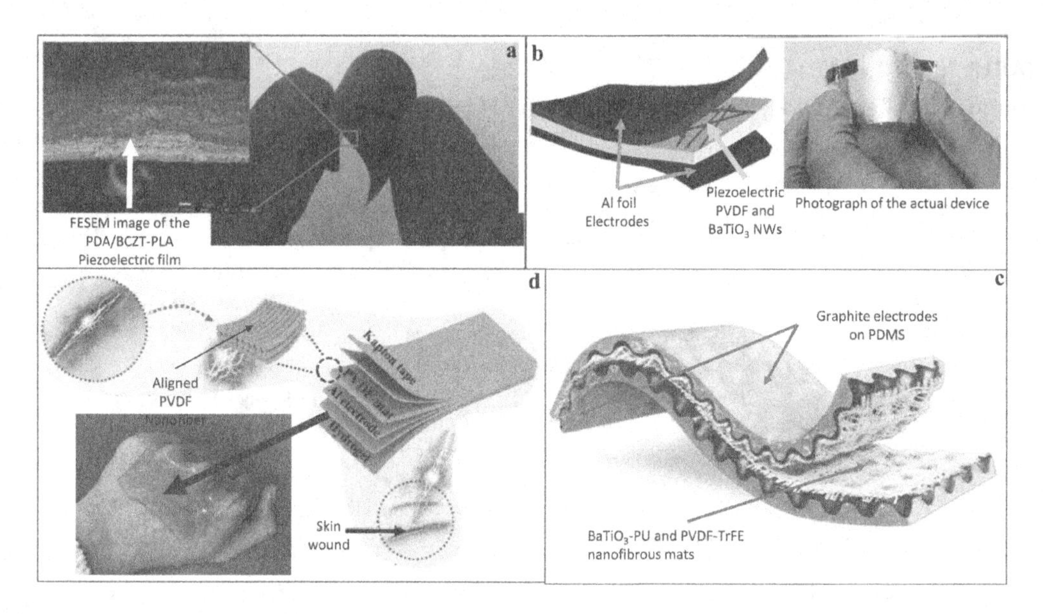

FIGURE 10.42 In vitro FS-WNGs based on piezoelectric effect. (a) Real and field-emission scanning electron microscope (FESEM) image of the BCZT-PLA nanocomposite film. (Reused with permission from Hanani et al. (2021). Copyright Elsevier 2020.) (b) Schematic and photograph of the PVDF-$BaTiO_3$ piezoelectric pressure sensor. (Reused with permission from Guo et al. (2018). Copyright Royal Society of Chemistry 2018.) (c) Schematic representation of the stretchable NF-SPENG with a stacked nanofiber mat and graphite electrodes. (Reused with permission from (Siddiqui et al. 2018). Copyright John Wiley and Sons 2017.) (d) Schematic illustration showing the fabrication of the bioinspired patch consisting of the self-adhesive matrix of PDA-PAAm hydrogel and the PENG based on aligned PVDF nanofibers mat. (Reused with permission from Du et al. (2020). Copyright Springer Nature 2020.)

from human body motion. Based on the size and thickness of aligned nanofibers mats, the output pulse voltage ranges from 0.3 to 1.8 V (Du et al. 2020). A similar type of piezoelectric skin patch of ZnO nanostructures dispersed in a PDMS matrix capable of regenerating cells via the piezoelectric effect was also reported (Bhang et al. 2017). In another study, Maity et al. demonstrated a method to enhance the piezoelectricity of the organic PENGs made of multi-layered electrospun PVDF nanofibrous mats. The FS-WPENG was capable of harvesting electrical output from human motions like foot strikes during walking. Herein, PEDOT-coated PVDF nanofibrous mats were utilized as the electrodes. Continuous pressing and releasing by human feet enable mechanical stress to be converted into electrical energy. When subjected to 8.3 kPa stress, the FS-WPENG generated V_{oc} of 48 V and I_{sc} of 5 μA (Maity and Mandal 2018). Table 10.13 summarizes the recently developed in Vitro FS-WPENGs.

10.4.2.1.3 In Vitro FS-WNGs Based on Thermoelectric Effect (FS-WTEGs)

A study conducted by Zhao et al. demonstrates the development of a self-powered FS-WTEG based on CNFs/Bi_2Te_3 thermoelectric composite film that can harness thermal energy directly from human skin and convert it into electrical output employing the Seebeck effect. The vapor filtration process prepared the superhydrophobic CNFs/Bi_2Te_3 film. CNFs/$Bi_{0.5}Sb_{1.5}Te_3$ and CNF/$Bi_2Se_{0.3}Te_{2.7}$ films were used as p and n-type materials of the main FS-WTEG. They had Seebeck coefficients of 154–130 μV/K, respectively. Ag wires and Ag paste were applied as conductive elements. With five p-n terminals, at a 60 DK temperature gradient, this FS-WTEG exerted a peak voltage of 22.34 mV, power of 561 nW, and PD of 8.93 nW cm^{-2} (Zhao et al. 2020). The schematics of FS-WTEG structure for human body energy harvesting are shown in Figure 10.43(a). Suarez et al. reported a flexible FS-WTEG capable of continuously harnessing heat from the human body as electrical energy. The device was based on Galium-Indium liquid metal interconnections and the flexible-bulky leg

TABLE 10.13
Summary of Recently Developed in Vitro FS-WPENGs

Active Materials	Electrodes	Output	Applied As	Special Attributes	Ref.
Selenium (Se) nanowires	Ag nanowires	At 0.74% compressive strain, V_{oc} and I_{sc} were recorded to be ~0.45 V and ~1.67 nA	Physiological monitoring	Converts time varied mechanical vibration (gesture, pulse, vocal movement) generated by the human body into electrical signals	Wu et al. (2019)
Molybdenum disulfide (MoS_2) nanosheet film	MoS_2 nanoflakes	Upon applying 0.12% bending strain, the open circuit voltages were ~37 and ~35 mV during forward and backward motions respectively	Conversion of strain in human feet during walking	Capable of excellent energy harnessing from human bending motion	Han et al. (2019)
BCST nanoparticles dispersed in PDMS matrix	Ti wire and Ag layer	When tested under 2 N force, the device generated V_{oc} of ~52 V, I_{sc} of ~600 nA, and the PD of ~0.8 mW m^{-2}	Human muscle applied sensor	Detecting motion of different body parts (spinal cord, elbow, foot, knee, jaw, palm, throat, breathe, etc.)	Alluri et al. (2018)
PVDF thin film	Ag layer	Under the condition, where input force was 4 N at the angles of 0°, 45°, and 90°, the maximum output voltages and power outputs were 1.75, 1.29, 0.98 V and 0.064, 0.026, 0.020 μW respectively	Mechanical strain from various human motion	This device is a strain power harvester capable of harnessing input energy in the form of strain at multiple directions (bending, sliding, pressing, stretching, etc.)	Kim et al. (2018)
PVDF films	Al	The resting voltage potential was recorded at 79.6 mV	Artificial skin	Can generate receptor potential of sensory organs emulating the human skin. Can also monitor the heart-rate, and measure the ballistocardiogram in real-time	Chun et al. (2018)
PZT thin film	Ultrathin PET substrate	In the case of a mid-aged healthy male, the average voltage of 65 and 81.5 mV for heartbeat 73 and 100 beats per minute (bpm) were recorded	Arterial pulse monitor	A real-time pulse monitoring system through wireless transmission of the detected signals to a smartphone	Park et al. (2017)
$BaTiO_3$	Ag NWs nanowires	The maximum voltage of 105 V, PD of 102 μW cm^{-2}, and current density of 6.5 μA cm^{-2} were reported	Physiological signal monitoring	It can detect human physiological signals (pulse, pronunciation, blinking, arm movement, etc.)	Chen et al. (2017)

Abbreviations: (BCZT) ($Ba_{0.85}Ca_{0.15}Zr_{0.10}Ti_{0.90}O_3$), PDMS (polydimethylsiloxane), PVDF (polyvinylidene fluoride), PET (polyethylene terephthalate), V_{oc} (open circuit voltage), I_{sc} (short circuit current), PD (power density), Pb [Zrx, Ti_{1-x}]O_3, lead zirconate titanate (PZT), Al (aluminum), Ag (silver), Ti (titanium), NWs (nanowires).

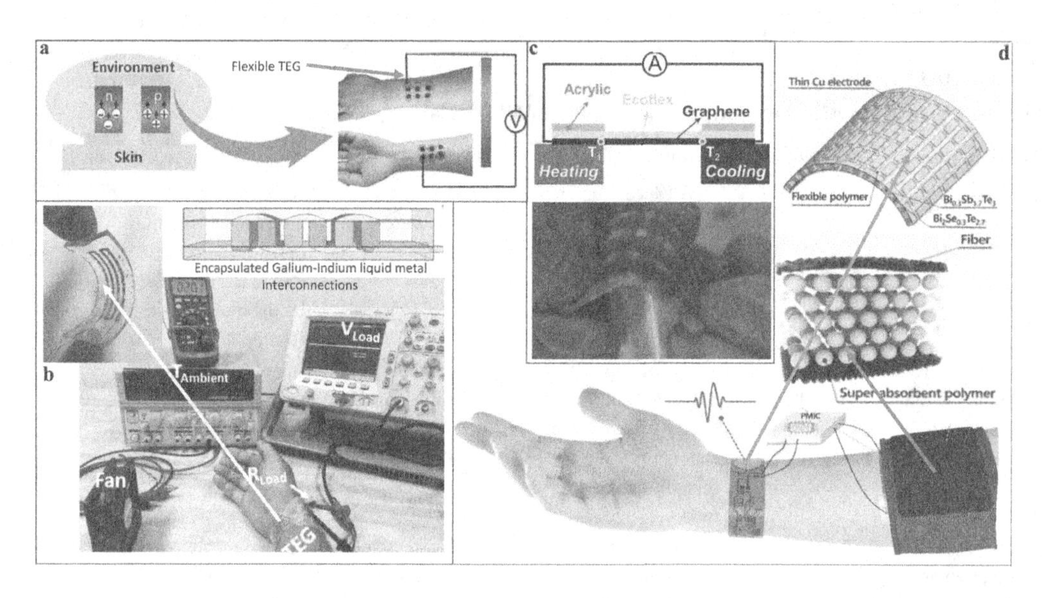

FIGURE 10.43 In vitro FS-WNGs based on thermoelectric effect. (a) Schematics of CNF-Bi_2Te_3-films-based FS-WTEG structure directly implantable on human skin, (Reused with permission from Zhao et al. (2020). Copyright Elsevier 2020.) (b) Bent flexible FS-WTEG; encapsulated final interconnections of FS-TEG; experiment setup of the FS-WTEG. (Reused with permission from Li et al. (2017). Copyright Elsevier 2017.) (c) Schematic diagram showing the structure of stretchable thermoelectric device based on graphene-silicone rubber nanocomposite film and its stretched condition. (Reused with permission from Zhang, Zhang et al. (2019). Copyright Elsevier 2018.) (d) Schematic diagram of an FS-WTEG-based self-powered wearable ECG system. (Reused with permission from Kim et al. (2018). Copyright American Chemical Society 2018.)

approach. $Bi_{0.5}$-$Sb_{1.5}Te_3$ was used as p-type, and $Bi_2Se_{0.3}Te_{2.7}$ was used as n-type flexible legs with Seebeck coefficient, electric conductivity, and thermal conductivity of μVK^{-1}, 900 S cm^{-1}, and 1.45 Wm^{-1} K^{-1}, respectively. The whole structure was fixed on an Al stencil and was encapsulated with PDMS. Figure 10.43(b) illustrates the flexible device and its power harnessing technique from the human body. At 35°C ambient temperature, the peak voltage and power were observed as 0.69 mV and 0.33 μW while for 24°C they were 2.96 mV and 1.48–6.0 μW, respectively (Suarez et al. 2016). Zhang et al. reported another self-powered and highly sensitive FS-WTEG (Figure 10.43(c)) strain sensor based on graphene-silicone rubber composite film. This stretchable device is capable of tracking human motion by generating thermoelectric effect-induced electrical output from the temperature difference that exists between human skin and surrounding temperature. Graphene and silicone rubber (Ecoflex) were combined to prepare thin film graphene/polymer nanocomposite strips. Al tapes were used as electrodes. Under a temperature difference of 14.5 DK the FS-WTEG generated a DC output of voltage ~0.48 mV and current of ~130 nA (Zhang, Zhang et al. 2019). Zhang et al. further conducted another study with a similar setup to harness photo-thermoelectric (PTE) induced electricity by illuminating one side of the graphene-silicone rubber composite film. Under induced light of 980 nm wavelength, the output signals were found to be 0.49 mV and 1.0 μA with a responsivity of 42 μA/W (Zhang, Song et al. 2019). A self-powered skin wearable ECG (Figure 10.43(d)) device based on thermoelectric effect was reported by Kim et al. The wearable FS-WTEG harnessed enough energy from the human body heat to continuously drive the ECG system. Furthermore, to induce a stable temperature difference, a flexible heatsink of water containing polymeric (sodium polyacrylate) particles was employed. Herein, the inner part of the heat sink contains water, whereas the outer part dissipates heat as water evaporates. The FS-WTEG was able to generate a power of 1.5 mW in the initial 10 minutes of operation, which corresponds to a PD of 38 μW cm^{-2} (Kim et al. 2018). Table 10.14 summarizes the recently developed in Vitro FS-WNGs based on the thermoelectric effect.

TABLE 10.14
Summary of Recently Developed in Vitro-FS-WNGs Based on the Thermoelectric Effect

Active Materials	Electrodes	Output	Applied As	Special Attributes	Ref.
Bi_2Te_3 (n-type) and Sb_2Te_3(p-type)	Cu wires	PD of 3 $\mu W\ cm^{-2}$ at 22 DK	A wearable, flexible heat energy harvester that harnesses energy from the temperature gradient between human skin (37°C) and ambient (15°C) wearable on skin	Fabrication of lateral Y-type device structure instead of a common π structure contributes to enhancing output performance	Huu, Nguyen Van, and Takahito (2018)
Bi_2Te_3 as both p-type and n-type	Al_2O_3 and Cu as ceramic substrate and metal conductor	The output power generation of the TEG was found to vary between 5 and 50 μW in between the resting stage and riding a bike respectively	A wearable TEG that is applied on human arms or legs while the subject is performing daily activities like jogging, sitting walking, or bike riding	Whereas maximum of the TEGs available reports about the application at the upper part of the body, this TEG generates output result acting on the lower limbs of the human body as well	Proto et al. (2018)
MoS_2/PU photothermal film and Te/ PEDOT thermoelectric layer	Ag and Cu wires	Upon application of IR laser ($\lambda = 808$ nm) with a PD of 2.625 $W\ cm^{-2}$, at 45 DK, the peak output voltage and current were 1.9 mV and 1.5 μA, respectively	A flexible photo thermal film-based wearable TEG for harnessing energy from IR and powering wearable devices	The transferability, flexibility, and photothermal characteristics can be easily modified by tuning MoS_2	He et al. (2018)
Bi_2Te_3 NWs as n-type and Te-PEDOT:PSS as p-type	Ag paste and Cu inlaid in ceramic	The output voltage of 56 mV and PD of 32 $\mu W\ cm^{-2}$ at 60 DK	A flexible, composite thin film-based TEG capable of harnessing electrical energy from wasted heat generated by the human body	–	Li et al. (2017)
Bi_2Te_3 as both p and n-type materials	Carbon film and Cu	Utilizing an optimized heat spreader the maximum power output was recorded ~6 and ~20 $\mu W\ cm^{-2}$ when the subject person was stationery and walking, respectively	A flexible self-powered TEG implanted in different body locations (wrist, upper arm, chest, etc.) to generate electrical power from body generated heat	For enhancing output power generation considerably optimized heat spreaders were used	Hyland et al. (2016)

Abbreviations: Te-PEDOT:PSS (tellurium-poly (3,4-ethylenedioxythiophene):poly(styrene sulfonate)), Bi_2Te_3 (bismuth telluride), Sb_2Te_3 (antimony telluride), PCE (power conversion efficiency), MoS_2 (molybdenum disulfide), Cu (copper), Ag (silver), PD (power density), IR (infrared light), TEG (thermoelectric nanogenerator).

10.4.2.1.4 In Vitro FS-WNGs Based on Pyroelectric Effect (FS-WPyNG)

Jiang et al. reported a self-powered biocompatible FS-WPyNG capable of harnessing energy from near-infrared radiation and powering wireless devices via generating electric pulses. The constituent elements include a conductive polymer (PEDOT:PSS) used as electrode and PVDF layer as pyroelectric material (pyroelectric coefficient 40 μC m^{-2} K^{-1}) that constructed PEDOT/PVDF/PEDOT sandwiched interface structure. This laminated structure showed the maximum output of 23.4 V in a combination of 6 such cells at a temperature increment of 7K. This device showed the potentiality to be utilized as an independently implantable medical device (IMD) or alternate power source for IMD reinforced by successful application as the nerve stimulant of a frog (Jiang et al. 2018). In another study, Zhang et al. reported an FS-WPyNG based on flexible PVDF thin film capable of powering LEDs and LCDs effectively. Herein, the PVDF film acted as the pyroelectric material, whereas two Al layers were incorporated as electrodes. Upon encountering a temperature fluctuation of 50K, the device generated peak V_{oc}, $I_{sc,}$ and power of 8.2V, 0.8 μA, and 2.2 μW, respectively, at a load impedance of 0.1-MΩ. Apart from scavenging wasted heat to generate electricity and working as a power source for low-powered electronics, it was capable of powering a thermo-sensor (Zhang et al. 2016). Another interesting study reports, application of the pyroelectric effect to power a computer mouse from the heat of the user's palm without any external power source like battery or USB connection. PZT, a composite of $PbTiO_3$ and $PbZrO_3$, was chosen as the pyroelectric material because of its high pyroelectric coefficient as well as availability. For different hand positions over the mouse with the pyroelectric substrate, the peak voltage and power observed were 10V and 150 μW, respectively (Potnuru and Tadesse 2014). Table 10.15 summarizes the recently developed in Vitro FS-WNGs based on the pyroelectric effect.

10.4.2.1.5 In Vitro FS-WNGs Based on Solar Energy

Li et al. demonstrated a rechargeable capacitor for constantly charging a wearable strain sensor (Figure 10.44(a)). It receives its power from a PSC-based energy harvester attached directly to the skin.

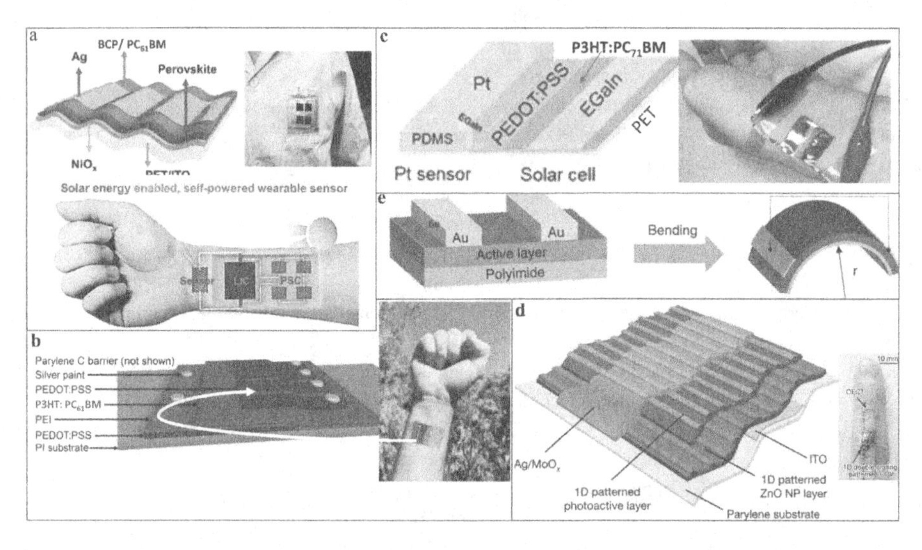

FIGURE 10.44 (a) Illustration of flexible PSC structure and solar energy enabled self-powered wearable sensor applied on textile and bare skin. (Reused with permission from Li et al. (2019). Copyright Elsevier 2019.) (b) Schematics and application of an OSC directly applicable on human skin, (Reused with permission from O'Connor et al. (2016). Copyright Elsevier 2015.) (c) Schematics of the coplanar device with the corresponding circuit and its application on the human hand. (Reused with permission from Hsieh, Hsu, and Chen (2018). Copyright American Chemical Society 2018.) (d) Illustration of SC device structure based on PBDTTT-OFT and $PC_{71}BM$. (Reused with permission from Park et al. (2018). Copyright Springer Nature 2018.) (e) Illustration of OSC blend film deposited on the flexible PI substrate bent under mechanical stress. (Reused under the terms of the Creative Commons CC BY license from Kim et al. (2015). Copyright Springer Nature 2015.)

TABLE 10.15
Summary of Recently Developed in Vitro FS-WNGs Based on the Pyroelectric Effect

Active Materials	Electrodes	Output	Applied As	Special Attributes	Ref.
DuraAct patch transducer	–	At 30 DK peak V_{oc}, I_{sc}, and PD of 1.5 V, 1.5 μA, and 0.034 μW cm^{-2}, respectively	The device can harvest energy via human finger tapping as well as from breathing	This device on human body application has the potential to harvest energy from dissipated heat as well as from human respiration	Sultana et al. (2018)
PVDF	Cu wires	At 5°C ambient temperature, the peak V_{oc}, $I_{sc,}$ and power were observed as 42 V, 2.5 μA, and 8.31 μW, respectively at 50-MΩ load resistance	Temperature difference generated due to human respiration when the device was attached to an N95 mask	The energy scavenged from this device can be used to drive electronics like LEDs and LCDs. It can also act as self-powered temperature and breathing sensors	Xue et al. (2017)
PVDF	Al	With 70% covered electrodes, the current and voltage was observed ~20 nA and ~42 V compared to fully covered electrodes 7 nA and 16 V, respectively	The flexible device can be mounted on the human body that can act as a power source for microelectronics	Micropatterning over the flexible PVDF film enhances V_{oc} and I_{sc} by 380% and 420%, respectively	Zabek et al. (2015)
PVDF	Ag-Pb	At a temperature difference of 1.5°C, the optimum output power was ~0.154 μW at 1.18 V	Mountable on the human forearm. Can generate electricity from the temperature difference between human forearm skin and ambient	Honeycombed-shaped PyNG electrode and protective layer model allows better penetration of human body generated heat into the pyroelectric materials	Kim et al. (2015)

Abbreviations: PVDF (polyvinylidene fluoride), V_{oc} (open circuit voltage), I_{sc} (short circuit current), PD (power density).

This PV unit was formed by coating PET flexible sheet with indium tin oxide (ITO/PET) in the form of a bendable ultrathin wearable substrate. The whole flexible PSC unit was composed of Ag/BCP/$PC_{61}BM/MA_{1-y}FAyPbI_{3-x}Cl_x$ (perovskite material)/NiO_x/ITO/PET. The V_{oc}, J_{sc}, and PCE were 3.95 V, 18.71 mA cm^{-2} and 14.01%, respectively (Li et al. 2019). Another study conducted by O'Connor et al. demonstrates an OSC directly applicable to human skin (Figure 10.44(b)). Both the anode and cathode of such an energy harvester were formed of PEDOT:PSS film. The device was constructed as PI substrate/PEDOT:PSS/polyethyleneimine (PEI)/poly(3-heptylthio-phene) (P3HpT): $PC_{61}BM$)/PEDOT:PSS. A PI tape enabled the device to stick to human skin. Additionally, a parylene layer was vapor-deposited on the whole SC to protect it from photochemical degradation. At 980 W m^{-2}, the device was capable of powering a digital watch and LED. When eight OSCs were connected in series, a voltage of 2.23 V was generated (O'Connor et al. 2016). In a similar study, Hsieh et al. designed a coplanar SC on a flexible elastic substrate capable of harnessing energy even from the indoor source of illumination as low as 2 mW cm^{-2}. The proposed coplanar device (Figure 10.44(c)) was made in combination with a Pt metal film, a strain sensor layered in the formation as PET/PDMS/Pt, and a flexible polymer part acting as the SC layered in the following order PET/PEDOT:PSS/P3HT:$PC_{61}BM$/eutectic Gallium indium (EGaIn). Here the SC functions as the power source to drive the Pt film strain sensor for continuous activity. The SC demonstrated 0.64 V_{oc} and 8.05 mA cm^{-2} J_{sc} (Hsieh, Hsu, and Chen 2018). A study performed by Park et al. reports an ultra-flexible self-powered generator that could be applied on the skin as a sensor to interpret biometric signs. A 1D-patterned OSC was affiliated with this device to provide required driving electrical power. As PV layer, a composite of poly[4,8-bis(5-(2-ethylhexyl)thiophen-2-yl) benzo[1,2-b;4,5-b′]dithiophene-2,6-diyl-alt-(4-octyl-3-fluorothieno[3,4-b]thiophene)-2-carboxylate-2-6-diyl] (PBDTTT-OFT) polymer with $PC_{71}BM$ was used. Herein, ITO performed as the bottom electrode with a ZnO NPs layer functioned as the electron supporting layer. To fabricate the top electrode, MoO_x and Ag were sequentially deposited. The whole device was mounted on a flexible parylene substrate. The schematic of such a device has been depicted below in Figure 10.44(d). The device generated a V_{oc} of 0.785 V, a current density of 19.17 mA cm^{-2}, and PCE of 10.5%. A higher power output per gram (11.46 W g^{-1}) was achieved (Park et al. 2018). An all polymer-based OSC-based on poly[4,8-bis(5-(2-ethylhexyl)thiophen-2-yl)benzo[1,2-b:4,5-b′]dithiophene-alt-1,3-bis(thiophen2-yl)-5-(2-hexyldecyl)-4*H*-thieno[3,4-c]pyrrole-4,6(5*H*)-dione] (PBDTTTPD) as the electron donor and poly[[*N*,*N*′-bis(2-hexyldecyl)-naphthalene-1,4,5,8-bis(dicarboximide)-2,6-diyl]-alt5,5′-thiophene] (P(NDI2HD-T)) as the electron acceptor were deposited on a flexible plastic PI film. Herein, a pair of Au electrodes were used to extract the energy. The structure schematics of the device are shown in Figure 10.44(e). The OSC demonstrated a V_{oc} of 1.06 V and PCE of 6.64% (Kim et al. 2015). Table 10.16 summarizes the recently developed in Vitro FS-WNGs based on solar energy.

TABLE 10.16
Summary of Recently Developed in Vitro FS-WNGs Based on Solar Energy

Active Materials	Electrodes	Output	Applied As	Special Attributes	Ref.
Carbazole dye	TiO_2 as photoanode and Pt-coated PET-ITO as photocathode	Under sunlight V_{oc} and current density of 0.69 V and 4.76 mA cm^{-2} at 2% PCE were achieved	This flexible device can be worn on the human body and can be used as a power source for microelectronics	DSSC contains electrolyte having self-healing properties. When performance is hindered upon application of mechanical stress, up to 94% of performance can be recovered upon a cooling treatment that regenerates the electrode/ electrolyte interface	Sonigara et al. (2018)

(Continued)

TABLE 10.16 *(Continued)*
Summary of Recently Developed in Vitro FS-WNGs Based on Solar Energy

Active Materials	Electrodes	Output	Applied As	Special Attributes	Ref.
DM as HTL on perovskite-based $(FAPbI_3)_{0.95}$ $(MAPbBr_3)_{0.05}$ thin film	FTO/TiO_2 and Au	V_{oc}, J_{sc}, and PCE of 1.14 V, 24.9 mA cm^{-2}, and 23.2% were achieved	–	A thermally stable (capable of showing 95% initial performance at 60°C) and durable PSC with high T_g and a high PCE of 23.2%	Jeon et al. (2018)
$CsPbI_2Br$ perovskite film	FTO/TiO_2 and Au	V_{oc}, J_{sc}, and PCE of 1.19 V, 14.6 mA cm^{-2}, and 11.7% were achieved	A flexible and inorganic thin film-based PSC, wearable on the skin for powering microelectronics	Inorganic perovskite thin film based on $CsPbI_2Br$ shows much better thermal, illumination stability	Jiang et al. (2018)
PBDB-T as the donor, F-M as the acceptor active layers	ITO/ZnO layer and Ag layer	V_{oc}, J_{sc}, and PCE of 1.642 V, 14.35 mA cm^{-2}, and 17.36% were achieved	A polymer active-layer-based lightweight SC wearable on the human body	–	Meng et al. (2018)
P3HT:PC_{61}BM or PBDTTT-C-T:PC_{71}BM	CNTs-doped PEDOT: PSS and Ag	V_{oc} and J_{sc}, of 0.58 + 0.03V V and 14.35 mA cm^{-2} were achieved	–	CNT-loaded optimized PEDOT:PSS shows a high amount of electrode conductivity 3264.27 s cm^{-1}	Hu et al. (2014)

Abbreviations: DM [N2, N2′, N7, N7′-tetrakis(9,9-dimethyl-9*H*-fluoren-2-yl)-N2, N2′, N7, N7′ tetrakis(4-methoxyphenyl)-9,9′-spirobi[fluorene]-2,2′,7,7′-tetraamine], Cesium lead bromide ($CsPbI_2Br$), PBDB-T [Poly[[4,8-bis[5-(2-ethylhexyl)-2-thienyl]benzo[1,2-b:4,5-b′]dithiophene-2,6-diyl]-2,5-thiophenediyl[5,7-bis(2-ethylhexyl)-4,8-dioxo-4*H*,8*H*-benzo[1,2-c:4,5-c′]dithiophene-1,3-diyl]]], HTL (hole transporting layer), P3HT:PC_{61}BM [(poly(3-hexylthiophene) (P3HT):(6,6)-phenyl-C61 butyric acid methyl ester (PC61BM)], PBDTTT-C-T:PC_{71}BM [poly[4,8-bis(2-ethyl-hexyl-thiophene-5-yl)-benzo[1,2-b:4,5-b]dithiophene-2,6-diyl]-alt-[2-(2-ethyl-hexanoyl)-thieno[3,4-b]thiophen-4,6-diyl] (PBDTTT-C-T)/(6,6)-phenyl-C71 butyric acid methyl ester (PC71BM)], FTO (Fluorine-doped tin-oxide), PET (polyethylene terephthalate (PET), ITO (indium tin oxide, Pt (platinum), Au (gold), ZnO (zinc oxide), CNTs (carbon nanotubes), PEDOT:PSS [poly(3,4-ethylenedioxythiophene) polystyrene sulfonate], V_{oc} (open circuit voltage), PCE (power conversion efficiency), J_{sc} (current density), T_g (glass transition).

10.4.2.1.6 In Vitro FS-WNGs Based on Biochemical Energy

BFCs application as a wearable substrate encompasses their activity as an individual sensor, self-powered harvester, or as a source for powering other wearable devices fueled by biological fluids (Jeerapan, Sempionatto, and Wang 2020). Feasibility of the concept for using BFCs in low power consumable drug delivery system, cardio-stimulators, glucose biosensors, etc. is a potential field of interest (Slaughter and Kulkarni 2015).

A very recent study conducted by Shitanda et al. reports a paper-based biofuel cell (PBFC) capable of generating power from lactate present in biofluid. The cell is in the form of a wearable bandage capable of generating power in the presence of lactate in the human sweat. LO_x treated anode, and BOD-treated cathode were the integral parts of the PBFC. An array of six PBFCs was capable of generating V_{oc} of 3.4 V when connected in series and peak power output of 4.3 nW (at 2.44 V) when arranged in a 6 × 6 array formation. The power generation intensity principally depends on the concentration of the lactate biofluid (Shitanda et al. 2021). The application on the human body has been shown in Figure 10.45(a). Such an energy harvester was capable of powering a low-energy Bluetooth device. Bandodkar et al. demonstrated the fabrication and real-life application of a soft and stretchable skin-based electronic BFC (E-BFC) capable of harvesting energy from sweat lactate generated by the human body. CNTs-NQ and CNTs-Ag_2O were applied as anode and cathode, respectively. The electrodes were organized in an iron bridge architecture where serpentine-shaped Au was used for bridging the electrodes. The device demonstrated V_{oc} and PD of 0.5 V and 1.2 mW cm^{-2}, respectively. When applied on human skin while exercising, a power of ~1 mW was observed. The generated power was enough to power LED or Bluetooth low energy (BLE) radio (Figure 10.45(b)) (Bandodkar et al. 2017). Another study performed by Reid et al. reports a based contact lens structured BFC that can generate power from lactate present in tears. Herein, the anode was composed of poly (methylene green) and a hydrogel matrix with lactate dehydrogenase and NAD^+. The cathode was composed of 1-pyrenemethyl anthracene-2-carboxylate and BOD. Both the anode and cathode were constructed onto Buckypaper electrodes. When tested with a synthetic tear at the 35°C, the device demonstrated V_{oc}, maximum current, and PD of 0.413 ± 0.06 V, 61.3 ± 2.9 μA cm^{-2}, and 8.01 ± 1.4 μW cm^{-2}, respectively (Reid, Minteer, and Gale 2015). The device is shown in Figure 10.45(c). Table 10.17 summarizes the recently developed in Vitro FS-WNGs based on biochemical energy.

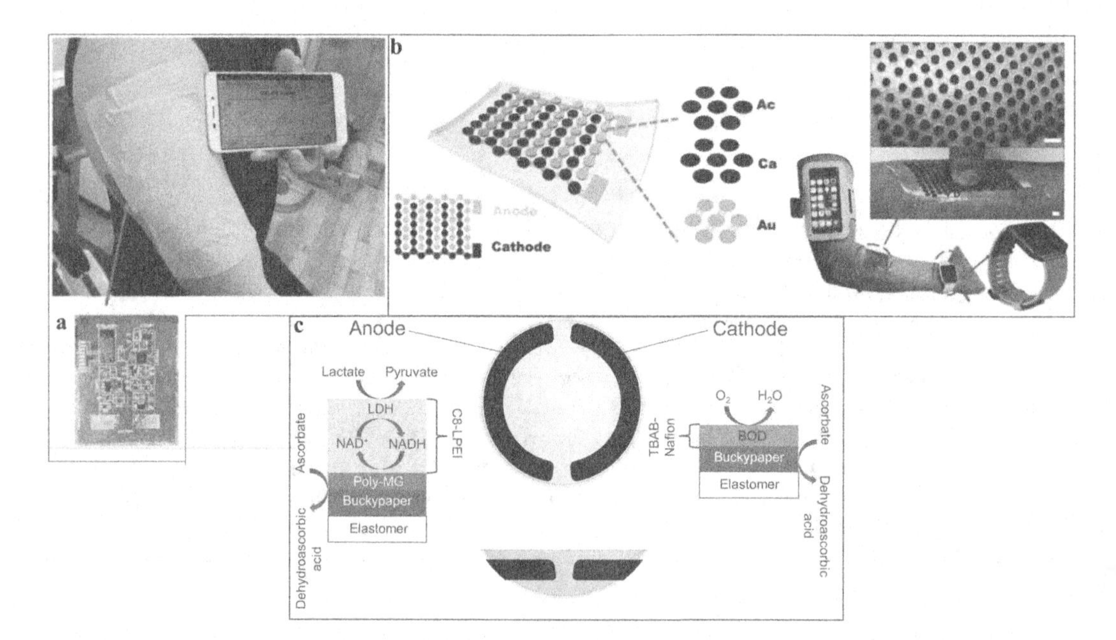

FIGURE 10.45 (a) Application of a PBFC capable of generating power from sweat lactate on a human arm. (Reused with permission from Shitanda et al. (2021). Copyright Elsevier 2021.) (b) Illustration of disintegrated E-BFC layers and pictures of the device worn on human skin, (Reused with permission from Bandodkar et al. (2017). Copyright Royal Society of Chemistry 2017.) (c) Illustration and mechanism of a wearable BFC lens capable of generating electricity from lactate present in tears. (Reused with permission from Reid, Minteer, and Gale (2015). Copyright Elsevier 2014.)

TABLE 10.17
Summary of Recently Developed in Vitro FS-WNGs Based on Biochemical Energy

Active Materials	Electrode	Output	Applied As	Special Attributes	Ref.
Lactate and glucose	CNTs coated with PVC and chitosan anode, Au coated with Pt black and Nafion cathode	This device demonstrated an output range ~1 V with a maximum current of ~2 μA	Attached to the skin for continuous monitoring concentration of lactate, glucose, pH, sweat rate, and loss	The device has an installed colorimetric system that provides the visual color output along with the numerical value	Bandodkar et al. (2019)
Lactate	LO_x-modified bioanode and BOD-treated biocathode	The peak V_{oc}, I_{sc}, and J_{sc} were 0.55 ± 0.03 V, 140 ± 4 μA, and 30 μA cm^{-2}, respectively	Skin wearable patch-type enzymatic BFC capable of generating electrical output from perspiration applied near-throat area	A wireless network can be incorporated to track real-time energy generation via a cell phone	Escalona-Villalpando et al. (2019)
Fructose	FDH-modified carbon fabric anode and BOD-modified carbon fabric cathode	The peak V_{oc} and current density were recorded 0.75 V and ~300 μA cm^{-2}	Skin-worn organic patch to induce iontophoresis	Capable of providing the necessary power for driving electrically controlled application of the iontophoresis patch	Ogawa, Kato et al. (2015)
Glucose	10% GO_x + 3% TTF + 87% carbon paste bioanode; Carbon paste + Pt black biocathode	At glucose concentration range of 5–10 mM, the PD range was 3–5 mW cm^{-2}	A self-powered microneedle-based BFC attached to the skin	In the presence of different protein in the endothermal locations, the microneedle array demonstrated high stability	Valdés-Ramírez et al. (2014)
Lactate	Pt as both electrodes	At lactate concentration 8 mM, the device demonstrated a power density of 25 mW cm^{-2}	Skin worn epidermal BFC based on temporary transfer tattoos	The electrodes in the form of tattooed epidermal designs can withstand repeated mechanical stress and deformations	Jia et al. (2013)

Abbreviations: CNTs (carbon nanotubes), PVC (polyvinyl chloride), Au (gold), Pt (platinum), LO_x (lactate oxidase), BOD (bilirubin oxidase), TTF (tetrathiafulvalene), GO_x (glucose oxidase), FDH (fructose dehydrogenese), V_{oc} (open circuit voltage), J_{sc} (short circuit current density), PD (power density).

10.4.2.1.7 In Vitro-FS-WNGs Based on Hybrid Energy

The main goal behind developing wearable hybrid nanogenerators (HNGs) is to be able to harness energy from multiple sources of the environment (Pan et al. 2011). Among various hybridized WNGs available, HNGs of PENG and TENG have gained much attention due to their low-cost structural conception as well as proficient mechanical adaptability (Shawon et al. 2021). Other devices like SCs, BFCs, and PyNGs are also observed to be integrated as HNGs for wearable applications. Table 10.18 summarizes the recently developed in Vitro FS-WNGs based on hybrid energy.

TABLE 10.18
Summary of Recently Developed in Vitro FS-WNGs Based on Hybrid Energy

Hybrid of	Active Materials	Electrodes	Output	Special Attributes	Ref.
SC + PyNG	Modified rGO-PEI (SC substrate), polarized PVDF film (pyroelectric layer)	Ag	Under single radiation of sunlight, PD of 21.3 mW m^{-2} was achieved at a light oscillation frequency of 25 mHz	Wearable bracelet capable of tracking fitness. The sunlight triggered PyNG can enhance the output performance ability of traditional PyNG	Li et al. (2020)
PENG + TENG + PyNG	PVDF as piezoelectric and pyroelectric substance; PDMS as a triboelectric layer	Ag	As a combination of three types of generators, it is capable of harnessing maximum of the human mechanical and thermal energies. A maximum V_{oc} of 55 and 86 V can be achieved for TENG-PENG and PyNG, respectively	Apart from self-powering function and powering other devices, it can continuously monitor heartbeat, respiration, and swallowing	Park et al. (2018)
PENG + TENG	ZnO as a constituent of PENG; PDMS as a constituent of TENG, copper wires, and silver paste for connecting the electrodes	PDMS/ZnO NFs/Au/PET as bottom electrode and Au/PDMS as top electrode	The 3×3 cm^2-sized device produced voltage, current density, and PD of ~470 V, ~60 μA cm^{-2}, and ~28.2 mW cm^{-2}, respectively	Without any battery or any other storage system, this self-powered device was capable of powering up LED	He et al. (2018)
PENG + PyNG	PVDF	Conductive PEDOT:PSS-PVP and CNTs	In case of mechanical impact, under 0.65, 1.2, and 3 Hz, the current peak outputs were, (positive/negative) 0.54 μA/–0.21 μA, 2.32 μA/–0.78 μA, and 3.76 μA/–0.94 μA, respectively. Electrical energy was 0.78, 2.39, and 2.85 μJ, respectively. Upon heating/cooling from 313 to 319K and vice versa, positive/negative current was 20 nA/–16 nA with a corresponding voltage of 0.01 V/–0.087 V	Can simultaneously generate energy under strain and thermal gradient	You et al. (2018)
TENG + PENG	PVDF-TrFE as a piezoelectric layer, PDMS as a triboelectric	PU nanofiber coated with CNTs and AgNWs	Upon simulation under compressive stress, the piezoelectric and triboelectric part generated the highest PD of 0.82 and 630 μW cm^{-2}, respectively	The device is highly sensitive and thus is capable of performing real-time tracking of physiological activities like radial artery pulse respiratory system, etc.	Chen et al. (2017)

(Continued)

TABLE 10.18 *(Continued)*
Summary of Recently Developed in Vitro FS-WNGs Based on Hybrid Energy

Hybrid of	Active Materials	Electrodes	Output	Special Attributes	Ref.
TEG + PyNG	$Bi_{0.5}Sb_{1.5}Te_3$ (p-type) and $Bi_2Sb_{0.3}Te_{2\cdot7}$ (n-type); PVDF as pyroelectric material	Conductive threads, Ag-Pd ink	Under a difference in temperature of ~6.5°C at the rate of ~0.62°C s^{-1}, the resultant PD and power were recorded ~1.5 $\mu W\ m^{-2}$ and ~ 4.5 nW, respectively	This WHNG can harness energy from the human body regardless of temperature change	Kim et al. (2017)
PENG + PyNG	Single crystal flexible PMN-PT film (both piezoelectric and pyroelectric material)	Au/Ti layer electrodes	Piezoelectric output: a voltage output of 1.5 V is recorded with the folding and release action of the wrist. A peak of 3.4 V is observed when the device was subjected to 1-Hz frequency. 20 mV was observed for a sound frequency of 90 dB Pyroelectric output: On a temperature difference of 8K the output voltage, current, and PD were 0.1V, 20 nA, 2 mW cm^{-3}, respectively	This device is capable of sensing pressure, strain as well as wind and sound waves. It can also sense surrounding temperature fluctuations induced by water or light	Chen et al. (2017)
BFC + PENG	Enzymes LO_x (lactate oxidase), GO_x (glucose oxidase), uricase, and urease for the BFC and ZnO nanowire arrays for the PENG	Ti interdigital electrodes	For LO_x, GO_x, uricase, Urease /ZnO NWs, (a) 72.4, 57.8, 72.8, and 57.6 mV, respectively, when human runs for 8 minutes at 6 km h^{-1} speed, (b) 28.2, 39.2, 40.4, and 46 mV, respectively, when a person runs for 30 minutes at 12 km h^{-1} speed	Can monitor the physical condition via analyzing human perspiration while running	Han et al. (2017)

Abbreviations: SC (solar cell), PyNG (pyroelectric nanogenerator), PENG (piezoelectric nanogenerator), TENG (triboelectric nanogenerator), TEG (thermoelectric nanogenerator), BFC (biofuel cell), rGO-PEI (reduce graphene oxide-polyethyleneimine), PVDF (polyvinylidene fluoride), PDMS (polydimethyl siloxane), ZnO (zinc oxide), PVDF-Tr-FE (PVDF-trifluoroethylene), PMN-PT (lead magnesium niobate-lead titanate, PU (polyurethane), CNTs (carbon nanotubes), LO_x (lactate oxidase), GO_x (glucose oxidase), PEDOT: PSS-PVP (poly(3,4-ethylenedioxythiophene):poly(styrene sulfonate)-polyvinyl pyrrolidone), Ag (silver), NFs (nanofibers), NWs (nanowires), Pd (palladium), Ti (titanium), Au (gold(), PD (power density), V_{oc} (open circuit voltage).

10.4.2.2 FS-WNGS Implantable Inside Human Body (In Vivo)

This section contains information on NGs which are generally implanted inside the human body. Over the past few decades, exploring the promising potential that WNGs, especially PENG, TENG, PyNG, TEG, SC, and any of their hybrid has helped to find a replacement to harness energy from low-frequency sources. The typical power sources have proven to be complicated to use efficiently in an in vivo application (Dagdeviren, Li, and Wang 2017; Hanani et al. 2021). If a system can utilize mechanical movement due to organ activities as a power source inside the human body, a self-powered system can be utilized (Liu et al. 2018). Moreover, a self-powered system is important in this regard as the hassle with replacement of exhausted power sources can be avoided. The applications are mostly in the realm of the biomedical domain. Therefore, these biomechanical and biochemical energy harvesters can be a permanent, sustainable source of power for implantable devices (Yu et al. 2016). In addition to that, they have the potentiality to become a new and self-powered way of sensing, driving, or monitoring physiological and pathological parameters like blood flow rate, blood pressure (BP), breathing pattern, pulse, heart rate, cardiac signals, body temperature, respiratory rhythm, and rate, etc. (Shawon et al. 2021; Zhao et al. 2020). Hence, implantable generators are becoming medical tools for enabling quick disease detection as well as real-time update of bodily functions (Liu, Li et al. 2019; Mahmud et al. 2018; Shi, Li, and Fan 2018; Shi et al. 2021; Wang, He, and Lee 2019).

The very first in vivo application utilizing piezoelectric material was demonstrated by Wang et al. back in 2010. The in vivo biomechanical energy harvesting application with triboelectric mechanism was reported by Zheng et al. back in 2014 for the first time (Zheng et al. 2014). For any NG to be implanted, flexibility is perhaps the most important characteristic (Li et al. 2020; Zhang et al. 2015). Moreover, the biocompatibility and toxicity of the material to be used are also crucial, which can be controlled by the encapsulation concept to separate the materials from internal body organs and fluids (Jiang et al. 2018; Li et al. 2010). The typical implantable locations include the diaphragm, liver, pericardium organs, heart, gastric organs, cochlear, etc. (Dagdeviren et al. 2014; Liu et al. 2018; Mostafalu and Sonkusale 2014; Shi et al. 2019). Some of the devices that can be used in vivo applications are human body implantable devices such as nerve stimulators, pacemakers, and pumps for drug delivery, etc. (Bandodkar and Wang 2016).

10.4.2.2.1. In Vivo FS-WNGs Based on Triboelectric Effect

A self-powered symbiotic pacemaker (SPM) based on implantable TENG (iTENG) was developed to harness energy from cardiac motion and to drive cardiac pacing as well as correct sinus arrhythmia. Nanostructured PTFE film and Al triboelectric layers with EVA/Ti spacer layer and Teflon/PDMS as the top and bottom encapsulation layers were assembled as illustrated in (Figure 10.47(a)) to fabricate the iTENG. The device generated a V_{oc} of 65.2 V and harnessed 0.495 μJ energy surpassing the pacing threshold energy of a human (0.377 μJ) (Ouyang et al. 2019). Liu et al. reported a research work that demonstrates an ultrasensitive self-powered TENG-based endocardial pressure sensor (SEPS). The triboelectric constituent material uses blood flow inside the heart chamber as the source of energy to generate electricity. The sensitivity of such a device was found to be excellent at 1.195 mV mmHg^{-1}. A PTFE film was incorporated as a triboelectric layer and a thin Au layer was used as an electrode (Figure 10.46(c)). A Kapton layer acted as a flexible base substrate, and an Al foil was used as both the second triboelectric layer and the second electrode. During testing, the corona discharge effect was also evaluated. The V_{oc} before and after corona discharge was recorded 1.2 and 6.2 V, respectively. Under the highest pressure of 350 mmHg, the V_{oc} and I_{sc} of 0.45 V and 0.2 μA were reported (Liu, Ma et al. 2019). Research work performed by Ma et al. demonstrated a flexible triboelectric self-powered implantable active sensor (iTEAS) (Figure 10.46(b)). This device showed the capability to scavenge biomechanical energy from the ambient, converting it into electrical output, and perform real-time monitoring of the heart rate, respiratory rate, blood flow pressure, etc. The WNG consisted of electrodes, spacers, and triboelectric layers contained

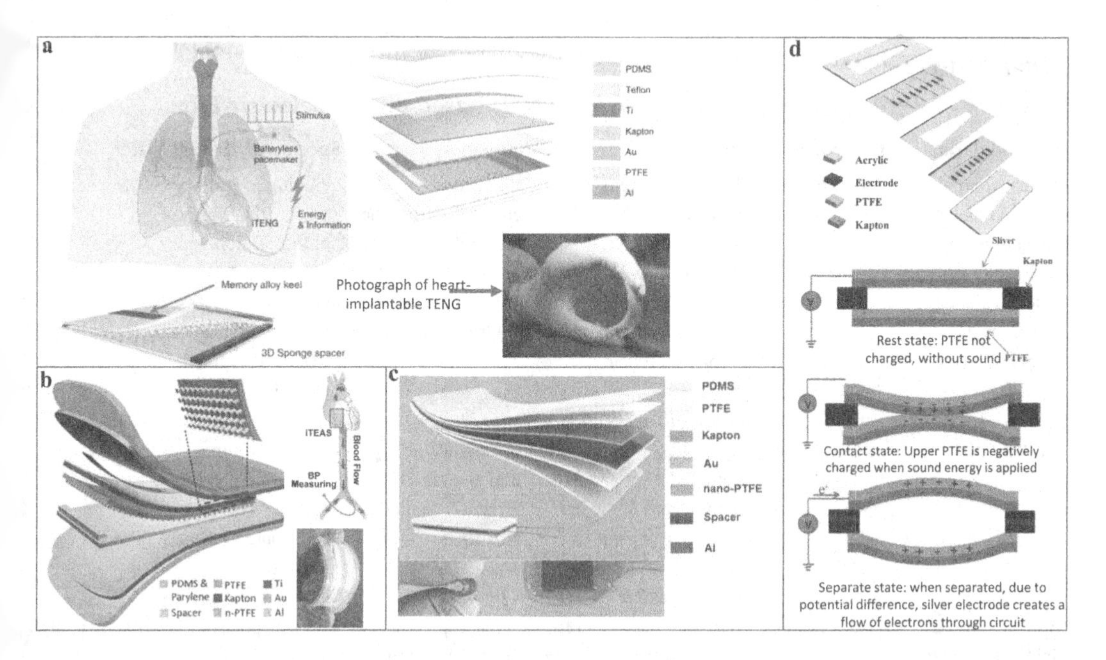

FIGURE 10.46 In vivo FS-WNGs based on triboelectric effect. (a) Illustration of symbiotic cardiac pacemaker system & photograph of iTENG under bending. (Reused under the terms of the Creative Commons CC BY license from Ouyang et al. (2019). Copyright Springer Nature 2019.) (b) 3D view of the iTEAS structure, mechanism for monitoring blood flow velocity, photograph of flexible TENG, (Reused with permission from Ma et al. (2016). Copyright American Chemical Society 2016.) (c) Schematic illustration of the structure and photograph under bending of the SEPS. (Reused with permission from Liu, Ma et al. (2019). Copyright John Wiley and Sons 2018.) (d) The 3D view of the major components of the TENG for frequency selectivity and its working principle. (Reused with permission from Liu et al. (2018). Copyright Springer Nature 2018.)

in a biocompatible multilayered encapsulation. A PTFE thin film was used as a triboelectric layer, Kapton film worked as the base substrate, Au layer served as one electrode, and an Al layer functioned as the second electrode and triboelectric layer. The test was conducted by implanting in the pericardial sack of a swine. For heartbeat and breathing, the resultant output signal was V_{oc} ~10 V and I_{sc}~4 μA (Ma et al. 2016). In a study conducted by Liu et al., a TENG was integrated with a novel cochlear membrane (typically implanted inside the human ear) to covert acoustic vibration into electrical output. A combination of nine silver electrodes with a PTFE membrane was incorporated as the basilar acoustic sensor bionic membrane. Additionally, a Kapton film and two acrylic plates with trapezoid-shaped slits were attached to the PTFE film (Figure 10.46(d)). The device can sense various acoustic sound waves from the triboelectric output voltage. Each of the nine electrodes can generate various triboelectric voltages from the same sound pressure (Liu et al. 2018). Table 10.19 summarizes the recently developed in Vivo FS-WNGs based on triboelectric effect.

10.4.2.2.2 In Vivo FS-WNGs Based on Piezoelectric Effect

Cheng et al. reported an implantable self-powered flexible BP monitor based on piezoelectric thin film. Harvesting biomechanical energy from the in vivo ambient, this device generates electricity for self-powered operation via the piezoelectric effect. The flexible device in the film structure is wrapped around the aorta to monitor BP continuously. The constituents include a very thin piezoelectric PVDF film sandwiched between Al electrodes. PI films were used to encapsulate the whole arrangement. (Figure 10.47(a)) shows the schematics and location of the installation of the device. This WNG can work in vivo situations, and when experimented on an average adult, it exhibited a sound sensitivity of 14.32 mV/mmHg and a maximum power of 40 nW (Cheng et al. 2016). A recent

TABLE 10.19
Summary of Recently Developed in Vivo FS-WNGs Based on Triboelectric Effect

Active Materials	Electrodes	Output	Applied As	Special Attributes	Ref.
Bioabsorbable polymers as two friction layers (cellulose, chitin, silk fibroin, egg white, or rice paper)	Thin Mg film	The observed range of voltage, current, and PD were 8–55 V, 0.08–0.6 μA, and ~21.6 mW m^{-2} for different combination of bioabsorbable polymers	Can generate electricity from the biomechanical movement of organs in vivo environment	Utilizing as a power source, the heartbeat of the cardiomyotic cluster can be made faster. Thus diseases like arrhythmia and bradycardia can be treated	Jiang et al. (2018)
POC vs PLGA or PLA or PCL	Au and Mg deposited layers	For in vitro application V_{oc} and I_{SC} were recorded 28 V and 220 nA whereas in the in vivo environment (planted inside a rat) the V_{oc} was 2 V	A tunable biodegradable implantable TENG (iTENG) having NIR light sensitivity	The output voltage in vivo can promote migration of cell and thus can contribute to wound healing	Li et al. (2018)
Nanostructured-PTFE and Al foil encapsulated by PDMS	Au layer and Al foil	The observed V_{oc} and I_{sc} were ~14 V and ~5 μA, respectively	A biocompatible iTENG for in vivo biomechanical energy harvester driven by the heartbeat	This self-powered device can be integrated with a wireless transmission for real-time cardiac monitoring	Zheng et al. (2016)
Kapton film encapsulated by PDMS film	Serpentine-shaped Cu wires	At 44-MΩ load resistance, PD of ~5 W m^{-2} was observed in a compressive mode. Recorded V_{oc} and I_{sc} were approximate to ~700 V and ~75 μA, respectively	Self-powered stretchable, shape adaptive, flexible biomedical monitoring TENG device directly applied to the human body	Can operate flexibly in both stretching and compressing mode. Can be attached to monitor motions of muscle, joints as well as Adam's apple	Yang et al. (2015)
Kapton substrate and Al foil encapsulated by PDMS film	Au layer and Al foil	The achieved V_{oc}, I_{sc}, and PD were found ~12 V, ~0.25 μA, and ~8.44 mW m^{-2}, respectively	An iTENG capable of harnessing energy from the in vivo biomechanical movement during breathing	To drive implanted pacemaker	Zheng et al. (2014)

Abbreviations: POC [(poly (1,8-octanediol-*co*-citric acid)], PLGA (polylactic-*co*-glycolic acid), PLA (polylactic acid), PCL (polycaprolactone), PTFE (polytetrafluoroethylene), Al (aluminum), PDMS (polydimethyl siloxane), Mg (magnesium), Au (gold), Cu (copper), PD (power density), V_{oc} (open circuit voltage), I_{sc} (short circuit current), NIR (near-field IR), iTENG (implantable TENG).

study conducted by Azimi et al. utilized a polymer composite-based PENG as a self-powered pacemaker. Non-toxic PVDF and ZnO-rGO (reduced graphene oxide) were used as piezoelectric materials. When tested, the device was capable of harnessing 0.487-μJ energy from the cardiac motion of the left ventricle. A PD of 138 ± 2.82 μW cm^{-3} at only 1-Hz frequency of mechanical frequency was recorded. The V_{oc} of 6.06 ± 0.08 V and I_{sc} of 3.46 ± 0.70 μA proved that it could provide enough energy to run a commercial pacemaker (Azimi et al. 2021). The schematic of the device is shown in

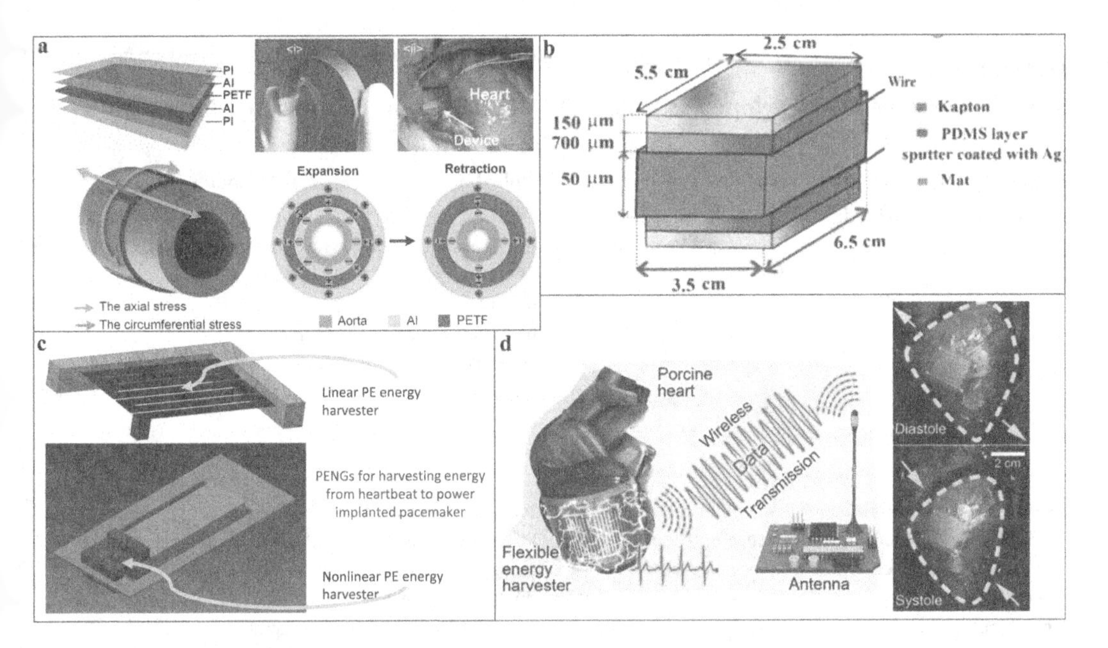

FIGURE 10.47 In vivo FS-WNGs based on piezoelectric effect. (a) The schematics and implantation location of installation of PTFE-based BP monitoring device. (Reused with permission from Cheng et al. (2016). Copyright Elsevier 2016.) (b) Schematics of PENG; photograph of PENG sutured on epicardium facing the lateral wall. (Reused with permission Azimi et al. (2021). Copyright Elsevier 2021.) (c) The linear and nonlinear fabricated PENG for powering a pacemaker. (Reused with permission Karami and Inman (2012). Copyright American Institute of Physics 2012.) (d) Conceptual illustration of self-powered in vivo PENG; Picture of the PENG when the heart expands in diastole & squeezes in systole. (Reused with permission Kim et al. (2017). Copyright John Wiley and Sons 2017.)

Figure 10.47(b). A study conducted by Karimi et al. demonstrated the development of a piezoelectric linear and nonlinear device that can convert the vibration from heartbeat to power pacemakers. The linear zigzag device was constructed of brass and polyepoxide. The nonlinear device was constructed of bimorph structured brass and magnets as tip and base. For linear and nonlinear PENG as shown in (Figure 10.47(c)), the recorded powers were 10 and 8 μW, respectively, at 39-Hz heartbeat oscillation (Karami and Inman 2012). A study conducted by Kim et al. demonstrates an in vivo self-powered PENG based on Mn-doped $(1-x)Pb(Mg_{1/3}Nb_{2/3})O_3-(x)Pb(Zr, Ti)O_3$ (PMN-PZT). The device could transfer data wirelessly. Adding to that, this biocompatible device could generate electrical output from heartbeats. Au-interdigitated electrodes were employed to harness energy. The device generated V_{oc} and I_{sc} of 17.8 V and 1.74 μA, respectively. This device could work not only as an independent sensor for cardiac movement and heart rate tracker but also as a power source for other implantable devices (Kim et al. 2017). The device and its operation during the functionality of the heart are shown in Figure 10.47(d). Table 10.20 summarizes the recently developed in Vivo FS-WNGs based on piezoelectric effect.

10.4.2.2.3 In Vivo FS-WNGs Based on Biochemical Energy

While considering BFC for in vivo application, some important points are needed to be kept in mind. The location of the implantable device is the most important aspect. The location should be such a place where there is ample availability of replenishing biofluids like glucose and lactase. The physiological surroundings should be supporting enough so that the electrocatalytic reactions are carried out with optimum efficiency. The devices must be biocompatible and protected from external hostile particles by sealed container membranes (Cosnier, Le Goff, and Holzinger 2014).

TABLE 10.20
Summary of Recently Developed in Vivo FS-WNGs Based on Piezoelectric Effect

Active Materials	Electrodes	Output	Applied As	Special Attributes	Ref.
Poly-L-lactide	Mo or Mg	Peak to peak voltage of 0.9 V at 23 N force	A self-sensing bionic force sensor capable of being integrated directly into the biological organs to detect internal pressure buildup	Has the potential to sense important bio-physiological (diaphragmatic contraction, intra-abdominal, intra-articular, intraocular, intracranial, etc.) pressure buildup	Curry et al. (2018)
PVDF nanofibers encapsulated by PI/PET thin film	Al foils and Cu wire electrodes	An output voltage in the range of 0.4–0.6 V cm^{-2} was observed	This self-powered PENG can harness energy from the rhythmic movement of the femoral and carotid artery	The device can function both in vivo and in vitro conditions to demonstrate incredibly good sensitivity even under very low frequency ~0.1 Hz	Li et al. (2018)
PVDF thin film encapsulated by PDMS	Au electrodes	At 20-Hz frequency and 6 N force, the peak V_{oc} and I_{sc} were recorded 3.8 V and 3.5 μA. Moreover, V_{oc} of ~200 mV was observed from the gentle movement of muscle in a rodent's body	An implantable structured film-based NG capable of harnessing power from in vivo environment due to muscle movement	Has the potentiality to operate over a long period (years)	Yu et al. (2016)
PZT film encapsulated by PDMS	Au layer	During cardiac action, the maximum peak to peak voltage of ~3 V was observed	An ultra-flexible device capable of harvesting energy from cardiac motion and is placed from the apex of the left ventricle to the right ventricle	–	Lu et al. (2015)
Flexible PZT layer encapsulated by PI	Ti/Pt as the bottom electrode, Cr/Au as the top electrode	The peak voltage observed from this device was 3.7 V. A PD of 0.18 μW cm^{-2} was observed when the device ribbon was mounted at a 45° angle	A flexible PENG capable of energy conversion from in vivo relaxation and contractile motion of heart, diaphragm, and lung	–	Dagdeviren et al. (2014)

Abbreviations: PLLA (poly-L-lactide polymer), PVDF (polyvinylidene fluoride), PI (polyimide), PET (polyethylene terephthalate), PDMS (polydimethylsiloxane), PZT (lead zirconate titanate), Mo (molybdenum), Mg (magnesium), Al (aluminum), Cu (copper), Au (gold), Ti (titanium), Pt (platinum), Cr (chromium), PENG (piezoelectric nanogenerator), V_{oc} (open circuit voltage), I_{sc} (short circuit current), PD (power density).

TABLE 10.21
Summary of Recently Developed in Vivo FS-WNGs Based on Biochemical Energy

Active Materials	Electrodes	Output	Applied As	Special Attributes	Ref.
Gastrointestinal gastric juice biofuel	Zinc as anode and palladium as cathode	For the 15 mm^2 contact area, the peak power of 1.25 mW and V_{oc} of 0.75 V were reported	Implanted in endoscopy application areas	To drive a wireless endoscopy capsule	Mostafalu and Sonkusale (2014)
Olive oil (binder) + activated charcoal (filler) = edible paste as biofuel	Ag (silver)/AgCl and carbon ink	The device demonstrated an improved sensitivity of 6.9 nA mM^{-1} compared to a traditional glucose sensor	Embedded into stomach and gastrointestinal areas	Prolonged longevity in extreme acidic condition (i.e., strong stomach acid); also it showed a linear response in the glucose concentration between 2 and 10 mM	Montiel et al. (2019)

A recent study by Bollella et al. reports a millimeter-scaled implantable BFC capable of utilizing glucose/O_2 solution mimicking human metabolites as biofuel. The device was implanted in the dorso-posterior part of an animal, between the heart and body wall to explore the possibility of human in vivo application. The device demonstrated an electrical output in the range 2–10 μW based on the concentration of the metabolites in the fluid. In this experiment, the BFC was used as the power source for a micro-electric temperature sensor with a wireless data transmission system. MWCNT was incorporated as the electrodes; in the case of anode and cathode, PQQ (pyrroloquinoline quinone)-GHD and BOD were used respectively as modification enzymes (Bollella et al. 2020). Southcott et al. designed an implantable BFC and tested it to harness energy from a serum solution and condition that emulates the human blood circulatory system. The device was developed to be able to use as a power source for a pacemaker. The constituents include Buckypaper catalytic electrodes, where the anode was modified with PQQ-GHD, and the cathode was treated with laccase enzymes. The designed biofuel, i.e., the human serum solution, contained a glucose concentration of 6.4 mM. Such a device could generate V_{oc}, I_{sc}, J_{sc} of 470 mV, 5 mA, and 0.83 mA cm^{-2}, respectively (Southcott et al. 2013). Table 10.21 summarizes the recently developed in Vivo FS-WNGs based on biochemical energy.

10.4.2.2.4 In Vivo FS-WNGs Based on Hybrid Energy

An HNG implantable inside the human body can provide the prospect for not only powering the small electronics but also as a wireless transmitter sensor to detect human body-related phenomena (Hansen et al. 2010). Although the application is limited because of the size, shape, and flexibility of such a device, a few of the research have been shown in Table 10.22.

10.5 LIMITATIONS AND CHALLENGES

It is without a doubt that NGs hold the key to potentially the future of sustainable energy harvesting and solution to the scarcity of renewable energy sources. This review covers most of the recent

TABLE 10.22
Summary of Recently Developed In Vivo-FS-WNGs Based on Hybrid Energy

Hybrid of	Active Materials	Electrodes	Output	Applied As	Special Attributes	Ref.
TENG + BFC	For TENG, Al foil and Kapton film acted as friction layers. For BFC, glucose was used. Both TENG and BFC were integrated into a flexible PET substrate	Thin Cu film for TENG, Au films were used as both anode and cathode for BFC	For TENG, the peak V_{oc} of 22 V and I_{sc} of 0.24 μA, respectively. The observed peak PD of 3.3 mW cm^{-2} at 70-MΩ load resistance. In the case of BFC for a 2 × 2 cm^{-2} sized device, the I_{sc} was recorded 6 μA	Highly efficient WHNG implantable inside human skin to harness superimposed electrical output from biomechanical and chemical energies simultaneously	Since both the separate harvesters were connected in parallel structure, it provided an amplified superimposed output	Li et al. (2020)
TENG + PENG	$BaTiO_3$-coated PDMS film as a piezoelectric material; Al foil and $BaTiO_3$-PDMS film acted as a contact separation layer for triboelectric effect	Au thin film electrode	Upon application of a linear motion of 1-Hz frequency, the TENGs generated peak V_{oc} and I_{sc} were 40 V and 0.5 μA, respectively. The combined V_{oc}, I_{sc}, and PD were 60 V, 1 μA, and 97.41 mW m^{-2}, respectively at 50-MΩ load resistance	Implantable in vivo environment to generate electricity from linear motion and act as a sustainable power source for implantable medical devices	The presence of piezoelectric elements helps to enhance the output of triboelectric outcomes via a synergistic effect. The flexible hybrid device can act as a power source, rectifier, and micro battery	Shi et al. (2016)
PENG + BFC	PVDF NFs as piezoelectric and glucose as biofuel	Ag paste for PENG; GO_x/CNTs and laccase/CNTs as anode and cathode respectively for BFC	For the PENG part, the peak V_{oc} and I_{sc} were found to be ~20 mV and ~0.3 nA, respectively. For the BFC part the V_{oc} and I_{sc} were ~ 60 mV and 13 nA, respectively	Can be used inside the body as it is capable of harvesting biomechanical motion of heartbeat and biochemical energy from glucose/O_2 of blood	The hybrid device is capable of harnessing energy simultaneously using both parts and also can harvest energy as a single unit using any of the harvesting components	Hansen et al. (2010)

Abbreviations: TENG (triboelectric nanogenerator), BFC (biofuel cell), PENG (piezoelectric nanogenerator), Al (aluminum foil), PET (polyethylene terephthalate), PDMS (polydimethylsiloxane), NFs (nanofibers), Cu (copper), Au (gold), Ag (silver), GO_x (glucose oxidase), CNTs (carbon nanotubes), PD (power density), V_{oc} (open circuit voltage), I_{sc} (short circuit current).

studies is a testament to the fact that we are certainly going in the right direction. However, some challenges and obstacles persist in almost all types of NGs mentioned in the review. In most cases, the NGs suffer from the inadequate density of generated energy resulting in low output voltages, making the devices inefficient (Li et al. 2019). In the case of implantable devices, encapsulation is a crucial part of keeping the devices separated from the inside body environment (Shi et al. 2016).

There are a lot of wearable applications regarding PENGs reported by researchers all over the world. However, the list of difficulties and limitations is not short. Although ceramic materials demonstrate remarkable piezoelectric properties, they are restricted in wearable applications due to their rigidity as well as toxicity (Baniasadi et al. 2015; Guo et al. 2018). The basis of the piezoelectric effect, i.e., the depolarization of active materials, can cause a substantial impact on the sensitivity (Yu et al. 2020). In case of PENGs made of polymeric materials can show poor sensitivity due to their inherent properties around the surrounding ambient noise (Park et al. 2017). Furthermore, with polymer-based PENGs, lower piezoelectric coefficient, elevated cost, poor flexibility, fragility, etc., may cause a significant amount of difficulties in wearable applications (Wang et al. 2017). Since organic material-based PENGs have an inhomogeneous structure of molecules, it is extremely difficult to introduce any piezoelectric performance improving measure (Li et al. 2018). Transition metal di-chalcogenide-based PENGs show poor performance due to their nano-scaled energy generation (Han et al. 2019). Lead (Pb)-based perovskite PENGs are toxic in nature and non-eco-friendly, so they are not suitable for wearable applications (Han et al. 2019).

In the case of TENGs in wearable applications, researchers worldwide have pointed out many difficulties where improvements are required. Some typical problems are lower output power, unstable electricity harnessing, poor utilization of energies from multiple sources (Shi et al. 2021). TENGs performance degradation can be commonly associated with vulnerability to humidity present in the atmosphere as well as damages caused by friction (Jung et al. 2014; Lee and Roh 2019). Since the constituent counter surfaces are required to stay in touch with each other, good abrasion resistance of the component materials plays a crucial part. Being composed of completely two different sets of materials, difficulties are common in operation and assembly during the contact separation mechanism (Chu et al. 2018). Adding to that, poor detection of biosignals may further damage their expected performance (Park et al. 2017). In the case of implantable applications, the biocompatibility of the materials, limited space, high internal impedance, complex operating environment, etc., possess some difficulties that are yet to be resolved (Shi et al. 2021).

With the wearable application of TEG, scientists have encountered numerous difficulties. The flexibility of the device has always been an issue when it covers a large area of the body (Nozariasbmarz et al. 2020). In the case of organic-material-based TEGs, high contact resistance results in confined output performance, with inorganic TEGs, the inherent brittleness and rigid tendency confines application as a complex structure. On the other hand, polymer-based TEGs inadequate biocompatibility, as well as their inclination toward rapid aging, circumscribe their applications (Li et al. 2017; Zhao et al. 2020). The temperature difference between the human body and ambient being not that great may contribute to its rather poor performance (Elsheikh et al. 2014). Also, most of the TEGs have demonstrated lower PCE within the range of 5–10% (Siddique, Mahmud, and Heyst 2017). When applied directly on the skin, great thermal resistance either in the skin/TEG or TEG/ambient interface may cause difficulties (Suarez et al. 2016). Furthermore, they have been reported to create a load imbalance between the internal and external portion of the device (Siddique, Mahmud, and Heyst 2017). Yap et al. highlighted in their study that TEGs are good in providing the necessary power to drive only low-power consumable devices or microelectronics rather than power high-energy consumable devices like cell phones (Yap, Naayagi, and Woo 2017).

With PyNG, frequent change in temperature results in better energy harnessing. In wearable applications, since the body temperature does not alter much over time, it is difficult to harness energy with high efficiency (Kim et al. 2017). Furthermore, H. Li et al. highlighted the output power generation is generally less than 5 mW m^{-2}. The bulky design further circumscribes the performance in terms of wearable applications (Li et al. 2020).

For WSCs applications, many difficulties have been reported over the past decade. A major drawback is associated with sunlight power-based SCs dependence on weather conditions puts unpredictability in their performance (Shi et al. 2021). Xu et al. mentioned the hardship of SCs with a thin bottom layer to be mounted on the curved surface of the fiber. The unstable behavior of matters at different temperatures is another problem. Evaporation of electrolyte in case of DSSC and deterioration of PSC forming materials impede their functionality by a great deal (Xu et al. 2020). One of the most challenging facts for DSSC as well as polymer-based SCs is, with an increase of cell length, the conversion efficiency decreases. Hence, the maximum size of the fiber-shaped SCs is forced to be kept in millimeters to ensure decent PCE (Pan et al. 2014). In their study, Kim et al. (2015) highlighted the limited power conversion efficiency (5–6%) of polymer-based SCs. Furthermore, with DSSC, additional enclosure or packing must contain volatile electrolytes that create difficulties in wearability (Yuan, Hu, and Chen 2020). The shorter life span causes the harvester to be replaced frequently (Sonigara et al. 2018). With OSC, the poor performance is mostly because of the lower amount of charge mobility by the constituent organic matters (Meng et al. 2018). Chao Li et al. mentioned, heavyweight, lack of flexibility as well as bulky shapes not only hinder comfortability but also limit wearability. Limited PD further adds to the challenge (Li et al. 2019).

With wearable applications of BFC, lower output and PD may be a result of limited access since biofuel active components are capsulated and barred inside the bulky membrane and thick cell walls. Moreover, cytotoxicity of the biofluid may heavily influence the wearable application (Bandodkar and Wang 2016). One of the very basic drawbacks of in vivo application is its insufficient voltage output, which is generally within the range of 0.5 V (Shi, Li, and Fan 2018). One of the common locations for the BFCs implanted inside the human body is blood vessels. Even so, Cosnier, Le Goff, and Holzinger (2014) mentioned the existence of an alien material may hinder blood circulation that can cause unwanted consequences. Furthermore, the slowness of oxygen reducing reaction in the electrode also affects the performance of the whole cell. Better current density is achievable with metallic electrodes. However, it adds a significant amount to the list of costs (Bandodkar and Wang 2016).

10.6 CONCLUSION

In this book chapter, the state-of-the-art mechanisms and techniques of harvesting different kinds of ambient energy by WNGs were summarized. Furthermore, major approaches undertaken to fabricate different kinds of traditional textiles and flexible wearable-sheet-based WNGs were disclosed. Demonstrations of application in the form of powering wearable electronics and self-powered systems have been reported. The reported electrical energy output and demonstration of numerous in vitro and in vivo applications exhibit the potentiality of such WNGs to be utilized in self-sustainable systems. However, the big challenges that need to be addressed for WNGs are stable output performance, large-scale production, true wearability, and washability. Throughout the chapter, we have seen heavy use of nontraditional textiles materials in the fabrication of WNGs. The utilization of such materials reduces the overall wearability and flexibility of the WNGs, which can create wearable discomfort. So, it is clear that there is still scope for improvement in the area of selecting materials with more promise as well as introducing more production-friendly fabrication methods for wearable applications. Moreover, the environmental dependency of SCs, TEGs, PyNGs, and BFCs has significantly reduced their energy output and practical efficiency. In that case, TENG- and PENG-based WNGs showed greater promise in higher energy output and power efficiency. Future studies can also be propelled toward this area. Additionally, the concept of HNGs was disclosed to compensate for environmental disadvantages among the different WNGs. In conclusion, WNGs are expected to play a significant role in the upcoming fourth industrial revolution (4IR) since they have shown enormous potential for IoTs-based applications especially for wireless communications, health care, personal care, safety, security, and human-machine interfaces. For this to work smoothly, WNGs must also become a self-sustained and self-powered system to provide the energy needed for a particular application. However, the energy generation of WNGs is intermittent in most cases. Although many studies have demonstrated that this part of the limitation is already being worked on via rectifying and providing necessary storage of energy, it is still not enough. Therefore, appropriate power processing and storing electronics and electrical circuitry must also be focused on with equal importance. To look forward to WNGs as an important component of the future, a look at the current situation is necessary without a doubt. This book chapter is hoped to put light on the present scenarios to help achieve that feat.

REFERENCES

Abate, Antonio, Juan P. Correa-Baena, Michael Saliba, Mohd Sukor Su'ait, and Federico Bella. 2018. "Perovskite Solar Cells: From the Laboratory to the Assembly Line." *Chemistry – A European Journal* 24 (13): 3083–100. https://doi.org/10.1002/chem.201704507.

Ahn, Yongho, Seunghwan Song, and Kwang S. Yun. 2015. "Woven Flexible Textile Structure for Wearable Power-Generating Tactile Sensor Array." *Smart Materials and Structures* 24 (7). https://doi.org/10.1088/0964-1726/24/7/075002.

Alanne, Kari, and Arto Saari. 2006. "Distributed Energy Generation and Sustainable Development." *Renewable and Sustainable Energy Reviews* 10 (6): 539–58. https://doi.org/10.1016/j.rser.2004.11.004.

Allison, Linden K., and Trisha L. Andrew. 2019. "A Wearable All-Fabric Thermoelectric Generator." *Advanced Materials Technologies* 4 (5): 1800615. https://doi.org/10.1002/admt.201800615.

Alluri, Nagamalleswara R., Venkateswaran Vivekananthan, Arunkumar Chandrasekhar, and Sang J. Kim. 2018. "Adaptable Piezoelectric Hemispherical Composite Strips Using a Scalable Groove Technique for a Self-Powered Muscle Monitoring System." *Nanoscale* 10 (3): 907–13. https://doi.org/10.1039/c7nr06674k.

Anja Lund, Christian Jonasson, Christer Johansson, Daniel Haagensen, Bengt Hagstr€om. 2012. "Piezoelectric Polymeric Bicomponent Fibers Produced by Melt Spinning." *Journal of Applied Polymer Science* 126 (5): 490–500. https://doi.org/10.1002/app.

Azimi, Sara, Allahyar Golabchi, Abdolhossein Nekookar, Shahram Rabbani, Morteza H. Amiri, Kamal Asadi, and Mohammad Mahdi Abolhasani. 2021. "Self-Powered Cardiac Pacemaker by Piezoelectric Polymer Nanogenerator Implant." *Nano Energy* 83 (May): 105781. https://doi.org/10.1016/j.nanoen.2021.105781.

Bai, Suo, Lu Zhang, Qi Xu, Youbin Zheng, Yong Qin, and Zhong L. Wang. 2013. "Two Dimensional Woven Nanogenerator." *Nano Energy* 2 (5): 749–53. https://doi.org/10.1016/j.nanoen.2013.01.001.

Bairagi, Satyaranjan, and S. Wazed Ali. 2019. "Nanorod-Based Nanocomposite Fi Bre." *European Polymer Journal* 116 (April): 554–61. https://doi.org/10.1016/j.eurpolymj.2019.04.043.

Bandodkar, Amay J. 2017. "Review—Wearable Biofuel Cells: Past, Present and Future." *Journal of the Electrochemical Society* 164 (3): H3007–14. https://doi.org/10.1149/2.0031703jes.

Bandodkar, Amay J., Philipp Gutruf, Jungil Choi, Kun H. Lee, Yurina Sekine, Jonathan T. Reeder, William J. Jeang, et al. 2019. "Battery-Free, Skin-Interfaced Microfluidic/Electronic Systems for Simultaneous Electrochemical, Colorimetric, and Volumetric Analysis of Sweat." *Science Advances* 5 (1). https://doi.org/10.1126/sciadv.aav3294.

Bandodkar, Amay J., Itthipon Jeerapan, Jung M. You, Rogelio Nuñez-Flores, and Joseph Wang. 2016. "Highly Stretchable Fully-Printed CNT-Based Electrochemical Sensors and Biofuel Cells: Combining Intrinsic and Design-Induced Stretchability." *Nano Letters* 16 (1): 721–27. https://doi.org/10.1021/acs.nanolett.5b04549.

Bandodkar, Amay J., and Joseph Wang. 2016. "Wearable Biofuel Cells: A Review." *Electroanalysis* 28 (6): 1188–200. https://doi.org/10.1002/elan.201600019.

Bandodkar, Amay J., Jung M. You, Nam H. Kim, Yue Gu, Rajan Kumar, A. M. Vinu Mohan, Jonas Kurniawan, et al. 2017. "Soft, Stretchable, High Power Density Electronic Skin-Based Biofuel Cells for Scavenging Energy from Human Sweat." *Energy and Environmental Science* 10 (7): 1581–89. https://doi.org/10.1039/c7ee00865a.

Baniasadi, Mahmoud, Jiacheng Huang, Zhe Xu, Salvador Moreno, Xi Yang, Jason Chang, Manuel A. Quevedo-Lopez, Mohammad Naraghi, and Majid Minary-Jolandan. 2015. "High-Performance Coils and Yarns of Polymeric Piezoelectric Nanofibers." *ACS Applied Materials and Interfaces* 7 (9): 5358–66. https://doi.org/10.1021/am508812a.

Bauer, Siegfried., and F. Bauer. 2008. "Piezoelectric Polymers and Their Applications." *Springer Series in Materials Science* 114: 157–77. https://doi.org/10.1007/978-3-540-68683-5_6.

Baxter, Jason B. 2012. "Commercialization of Dye Sensitized Solar Cells: Present Status and Future Research Needs to Improve Efficiency, Stability, and Manufacturing." *Journal of Vacuum Science & Technology A: Vacuum, Surfaces, and Films* 30 (2): 020801. https://doi.org/10.1116/1.3676433.

Beeby, Stephen P., Russel N. Torah, M. John Tudor, Peter Glynne-Jones, Terence O'Donnell, Chitta R. Saha, and Saibal Roy. 2007. "A Micro Electromagnetic Generator for Vibration Energy Harvesting." *Journal of Micromechanics and Microengineering* 17 (7): 1257–65. https://doi.org/10.1088/0960-1317/17/7/007.

Bhang, Suk H., Woo S. Jang, Jin Han, Jeong-Kee Yoon, Wan-Geun La, Eungkyu Lee, Youn S. Kim, et al. 2017. "Zinc Oxide Nanorod-Based Piezoelectric Dermal Patch for Wound Healing." *Advanced Functional Materials* 27 (1): 1603497. https://doi.org/10.1002/adfm.201603497.

Bhatnagar, Vikrant, and Philip Owende. 2015. "Energy Harvesting for Assistive and Mobile Applications." *Energy Science and Engineering* 3 (3): 153–73. https://doi.org/10.1002/ese3.63.

Bollella, Paolo, Inhee Lee, David Blaauw, and Evgeny Katz. 2020. "A Microelectronic Sensor Device Powered by a Small Implantable Biofuel Cell." *ChemPhysChem* 21 (1): 120–28. https://doi.org/10.1002/cphc.201900700.

Bolt, Robin, Michele Magno, Thomas Burger, Aldo Romani, and Luca Benini. 2017. "Kinetic AC/DC Converter for Electromagnetic Energy Harvesting in Autonomous Wearable Devices." *IEEE Transactions on Circuits and Systems II: Express Briefs* 64 (12): 1422–26. https://doi.org/10.1109/TCSII.2017.2768391.

Bowen, Chris R., John Taylor, Emmanuel Leboulbar, Daniel Zabek, Aditya Chauhan, and Rahul Vaish. 2014. "Pyroelectric Materials and Devices for Energy Harvesting Applications." *Energy and Environmental Science* 7 (12): 3836–56. https://doi.org/10.1039/c4ee01759e.

Cao, Ran, Xianjie Pu, Xinyu Du, Wei Yang, Jiaona Wang, Hengyu Guo, Shuyu Zhao, et al. 2018. "Screen-Printed Washable Electronic Textiles as Self-Powered Touch/Gesture Tribo-Sensors for Intelligent Human–Machine Interaction." *ACS Nano* 12: 5190–6 https://doi.org/10.1021/acsnano.8b02477.

Cao, Zhuo, Michael J. Tudor, Russel N. Torah, and Steve P. Beeby. 2016. "Screen Printable Flexible BiTe-SbTe-Based Composite Thermoelectric Materials on Textiles for Wearable Applications." *IEEE Transactions on Electron Devices* 63 (10): 4024–30. https://doi.org/10.1109/TED.2016.2603071.

Chai, Zhisheng, Nannan Zhang, Peng Sun, Yi Huang, Chuanxi Zhao, Hong J. Fan, Xing Fan, and Wenjie Mai. 2016. "Tailorable and Wearable Textile Devices for Solar Energy Harvesting and Simultaneous Storage." *ACS Nano* 10 (10): 9201–7. https://doi.org/10.1021/acsnano.6b05293.

Champier, Daniel. 2017. "Thermoelectric Generators: A Review of Applications." *Energy Conversion and Management.* https://doi.org/10.1016/j.enconman.2017.02.070.

Chen, Baodong, Wei Tang, Tao Jiang, Laipan Zhu, Xiangyu Chen, and Chuan He. 2018. "Three-Dimensional Ultra Flexible Triboelectric Nanogenerator Made by 3D Printing." *Nano Energy* 45 (December): 380–89. https://doi.org/10.1016/j.nanoen.2017.12.049.

Chen, Guorui, Yongzhong Li, Michael Bick, and Jun Chen. 2020. "Smart Textiles for Electricity Generation." *Chemical Reviews.* https://doi.org/10.1021/acs.chemrev.9b00821.

Chen, Jun, Yi Huang, Nannan Zhang, Haiyang Zou, Ruiyuan Liu, Changyuan Tao, Xing Fan, and Zhong L. Wang. 2016. "Micro-Cable Structured Textile for Simultaneously Harvesting Solar and Mechanical Energy." *Nature Energy* 1 (10): 1–8. https://doi.org/10.1038/nenergy.2016.138.

Chen, Ru-Jun, Yi-Bo Zhang, Ting Liu, Bing-Qing Xu, Yuan-Hua Lin, Ce-Wen Nan, and Yang Shen. 2017. "Addressing the Interface Issues in All-Solid-State Bulk-Type Lithium Ion Battery via an All-Composite Approach." *ACS Applied Materials & Interfaces* 9 (11): 9654–61. https://doi.org/10.1021/acsami.6b16304.

Chen, Xiaoliang, Kaushik Parida, Jiangxin Wang, Jiaqing Xiong, Meng F. Lin, Jinyou Shao, and Pooi S. Lee. 2017. "A Stretchable and Transparent Nanocomposite Nanogenerator for Self-Powered Physiological Monitoring." *ACS Applied Materials and Interfaces* 9 (48): 42200–209. https://doi.org/10.1021/acsami.7b13767.

Chen, Xuexian, Yu Song, Zongming Su, Haotian Chen, Xiaoliang Cheng, Jinxin Zhang, Mengdi Han, and Haixia Zhang. 2017. "Flexible Fiber-Based Hybrid Nanogenerator for Biomechanical Energy Harvesting and Physiological Monitoring." *Nano Energy* 38 (August): 43–50. https://doi.org/10.1016/j.nanoen.2017.05.047.

Chen, Yan, Yang Zhang, Feifei Yuan, Fei Ding, and Oliver G. Schmidt. 2017. "A Flexible PMN-PT Ribbon-Based Piezoelectric-Pyroelectric Hybrid Generator for Human-Activity Energy Harvesting and Monitoring." *Advanced Electronic Materials* 3 (3): 1600540. https://doi.org/10.1002/aelm.201600540.

Cheng, Xiaoliang, Xiang Xue, Ye Ma, Mengdi Han, Wei Zhang, Zhiyun Xu, Hao Zhang, and Haixia Zhang. 2016. "Implantable and Self-Powered Blood Pressure Monitoring Based on a Piezoelectric Thinfilm: Simulated, in Vitro and in Vivo Studies." *Nano Energy* 22 (April): 453–60. https://doi.org/10.1016/j.nanoen.2016.02.037.

Cheng, Yin, Xin Lu, Kwok H. Chan, Ranran Wang, Zherui Cao, Jing Sun, and Ghim W. Ho. 2017. "A Stretchable Fiber Nanogenerator for Versatile Mechanical Energy Harvesting and Self-Powered Full-Range Personal Healthcare Monitoring." *Nano Energy* 41: 511–18. https://doi.org/10.1016/j.nanoen.2017.10.010.

Cheng, Yin, Xin Lu, Kwok Hoe, Ranran Wang, Zherui Cao, Jing Sun, and Ghim Wei. 2017. "A Stretchable Fiber Nanogenerator for Versatile Mechanical Energy Harvesting and Self-Powered Full-Range Personal Healthcare Monitoring." *Nano Energy* 41 (July): 511–18. https://doi.org/10.1016/j.nanoen.2017.10.010.

Cho, Seok H., Jaegab Lee, Mi J. Lee, Hyo J. Kim, Sung M. Lee, and Kyung C. Choi. 2019. "Plasmonically Engineered Textile Polymer Solar Cells for High-Performance, Wearable Photovoltaics." *ACS Applied Materials and Interfaces* 11 (23): 20864–72. https://doi.org/10.1021/acsami.9b05048.

Chou, Xiujian, Jie Zhu, Shuo Qian, Xushi Niu, Jichao Qian, Xiaojuan Hou, Jiliang Mu, et al. 2018. "All-in-One Filler-Elastomer-Based High-Performance Stretchable Piezoelectric Nanogenerator for Kinetic Energy Harvesting and Self-Powered Motion Monitoring." *Nano Energy* 53 (June): 550–58. https://doi.org/10.1016/j.nanoen.2018.09.006.

Chu, Yao, Junwen Zhong, Huiliang Liu, Yuan Ma, Nathaniel Liu, Yu Song, Jiaming Liang, et al. 2018. "Human Pulse Diagnosis for Medical Assessments Using a Wearable Piezoelectret Sensing System." *Advanced Functional Materials* 28 (40): 1803413. https://doi.org/10.1002/adfm.201803413.

Chun, Kyoung-Yong, Young J. Son, Eun-Seok Jeon, Sehan Lee, and Chang-Soo Han. 2018. "A Self-Powered Sensor Mimicking Slow- and Fast-Adapting Cutaneous Mechanoreceptors." *Advanced Materials* 30 (12): 1706299. https://doi.org/10.1002/adma.201706299.

Cosnier, Serge, Alan L. Goff, and Michael Holzinger. 2014. "Towards Glucose Biofuel Cells Implanted in Human Body for Powering Artificial Organs: Review." *Electrochemistry Communications*. https://doi.org/10.1016/j.elecom.2013.09.021.

Covaci, Corina, and Aurel Gontean. 2020. "Piezoelectric Energy Harvesting Solutions: A Review." *Sensors (Switzerland)* 20 (12): 1–37. https://doi.org/10.3390/s20123512.

Cui, Nuanyang, Jinmei Liu, Long Gu, Suo Bai, Xiaobo Chen, and Yong Qin. 2015. "Wearable Triboelectric Generator for Powering the Portable Electronic Devices." *ACS Applied Materials & Interfaces* 7: 18225–30. https://doi.org/10.1021/am5071688.

Curry, Eli J., Kai Ke, Meysam T. Chorsi, Kinga S. Wrobel, Albert N. Miller, Avi Patel, Insoo Kim, et al. 2018. "Biodegradable Piezoelectric Force Sensor." *Proceedings of the National Academy of Sciences of the United States of America* 115 (5): 909–14. https://doi.org/10.1073/pnas.1710874115.

Dagdeviren, Canan, Zhou Li, and Zhong L. Wang. 2017. "Energy Harvesting from the Animal-Human Body for Self-Powered Electronics." *Annual Review of Biomedical Engineering* 19 (June): 85–108. https://doi.org/10.1146/annurev-bioeng-071516-044517.

Dagdeviren, Canan, Byung D. Yang, Yewang Su, Phat L. Tran, Pauline Joe, Eric Anderson, Jing Xia, et al. 2014. "Conformal Piezoelectric Energy Harvesting and Storage from Motions of the Heart, Lung, and Diaphragm." *Proceedings of the National Academy of Sciences of the United States of America* 111 (5): 1927–32. https://doi.org/10.1073/pnas.1317233111.

Deng, Fang, Huangbin Qiu, Jie Chen, Lu Wang, and Bo Wang. 2017. "Wearable Thermoelectric Power Generators Combined with Flexible Supercapacitor for Low-Power Human Diagnosis Devices." *IEEE Transactions on Industrial Electronics* 64 (2): 1477–85. https://doi.org/10.1109/TIE.2016.2613063.

Giacomo, Francesco D., Azhar Fakharuddin, Rajan Jose, and Thomas M. Brown. 2016. "Progress, Challenges and Perspectives in Flexible Perovskite Solar Cells." *Energy and Environmental Science*. https://doi.org/10.1039/c6ee01137c.

Dong, Kai, Jianan Deng, Wenbo Ding, Aurelia C. Wang, Peihong Wang, Chaoyu Cheng, Yi-cheng Wang, et al. 2018. "Versatile Core – Sheath Yarn for Sustainable Biomechanical Energy Harvesting and Real-Time Human-Interactive Sensing." *Advanced Energy Materials* 8: 1801114. https://doi.org/10.1002/aenm.201801114.

Dong, Kai, Jianan Deng, Yunlong Zi, Yi C. Wang, Cheng Xu, Haiyang Zou, Wenbo Ding, et al. 2017. "3D Orthogonal Woven Triboelectric Nanogenerator for Effective Biomechanical Energy Harvesting and as Self-Powered Active Motion Sensors." *Advanced Materials* 29 (38): 1–11. https://doi.org/10.1002/adma.201702648.

Dong, Kai, Xiao Peng, and Zhong L. Wang. 2020a. "Fiber/Fabric-Based Piezoelectric and Triboelectric Nanogenerators for Flexible/Stretchable and Wearable Electronics and Artificial Intelligence." *Advanced Materials* 32 (5): 1–43. https://doi.org/10.1002/adma.201902549.

Dong, Kai, Xiao Peng, and Zhong L. Wang. 2020b. "Fiber/Fabric-Based Piezoelectric and Triboelectric Nanogenerators for Flexible/Stretchable and Wearable Electronics and Artificial Intelligence." *Advanced Materials* 32 (5): 1–43. https://doi.org/10.1002/adma.201902549.

Dong, Kai, Yi C. Wang, Jianan Deng, Yejing Dai, Steven L. Zhang, Haiyang Zou, Bohong Gu, Baozhong Sun, and Zhong L. Wang. 2017. "A Highly Stretchable and Washable All-Yarn-Based Self-Charging Knitting Power Textile Composed of Fiber Triboelectric Nanogenerators and Supercapacitors." *ACS Nano* 11 (9): 9490–99. https://doi.org/10.1021/acsnano.7b05317.

Du, Shuo, Nuoya Zhou, Yujie Gao, Ge Xie, Hongyao Du, Hao Jiang, Lianbin Zhang, Juan Tao, and Jintao Zhu. 2020. "Bioinspired Hybrid Patches with Self-Adhesive Hydrogel and Piezoelectric Nanogenerator for Promoting Skin Wound Healing." *Nano Research* 13 (9): 2525–33. https://doi.org/10.1007/s12274-020-2891-9.

Du, Yueyu, K. F. Cai, Shirley Z. Shen, R. Donelsonand, J. Y. Xu, H. X. Wang, and Tong Lin. 2017. "Multifold Enhancement of the Output Power of Flexible Thermoelectric Generators Made from Cotton Fabrics Coated with Conducting Polymer." *RSC Advances* 7 (69): 43737–42. https://doi.org/10.1039/c7ra08663f.

Du, Yong, Kefeng Cai, Song Chen, Hongxia Wang, Shirley Z. Shen, Richard Donelson, and Tong Lin. 2015. "Thermoelectric Fabrics: Toward Power Generating Clothing." *Scientific Reports* 5 (1): 1–6. https://doi.org/10.1038/srep06411.

Elahi, Hassan, Khushboo Munir, Marco Eugeni, Sofiane Atek, and Paolo Gaudenzi. 2020. "Energy Harvesting towards Self-Powered Iot Devices." *Energies* 13 (21): 1–31. https://doi.org/10.3390/en13215528.

Escalona-Villalpando, Ricardo Antonio, E. Ortiz-Ortega, J. P. Bocanegra-Ugalde, Shelley D. Minteer, J. Ledesma-García, and L. G. Arriaga. 2019. "Clean Energy from Human Sweat Using an Enzymatic Patch." *Journal of Power Sources* 412 (February): 496–504. https://doi.org/10.1016/j.jpowsour.2018.11.076.

Falk, Magnus, Viktor Andoralov, Zoltan Blum, Javier Sotres, Dmitry B. Suyatin, Tautgirdas Ruzgas, Thomas Arnebrant, and Sergey Shleev. 2012. "Biofuel Cell as a Power Source for Electronic Contact Lenses." *Biosensors and Bioelectronics* 37 (1): 38–45. https://doi.org/10.1016/j.bios.2012.04.030.

Fan, Feng R., Zhong Q. Tian, and Zhong L. Wang. 2012. "Flexible Triboelectric Generator." *Nano Energy* 1 (2): 328–34. https://doi.org/10.1016/j.nanoen.2012.01.004.

Fatet, Jérôme. 2005. "Edmond Becquerel's Electrochemical Actinometer." *Archives Des Sciences* 58 (January): 147–56.

Fu, Xuemei, Limin Xu, Jiaxin Li, Xuemei Sun, and Huisheng Peng. 2018. "Flexible Solar Cells Based on Carbon Nanomaterials." *Carbon* 139: 1063–73. https://doi.org/10.1016/j.carbon.2018.08.017.

Gagge, A. Pharo, and Richard R. Gonzalez. 2011. "Mechanisms of Heat Exchange: Biophysics and Physiology." In *Comprehensive Physiology*, 45–84. Hoboken, NJ: John Wiley & Sons, Inc. https://doi.org/10.1002/cphy.cp040104.

Gao, Huipu, Pham T. Minh, Hong Wang, and Sergiy Minko. 2018. "High-Performance Flexible Yarn for Wearable Piezoelectric Nanogenerators." *Smart Materials and Structures,* 27: 095018. https://doi.org/10.1088/1361-665X/AAD718.

Gao, Huipu, Darya Asheghali, Nataraja S. Yadavalli, Minh T. Pham, Tho D. Nguyen, Sergiy Minko, and Suraj Sharma. 2019. "Fabrication of Core-Sheath Nanoyarn via Touchspinning and Its Application in Wearable Piezoelectric Nanogenerator." *Journal of the Textile Institute* 119: 1–10. https://doi.org/10.1080/00405000.2019.1678558.

Gao, Zhen, Peng Liu, Xuemei Fu, Limin Xu, Yong Zuo, Bo Zhang, Xuemei Sun, and Huisheng Peng. 2019. "Flexible Self-Powered Textile Formed by Bridging Photoactive and Electrochemically Active Fiber Electrodes." *Journal of Materials Chemistry A* 7 (24): 1447–54. https://doi.org/10.1039/c9ta04178h.

Gljušćić, Petar, Saša Zelenika, David Blažević, and Ervin Kamenar. 2019. "Kinetic Energy Harvesting for Wearable Medical Sensors." *Sensors (Switzerland)* 19 (22). https://doi.org/10.3390/s19224922.

Gogurla, Narendar, Biswajit Roy, Ji Y. Park, and Sunghwan Kim. 2019. "Skin-Contact Actuated Single-Electrode Protein Triboelectric Nanogenerator and Strain Sensor for Biomechanical Energy Harvesting and Motion Sensing." *Nano Energy* 62 (May): 674–81. https://doi.org/10.1016/j.nanoen.2019.05.082.

Gong, Jianliang, Bingang Xu, Xiaoyang Guan, Yuejiao Chen, Shengyan Li, and Jie Feng. 2019. "Towards Truly Wearable Energy Harvesters with Full Structural Integrity of Fiber Materials." *Nano Energy* 58 (November): 365–74. https://doi.org/10.1016/j.nanoen.2019.01.056.

Gong, Wei, Chengyi Hou, Yinben Guo, Jie Zhou, Jiuke Mu, Yaogang Li, Qinghong Zhang, and Hongzhi Wang. 2017a. "A Wearable, Fibroid, Self-Powered Active Kinematic Sensor Based on Stretchable Sheath-Core Structural Triboelectric Fiber." *Nano Energy* 39 (July): 673–83. https://doi.org/10.1016/j.nanoen.2017.08.003.

Gong, Wei, Chengyi Hou, Yinben Guo. 2017. "A Wearable, Fibroid, Self-Powered Active Kinematic Sensor Based on Stretchable Sheath-Core Structural Triboelectric Fibers." *Nano Energy* 39: 673–83. https://doi.org/10.1016/j.nanoen.2017.08.003.

Gong, Wei, Chengyi Hou, Jie Zhou, Yinben Guo, Wei Zhang, Yaogang Li, Qinghong Zhang, and Hongzhi Wang. 2019. "Continuous and Scalable Manufacture of Amphibious Energy Yarns and Textiles." *Nature Communications* 10 (1): 1–8. https://doi.org/10.1038/s41467-019-08846-2.

Gould, Christopher A., Noel Y. A. Shammas, Steve Grainger, and Ian Taylor. 2008. "A Comprehensive Review of Thermoelectric Technology, Micro-Electrical and Power Generation Properties." In *2008 26th International Conference on Microelectronics, Proceedings, MIEL* 2008, 329–32. https://doi.org/10.1109/ICMEL.2008.4559288.

Gu, Yuhan, Weiqun Liu, Caiyou Zhao, and Ping Wang. 2020. "A Goblet-like Non-Linear Electromagnetic Generator for Planar Multi-Directional Vibration Energy Harvesting." *Applied Energy* 266 (May): 114846. https://doi.org/10.1016/j.apenergy.2020.114846.

Guo, Jiajun, Min Nie, and Qi Wang. 2020. "A Piezoelectric Poly(Vinylidene Fluoride) Tube Featuring Highly-Sensitive and Isotropic Piezoelectric Output for Compression." *RSC Advances* 11 (2): 1182–86. https://doi.org/10.1039/d0ra09131f.

Guo, Wenzhe, Cenxiao Tan, Kunming Shi, Junwen Li, Xiao X. Wang, Bin Sun, Xingyi Huang, Yun Z. Long, and Pingkai Jiang. 2018. "Wireless Piezoelectric Devices Based on Electrospun PVDF/BaTiO3 NW Nanocomposite Fibers for Human Motion Monitoring." *Nanoscale* 10 (37): 17751–60. https://doi.org/10.1039/c8nr05292a.

Guo, Yinben, Xiao S. Zhang, Ya Wang, Wi Gong, Qinghong Zhang, Hongzhi Wang, and Juergen Brugger. 2018. "All-Fiber Hybrid Piezoelectric-Enhanced Triboelectric Nanogenerator for Wearable Gesture Monitoring." *Nano Energy* 48 (February): 152–60. https://doi.org/10.1016/j.nanoen.2018.03.033.

Gupta, Mukti N., Suman, and Surendra Kumar Yadav. 2004. "Electricity Generation Due to Vibration of Moving Vehicles Using Piezoelectric Effect." *Advance in Electronic and Electric Engineering* 4 (3): 313–18. https://doi.org/10.1088/0964-1726/13/5/018.

Halim, Miah A., Hyunok Cho, Md Salauddin, and Jae Y. Park. 2016. "A Miniaturized Electromagnetic Vibration Energy Harvester Using Flux-Guided Magnet Stacks for Human-Body-Induced Motion." *Sensors and Actuators, A: Physical* 249 (October): 23–31. https://doi.org/10.1016/j.sna.2016.08.008.

Elsheikh, Mohamed H., Dhafer A. Shnawah, Mohd Faizul M. Sabri, Suhana B. M. Said, Masjuki H. Hassan, Mohamed Bashir Ali Bashir, and Mahazani Mohamad. 2014. "A Review on Thermoelectric Renewable Energy: Principle Parameters That Affect Their Performance." *Renewable and Sustainable Energy Reviews. Pergamon*. https://doi.org/10.1016/j.rser.2013.10.027.

Han, Jin K., Min A. Kang, Chong Y. Park, Minbaek Lee, Sung Myung, Wooseok Song, Sun S. Lee, Jongsun Lim, and Ki S. An. 2019. "Attachable Piezoelectric Nanogenerators Using Collision-Induced Strain of Vertically Grown Hollow MoS_2 Nanoflakes." *Nanotechnology* 30 (33). https://doi.org/10.1088/1361-6528/ab1d06.

Han, Wuxiao, Haoxuan He, Linlin Zhang, Chuanyi Dong, Hui Zeng, Yitong Dai, Lili Xing, Yan Zhang, and Xinyu Xue. 2017. "A Self-Powered Wearable Noninvasive Electronic-Skin for Perspiration Analysis Based on Piezo-Biosensing Unit Matrix of Enzyme/ZnO Nanoarrays." *ACS Applied Materials and Interfaces* 9 (35): 29526–37. https://doi.org/10.1021/acsami.7b07990.

Hanani, Zouhair, Ilyasse Izanzar, M'barek Amjoud, Daoud Mezzane, Mohammed Lahcini, Hana Uršič, Uroš Prah, et al. 2021. "Lead-Free Nanocomposite Piezoelectric Nanogenerator Film for Biomechanical Energy Harvesting." *Nano Energy* 81 (March): 105661. https://doi.org/10.1016/j.nanoen.2020.105661.

Hansen, Benjamin J., Ying Liu, Rusen Yang, and Zhong L. Wang. 2010. "Hybrid Nanogenerator for Concurrently Harvesting Biomechanical and Biochemical Energy." *ACS Nano* 4 (7): 3647–52. https://doi.org/10.1021/nn100845b.

Hashemi, Seyyed A., Seeram Ramakrishna, and Armin G. Aberle. 2020. "Recent Progress in Flexible-Wearable Solar Cells for Self-Powered Electronic Devices." *Energy and Environmental Science* 13 (3): 685–743. https://doi.org/10.1039/c9ee03046h.

He, Minghui, Yu J. Lin, Che M. Chiu, Weifeng Yang, Binbin Zhang, Daqin Yun, Yannan Xie, and Zong H. Lin. 2018. "A Flexible Photo-Thermoelectric Nanogenerator Based on MoS2/PU Photothermal Layer for Infrared Light Harvesting." *Nano Energy* 49 (July): 588–95. https://doi.org/10.1016/j.nanoen.2018.04.072.

He, Wei, Gan Zhang, Xingxing Zhang, Jie Ji, Guiqiang Li, and Xudong Zhao. 2015. "Recent Development and Application of Thermoelectric Generator and Cooler." *Applied Energy* 143: 1–25. https://doi.org/10.1016/j.apenergy.2014.12.075.

He, Wen, Yongteng Qian, Byeok Song Lee, Fangfang Zhang, Aamir Rasheed, Jae E. Jung, and Dae J. Kang. 2018. "Ultrahigh Output Piezoelectric and Triboelectric Hybrid Nanogenerators Based on ZnO Nanoflakes/Polydimethylsiloxane Composite Films." *ACS Applied Materials and Interfaces* 10 (51): 44415–20. https://doi.org/10.1021/acsami.8b15410.

He, Xu, Yunlong Zi, Hengyu Guo, Haiwu Zheng, Yi Xi, Changsheng Wu, Jie Wang, Wei Zhang, Canhui Lu, and Zhong L. Wang. 2017. "A Highly Stretchable Fiber-Based Triboelectric Nanogenerator for Self-Powered Wearable Electronics." *Advanced Functional Materials* 27: 1604378. https://doi.org/10.1002/adfm.201604378.

He, Xu, Haiyang Zou, Zhishuai Geng, Xingfu Wang, Wenbo Ding, Fei Hu, Yunlong Zi, et al. 2018. "A Hierarchically Nanostructured Cellulose Fiber-Based Triboelectric Nanogenerator for Self-Powered Healthcare Products." *Advanced Functional Materials* 28 (45): 1–8. https://doi.org/10.1002/adfm.201805540.

Herzog, Antonia V., Timothy E. Lipman, and Daniel M. Kammen. 2001. "Renewable Energy Sources: A Variable Choice." *Environment: Science and Policy for Sustainable Development* 43 (10): 8–20. https://doi.org/10.1080/00139150109605150.

Hsieh, Hsing H., Fang C. Hsu, and Yang F. Chen. 2018. "Energetically Autonomous, Wearable, and Multifunctional Sensor." *ACS Sensors* 3 (1): 113–20. https://doi.org/10.1021/acssensors.7b00690.

Hu, Xiaotian, Lie Chen, Yong Zhang, Qiao Hu, Junliang Yang, and Yiwang Chen. 2014. "Large-Scale Flexible and Highly Conductive Carbon Transparent Electrodes via Roll-to-Roll Process and Its High Performance Lab-Scale Indium Tin Oxide-Free Polymer Solar Cells." *Chemistry of Materials* 26 (21): 6293–302. https://doi.org/10.1021/cm5033942.

Hu, Youfan, and Zhong L. Wang. 2014. "Recent Progress in Piezoelectric Nanogenerators as a Sustainable Power Source in Self-Powered Systems and Active Sensors." *Nano Energy* 14: 3–14. https://doi.org/10.1016/j.nanoen.2014.11.038.

Huang, Tao, Jing Zhang, Bin Yu, Hao Yu, Hairu Long, Hongzhi Wang, Qinghua Zhang, and Meifang Zhu. 2019. "Fabric Texture Design for Boosting the Performance of a Knitted Washable Textile Triboelectric Nanogenerator as Wearable Power." *Nano Energy* 58 (January): 375–83. https://doi.org/10.1016/j.nanoen.2019.01.038.

Hussain, Aftab M., Farhan A. Ghaffar, Sung I. Park, John A. Rogers, Atif Shamim, and Muhammad M. Hussain. 2015. "Metal/Polymer Based Stretchable Antenna for Constant Frequency Far-Field Communication in Wearable Electronics." *Advanced Functional Materials* 25 (42): 6565–75. https://doi.org/10.1002/adfm.201503277.

Huu, Trung N., Toan N. Van, and Ono Takahito. 2018. "Flexible Thermoelectric Power Generator with Y-Type Structure Using Electrochemical Deposition Process." *Applied Energy* 210 (January): 467–76. https://doi.org/10.1016/j.apenergy.2017.05.005.

Hyland, Melissa, Haywood Hunter, Jie Liu, Elena Veety, and Daryoosh Vashaee. 2016. "Wearable Thermoelectric Generators for Human Body Heat Harvesting." *Applied Energy* 182 (November): 518–24. https://doi.org/10.1016/j.apenergy.2016.08.150.

Indira, Sridhar S., Chockalingam A. Vaithilingam, Kameswara S. P. Oruganti, Faizal Mohd, and Saidur Rahman. 2019. "Nanogenerators as a Sustainable Power Source: State of Art, Applications, and Challenges." *Nanomaterials*. https://doi.org/10.3390/nano9050773.

Ito, Mitsuhiro, Takuya Koizumi, Hirotaka Kojima, Takeshi Saito, and Masakazu Nakamura. 2017. "From Materials to Device Design of a Thermoelectric Fabric for Wearable Energy Harvesters." *Journal of Materials Chemistry A* 5 (24): 12068–72. https://doi.org/10.1039/c7ta00304h.

Jeerapan, Itthipon, Juliane R. Sempionatto, Adriana Pavinatto, Jung M. You, and Joseph Wang. 2016. "Stretchable Biofuel Cells as Wearable Textile-Based Self-Powered Sensors." *Journal of Materials Chemistry A* 4 (47): 18342–53. https://doi.org/10.1039/C6TA08358G.

Jeerapan, Itthipon, Juliane R. Sempionatto, and Joseph Wang. 2020. "On-Body Bioelectronics: Wearable Biofuel Cells for Bioenergy Harvesting and Self-Powered Biosensing." *Advanced Functional Materials* 30 (29): 1906243. https://doi.org/10.1002/adfm.201906243.

Jeon, Nam J., Hyejin Na, Eui H. Jung, Tae Y. Yang, Yong G. Lee, Geunjin Kim, Hee W. Shin, Sang I. Seok, Jaemin Lee, and Jangwon Seo. 2018. "A Fluorene-Terminated Hole-Transporting Material for Highly Efficient and Stable Perovskite Solar Cells." *Nature Energy* 3 (8): 682–89. https://doi.org/10.1038/s41560-018-0200-6.

Jeong, Eun G., Yongmin Jeon, Seok H. Cho, and Kyung C. Choi. 2019. "Textile-Based Washable Polymer Solar Cells for Optoelectronic Modules: Toward Self-Powered Smart Clothing." *Energy and Environmental Science* 12 (6): 1878–89. https://doi.org/10.1039/c8ee03271h.

Jia, Wenzhao, Gabriela Valdés-Ramírez, Amay J. Bandodkar, Joshua R. Windmiller, and Joseph Wang. 2013. "Epidermal Biofuel Cells: Energy Harvesting from Human Perspiration." *Angewandte Chemie International Edition* 52 (28): 7233–36. https://doi.org/10.1002/anie.201302922.

Jia, Wenzhao, Xuan Wang, Somayeh Imani, Amay J. Bandodkar, Julian Ramírez, Patrick P. Mercier, and Joseph Wang. 2014. "Wearable Textile Biofuel Cells for Powering Electronics." *Journal of Materials Chemistry A* 2 (43): 18184–89. https://doi.org/10.1039/c4ta04796f.

Jiang, Hong, Jiangshan Feng, Huan Zhao, Guijun Li, Guannan Yin, Yu Han, Feng Yan, Zhike Liu, and Shengzhong F. Liu. 2018. "Low Temperature Fabrication for High Performance Flexible $CsPbI_2Br$ Perovskite Solar Cells." *Advanced Science* 5 (11): 1801117. https://doi.org/10.1002/advs.201801117.

Jiang, Weitao, Tingting Zhao, Hongzhong Liu, Rui Jia, Dong Niu, Bangdao Chen, Yongsheng Shi, Lei Yin, and Bingheng Lu. 2018. "Laminated Pyroelectric Generator with Spin Coated Transparent Poly(3,4-Ethylenedioxythiophene) Polystyrene Sulfonate (PEDOT:PSS) Electrodes for a Flexible Self-Powered Stimulator." *RSC Advances* 8 (27): 15134–40. https://doi.org/10.1039/c8ra00491a.

Jiang, Wen, Hu Li, Zhuo Liu, Zhe Li, Jingjing Tian, Bojing Shi, Yang Zou, et al. 2018. "Fully Bioabsorbable Natural-Materials-Based Triboelectric Nanogenerators." *Advanced Materials* 30 (32): 1801895. https://doi.org/10.1002/adma.201801895.

Jin, Long, Songyuan Ma, Weili Deng, Cheng Yan, Tao Yang, Xiang Chu, Guo Tian, Da Xiong, Jun Lu, and Weiqing Yang. 2018. "Polarization-Free High-Crystallization β-PVDF Piezoelectric Nanogenerator toward Self-Powered 3D Acceleration Sensor." *Nano Energy* 50 (May): 632–38. https://doi.org/10.1016/j.nanoen.2018.05.068.

Jinno, Hiroaki, Kenjiro Fukuda, Xiaomin Xu, Sungjun Park, Yasuhito Suzuki, Mari Koizumi, Tomoyuki Yokota, Itaru Osaka, Kazuo Takimiya, and Takao Someya. 2017. "Stretchable and Waterproof Elastomer-Coated Organic Photovoltaics for Washable Electronic Textile Applications." *Nature Energy* 2 (10): 780–85. https://doi.org/10.1038/s41560-017-0001-3.

Jung, Sungmook, Jongsu Lee, Taeghwan Hyeon, Minbaek Lee, and Dae H. Kim. 2014. "Fabric-Based Integrated Energy Devices for Wearable Activity Monitors." *Advanced Materials (Deerfield Beach, Fla.)* 26 (36): 6329–34. https://doi.org/10.1002/adma.201402439.

Jung, Woo S., Min G. Kang, Hi G. Moon, Seung H. Baek, Seok J. Yoon, Zhong L. Wang, Sang W. Kim, and Chong Y. Kang. 2015. "High Output Piezo/Triboelectric Hybrid Generator." *Scientific Reports* 5: 16. https://doi.org/10.1038/srep09309.

Kannan, Nadarajah, and Divagar Vakeesan. 2016. "Solar Energy for Future World – A Review." *Renewable and Sustainable Energy Reviews.* https://doi.org/10.1016/j.rser.2016.05.022.

Karami, M. Amin, and Daniel J. Inman. 2012. "Powering Pacemakers from Heartbeat Vibrations Using Linear and Nonlinear Energy Harvesters." *Applied Physics Letters* 100 (4): 042901. https://doi.org/10.1063/1.3679102.

Khalid, Salman, Izaz Raouf, Asif Khan, Nayeon Kim, and Heung S. Kim. 2019. "A Review of Human-Powered Energy Harvesting for Smart Electronics: Recent Progress and Challenges." *International Journal of Precision Engineering and Manufacturing-Green Technology* 6 (4): 821–51. https://doi.org/10.1007/s40684-019-00144-y.

Kim, Choong S., Hyeong M. Yang, Jinseok Lee, Gyu S. Lee, Hyeongdo Choi, Yong J. Kim, Se H. Lim, Seong H. Cho, and Byung J. Cho. 2018. "Self-Powered Wearable Electrocardiography Using a Wearable Thermoelectric Power Generator." *ACS Energy Letters* 3 (3): 501–7. https://doi.org/10.1021/acsenergylett.7b01237.

Kim, Dong H., Hong J. Shin, Hyunseung Lee, Chang K. Jeong, Hyewon Park, Geon-Tae Hwang, Ho-Yong Lee, et al. 2017. "In Vivo Self-Powered Wireless Transmission Using Biocompatible Flexible Energy Harvesters." *Advanced Functional Materials* 27 (25): 1700341. https://doi.org/10.1002/adfm.201700341.

Kim, Hee S., Weishu Liu, and Zhifeng Ren. 2017. "The Bridge between the Materials and Devices of Thermoelectric Power Generators." *Energy and Environmental Science* 10 (1): 69–85. https://doi.org/10.1039/c6ee02488b.

Kim, Hyunjin, Seong M. Kim, Hyungbin Son, Hyeok Kim, Boongik Park, Jiyeon Ku, Jung I. Sohn, et al. 2012. "Enhancement of Piezoelectricity via Electrostatic Effects on a Textile Platform." *Energy and Environmental Science* 5 (10): 8932–36. https://doi.org/10.1039/c2ee22744d.

Kim, Joo G., Byungrak Son, Santanu Mukherjee, Nicholas Schuppert, Alex Bates, Osung Kwon, Moon J. Choi, Hyun Y. Chung, and Sam Park. 2015. "A Review of Lithium and Non-Lithium Based Solid State Batteries." *Journal of Power Sources* 282: 299–322. https://doi.org/10.1016/j.jpowsour.2015.02.054.

Kim, Kyeong N., Jinsung Chun, Jin. Kim, Keun Y. Lee, Jang-Ung Park, Sang-Woo Kim, Zhong L. Wang, Jeong M. Baik, and Jeong M. Baik. 2015. "Highly Stretchable 2D Fabrics for Wearable Triboelectric Nanogenerator under Harsh Environments." *ACS Nano* 9 (6): 6394–400. https://doi.org/10.1021/acsnano.5b02010.

Kim, Min K., Myoung S. Kim, Seok Lee, Chulki Kim, and Yong J. Kim. 2014. "Wearable Thermoelectric Generator for Harvesting Human Body Heat Energy." *Smart Materials and Structures* 23 (10). https://doi.org/10.1088/0964-1726/23/10/105002.

Kim, Min O., Soonjae Pyo, Yongkeun Oh, Yunsung Kang, Kyung H. Cho, Jungwook Choi, and Jongbaeg Kim. 2018. "Flexible and Multi-Directional Piezoelectric Energy Harvester for Self-Powered Human Motion Sensor." *Smart Materials and Structures* 27 (3): 035001. https://doi.org/10.1088/1361-665X/aaa722.

Kim, Minji, Yuen S. Wu, Edwin C. Kan, and Jintu Fan. 2018. "Breathable and Flexible Piezoelectric ZnO@ PVDF Fibrous Nanogenerator for Wearable Applications." *Polymers* 10 (7). https://doi.org/10.3390/polym10070745.

Kim, Myoung S., Sung E. Jo, Hye R. Ahn, and Yong J. Kim. 2015. "Modeling of a Honeycomb-Shaped Pyroelectric Energy Harvester for Human Body Heat Harvesting." *Smart Materials and Structures* 24 (6): 065032. https://doi.org/10.1088/0964-1726/24/6/065032.

Kim, Myoung S., Min K. Kim, Kyongtae Kim, and Yong J. Kim. 2017. "Design of Wearable Hybrid Generator for Harvesting Heat Energy from Human Body Depending on Physiological Activity." *Smart Materials and Structures* 26 (9): 095046. https://doi.org/10.1088/1361-665X/aa82d5.

Kim, Suk L., Kyungwho Choi, Abdullah Tazebay, and Choongho Yu. 2014. "Flexible Power Fabrics Made of Carbon Nanotubes for Harvesting Thermoelectricity." *ACS Nano* 8 (3): 2377–86. https://doi.org/10.1021/nn405893t.

Kim, Sun J., Ju H. We, and Byung J. Cho. 2014. "A Wearable Thermoelectric Generator Fabricated on a Glass Fabric." *Energy and Environmental Science* 7 (6): 1959–65. https://doi.org/10.1039/c4ee00242c.

Kim, Taesu, Jae H. Kim, Tae E. Kang, Changyeon Lee, Hyunbum Kang, Minkwan Shin, Cheng Wang, et al. 2015. "Flexible, Highly Efficient All-Polymer Solar Cells." *Nature Communications* 6 (May): 17. https://doi.org/10.1038/ncomms9547.

Korkmaz, Satiye, and Afşin Kariper. 2021. "Pyroelectric Nanogenerators (PyNGs) in Converting Thermal Energy into Electrical Energy: Fundamentals and Current Status." *Nano Energy* 84 (January). https://doi.org/10.1016/j.nanoen.2021.105888.

Kraemer, Daniel, Bed Poudel, Hsien P. Feng, J. Christopher Caylor, Bo Yu, Xiao Yan, Yi Ma, et al. 2011. "High-Performance Flat-Panel Solar Thermoelectric Generators with High Thermal Concentration." *Nature Materials* 10 (7): 532–8. https://doi.org/10.1038/nmat3013.

Kumavat, Priyanka P., Prashant Sonar, and Dipak S. Dalal. 2017. "An Overview on Basics of Organic and Dye Sensitized Solar Cells, Their Mechanism and Recent Improvements." *Renewable and Sustainable Energy Reviews* 78 (April): 1262–87. https://doi.org/10.1016/j.rser.2017.05.011.

Kwak, Sung S., Han Kim, Wanchul Seung, Jihye Kim, Ronan Hinchet, and Sang W. Kim. 2017. "Fully Stretchable Textile Triboelectric Nanogenerator with Knitted Fabric Structures." *ACS Nano* 11 (11): 10733–41. https://doi.org/10.1021/acsnano.7b05203.

Kwon, Cheong H., Yongmin Ko, Dongyeeb Shin, Seung W. Lee, and Jinhan Cho. 2019. "Highly Conductive Electrocatalytic Gold Nanoparticle-Assembled Carbon Fiber Electrode for High-Performance Glucose-Based Biofuel Cells." *Journal of Materials Chemistry A* 7 (22): 13495–505. https://doi.org/10.1039/c8ta12342j.

Kwon, Cheong H., Sung H. Lee, Young B. Choi, Jae A. Lee, Shi H. Kim, Hyug H. Kim, Geoffrey M. Spinks, et al. 2014. "High-Power Biofuel Cell Textiles from Woven Biscrolled Carbon Nanotube Yarns." *Nature Communications* 5 (1): 1–7. https://doi.org/10.1038/ncomms4928.

Lai, Ying-chih, Jianan Deng, Steven L. Zhang, Simiao Niu, Hengyu Guo, and Zhong L. Wang. 2017. "Single-Thread-Based Wearable and Highly Stretchable Triboelectric Nanogenerators and Their Applications in Cloth-Based Self-Powered Human-Interactive and Biomedical Sensing." *Advanced Functional Materials* 27: 1604462. https://doi.org/10.1002/adfm.201604462.

Lang, Sidney B. 2005. "Pyroelectricity: From Ancient Curiosity to Modern Imaging Tool." *Physics Today* 58 (8): 31–6. https://doi.org/10.1063/1.2062916.

Lee, Hyewon, and Jung S. Roh. 2019. "Wearable Electromagnetic Energy-Harvesting Textiles Based on Human Walking." *Textile Research Journal* 89 (13): 2532–41. https://doi.org/10.1177/0040517518797349.

Lee, In, and Kyoochun Lee. 2015. "The Internet of Things (IoT): Applications, Investments, and Challenges for Enterprises." *Business Horizons* 58 (4): 431–40. https://doi.org/10.1016/j.bushor.2015.03.008.

Lee, Jae A., Ali E. Aliev, Julia S. Bykova, Mônica J. de Andrade, Daeyoung Kim, Hyeon J. Sim, Xavier Lepró, et al. 2016. "Woven-Yarn Thermoelectric Textiles." *Advanced Materials* 28 (25): 5038–44. https://doi.org/10.1002/adma.201600709.

Leng, Qiang, Lin Chen, Hengyu Guo, Jianlin Liu, Guanlin Liu, Chenguo Hu, and Yi Xi. 2014. "Harvesting Heat Energy from Hot/Cold Water with a Pyroelectric Generator." *Journal of Materials Chemistry A* 2 (30): 11940–7. https://doi.org/10.1039/c4ta01782j.

Leonov, Vladimir, and Ruud J. M. Vullers. 2009. "Wearable Electronics Self-Powered by Using Human Body Heat: The State of the Art and the Perspective." *Journal of Renewable and Sustainable Energy* 1 (6): 062701. https://doi.org/10.1063/1.3255465.

Lewis, Nathan S. 2007. "Toward Cost-Effective Solar Energy Use." *Science* 315 (5813): 798–801. https://doi.org/10.1126/science.1137014.

Li, Changcun, Fengxing Jiang, Congcong Liu, Wenfang Wang, Xuejing Li, Tongzhou Wang, and Jingkun Xu. 2017. "A Simple Thermoelectric Device Based on Inorganic/Organic Composite Thin Film for Energy Harvesting." *Chemical Engineering Journal* 320 (July): 201–10. https://doi.org/10.1016/j.cej.2017.03.023.

Li, Chao, Shan Cong, Zhengnan Tian, Yingze Song, Lianghao Yu, Chen Lu, Yuanlong Shao, et al. 2019. "Flexible Perovskite Solar Cell-Driven Photo-Rechargeable Lithium-Ion Capacitor for Self-Powered Wearable Strain Sensors." *Nano Energy* 60 (March): 247–56. https://doi.org/10.1016/j.nanoen.2019.03.061.

Li, Haitao, Charlynn S. L. Koh, Yih H. Lee, Yihe Zhang, Gia C. Phan-Quang, Chao Zhu, Zheng Liu, et al. 2020. "A Wearable Solar-Thermal-Pyroelectric Harvester: Achieving High Power Output Using Modified RGO-PEI and Polarized PVDF." *Nano Energy* 73 (January): 104723. https://doi.org/10.1016/j.nanoen.2020.104723.

Li, Hu, Xiao Zhang, Luming Zhao, Dongjie Jiang, Lingling Xu, Zhuo Liu, Yuxiang Wu, et al. 2020. "A Hybrid Biofuel and Triboelectric Nanogenerator for Bioenergy Harvesting." *Nano-Micro Letters* 12 (1): 1–12. https://doi.org/10.1007/s40820-020-0376-8.

Li, Hui, Shuyu Zhao, Xinyu Du, Jiaona Wang, Ran Cao, Yi Xing, and Congju Li. 2018. "A Compound Yarn Based Wearable Triboelectric Nanogenerator for Self-Powered Wearable Electronics." 3: 1800065. https://doi.org/10.1002/admt.201800065.

Li, Peng, Yang Guo, Jiuke Mu, Hongzhi Wang, Qinghong Zhang, and Yaogang Li. 2016. "Single-Walled Carbon Nanotubes/Polyaniline-Coated Polyester Thermoelectric Textile with Good Interface Stability Prepared by Ultrasonic Induction." *RSC Advances* 6 (93): 90347–53. https://doi.org/10.1039/c6ra16532j.

Li, Tong, Zhang Q. Feng, Ke Yan, Tao Yuan, Wuting Wei, Xu Yuan, Chao Wang, Ting Wang, Wei Dong, and Jie Zheng. 2018. "Pure OPM Nanofibers with High Piezoelectricity Designed for Energy Harvesting in Vitro and in Vivo." *Journal of Materials Chemistry B* 6 (33): 5343–52. https://doi.org/10.1039/c8tb01702f.

Li, Yaowen, Guiying Xu, Chaohua Cui, and Yongfang Li. 2018. "Flexible and Semitransparent Organic Solar Cells." *Advanced Energy Materials* 8 (7): 1–28. https://doi.org/10.1002/aenm.201701791.

Li, Zhaoling, Miaomiao Zhu, Qian Qiu, Jianyong Yu, and Bin Ding. 2018. "Multilayered Fiber-Based Triboelectric Nanogenerator with High Performance for Biomechanical Energy Harvesting." *Nano Energy* 53 (August): 726–33. https://doi.org/10.1016/j.nanoen.2018.09.039.

Li, Zhe, Hongqing Feng, Qiang Zheng, Hu Li, Chaochao Zhao, Han Ouyang, Sehrish Noreen, et al. 2018. "Photothermally Tunable Biodegradation of Implantable Triboelectric Nanogenerators for Tissue Repairing." *Nano Energy* 54 (December): 390–9. https://doi.org/10.1016/j.nanoen.2018.10.020.

Li, Zhe, Qiang Zheng, Zhong L. Wang, and Zhou Li. 2020. "Nanogenerator-Based Self-Powered Sensors for Wearable and Implantable Electronics." *Research* 2020: 1–25. https://doi.org/10.34133/2020/8710686.

Li, Zhou, Guang Zhu, Rusen Yang, Aurelia C. Wang, and Zhong L. Wang. 2010. "Muscle-Driven in Vivo Nanogenerator." *Advanced Materials* 22 (23): 2534–37. https://doi.org/10.1002/adma.200904355.

Liang, Daxin, Haoran Yang, Scott W. Finefrock, and Yue Wu. 2012. "Flexible Nanocrystal-Coated Glass Fibers for High-Performance Thermoelectric Energy Harvesting." *Nano Letters* 12 (4): 2140–5. https://doi.org/10.1021/nl300524j.

Lin, Zhiming, Jun Yang, Xiaoshi Li, Yufen Wu, Wei Wei, Jun Liu, Jun Chen, and Jin Yang. 2018. "Large-Scale and Washable Smart Textiles Based on Triboelectric Nanogenerator Arrays for Self-Powered Sleeping Monitoring." *Advanced Functional Materials* 28: 1704112. https://doi.org/10.1002/adfm.201704112.

Liu, Huicong, Zhangping Ji, Tao Chen, Lining Sun, Suchith C. Menon, and Chengkuo Lee. 2015. "An Intermittent Self-Powered Energy Harvesting System from Low-Frequency Hand Shaking." *IEEE Sensors Journal* 15 (9): 4782–90. https://doi.org/10.1109/JSEN.2015.2411313.

Liu, Mengmeng, Zifeng Cong, Xiong Pu, Wenbin Guo, Ting Liu, Meng Li, Yang Zhang, Weiguo Hu, and Zhong L. Wang. 2019a. "High-Energy Asymmetric Supercapacitor Yarns for Self-Charging Power Textiles" 1806298: 1–12. https://doi.org/10.1002/adfm.201806298.

Liu, Mengmeng, Zifeng Cong, Xiong Pu, Wenbin Guo, Ting Liu, Meng Li, Yang Zhang, Weiguo Hu, and Zhong L. Wang. 2019b. "High-Energy Asymmetric Supercapacitor Yarns for Self-Charging Power Textiles." *Advanced Functional Materials* 29 (41): 1–12. https://doi.org/10.1002/adfm.201806298.

Liu, Peng, Zhen Gao, Limin Xu, Xiang Shi, Xuemei Fu, Ke Li, Bo Zhang, Xuemei Sun, and Huisheng Peng. 2018. "Polymer Solar Cell Textiles with Interlaced Cathode and Anode Fibers." *Journal of Materials Chemistry A* 6 (41): 19947–53. https://doi.org/10.1039/c8ta06510a.

Liu, Yudong, Yaxing Zhu, Jingyu Liu, Yang Zhang, Juan Liu, and Junyi Zhai. 2018. "Design of Bionic Cochlear Basilar Membrane Acoustic Sensor for Frequency Selectivity Based on Film Triboelectric Nanogenerator." *Nanoscale Research Letters* 13 (1): 191. https://doi.org/10.1186/s11671-018-2593-3.

Liu, Zhaoxian, Zhizhen Zhao, Xiangwen Zeng, Xiuli Fu, and Youfan Hu. 2019. "Expandable Microsphere-Based Triboelectric Nanogenerators as Ultrasensitive Pressure Sensors for Respiratory and Pulse Monitoring." *Nano Energy* 59 (May): 295–301. https://doi.org/10.1016/j.nanoen.2019.02.057.

Liu, Zhuo, Hu Li, Bojing Shi, Yubo Fan, Zhong L. Wang, and Zhou Li. 2019. "Wearable and Implantable Triboelectric Nanogenerators." *Advanced Functional Materials* 29 (20): 1808820. https://doi.org/10.1002/adfm.201808820.

Liu, Zhuo, Ye Ma, Han Ouyang, Bojing Shi, Ning Li, Dongjie Jiang, Feng Xie, et al. 2019. "Transcatheter Self-Powered Ultrasensitive Endocardial Pressure Sensor." *Advanced Functional Materials* 29 (3): 1807560. https://doi.org/10.1002/adfm.201807560.

Löf, George O. G., John A. Duffie, and C. O. Smith. 1966. "World Distribution of Solar Radiation." *Solar Energy* 10 (1): 27–37. https://doi.org/10.1016/0038-092X(66)90069-7.

Lu, Bingwei, Ying Chen, Dapeng Ou, Hang Chen, Liwei Diao, Wei Zhang, Jun Zheng, Weiguo Ma, Lizhong Sun, and Xue Feng. 2015. "Ultra-Flexible Piezoelectric Devices Integrated with Heart to Harvest the Biomechanical Energy." *Scientific Reports* 5 (November). https://doi.org/10.1038/srep16065.

Lu, Zhisong, Huihui Zhang, Cuiping Mao, and Chang M. Li. 2016. "Silk Fabric-Based Wearable Thermoelectric Generator for Energy Harvesting from the Human Body." *Applied Energy* 164: 57–63. https://doi.org/10.1016/j.apenergy.2015.11.038.

Lv, Jian, Itthipon Jeerapan, Farshad Tehrani, Lu Yin, Cristian A. Silva-Lopez, Ji H. Jang, Davina Joshuia, et al. 2018. "Sweat-Based Wearable Energy Harvesting-Storage Hybrid Textile Devices." *Energy and Environmental Science* 11 (12): 3431–42. https://doi.org/10.1039/c8ee02792g.

Ma, Ye, Qiang Zheng, Yang Liu, Bojin Shi, Xiang Xue, Weiping Ji, Zhuo Liu, et al. 2016. "Self-Powered, One-Stop, and Multifunctional Implantable Triboelectric Active Sensor for Real-Time Biomedical Monitoring." *Nano Letters* 16 (10): 6042–51. https://doi.org/10.1021/acs.nanolett.6b01968.

Ma, Yuan, Wangshu Tong, Wenjiang Wang, Qi An, and Yihe Zhang. 2018. "Montmorillonite/PVDF-HFP-Based Energy Conversion and Storage Films with Enhanced Piezoelectric and Dielectric Properties." *Composites Science and Technology* 168 (June): 397–403. https://doi.org/10.1016/j.compscitech.2018.10.009.

Mahmud, M. A. Parvez, Nazmul Huda, Shahjadi H. Farjana, Mohsen Asadnia, and Candace Lang. 2018. "Recent Advances in Nanogenerator-Driven Self-Powered Implantable Biomedical Devices." *Advanced Energy Materials* 8 (2): 1701210. https://doi.org/10.1002/aenm.201701210.

Maity, Kuntal, and Dipankar Mandal. 2018. "All-Organic High-Performance Piezoelectric Nanogenerator with Multilayer Assembled Electrospun Nanofiber Mats for Self-Powered Multifunctional Sensors." *ACS Applied Materials and Interfaces* 10 (21): 18257–69. https://doi.org/10.1021/acsami.8b01862.

Martin, Thomas, Mark Jones, Josh Edmison, and Ravi Shenoy. 2003. "Towards a Design Framework for Wearable Electronic Textiles." In *Proceedings – International Symposium on Wearable Computers, ISWC*, 190–9. https://doi.org/10.1109/iswc.2003.1241411.

Meng, Lingxian, Yamin Zhang, Xiangjian Wan, Chenxi Li, Xin Zhang, Yanbo Wang, Xin Ke, et al. 2018. "Organic and Solution-Processed Tandem Solar Cells with 17.3% Efficiency." *Science* 361 (6407): 1094–8. https://doi.org/10.1126/science.aat2612.

Minteer, Shelley D., Bor Y. Liaw, and Michael J. Cooney. 2007. "Enzyme-Based Biofuel Cells." *Current Opinion in Biotechnology* 18 (3): 228–34. https://doi.org/10.1016/j.copbio.2007.03.007.

Miyake, Takeo, Keigo Haneda, Syuhei Yoshino, and Matsuhiko Nishizawa. 2013. "Flexible, Layered Biofuel Cells." *Biosensors and Bioelectronics* 40 (1): 45–9. https://doi.org/10.1016/j.bios.2012.05.041.

Mokhtari, Fatemeh, Javad Foroughi, Tian Zheng, Zhenxiang Cheng, and Geoffrey M. Spinks. 2019. "Triaxial Braided Piezo Fiber Energy Harvesters for Self-Powered Wearable Technologies." *Journal of Materials Chemistry A* 7 (14): 8245–57. https://doi.org/10.1039/c8ta10964h.

Montiel, Víctor R.-V., Juliane R. Sempionatto, Susana Campuzano, José M. Pingarrón, Berta E. F. de Ávila, and Joseph Wang. 2019. "Direct Electrochemical Biosensing in Gastrointestinal Fluids." *Analytical and Bioanalytical Chemistry* 411 (19): 4597–604. https://doi.org/10.1007/s00216-018-1528-2.

Mostafalu, Pooria, and Sameer Sonkusale. 2014. "Flexible and Transparent Gastric Battery: Energy Harvesting from Gastric Acid for Endoscopy Application." *Biosensors and Bioelectronics* 54 (April): 292–6. https://doi.org/10.1016/j.bios.2013.10.040.

Ning, Chuan, Lan Tian, Xinya Zhao, Shengxin Xiang, Yingjie Tang, Erjun Liang, and Yanchao Mao. 2018. "Washable Textile-Structured Single-Electrode Triboelectric Nanogenerator for Self-Powered Wearable Electronics." *Journal of Materials Chemistry A* 6: 19143–50. https://doi.org/10.1039/c8ta07784c.

Niu, Simiao, Xiaofeng Wang, Fang Yi, Yu S. Zhou, and Zhong L. Wang. 2015. "A Universal Self-Charging System Driven by Random Biomechanical Energy for Sustainable Operation of Mobile Electronics." *Nature Communications* 6 (1): 1–8. https://doi.org/10.1038/ncomms9975.

Nolas, George S., Jeffrey Sharp, and H. Julian Goldsmid. 2001. "Thermoelectrics". In *Springer Series in Materials Science*, Vol. 45. Berlin, Heidelberg: Springer Berlin Heidelberg. https://doi.org/10.1007/978-3-662-04569-5.

Nozariasbmarz, Amin, Henry Collins, Kelvin Dsouza, Mobarak H. Polash, Mahshid Hosseini, Melissa Hyland, Jie Liu, et al. 2020. "Review of Wearable Thermoelectric Energy Harvesting: From Body Temperature to Electronic Systems." *Applied Energy*. https://doi.org/10.1016/j.apenergy.2019.114069.

O'Connor, Timothy F., Aliaksandr V. Zaretski, Suchol Savagatrup, Adam D. Printz, Cameron D. Wilkes, Mare I. Diaz, Eric J. Sawyer, and Darren J. Lipomi. 2016. "Wearable Organic Solar Cells with High Cyclic Bending Stability: Materials Selection Criteria." *Solar Energy Materials and Solar Cells* 144: 438–44. https://doi.org/10.1016/j.solmat.2015.09.049.

Ogawa, Yudai, Koichiro Kato, Takeo Miyake, Kuniaki Nagamine, Takuya Ofuji, Syuhei Yoshino, and Matsuhiko Nishizawa. 2015. "Organic Transdermal Iontophoresis Patch with Built-in Biofuel Cell." *Advanced Healthcare Materials* 4 (4): 506–10. https://doi.org/10.1002/adhm.201400457.

Ogawa, Yudai, Yuki Takai, Yuto Kato, Hiroyuki Kai, Takeo Miyake, and Matsuhiko Nishizawa. 2015. "Stretchable Biofuel Cell with Enzyme-Modified Conductive Textiles." *Biosensors and Bioelectronics* 74: 947–52. https://doi.org/10.1016/j.bios.2015.07.063.

Ouyang, Han, Zhuo Liu, Ning Li, Bojing Shi, Yang Zou, Feng Xie, Ye Ma, et al. 2019. "Symbiotic Cardiac Pacemaker." *Nature Communications* 10 (1): 1–10. https://doi.org/10.1038/s41467-019-09851-1.

Ouyang, Han, Jingjing Tian, Guanglong Sun, Yang Zou, Zhuo Liu, Hu Li, Luming Zhao, et al. 2017. "Self-Powered Pulse Sensor for Antidiastole of Cardiovascular Disease." *Advanced Materials* 29 (40): 1–10. https://doi.org/10.1002/adma.201703456.

Pan, Caofeng, Zetang Li, Wenxi Guo, Jing Zhu, and Zhong L. Wang. 2011. "Fiber-Based Hybrid Nanogenerators for/as Self-Powered Systems in Biological Liquid." *Angewandte Chemie – International Edition* 50 (47): 11192–6. https://doi.org/10.1002/anie.201104197.

Pan, Shaowu, Zhibin Yang, Peining Chen, Jue Deng, Houpu Li, and Huisheng Peng. 2014. "Wearable Solar Cells by Stacking Textile Electrodes." *Angewandte Chemie* 126 (24): 6224–8. https://doi.org/10.1002/ange.201402561.

Papagiannakis, Athanassios T., Samer H. Dessouky, Arturo H Montoya, and HosseinRoshani. 2016. "Energy Harvesting from Roadways." *Procedia Computer Science* 83 (Seit): 758–65. https://doi.org/10.1016/j.procs.2016.04.164.

Park, Dae Y., Daniel J. Joe, Dong H. Kim, Hyewon Park, Jae H. Han, Chang K. Jeong, Hyelim Park, Jung G. Park, Boyoung Joung, and Keon J. Lee. 2017. "Self-Powered Real-Time Arterial Pulse Monitoring Using Ultrathin Epidermal Piezoelectric Sensors." *Advanced Materials* 29 (37): 1702308. https://doi.org/10.1002/adma.201702308.

Park, Jiwon, A. Young Choi, Chang J. Lee, Dogyun Kim, and Youn T. Kim. 2017. "Highly Stretchable Fiber-Based Single-Electrode Triboelectric Nanogenerator for Wearable Devices." *RSC Advances* 7 (86): 54829–34. https://doi.org/10.1039/c7ra10285b.

Park, Jiwon, Dogyun Kim, A. Young Choi, and Youn T. Kim. 2018. "Flexible Single-Strand Fiber-Based Woven- Structured Triboelectric Nanogenerator for Self-Powered Electronics." *APL Materials* 6: 101106. https://doi.org/10.1063/1.5048553.

Park, Sungjun, Soo W. Heo, Wonryung Lee, Daishi Inoue, Zhi Jiang, Kilho Yu, Hiroaki Jinno, et al. 2018. "Self-Powered Ultra-Flexible Electronics via Nano-Grating-Patterned Organic Photovoltaics." *Nature* 561 (7724): 516–21. https://doi.org/10.1038/s41586-018-0536-x.

Patel, Divyesh, Shruti B. Mehta, and Pratik Shah. 2015. "Review of Thermoelectricity to Improve Energy Quality." *International Journal of Emerging Technologies and Innovative Research* 2 (3): 847–50.

Potnuru, Akshay, and Yonas Tadesse. 2014. "Characterization of Pyroelectric Materials for Energy Harvesting from Human Body." *Integrated Ferroelectrics* 150 (1): 23–50. https://doi.org/10.1080/10584587.2014.873319.

Proto, Antonino, Daniele Bibbo, Martin Cerny, David Vala, Vladimir Kasik, Lukas Peter, Silvia Conforto, Maurizio Schmid, and Marek Penhaker. 2018. "Thermal Energy Harvesting on the Bodily Surfaces of Arms and Legs through a Wearable Thermo-Electric Generator." *Sensors (Switzerland)* 18 (6). https://doi.org/10.3390/s18061927.

Pu, Xiong, Linxuan Li, Huanqiao Song. 2015. "A Self-Charging Power Unit by Integration of a Textile Triboelectric Nanogenerator and a Flexible Lithium-Ion Battery for Wearable Electronics." Advanced Materials 27, 2472–2478. https://doi.org/10.1002/adma.201500311.

Pu, Xiong, Linxuan Li, Mengmeng Liu. 2016. "Wearable Self-Charging Power Textile Based on Flexible Yarn Supercapacitors and Fabric Nanogenerators." 98–105. https://doi.org/10.1002/adma.201504403.

Pu, Xiong, Linxuan Li, Huanqiao Song, Chunhua Du, Zhengfu Zhao, Chunyan Jiang, Guozhong Cao, Weiguo Hu, and Zhong L. Wang. 2015. "A Self-Charging Power Unit by Integration of a Textile Triboelectric Nanogenerator and a Flexible Lithium-Ion Battery for Wearable Electronics." 2472–8. https://doi.org/10.1002/adma.201500311.

Pu, Xiong, Weixing Song, Mengmeng Liu, Chunwen Sun, Chunhua Du, Chunyan Jiang, Xin Huang, Dechun Zou, Weiguo Hu, and Zhong L. Wang. 2016. "Wearable Power-Textiles by Integrating Fabric Triboelectric Nanogenerators and Fiber-Shaped Dye-Sensitized Solar Cells." *Advanced Energy Materials* 6 (20). https://doi.org/10.1002/aenm.201601048.

Qin, Weiwei, Tao Li, Yutong Li, Junwen Qiu, Xianjun Ma, Xiaoqiang Chen, Xuefeng Hu, and Wei Zhang. 2016. "A High Power ZnO Thin Film Piezoelectric Generator." *Applied Surface Science* 364: 670–75. https://doi.org/10.1016/j.apsusc.2015.12.178.

Qin, Zhen, Yingying Yin, Wenzheng Zhang, Congju Li, and Kai Pan. 2019. "Wearable and Stretchable Triboelectric Nanogenerator Based on Crumpled Nanofibrous Membranes." *ACS Applied Materials and Interfaces* 11 (13): 12452–59. https://doi.org/10.1021/acsami.8b21487.

Qiu, Qian, Miaomiao Zhu, Zhaoling Li, Kaili Qiu, Xiaoyan Liu, and Jianyong Yu. 2019. "Highly Flexible, Breathable, Tailorable and Washable Power Generation Fabrics for Wearable Electronics." *Nano Energy* 58 (February): 750–58. https://doi.org/10.1016/j.nanoen.2019.02.010.

Qu, Sanyin, Yanling Chen, Wei Shi, Mengdi Wang, Qin Yao, and Lidong Chen. 2018. "Cotton-Based Wearable Poly(3-Hexylthiophene) Electronic Device for Thermoelectric Application with Cross-Plane Temperature Gradient." *Thin Solid Films* 667 (April): 59–63. https://doi.org/10.1016/j.tsf.2018.09.046.

Rabaey, Korneel, and Willy Verstraete. 2005. "Microbial Fuel Cells: Novel Biotechnology for Energy Generation." *Trends in Biotechnology* 23: 291–8: https://doi.org/10.1016/j.tibtech.2005.04.008.

Ramadan, Khaled S., Dan Sameoto, and Stephane Evoy. 2014. "A Review of Piezoelectric Polymers as Functional Materials for Electromechanical Transducers." *Smart Materials and Structures* 23 (3). https://doi.org/10.1088/0964-1726/23/3/033001.

Rasmussen, Michelle, Sofiene Abdellaoui, and Shelley D. Minteer. 2016. "Enzymatic Biofuel Cells: 30 Years of Critical Advancements." *Biosensors and Bioelectronics* 76: 91–102. https://doi.org/10.1016/j.bios.2015.06.029.

Razavi, Seyedalireza, Lorenzo Iannucci, and Emile S. Greenhalgh. 2020. "A Piezo Smart-Braid Harvester and Damper for Multifunctional Fiber Reinforced Polymer Composites." *Energy Technology* 8: 2000777. https://doi.org/10.1002/ente.202000777.

Reid, Russell C., Shelley D. Minteer, and Bruce K. Gale. 2015. "Contact Lens Biofuel Cell Tested in a Synthetic Tear Solution." *Biosensors and Bioelectronics* 68 (June): 142–8. https://doi.org/10.1016/j.bios.2014.12.034.

Roldán-Carmona, Cristina, Olga Malinkiewicz, Alejandra Soriano, Guillermo M. Espallargas, Ana Garcia, Patrick Reinecke, Thomas Kroyer, M. Ibrahim Dar, Mohammad Khaja Nazeeruddin, and Henk J. Bolink. 2014. "Flexible High Efficiency Perovskite Solar Cells." *Energy and Environmental Science* 7 (3): 994–7. https://doi.org/10.1039/c3ee43619e.

Ruiz, Olivia, Mark Cochrane, Manni Li, Yan Yan, Ke Ma, Jintao Fu, Zeyu Wang, Sarah H. Tolbert, Vivek B. Shenoy, and Eric Detsi. 2018. "Enhanced Cycling Stability of Macroporous Bulk Antimony-Based Sodium-Ion Battery Anodes Enabled through Active/Inactive Composites." *Advanced Energy Materials* 8 (31): 1801781. https://doi.org/10.1002/aenm.201801781.

Ryan, Jason D., Anja Lund, Anna I. Hofmann, Renee Kroon, Ruben Sarabia-Riquelme, Matthew C. Weisenberger, and Christian Müller. 2018. "All-Organic Textile Thermoelectrics with Carbon-Nanotube-Coated n-Type Yarns." *ACS Applied Energy Materials* 1 (6): 2934–41. https://doi.org/10.1021/acsaem.8b00617.

Ryu, Hanjun, and Sang W. Kim. 2019. "Emerging Pyroelectric Nanogenerators to Convert Thermal Energy into Electrical Energy." *Small* 17: 1903469. https://doi.org/10.1002/smll.201903469.

Sakai, Hideki, Takaaki Nakagawa, Yuichi Tokita, Tsuyonobu Hatazawa, Tokuji Ikeda, Seiya Tsujimura, and Kenji Kano. 2009. "A High-Power Glucose/Oxygen Biofuel Cell Operating under Quiescent Conditions." *Energy and Environmental Science* 2 (1): 133–38. https://doi.org/10.1039/b809841g.

Satharasinghe, Achala, Theodore Hughes-Riley, and Tilak Dias. 2020. "A Review of Solar Energy Harvesting Electronic Textiles." *Sensors (Switzerland)* 20 (20): 1–39. https://doi.org/10.3390/s20205938.

Seung, Wanchul, Manoj K. Gupta, Keun Y. Lee, Kyung S. Shin, Ju H. Lee, Tae Y. Kim, Sanghyun Kim, Jianjian Lin, Jung H. Kim, and Sang W. Kim. 2015. "Nanopatterned Textile-Based Wearable Triboelectric Nanogenerator." *ACS Nano* 9 (4): 3501–9. https://doi.org/10.1021/nn507221f.

Shawon, SK Md Ali Zaker, Andrew X. Sun, Valeria S. Vega, Brishty D. Chowdhury, Phong Tran, Zaida D. Carballo, Jim A. Tolentino, et al. 2021. "Piezo-Tribo Dual Effect Hybrid Nanogenerators for Health Monitoring." *Nano Energy*. https://doi.org/10.1016/j.nanoen.2020.105691.

Shi, Bojing, Zhou Li, and Yubo Fan. 2018. "Implantable Energy-Harvesting Devices." *Advanced Materials* 30 (44): 1801511. https://doi.org/10.1002/adma.201801511.

Shi, Bojing, Zhuo Liu, Qiang Zheng, Jianping Meng, Han Ouyang, Yang Zou, Dongjie Jiang, et al. 2019. "Body-Integrated Self-Powered System for Wearable and Implantable Applications." *ACS Nano* 13 (5): 6017–24. https://doi.org/10.1021/acsnano.9b02233.

Shi, Bojing, Qiang Zheng, Wen Jiang, Ling Yan, Xinxin Wang, Hong Liu, Yan Yao, Zhou Li, and Zhong L. Wang. 2016. "A Packaged Self-Powered System with Universal Connectors Based on Hybridized Nanogenerators." *Advanced Materials* 28 (5): 846–52. https://doi.org/10.1002/adma.201503356.

Shi, Kunming, Bin Sun, Xingyi Huang, and Pingkai Jiang. 2018. "Synergistic Effect of Graphene Nanosheet and BaTiO3 Nanoparticles on Performance Enhancement of Electrospun PVDF Nanofiber Mat for Flexible Piezoelectric Nanogenerators." *Nano Energy* 52 (July): 153–62. https://doi.org/10.1016/j.nanoen.2018.07.053.

Shi, Qiongfeng, Zhongda Sun, Zixuan Zhang, and Chengkuo Lee. 2021. "Triboelectric Nanogenerators and Hybridized Systems for Enabling Next-Generation IoT Applications." *Research* 2021: 1–30. https://doi.org/10.34133/2021/6849171.

Shinjo, Teruya. 2013. *Nanomagnetism and Spintronics*, Second Edition. Oxford, UK: Elsevier Inc. https://doi.org/10.1016/C2013-0-00584-1.

Shitanda, Isao, Yukiya Morigayama, Risa Iwashita, Himeka Goto, Tatsuo Aikawa, Tsutomu Mikawa, Yoshinao Hoshi, et al. 2021. "Paper-Based Lactate Biofuel Cell Array with High Power Output." *Journal of Power Sources* 489 (March): 229533. https://doi.org/10.1016/j.jpowsour.2021.229533.

Shuvo, Ikra I., Justine Decaens, Dominic Lachapelle, and Patricia I. Dolez. 2021. "Smart Textiles Testing: A Roadmap to Standardized Test Methods for Safety and Quality-Control." https://doi.org/ 10.5772/intechopen.96500.

Siddique, Abu R. M., Shohel Mahmud, and Bill V. Heyst. 2017. "A Review of the State of the Science on Wearable Thermoelectric Power Generators (TEGs) and Their Existing Challenges." *Renewable and Sustainable Energy Reviews*. https://doi.org/10.1016/j.rser.2017.01.177.

Siddique, Abu R. M., Ronil Rabari, Shohel Mahmud, and Bill V. Heyst. 2016. "Thermal Energy Harvesting from the Human Body Using Flexible Thermoelectric Generator (FTEG) Fabricated by a Dispenser Printing Technique." *Energy* 115: 1081–91. https://doi.org/10.1016/j.energy.2016.09.087.

Siddiqui, Saqib, Do I. Kim, Le T. Duy, Minh T. Nguyen, Shoaib Muhammad, Won S. Yoon, and Nae E. Lee. 2015. "High-Performance Flexible Lead-Free Nanocomposite Piezoelectric Nanogenerator for Biomechanical Energy Harvesting and Storage." *Nano Energy* 15: 177–85. https://doi.org/10.1016/j.nanoen.2015.04.030.

Siddiqui, Saqib, Han B. Lee, Do-I. Kim, Le T. Duy, Adeela Hanif, and Nae-Eung Lee. 2018. "An Omnidirectionally Stretchable Piezoelectric Nanogenerator Based on Hybrid Nanofibers and Carbon Electrodes for Multimodal Straining and Human Kinematics Energy Harvesting." *Advanced Energy Materials* 8 (2): 1701520. https://doi.org/10.1002/aenm.201701520.

Sim, Hyeon J., Changsoon Choi, Chang J. Lee, Youn T. Kim, Geoffrey M. Spinks, Marcio D. Lima, Ray H. Baughman, and Seon J. Kim. 2015. "Flexible, Stretchable and Weavable Piezoelectric Fiber." *Advanced Engineering Materials* 17 (9): 1270–5. https://doi.org/10.1002/adem.201500018.

Sim, Hyeon J., Changsoon Choi, Chang J. Lee, Youn T. Kim, and Seon J. Kim. 2015. "Flexible Two-Ply Piezoelectric Yarn Energy Harvester." *Current Nanoscience* 11 (4): 539–44. https://doi.org/10.2174/157341371166150225231434.

Sim, Hyeon J., Dong Y. Lee, Hyunsoo Kim, Young B. Choi, Hyug H. Kim, Ray H. Baughman, and Seon J. Kim. 2018. "Stretchable Fiber Biofuel Cell by Rewrapping Multiwalled Carbon Nanotube Sheets." *Nano Letters* 18 (8): 5272–8. https://doi.org/10.1021/acs.nanolett.8b02256.

Slaughter, Gymama, and Tanmay Kulkarni. 2015. "Enzymatic Glucose Biofuel Cell and Its Application." *Journal of Biochips & Tissue Chips* 05 (01): 1–10. https://doi.org/10.4172/2153-0777.1000110.

Soin, Navneet, Tahir H. Shah, Subhash C. Anand, Junfeng Geng, Wiwat Pornwannachai, Pranab Mandal, David Reid, et al. 2014. "Novel '3-D Spacer' All Fibre Piezoelectric Textiles for Energy Harvesting Applications." *Energy and Environmental Science* 7 (5): 1670–9. https://doi.org/10.1039/c3ee43987a.

Song, Seunghwan, and Kwang S. Yun. 2015. "Design and Characterization of Scalable Woven Piezoelectric Energy Harvester for Wearable Applications." *Smart Materials and Structures* 24 (4). https://doi.org/10.1088/0964-1726/24/4/045008.

Song, Weixing, Baoheng Gan, Tao Jiang, Yue Zhang, Aifang Yu, Hongtao Yuan, Ning Chen, Chunwen Sun, and Zhong L. Wang. 2016. "Nanopillar Arrayed Triboelectric Nanogenerator as a Self-Powered Sensitive Sensor for a Sleep Monitoring System." *ACS Nano* 10 (8): 8097–103. https://doi.org/10.1021/acsnano.6b04344.

Sonigara, Keval K., Hiren K. Machhi, Jayraj V. Vaghasiya, Alain Gibaud, Swee C. Tan, and Saurabh S. Soni. 2018. "A Smart Flexible Solid State Photovoltaic Device with Interfacial Cooling Recovery Feature through Thermoreversible Polymer Gel Electrolyte." *Small* 14 (36). https://doi.org/10.1002/smll.201800842.

Southcott, Mark, Kevin MacVittie, Jan Halámek, Lenka Halámková, William D. Jemison, Robert Lobel, and Evgeny Katz. 2013. "A Pacemaker Powered by an Implantable Biofuel Cell Operating under Conditions Mimicking the Human Blood Circulatory System-Battery Not Included." *Physical Chemistry Chemical Physics* 15 (17): 6278–83. https://doi.org/10.1039/c3cp50929j.

Suarez, Francisco, Amin Nozariasbmarz, Daryoosh Vashaee, and Mehmet C. Öztürk. 2016. "Designing Thermoelectric Generators for Self-Powered Wearable Electronics." *Energy and Environmental Science* 9 (6): 2099–113. https://doi.org/10.1039/c6ee00456c.

Sultana, Ayesha, Md Mehebub Alam, Sujoy K. Ghosh, Tapas R. Middya, and Dipankar Mandal. 2019. "Energy Harvesting and Self-Powered Microphone Application on Multifunctional Inorganic-Organic Hybrid Nanogenerator." *Energy* 166: 963–71. https://doi.org/10.1016/j.energy.2018.10.124.

Sultana, Ayesha, Md Mehebub Alam, Tapas R. Middya, and Dipankar Mandal. 2018. "A Pyroelectric Generator as a Self-Powered Temperature Sensor for Sustainable Thermal Energy Harvesting from Waste Heat and Human Body Heat." *Applied Energy* 221 (July): 299–307. https://doi.org/10.1016/j.apenergy.2018.04.003.

Sun, Tingting, Beiying Zhou, Qi Zheng, Lianjun Wang, Wan Jiang, and Gerald J. Snyder. 2020. "Stretchable Fabric Generates Electric Power from Woven Thermoelectric Fibers." *Nature Communications* 11 (1): 1–10. https://doi.org/10.1038/s41467-020-14399-6.

Talbourdet, Anaëlle, François Rault, Guillaume Lemort, Cédric Cochrane, Eric Devaux, and Christine Campagne. 2018. "3D Interlock Design 100% PVDF Piezoelectric to Improve Energy Harvesting." *Smart Materials and Structures* 27 (7). https://doi.org/10.1088/1361-665X/aab865.

Tao, Xiaoming. 2019. "Study of Fiber-Based Wearable Energy Systems." *Accounts of Chemical Research* 5. https://doi.org/10.1021/acs.accounts.8b00502.

Tian, Zhumei, Jian He, Xi Chen, Zengxing Zhang, Tao Wen, Cong Zhai, and Jianqiang Han. 2017. "Performance-Boosted Triboelectric Textile for Harvesting Human Motion Energy." *Nano Energy* 39 (May): 562–70. https://doi.org/10.1016/j.nanoen.2017.06.018.

Valdés-Ramírez, Gabriela, Ya C. Li, Jayoung Kim, Wenzhao Jia, Amay J. Bandodkar, Rogelio Nuñez-Flores, Philip R. Miller, et al. 2014. "Microneedle-Based Self-Powered Glucose Sensor." *Electrochemistry Communications* 47 (October): 58–62. https://doi.org/10.1016/j.elecom.2014.07.014.

Varma, Sreekanth J., Kowsik S. Kumar, Sudipta Seal, Swaminathan Rajaraman, and Jayan Thomas. 2018. "Fiber-Type Solar Cells, Nanogenerators, Batteries, and Supercapacitors for Wearable Applications." *Advanced Science* 5 (9): 1800340. https://doi.org/10.1002/advs.201800340.

Wang, Bo, Chen Liu, Yongjun Xiao, Junwen Zhong, Wenbo Li, Yongliang Cheng, Bin Hu, Liang Huang, and Jun Zhou. 2017. "Ultrasensitive Cellular Fluorocarbon Piezoelectret Pressure Sensor for Self-Powered Human Physiological Monitoring." *Nano Energy* 32 (February): 42–9. https://doi.org/10.1016/j.nanoen.2016.12.025.

Wang, Jiahui, Tianyiyi He, and Chengkuo Lee. 2019. "Development of Neural Interfaces and Energy Harvesters towards Self-Powered Implantable Systems for Healthcare Monitoring and Rehabilitation Purposes." *Nano Energy*. https://doi.org/10.1016/j.nanoen.2019.104039.

Wang, Jie, Shengming Li, Fang Yi, Yunlong Zi, Jun Lin, Xiaofeng Wang, Youlong Xu, and Zhong Lin Wang. 2016. "Sustainably Powering Wearable Electronics Solely by Biomechanical Energy." *Nature Communications* 7:12744. https://doi.org/10.1038/ncomms12744.

Wang, Ruoxing, Shengjie Gao, Zhen Yang, Yule Li, Weinong Chen, Benxin Wu, and Wenzhuo Wu. 2018. "Engineered and Laser-Processed Chitosan Biopolymers for Sustainable and Biodegradable Triboelectric Power Generation." *Advanced Materials* 30 (11): 1706267. https://doi.org/10.1002/adma.201706267.

Wang, Ruoxing, Liwen Mu, Yukai Bao, Han Lin, Tuo Ji, Yijun Shi, Jiahua Zhu, and Wenzhuo Wu. 2020. "Holistically Engineered Polymer–Polymer and Polymer–Ion Interactions in Biocompatible Polyvinyl Alcohol Blends for High-Performance Triboelectric Devices in Self-Powered Wearable Cardiovascular Monitorings." *Advanced Materials* 32 (32): 2002878. https://doi.org/10.1002/adma.202002878.

Wang, Z. 2011. *Nanogenerators for Self-Powered Devices and Systems*. Atlanta GA: School of Materials Science and Engineering, Georgia Institute of Technology.

Wang, Zhong L. 2017. "On Maxwell's Displacement Current for Energy and Sensors: The Origin of Nanogenerators." *Materials Today* 20 (2): 74–82. https://doi.org/10.1016/j.mattod.2016.12.001.
Wang, Zhong L., and Jinhui Song. 2006. "Piezoelectric Nanogenerators Based on Zinc Oxide Nanowire Arrays." *Science* 312 (5771): 242–6. https://doi.org/10.1126/science.1124005.
Wen, Zhen, Min H. Yeh, Hengyu Guo, Jie Wang, Yunlong Zi, Weidong Xu, Jianan Deng, et al. 2016. "Self-Powered Textile for Wearable Electronics by Hybridizing Fiber-Shaped Nanogenerators, Solar Cells, and Supercapacitors." *Science Advances* 2 (10): e1600097. https://doi.org/10.1126/sciadv.1600097.
Williams, David, Nanda G. Mandal, and Anand Sharma. 2011. "Electricity and Magnetism." *Anaesthesia and Intensive Care Medicine* 12 (9): 423–5. https://doi.org/10.1016/j.mpaic.2011.06.015.
Wu, Changsheng, Aurelia C. Wang, Wenbo Ding, Hengyu Guo, and Zhong L. Wang. 2019. "Triboelectric Nanogenerator: A Foundation of the Energy for the New Era." *Advanced Energy Materials* 9 (1). https://doi.org/10.1002/aenm.201802906.
Wu, Chaoxing, Tae W. Kim, Tailiang Guo, and Fushan Li. 2017. "Wearable Ultra-Lightweight Solar Textiles Based on Transparent Electronic Fabrics." *Nano Energy* 32 (November): 367–73. https://doi.org/10.1016/j.nanoen.2016.12.040.
Wu, Min, Yixiu Wang, Shengjie Gao, Ruoxing Wang, Chenxiang Ma, Zhiyuan Tang, Ning Bao, Wenxuan Wu, Fengru Fan, and Wenzhuo Wu. 2019. "Solution-Synthesized Chiral Piezoelectric Selenium Nanowires for Wearable Self-Powered Human-Integrated Monitoring." *Nano Energy* 56 (February): 693–9. https://doi.org/10.1016/j.nanoen.2018.12.003.
Wu, Qian, and Jinlian Hu. 2016. "Waterborne Polyurethane Based Thermoelectric Composites and Their Application Potential in Wearable Thermoelectric Textiles." *Composites Part B* 107: 59–66. https://doi.org/10.1016/j.compositesb.2016.09.068.
Wu, Qian, and Jinlian Hu. 2017. "A Novel Design for a Wearable Thermoelectric Generator Based on 3D Fabric Structure." *Smart Materials and Structures* 26 (4): 045037. https://doi.org/10.1088/1361-665X/aa5694.
Xiao, Xinxin, Hong Q. Xia, Ranran Wu, Lu Bai, Lu Yan, Edmond Magner, Serge Cosnier, Elisabeth Lojou, Zhiguang Zhu, and Aihua Liu. 2019. "Tackling the Challenges of Enzymatic (Bio)Fuel Cells." *Chemical Reviews* 119 (16): 9509–58. https://doi.org/10.1021/acs.chemrev.9b00115.
Xie, Xing, Liangbing Hu, Mauro Pasta, George F. Wells, Desheng Kong, Craig S. Criddle, and Yi Cui. 2011. "Three-Dimensional Carbon Nanotube-Textile Anode for High-Performance Microbial Fuel Cells." *Nano Letters* 11 (1): 291–6. https://doi.org/10.1021/nl103905t.
Xie, Xing, Meng Ye, Liangbing Hu, Nian Liu, James R. McDonough, Wei Chen, H. N. Alshareef, Craig S. Criddle, and Yi Cui. 2012. "Carbon Nanotube-Coated Macroporous Sponge for Microbial Fuel Cell Electrodes." *Energy and Environmental Science* 5 (1): 5265–70. https://doi.org/10.1039/c1ee02122b.
Xiong, Jiaqing, Peng Cui, Xiaoliang Chen, Jiangxin Wang, Kaushik Parida, Meng-fang Lin, and Pooi S. Lee. 2018. "Skin-Touch-Actuated Textile-Based Triboelectric Nanogenerator with Black Phosphorus for Durable Biomechanical Energy Harvesting." *Nature Communications* 9: 4280. https://doi.org/10.1038/s41467-018-06759-0.
Xu, Limin, Xuemei Fu, Fei Liu, Xiang Shi, Xufeng Zhou, Meng Liao, Chuanrui Chen, et al. 2020. "A Perovskite Solar Cell Textile That Works at −40 to 160°C." *Journal of Materials Chemistry A* 8 (11): 5476–83. https://doi.org/10.1039/c9ta13785h.
Xue, Hao, Quan Yang, Dingyi Wang, Weijian Luo, Wenqian Wang, Mushun Lin, Dingli Liang, and Qiming Luo. 2017. "A Wearable Pyroelectric Nanogenerator and Self-Powered Breathing Sensor." *Nano Energy* 38 (August): 147–54. https://doi.org/10.1016/j.nanoen.2017.05.056.
Yadav, Abhishek, Kevin P. Pipe, and Max Shtein. 2008. "Fiber-Based Flexible Thermoelectric Power Generator." *Journal of Power Sources* 175 (2): 909–13. https://doi.org/10.1016/j.jpowsour.2007.09.096.
Yang, Enlong, Zhe Xu, Lucas K. Chur, Ali Behroozfar, Mahmoud Baniasadi, Salvador Moreno, Jiacheng Huang, Jules Gilligan, and Majid Minary-Jolandan. 2017. "Nanofibrous Smart Fabrics from Twisted Yarns of Electrospun Piezopolymer." *ACS Applied Materials and Interfaces* 9 (28): 24220–9. https://doi.org/10.1021/acsami.7b06032.
Yang, Lei, Zhi-Gang Chen, Matthew S. Dargusch, and Jin Zou. 2018. "High Performance Thermoelectric Materials: Progress and Their Applications." *Advanced Energy Materials* 8 (6): 1701797. https://doi.org/10.1002/aenm.201701797.
Yang, Po K., Long Lin, Fang Yi, Xiuhan Li, Ken C. Pradel, Yunlong Zi, Chih I. Wu, Jr Hau He, Yue Zhang, and Zhong L. Wang. 2015. "A Flexible, Stretchable and Shape-Adaptive Approach for Versatile Energy Conversion and Self-Powered Biomedical Monitoring." *Advanced Materials* 27 (25): 3817–24. https://doi.org/10.1002/adma.201500652.

Yang, Rusen, Yong Qin, Liming Dai, and Zhong L. Wang. 2009. "Power Generation with Laterally Packaged Piezoelectric Fine Wires." *Nature Nanotechnology* 4 (1): 34–9.

Yang, Ya, Wenxi Guo, Ken C. Pradel, Guang Zhu, Yusheng Zhou, Yan Zhang, Youfan Hu, Long Lin, and Zhong L. Wang. 2012. "Pyroelectric Nanogenerators for Harvesting Thermoelectric Energy." *Nano Letters* 12 (6): 2833–8. https://doi.org/10.1021/nl3003039.

Yang, Ya, Ken C. Pradel, Qingshen Jing, Jyh M. Wu, Fang Zhang, Yusheng Zhou, Yue Zhang, and Zhong L. Wang. 2012. "Thermoelectric Nanogenerators Based on Single Sb-Doped ZnO Micro/Nanobelts." *ACS Nano* 6 (8): 6984–9. https://doi.org/10.1021/nn302481p.

Yang, Ya, Sihong Wang, Yan Zhang, and Zhong L. Wang. 2012. "Pyroelectric Nanogenerators for Driving Wireless Sensors." *Nano Letters* 12 (12): 6408–13. https://doi.org/10.1021/nl303755m.

Yang, Yanqin, Na Sun, Zhen Wen, Ping Cheng, Hechuang Zheng, Huiyun Shao, and Yujian Xia. 2018. "Liquid-Metal-Based Super-Stretchable and Structure-Designable Triboelectric Nanogenerator for Wearable Electronics." *ACS Nano* 12: 2027–34. https://doi.org/10.1021/acsnano.8b00147.

Yang, Yanqin, Lingjie Xie, Zhen Wen, Chen Chen, Xiaoping Chen, Aiming Wei, Ping Cheng, Xinkai Xie, and Xuhui Sun. 2018. "Coaxial Triboelectric Nanogenerator and Supercapacitor Fiber-Based Self-Charging Power Fabric." *ACS Applied Materials and Interfaces* 10 (49): 42356–62. https://doi.org/10.1021/acsami.8b15104.

Yang, Zhibin, Jue Deng, Xuemei Sun, Houpu Li, and Huisheng Peng. 2014. "Stretchable, Wearable Dye-Sensitized Solar Cells." *Advanced Materials* 26 (17): 2643–47. https://doi.org/10.1002/adma.201400152.

Yap, Y. Z., Ramasamy T. Naayagi, and Wai L. Woo. 2017. "Thermoelectric Energy Harvesting for Mobile Phone Charging Application." In *IEEE Region 10 Annual International Conference, Proceedings/TENCON*, 3241–45. Institute of Electrical and Electronics Engineers Inc. https://doi.org/10.1109/TENCON.2016.7848649.

Ye, Meidan, Xiaoru Wen, Mengye Wang, James Iocozzia, Nan Zhang, Changjian Lin, and Zhiqun Lin. 2015. "Recent Advances in Dye-Sensitized Solar Cells: From Photoanodes, Sensitizers and Electrolytes to Counter Electrodes." *Materials Today* 18 (3): 155–62. https://doi.org/10.1016/j.mattod.2014.09.001.

Yin, Sijie, Zewen Jin, and Takeo Miyake. 2019. "Wearable High-Powered Biofuel Cells Using Enzyme/Carbon Nanotube Composite Fibers on Textile Cloth." *Biosensors and Bioelectronics* 141 (September): 111471. https://doi.org/10.1016/j.bios.2019.111471.

You, Ming H., Xiao X. Wang, Xu Yan, Jun Zhang, Wei Z. Song, Miao Yu, Zhi Y. Fan, Seeram Ramakrishna, and Yun Z. Long. 2018. "A Self-Powered Flexible Hybrid Piezoelectric-Pyroelectric Nanogenerator Based on Non-Woven Nanofiber Membranes." *Journal of Materials Chemistry A* 6 (8): 3500–09. https://doi.org/10.1039/c7ta10175a.

Yu, Aifang, Xiong Pu, Rongmei Wen, Mengmeng Liu, Tao Zhou, Ke Zhang, Yang Zhang, Junyi Zhai, Weiguo Hu, and Zhong L. Wang. 2017. "Core-Shell-Yarn-Based Triboelectric Nanogenerator Textiles as Power Cloths." *ACS Nano* 11 (12): 12764–71. https://doi.org/10.1021/acsnano.7b07534.

Yu, Junbin, Xiaojuan Hou, Jian He, Min Cui, Chao Wang, Wenping Geng, Jiliang Mu, Bing Han, and Xiujian Chou. 2020. "Ultra-Flexible and High-Sensitive Triboelectric Nanogenerator as Electronic Skin for Self-Powered Human Physiological Signal Monitoring." *Nano Energy* 69 (March): 104437. https://doi.org/10.1016/j.nanoen.2019.104437.

Yu, Xinghai, Jian Pan, Jing Zhang, Hao Sun, Sisi He, Longbin Qiu, Huiqing Lou, Xuemei Sun, and Huisheng Peng. 2017. "A Coaxial Triboelectric Nanogenerator Fiber for Energy Harvesting and Sensing under Deformation." *Journal of Materials Chemistry A* 5 (13): 6032–37. https://doi.org/10.1039/C7TA00248C.

Yu, Yanhao, Haiyan Sun, Hakan Orbay, Feng Chen, Christopher G. England, Weibo Cai, and Xudong Wang. 2016. "Biocompatibility and in Vivo Operation of Implantable Mesoporous PVDF-Based Nanogenerators." *Nano Energy* 27 (September): 275–81. https://doi.org/10.1016/j.nanoen.2016.07.015.

Yuan, Kai, Ting Hu, and Yiwang Chen. 2020. "Flexible and Wearable Solar Cells and Supercapacitors." In *Flexible and Wearable Electronics for Smart Clothing*, 87–129. Germany: Wiley-VCH. https://doi.org/10.1002/9783527818556.ch5.

Zabek, Daniel, John Taylor, Emmanuel L. Boulbar, and Chris R. Bowen. 2015. "Micropatterning of Flexible and Free Standing Polyvinylidene Difluoride (PVDF) Films for Enhanced Pyroelectric Energy Transformation." *Advanced Energy Materials* 5 (8): 1401891. https://doi.org/10.1002/aenm.201401891.

Zeng, Wei, Xiao M. Tao, Song Chen, Songmin Shang, Helen L. W. Chan, and Siu H. Choy. 2013. "Highly Durable All-Fiber Nanogenerator for Mechanical Energy Harvesting." *Energy and Environmental Science* 6 (9): 2631–8. https://doi.org/10.1039/c3ee41063c.

Zhang, Ding, Yiding Song, Lu Ping, Suwen Xu, De Yang, Yuanhao Wang, and Ya Yang. 2019. "Photo-Thermoelectric Effect Induced Electricity in Stretchable Graphene-Polymer Nanocomposites for Ultrasensitive Strain Sensing." *Nano Research* 12 (12): 2982–87. https://doi.org/10.1007/s12274-019-2541-2.

Zhang, Ding, Yuanhao Wang, and Ya Yang. 2019. "Design, Performance, and Application of Thermoelectric Nanogenerators." *Small* 15 (32): 1805241. https://doi.org/10.1002/smll.201805241.

Zhang, Ding, Kewei Zhang, Yuanming Wang, Yuanhao Wang, and Ya Yang. 2019. "Thermoelectric Effect Induced Electricity in Stretchable Graphene-Polymer Nanocomposites for Ultrasensitive Self-Powered Strain Sensor System." *Nano Energy* 56 (February): 25–32. https://doi.org/10.1016/j.nanoen.2018.11.026.

Zhang, Hao, Xiao S. Zhang, Xiaoliang Cheng, Yang Liu, Mengdi Han, Xiang Xue, Shuaifei Wang, et al. 2015. "A Flexible and Implantable Piezoelectric Generator Harvesting Energy from the Pulsation of Ascending Aorta: In Vitro and in Vivo Studies." *Nano Energy* 12 (March): 296–304. https://doi.org/10.1016/j.nanoen.2014.12.038.

Zhang, Hong, Jiaqi Cheng, Francis Lin, Hexiang He, Jian Mao, Kam S. Wong, Alex K. Y. Jen, and Wallace C. H. Choy. 2016. "Pinhole-Free and Surface-Nanostructured Niox Film by Room-Temperature Solution Process for High-Performance Flexible Perovskite Solar Cells with Good Stability and Reproducibility." *ACS Nano* 10 (1): 1503–11. https://doi.org/10.1021/acsnano.5b07043.

Zhang, Hulin, Yuhang Xie, Xiaomei Li, Zhenlong Huang, Shangjie Zhang, Yuanjie Su, Bo Wu, Long He, Weiqing Yang, and Yuan Lin. 2016. "Flexible Pyroelectric Generators for Scavenging Ambient Thermal Energy and as Self-Powered Thermosensors." *Energy* 101 (April): 202–10. https://doi.org/10.1016/j.energy.2016.02.002.

Zhang, Nannan, Jun Chen, Yi Huang, Wanwan Guo, Jin Yang, Jun Du, Xing Fan, and Changyuan Tao. 2016. "A Wearable All-Solid Photovoltaic Textile." *Advanced Materials* 28 (2): 263–9. https://doi.org/10.1002/adma.201504137.

Zhang, Ting, Kaiwei Li, Jing Zhang, Ming Chen, Zhe Wang, Shaoyang Ma, Nan Zhang, and Lei Wei. 2017. "High-Performance, Flexible, and Ultralong Crystalline Thermoelectric Fibers." *Nano Energy* 41 (September): 35–42. https://doi.org/10.1016/j.nanoen.2017.09.019.

Zhang, Zhi, Ying Chen, and Jiansheng Guo. 2019. "ZnO Nanorods Patterned-Textile Using a Novel Hydrothermal Method for Sandwich Structured-Piezoelectric Nanogenerator for Human Energy Harvesting." *Physica E: Low-Dimensional Systems and Nanostructures* 105 (June 2018): 212–8. https://doi.org/10.1016/j.physe.2018.09.007.

Zhang, Zhitao, Xueyi Li, Guozhen Guan, Shaowu Pan, Zhengju Zhu, Dayong Ren, and Huisheng Peng. 2014. "A Lightweight Polymer Solar Cell Textile That Functions When Illuminated from Either Side." *Angewandte Chemie – International Edition* 53 (43): 11571–4. https://doi.org/10.1002/anie.201407688.

Zhang, Zixuan, Tianyiyi He, Minglu Zhu, Zhongda Sun, Qiongfeng Shi, Jianxiong Zhu, Bowei Dong, Mehmet R. Yuce, and Chengkuo Lee. 2020. "Deep Learning-Enabled Triboelectric Smart Socks for IoT-Based Gait Analysis and VR Applications." *npj Flexible Electronics* 4 (1): 1–12. https://doi.org/10.1038/s41528-020-00092-7.

Zhao, Gengrui, Yawen Zhang, Nan Shi, Zhirong Liu, Xiaodi Zhang, Mengqi Wu, Caofeng Pan, Hongliang Liu, Linlin Li, and Zhong L. Wang. 2019. "Transparent and Stretchable Triboelectric Nanogenerator for Self-Powered Tactile Sensing." *Nano Energy* 59 (January): 302–10. https://doi.org/10.1016/j.nanoen.2019.02.054.

Zhao, Luming, Hu Li, Jianping Meng, and Zhou Li. 2020. "The Recent Advances in Self-powered Medical Information Sensors." *InfoMat* 2 (1): 212–34. https://doi.org/10.1002/inf2.12064.

Zhao, Tianming, Junlang Li, Hui Zeng, Yongming Fu, and Haoxuan He. 2018. "Self-Powered Wearable Sensing-Textiles for Real-Time Detecting Environmental Atmosphere and Body Motion Based on Surface-Triboelectric Coupling Effect." *Nanotechnology* 29 (40): 405504. https://doi.org/10.1088/1361-6528/aad3fc.

Zhao, Xuan, Chuanshan Zhao, Yifei Jiang, Xingxiang Ji, Fangong Kong, Tong Lin, Hao Shao, and Wenjia Han. 2020. "Flexible Cellulose Nanofiber/Bi_2Te_3 Composite Film for Wearable Thermoelectric Devices." *Journal of Power Sources* 479 (October): 229044. https://doi.org/10.1016/j.jpowsour.2020.229044.

Zhao, Zhenfu, Xiong Pu, Chunhua Du, Linxuan Li, Chunyan Jiang, and Weiguo Hu. 2016. "Freestanding Flag-Type Triboelectric Nanogenerator for Harvesting High-Altitude Wind Energy from Arbitrary Directions." *ACS Nano* 10: 1780–7. https://doi.org/10.1021/acsnano.5b07157.

Zhao, Zhizhen, Casey Yan, Zhaoxian Liu, Xiuli Fu, Lian-mao Peng, and Youfan Hu. 2016. "Machine-Washable Textile Triboelectric Nanogenerators for Effective Human Respiratory Monitoring through Loom Weaving of Metallic Yarns." *Advanced Materials* 28: 10267–74. https://doi.org/10.1002/adma.201603679.

Zheng, Qiang, Bojing Shi, Fengru Fan, Xinxin Wang, Ling Yan, Weiwei Yuan, Sihong Wang, Hong Liu, Zhou Li, and Zhong L. Wang. 2014. "In Vivo Powering of Pacemaker by Breathing-Driven Implanted Triboelectric Nanogenerator." *Advanced Materials* 26 (33): 5851–56. https://doi.org/10.1002/adma.201402064.

Zheng, Qiang, Hao Zhang, Bojing Shi, Xiang Xue, Zhuo Liu, Yiming Jin, Ye Ma, et al. 2016. "In Vivo Self-Powered Wireless Cardiac Monitoring via Implantable Triboelectric Nanogenerator." *ACS Nano* 10 (7): 6510–18. https://doi.org/10.1021/acsnano.6b02693.

Zheng, Zhuang H., Ping Fan, Jing T. Luo, Guang X. Liang, and Dong P. Zhang. 2013. "Enhanced Thermoelectric Properties of Antimony Telluride Thin Films with Preferred Orientation Prepared by Sputtering a Fan-Shaped Binary Composite Target." *Journal of Electronic Materials* 42 (12): 3421–5. https://doi.org/10.1007/s11664-013-2779-5.

Zhong, Junwen, Yan Zhang, Qize Zhong, Qiyi Hu, Bin Hu, Zhong L. Wang, and Jun Zhou. 2014. "Fiber-Based Generator for Wearable Electronics and Mobile Medication." *ACS Nano* 6: 6273–80.

Zhou, Tao, Chi Zhang, Chang B. Han, Feng R. Fan, Wei Tang, and Zhong L. Wang. 2014. "Woven Structured Triboelectric Nanogenerator for Wearable Devices." *Appl. Mater. Interfaces* 6 (16): 14695–701. https://doi.org/10.1021/am504110u.

Zhou, Yuman, Jianxin He, Hongbo Wang, Kun Qi, Nan Nan, Xiaolu You, Weili Shao, Lidan Wang, Bin Ding, and Shizhong Cui. 2017. "Highly Sensitive, Self-Powered and Wearable Electronic Skin Based on Pressure-Sensitive Nanofiber Woven Fabric Sensor." *Scientific Reports* 7 (1): 1–9. https://doi.org/10.1038/s41598-017-13281-8.

Zhu, Minshen, Yang Huang, Wing S. Ng, Junyi Liu, Zifeng Wang, Zhengyue Wang, Hong Hu, and Chunyi Zhi. 2016. "3D Spacer Fabric Based Multifunctional Triboelectric Nanogenerator with Great Feasibility for Mechanized Large-Scale Production." *Nano Energy* 27: 439–46. https://doi.org/10.1016/j.nanoen.2016.07.016.

Zou, Haiyang, Ying Zhang, Litong Guo, Peihong Wang, Xu He, Guozhang Dai, Haiwu Zheng, et al. 2019. "Quantifying the Triboelectric Series." *Nature Communications* 10 (1): 1–9. https://doi.org/10.1038/s41467-019-09461-x.

Zou, Yang, Puchuan Tan, Bojing Shi, Han Ouyang, Dongjie Jiang, Zhuo Liu, Hu Li, et al. 2019. "A Bionic Stretchable Nanogenerator for Underwater Sensing and Energy Harvesting." *Nature Communications* 10 (1): 1–10. https://doi.org/10.1038/s41467-019-10433-4.

11 Flexible Nanogenerators
A Promising Route of Harvesting Mechanical Energy

Syed Wazed Ali, Satyaranjan Bairagi, and Swagata Banerjee

CONTENTS

11.1 INTRODUCTION

Nowadays, continuous supply of energy to run the wearable and portable electronic devices is a challenging task. In maximum cases, fossil fuels have been utilized to omit this energy problem in the electronic markets. But the consumption of fossil fuels has different environmental threats as is known for us. From the past, researchers have tried to establish ecofriendly energy supplying technologies to the electronic goods so that environmental threats are omitted. And interestingly it has been found that there are many technologies by which renewable energy can be utilized to provide continuous power to the wearable and portable electronic goods by maintaining the clean environment. Moreover, our quality life will also get enhanced. There are many inexhaustive energy sources such as thermal, solar, mechanical, chemical and biological sources of energy. The mechanical energy is one of the most usable energy sources among others due to its abundant availability. It exists in different forms in our surrounding environment such as wind energy, human body movement, vibration energy and many more. It can also be harvested from sources such as human motion and walking that serve as a green energy source [1–4]. Two main mechanical energy harvesting technologies are piezoelectric and triboelectric. These two technologies can convert the mechanical energy into useful electrical energy in a systematic way. The piezoelectric energy harvesting can be defined as the generation of electrical energy when materials are subjected to the mechanical stress and vice versa. Similarly, triboelectric energy harvesting is nothing but the technology to harvest electrical energy from the mechanical energy when two dissimilar surfaces are rubbed together [5]. There are many piezoelectric and triboelectric materials such as

DOI: 10.1201/9781003187615-11

poly(vinylidene difluoride) (PVDF) and its copolymers (piezoelectric materials), and polytetrafluroethylene (PTFE), polypropylene (PP), silk, cotton (triboelectric materials) [6, 7]. To date, many advanced piezoelectric and triboelectric materials have been explored by different research groups for mechanical energy harvesting. In addition, in the recent days, researchers are trying to develop different flexible piezoelectric and triboelectric nanogenerators (TENGs) using different polymeric and other filler materials. As yet, the maximum power densities of 500 W/m^2 (for a single nanogenerator) and 15 MW/m^2 (for the arrays of nanogenerators) have been reported for the TENG. Thus, TENG has been used not only in electronic applications (to supply the power) but also to harvest lump of energy from the wind and other mechanical energy sources [8, 9].

Other outmost applications of the nanogenerators are bendable electronic and biomedical. This is because their inherent flexibility, bendability, twistability and so on. For this reason, the utilization of flexible electronic gadgets has increased in daily life. Therefore, the selection of suitable materials having enough flexibility and mechanical stability is a challenging task [10–13]. In this chapter, two mechanical energy harvesting technologies, including flexible piezoelectric and TENGs, have been discussed in a systematic way. In addition, applications of flexible nanogenerators have also been included as a separate section of this chapter.

11.2 ENERGY HARVESTING

Energy harvesting is defined as the harvesting or scavenging of the energy from surrounding sources such as solar energy, wind energy, thermal energy and so forth and stores this energy in the form of batteries, capacitors, etc. The energy harvesting process has been started more than 50 years back. It was started from harvesting the energy from the waterwheels or windmills. But very recently, many technologies have been used to harvest different small-scale waste energies as well. And researchers are trying to use modern energy harvesting devices as power sources in the small-scale electronic goods instead of battery [14]. Amongst various technical approaches of harvesting energy, triboelectric and piezoelectric technologies are the most promising electromechanical energy harvesting technology. They can harvest mechanical energy available in our surrounding environment in a simple way [15]. In this chapter, mainly these two technologies have been elaborated in detail.

11.3 FLEXIBLE PIEZOELECTRIC NANOGENERATORS

11.3.1 Piezoelectric Materials

The piezoelectric effect was first explored in quartz and Rochelle salt in 1880 by Curie brothers [16]. But in that time, they were not able to discover indirect piezoelectric effect which was first discovered by Lippmann in 1881 [17]. There are more than two hundred piezoelectric materials which can be used to harvest the energy, including quartz, lead titanate (PT), cadmium sulphide (CdS), barium titanate (BT), lead zirconate titanate (PZT), lead magnesium niobate (PMN), lead lanthanum zirconate titanate (PLZT), PVDF, poly(vinylidene difluoride-trifluroethane) (PVDF-TrFE) and many more. Barium titanate (BT) was first exposed as a piezoelectric material and PZT is the most usable piezoelectric material. BT and PZT all are the ceramic-based piezoelectric materials. Ceramic-based materials are brittle in nature having higher piezoelectric properties than polymeric materials [18]. Specially, PZT is the most preferred piezoelectric materials with piezoelectric constant (d_{33}) of >400 pC/N and relative permittivity (ε) of >1800. However, its lead content is above 60%, which makes it toxic in nature. Therefore, PZT is considered a non-environment-friendly piezoelectric material. Researchers have synthesized potassium sodium niobate (KNN) as an alternative of PZT materials which has superior electrical, dielectric, ferroelectric and piezoelectric properties. It shows a higher piezoelectric constant (d_{33}) of 100–400 pC/N and Curie temperature in the range of 217–304°C. Moreover, KNN can reveal maximum piezoelectric properties for its crystal structure with morphotropic phase boundary (MPB) [19]. It has also shown a higher dielectric constant (ε_r) (~290) and a remnant polarization value (~33 μC/cm^2). On the other hand, the most usable piezoelectric polymer is PVDF because of its higher dielectric constant (ε_r) (~10) and flexibility as compared to other piezoelectric polymers (polypropylene, Nylon 11, polylactic acid). PVDF is a semicrystalline (~60% crystallinity) polymer having five different crystalline structures (α, β, γ, δ and ε). The β-crystalline phase of a PVDF polymer is responsible for its piezoelectric and ferroelectric properties. However, an α-crystalline phase is the most stable state of PVDF polymers and therefore, for getting efficient piezoelectric effect from PVDF polymer, the α-crystalline phase must be transformed into the β-crystalline phase. There are various methods by which the α-crystalline phase of PVDF can be converted into the β-crystalline phase. For instance, by mechanical stretching, the inclusion of filler is made in the PVDF matrix, by electrospinning and many more [19]. But a lower electromechanical coupling coefficient (~0.30) and the Curie temperature (~90°C) of PVDF limit its piezoelectric applications. In summary, ceramic-based piezometric materials have higher piezoelectric properties, but brittleness limits their applications in the different remote places where flexibility is the prime requirement. In contrast, polymer-based materials are flexible in nature and can be used to prepare a flexible piezoelectric device. However, their piezoelectric properties are lower than ceramic-based materials. Therefore, composite-based piezoelectric materials are one of the best solutions as compared to ceramic and polymeric materials as a single component. In composite-based piezoelectric materials, the best properties from both the ceramic and polymer parts can effectively be tuned.

11.3.2 Terminology Related to Piezoelectric Energy Harvesting

Piezoelectricity can be summarized into the following two linear constitutive piezoelectric equations [20]:

$$S_p = s^E_{pq} T_q + d_{pk} E_k \tag{11.1}$$

$$D_i = d_{iq} T_q + \varepsilon^T_{ik} + E_k \tag{11.2}$$

where s^E_{pq} is an elastic compliance tensor at constant electric field, ε^T_{ik} is a dielectric constant tensor under constant stress, d_{pk} is a piezoelectric constant tensor, S_p is the mechanical strain in the p direction, D_i is electric displacement in the i direction, T_q is mechanical stress in the q direction, and E_k is the electric field in the k direction.

Piezoelectricity happens due to the cross-coupling of the mechanical variables such as stress and strain and electrical variables such as electrical charge density and electric field, and two methods of coupling modes are available: the 31 and 33 modes.

In piezoelectric coefficient designation, d_{ij}, the first subscript is the direction of the electric field, while the second subscript denotes the direction of the mechanical deformation or stress. So, in the 31 coupling mode, the deformation force applied is perpendicular to the direction of the poled molecules, whereas in the 33 mode, the force and the poling direction are in the same direction. Just to understand this in a better way, if we compress a piezoelectric material which is put in the sole of a shoe, it would be 33 mode, but the bending movement of PVDF sheet is 31 mode [21].

To measure the piezoelectric response of a material, it would be possible to use the direct effect or the converse effect so as to get the piezoelectric coefficient or the electromechanical coupling coefficient [22].

Definitions of the terminologies such as piezoelectric charge constant (d_{ij}), piezoelectric voltage constant (g_{ij}), dielectric constant and loss, coupling coefficient (k_p), compliance (s) along with their formula are given in the following sections.

11.3.2.1 Piezoelectric Coefficient

Piezoelectric coefficient, d_{ij}, can be measured by applying a deformation stress and measuring the generated voltage using an electrometer. The directions of the force and voltage can be perpendicular to get d_{31} or parallel to get d_{33}.

And it is the ratio of the charge generated to the applied stress in its direct effect or the strain developed due to the applied voltage in the converse effect determines the coefficient d_{ij} as shown in Equation 11.1:

$$D = dD/dX \text{ or } dS/dE, \tag{11.3}$$

where D is the energy density and S is the strain.

The units of measurement would be C/N or m/V for direct and converse effect, respectively.

Therefore, the higher the piezoelectric coefficient, the better the material is at converting mechanical stress into electrical charge [21, 22].

11.3.2.2 Dielectric Constant

Permittivity is the ability of a material to polarize under an electric field which is related to dielectric constant. Permittivity of a material to permittivity of a vacuum is relative permittivity. When an electric field is applied, electronic, atomic and orientation polarizations occur due to the separation of positive and negative charges to behave like electric dipoles. Relative permittivity consists of real and imaginary parts which are dielectric constant and dielectric loss ($\tan\delta$). The relaxation time in a polymer is the cause for the inability of polarization process to follow the rate of change of oscillating-applied electric field which is dielectric loss.

11.3.2.3 Electromechanical Coupling Coefficient (k_{ij})

The capacity of a piezo material to convert electrical energy into mechanical energy or vice versa is characterized by the electromechanical coupling coefficient k_{ij}. The higher the coefficient the more appropriate for energy harvesting devices:

$$K = \frac{\text{Electrical energy converted to mechanical energy}}{\text{Input electrical energy}} \tag{11.4}$$

$$K = \frac{\text{Mechanical energy converted to electrical energy}}{\text{Input mechanical energy}} \tag{11.5}$$

11.3.3 Mechanical Energy Harvesting by Flexible Piezoelectric Nanogenerator

There are different technologies as mentioned in the previous section, by which mechanical energy can be harvested. The piezoelectric energy harvesting technology has recently gained great attention than other technologies because of its cost-effectiveness, easy processability and higher energy generation capability. With the time, flexible piezoelectric nanogenerators have drawn a great interest in different applications such as energy harvesting, robotics, metrology, soft and integrated electronic skin. This is obviously due to their desirable flexibility, stretchability and lightweight. Considering the flexibility and lightweight characteristics of PVDF polymers, from the very past advanced flexible piezoelectric devices have been fabricated out of this material for wearables and implantable biomedical applications. Recently, flexible piezoelectric nanogenerators have been documented as an inexpensive energy harvesting tactic due to its various advantages as mentioned earlier [23]. In this section, flexible piezoelectric technology–based nanogenerators made of different materials have been conferred in a compact way.

The ceramic or metal oxide–based piezoelectric materials are brittle in nature, and therefore, it is an impossible task to develop flexible piezoelectric nanogenerators using those brittle materials alone. Those materials can be implemented along with flexible substrates or polymer matrices. The researchers have already tried to develop flexible piezoelectric nanogenerators based on ceramic or metal oxide growth flexible substrates. In one such study, Yang *et al.* [24] have developed a flexible piezoelectric energy harvester where they have used ZnO piezoelectric fine wire (PFW) as a piezoelectric material. In this regard, ZnO fine wire is lied laterally on the Kapton polyimide flexible substrate followed by pasting of silver paste on two ends of the PFW by taking help of a microscope. The entire assembly is then encapsulated into the flexible polymer which serves as a protecting cover of the device from outside environment. This flexible nanogenerator has shown an output voltage of ~50 mV and an output current of 400–700 pA against a strain in the range of 0.05%–0.1%. Since piezoelectric performance is lower with a single ZnO nanowire (NW)-based flexible nanogenerator, researchers have tried to develop NWs array–based nanogenerators. Xu *et al.* [25] have fabricated a flexible piezoelectric nanogenerator based on the array of ZnO NWs. ZnO NWs have been integrated on the Kapton film in two directions (i.e. lateral and vertical). Each nanogenerator contains 700 rows of ZnO NWs. The lateral integrated ZnO NWs-based piezoelectric nanogenerator has shown a maximum output voltage of 1.26 V and a current of 28.8 nA, consequently at a mechanical strain of ~0.19%, the vertically integrated ZnO NWs-based nanogenerator has shown a power density of ~2.7 mW/cm^3. Finally, authors have concluded that these flexible piezoelectric nanogenerators are capable of charging the battery and capacitor as well. Zhu *et al.* [26] have also reported the piezoelectric performance of ZnO NWs arrays–based flexible nanogenerators. It has been found that this flexible nanogenerator can generate an open-circuit output voltage of ~2.03 V and ~11 mW/cm^3 power density, respectively. PZT is the preferred piezoelectric materials since it has a very high dielectric and piezoelectric constant as mentioned earlier. However, their brittle nature and toxicity problems limit their applications, although PZT-based flexible piezoelectric nanogenerators have been developed by the different research groups. For example, there are various piezoelectric polymers such as PVDF, polylactic acid (PLA) and Nylon 11. PVDF polymers have drawn great interest among them as it has higher piezoelectric properties as compared to other polymers. Keeping the advantages of PVDF polymers in mind, researchers have tried to develop flexible nanogenerators based on PVDF polymers. In one such research, Bairagi *et al.* [27] have developed a piezoelectric energy harvester, which is flexible, cost-effective and exhibit superior piezoelectric effect in terms of voltage and power density. Here, a unique nanogenerator has been engineered comprising nanorods with an average aspect ratio of 8.5 and made up of lead-free KNN and PVDF without using separate electrical poling process. Interestingly, the authors have mentioned that KNN nanorods influence a negative effect on the growth of beta nucleation in a PVDF matrix but, on the other hand, significantly help in generation of enhanced piezoelectric response in the nanocomposite sample by the action of self-orientation of the dipoles during the preparation of nano-fibrous strand in electrospinning process. Therefore, such an approach can resolve the problem of poling the polymer and filler dipoles separately in a subsequent process. The results exhibited by the nanodevice demonstrated high open-circuit output voltage and current of 17.5 V and 0.522 μA, respectively, along with a current density of 0.13 μA/cm^2 by finger tapping repeatedly. Also, the device can power an LED of 2 V practically. So, from this study, it can be emphasized that a flexible piezoelectric nanogenerator shows potential applications in the field of electronic wearables. In the above-mentioned study, the researchers have clearly highlighted the use of lead-free KNN for making nanogenerators to light LEDs. In that context, the same group has reported the use of silane–modified KNN along with PVDF electrospun fibrous web-based nanogenerators. The authors stated that by the surface modification of KNN and its incorporation into PVDF nanoweb can enhance the performance towards piezoelectricity notably because of the deagglomeration of the nano fillers within the PVDF matrix and better interaction between the PVDF polymeric chains and KNN nanorods. Results proved that the PVDF containing 3% surface-modified KNN nanocomposites showed remarkable rise in output voltage (21 V), output current (22 mA), current density (5.5 mA/cm^2)

and power density (115.5 mW/cm²). The images of output current, current density and power density of the pure Poly vinylidene difluoride (PDVF) and surface-modified KNN (at different concentrations) incorporated PVDF-based nanogenerator are shown in Figure 11.1 [28].

In the previous work, the researchers have used electrospinning process to manufacture the nanocomposite. But in this trailing work to be mentioned next, the authors have used twin-screw melt mixing process to prepare PVDF-incorporated KNN filaments-based composites followed by

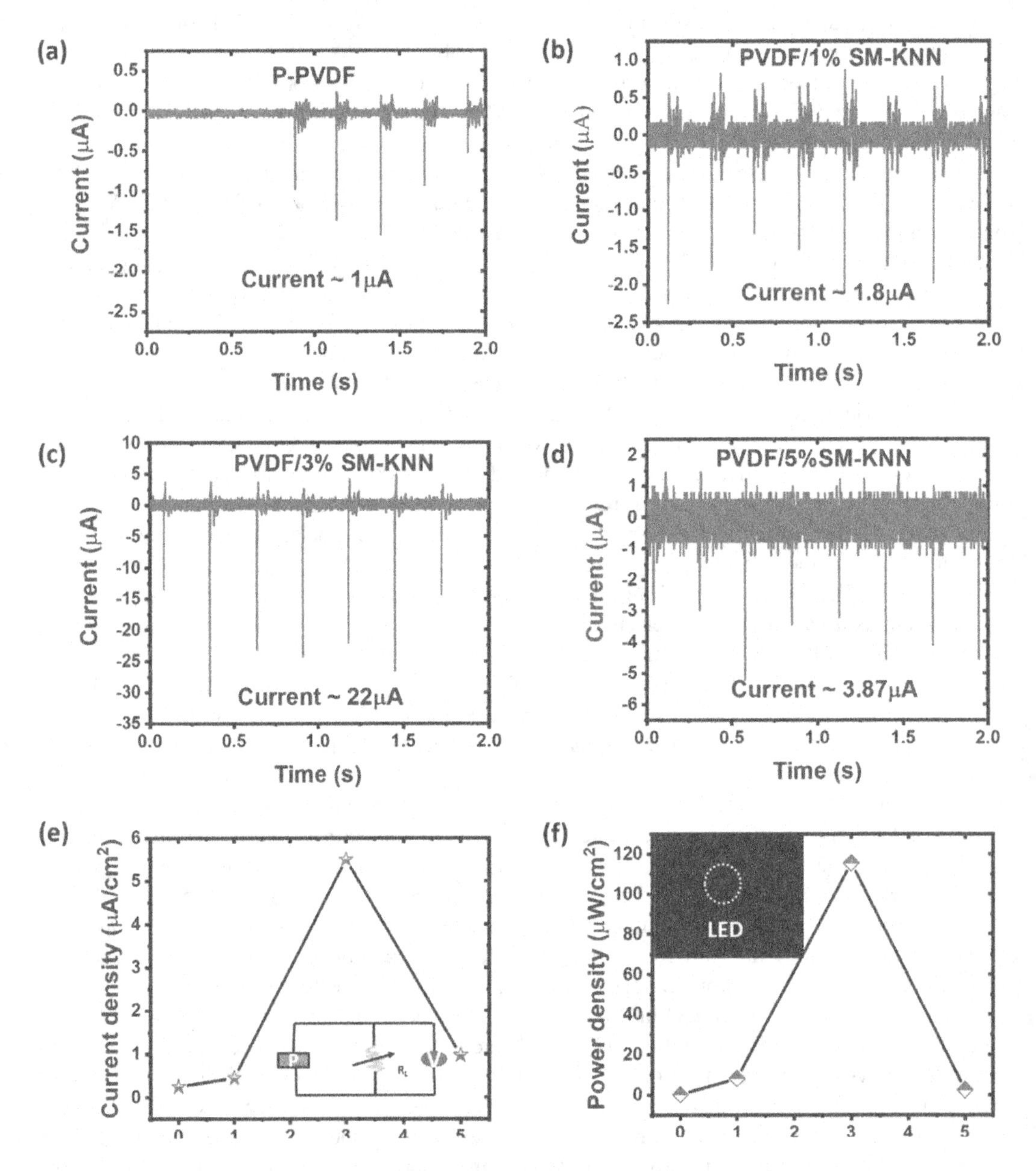

FIGURE 11.1 Image portraying the output current of (a) pure PVDF nanogenerators, (b) PVDF with 1% surface-modified KNN-based nanogenerators, (c) PVDF with 3% surface-modified KNN-based nanogenerators, (d) PVDF with 5% surface-modified KNN nanogenerators, (e) current densities shown by different nanogenerators, including varying proportions of surface-modified-KNN nanorods (inside: circuit diagram for current measurement) and (f) resultant power densities of different nanogenerators with various proportions of surface-modified-KNN nanorods inside the matrix of PVDF [28].

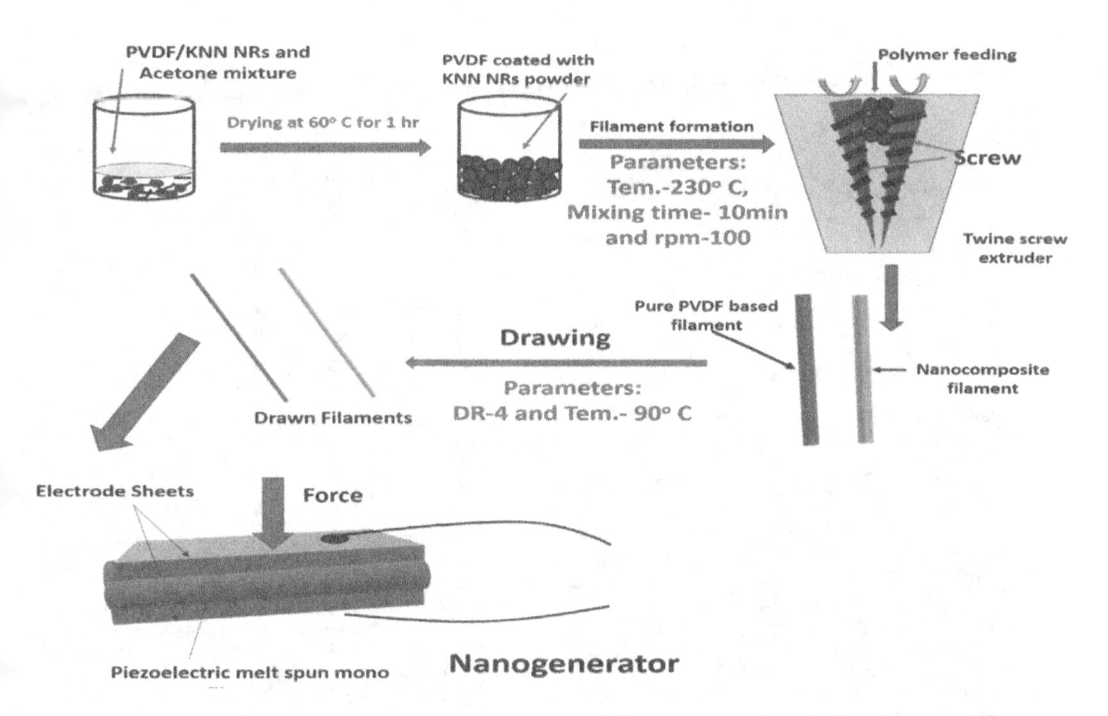

FIGURE 11.2 Flow chart displaying the production of melt-spun nanorod-based filaments along with the preparation of the nanogenerator [29].

the process of melt spinning and the set-up of the same is shown in Figure 11.2. In this method, different loading percentages of KNN nanorods were varied on the polymer weight. The nanogenerator fabricated from PVDF/4% KNN filament showed better beta fraction values of around 26% compared to others with low level of KNN nanorods loading. The authors reported that 4% PVDF/KNN nanorods–based filaments exhibited a voltage of 3.7 V and a current of 0.326 μA. It was concluded by them that the performance shown by the nanogenerator is powerful enough to use them in portable and wearable electronic goods in textiles [29].

In all the above-mentioned works the authors have used KNN as a filler material in PVDF polymers to enhance its properties, while making a nanocomposite material. For better understanding about the influence of KNN in the polymer-based composite, the same group of authors has revealed a designed nanocomposite film with the combination of KNN nanorods and PVDF matrix [30]. Having varied the KNN loading in PVDF matrix, it was found that the nanocomposite containing 10% KNN in a PVDF film exhibited the best results in terms of a high beta fraction value of approximately 98%. The enhanced beta fraction in the nanocomposite sample is attributed to the KNN fillers acting as a nucleating agent of beta phase of the PVDF matrix. Figure 11.3 (scanning electron microscopy [SEM] images) illustrates the surface morphology of both the pure PVDF and 10% KNN nanorods–loaded PVDF sample. Furthermore, 10% KNN nanorods containing a PVDF film exhibited maximized effects like a dielectric constant (ε) value of around 23 and a remnant polarization (Pr) value of 145.0×10^{-3} μC/cm^2 as shown in Figure 11.4. Similarly, the output voltage and current of about 3.4 V and 0.100 μA, respectively, are exhibited by PVDF/10% KNN nanogenerators. The values for current and voltage for a pure PVDF film were 1.2 V and 0.051 μA, respectively. The current density was found to be increased with 10% KNN loading, i.e. 0.025 μA/cm^2, whereas the other loading percentages of KNN nanorods gave lower current densities of 0.020 μA/cm^2 for 15%, 0.018 μA/cm^2 for 20%, 0.015 μA/cm^2 for 5% and finally 0.012 μA/cm^2 for 0%. Hence, a loading of 10% KNN nanorods in the PVDF matrix when engineered into a nanogenerator gave optimal piezoelectric outputs.

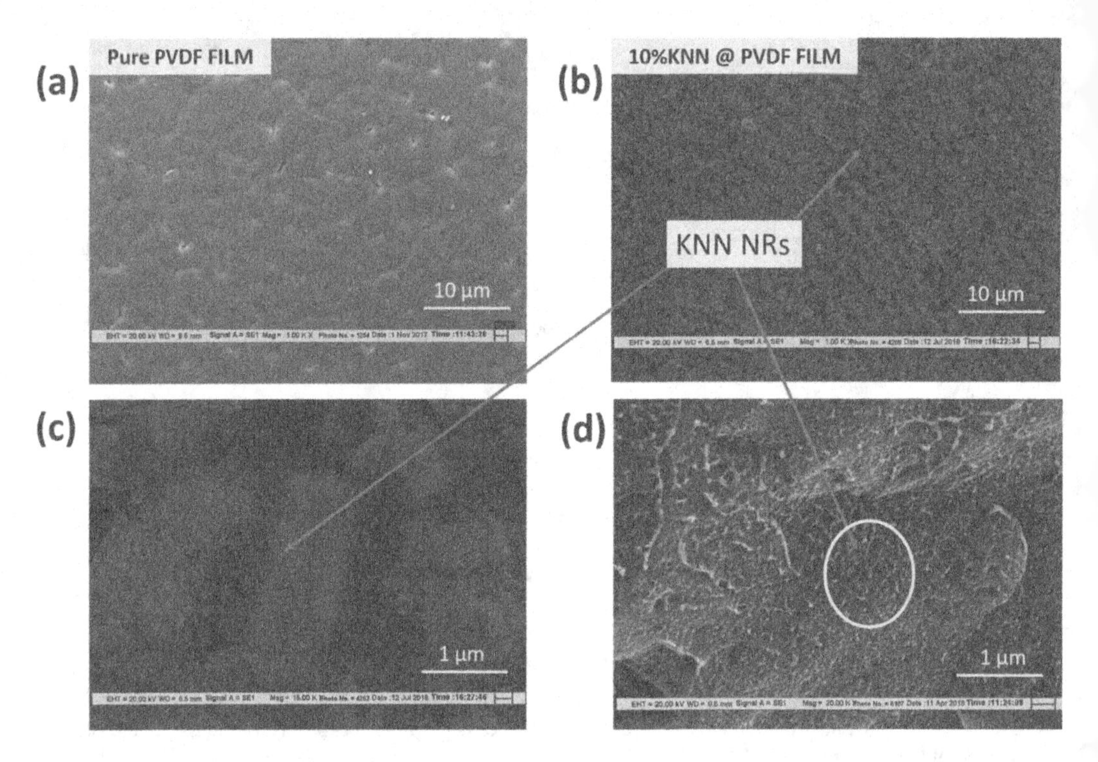

FIGURE 11.3 SEM images of (a) a pure and (b) a 10% KNN-loaded PVDF film, (c) SEM image with a higher magnification of 10% KNN nanorods inside the PVDF matrix, and (d) SEM image showing the cross-section of 10% KNN nanorods inside PVDF matrix [30].

Furthermore, the same group of researchers has also carried out a novel work based on the effect of different percentage loadings of carbon nanotubes on the piezoelectricity of PVDF/KNN-based electrospun nanocomposite. The inclusion of 0.1% carbon nanotubes in PVDF/KNN nanocomposite showed a noticeable enhancement in the performance of the nanogenerator, with an output voltage, current and power density of 23.24 V, 9 mA and 52.29 mW/cm^2, respectively. Also, the authors have stated the values of voltage, current and power density as 12 V, 18 mA and 54 mW/cm^2 when

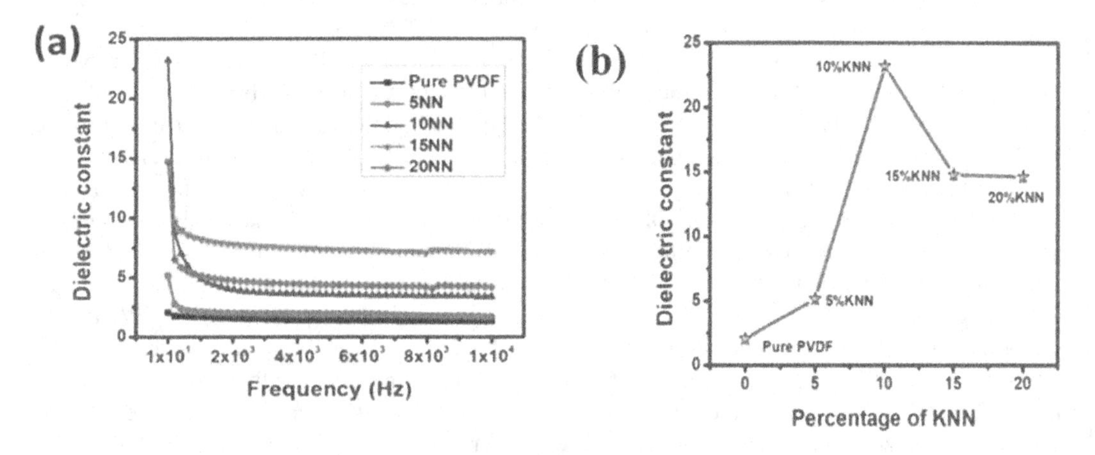

FIGURE 11.4 Dielectric constants of (a) different films, and (b) various KNN nanorods loaded with different percentages [30].

a compressive force was applied. The incorporation of carbon nanotubes into the nanocomposite sample enhanced the beta-fraction value. The reason as reported by the researchers is that carbon nanotubes can create a conductive path when loaded as a nanofiller in the polymer jet, and this results in improved mechanical stretching of as-prepared electrospun fibres. Therefore, it can be concluded that such a lead-free nanogenerator could act as a promising candidate in the case of a self-powered portable and wearable electronic goods, which require a very minimal amount of power [31].

11.4 FLEXIBLE TRIBOELECTRIC NANOGENERATORS

11.4.1 Triboelectric Materials

The triboelectric effect is one of those scientific phenomena that we experience in our daily activities. It is a contact-induced electrification that is caused because of the difference in the relative polarities of the materials in contact. It is an everyday electrostatics phenomenon. When two different material surfaces come in contact, charge transfer takes place between them in order to equalize their electrochemical potential. On separation, one of the surfaces retains the electrons, while the other let go of them. This results in a generation of surface charges on the two materials. This is the basic mechanism behind the triboelectrification [32]. The triboelectric materials are arranged in a series as shown in Figure 11.5, called the triboelectric series. This helps one to know about the relative charges developed on the material surfaces when they come in contact.

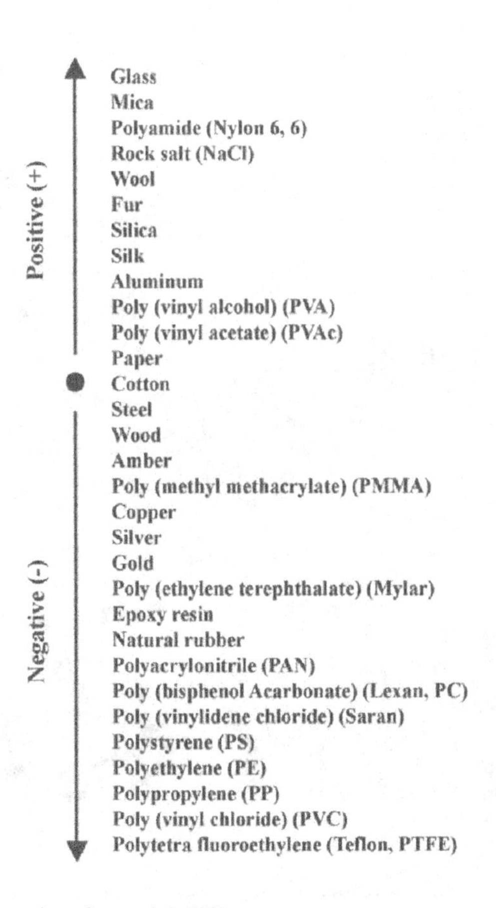

FIGURE 11.5 Triboelectric series of materials [33].

11.4.2 Mechanical Energy Harvesting by Flexible Triboelectric Nanogenerators

The triboelectric effect has been explored immensely in the field of energy harvesting [34–36]. The nanogenerators formed out of the triboelectric materials have shown appreciable energy output. These nanogenerators can be of the following types based on their contact modes [35]:

a. Vertical contact: When the two triboelectric materials come in close contact, they generate charges of opposite polarity. Electrodes are placed on one side of the two-poled dielectric film. Flow of electrons takes place when the surfaces come in contact, while the charge gets accumulated when the contact is cut off.
b. Lateral slip: When the two triboelectric surfaces laterally slide past each other, they create charges on their surfaces. The polarization induced is in the direction of friction, which causes the flow of electron resulting in balancing out of charges.
c. Single electrode: The movement of the TENG towards and away from the grounded dielectric film results in changes of the local electric field distribution. Electron flow takes place between the bottom electrode and the ground in order to maintain the potential of the film.
d. Non-connected electrodes: A pair of electrodes is placed below the dielectric. The working is illustrated in Figure 11.6.

An interesting TENG was prepared with the aid of 3D-printing technology (Figure 11.7). The distinguishing feature of this nanogenerator is its printed resin components that impart it ultra-flexibility. The triboelectric layer consists of the photopolymer resins and the ionic hydrogel forms

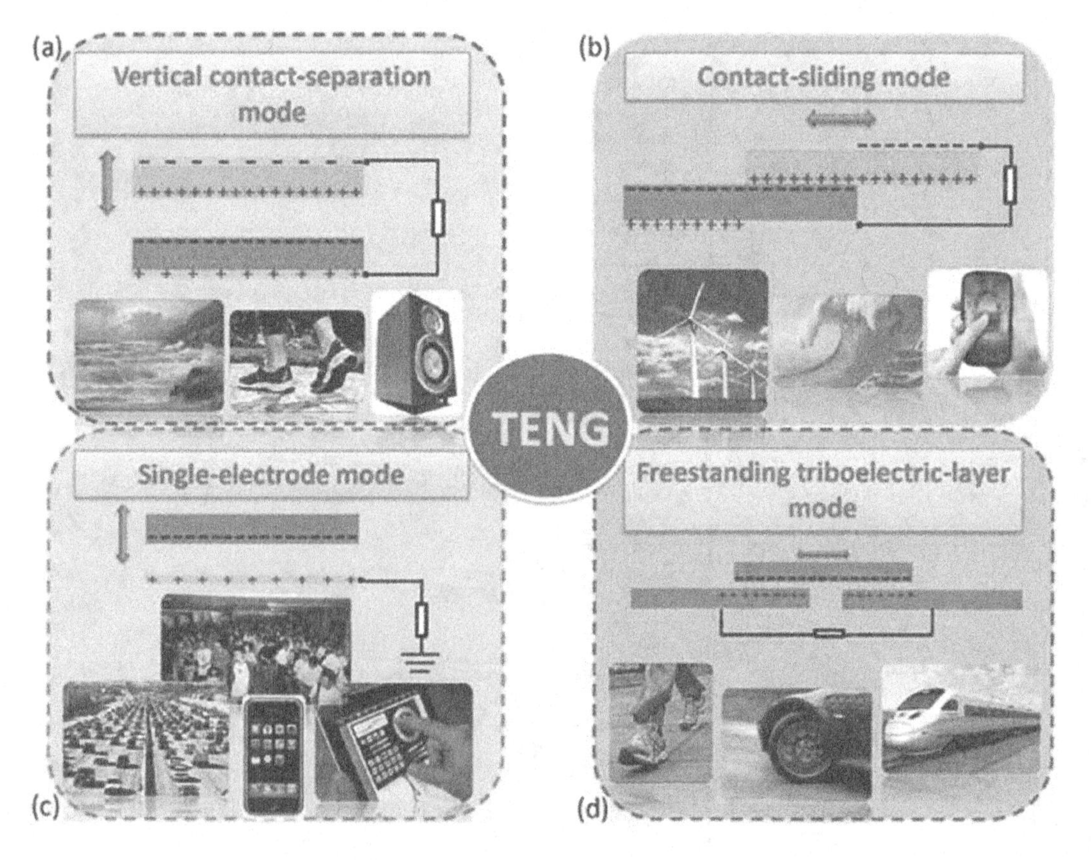

FIGURE 11.6 (a) Contact-separation mode, (b) contact-sliding mode, (c) single-electrode mode, and (d) freestanding triboelectric-layer mode of TENGs [35].

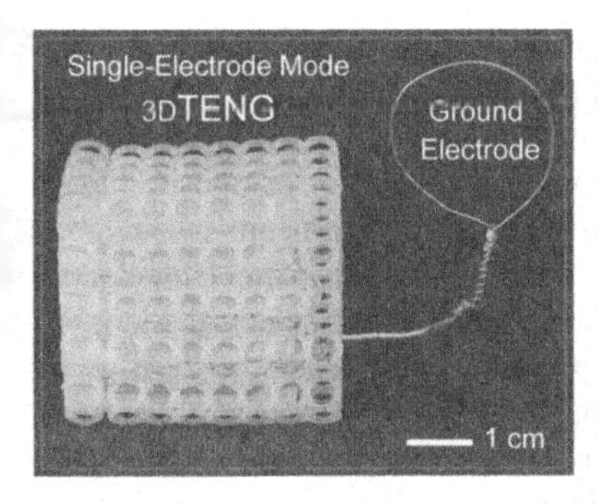

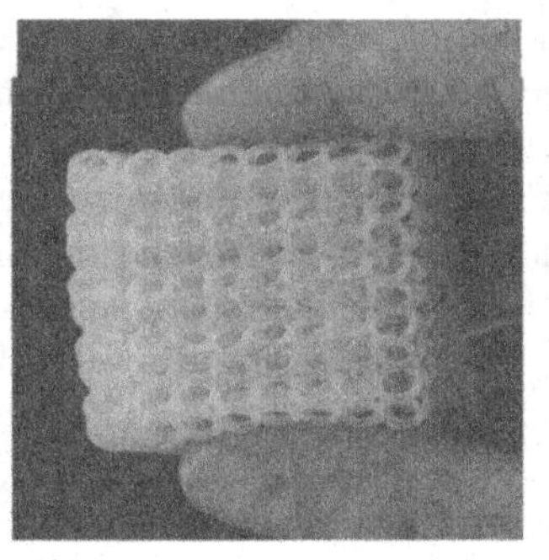
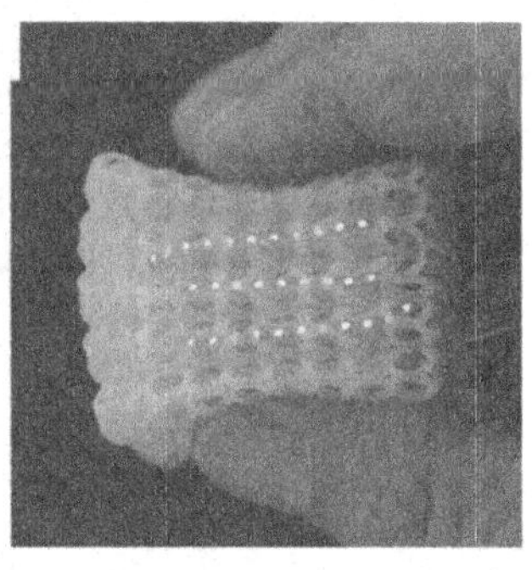

FIGURE 11.7 Ultra-flexible 3D-printed TENG [37].

the electrode. The TENG works in a single-electrode mode with the electrode being grounded. Under fully compressed mode, the resin approaches each other and develop charges of opposite polarity due to their difference in surface electron affinities. At such a juncture, the two opposite charges almost coincide with each other, thus producing zero net potential. When the layers separate, the static charge on one resin layer induces positive charge at the ionic hydrogel interface. This creates a negative potential between the electrode and ground that causes the current to flow until maximum separation. This innovative TENG is soft, elastic and hence can be incorporated into textile structures. This is also claimed to be a self-powered source that may find applications in electronic wearables [37].

The cellulose nanofibres (CNFs) have a number of hydroxyl groups on their surface. This gives an opportunity to surface functionalize the material and improve the triboelectric properties. The methyl group is electron donating group, while the nitro group is the electron withdrawing group. The triboelectric properties of the CNFs were tuned by the introduction of the nitro and methyl groups through chemical reactions. This demonstrated a low cost eco-friendly application of CNFs in TENG applications [38]. A TENG was made out of knitted fabric structures that had high stretchability. Silver and PTFE were used as the triboelectric materials for developing the TENG. The plain, double and rib-knitted structures were formed as the TENG material. Out of them, the rib-knitted structure showed the highest stretchability. Due to the difference in electron affinities, PTFE acquires a negative charge and silver has a positive charge. The gap variation between the PTFE and silver fabrics induces a flow of electrons that generate AC signals. The rib-knitted triboelectric structure gives a higher triboelectric performance because of the higher contact surface area that leads to more tribo-charge generation and hence higher triboelectric voltage [39]. A novel snow TENG was fabricated to harvest energy from the friction, while sliding on snow. It operates on a single electrode mode utilizing both contact electrification and electrostatic induction. The layers of the device are transparent, and the device as a whole is flexible. It utilizes the phenomenon of snow triboelectrification as the airborne snow particles slide past the device layers [40].

11.5 CONCLUSIONS AND FUTURE OUTLOOK

Harvesting energy from inexhaustive non-conventional sources is a budding area of research. Apprehending an energy crisis in the near future, the non-renewable energy sources are being explored for their potential to produce energy. The mechanical energy harvesting is gaining attention because of a wide scope of applications. The nanogenerators help to contribute in such mechanical energy harvesting. The piezoelectric or triboelectric-based nanogenerators help to harvest energy from the otherwise wasted mechanical form of energy. These nanogenerators are also used in biomechanical energy harvesting which includes energy harvesting based on human motions. These nanogenerators are being investigated because of their high conversion efficiency. Moreover, the power generated is enough to meet the energy requirements of the small power electronics.

With the demand of wearable electronics increasing, the flexibility of the nanogenerators is also being considered. The flexible and stretchable nanogenerators find applications in the electronic wearables. Different structures of the nanogenerators contribute to different requirements of the electronic wearables. They may sometimes be woven, knitted or composite-based. Different composite preparation techniques also have their respective pros and cons in the fabrication of a nanogenerator. The different classes of materials in the piezoelectric and triboelectric category are also selected based on the end use. However, in the case of piezoelectric ceramics, a trend towards lead-free approach is observed to make the products eco-friendly. A lot is going on in this field to widen the application perspective of nanogenerators. New materials are being synthesized through processes like doping and hybridization that would further enhance the output of such nanogenerators. There is much being done and a lot that remains to be explored in this field of nanogenerators.

ACKNOWLEDGEMENT

The authors are thankful to Device Development Programme (DDP), Department of Science and Technology (DST), Govt. of India (Sanction Letter: DST/TDT/DDP-05/2018 (G)) and also to Science and Engineering Research Board (SERB), The Govt. of India (File No. YSS/2014/000964) for financial support to execute the research activities on flexible nanogenerators at IIT Delhi.

REFERENCES

1. L.S. Nathan, Toward cost-effective solar energy use, Science. 315 (2007) 798–802.
2. M.S. Dresselhaus, I.L. Thomas, Alternative energy technologies, Nature. 414 (2001) 332–337. https://doi.org/10.1038/35104599.
3. Z.L. Wang, W. Wu, Nanotechnology-enabled energy harvesting for self-powered micro-/nanosystems, Angew. Chem. Int. Ed. 51 (2012) 11700–11721. https://doi.org/10.1002/anie.201201656.
4. X. Chen, C. Li, M. Grätzel, R. Kostecki, S.S. Mao, Nanomaterials for renewable energy production and storage, Chem. Soc. Rev. 41 (2012) 7909–7937. https://doi.org/10.1039/c2cs35230c.
5. Z.-L. Wang, Self-powered nano tech, Sci. Am. 298 (2013) 82–87.
6. Z.L. Wang, J. Song, Piezoelectric nanogenerators based on zinc oxide nanowire arrays, Science. 312 (2006) 242–246.
7. Z.L. Wang, Towards self-powered nanosystems: from nanogenerators to nanopiezotronics, Adv. Funct. Mater. 18 (2008) 3553–3567. https://doi.org/10.1002/adfm.200800541.
8. Z.L. Wang, R. Yang, J. Zhou, Y. Qin, C. Xu, Y. Hu, S. Xu, Lateral nanowire/nanobelt based nanogenerators, piezotronics and piezo-phototronics, Mater. Sci. Eng. R Rep. 70 (2010) 320–329. https://doi.org/10.1016/j.mser.2010.06.015.
9. L. Jin, J. Chen, B. Zhang, W. Deng, L. Zhang, H. Zhang, X. Huang, M. Zhu, W. Yang, Z.L. Wang, Self-powered safety helmet based on hybridized nanogenerator for emergency, ACS Nano. 10 (2016) 7874–7881. https://doi.org/10.1021/acsnano.6b03760.
10. L. Nyholm, G. Nyström, A. Mihranyan, M. Strømme, Toward flexible polymer and paper-based energy storage devices, Adv. Mater. 23 (2011) 3751–3769. https://doi.org/10.1002/adma.201004134.
11. X. Wang, X. Lu, B. Liu, D. Chen, Y. Tong, G. Shen, Flexible energy-storage devices: design consideration and recent progress, Adv. Mater. 26 (2014) 4763–4782. https://doi.org/10.1002/adma.201400910.
12. K. Xie, B. Wei, Materials and structures for stretchable energy storage and conversion devices, Adv. Mater. 26 (2014) 3592–3617. https://doi.org/10.1002/adma.201305919.
13. M. Koo, K. Il Park, S.H. Lee, M. Suh, D.Y. Jeon, J.W. Choi, K. Kang, K.J. Lee, Bendable inorganic thin-film battery for fully flexible electronic systems, Nano Lett. 12 (2012) 4810–4816. https://doi.org/10.1021/nl302254v.
14. S. Garain, S. Jana, T.K. Sinha, D. Mandal, Design of in situ poled Ce^{3+}-doped electrospun PVDF/graphene composite nanofibers for fabrication of nanopressure sensor and ultrasensitive acoustic nanogenerator, ACS Appl. Mater. Interfaces. 8 (2016) 4532–4540. https://doi.org/10.1021/acsami.5b11356.
15. S.R. Anton, H.A. Sodano, A review of power harvesting using piezoelectric materials, Smart Mater. Struct. 16 (2007) 1–21. https://doi.org/10.1088/0964-1726/16/3/R01.

16. J. Curie, P. Curie, Développement par compression de l'électricité polaire dans les cristaux hémièdres à faces inclinées, Bull. La Société Minéralogique Fr. 3 (1880) 90–93. https://doi.org/10.3406/bulmi.1880.1564.
17. G. Lippmann, Principe de la conservation de l'elecricit´e, Ann. Chem. Phys. 10 (1881) 381–394. https://doi.org/10.1051/jphystap:0188100100038100
18. A. Teka, S. Bairagi, M. Shahadat, M. Joshi, S. Ziauddin Ahammad, S. Wazed Ali, Poly(vinylidene fluoride) (PVDF)/potassium sodium niobate (KNN)–based nanofibrous web: a unique nanogenerator for renewable energy harvesting and investigating the role of KNN nanostructures, Polym. Adv. Technol. 29 (2018) 2537–2544. https://doi.org/10.1002/pat.4365.
19. S. Bairagi, S.W. Ali, Effects of surface modification on electrical properties of KNN nanorod-incorporated PVDF composites, J. Mater. Sci. 54 (2019) 11462–11484. https://doi.org/10.1007/s10853-019-03719-x.
20. S. Banerjee, S. Bairagi, S. Wazed Ali, A critical review on lead-free hybrid materials for next generation piezoelectric energy harvesting and conversion, Ceram. Int. (2021). https://doi.org/10.1016/j.ceramint.2021.03.054.
21. J.S. Harrison, Z. Ounaies, Piezoelectric, 2001.
22. T.R.T. Dargaville, M.C. Celina, J. Elliot, P.M. Chaplya, J.M. Elliott, G.D. Jones, D.M. Mowery, R.A. Assink, R.L. Clough, J.W. Martin, Characterization, performance and optimization of PVDF as a piezoelectric film for advanced space mirror concepts, Optimization. (2005). https://doi.org/SAND2005-6846.
23. Y. Liu, L. Wang, L. Zhao, X. Yu, Y. Zi, Recent progress on flexible nanogenerators toward self-powered systems, InfoMat. 2 (2020) 318–340. https://doi.org/10.1002/inf2.12079.
24. R. Yang, Y. Qin, L. Dai, Z.L. Wang, Power generation with laterally packaged piezoelectric fine wires, Nat. Nanotechnol. 4 (2009) 34–39. https://doi.org/10.1038/nnano.2008.314.
25. S. Xu, Y. Qin, C. Xu, Y. Wei, R. Yang, Z.L. Wang, Self-powered nanowire devices, Nat. Nanotechnol. 5 (2010) 366–373. https://doi.org/10.1038/nnano.2010.46.
26. G. Zhu, R. Yang, S. Wang, Z.L. Wang, Flexible high-output nanogenerator based on lateral ZnO nanowire array, Nano Lett. 10 (2010) 3151–3155. https://doi.org/10.1021/nl101973h.
27. S. Bairagi, S.W. Ali, Influence of high aspect ratio lead-free piezoelectric fillers in designing flexible fibrous nanogenerators: demonstration of significant high output voltage, Energy Technol. 7 (2019) 1–10. https://doi.org/10.1002/ente.201900538.
28. S. Bairagi, S.W. Ali, Flexible lead-free PVDF/SM-KNN electrospun nanocomposite based piezoelectric materials: significant enhancement of energy harvesting efficiency of the nanogenerator, Energy. 198 (2020) 117385. https://doi.org/10.1016/j.energy.2020.117385.
29. S. Bairagi, S.W. Ali, A unique piezoelectric nanogenerator composed of melt-spun PVDF/KNN nanorod-based nanocomposite fibre, Eur. Polym. J. 116 (2019) 554–561. https://doi.org/10.1016/j.eurpolymj.2019.04.043.
30. S. Bairagi, S.W. Ali, Poly (vinylidine fluoride) (PVDF)/potassium sodium niobate (KNN) nanorods based flexible nanocomposite film: influence of KNN concentration in the performance of nanogenerator, Org. Electron. 78 (2020) 105547. https://doi.org/10.1016/j.orgel.2019.105547.
31. S. Bairagi, S.W. Ali, Investigating the role of carbon nanotubes (CNTs) in the piezoelectric performance of a PVDF/KNN-based electrospun nanogenerator, Soft Matter. 16 (2020) 4876–4886. https://doi.org/10.1039/d0sm00438c.
32. Z.L. Wang, Triboelectric nanogenerators as new energy technology and self-powered sensors – principles, problems and perspectives, Faraday Discuss. 176 (2014) 447–458. https://doi.org/10.1039/c4fd00159a.
33. S. Liu, T. Hua, X. Luo, N. Yi Lam, X. Ming Tao, L. Li, A novel approach to improving the quality of chitosan blended yarns using static theory, Text. Res. J. 85 (2015) 1022–1034. https://doi.org/10.1177/0040517514559576.
34. S. Shin, Y.E. Bae, H.K. Moon, J. Kim, S. Choi, Y. Kim, Formation of triboelectric series via atomic-level surface functionalization for triboelectric energy harvesting, 2017. https://doi.org/10.1021/acsnano.7b02156.
35. Ö. Faruk Ünsal, A. Çelik Bedeloğlu, Recent trends in flexible nanogenerators: a review, Mater. Sci. Res. India. 15 (2018) 114–130. https://doi.org/10.13005/msri/150202.
36. Y. Jao, P. Yang, C. Chiu, J. Lin, S. Chen, D. Choi, H. Lin, Author's accepted manuscript, Nano Energy. (2018). https://doi.org/10.1016/j.nanoen.2018.05.071.
37. B. Chen, W. Tang, T. Jiang, L. Zhu, X. Chen, C. He, L. Xu, H. Guo, P. Lin, D. Li, J. Shao, Z.L. Wang, Three-dimensional ultraflexible triboelectric nanogenerator made by 3D printing, Nano Energy. 45 (2018) 380–389. https://doi.org/10.1016/j.nanoen.2017.12.049.

38. C. Yao, X. Yin, Y. Yu, Z. Cai, X. Wang, Chemically functionalized natural cellulose materials for effective triboelectric nanogenerator development, Adv. Funct. Mater. 27 (2017) 1700794. https://doi.org/10.1002/adfm.201700794.
39. S.S. Kwak, H. Kim, W. Seung, J. Kim, R. Hinchet, S.W. Kim, Fully stretchable textile triboelectric nanogenerator with knitted fabric structures, ACS Nano. 11 (2017) 10733–10741. https://doi.org/10.1021/acsnano.7b05203.
40. A. Ahmed, I. Hassan, I.M. Mosa, E. Elsanadidy, G.S. Phadke, M.F. El-Kady, J.F. Rusling, P. Ravi, R.B. Kaner, Nano energy all printable snow-based triboelectric nanogenerator, Nano Energy. 60 (2019) 17–25. https://doi.org/10.1016/j.nanoen.2019.03.032.

12 Nanogenerators as a Sustainable Power Source

Muhammad Mudassir Iqbal, Gulzar Muhammad, Tania Saif, Muhammad Shahbaz Aslam, Muhammad Arshad Raza, Muhammad Ajaz Hussain, and Muhammad Tahir Haseeb

CONTENTS

12.1 INTRODUCTION

Nanogenerators (NGs) transform mechanical energy into useful electrical energy and are employed as sustainable power sources in a variety of devices. The sustainable power from NGs shows a wide range of uses in the fields of energy sciences, the safety of the environment, wearable gadgets, self-powered sensors, medical sciences, robot science, artificial intelligence, human-machine interface, smooth traffic, smart towns, and fiber and textile sensors (Wang and Song 2006; Wang 2020; Wu et al. 2020). Internet of things demands dispersed power generation from sustainable energy sources like solar, thermal, wind, and mechanical triggering/vibration (Wu et al. 2019). Wearable electronic devices used under water for human motion monitoring and rescue systems rely on sustainable energy sources for power generation (Zou et al. 2019). Self-powered sensing networks drive industry worldwide through technology and cost-effectiveness. NGs designed by combining the piezoelectric or triboelectric effects with semiconducting properties of gases enable us to sense and characterize gases and do not require any external aid of power. Additional advantages of self-powered sensors are decrement in power demand and prerequisite space for installation (Wen et al. 2017). The hybrid NG shows the capability of applications in the arena of logic devices, power supplies, prosthetics, antistatic protection, and self-powered sensors in a network (Rasel et al. 2019). Energy harvesting from our living environment is the method of choice to meet the energy demand. The advantage is sustainable power generation without polluting the environment. However, the existence of energy in living systems at lower incidence makes it difficult to harvest energy from the living environment

DOI: 10.1201/9781003187615-12

(Zhang et al. 2013). NGs are mainly divided into three types, piezoelectric, triboelectric, and pyroelectric NGs, based on phenomena resulting in the production of output power. NGs harvest energy through the piezoelectric, triboelectric, or pyroelectric effects to produce electric current (Wang 2020).

The driving force that converts the mechanical energy into electrical energy in the NGs is Maxwell's displacement current. Maxwell's equation extended by the addition of a new term medium polarization due to non-electric field (like strain) *Ps in* the displacement vector *D* explains medium polarization initiated by surface charges through piezoelectric and triboelectric effects in NGs. Medium polarization due to non-electric field remains even if E=0. The theory enables quantification of performance and electromagnetic behavior of NGs (Wang 2020).

Figures of merit help to quantify the output performance of NGs. The breakdown effect greatly poses an effect on output power. Figures of merit that do not consider breakdown effect are less effective in output power quantification. A standard method has been proposed to quantify the output power of the triboelectric NGs (TENGs). The results of the experiments show consistency with Paschen's law. The method is universal as it also has been used for piezoelectric NGs (Xia et al. 2019). A figure of merits to explain the efficiency of TENG is devised from the structural and material figures of merit. After derivation and simulation, the structural figure of merit is compared with different structures of TENG for the performance (Zi et al. 2015).

In this chapter, NGs are discussed as a sustainable source of energy. The main focus is on construction, working, and applications of NGs. The characterization and evaluation methods used to characterize fabricated dielectric and conducting materials and constructed NGs are also described in this chapter. The chapter starts with an introduction to NGs. Later, TENGs are described in detail. Four application categories of TENGs are explained in detail. Then applications of TENGs are discussed followed by a detailed explanation of triboelectric materials. Here, two special types of TENGs, ionic-skin TENG (IS-TENG) and bio-TENG, are explained. In the following part of the chapter, other types of NGs are discussed one by one. In the last, hybrid NGs are described.

12.2 TRIBOELECTRIC NANOGENERATORS (TENGs)

The sustainable development of human society, artificial intelligence, portable electronics, and widely distributed sensing networks call for renewable and eco-friendly energy resources (Yick et al. 2008). Mechanical energy is a more widely distributed form of energy in the environment as compared to other renewable energy sources and is independent of weather and working environmental conditions. TENGs have a great ability for harvesting mechanical energy (Cheng et al. 2019). Methods and models have been designed to calculate the minimum electrostatic force required to drive a TENG. The models also help in calculating the efficiency of TENG to convert the energy during a single cycle and the minimum amount of the mechanical input energy to drive the TENG (Xu et al. 2019). Direct powering from TENGs to small electronic gadgets is difficult due to high resistance. Output power from TENGs is further decreased by the high resistance of the circuits. Pulse controllable voltage sources provide consistent maximum output energy. An inducer, a diode, a capacitor, and a unidirectional switch are combined with TENG to construct a novel power management system. The unidirectional switch delivers maximum output energy. The resulting device is capable of storing energy with high efficiency and shows potential to power electronic watch and high-brightness quantum dot light-emitting diode (LED) (Qin et al. 2018; Xia et al. 2020). The charging of capacitors by TENGs is well analyzed to explore the output current of the power characteristics of TENGs (Shao et al. 2018a,b, 2019).

Parasitic capacitance is damaging and arises due to the phenomena of capacitance taking place in the components in proximity. TENGs with a different arrangement of structural parts and dimensions designed to study the parasitic capacitance of the layered TENG have been evaluated in a vacuum. The dielectric material is arranged in two different patterns as XYYX or XYXY forming the symmetrical and alternating pattern. Edge effect is prominent in small layered TENGs, which enhances parasitic capacitance and reduces the output power of the TENG (Yin et al. 2018). The programed TENG uses mechanical switches to regulate the energy operation. Combining three unit processes in a particular sequence results in the increment of power from the TENG. Numerous programs are possible to power the particular TENG-based devices. Three programs, current amplifier, Bennet doubler, and charge oscillator, show achievement of output voltage up to kV level. The output of the programed TENG is not dependent upon the triboelectric and nature of the conducting material. The programed TENG also overcomes the problem faced with high-k coating materials and high-voltage breakdown. The novel approach to regulate mechanical switches through programing reveals new research horizons in the research on the TENGs as the sustainable source of power (Wang et al. 2020).

The performance of TENGs decreases with the deforming of dielectric layers. To increase the durability of TENG, steel wire has been introduced as an electrode in the fiber-shaped TENG. The resulting TENG is highly stable, stretchable, and tailorable. The TENG can easily sense human motion and is suitable to power wearable devices (Xie et al. 2019). Self-charge pumping TENGs use floating layer structures to accumulate charges for electrostatic induction and a charge pump that pumps charges into the floating layer. The novel design increments charge density and hence the power output (Xu et al. 2018). Flipping of charges on the conducting material by use of the LC oscillating circuit manages power output from the TENG. The LC oscillating circuit consists of a diode, a switch, an inductor, and a capacitor. The increment in the power is due to the increased level of the equivalent charge density than that of saturated charge density (Xu et al. 2019).

Wearable bioelectronics has seen a marvelous advancement in size reduction and multifunction performance. Conventional powering schemes are not capable of delivering sustainable power supply to the miniaturized and multifunctional wearable bioelectronics. TENGs are shown to meet the challenge by converting the biochemical energy associated with human motions into the electric current competently (Zou et al. 2020).

12.2.1 Major Applications of TENGs

So far, TENG can be divided into four categories based on its applications, including self-operating/self-powered sensors, direct high-voltage power sources, blue energy, and nano/micropower source (Li et al. 2019). Figure 12.1 illustrates major applications of TENGs as a sustainable source of energy (Luo and Wang 2020).

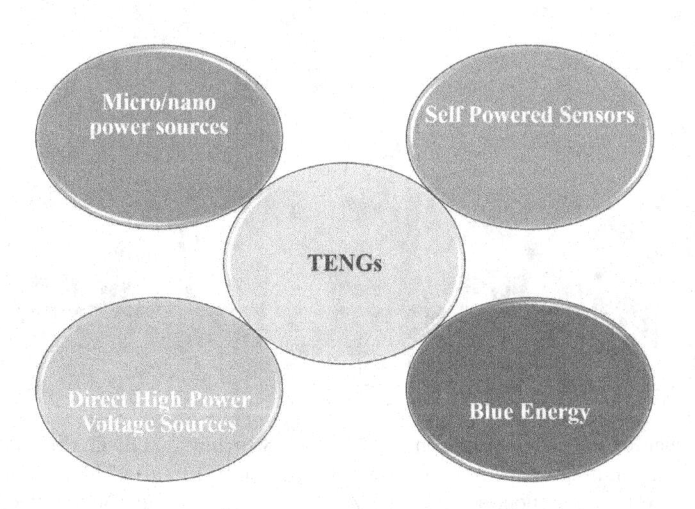

FIGURE 12.1 Four uses of TENGs.

12.2.1.1 Self-Powered Sensors

Mechanical stimuli can be easily transformed into electric signals without requiring any transducers with the help of a TENG, which have the potential to develop self-powered sensors in the absence of any external power source, like tactile, acceleration, photoelectric, chemical, pressure and acoustic sensors (Li et al. 2015; Pang et al. 2016). In the meantime, the human-machine interface is a field that makes the interface between machines and users using different techniques in applications like healthcare, security, and information communication (Guo et al. 2018). Hence, TENG-based human-machine interface systems have been developed that can translate eye blink motion into controllable commands. The photograph and design of this mechanosensation TENG are shown in Figure 12.2. The design of the circular mechanosensation TENG is based on the single-electrode mode having multilayered structures that monitor eye motion by flexibly mounting on an eyeglass (Pu et al. 2017).

Electronic skin, a kind of artificial skin, is another example of a TENG-powered system (Fan et al. 2016). The system replicates sensing functionalities of human skin through electronic components and flexible materials making it useful for medical diagnostics and other applications (Hammock et al. 2013). Flexible tactile sensors are requisite components. Different approaches such as ultra-sensitivity, flexibility, transparency, wide range, and stretchability are effective regarding tactile sensors for electronic skin (Someya et al. 2004). Polyethylene terephthalate (PET) film at the top acts as a flexible substrate and the top layer is spin-coated by polydimethylsiloxane (PDMS) and used as a triboelectrification layer. The bottom of the substrate is enclosed by a PET film containing a copolymer of ethylene-vinyl acetate (EVA) for electrode protection. A triboelectric sensor matrix (16×16 pixelated) was created having 5-dpi resolution with a pixel size of 2.5×2.5 mm^2 in a laboratory environment. By adopting the microfabrication technique, pixel size can be minimized to micron-meter and the smallest pixel size obtained was 500×500 m^2 (Wang et al. 2016; Liu et al. 2020).

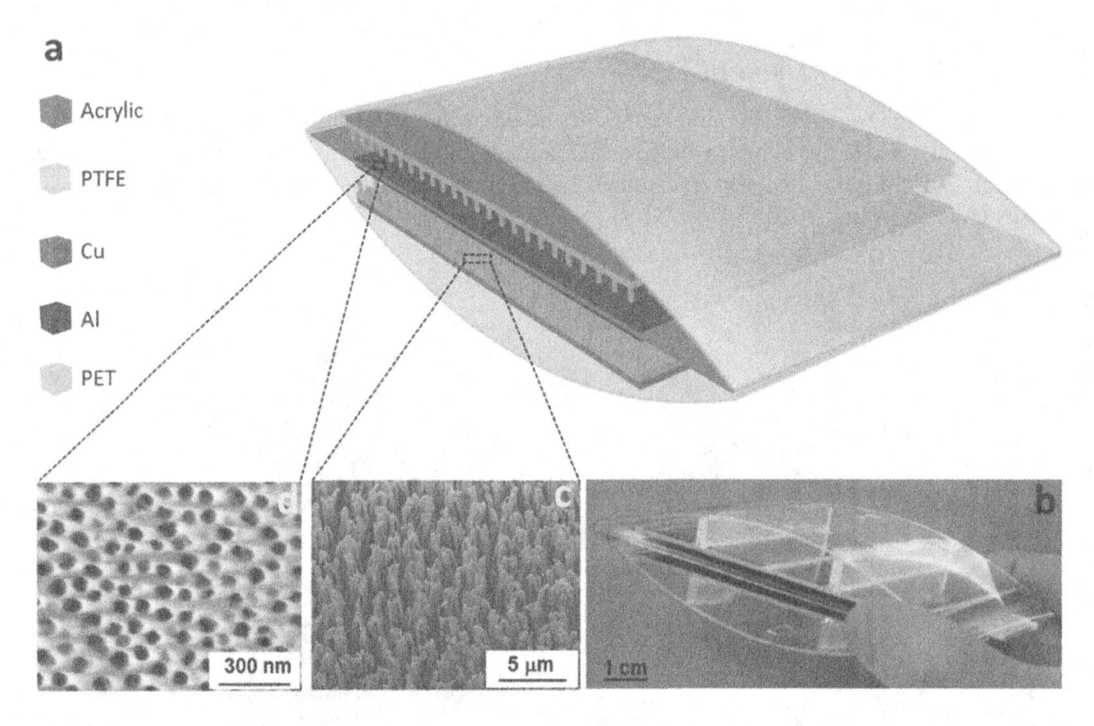

FIGURE 12.2 Structure and characterization of TENG. (a) Structure of TENG. (b) A complete functional unit. (c) Scanning electron microscopy image shows nanowire on PTFE. (d) SEM photograph of conducting electrode of aluminum having nanopores. PET: polyethylene terephthalate; polytetrafluoroethylene: PTFE. (Reprinted with permission from Chen et al. (2015). Copyright (2015) American Chemical Society.)

TABLE 12.1
Various Types of NGs Harvest Energy from the Human Body

NG Types	NG Active Materials	Electrodes	Substrate	Dimension of NGs	Applicability on Human Body	Voltage
Triboelectric	PTFE and Aluminum	Al	PI	Thickness 0.95 mm Area 38 × 38 mm^2	Hand palm tapping	215 V
	FEP and Aluminum	Al, Cu	PI	Area 52 × 58 mm^2	Heel strike	700 V
	PTFE and Al	Al, Cu	PET	Thickness 0.6 mm Area 50 × 50 mm^2	Walking with 2-kg pack load	400 V
Piezoelectric	Zinc oxide film	Al	PMMA and Al foil	Thickness 18 μm Area 13.5 × 5 mm^2	Blinking motion	0.2 V
	PZT film	Au/Cr and Au/Ti	PET	Thickness 225 μm Area 21×50 mm^2	Wrist movement	2 μA
Thermoelectric	PEDOT: PSS and p: Sb_2Te_3, n: Bi_2Te_3	Silver film	PI	Thermocouples 15 Area 30 × 50 mm^2	Worn on wrist	12 mV

Source: Reprinted with permission of Elsevier from Proto et al. (2017).
Abbreviations: Al, Aluminum; PTFE, polytetrafluoroethylene; PI, polyimide; FEP, fluorinated ethylene propylene; PET, polyethylene terephthalate; PZT, lead zirconate titanate; Cr, Chromium; Au, Gold; PMMA, poly (methyl methacrylate); PEDOT: PSS, poly (3,4-ethylene dioxythiophene)-poly (styrene sulfonate); Sb_2Te_3, antimony telluride; Bi_2Te_3, bismuth telluride.

An important feature of NG to wear is the fixation on the human body by comfort and impalpable for the person wearing NG just like clothing that can easily harvest the energy from the body. Table 12.1 shows triboelectric, piezoelectric, and thermoelectric NGs employed for harvesting energy from the surface of a human body (Proto et al. 2017).

Fiber-based hybrid NGs consisting of fiber NGs and fiber biofuel cell designed on a carbon fiber are used as self-powered nano-system in bio-liquid for conversion of biochemical and mechanical energy into electrical energy for applications in monitoring blood pressure and operations of oil/gas/water pipes (Li and Wang 2011; Pan et al. 2011).

12.2.1.2 Direct High-Voltage Power Sources

High voltage from TENG can effectively derive or control electrically responsive devices like ferroelectric materials, electrostatic air cleaners, dielectric elastomers, field emitters, and piezoelectric materials (Zi et al. 2018). TENG-based high-voltage power sources do not need power converters and output a small current as compared to conventional high-voltage power sources, making them have various benefits such as cost-effectiveness, portability, simple structure, and safety. An agile piezoelectric micro-actuator by combining piezoelectric ceramics with planar sliding TENG for 2D optical direction modulation is illustrated in Figure 12.4 (Zhang et al. 2015).

12.2.1.3 Micro/Nano-Power Sources

TENGs have major applications as micro/nano-power sources on account of their low cost, vast material, structural choices, and lightweight. TENGs show the ability to harvest the nanomechanical energy from its environment, i.e., heartbeat, vibration, human walk, and wind (Yang et al. 2013). Many researchers have struggled to utilize low-frequency biomechanical energy that otherwise gets wasted by using TENG. The TENG that can adapt the shape contains a liquid with conductive property and elastic polymer cover (Yi et al. 2016). Due to the greater flexibility of the rubber cover and deformability of the liquid electrode, stretchable TENG can be used on any curvilinear surface (Ma et al. 2018). A shape adaptive TENG is illustrated in Figure 12.5.

A wind-driven TENG contains fluorinated ethylene propylene and silver nanowire (Ag NWs) acting as a triboelectric material to power wearable electronics in place of batteries. This fabricated TENG could produce power, voltage, and current of 0.18 mW, 150 V, and 7.5 μA, respectively, under a wind speed of 20 m s^{-1}. By integrating wind-driven TENG into self-powered devices (i.e., mask, shoe, and bracelet) as an energy source could offer advanced approaches to protect the environment and improve the quality of human life (Jiang et al. 2018).

12.2.2 Triboelectric Materials

Recently, based on triboelectricity, a new kind of generator has been invented that utilizes conventional organic materials, thus making it cost-effective, light, and easy to scale up and fabricate (Wang 2013). The energy conversion efficiency obtained through the TENG is 55% (Lin et al. 2015). TENG has arch-shaped upper and lower plates made up of PET with multilayer core, bent naturally by heat treatment as shown in Figure 12.2(a), which helps to effectively separate charge and contact with help of film elasticity. The fabrication of the unit is shown in Figure 12.2(b). Forming an upper and a lower layer of the functional core, holding a sandwiched structure, is a polytetrafluoroethylene (PTFE) film having copper deposition as back electrodes. An exposed PTFE surface was used to create PTFE nanowire arrays through a top-down method by reactive ion etching, which helps to increase the charge density of contact esterification (Fang et al. 2009; Yang et al. 2013). Figure 12.2(c) reveals scanning electron microscopy results in the form of image-aligned PTFE nanowires, indicating that the average length of PTFE nanowire is 1.5 ± 0.5 μm and the average clustering diameter is 54 ± 3 nm. Functional core with top and bottom layers has a thin film of aluminum with a nanoporous surface, sandwiched between layers as shown in Figure 12.2(d), works as both contact surface and electrode. The depth and average diameter of aluminum nanopores are 0.8 ± 0.2 μm and 57 ± 5 nm, respectively, having a distribution density of 213 μm^{-2}. Acrylic, on account of its lightweight, low cost, decent strength, and good machinability was chosen as structural supporting material (Chen et al. 2015).

The basic working mechanism of TENG is established on a coupling between electrostatic induction and contact electrification (Chen et al. 2013; Su et al. 2014) as shown in Figure 12.3 illustrating both charge delivery (up) and potential supply by analysis through COMSOL (down) for understanding. By applying an external force with the help of a rolling ball, on the uppermost surface of the functional unit, the two layers of PTFE are brought closer touches the conducting layer of aluminum located in the middle at once, transferring charge at contact interfaces (Wang et al. 2012; Yang et al. 2015). In consonance with triboelectric series, injection of electrons occur into PTFE from aluminum as PTFE is more triboelectrically negative than aluminum, producing negative triboelectric charges on PTFE and positive charges on aluminum (Figure 12.3(a)). Thereafter, if external force vanishes, the elasticity of the arch-shaped plates will result in a separation between the aluminum and PTFE following no longer coincidences between negative and positive triboelectric charges on the same plate and generating an inner dipole moment between two sets of contact surfaces. This dipole moment helps in balancing the electric field by moving free electrons from the copper electrode to the aluminum electrode, generating positive charges on a copper electrode (Figure 12.3(b)). When the corresponding separation is maximum and the upper plate achieves the highest point, the flow of electrons stops (Figure 12.3(c)). Persistently, dipole moment weakens due to reduced separation between contact surfaces, allowing backflow of free electrons to copper electrode unless both contact surfaces contact with each other (Figure 12.3(d)), resulting in a cycle of electricity generating process.

Innovatively exploiting the surface charge effect between a thin layer of metal as the electrode and conventional polymers, triboelectrification-based TENG networks are cost-effective, eco-friendly, lightweight, anticorrosive to the marine environment, and efficient to float on the surface

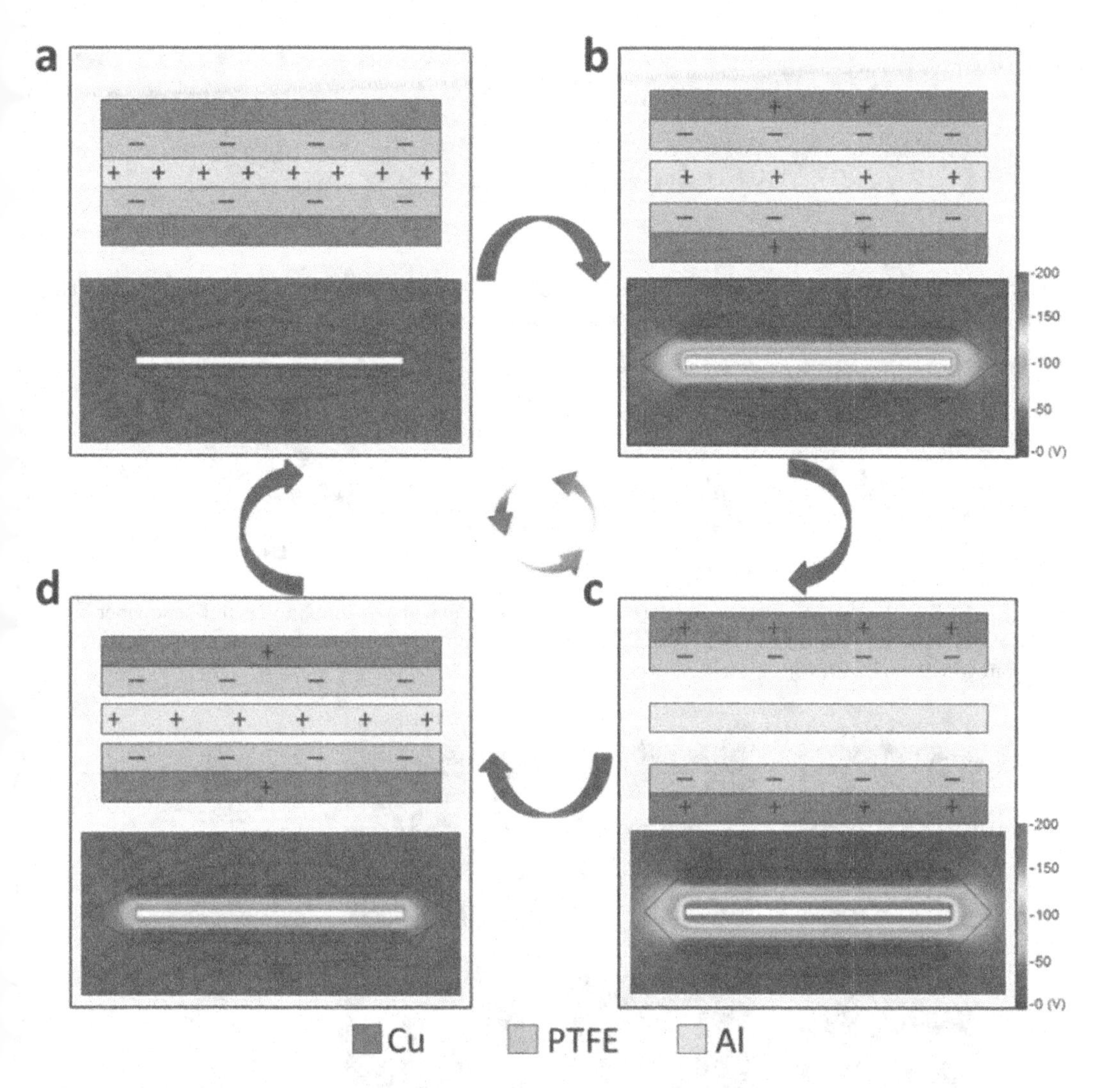

FIGURE 12.3 Production of an electrical signal by TENG. Both illustration of charge distribution (up) and potential distribution by COMSOL (down) are shown to explain the operating principle of an aluminum functional core. (a) The initial state of negatively charged PTFE after coming in contact with aluminum. (b) Separation of aluminum and PTFE. (c) Increase in separation. (d) Reduction in separation between contact surfaces. (Reprinted with permission from Chen et al. (2015). Copyright (2015) American Chemical Society.)

of the water. Thus, TENG network represents a sustainable and efficient way to gain water wave energy. Besides, integrating TENG network with wind energy can act as 'substrates' to harvest different kinds of energies (Chen et al. 2015).

Graphite is coated with the help of a brush on sandpaper and hot-pressed to implant a protective coating as illustrated in Figure 12.4. The sandpaper coated with graphite is placed between PTFE sheet and aluminum layer to construct a TENG illustrated in Figure 12.5. Graphite-coated sandpaper and aluminum serve as electrodes. Aluminum also acts as triboelectric material. The NG is capable of powering portable and wearable devices. The NG powers 120 blue LEDs in series and liquid display. The sandpaper surface morphology is characterized by SEM (Figure 12.6) and the working mechanism of TENG is evaluated by COMSOL simulation (Figure 12.7) (Ankanahalli Shankaregowda et al. 2019).

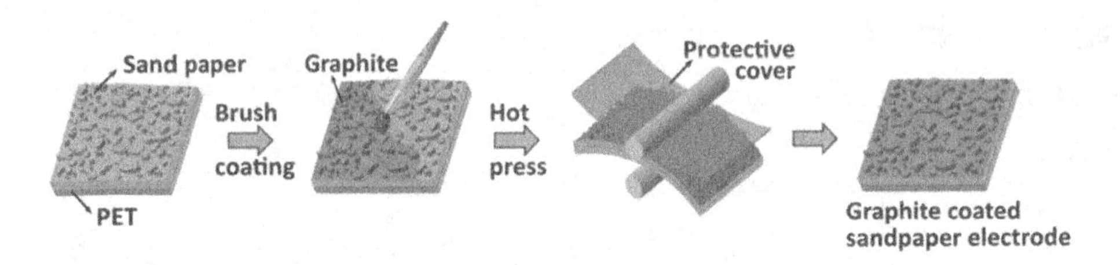

FIGURE 12.4 Illustration of the scheme of graphite coating on sandpaper. (Reprinted with permission under Creative Commons License from Ankanahalli Shankaregowda et al. (2019).)

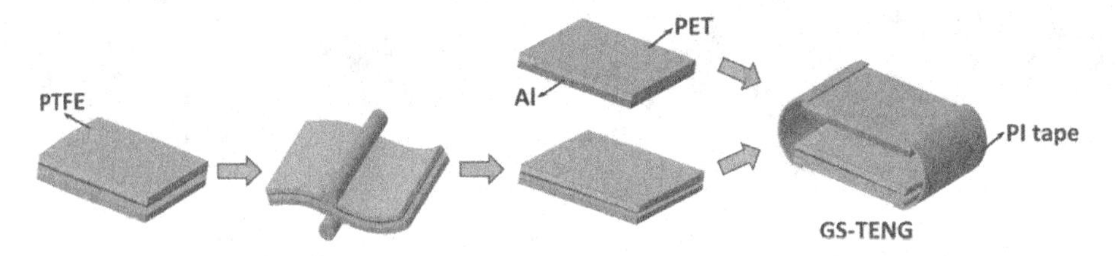

FIGURE 12.5 Illustration of steps involved in the construction of the graphite-coated sandpaper-based TENG (GS-TENG). (Reprinted with permission under Creative Commons License from Ankanahalli Shankaregowda et al. (2019).)

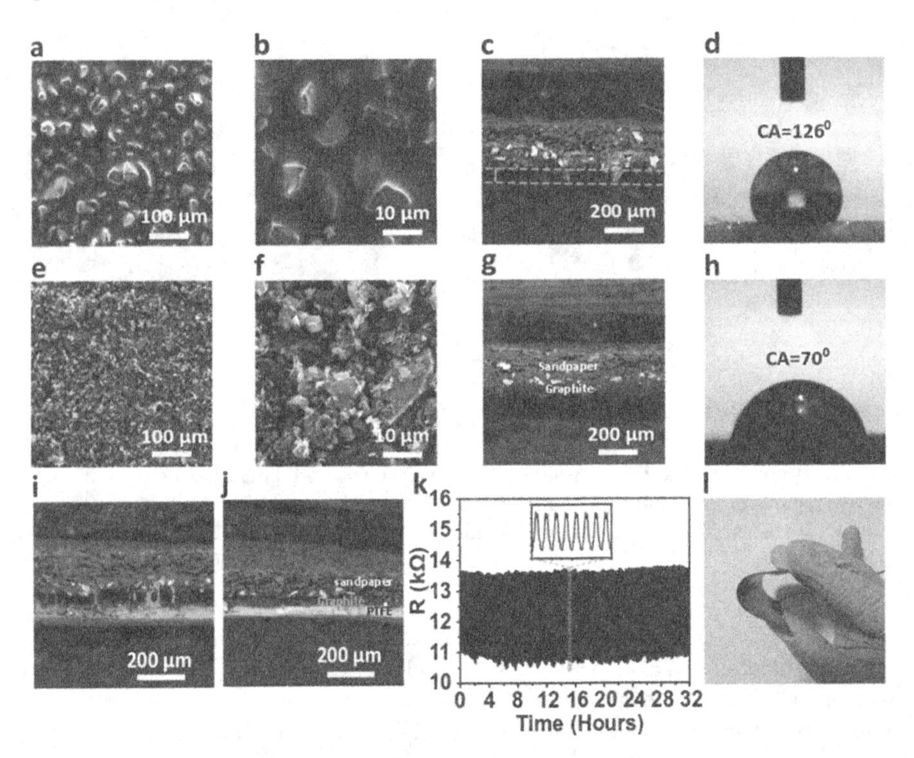

FIGURE 12.6 Characterization of graphite-coated sandpaper-based TENG. SEM images of sandpaper without coating: (a) high magnification, (b) low magnification, (c) side view, (d) image revealing the contact angle. SEM images of sandpaper after graphite coating: (e) high magnification, (f) low magnification, (g) side view, (h) image revealing the contact angle. SEM images of graphite-coated sandpaper combined with the polytetrafluoroethylene (PTFE) film: (i) before hot rolling, (j) after hot rolling, (k) elasticity assessment of the graphite-coated sandpaper/polyethylene terephthalate (PET) with PTFE film tape, (l) image of GS-TENG revealing flexibility. (Reprinted with permission under Creative Commons License from Ankanahalli Shankaregowda et al. (2019).)

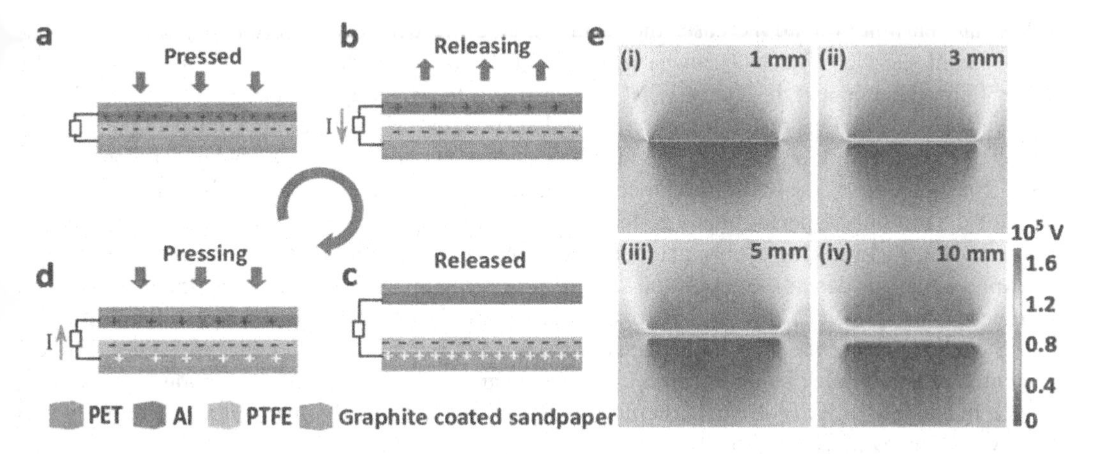

FIGURE 12.7 Scheme illustrating the mechanism during the production of an electric signal by GS-TENG. When pressed, the aluminum comes in contact with PTFE (a). When released, the Al electrode separates from the PTFE dielectric layer and current flows from electrode to dielectric material via external circuit (b). After releasing current, the PTFE and graphite layers come in the equilibrium of charges (c). When an external force is reapplied, the current flows in the reverse direction (d). (e) COMSOL simulation results at each of the four steps during the working of GS-TEND. (Reprinted with permission under Creative Commons License from Ankanahalli Shankaregowda et al. (2019).)

Segmentation and multilayer cylindrical design of TENG increments the output power. The contact area of the PTFE and Al petals affects the output power by TENG. The power generation by the TENG is divided into four steps. PTFE and Al petals are in (a) complete contact (b) separating, (c) fully separated, and (d) contacting (Tang et al. 2014). In a novel design of TENG, polyimide surfaces in contact slide on PTFE films in a planar manner. Nanowires are imprinted on polyimide and PTFE layers. The TENG creates approximately 1300 V and can be used to power hundreds of LEDs connected in series directly. Investigations reveal the relation between output power and the sliding motion of layers. The planar sliding TENG shows high efficiency and is suitable for converting mechanical energy into electrical energy (Wang et al. 2013). In a novel TENG, the triboelectric layer slides between two fixed electrodes in the same plane and harvests energy from the random motion of entities like a human. Electrostatic induction generates electric current efficiently when triboelectric film approaches electrodes alternatively (Wang et al. 2014). TENG aluminum electrode is sandwiched between PTFE film and paper. The TENG is folded in a slinky way. Paper is recyclable and decreases cost. The TENG is lightweight and shows flexibility due to paper. The TENG can harvest energy from human movements like stretching, lifting, and twisting (Yang et al. 2015).

The metal electrodes used conventionally in TENGs corrode easily in tough environments. A novel TENG uses plastic, polypyrrole as conducting material instead of metal electrodes. The polypyrrole-based TENG shows increased charge density due to increased surface area in nanostructures. A cost-effective self-power system designed by combining the polypyrrole-based TENG and polypyrrole-based capacitor is capable of self-charging (Wang et al. 2016).

PDMS-based TENG converts irregular mechanical energy into electrical energy efficiently. The NG drives 10 LEDs in series at a time (Shin et al. 2016).

TENGs consisting of poly(3,4-ethylene dioxythiophene) wired with silver interceding between layers of PDMS are self-healing. The mechanical damage is repaired by imine bonds in PDMS network. The NGs are elastic and transparent. The NG is usable in soft power sources (Sun et al. 2018). An increase in surface area of triboelectric material increases the output power of the TENG by increment in charge density. The porous PDMS fabricated on nanogram silicon shows increased voltage and current, can drive 400 LEDs connected in series, and can power a wireless transmitter through a capacitor (Tantraviwat et al. 2020). Modification of PDMS with $K_{0.5}Na_{0.5}NbO_3$

(KNN) and barium titanate increases the surface area of triboelectric material and hence the output power of the NG. The NG can be integrated into small devices as a sustainable source of energy (Vivekananthan et al. 2020). TENG consisting of PDMS-graphene oxide film and conducting electrodes of fabric is tested for experimental verification of proposed model based on charge conservation and zero loop voltage. The model evaluates the output current, voltage, and power of the TENG for resistance offered by the connected device. The model explains the effect of three types of motion on the performance of the TENG. Three types of motion studied are sinusoidal motion cycles, real walking cycles, and constant speed. The verification of the proposed model by the experiment provides an escort for the design of the structure and helps to select the material for the device construction (Yang et al. 2016). The aluminum layer sandwiched between two films of PDMS converts external energy into electric current and provides double output frequency. Finite element simulation helps us understand the mechanism of the electric current production by the TENG. In addition, micro- or nanostructures on PDMS surface greatly increment the efficiency of the novel TENG (Zhang et al. 2013).

The TENG fabricated from porous cellulose paper and PDMS acts as a sensor power source. As a sensor, the NG monitors the operation of a remote machine through a wireless transmission system. The NG sensor sends messages associated with the movement of fingers. The NG is also capable of acting as a power source during electropolymerization of polyaniline on the electrode of carbon nanotube. Barium titanate and silver nanoparticles are used for dielectric modulation of the NG by changing the permittivity of friction material (Shi et al. 2020). Cellulose sheets are used along with titanium oxide and silver nanoparticles to construct the lightweight and elastic NGs. The titanium oxide sheets are coated with Ag nanoparticles by the dip-coating technique. Ag nanoparticles help in charge transportation. The NG is capable of detecting simple human motions like finger tapping, rubbing, and foot trampling (Sriphan et al. 2020). Cellulose is found in abundance and is advantageous as being a renewable source and biodegradable material. The cellulose nano-fibrils are transparent and flexible materials with a rough surface. The cellulose nano-fibril film and fluorinated ethylene propylene are combined in the TENG assembly. The resulting TENG is further combined with cardboard fibers by the cold-pressed technique. The mechanical energy from a normal human step results in approximately 30 V and 90 μA output by the TENG (Yao et al. 2016). The TENG-containing liquid electrode consists of poly(3,4-ethylene dioxythiophene), polystyrene sulfonate, and silicone rubber. Silicone rubber provides high elasticity and triboelectric property to the NG. The TENG retains stable output performance even upon deformation. The TENG is capable of powering wearable electronic devices (Shi et al. 2019).

The laser-induced graphene hybrid material is obtained from polyimide and cork by irradiating with the infrared laser. The hybrid material increases the conductivity and the triboelectric ability of the NGs. The laser-induced graphene is applicable to prepare different geometries of the open circuit with high voltage and utilize the mechanical energy produced from the skin contact (Stanford et al. 2019). The waste plastic bottle made of PET is used to device a novel TENG. Polybutylene adipate terephthalate films on the sandpaper make active layers of the TENG. The gadget looks like eye shape and serves as a sensor for the vehicle security alarming system and can be planted in the tire of the vehicle. The sensor sends the information through the Bluetooth. The gadget also scavenges the biomechanical energy from small motions (Sukumaran et al. 2020).

The polyvinyl-alcohol-based hydrogel TENGs are flexible, recyclable, and environment-friendly. The tube-shaped hydrogel-based TENG shows the ability to harvest mechanical energy from human motions. The TENG is capable of harvesting the energy from bending, twisting, and stretching (Xu et al. 2017). The temperature ranges of hydrogel-based TENGs are narrow due to freezing and loss of water by evaporation. The high stretching ability and the ionic conduction exhibited by the ionogel-based TENGs are due to the dipole-dipole and ion-dipole forces in the ionogel network. The ionogel-based NGs are transparent, stretchable, and used in a variety of applications like wearable electronic devices, electronic skin, and artificially intelligent devices (Sun et al. 2019).

The TENGs are used in four different manners to power nano-systems; however, the multilayer structure of films in the TENGs poses the complication of the practical use in the moveable structural parts. A single layer consisting of the natural nanofibers implanted with the ions is used in the single-layer TENG. The single-layer TENG design also eliminates the use of the dielectric as an electrode. The investigations show that the ions efficiently transfer the charge in a single layer (Ba et al. 2020). The natural polymers containing the ions are used to construct the TENG with one layer. The resulting TENG is simple and lacks the dielectric film. The investigations analyze the effect of the ions on the performance of the biodegradable TENG (Yang et al. 2016).

Barium titanate is a ceramic material with high dielectric property and possesses the ability to trap the charge through the property of ferroelectric character. A novel TENG consists of barium titanate and the copolymer of vinylidene fluoride with trifluoroethylene. The introduction of the polarization by the electricity in the NG results in the high triboelectric charge transfer. The NG is applicable in the self-powered watches without external power aid (Seung et al. 2017).

Cesium lead tribromide ($CsPbBr_3$) is not only stable but also shows good optical and dielectric properties. The perovskites are obtained by the introduction of the barium ions into the cesium lead tribromide crystals. The perovskites increase the output of the TENG due to the increased space charge polarization and work function (Wang et al. 2020).

12.2.3 Ionic-Skin TENGs

Nowadays, the need for the development of self-healable, deformable and transparent electronic devices like strain sensors, LEDs, transistors, and energy storage devices has increased (Oh et al. 2016). Transparency is needed to visually transmit the information, which is used in biomedical imaging, touch screens, user-interactive displays, and therapeutic optogenetics (Sun et al. 2014). The self-healing ability of the devices to regain their functionality after the mechanical damage is also crucial (Chortos et al. 2016). Additionally, the deformable devices that can perform without any changes are also high in demand (Sekitani et al. 2008).

Recently, the TENGs have been used as a source of power for these devices owing to the shown capacity to work at the irregular and low mechanical impulses, the environmental friendliness, the high-voltage output, and the low processing cost (Sekitani et al. 2008; Askari et al. 2018). However, the TENGs have metallic electrodes that lack stretchability, transparency, and self-healing properties. Therefore, efforts were made to develop highly transparent, stretchable, and self-healing IS-TENG by using the slime-based ionic conductors for collecting the electricity to power the electronic devices (Wang et al. 2016).

The IS-TENG consists of a silicone rubber layer used as the triboelectrically negative material, with a thickness of $100 \pm 10\ \mu m$, a cross-linked polyvinyl alcohol gel containing the slime layer working as an ionic current collector with a thickness of 1 mm, and a double-sided foam tape acting as the substrate having a thickness of 1 mm. According to the triboelectric series, energy is harvested due to coupling between the electrostatic induction and the triboelectric effect (Wang 2015). The human skin due to its low electronegativity acts as the triboelectrically positive material, while the silicone rubber due to its high electronegativity acts as the triboelectrically negative material (Lai et al. 2016). At first, no surface charge is produced as the human skin and silicone rubber surfaces are not in contact; consequently, the distribution of the positive (Na^+) and the negative ($B(OH)^{4-}$) ions of ionic collector occurs across the polyvinyl alcohol matrix. The transfer of the charge from the human skin to the silicone rubber layer occurs when both the surfaces come in contact with each other, making the human skin positively charged and the silicone rubber negatively charged (Lai et al. 2016). When a finger is moved away from the surface, the separation of two oppositely charged surfaces in the space takes place creating a potential difference. Thus, an electrical double layer is formed when the accumulation of the negative charge ($B(OH)^{4-}$) at the slime/copper interface and the positive ions (Na^+) at the slime/silicone rubber interface is induced due to an unscreened negative charge on silicone rubber surface (Ye et al. 2015). This results in the generation of the voltage output due to the flow of the charge from a copper electrode to the ground. When the maximum separation distance is created between the two oppositely charged surfaces, the electrostatic equilibrium is attained. After applying an external force, the potential difference, as well as the distance between the silicone rubber layer and the human skin, decreases, allowing the flow of the charge in the opposite direction (from ground to slime/copper interface), thus generating the voltage in a reverse direction. Hence, repeating the earlier process generates multiple voltage outputs (Chun et al. 2015).

The application of the designed IS-TENG as an energy source for electronic devices has been displayed by powering digital watches, commercial LEDs, and skin-attachable touch screens. Hence, the disruptive technology with the innovative concept exhibits the new prospects for the utilization of the TENG in a wide range of applications like sports, implantable electronics, health monitoring, and smart robotics empowering next-generation electronics.

12.2.4 Bio-TENGs

The bio-TENGs based on environment-friendly and degradable biomaterials are becoming a new energy source owing to their great potential in implantable devices and the field of the Internet of things. The biomaterials utilized in bio-TENGs are categorized into proteins, lipids, and polysaccharides. Just like the conventional TENGs, the design of the bio-TENGs has three basic kinds: the vertical contact-separation structure (Niu et al. 2013), the single-electrode structure (Niu and Wang 2015), and the freestanding triboelectric layer structure. Out of all the three, the vertical contact-separation structure is the most commonly used, followed by the single-electrode structure and the last one is the freestanding triboelectric structure.

The vertical contact-separation structure contains two dielectric films having the opposite triboelectricity polarity along with two electrodes fixed to the back of the dielectric film. When the vertical force combines the dielectric films, the formation of the opposite charges takes place in the contact surfaces on account of different electron affinities of both dielectric films (Hinchet et al. 2015). By applying an external force, an inductive potential difference is created between electrodes on the back sides of the two films. When two electrodes are connected through an external circuit, a reverse potential difference is formed due to the movement of electrons to balance the electrostatic field. Furthermore, when the two films are again in contact, the potential difference vanishes that was created by the triboelectric charge and backflow of electrons. During the contact-separation process, this oscillatory motion of the electrons between the two membranes produces an alternating current that is utilized for the external loads. During the bio-TENG designing, based on the vertical contact-separation structure, the biomaterials are utilized in the fabrication of the electron-donating layer. A chitosan-based bio-TENG with the vertical contact-separation structure is prepared in which the triboelectric negative layer is formed from polyimide film, the triboelectric positive layer from the chitosan-based film, and two same-size aluminum foils as the induction electrodes fixed to the backside of the triboelectric films (Chen et al. 2019). When the external force separates or combines the triboelectric films, a change occurs in the potential of the two electrodes that results in the motion of the electrons between two electrodes, thus producing the current. The flexible few-layer graphene-based electrodes by replacing the gold electrodes in the TENGs operating in vertical contact mode achieve a 26-fold increase in the power density (Pace et al. 2020).

The working principle of the single-electrode current is the same, the coupling of electrostatic induction and the triboelectricity, just like bio-TENG with vertical contact-separation structure. When the triboelectrification occurs between the dielectric material and the electrode, the charge is produced on the surface of the dielectric material. Due to the electrostatic charge on its surface, the electrons are pulled or pushed by the dielectric material due to a change in the potential during the contact with and the separation from the electrode (Jiang et al. 2019). A completely stretchable bio-TENG possessing the single-electrode structure is designed to harvest the energy of a leaf, where the leaf acts as an electron-donating layer. Due to the flexibility, bio-TENG easily attaches to the leaf surface and harvest the energy through the triboelectric effect (Pu et al. 2017; Ahmed et al. 2019).

The freestanding model consists of moving objects and patterned electrodes under the dielectric material possessing the same size as that of a moving object. When the charge approaches and moves away after triboelectrification between the moving object and the triboelectric material, the charge distribution on both electrodes occurs leading to the motion of the electrons through an external circuit (Lin et al. 2016; Kwak et al. 2019). The bio-TENG shows various potential applications such as in health-care monitoring, agricultural self-powered sensing, and implantable medical equipment.

- As passive sensing devices, bio-TENGs are attached to the human body to monitor the muscle status, pulse, and sweat without any side effects. Bio-TENGs are used as battery-free sensors.
- In comparison to conventional energy sources, the bio-TENGs are controllable, environment-friendly, and compatible with power sensing in agriculture. Moreover, if the bio-TENGs are tossing out, the biomaterials present in bio-TENG can be used as food for livestock.
- The bio-TENGs get enough energy from the blood flow, contraction, expansion, and heartbeat of the human body as compared to the traditional energy to work as an electrical stimulation device. Furthermore, bio-TENGs enable implantable devices to reduce the risk of a second surgery and a patient's pain (Li et al. 2020).

12.2.5 Network TENGs

The demand for sustainable energy with fewer carbon emissions is essential for the sustainable growth of human civilization. Wind, solar energy, and water waves are sources of renewable and clean energy with high potential. However, the water waves in the ocean show greater advantages over solar and wind energy (Khaligh and Onar 2017; Sripadmanabhan Indira et al. 2019). Energy generated through water waves depends less on the duration of day and night, weather, temperature, and season. Mainly, the mechanical energy is harvested based on the principle of electromagnetic effects, but it shows certain limitations. First, an electromagnetic generator produces energy from a flowing stream, thus making it difficult to effectively harvest wave energy of water. Second, high-quality material is required for the fabrication of the electromagnetic generator, making it costly for practice on a large surface area. Lastly, the electromagnetic generator is heavy with a larger mass density because it contains metal coils and magnets, making it less efficient to float on the water surfaces unless support is provided by a buoy platform or floater. In the ocean, the surface of the water possesses the most dynamic energy (Salter 1974; Chen 2016).

12.3 PIEZOELECTRIC NGs

Piezoelectric NGs can harvest mechanical energy from the living environment. Piezoelectric NGs are used to power nano-systems and portable electronic devices. Piezoelectric NG is constructed from electrospun barium titanate ($BaTiO_3$) nanofibers followed by calcination. Barium titanate nanofibers are aligned in a matrix of PDMS in different patterns and placed between electrodes. The morphology of nanofibers is characterized by SEM and crystal structure by X-ray diffraction (Yan and Jeong 2016). Piezoelectric nanowire array consisting of $PbZr_xTi_{x-1}O_3$ (PZT) shows an ability to provide high output power and harvests energy from random motion. The NG can be combined with a rectifier and capacitor to store energy. The NG is used to power LEDs and mobile phones (Xu et al. 2010).

12.4 THERMOELECTRIC NGs

The thermoelectric NGs were constructed by placing a SiN thin membrane over a silicon substrate. The n-type and p-type materials are structured over SiN film by the lift-off process. Ni or Au provides the metallic contact between the n-type and p-type material at the top of SiN membrane. At the top, a photoresist is placed to protect the electrical contact. The SiN layer is etched to suspend the thermoelectric junction in the air and allow the flow of cold air through the microstructure. The assembly allows a decreased thermal coupling between the metallic conductor and the silicone frame and increased thermal exchange with the surrounding air or radiations. Second, the electrical resistance of the device is decreased to allow its use as a generator. The NGs have connected in series and/or in parallel to power a device. The type of connection is determined by the final resistance of the load. The NG provides the power upon a small change in the temperature and may be used in the temperature or air sensing after the optimization (Tainoff et al. 2019). The antimony-doped zinc oxide (ZnO) nanobelt is connected to the glass at its ends to construct a thermoelectric NG. The temperature difference between the two electrodes is kept at 30 K. The thermoelectric NG may be used to power a self-power temperature sensor (Yang et al. 2012).

12.5 PYROELECTRIC NGs

The pyroelectric NGs are noticed for the prospective applications in power generation. The pyroelectric effect converts the thermal energy into the electric current. The pyroelectric effect has been employed to sense infrared radiations. The miniaturized sensor technology utilizes the pyroelectric NGs as a sustainable power source. Pyroelectric NGs harvest energy from the sun, body heat, and heaters. A variety of polymers and ceramic materials are used in the pyroelectric NGs to harvest the thermal energy through the pyroelectric effect (Ryu and Kim 2019).

12.6 HYBRID NGs

The hybrid electromagnetic TENG uses wind energy for power generation. Using the novel power management system with TENG increments the power storage in a capacitor. The hybrid NG shows an increased ability to power the temperature sensor (Wang and Yang 2017). Bionic NG for application in underwater usable devices uses combined triboelectric effect and electrostatic induction for sustainable power generation (Zou et al. 2019). Self-powered NG capable of deriving the energy from human motion comprises electromagnetic and triboelectric generators. The construction of hybrid NG and surface morphology of triboelectric layer as assessed by FESEM is illustrated in Figure 12.8. The use of the Halbach magnet array in electromagnetic generators increases the magnetic flux density and output power. Incorporation of nanowire structure in the PTFE film and amendment of the polyamide 11 through the phase inversion method increments the energy deriving capability of the triboelectric generators. The working mechanism of hybrid NG is illustrated in Figure 12.9. The hybrid NG is capable of deriving the biochemical energy to power Bluetooth computer mouse, smartwatches, and smartphones. The COMSOL simulation is used to study the distribution of magnetic and electric fields in the hybrid NG as illustrated in Figure 12.10 (Rana et al. 2020).

Continuous energy harvesting becomes challenging in the harsh conditions posed by some environments. Hybrid all-in-one power TENGs are spherical and designed to adapt to such environments for uninterrupted output power. The spherical TENGs can harvest energy from different energy sources like the sun, wind, and rain. The spherical TENGs contain solar cells and can harvest the energy from a fluid as well. The TENGs are used in the self-powered humidity control system, forest fire alarms, and pipeline monitors (Xu et al. 2020).

The triboelectric and the piezoelectric effects are combined in a hybrid NG to harvest the biomechanical energy from the skin upon clapping of the hand. The hybrid NGs are wearable and capable

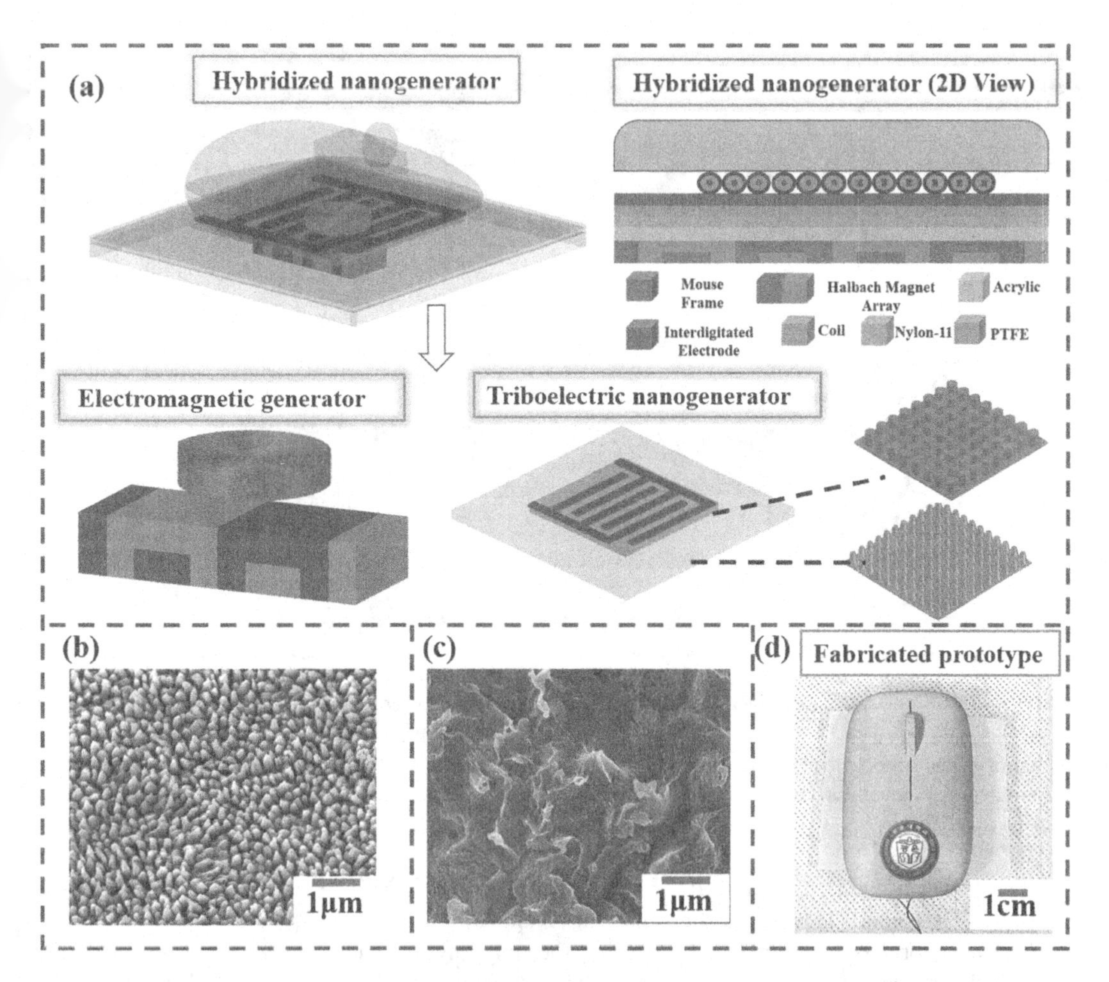

FIGURE 12.8 Illustration of the structure of human interactive hybrid NG consisting of electromagnetic NG and TENG (a) and characterization of surface morphology by FESEM (b) nylon/11 and (c) PTFE surface. (d) Device taking power from hybrid NG. (Reprinted with permission of Elsevier from Rana et al. (2020).)

of powering portable devices and eliminating the requirement to use a battery. The hybrid NG contains polyvinylidene fluoride enclosed in polyamide between two layers of nanopillar PDMS. One clap can deliver two triboelectric and one piezoelectric output instantaneously. The hybrid NG is capable of powering the pedometer and recharges a trimmer, a personal Wi-Fi router, and a smartphone one at a time (Rasel et al. 2019).

The fast-growing development of sensor technology and the Internet of things demands a sustainable supply of power sources for their operation. A self-powered helmet is powered by changing the mechanical vibration through the hybridized NGs formed by the integration of an electromagnetic generator and a TENG in a sandwiched structure. The self-powered helmet can power the electronics when a person is running, walking, shaking hands, or jumping (Jin et al. 2016).

Human life has become intelligent and efficient due to technological advances (Pu et al. 2017). The advances include wearable electronics to monitor the human position, motion, and heartbeat, etc. (Yamamoto et al. 2016). Nonetheless, the battery life span makes the applications limited. The facts pose an urgent need for the development of self-powered systems to harvest the energy to derive the system. The fabric-based TENGs have been used to convert human motion energy into electrical energy (Liu et al. 2019). Among various types of energy, mechanical energy is the most suitable form for wearable electronics. Thus, a small-size high-power NG is incorporated into shoes

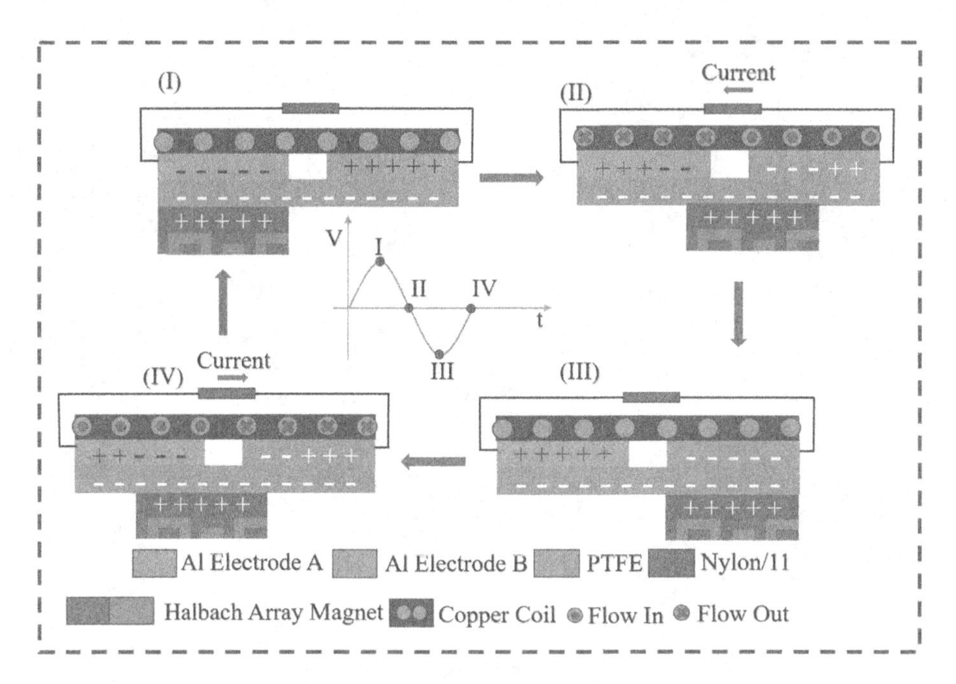

FIGURE 12.9 Internal layer structure and design of the HMI-HBNG showing charge distributions and electricity generation (I–IV). The figure shows the process when the computer mouse moves during sliding operation of the triboelectric nanogenerator (TENG), and the current flow inside the coils when the coil of the electromagnetic generator (EMG) moves concerning magnet. (Reprinted with permission of Elsevier from Rana et al. (2020).)

for high power usage (Jing et al. 2015; Li et al. 2017). A hybrid NG, due to its high-power output, combines TENG and electromagnetic generator (Wang and Yang 2017).

A novel hybrid NG is prepared by treating the polyvinyl chloride film with CF_4 by using an inductively coupled plasma. The treated polyvinyl chloride film is cut into circles along with conductive fabric film, adhering the ends to the substrate to form joint TENG. The mirror symmetry is shown by the top and bottom part of TENG, while every part possesses two round electrodes combined with a substrate. The packaging of the device is achieved in size with a height of 1.2 cm and a radius of 2 cm. Upon embedding the mini-sized device into the shoe, the device operates as an energy cell to supply power as a lithium battery (3.7 V) shown to charge about 2.7–3.0 V after walking for 30 minutes and a capacitor (1000 μF) is charged 5.0 V after 100 cycles of vibrations. Besides, the device helps to detect the user's location through a GPS or can charge a cell phone (Liu et al. 2018).

12.7 FUTURE HORIZONS

Since their discovery in 2006, NGs have experienced tremendous development and advancement in size structure and dielectric material. NGs have been used extensively and innovatively in a wide range of applications. Still, there is much to explore in the field of NGs. New dielectric materials able to harvest a minimum amount of energy and offering maximum and stable output are being explored. Scientists have developed electrodeless NGs and are looking for more improvement. NGs have great potential in blue energy. NGs are connected to a network to increase output power in the future. NGs may be used as a high-voltage source. Simulation software helps us improve the output performance and understand the behavior of electric current produced by NGs. Scientists are working to achieve maximum output current, voltage, and performance of NGs in minimum size and cost. Ultimately, this will improve the human lifestyle and the horizon is open.

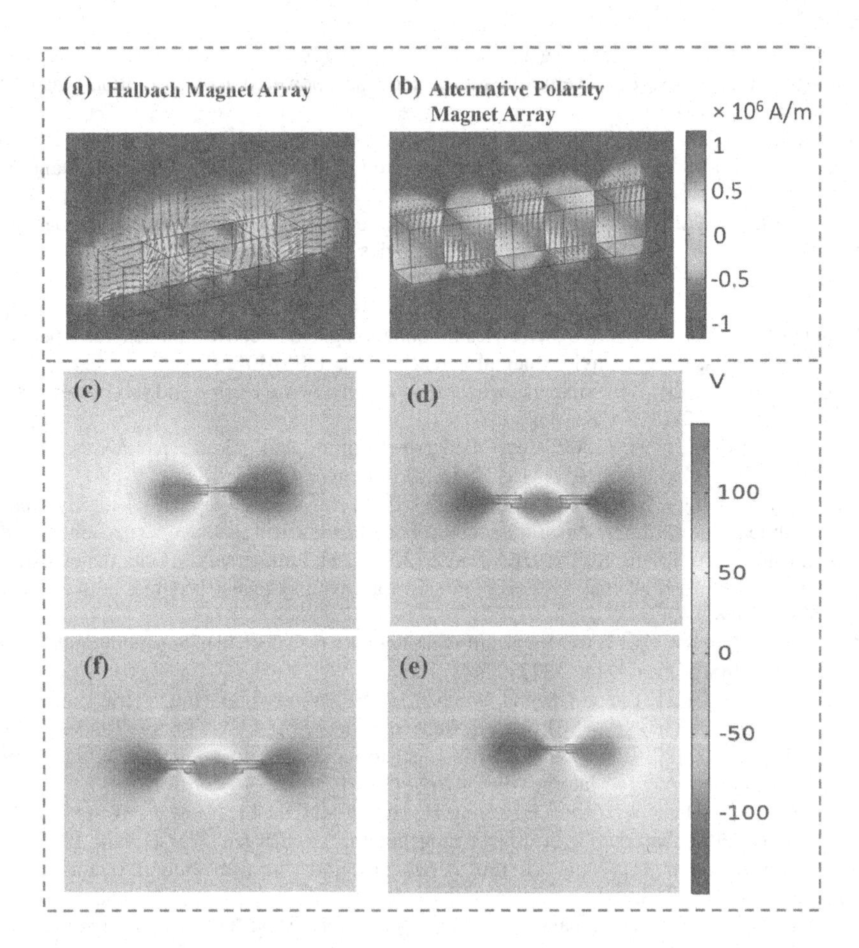

FIGURE 12.10 COMSOL software simulation analysis of the EMG and TENG. (a) Distribution of magnetic field strength of Halbach magnet array of cross-sectional 3D schematic view. (b) Magnetic field strength distribution of the alternative polarity magnet array. Potential distribution across the interdigitated electrode of the TENG in an open circuit condition in four states: (c) initial state (d) first intermediate state, (e) final state, and (f) second intermediate state. (Reprinted with permission of Elsevier from Rana et al. (2020).)

REFERENCES

Ahmed A, Hassan I, El-Kady MF, Radhi A, Jeong CK, Selvaganapathy PR, Zu J, Ren S, Wang Q, Kaner RB (2019) Integrated triboelectric nanogenerators in the era of the internet of things. Advanced Science 6(24): 1802230.

Ankanahalli Shankaregowda S, Sagade Muktar Ahmed RF, Liu Y, Bananakere Nanjegowda C, Cheng X, Shivanna S, Ramakrishna S, Yu Z, Zhang X, Sannathammegowda K (2019) Dry-coated graphite onto sandpaper for triboelectric nanogenerator as an active power source for portable electronics. Nanomaterials 9(11): 1585.

Askari H, Khajepour A, Khamesee MB, Saadatnia Z, Wang ZL (2018) Piezoelectric and triboelectric nanogenerators: trends and impacts. Nano Today 22: 10–13.

Ba Y-Y, Bao J-F, Deng H-T, Wang Z-Y, Li X-W, Gong T, Huang W, Zhang X-S (2020) Single-layer triboelectric nanogenerators based on ion-doped natural nanofibrils. ACS Applied Materials & Interfaces 12(38): 42859–42867.

Chen J (2016). Triboelectric nanogenerators. (Doctor of Philosophy), Georgia Institute of Technology, Atlanta, GA.

Chen J, Yang J, Li Z, Fan X, Zi Y, Jing Q, Guo H, Wen Z, Pradel KC, Niu S, Wang ZL (2015) Networks of triboelectric nanogenerators for harvesting water wave energy: a potential approach toward blue energy. ACS Nano 9(3): 3324–3331. doi:10.1021/acsnano.5b00534

Chen J, Zhu G, Yang W, Jing Q, Bai P, Yang Y, Hou TC, Wang ZL (2013) Harmonic-resonator-based triboelectric nanogenerator as a sustainable power source and a self-powered active vibration sensor. Advanced Materials 25(42): 6094–6099.

Chen S, Jiang J, Xu F, Gong S (2019) Crepe cellulose paper and nitrocellulose membrane-based triboelectric nanogenerators for energy harvesting and self-powered human-machine interaction. Nano Energy 61: 69–77.

Cheng T, Gao Q, Wang ZL (2019) The current development and future outlook of triboelectric nanogenerators: a survey of literature. Advanced Materials Technologies 4(3): 1800588.

Chortos A, Liu J, Bao Z (2016) Pursuing prosthetic electronic skin. Nature Materials 15(9): 937–950.

Chun J, Kim JW, Jung W-S, Kang C-Y, Kim S-W, Wang ZL, Baik JM (2015) Mesoporous pores impregnated with Au nanoparticles as effective dielectrics for enhancing triboelectric nanogenerator performance in harsh environments. Energy Environmental Science 8(10): 3006–3012.

Fan FR, Tang W, Wang ZL (2016) Flexible nanogenerators for energy harvesting and self-powered electronics. Advanced Materials 28(22): 4283–4305.

Fang H, Wu W, Song J, Wang ZL (2009) Controlled growth of aligned polymer nanowires. The Journal of Physical Chemistry C 113(38): 16571–16574. doi:10.1021/jp907072z

Guo H, Pu X, Chen J, Meng Y, Yeh M-H, Liu G, Tang Q, Chen B, Liu D, Qi S (2018) A highly sensitive, self-powered triboelectric auditory sensor for social robotics and hearing aids. Science Robotics 3(20): 1-9.

Hammock ML, Chortos A, Tee BCK, Tok JBH, Bao Z (2013) 25th anniversary article: the evolution of electronic skin (e-skin): a brief history, design considerations, and recent progress. Advanced Materials 25(42): 5997–6038.

Hinchet R, Seung W, Kim SW (2015) Recent progress on flexible triboelectric nanogenerators for selfpowered electronics. ChemSusChem 8(14): 2327–2344.

Jiang D, Xu M, Dong M, Guo F, Liu X, Chen G, Wang ZL (2019) Water-solid triboelectric nanogenerators: an alternative means for harvesting hydropower. Renewable and Sustainable Energy Reviews 115: 109366.

Jiang Q, Chen B, Yang Y (2018) Wind-driven triboelectric nanogenerators for scavenging biomechanical energy. ACS Applied Energy Materials 1(8): 4269–4276.

Jin L, Chen J, Zhang B, Deng W, Zhang L, Zhang H, Huang X, Zhu M, Yang W, Wang ZL (2016) Self-powered safety helmet based on hybridized nanogenerator for emergency. ACS Nano 10(8): 7874–7881.

Jing Q, Xie Y, Zhu G, Han RP, Wang ZL (2015) Self-powered thin-film motion vector sensor. Nature Communications 6(1): 1–8.

Khaligh A, Onar OC (2017). Energy harvesting: solar, wind, and ocean energy conversion systems, CRC Press, USA.

Kwak SS, Yoon HJ, Kim SW (2019) Textile-based triboelectric nanogenerators for self-powered wearable electronics. Advanced Functional Materials 29(2): 1804533.

Lai Y-C, Deng J, Niu S, Peng W, Wu C, Liu R, Wen Z, Wang ZL (2016) Electric eel-skin-inspired mechanically durable and super-stretchable nanogenerator for deformable power source and fully autonomous conformable electronic-skin applications. Advanced Materials 28(45): 10024–10032. doi:10.1002/adma.201603527

Li S, Wang J, Peng W, Lin L, Zi Y, Wang S, Zhang G, Wang ZL (2017) Sustainable energy source for wearable electronics based on multilayer elastomeric triboelectric nanogenerators. Advanced Energy Materials 7(13): 1602832.

Li X, Jiang C, Ying Y, Ping J (2020) Biotriboelectric nanogenerators: materials, structures, and applications. Advanced Energy Materials 10(44): 2002001.

Li X, Xu G, Xia X, Fu J, Huang L, Zi Y (2019) Standardization of triboelectric nanogenerators: progress and perspectives. Nano Energy 56: 40–55.

Li Z, Chen J, Yang J, Su Y, Fan X, Wu Y, Yu C, Wang ZL (2015) β-Cyclodextrin enhanced triboelectrification for self-powered phenol detection and electrochemical degradation. Energy Environmental Science 8(3): 887–896.

Li Z, Wang ZL (2011) Air/liquid-pressure and heartbeat-driven flexible fiber nanogenerators as a micro/nano-power source or diagnostic sensor. Advanced Materials 23(1): 84–89.

Lin L, Xie Y, Niu S, Wang S, Yang P-K, Wang ZL (2015) Robust triboelectric nanogenerator based on rolling electrification and electrostatic induction at an instantaneous energy conversion efficiency of ~55%. ACS Nano 9(1): 922–930.

Lin Z, Chen J, Yang J (2016) Recent progress in triboelectric nanogenerators as a renewable and sustainable power source. Journal of Nanomaterials 2016: 5651613.

Liu J, Gu L, Cui N, Xu Q, Qin Y, Yang R (2019) Fabric-based triboelectric nanogenerators. Research 2019: 1091632.

Liu L, Tang W, Deng C, Chen B, Han K, Zhong W, Wang ZL (2018) Self-powered versatile shoes based on hybrid nanogenerators. Nano Research 11(8): 3972–3978.
Liu Y, Wang L, Zhao L, Yu X, Zi Y (2020) Recent progress on flexible nanogenerators toward self-powered systems. InfoMat 2(2): 318–340.
Luo J, Wang ZL (2020) Recent progress of triboelectric nanogenerators: from fundamental theory to practical applications. EcoMat 2(4): e12059.
Ma M, Kang Z, Liao Q, Zhang Q, Gao F, Zhao X, Zhang Z, Zhang Y (2018) Development, applications, and future directions of triboelectric nanogenerators. Nano Research 11(6): 2951–2969.
Niu S, Wang S, Lin L, Liu Y, Zhou YS, Hu Y, Wang ZL (2013) Theoretical study of contact-mode triboelectric nanogenerators as an effective power source. Energy & Environmental Science 6(12): 3576–3583.
Niu S, Wang ZL (2015) Theoretical systems of triboelectric nanogenerators. Nano Energy 14: 161–192.
Oh JY, Rondeau-Gagné S, Chiu Y-C, Chortos A, Lissel F, Wang G-JN, Schroeder BC, Kurosawa T, Lopez J, Katsumata T (2016) Intrinsically stretchable and healable semiconducting polymer for organic transistors. Nature 539(7629): 411–415.
Pace G, Ansaldo A, Serri M, Lauciello S, Bonaccorso F (2020) Electrode selection rules for enhancing the performance of triboelectric nanogenerators and the role of few-layers graphene. Nano Energy 76: 104989.
Pan C, Li Z, Guo W, Zhu J, Wang ZL (2011) Fiber-based hybrid nanogenerators for/as self-powered systems in biological liquid. Angewandte Chemie International Edition 50(47): 11192–11196.
Pang Y, Xue F, Wang L, Chen J, Luo J, Jiang T, Zhang C, Wang ZL (2016) Tribotronic enhanced photoresponsivity of a MoS_2 phototransistor. Advanced Science 3(6): 1500419.
Proto A, Penhaker M, Conforto S, Schmid M (2017) Nanogenerators for human body energy harvesting. Trends in Biotechnology 35(7): 610–624.
Pu X, Guo H, Chen J, Wang X, Xi Y, Hu C, Wang ZL (2017) Eye motion triggered self-powered mechnosensational communication system using triboelectric nanogenerator. Science Advances 3(7): e1700694.
Pu X, Liu M, Chen X, Sun J, Du C, Zhang Y, Zhai J, Hu W, Wang ZL (2017) Ultrastretchable, transparent triboelectric nanogenerator as electronic skin for biomechanical energy harvesting and tactile sensing. Science Advances 3(5): e1700015.
Qin H, Cheng G, Zi Y, Gu G, Zhang B, Shang W, Yang F, Yang J, Du Z, Wang ZL (2018) High energy storage efficiency triboelectric nanogenerators with unidirectional switches and passive power management circuits. Advanced Functional Materials 28(51): 1805216. doi:10.1002/adfm.201805216
Rana SMS, Rahman MT, Salauddin M, Maharjan P, Bhatta T, Cho H, Park JY (2020) A human-machine interactive hybridized biomechanical nanogenerator as a self-sustainable power source for multifunctional smart electronics applications. Nano Energy 76: 105025.
Rasel MS, Maharjan P, Park JY (2019) Hand clapping inspired integrated multilayer hybrid nanogenerator as a wearable and universal power source for portable electronics. Nano Energy 63: 103816.
Ryu H, Kim SW (2019) Emerging pyroelectric nanogenerators to convert thermal energy into electrical energy. Small 17(9): 1903469.
Salter SH (1974) Wave power. Nature 249(5459): 720–724.
Sekitani T, Noguchi Y, Hata K, Fukushima T, Aida T, Someya T (2008) A rubberlike stretchable active matrix using elastic conductors. Science 321(5895): 1468–1472.
Seung W, Yoon HJ, Kim TY, Ryu H, Kim J, Lee JH, Lee JH, Kim S, Park YK, Park YJ (2017) Boosting power-generating performance of triboelectric nanogenerators via artificial control of ferroelectric polarization and dielectric properties. Advanced Energy Materials 7(2): 1600988.
Shao J, Jiang T, Tang W, Chen X, Xu L, Wang ZL (2018a) Structural figure-of-merits of triboelectric nanogenerators at powering loads. Nano Energy 51: 688–697.
Shao J, Jiang T, Tang W, Xu L, Kim TW, Wu C, Chen X, Chen B, Xiao T, Bai Y (2018b) Studying about applied force and the output performance of sliding-mode triboelectric nanogenerators. Nano Energy 48: 292–300.
Shao J, Willatzen M, Jiang T, Tang W, Chen X, Wang J, Wang ZL (2019) Quantifying the power output and structural figure-of-merits of triboelectric nanogenerators in a charging system starting from the Maxwell's displacement current. Nano Energy 59: 380–389.
Shi J, Chen X, Li G, Sun N, Jiang H, Bao D, Xie L, Peng M, Liu Y, Wen Z (2019) A liquid PEDOT: PSS electrode-based stretchable triboelectric nanogenerator for a portable self-charging power source. Nanoscale 11(15): 7513–7519.
Shi K, Zou H, Sun B, Jiang P, He J, Huang X (2020) Dielectric modulated cellulose paper/PDMS-based triboelectric nanogenerators for wireless transmission and electropolymerization applications. Advanced Functional Materials 30(4): 1904536.

Shin SY, Saravanakumar B, Ramadoss A, Kim SJ (2016) Fabrication of PDMS-based triboelectric nanogenerator for self-sustained power source application. International Journal of Energy Research 40(3): 288–297.
Someya T, Sekitani T, Iba S, Kato Y, Kawaguchi H, Sakurai T (2004) A large-area, flexible pressure sensor matrix with organic field-effect transistors for artificial skin applications. Proceedings of the National Academy of Sciences 101(27): 9966–9970.
Sripadmanabhan Indira S, Aravind Vaithilingam C, Oruganti KSP, Mohd F, Rahman S (2019) Nanogenerators as a sustainable power source: state of art, applications, and challenges. Nanomaterials 9(5): 773.
Sriphan S, Charoonsuk T, Maluangnont T, Pakawanit P, Rojviriya C, Vittayakorn N (2020) Multifunctional nanomaterials modification of cellulose paper for efficient triboelectric nanogenerators. Advanced Materials Technologies 5(5): 2000001.
Stanford MG, Li JT, Chyan Y, Wang Z, Wang W, Tour JM (2019) Laser-induced graphene triboelectric nanogenerators. ACS Nano 13(6): 7166–7174.
Su Y, Zhu G, Yang W, Yang J, Chen J, Jing Q, Wu Z, Jiang Y, Wang ZL (2014) Triboelectric sensor for self-powered tracking of object motion inside tubing. ACS Nano 8(4): 3843–3850.
Sukumaran C, Vivekananthan V, Mohan V, Alex ZC, Chandrasekhar A, Kim S-J (2020) Triboelectric nanogenerators from reused plastic: an approach for vehicle security alarming and tire motion monitoring in rover. Applied Materials Today 19: 100625.
Sun J, Pu X, Liu M, Yu A, Du C, Zhai J, Hu W, Wang ZL (2018) Self-healable, stretchable, transparent triboelectric nanogenerators as soft power sources. ACS Nano 12(6): 6147–6155.
Sun JY, Keplinger C, Whitesides GM, Suo Z (2014) Ionic skin. Advanced Materials 26(45): 7608–7614.
Sun L, Chen S, Guo Y, Song J, Zhang L, Xiao L, Guan Q, You Z (2019) Ionogel-based, highly stretchable, transparent, durable triboelectric nanogenerators for energy harvesting and motion sensing over a wide temperature range. Nano Energy 63: 103847. doi:10.1016/j.nanoen.2019.06.043
Tainoff D, Proudhom A, Tur C, Crozes T, Dufresnes S, Dumont S, Bourgault D, Bourgeois O (2019) Network of thermoelectric nanogenerators for low power energy harvesting. Nano Energy 57: 804–810.
Tang W, Zhang C, Han CB, Wang ZL (2014) Enhancing output power of cylindrical triboelectric nanogenerators by segmentation design and multilayer integration. Advanced Functional Materials 24(42): 6684–6690.
Tantraviwat D, Buarin P, Suntalelat S, Sripumkhai W, Pattamang P, Rujijanagul G, Inceesungvorn B (2020) Highly dispersed porous polydimethylsiloxane for boosting power-generating performance of triboelectric nanogenerators. Nano Energy 67: 104214.
Vivekananthan V, Raj NPMJ, Alluri NR, Purusothaman Y, Chandrasekhar A, Kim S-J (2020) Substantial improvement on electrical energy harvesting by chemically modified/sandpaper-based surface modification in micro-scale for hybrid nanogenerators. Applied Surface Science 514: 145904.
Wang H, Zhu J, He T, Zhang Z, Lee C (2020) Programmed-triboelectric nanogenerators—a multi-switch regulation methodology for energy manipulation. Nano Energy 78: 105241.
Wang J, Li S, Yi F, Zi Y, Lin J, Wang X, Xu Y, Wang ZL (2016) Sustainably powering wearable electronics solely by biomechanical energy. Nature Communications 7(1): 1–8.
Wang J, Wen Z, Zi Y, Zhou P, Lin J, Guo H, Xu Y, Wang ZL (2016) All-plastic-materials based self-charging power system composed of triboelectric nanogenerators and supercapacitors. Advanced Functional Materials 26(7): 1070–1076.
Wang S, Lin L, Wang ZL (2012) Nanoscale triboelectric-effect-enabled energy conversion for sustainably powering portable electronics. Nano Letters 12(12): 6339–6346.
Wang S, Lin L, Xie Y, Jing Q, Niu S, Wang ZL (2013) Sliding-triboelectric nanogenerators based on in-plane charge-separation mechanism. Nano Letters 13(5): 2226–2233.
Wang S, Xie Y, Niu S, Lin L, Wang ZL (2014) Freestanding triboelectric-layer-based nanogenerators for harvesting energy from a moving object or human motion in contact and non-contact modes. Advanced Materials 26(18): 2818–2824.
Wang X, Yang Y (2017) Effective energy storage from a hybridized electromagnetic-triboelectric nanogenerator. Nano Energy 32: 36–41.
Wang X, Zhang H, Dong L, Han X, Du W, Zhai J, Pan C, Wang ZL (2016) Self-powered high-resolution and pressure-sensitive triboelectric sensor matrix for real-time tactile mapping. Advanced Materials 28(15): 2896–2903.
Wang Y, Duan J, Yang X, Liu L, Zhao L, Tang Q (2020) The unique dielectricity of inorganic perovskites toward high-performance triboelectric nanogenerators. Nano Energy 69: 104418.
Wang ZL (2013) Triboelectric nanogenerators as new energy technology for self-powered systems and as active mechanical and chemical sensors. ACS Nano 7(11): 9533–9557.

Wang ZL (2015) Triboelectric nanogenerators as new energy technology and self-powered sensors–Principles, problems and perspectives. Faraday Discussions 176: 447–458.
Wang ZL (2020) On the first principle theory of nanogenerators from Maxwell's equations. Nano Energy 68: 104272.
Wang ZL, Song J (2006) Piezoelectric nanogenerators based on zinc oxide nanowire arrays. Science 312(5771): 242–246.
Wen Z, Shen Q, Sun X (2017) Nanogenerators for self-powered gas sensing. Nano-Micro Letters 9(4): 45.
Wu C, Wang AC, Ding W, Guo H, Wang ZL (2019) Triboelectric nanogenerator: a foundation of the energy for the new era. Advanced Energy Materials 9(1): 1802906.
Wu Z, Cheng T, Wang ZL (2020) Self-powered sensors and systems based on Nanogenerators. Sensors 20(10): 2925.
Xia K, Wu D, Fu J, Xu Z (2020) A pulse controllable voltage source based on triboelectric nanogenerator. Nano Energy 77: 105112.
Xia X, Fu J, Zi Y (2019) A universal standardized method for output capability assessment of nanogenerators. Nature Communications 10(1): 1–9.
Xie L, Chen X, Wen Z, Yang Y, Shi J, Chen C, Peng M, Liu Y, Sun X (2019) Spiral steel wire based fiber-shaped stretchable and tailorable triboelectric nanogenerator for wearable power source and active gesture sensor. Nano-Micro Letters 11(1): 39.
Xu G, Li X, Xia X, Fu J, Ding W, Zi Y (2019) On the force and energy conversion in triboelectric nanogenerators. Nano Energy 59: 154–161.
Xu L, Bu TZ, Yang XD, Zhang C, Wang ZL (2018) Ultrahigh charge density realized by charge pumping at ambient conditions for triboelectric nanogenerators. Nano Energy 49: 625–633.
Xu L, Xu L, Luo J, Yan Y, Jia BE, Yang X, Gao Y, Wang ZL (2020) Hybrid All-in-One Power Source Based on High-Performance Spherical Triboelectric Nanogenerators for Harvesting Environmental Energy. Advanced Energy Materials 10(36): 2001669.
Xu S, Ding W, Guo H, Wang X, Wang ZL (2019) Boost the performance of triboelectric nanogenerators through circuit oscillation. Advanced Energy Materials 9(30): 1900772.
Xu S, Hansen BJ, Wang ZL (2010) Piezoelectric-nanowire-enabled power source for driving wireless microelectronics. Nature Communications 1(1): 1–5.
Xu W, Huang LB, Wong MC, Chen L, Bai G, Hao J (2017) Environmentally friendly hydrogel-based triboelectric nanogenerators for versatile energy harvesting and self-powered sensors. Advanced Energy Materials 7(1): 1601529.
Yamamoto Y, Harada S, Yamamoto D, Honda W, Arie T, Akita S, Takei K (2016) Printed multifunctional flexible device with an integrated motion sensor for health care monitoring. Science Advances 2(11): e1601473.
Yan J, Jeong YG (2016) High performance flexible piezoelectric nanogenerators based on BaTiO3 nanofibers in different alignment modes. ACS Applied Materials & Interfaces 8(24): 15700–15709.
Yang B, Zeng W, Peng ZH, Liu SR, Chen K, Tao XM (2016) A fully verified theoretical analysis of contact-mode triboelectric nanogenerators as a wearable power source. Advanced Energy Materials 6(16): 1600505.
Yang J, Chen J, Su Y, Jing Q, Li Z, Yi F, Wen X, Wang Z, Wang ZL (2015) Eardrum-Inspired active sensors for self-powered cardiovascular system characterization and throat-attached anti-Interference voice recognition. Advanced Materials 27(8): 1316–1326.
Yang P-K, Lin Z-H, Pradel KC, Lin L, Li X, Wen X, He J-H, Wang ZL (2015) based origami triboelectric nanogenerators and self-powered pressure sensors. ACS Nano 9(1): 901–907.
Yang W, Chen J, Zhu G, Yang J, Bai P, Su Y, Jing Q, Cao X, Wang ZL (2013) Harvesting energy from the natural vibration of human walking. ACS Nano 7(12): 11317–11324.
Yang Y, Pradel KC, Jing Q, Wu JM, Zhang F, Zhou Y, Zhang Y, Wang ZL (2012) Thermoelectric nanogenerators based on single Sb-doped ZnO micro/nanobelts. ACS Nano 6(8): 6984–6989.
Yao C, Hernandez A, Yu Y, Cai Z, Wang X (2016) Triboelectric nanogenerators and power-boards from cellulose nanofibrils and recycled materials. Nano Energy 30: 103–108.
Ye BU, Kim B-J, Ryu J, Lee JY, Baik JM, Hong K (2015) Electrospun ion gel nanofibers for flexible triboelectric nanogenerator: electrochemical effect on output power. Nanoscale 7(39): 16189–16194.
Yi F, Wang X, Niu S, Li S, Yin Y, Dai K, Zhang G, Lin L, Wen Z, Guo H (2016) A highly shape-adaptive, stretchable design based on conductive liquid for energy harvesting and self-powered biomechanical monitoring. Science Advances 2(6): e1501624.
Yick J, Mukherjee B, Ghosal D (2008) Wireless sensor network survey. Computer Networks 52(12): 2292–2330.

Yin X, Liu D, Zhou L, Li X, Zhang C, Cheng P, Guo H, Song W, Wang J, Wang ZL (2018) Structure and dimension effects on the performance of layered triboelectric nanogenerators in contact-separation mode. ACS Nano 13(1): 698–705.

Zhang C, Tang W, Pang Y, Han C, Wang ZL (2015) Active micro-actuators for optical modulation based on a planar sliding triboelectric nanogenerator. Advanced Materials 27(4): 719–726.

Zhang X-S, Han M-D, Wang R-X, Zhu F-Y, Li Z-H, Wang W, Zhang H-X (2013) Frequency-multiplication high-output triboelectric nanogenerator for sustainably powering biomedical microsystems. Nano Letters 13(3): 1168–1172.

Zi Y, Niu S, Wang J, Wen Z, Tang W, Wang ZL (2015) Standards and figure-of-merits for quantifying the performance of triboelectric nanogenerators. Nature Communications 6(1): 1–8.

Zi Y, Wu C, Ding W, Wang X, Dai Y, Cheng J, Wang J, Wang Z, Wang ZL (2018) Field emission of electrons powered by a triboelectric nanogenerator. Advanced Functional Materials 28(21): 1800610.

Zou Y, Raveendran V, Chen J (2020) Wearable triboelectric nanogenerators for biomechanical energy harvesting. Nano Energy 77: 105303.

Zou Y, Tan P, Shi B, Ouyang H, Jiang D, Liu Z, Li H, Yu M, Wang C, Qu X (2019) A bionic stretchable nanogenerator for underwater sensing and energy harvesting. Nature Communications 10(1): 1–10.

13 Recent Advances and Applications of Nanogenerators in Sensors, Photovoltaic, Wind, and Blue Energy

Pallavi Jain, Sapna Raghav, and Madhur Babu Singh

CONTENTS

13.1 INTRODUCTION

Energy harvesting is a technique for transforming various sources of renewable energy, such as thermal energy (TE), solar energy, kinetic energy (KE), and electrostatic energy, into electrical energy (EE) (Figure 13.1). The upcoming Internet of Things (IOT) will create a network defining bodies and objects individually and linked via dataflows and will agree for incessant monitoring of manifold activities and applications, allowing for better decision-making in innovative defense, health, and environmental rules, as well as fresh powering system requirements. The reconciliation of outlines taking into consideration energy harvesting from the atmosphere or from nearby moving items can address a justifiable choice for openly powering wearable electronic devices, devices to monitor health, and remote sensors in the IOT networks [1].

Nanogenerators (NGs) are the simple design of devices, which convert mechanical energy (ME) into EE. NGs do not require mechanical input of high frequency. Wang and Song displayed the

DOI: 10.1201/9781003187615-13

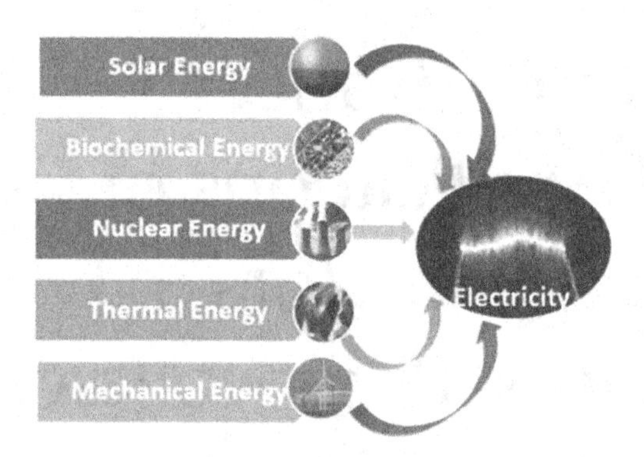

FIGURE 13.1 Different energy harvesting techniques.

very first NG in 2006 [2]. The system took advantage of the piezoelectric effect, produced by the escape of a modified variety of zinc oxide (ZnO) nanowires, undeniably proving the probability of converting ME into electricity for powering nanodevices. This revolutionary development was accompanied by a flurry of study that resulted in the creation of many NGs with an array of substances, system designs, and functionalities. Specifically, nanomaterials revealed the possibility to improve the productivity as well as to extend the scope of the device structures, while thrusts in accessible manufacture methods empowered the joining of inorganic nanomaterials on steady substrates, consequently setting out open doors for adjustable and stretchable gadgets impossible with usual methodologies [3].

13.1.1 History and Development of NGs

NGs are devices for energy harvesting that produce electricity from surrounding residual ME. Before 17th century over, researchers formed workable ways to produce static electricity. Around 1663, the most primitive type of friction machine was designed [4]. Afterward, a few scientists fragmented away at this device to build up the exhibition. The electromagnetic generator was invented in 1831 and employed extensively in thermal plants at present. The Wimshurst machine (1878) used static electricity for its working; it was used as a typical static electricity generator. Van de Graaff generator was invented in 1929; it generates direct current of high voltage. It can achieve the extremely high voltage of up to 25 MV. Afterward, nanotechnology was used to produce electricity using triboelectric, pyroelectric, and piezoelectric effects. High surface region and adjustable chemical and physical properties at nanodimensions insure an essentially extra productive improvement that catch, renovates, and stores various energies like EE, ME, TE, and radiant and chemical energy [5]. ME can be converted to EE using many ambient energy harvesting methods, viz., electromagnetic induction, piezoelectric effect, pyroelectric effect, and triboelectric effect. Different classes of MEs like human movement (e.g., strolling, heartbeat, running, and breathing), vibration, streaming water, raindrops, and wind, as well as residual heat could generate electricity due to pyroelectric effect. The premier NG, based on ZnO nanowires that worked on the piezoelectric effect, had 17–30% conversion efficiency, and was developed around 2006 [6].

Some customary methods like resonators and transducers are based on cantilever exists, which convert ME to electricity and work sufficiently on vibrations of high frequency; moreover, low-frequency mechanical vibrations can also be converted to electricity using NGs. ZnO nanowires are much more flexible and can sustain much more strain than a bulk of ZnO; thus, more electricity can be generated using ZnO nanowires [7]. ME harvesting in self-powered sensing and monitoring gadgets have potentially used NGs. The applications of NGs are expected to show sharp expansion as modernization unveils. Figure 13.2 depicts a variety of NGs.

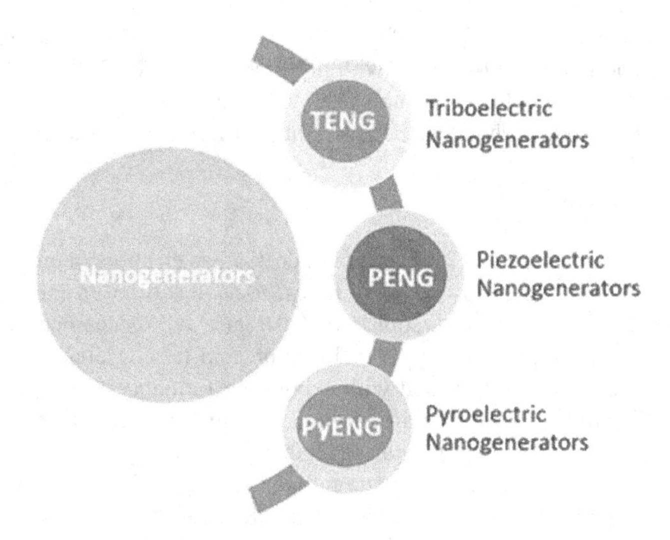

FIGURE 13.2 Types of nanogenerators.

This chapter intends to give an exhaustive survey on tribo-, pyro-, and piezo-based NGs, i.e., triboelectric NGs (TENGs), pyroelectric nanogenerators (PyENGs), and piezoelectric nanogenerators (PENGs) as sustainable energy-harvesting gadgets that includes the progression, enhancement, and utilizations in coming decades of NGs. Additionally, other applications of NGs are in sensors, photovoltaic, harvesting of wind, and blue energy.

13.2 PIEZOELECTRIC NANOGENERATORS

"Piezoelectricity" is Greek derivative of "piezein" as in "to press" and "elektron" as in "amber", implies electricity generated from pressure. The ability of certain materials to form isolated charges of opposite sign when an external mechanical deformation caused is known as piezoelectric effect. A piezo-potential is formed when both ends of the material has charges of opposite charge. Polymers and organic structures, which are non-centrosymmetric crystals, show piezoelectric effect [8]. Free electrons move through external circuit to adjust the polarization charges and to maintain the equilibrium when non-short-circuited crystal is subjected to external mechanical load.

One of the most prominent applications of the principles of piezoelectric effect is found in PENGs. A piezoelectric potential is generated between the internal and external Fermi levels at the contacts, when an external strain is posed on electrodes that have adjusted Fermi levels [9–11]. A balanced electrostatic level is attained for balancing Fermi levels difference by coursing charge carrier via the external force. Also, an applied electric field can also bring about mechanical strain in the material.

PENGs could be of two types [12]: in one type of PENGs, an electric field is generated when a strain is posed perpendicular to the length of the nanostructure. One bit of the nanostructure is extended, and the other is compressed causing piezo-potential difference when a strain is applied perpendicular to the axis of nanostructure. This situation was found to that of an n-type semiconductor, whereas in p-type, a reversed phenomenon was observed. The second type of PENGs is, in which when an external strain is posed perpendicular to the axis of nanowire placed between ohmic and Schottky contact, it gets compressed uniaxially. A negative potential is generated at the wire tip and a Fermi level is generated due to piezoelectric effect. A positive potential is generated when electrons move from tip of the wire to the base through the outer circuit. The progression of electron is impeded by the Schottky contact, as it moves via an external circuit. At the point when the force is eliminated, a prompt reduction in the piezoelectric effect and the neutralization of +ve potential at the tip was observed, resulting in the peak of voltage in reverse direction. Because of the in situ rectifying effect bias, Schottky diode is formed with minimal current. Once the probe comes in touch with the side of nanostructure, which is compressed with –ve potential, a +ve potential is generated at one side (maximum current), which can be due to huge difference of potential amid the sides.

13.2.1 Polymeric PENGs

Polymer-based NGs are made up of a layer of polymer that has electric contacts with the substrate that is adaptable/stretchable. Prior to the resent times, focus has been on materials/gadgets to (a) gather various energies like vibrational, sonic, water stream, and mechanical pressing factor; (b) enhance the accuracy of gadgets, life span, and output; and (c) coordinate the gadget mechanics with underlying prerequisites identified with explicit performance. Copolymers of polyvinylidene fluoride (PVDF) are by a long shot are prominently utilized to produce piezoelectricity due to favorable properties as far as fundamental adaptability, simplicity of handling, great chemical resistance from acids and bases, mechanical strength, and solvents. They have a very high fracture strain (~2%) due to which they are very flexible, adaptable, and stretchable energy harvesting gadgets, when compared to inorganic earthenware production and oxides of metal, for example, ZnO, lead zirconate/titanate ($O_3PbZr/PbTiO_3$), and $BaTiO_3$ that are portrayed by a natural crack strain <1% [13]. The conduct of plastic permits automated preparing depending on embellishment, drawing, projecting, and turning what's more, biocompatibility, they can be securely utilized in organic frameworks to make self-charging implantable gadgets and sensors. Nonetheless, accomplishing great execution needs electric poling to adjust both poles of the polar biphase toward path symmetrical to the plane of film [14] in this manner, restricting designing plan alternatives and, at the same time, requiring numerous preliminary advances and extra expense. Arising procedures of nano- and microfabrication show the possibility of improvement of the piezoelectric reaction and grow the scope of materialistic calculations and gadget structures that can be thought of.

13.2.2 PENGs: Applications

PENGs come out as a constant force source for different archaic use such as auto-controlled nano/micro sensors, self-powered hardware, wearable/adaptable gadgets, as well as biomedical uses [15, 16]. Hu et al. in 2011 [17] coordinated PENGs based on ZnO nanowires onto a tire's insides; the tire's disfigurement through the revolution produces a current of 1.5 V and 25 nA with a greatest force thickness 70 W/cm^3. The nanofibers of lead zirconate titanate (PZT) have greater piezoelectric voltage consistent and steady in dielectric, which makes it suitable for nanobattery and NG uses [18, 19]. PENGs that have outstanding adaptability as well as maximum yield potential difference have excellent usage in power hardware.

Different mixture energy harvester that coordinates PENG with different sorts of energy collectors like TENGs and PyENGs were produced for universal force and improved force change proficiency. A pair NG was constructed after the integration of PVDF NG with Si nanopillar-based solar cells. The developed device could gather energy from solar as well as sonic sources [20]. Zhu et al. in 2017 developed a piezo-phototronic impact-based Si-nano-hetero structure photovoltaic device. The solar cells showed an efficiency increase of 8.9%–9.5% [21]. Artificially supported PENGs based on composite produce a peak voltage of 65 V, which is changed over into DC yield and utilizes rectifier for charging power of light emitting diodes (LEDs) and capacitors [22]. A natural PENG dependent on multilayered construction of PVDF-nanofiber mats, trailed by poly(3,4-ethylenedioxythiphene) (PEDOT) covering, which shows an open-circuit voltage of 48 V under the pressure of 8.3 kPa was developed [23]. The yield indicated its applications infield of convenient hardware and self-fueled wearable. A PENG of inorganic-natural mixture dependent on elecrospun PVDF and zinc sulfide (ZnS) nanorods has a reverberation recurrence of 836 Hz, an acoustic affectability of ~58% [24]. As a result, it can

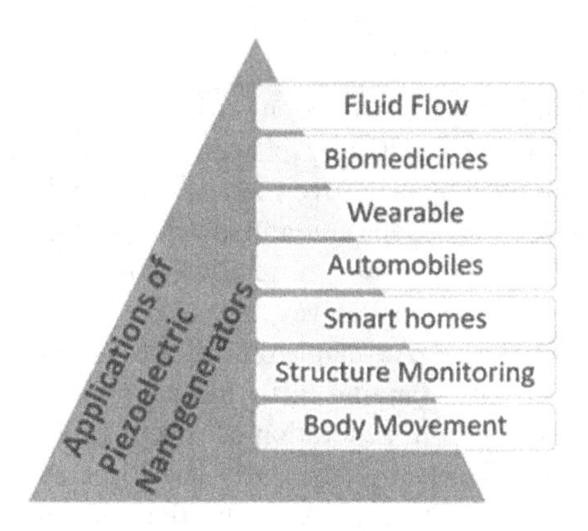

FIGURE 13.3 Some applications of PENGs.

detect noise, track security, harvest wind energy, and be used in sensors (self-powered). PENGs are used in a variety of protection systems sensors, including transportation monitoring [25], biomedical [26], wireless [27], and remote sensors (Figure 13.3).

13.2.3 Perspectives of PENGs

PENGs are much more efficient in performance than any other energy harvesting procedures (piezoelectric). The power yield of ZnO-based NGs can be 11–22 times that of PZT-based mass cantilever energy collectors. [28]. PENGs are multipurpose power sources that can be used in implantable and wearable applications. For this incorporation, future improvements are needed such as adaptability, solidness, and security. Improved semiconducting nanowires are essential to improve the performance and application of PENGs. ZnO-nanowires-dependent PENGs are most prominent, and its improved models are also available. Certain 1D nanomaterials such as ZnS, cadmium sulfide (CdS), gallium nitride (GaN), cadmium selenide (CdSe), indium nitride (InN), indium arsenide (InAs), and 2D material like molybdenum disulfide (MoS_2) exist, which have fabulous piezoelectric applicability [29, 30]. Improvement of foundational layout, joining, and pressing of NGs for self-powered sensors are a portion of things to come prerequisites that must be tended to for effective electro ME change.

13.3 TRIBOELECTRIC NGs

TENGs use a combination of electrostatic induction and triboelectrification to generate electricity. TENG was discovered accidently during the fabrication of PENG having small gap by Zhong and team. This fabricated NG produced a large output voltage that was later revealed to be caused by the triboelectric effect [31]. The charge generated due to the contact of two different materials is referred as triboelectrification, whereas the electricity is generated due to the movement of electrons from one electrode to another via an externally applied load to normalize the potential differences [32, 33]. The electron affinity brought into contact due to the dissimilar surfaces of two different materials in TENGs and the generation of tribolectric charges (time-dependent) occurs on the surfaces. An electric potential develops between them during the separation. The mechanism that causes charge to develop on the surface has not been clearly identified and is still a topic of discussion [34]. Insulators can hold the transferred charges for an extended time due to triboelectrification, and as a result, there exist several negative consequences in industrial production, aviation, transportation, etc. The first TENG was invented in 2012 that harvests atmospheric vibration or ME [35].

13.3.1 Modes of Operation of TENG

Electricity induction effects and triboelectrification are the underlying mechanism behind TENGs; elementary modes of operation of TENGs have been long investigated, as the field advances dynamically, and they operate on four primitive modes, viz., single-electrode mode (SEM), free-standing mode (FSM), vertical contact separation mode (VCSM), and linear sliding mode (LSM). Two entirely different triboelectric materials are required for these modes, and with correct conductor-insulator connection between each layer, either dielectric-dielectric or metal-dielectric arrangement form the mix. The intrinsic principle on which these modes work is the electric charge movement that breaks the electricity standing gift whenever there is a displacement in any of the triboelectric layer (TEL). This results in generation of potential drop amid electrodes. TELs get a forward and reverse potential between the electrodes due to perennial mechanical feat of layers with forward and reverse directions. This creates the positive and negative signals within the TENG output generating the AC signal.

13.3.2 Utility of TENGs

13.3.2.1. Self-Powered Sensors

TENGs can be utilized in home-based appliances such as windows, lights, doors, alarms, temperature sensors, etc. Advent of TENGs have promoted artificial intelligence (AI), IOT, human-machine interfacing scheme, and video games [36]. Intelligent keyboards, based on TENGs, E-skin [37], touch screen [38], acoustic sensors [39], and an artificial muscle [40], stand as a paradigm shift from human interface existing already. Polymer-based TENGs of polyamide (PA) film or Teflon (PTFE) film are applicable to energy-sensing element for detecting alcohol/water present in gas or liquid [41].

A self-powered measuring sensor and a versatile cylindrical spiral sort TENG (S-TENG) has been developed by researchers [42], that supported the TENG generating 250 V circuit voltage. The devices and sensors could utilize the incredible high-voltage output brought about by S-TENG, thus eliminating battery requirements. The residual KE of balls (football, basketball, and baseball), while one plays with ball, has been harvested using 3D stacked TENGs [43]. A relentless voltage of 5 V even at 3000 operating cycles was observed during experiments and, thus, proved useful in self-charging systems.

The energy of walking has also been reaped through shoes and backpacks using TENGs [44]. Liu and the team were able to develop a hybridized NG upon combining TENGs with magnetic NGs. This hybridized NG could be used by embedding it in trooper shoes, allowing it to harness the vibrational energy of walking [45] and allowing researchers and engineers to have a constant power supply. Zhong and coworkers, in 2012, were able to develop arc-shaped versatile NG that was able to harvest energy from writing. This device gave an incredible output of voltage as high as ~4.125 mV. It could be utilized in self-powered system and sensors [46]. Kinesio tape supported on elastic and wearable TENGs, showed a linear relationship with the elongated displacement and human gestures [47]. TENGs also harvest EE from the movement of eyeball [48], changing body posture [49], blinking of eyes [50], and movements of body parts among other things. Figure 13.4 depicts some of the applications of TENGs.

13.3.2.2 Blue Energy

The potential and common renewable energy source is water wave energy (WWE) and harvested traditionally utilizing electromagnetic generators (EMG). TENGs produced recently successfully harvest blue energy. TENGs could harvest energy even in irregular surroundings and could give constant frequency, about five cycles per second, which is considerably an advantage over EMG [51].

WWE harvesting TENGs with water, acting as a triboelectric material, have been reported [52]. Negatron flow is generated across the external circuits when there is a perennial contact among

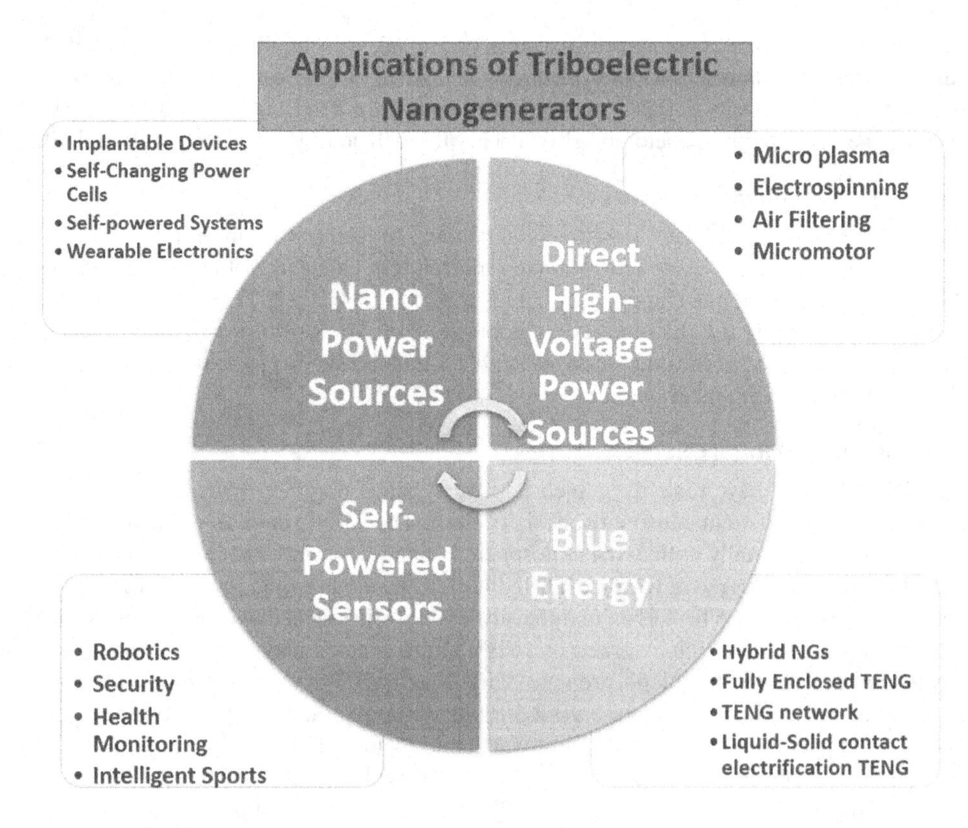

FIGURE 13.4 Some applications of TENGs.

water and the nonconductor material i.e., polydimethylsiloxane (PDMS) [53]. Deliquescent materials [54] and metals [55] tend to build up charge upon meeting water; this charge is sustained for a long duration [56]. 3D built-up structure of TENGs in the buoy ball harvested energy from wave to provide an output voltage of 110 V and 15 A current. These TENGs could be utilized in marine sciences and biology for fueling sensors. Water wave could rotate a spherical TENG (diameter 6 cm) generates 1-A current and 10-mW power instantly [57]. The TENGs developed by using Nylon 6.6 and Kapton possesses a basic design [58], lightweight, and could rock within water and harvest wave energy of lakes and oceans at a large scale. TENGs can float on the surface of water and harvest energy from feeble-frequency-wave [59]. These TENGs produced an output of 1.15 mW from 1 km^2 expanse. Electrical grid could be powered using wind turbines through TENGs, associated at the side of star panels. When combined with the power control unit, the TENGs network can generate a steady DC that could be stored in capacitors. TENGs network was merged with power control module that resulted in a 96-fold increase in stored energy when charging a capacitor [60]. B-TENG, a butterfly-inspired TENG, was developed that can harness the WWE of the ocean in all dimensions [61]. Another TENG with very low toxicity, high biodegradability, and wonderful biocompatibility was developed by utilizing alginate TEL [62]. TENG (2019) is newly reported that absorbs WWE and uses it to power a self-activated anti-biofouling device in water systems [63].

13.3.2.3 Harvesting Energy of Wind

A fluorine-doped tin compound (FTO) conductor and a PTFE film were used to create an extremely transparent TENG for extracting electrostatic energy from moving water. The PTFE film was rendered with a thickness >1 m to achieve better transparency. To extend the transmission of the TENG, an anti-reflection coating is used, allowing them to be inserted into solar cells, vehicles, and building glasses. When in contact with water, this TENG can produce 10 V voltage and 2 A/

cm^2 current density. When coupled with resistors (0.5 mW), the output density approached up to 11.6 mW/m^2. The transparent TENG was effectively integrated in solar cells by removing the existing security layer later in 2014 [64]. TENG was also combined with a silicon solar cell in another study, which will serve as a shared conductor to collect energy through both sunlight and raindrops [65].

13.3.2.4 TENG Application in Toys

TENG was recently changed into a commercial product by using it in an entirely different way. TENGs have designed a toy intended to make children less engaged in their electronic devices. TENGs made various toys like clapping toys, duck toys, puzzles, etc. A TEL was inserted between the layers of the clapping toy that creates a clapping sound. Similar idea was used in a duck toy. A TENG device was placed into the duck toy with 2 LEDs in the eye part, so when you click the middle half of the duck, it makes a noise.

13.3.2.5 Biodegradable TENG

Biodegradable TENGs have recently gained popularity since they are made of edible materials. They are implantable without causing cyanosis of flesh. The TENG devices are made from triboelectric materials that usually pollute the atmosphere in many ways in which properties of the material used determine the degree of toxicity. The edible TENG made from seaweed TEL is combined with silver foil (conductor). The TENG and the silver foil were attached to a rice sheet that acted as an active layer's base. The TENG worked in a SEM and contacted materials such as paper, tissue, vinyl polymer, and fluorinated olefin propene daily. The TENG device's toxicity level was determined using a 3-(4,5-dimethylthiazol-2-yl)-2,5-diphenyl tetrazolium bromide, cell imaging, and 4,6-diamidino-2-phenylindole. After which it was found that the TEL was biocompatible allowing cells to expand. Under a confocal magnifier, cell growth was observed with a blue color pattern. The electrical response and durability of the edible TENG device were its highest, demonstrating that edible TENG were very well for medical implants.

13.3.2.6 Waterproof TENG

TENG systems have several limitations that can be solved by using a variety of solutions, strategies, and engineering techniques. Since its inception, the main disadvantage of TENGs was the reduction of electrical efficiency due to humidity. This problem was solved with waterproof TENG (WP-TENG) system that can work in high humidity. WP-TENG can even work underwater. The positive TEL was made of nickel foam and the negative TEL was made of PDMS. This device was very lightweight, i.e., 1.9 g and by submerging in water, the waterproof potential was determined. The device had a significantly faster electrical response and power density, i.e., 80 V and 4 mW/m^2. Its water resistance potential was measured after dipping it into water for 24 hours, and then the result was measured. Humidity of various percentages of ratios was applied. The output was between 1/3 RH and 19 RH that was stable and suggested to use in harsh and rainy weather.

13.4 PYROELECTRIC NANOGENERATORS

For future generation, there is a need of great and efficient energy harvesting devices that have remarkable ability of transforming TE to EE. PyENGs have this ability to transform by employing pyroelectric material of nano-size. Some applications of PyENGs are shown in Figure 13.5. The temperature depending on time is one of the main differences between Seebeck effect-based TE harvesters and PyENGs. Since there is no chance of spontaneous polarization in the Seebeck effect, power generation from time-dependent temperature fluctuations is not achievable. In 2012, the Wang group has made the first pyroelectric nanogenerator that was based on ZnO nanowire arrays.

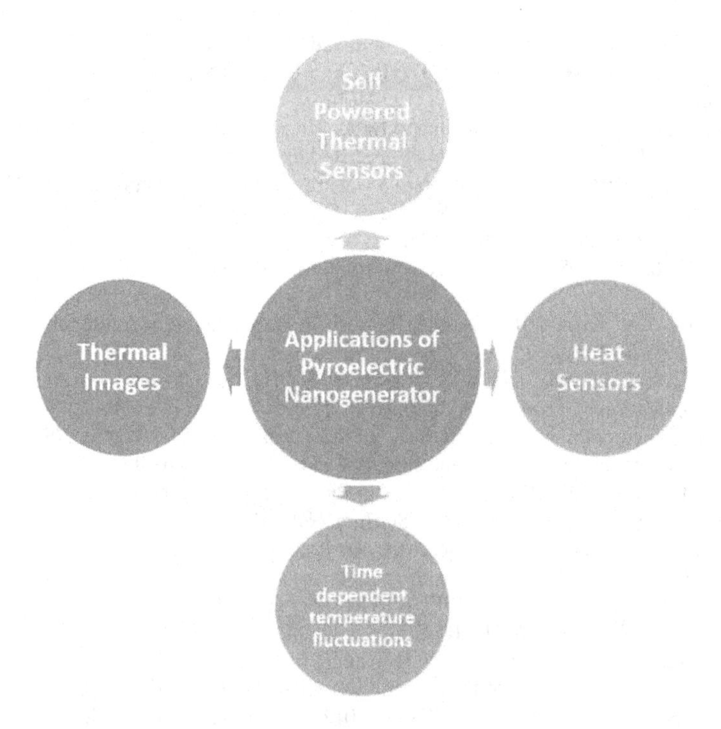

FIGURE 13.5 Applications of PyENGs.

13.5 CONCLUSIONS

In today's current scenario, the increasing demand of energy has become a challenge, and the invention of NGs has put a milestone in outcoming from these energy crises. As they step to optimization, flexibility, and efficiency, it is discovered that harvesting ambient ME for electronic and electrical systems is sustainable. These NGs are light, made of low-cost materials, simple to configure, compact, and, most of all, produce high-performance intensity, making them a sustainable technology. In the implementation of self-powered sensors, NGs can be an effective substitute for batteries. As compared to the output current, NGs have a very high output voltage. Rectifiers and voltage transformers can help with this high-voltage low-current dilemma by increasing current while lowering the voltage. Since the NGs power source is unsteady, capacitors and advanced power management circuits can be utilized to provide stable power to devices. The output power characteristics can be improved by transformers, inductors, and electronic logic control switches. As compared to low-frequency EMGs, TENG also has a higher power density. When it comes to the harvesting of blue energy, it is important to consider how well the device integrates with the power management module. The reliability of NGs in the harvesting of blue energy has been one of their main issues. NGs may also be combined with different harvesting systems like EMGs, turbines, solar PV systems, etc. to achieve better energy conversion performance. Using PyENGs, the issue of higher temperature control in concentrated PV can be solved. The innovation and implementations of NGs have expanded rapidly over the last decade, indicating that this technique could be extensively commercialized in the future.

REFERENCES

1. Wang, A.C., C. Wu, D. Pisignano, Z.L. Wang, and L. Persano. 2018. Polymer Nanogenerators: Opportunities and Challenges for Large-Scale Applications. J. Appl. Polym. Sci. 135: 45674.
2. Wang, Z.L., and J. Song. 2006. Piezoelectric Nanogenerators Based on Zinc Oxide Nanowire Arrays. Science 312: 242–246.

3. Qi, Y., J. Kim, T.D. Nguyen, B. Lisko, P.K. Purohit, and M.C. McAlpine. 2011. Enhanced Piezoelectricity and Stretchability in Energy Harvesting Devices Fabricated from Buckled PZT Ribbons. Nano Lett. 3: 1331–1336.
4. Schier, M.B. 2006. Draw the Lighting Down: Benjamin Franklin and Electrical Technology in the Age of Enlightenment; University of California Press: Berkeley, CA.
5. Kagan, C.R., L.E. Fernandez, Y. Gogotsi, P.T. Hammond, M.C. Hersam, A.E. Nel, R.C. Penner, C.G. Willson, and P.S. Weiss. 2016. Nano day: Celebrating the Next Decade of Nanoscience and Nanotechnology. ACS Nano 10: 9093–9103.
6. Wang, Z.L. 2006. Piezoelectric Nanogenerators Based on Zinc Oxide Nanowire Arrays. Science 312: 242–246.
7. Kumar, B., and S.W. Kim. 2012. Energy Harvesting Based on Semiconducting Piezoelectric ZnO Nanostructures. Nano Energy 1: 342–355.
8. Kholkin, A.L., N.A. Pertsev, and A.V. Goltsev. 2008. In Piezoelectric and Acoustic Materials for Transducer Applications; Safari, A., Akdoğan, E.K., Eds.; Springer: New York, NY; Chapter 2, p. 17.
9. Askari, H., A. Khajepour, M.B. Khamesee, Z. Saadatnia, and Z.L. Wang. 2018. Piezoelectric and Triboelectric Nanogenerators: Trends and Impacts. Nano Today 22: 10–13.
10. Hu, Y., and Z.L. Wang. 2014. Recent Progress in Piezoelectric Nanogenerators as a Sustainable Power Source Inself-Powered Systems and Active Sensors. Nano Energy 14: 3–14.
11. Zi, Y., and Z.L. Wang. 2017. Nanogenerators: An Emerging Technology Towards Nanoenergy. APL Mater. 5: 074103.
12. Roji, A.M.M., G. Jiji, and A.B.T. Raj. 2017. A Retrospect on the Role of Piezoelectric Nanogenerators in the Development of the Green World. RSC Adv. 7: 33642–33670.
13. Xue, X., S. Wang, W. Guo, Y. Zhang, and Z.L. Wang. 2012. Hybridizing Energy Conversion and Storage in a Mechanical to Electrochemical Process for Self-Charging Power Cell. Nano Lett. 12: 5048–5054.
14. Sharma, T., S.-S. Je, B. Gill, and J.X.J. Zhang. 2012. Patterning Piezoelectric Thin Film PVDF–TrFE Based Pressure Sensor for Catheter Application. Sens. Actuators, A 177: 87–92.
15. Jellaa, V., S. Ippili, J.-H. Eom, S.V.N. Pammi, J.-S. Jung, V.-D. Tran, V.H. Nguyen, A. Kirakosyan, S. Yun, D. Kim, M.R. Sihn, J. Choi, Y.-J. Kim, H.-J. Kim, and S.-G. Yoon. 2019. A Comprehensive Review of Flexible Piezoelectric Generators Based on Organic-Inorganic Metal Halide Perovskites. Nano Energy 57: 74–93.
16. Bera, B. 2016. Recent Advances in Piezoelectric Nano generators in Energy Harvesting Applications. Imp. J. Interdiscip. Res. 2: 1274–1291.
17. Hu, Y., C. Xu, Y. Zhang, L. Lin, R.L. Snyder, and Z.L. Wang. 2011. A Nanogenerator for Energy Harvesting from a Rotating Tire and its Application as a Self-Powered Pressure/Speed Sensor. Adv. Mater. 23: 4068–4071.
18. Chen, X., S. Xu, N. Yao, and Y. Shi. 2010. 1.6 V Nanogenerator for Mechanical Energy Harvesting Using PZT Nanofibers. Nano Lett. 10: 2133–2137.
19. Park, K.I., J.H. Son, G.T. Hwang, C.K. Jeong, J. Ryu, M. Koo, I. Choi, S.H. Lee, M. Byun, Z.L. Wang, and K.L. Jee. 2014. Highly-Efficient, Flexible Piezoelectric PZT Thin Film Nanogenerator on Plastic Substrates. Adv. Mater. 26: 2514–2520.
20. Lee, D.-Y., H. Kim, H.-M. Li, A.-R. Jang, Y.-D. Lim, S.N. Cha, Y.J. Park, D.J. Kang, and W.J. Yoo. 2013. Hybrid Energy Harvester Based on Nanopillar Solar Cells and PVDF Nanogenerator. Nanotechnology 24: 175402.
21. Zhu, L., L. Wang, C. Pan, L. Chen, F. Xue, B. Chen, L. Yang, L. Su, and Z.L. Wang. 2017. Enhancing the Efficiency of Silicon-Based Solar Cells by the Piezo-Phototronic Effect. ACS Nano 11: 1894–1900.
22. Lee, E.J., T.Y. Kim, S.W. Kim, S. Jeong, Y. Choi, and S.Y. Lee. 2018. High-Performance Piezoelectric Nanogenerators Based on Chemically-Reinforced Composites. Energy Environ. Sci. 11: 1425–1430.
23. Maity, K., and D. Mandal. 2018. All-Organic High-Performance Piezoelectric Nanogenerator with Multilayer Assembled Electrospun Nanofiber Mats for Self-Powered Multifunctional Sensors. ACS Appl. Mater. Interfaces 10: 18257–18269.
24. Sultana, A., M.M. Alam, S.K. Ghosh, T.R. Middya, and D. Mandal. 2018. Energy Harvesting and Self-Powered Microphone Application on Multifunctional Inorganic-Organic Hybrid Nanogenerator. Energy 166: 963–971.
25. Lin, L., Y. Hu, C. Xu, Y. Zhang, R. Zhang, X. Wen, and Z.L. Wang. 2013. Transparent Flexible Nanogenerator as Self-Powered Sensor for Transportation Monitoring. Nano Energy 2: 75–81.
26. Ali, F., W. Raza, X. Li, H. Gul, and K.-H. Kim. 2019. Piezoelectric Energy Harvesters for Biomedical Applications. Nano Energy 57: 879–902.

27. Yang, Y., S. Wang, Y. Zhang, and Z.L. Wang. 2012. Supplementary Materials Pyroelectric Nanogenerators for Driving Wireless Sensors. Nano Lett. Support. Inf. 12: 6408–6413.
28. Xu, S., Y. Qin, C. Xu, Y. Wei, R. Yang, and Z.L. Wang. 2010. Self-Powered Nanowire Devices. Nat. Nanotechnol. 5: 366–373.
29. Li, X., M. Sun, X. Wei, C. Shan, and Q. Chen. 2018. 1D Piezoelectric Material Based Nanogenerators: Methods, Materials and Property Optimization. Nanomaterials 8: 188.
30. Johar, M., M. Hassan, A. Waseem, J.-S. Ha, J. Lee, and S.-W. Ryu. 2018. Stable and High Piezoelectric Output of GaN Nanowire-Based Lead-Free Piezoelectric Nanogenerator by Suppression of Internal Screening. Nanomaterials 8: 437.
31. Weiss, P.S. 2015. A Conversation with Prof. Zhong Lin Wang, Energy Harvester. ACS Nano 9: 2221–2226.
32. Wang, Z.L., J. Chen, and L.Lin. 2015. Progress in Triboelectric Nanogenerators as a New Energy Technology and Self-Powered Sensors. Energy Environ. Sci. 8: 2250–2282.
33. Zi, Y., S. Niu, J. Wang, Z. Wen, W. Tang, and Z.L. Wang. 2015. Standards and Figure-of-Merits for Quantifying the performance of Triboelectric Nanogenerators. Nat. Commun. 6: 8376.
34. Wang, Z.L. 2013. Triboelectric Nanogenerators as New Energy Technology for Self-Powered Systems and as Active Mechanical and Chemical Sensors. ACS Nano 7: 9533–9557.
35. Niu, S., and Z.L. Wang. 2014. Theoretical Systems of Triboelectric Nanogenerators. Nano Energy 14: 161–192.
36. Ding, W., A.C. Wang, C. Wu, H. Guo, and Z.L. Wang. 2019. Human-Machine Interfacing Enabled by Triboelectric Nanogenerators and Tribotronics. Adv. Mater. Technol. 4: 1800487.
37. Wang, P., R. Liu, W. Ding, P. Zhang, L. Pan, G. Dai, H. Zou, K. Dong, C. Xu, and Z.L. Wang. 2018. Complementary Electromagnetic-Triboelectric Active Sensor for Detecting Multiple Mechanical Triggering. Adv. Funct. Mater. 28: 1705808.
38. Sun, J., X. Pu, M. Liu, A. Yu, C. Du, J. Zhai, W. Hu, and Z.L. Wang. 2018. Self-Healable, Stretchable, Transparent Triboelectric Nanogenerators as Soft Power Sources. ACS Nano 12: 6147–6155.
39. Yang, J., J. Chen, Y. Liu, W. Yang, Y. Su, and Z.L. Wang. 2014. Triboelectrification-Based Organic Film Nanogenerator for Acoustic Energy Harvesting and Self-Powered Active Acoustic Sensing. ACS Nano 8: 2649–2657.
40. Chen, X., T. Jiang, Y. Yao, L. Xu, Z. Zhao, and Z.L. Wang. 2016. Stimulating Acrylic Elastomers by a Triboelectric Nanogenerator toward Self-Powered Electronic Skin and Artificial Muscle. Adv. Funct. Mater. 26: 4906–4913.
41. Zhang, H., Y. Yang, Y. Su, J. Chen, C. Hu, Z. Wu, Y. Liu, C.P. Wong, Y. Bando, and Z.L. Wang. 2013. Triboelectricnanogenerator as Self-Powered Active Sensors for Detecting Liquid/Gaseous Water/Ethanol. Nano Energy 2: 693–701.
42. Li, X.H., C.B. Han, L.M. Zhang, and Z.L. Wang. 2015. Cylindrical Spiral Triboelectric Nanogenerator. Nano Res. 8: 3197–3204.
43. Qian, J., D.S. Kim, and D.W. Lee. 2018. On-Vehicle Triboelectric Nanogenerator Enabled Self-Powered Sensor for Tire Pressure Monitoring. Nano Energy 49: 126–136.
44. Hou, T.C., Y. Yang, H. Zhang, J. Chen, L.J. Chen, and Z.L. Wang. 2013. Triboelectric Nanogenerator Built Inside Shoe Insole for Harvesting Walking Energy. Nano Energy 2: 856–862.
45. Yang, W., J. Chen, G. Zhu, J. Yang, P. Bai, Y. Su, Q. Jing, X. Cao, and Z.L. Wang. 2013. Harvesting Energy from the Natural Vibration of Human Walking. ACS Nano 7: 11317–11324.
46. Niu, S., S. Wang, L. Lin, Y. Liu, Y.S. Zhou, Y. Hu, and Z.L. Wang. 2013. Theoretical Study of Contact-Mode Triboelectric Nanogenerators as an Effective Power Source. Energy Environ. Sci. 6: 3576–3583.
47. Wang, S., M. He, B. Weng, L. Gan, Y. Zhao, N. Li, and Y. Xie. 2018. Stretchable and Wearable Triboelectric Nanogenerator Based on Kinesio Tape for Self-Powered Human Motion Sensing. Nanomaterials 8: 657.
48. Lee, S., R. Hinchet, Y. Lee, Y. Yang, Z.H. Lin, G. Ardila, L. Montès, M. Mouis, and Z.L. Wang. 2014. Ultrathin Nanogenerators as Self-Powered Active Skin Sensors for Tracking Eye Ball Motion. Adv. Funct. Mater. 24: 1163–1168.
49. Pu, X., H. Guo, Q. Tang, L. Chen, L. Feng, G. Liu, X. Wang, Y. Xi, C. Hu, and Z.L. Wang. 2018. Rotation Sensing and Gesture Control of a Robot Joint via Triboelectric Quantization Sensor. Nano Energy 54: 453–460.
50. Pu, X., H. Guo, J. Chen, X. Wang, Y. Xi, C. Hu, and Z.L. Wang. 2017. Eye Motion Triggered Self-Powered Mechnosensational Communication System Using Triboelectric Nanogenerator. Sci. Adv. 3: e1700694.

51. Liang, X., T. Jiang, G. Liu, T. Xiao, L. Xu, W. Li, F. Xi, C. Zhang, and Z.L. Wang. 2019. Triboelectric Nanogenerator Networks Integrated with Power Management Module for Water Wave Energy Harvesting. Adv. Funct. Mater. 29: 1807241.
52. Jurado, U.T., S.H. Pu, and N.M. White. 2018. Water-Dielectric Single Electrode Mode Triboelectric Nanogenerators for Ocean Wave Impact Energy Harvesting. Proceedings 2: 714.
53. Lin, Z.H., G. Cheng, L. Lin, S. Lee, and Z.L. Wang. 2013. Water-Solid Surface Contact Electrification and its Use for Harvesting Liquid-Wave Energy. Angew. Chem. Int. Ed. 52: 12545–12549.
54. Gouveia, R.F., and F. Galembeck. 2009. Electrostatic Charging of Hydrophilic Particles Due to Water Adsorption. J. Am. Chem. Soc. 131: 11381–11386.
55. Ducati, T.R.D., L.H. Simoes, and F. Galembeck. 2010. Charge Partitioning at Gas-Solid Interfaces: Humidity Causes Electricity Buildup on Metals. Langmuir 26: 13763–13766.
56. Ovchinnikova, K., and G.H, Pollack. 2009. Can water store charge? Langmuir 25: 542–547.
57. Wang, X., S. Niu, Y. Yin, F. Yi, Z. You, and Z.L. Wang. 2015. Triboelectric Nanogenerator Based on Fully Enclosed Rolling Spherical Structure for Harvesting Low-Frequency Water Wave Energy. Adv. Energy. Mater. 5: 1501467.
58. Su, Y., Y. Yang, X. Zhong, H. Zhang, Z. Wu, Y. Jiang, and Z.L. Wang. 2014. Fully Enclosed Cylindrical Single-Electrode-Based Triboelectric Nanogenerator. ACS Appl. Mater. Interfaces 6: 553–559.
59. Chen, J., J. Yang, Z. Li, X. Fan, Y. Zi, Q. Jing, H. Guo, Z. Wen, K.C. Pradel, and S. Niu, and Z.L. Wang. 2015. Networks of Triboelectric Nanogenerators for Harvesting Water Wave Energy: A Potential Approach toward Blue Energy. ACS Nano 9: 3324–3331.
60. Wang, Z.L. 2017. Catch Wave Power in Floating Nets. Nature 542: 159–160.
61. Lei, R., H. Zhai, J. Nie, W. Zhong, Y. Bai, X. Liang, L. Xu, T. Jiang, X. Chen, and Z.L. Wang. 2018. Butterfly-Inspired Triboelectric Nanogenerators with Spring-Assisted Linkage Structure for Water Wave Energy Harvesting. Adv. Mater. Technol. 4: 1800514.
62. Pang, Y., F. Xi, J. Luo, G. Liu, T. Guo, and C. Zhang. 2018. An Alginate Film-Base Degradable Triboelectric Nanogenerator. RSC Adv. 8: 6719–6726.
63. Long, Y., Y. Yu, X. Yin, J. Li, C. Carlos, X. Dux, Y. Jiang, and X.E. Wang. 2019. Effective Anti-Biofouling Enabled by Surface electric Disturbance from Water Wave-Driven Nanogenerator. Nano Energy 57: 558–565.
64. Zheng, L., Z.H. Lin, G. Cheng, W. Wu, X. Wen, S. Lee, and Z.L. Wang. 2014. Silicon-Based Hybrid Cell for Harvesting Solar Energy and Raindrop Electrostatic Energy. Nano Energy 9: 291–300.
65. Liu, Y., N. Sun, J. Liu, Z. Wen, X. Sun, S.T. Lee, and B. Sun. 2018. Integrating a Silicon Solar Cell with a Triboelectric Nanogenerator via a Mutual Electrode for Harvesting Energy from Sunlight and Raindrops. ACS Nano 12: 2893–2899.

14 Large-Scale Applications of Triboelectric, Piezoelectric, and Pyroelectric Nanogenerators

Vishal Panwar, Atif Suhail, and Indranil Lahiri

CONTENTS

14.1 INTRODUCTION

Energy is one of the most life-threatening prerequisites of society, and non-renewable resources such as fossil fuels are depleting at an alarming rate, posing a significant threat to environmental conditions. Because of the severe effects of global energy crisis and battery disposal, there is a strong need to develop eco-friendly and independent nanoscale sources as alternative renewable energy sources. Renewable energy is considered as the fastest rising energy resource, accounting approx. 40% of total energy, according to the 2018 BP (British Petroleum) Energy Outlook (BP Energy Outlook 2018). Well-known renewable sources such as, bio, hydro, wind, and solar energy can help to meet the requirement for energy on a megawatt to gigawatt scale energy-harvesting techniques have risen to prominence in recent decades as Internet of Things (IoT), automated protection organizations, sensors, safety technology, wearable, cell phones, and portable devices. Also, other technologies have grown in popularity, requiring power resources which are perpetual and

DOI: 10.1201/9781003187615-14

maintenance-free. Typically, these devices use batteries as a source of power, having a finite lifespan and pose ecofriendly risks. All these devices may be powered by harvesting ambient mechanical energy. Among various sources of energy, ambient mechanical energy is plentiful and to be considered as the most necessary renewable form of energy in our atmosphere (Batra and Alomari 2017; Beeby, Tudor, and White 2006; Hu and Wang 2014).

Nanogenerators (NGs) are energy scavenging devices which use wasted mechanical energy available in the environment to produce electricity. Scientists had previously developed practical methods for generating electricity from friction by the end of the 17th century. In 1663, a primitive type of friction device was invented. Several researchers later worked on the machine and invented an electromagnetic generator (EMG) in 1831, and it is still the most prominent in various thermoelectric stations. In 1878, Wimshurst discovered an apparatus, which works on the static electricity also known as "static electricity generator." Van de Graaff produced exceptionally high-voltage DC generator in 1929. Modern Van de Graaff generators will generate up to 25 megavolts of electricity. Later, nanotechnology developed technologies to produce electricity using piezoelectric, triboelectric, and pyroelectric principles. Nanoscale structures' large surface area and their tunable physical as well as chemical properties offer even more appropriate technology for absorbing, altering, and storing energy from various sources, including thermal, electrical, radiant, chemical, and mechanical (Kagan et al. 2016). Figure 14.1(a) depicts the pivotal mechanical energy scavenging

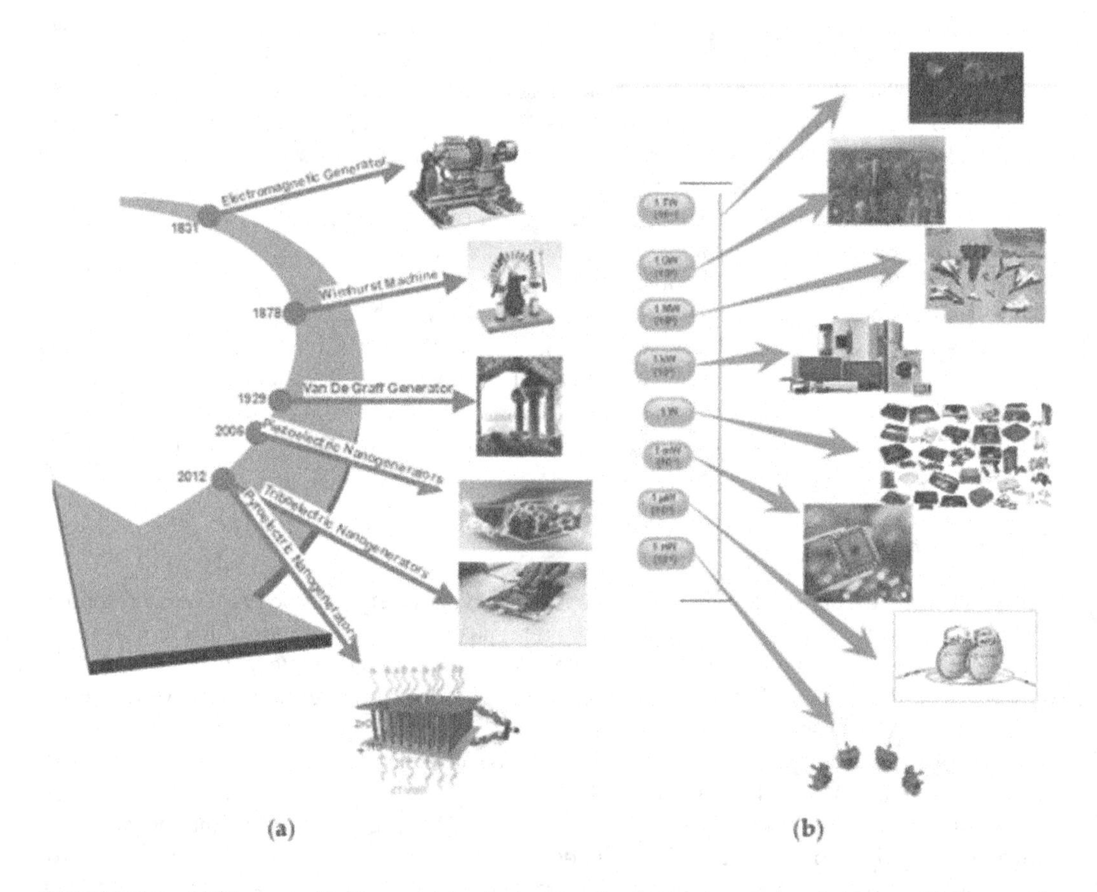

FIGURE 14.1 (a) The most significant mechanical energy-harvesting innovations in history. (b) The amount of energy required by various devices at different power levels. (Reprinted with permission from Indira et al. 2019.)

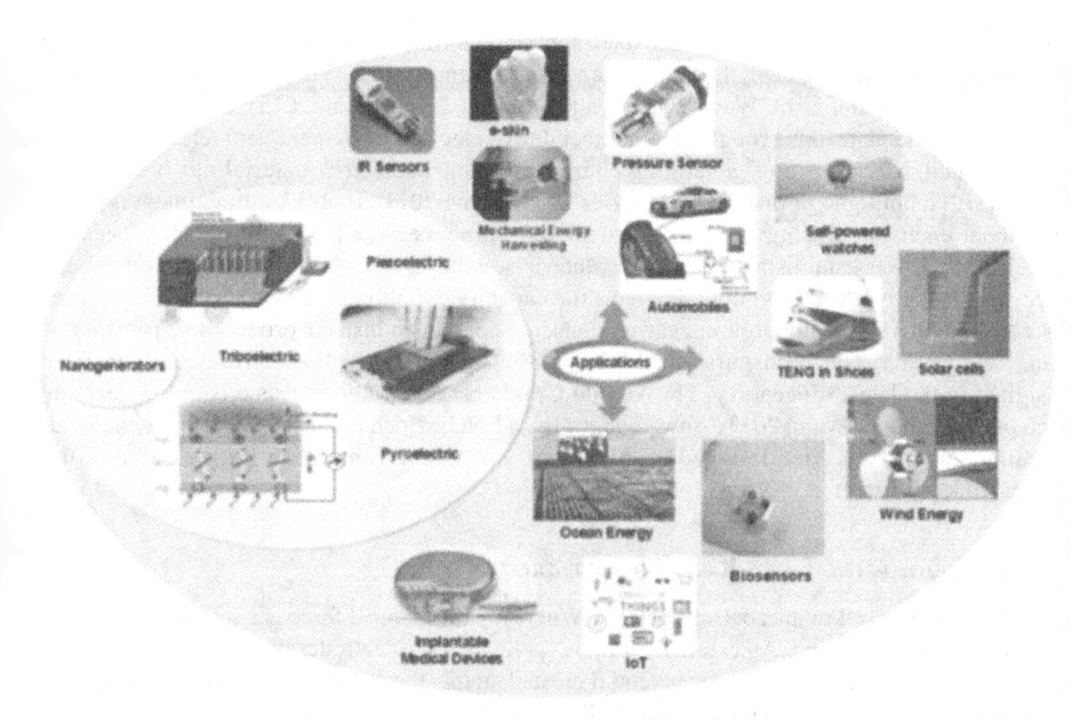

FIGURE 14.2 Nanogenerators of various types and their implementations in the Internet of Things (IoT) era. (Reprinted with permission from Indira et al. 2019.)

developments. Figure 14.1(b) represents the requirements of energy at various levels in different devices (Indira et al. 2019).

NGs are a new energy-harvesting technology that can absorb mechanical energy from various sources, including human motion (such as walking, heartbeat, running, and breathing), running water, waste heat, raindrops, wind, and vibration. In 2006, Wang and Song (2006) generated the foremost NG, which used nanowires of ZnO and the piezoelectric effect to achieve a 17–30% power conversion efficiency.

Some existing techniques, such as resonators and transducers, which were based on conventional cantilever are utilized to obtained electricity from mechanical vibrations. But these are applicable only for vibration, which have high frequency. NGs are capable to extract low-frequency mechanical vibrations effectively. However, ZnO-based nanowires twisted more than bulk ZnO causing no harm, allowing them to withstand more strain and thus produce more electricity (Kumar and Kim 2012; Wang and Song 2006). Figure 14.2 depicts the various forms of NGs and their applications. NGs have created novel platform and significant possibilities for self-power sensing, mechanical energy scavenging, and monitoring (Indira et al. 2019).

NGs produced using three different methods: triboelectric (Zhu, Lin et al. 2013), piezoelectric (Lu et al. 2009), and pyroelectric (Gao et al. 2016).

14.1.1 Triboelectric Effect-Based Nanogenerators (TENGs)

Triboelectric effect-based nanogenerators (TENGs) use a combination of triboelectrification and electrostatic induction to generate electricity. Zhong with the help of his colleagues invented a piezoelectric effect-based nanogenerator (PENG) that contained small gap, causing excessive output voltage. Belated, they reveal increase in output of voltage obtained as a result of triboelectric effect (Yang, Zhu et al. 2013). Charge production on the surface of two dissimilar materials referred to as triboelectrification as they come into touch. In electrostatic induction phenomenon, when external load applied electrons movement take place from one to another electrode to established the potential differences (Fan, Tian, and Wang 2012; Niu and Wang 2014; Wang 2013; Wang, Chen, and Lin 2015; Zi, Niu et al. 2015). In TENGs, as two unlike dielectric materials brought into contact due to electron movement, triboelectric charges are generated. Mechanism of generating charges in the triboelectric material has been vague. So, it is still a hot topic of discussion (Davies 1969; Wang 2013). In 2017, a fundamental physical model emerged, tracing the triboelectrification process back to Maxwell's Displacement current, which was established years ago. Materials, exhibiting the triboelectrification experience, are typically those insulators that have the capacity of holding the charges for a long span. As a result, there are numerous negative consequences in industrial production, transportation, and aviation. The Wimshurst machine (~1880) was the first triboelectric-based generator designed in the late 18th century. The Van de Graaff generator (1929) was created to generate high output voltage (Wang 2013). Subsequentially, TENG, which harvests ambient mechanical/vibration energy, was first designed in 2012 (Lin, Chen, and Yang 2016; Rathore et al. 2018; Wang 2013; Zi, Lin et al. 2015).

14.1.2 Piezoelectric Effect-Based Nanogenerators (PENGs)

PENG theory is based on piezoelectric effect. When any mechanical force is applied on the material, electricity is produced. An external strain is applied to two electrodes on a piezoelectric material near, culminating a piezoelectric potential created among the external and interior Fermi levels in PENG (Askari et al. 2018; Gao and Wang 2007; Hu and Wang 2014; Nechibvute, Chawanda, and Luhanga 2012; Noliac, n.d.; Wang 2012; Wang and Song 2006; Zi and Wang 2017). PENG works on the two different modes (Roji, Jiji, and Raj 2017; Zi and Wang 2017). When strain is applied to the perpendicular of individual structure of (nanowire/nanorod), resulting in an electric field. PENG's action is shown in Figure 14.3. When an atomic force microscopy probe introduced to apply a force perpendicular to the nanostructure's axis (Wang and Song 2006), the probe first contacted the stretched surface with positive potential, and the bias voltage at this interface is negative. With little current, a reversed bias Schottky diode is formed. A biased positive voltage with a pointy peak output current is produced at the interface when a probe with negative potential contacts the compressed side of the nanostructure. The electric field produced at the tip is eventually rebalanced by

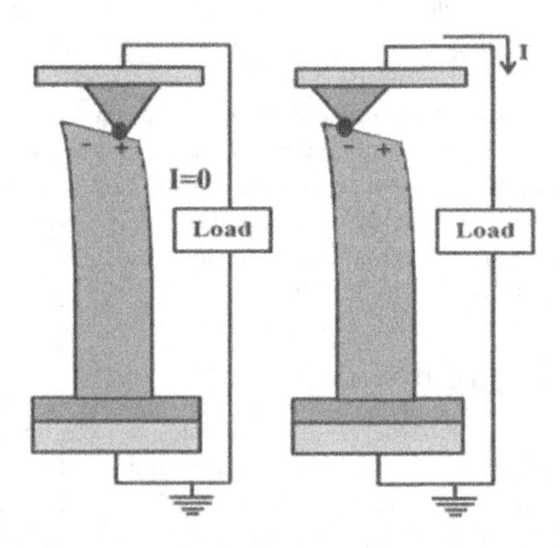

FIGURE 14.3 PENG's working principle and creating a Schottky contact when a perpendicular force is applied. (Reprinted with permission from Roji et al. 2017.)

the current flow caused by the ohmic contact created at the bottommost portion of the nanostructure (Wang and Song 2006; Zi and Wang 2017). Current is generated only at the top electrode touch with the negative potential, and no current is induced when the uppermost electrode connect to the positive potential. The external strain applied parallel to nanostructures' growing path in the other case (Figure 14.4(a and b)).

When any external force is introduced at the apex of the laterally developed nanowires, in-plane compression is generated within the nanowires. Due to piezoelectric effect, apex of the nanowires experienced the negative potential which raises the Fermi level frequency. Schottky touch restricted the movement of electrons within nanowires and thus directs them to an external circuit. The piezoelectricity fades because the applied force is removed. The detected performance has direct current characteristics due to the Schottky contact's in-situ rectifying effect.

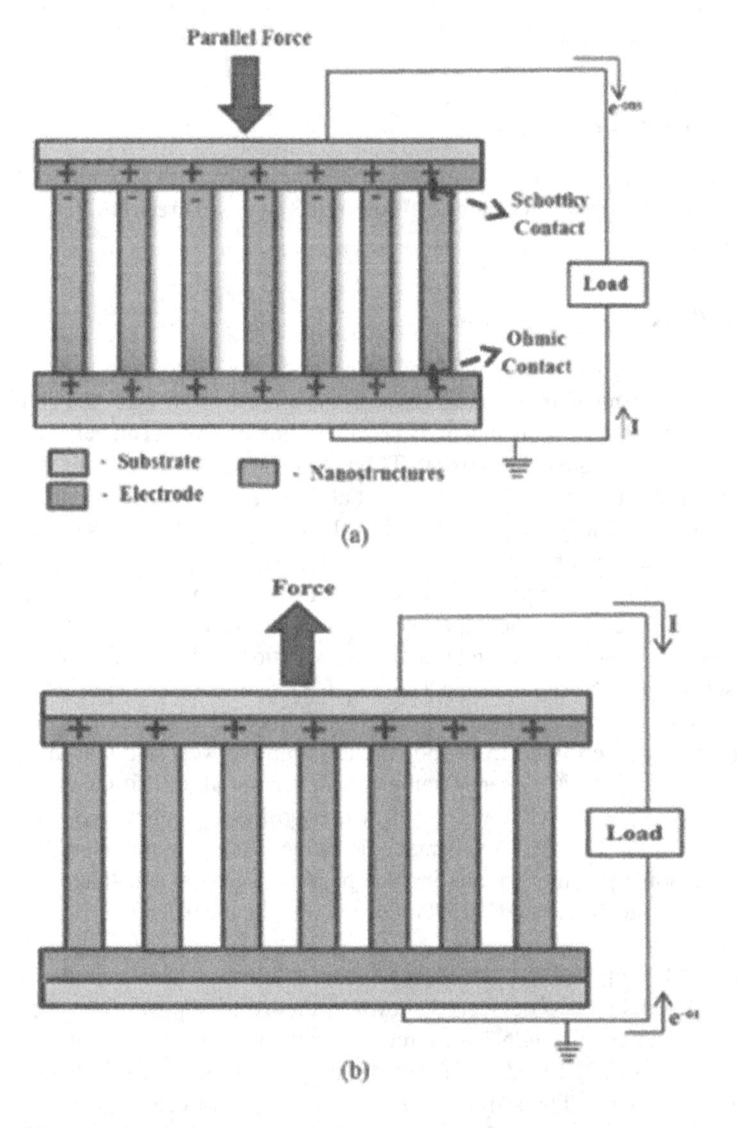

FIGURE 14.4 PENG's schematic. (a) After a force applied parallel to the axis. (b) Force has been removed. (Reprinted with permission from Roji et al. 2017.)

14.1.3 Pyroelectric Effect-Based Nanogenerator (PyENG)

Pyroelectric effect-based nanogenerators (PyENGs) are fantastic energy scavenging devices use nano-sized pyroelectric materials to transform thermal energy into electric energy. The temperature is time-dependent. The pyroelectric effect is based on the temperature induced change in spontaneous polarization in some anisotropic solids, which can be utilized to harvest thermal energy from temperature changes. However, the Seebeck effect is used to collect thermal energy by driving charge carrier diffusion with a temperature difference between the two ends of the device (Yang, Guo et al. 2012). The Wang group created the first PyENG in 2012, based on ZnO nanowire arrays (Yang, Guo et al. 2012). They are less study available on PyENGs.

This chapter will discuss triboelectric, piezoelectric, and PyENGs additionally NGs' brief introduction, progress, design, and a broad range of applications over the last decade. Scope of this chapter is restricted the use of NGs for scavenging numerous energies. Section 14.1 introduced the evolution of NGs and provides a broad overview of their development. Sections 14.2–14.4 outline the applications of triboelectric, piezoelectric, and PyENGs, respectively. Conclusions of the chapter are abridged in Section 14.5.

14.2 APPLICATIONS OF TENG

TENGs have vast area of application. However, applications are found broadly in following areas: Micropower Sources; Active Sensors for Self-Powered Systems; Direct HV (High Voltage) Power Sources and Blue Energy.

14.2.1 Micro/Nanoenergy Sources

TENGs have applications in various field including micropower sources used for self-propelled systems through scavenging biomechanical, like human movement, heartbeats, computer vibration, and wind. Biomechanical harvesting power of TENG is vital because it can generate splendid output even at very low range of frequencies. Earlier, TENG has large production in the field such as smart backpack (Chandrasekhar et al. 2017; Zhu, Chen et al. 2013), self-powered watch (Zhao, Wang, and Yang 2016), and acoustic energy harvester (Fan et al. 2015; Yang et al. 2014). On the basis of fabrication method, this field can be divided into two categories: shape-adjustable TENG and TENG based on fiber used for wearable power sources. It's important to note both these approaches are complementary instead mutually incompatible. Shape-adaptive devices use flexible and distorted materials to allow for a more natural and practical application to the human body. Successively, Yi et al. (2016) described a TENG comprising a conductive fluid and soft rubber. The conductive fluid, like a solution of common salt, was confined within the rubber. Such a fabricated system withstand tensile strain approx. 300 refrained from showing any signs of wear and tear after ≈55k deformation cycles. In addition to in vitro energy harvesting, Zheng et al. (2016) identified a versatile and biodegradable TENG (BD-TENG) for short-term in vivo biomechanical energy conversion. The multilayer structure of the BD-TENG was made up of biodegradable polymers (BDPs) and absorb metal. Further functional polymers, including self-healing polymers and shape memory polymers, have also been utilized to create new TENG devices with beneficial properties (Deng et al. 2018; Liu et al. 2018).

Zhong et al. (2014) proposed the primary fiber-based TENG, which could be spun into a piece of fabric in 2014. The fabricated device, as shown in Figure 14.5(a) was made up of two types of fibers, one with carbon nanotube (CNT) covered cotton yarn and the other with top layer of CNT is polytetrafluoroethylene (PTFE)-coated. The electricity generation mechanism relied on the separation of PTFE and CNT, caused by a shift in the interfiber distance caused by mechanical deformation of the woven fabric (Figure 14.5(b)). These TENGs were used to power the energy-storing devices, such as capacitor and also to activate a wireless human body temperature sensor in "power

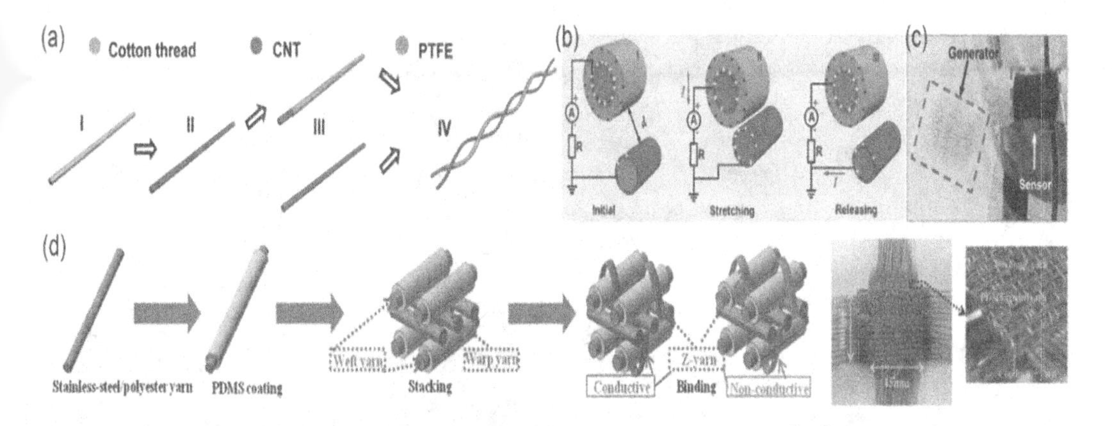

FIGURE 14.5 TENG-assisted fiber: (a) working mechanism, (b) construction procedure, (c) "power shirt" representation, and (d) three-dimensional picture of woven TENG. (Reprinted with permission from: (a–c) Zhong et al. 2014 and (d) Dong et al. 2017.)

shirt" (Figure 14.5(c)). 3D (three-dimensional) woven structures added to improve the production efficiency of the fiber-based TENG design (Dong et al. 2017). The construction process of such TENG is shown in Figure 14.5(d). The device concept proposed here can easily be used for large-area wearable clothing. Furthermore, by altering the primary fiber content to silicone rubber, the TENG based on fiber rendered to be highly flexible (He et al. 2017; Park et al. 2017).

Hybrid electricity production offers advanced peak power and the ability to harvest many forms of energy at the same time. By combining TENG and the PENG with a multiple layered structure transformed into a single fiber, Li et al. (2014b) described some basic principle about fiber-assisted hybrid NG. Chen et al. developed a hybrid electricity textile with the spinning pattern and material composition of a fabric TENG and a photovoltaic textile (Park et al. 2018). Combining fiber-shaped TENG with photovoltaic cells, and ultracapacitors turned into hybrid self-propelled textile (Wen et al. 2016).

14.2.2 Active Sensors for Self-Powered System

Tactile touch sensors (Lin, Xie et al. 2013; Meng et al. 2014; Zhu et al. 2014), acoustic sensors (Fan et al. 2015; Yang et al. 2014; Yu et al. 2015), motion and accelerated sensors (Chen et al. 2018; Wu et al. 2016; Zhou et al. 2014), and sensors based on chemical are all examples of innovative work (Li et al. 2015; Su et al. 2015; Wang, Chen, and Lin 2015). Meanwhile, as the IoT grows in popularity (Atzori, Iera, and Morabito 2010; Ornes 2016), advanced human-machine interfacing (HMI) is becoming more important to allow easy, secure, and human and external systems have never communicated in this way before. TENGs have been thoroughly tested for active sensing applications (Wang, Chen, and Lin 2015). The advantages of dynamics, a communicative biometric that measures people's typing habits, improves security without interfering with regular user activity (Monrose and Rubin 2009; Wu et al. 2018). Chen et al. (2015) suggested the idea of understanding keystroke dynamics by means of TENG-based sensing arrays in 2015. Their system operated on a single-electrode interaction isolation mode and suffered from working environment interferences. We then created a two-factor, pressure-enhanced keystroke-dynamics-based security framework and groundbreaking effort on triboelectric keyboards (Li et al. 2016), proficient of authenticating and recognizing operators via specific typing actions with an accuracy of up to 98.7% (Wu et al. 2019). Furthermore, optical devices like LEDs and a SE-mode-TENG-based keypad applicable to created touchscreens for security validation (Xiong et al. 2019).

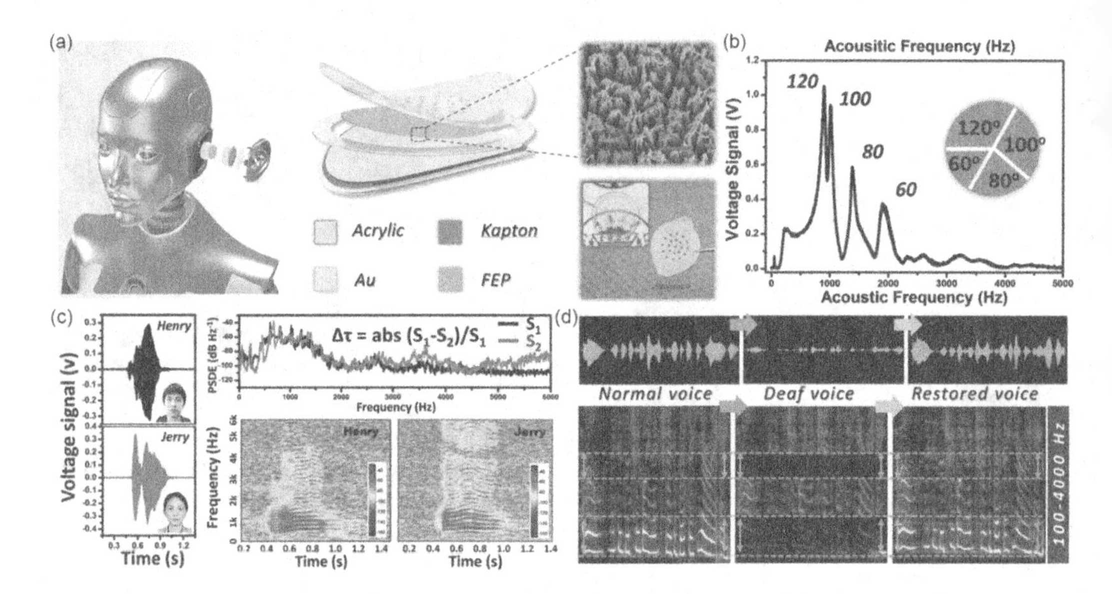

FIGURE 14.6 A TAS (triboelectric auditory sensor): (a) diagrammatic presentation and pictures, (b) spectrum of four divisional margins, (c) representation in audio identification, and (d) in medical field to assist restoration impede spectrums of hearing. (Reprinted with permission from Guo et al. 2018.)

TENG-assisted sensors also useful to create an innovative, hand-freely human machine interface to converted eye blinking micromotions into control commands (Pu et al. 2017), applicable for traditional HMI systems such as keyboards. A smart home control system was created by combining the mechanosensational TENG (msTENG) consist of signal processor circuit. In addition, the msTENG was combined with a wireless transceiver module to create a hands-free typing device. This TENG-based sensor allows individuals to monitor and also interact like a hands-free, intelligent manner.

Yang et al. (2014) created the first TENG-based auditory system that worked like a renewable energy cultivator and sensitive sensor at the same time in 2014. Following that, the idea of a paper-based TENG carrying the similar proficiencies was suggested, with more benefits like ultra-thin and effectively rolled (Fan et al. 2015). Such devices are used as an audible resource sensor and also self-propelled mike (Yu et al. 2015). It is proposed that self-propelled triboelectric audible sensor (TAS) used vastly for global automation and assisted auditory (Zhou et al. 2020). The multilayer TAS construction is shown in Figure 14.6(a). A multi-resonant spectrum also obtained using TAS, as shown in Figure 14.6(b). It is useful in voice-activated security systems. The TAS system captured the electrical signals of two people speaking the same word “Hello”, as shown in Figure 14.6(c).

The restoration impede spectrums of hearing are shown in Figure 14.6(d). A TAS-based hearing aid system also detected voices, showing that affected spectrum's acoustic information were repaired. The complication and expense of the signal processor circuits in traditional hearing supports are significantly reduced using a TAS-based method.

14.2.3 Blue Energy

Many researchers are interested in using TENGs to harness natural mechanical energy such as wind energy (Bae et al. 2014; Chen, Yang, and Wang 2018; Wang et al. 2018; Yang, Zhu et al. 2013), raindrop energy (Yeh et al. 2015; Zheng et al. 2014; Zhu et al. 2015), ultrasonic energy (Wang, Jiang, and Xu 2017; Yang, Zhang, Liu et al. 2013), and water wave energy (Wang 2017; Wen et al. 2014; Zhou et al. 2016). The idea of harvesting the kinetic energy of flowing water in huge oceans with a TENG, also known as blue energy (Chen et al. 2015; Wang, Jiang, and Xu 2017), especially interesting, as current studies indicate TENG is more efficient than an EMG at low-frequency vibration energy harvesting (Zi et al. 2016). The completely enclosed sphere structure (Chen et al. 2015; Wang, Jiang, and Xu 2017) and arch like sphere ball (Chen et al. 2015) are the most encouraging for the blue light among various types of TENG. Hence, energy can be harvested effectively using this approach (Wang et al. 2015) (Figure 14.7(a)), shows a freestanding triboelectric-layer-mode TENG (RF-TENG), used for blue energy harvesting in a spherical container. When it triggered by a wave generator, glow 70 green LEDs (Figure 14.7(b)) depict the concept of operation (Figure 14.7(c)) and show output of electric current obtained at an amplitude of 1.43 Hz. The relation between output and frequency is also shown in Figure 14.7(d). Material design improved the performance of the rolling sphere model. Soft synthetic material used to enhanced area, as shown in Figure 14.7(e) (Xu et al. 2018). The comparison analysis shown in Figure 14.7(f). It also revealed that ultraviolet treatment improved the production efficiency.

Three different types of connections have been suggested and studied, displayed in Figure 14.7(g), stiff plate connections fixed, flexible strip connections are deformable to some extent, and thread connections that can easily bend. Figure 14.7(h) shows that flexible connections by either the string or elastic strip in a 4×4 network, whether made with a string or an elastic

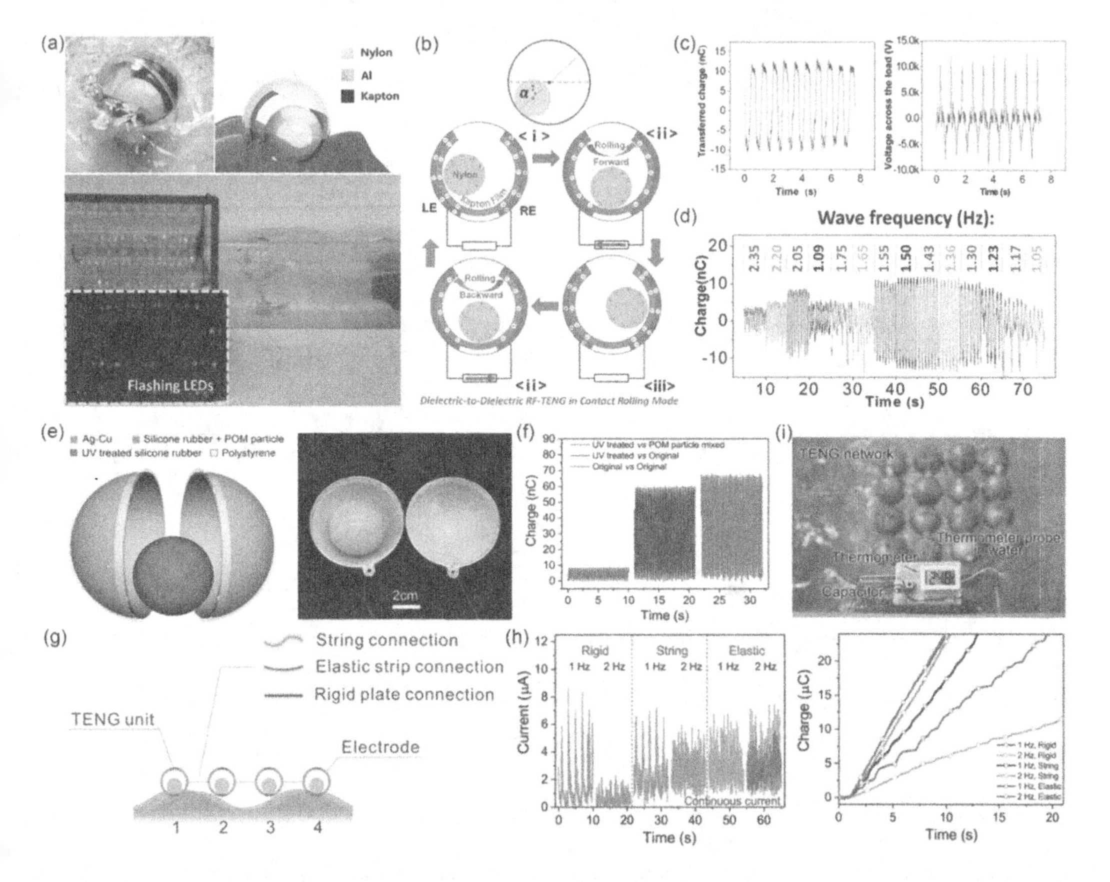

FIGURE 14.7 Radio frequency–based TENG: (a) pictures and diagram, (b) operational mechanism, (c) current output, (d) relation between output and frequency, (e) pictures and diagram of Si-assisted ball-shell TENG, (f) effect of changing the silicone material on the output is compared, (g) diagram shows TENG has a variety of connections, (h) output performance of various connections is compared, and (i) presentation of wire-connected TENG system used to control a thermostat. (Reprinted with permission from: (a–d) Wang et al. 2015 and (e–i) Xu et al. 2018.)

strip, outperformed rigid connections in terms transfer of charge, allocated to non-essential interior restraints among elements of the approach. This network connected to a thermometer to obtain the reading of temperature of water as demonstrated (Figure 14.7(i)). Jiang et al. proposed a TENG assisted with spring for blue energy which are designed with the help of acrylic box and spring, as shown in Figure 14.8(a). Copper-coated PTFE films were mounted to the blocks and electrodes fully covered with copper, fixed at the inner two walls of the box, resulting in PTFE and copper contact and separation due to mechanical stress applied externally. The spring supported and a standard TENGs were also compared in terms of electric outputs, as shown in Figure 14.8(b). Electric current increased from 351.03 to 747.84 nC, spring-assisted TENG generated several more current peaks. This is because some portion of instantaneous energy stored as elastic energy by the link spring. Later, this energy utilized further to initiate TENG. As a result, when the spring was used, the energy production per cycle increased by 150%. Several of these systems were combined in a single external box to boost a single device's output current, significantly improving its ability to

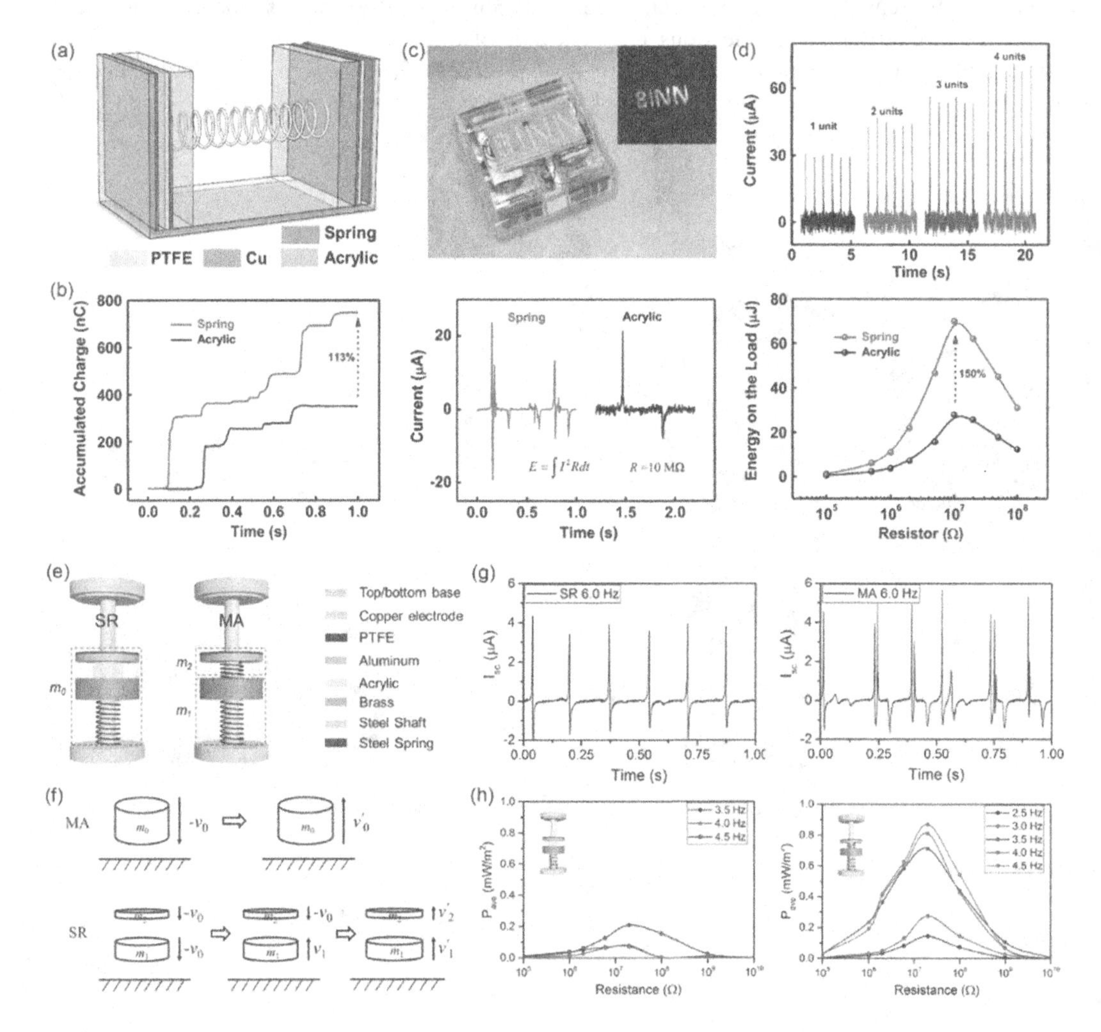

FIGURE 14.8 Spring-supported TENG: (a) image of the assembly, (b) the spring supported and a standard TENGs were compared in terms of electric outputs, (c) image of the bounded system with four component, (d) the output of the electric current of apparatus is compared to the number of component, (e) diagram of the SR- and MA-TENG, (f) depict the sequential collisions in the TENGs, (g and h) comparability of the output of current obtained from two devices, at 6- and 4-Hz vibration frequency, respectively. (Reprinted with permission from: (a–d) Jiang et al. 2017 and (e–h) Wu et al. 2017.)

initiative electronic devices like LEDs (Figure 14.8(c and d)). Additionally, a mechanical-based amplifier used to couple all the springs to enhance the resulting output of a single-spring resonator-assisted TENG (SR-TENG) (Chen and Wang 2017). Figure 14.8(e) shows the diagrammatic of a mechanical amplified-assisted TENG (MA-TENG) and a triboelectric single electrode rotational NG (SR-TENG). The main aspect is breaking of (m0) into two weights (m1 and m2) linked with the spring. The collision analysis of these designs revealed, in MA-TENG, because of successive collisions of the bottom base, m1, and m2 then m2 obtained the high velocity than m0, leading the PTFE film to reach the top copper electrode more quickly and regularly (Figure 14.8f). Using 6- and 4-Hz vibration frequency, the output of current obtained from two devices validated MA higher TENG's output frequency as shown in Figure 14.8(g and h).

These studies show a lot of room for improvement in TENG efficiency, and it has a bright future extensively in the area of blue energy. Also, wind energy can be harvested by stringing sphere TENGs and arranging them row by row, particularly in mild wind. It could have been an excellent alternative to conventional wind turbines, which only work when there's a lot of wind.

14.2.4 Direct HV Power Sources

TENG's inherent feature, HV, and low current make it a unique alternative to traditional HV power sources, providing unparalleled portability and protection. TENG-based HV sources hardly need refined power transformer since they operate at 1–10 kV, which reduces system complication and expense. Furthermore, since HV may not be sustained until the restricted charge is shifted, the lower current presents much less risk to staff and instruments. TENG is being used to produce nanoelectrospray ionization for extremely sensitive nano-coulomb molecular mass spectrometry (MS) (Li et al. 2017), additional advantage of TENG's HV performance also turns the charge transmission control with the gain of unprecedential ion generation control. Chen et al. are the first to propose air cleaner, driven by TENG (He and Wang 2018). In comparison of traditional electrostatic precipitators, it was self-powered does not generate ozone. The idea was earliest established by utilize a wind-driven rotating TENG with two parallel electrodes receiving its rectified DC output (He and Wang 2018). It also exposed using a rotating TENG, the particulate matter (PM) extracted productivity of traditional filters improved (Gu et al. 2017). Furthermore, using self-vibration of the tailpipe and a self-propelled triboelectric filter productivity obtained approx. 95.5% for PM2.5, shown in Figure 14.9(a) (Han et al. 2015). Recently, more advanced multiplayer triboelectric air filter (TAF) is created (Bai et al. 2018). Contrasting traditional masks, the TAF's extracted productivity hardly affected, afterward cleaning with a commercial detergent, suggesting that it can be used to create a reusable and effective face mask (Figure 14.9(b)). TENGs may also be used for electrostatic actuation due to their HV. Chen et al. demonstrated a dielectric elastomer actuator-based TENG-driven actuation device that would modify the spacing of tunable optical gratings or on/off status of an intelligent switch (Chen et al. 2016; Nie et al. 2018). The output-voltage obtained from the TENG controlled the movement of minute items like solid pellets and water droplets (Zheng et al. 2017) reasonably engineered electrodes on a substrate. On the basis of working mechanism, a self-controlled microfluidic conveyance unit powered, as shown in Figure 14.9(c) (Nie et al. 2018).

TENG can cross above 10 kV by introducing a voltage enhancing circuits, and it can be used to drive a variety of other high-voltage methods including electrospinning (Wu et al. 2019), field emission (Wu et al. 2019), and microplasma (Cheng et al. 2018b). Cheng et al. recently realized the idea of triboelectric microplasma by combining TENG and a plasma resource, resulting plasma powered solely through mechanical stimuli (Figure 14.9(d)). The triboelectric microplasma was successfully used for surface treatment and patterned luminescence. The findings are expected to broaden the field of plasma research by offering a clear and compact alternative to traditional plasma sources. TENG-based power sources have a wide range of applications in many fields, including high-voltage devices, actuators, environmental protection, and human health, due to their portability, controllability, safety, and high performance.

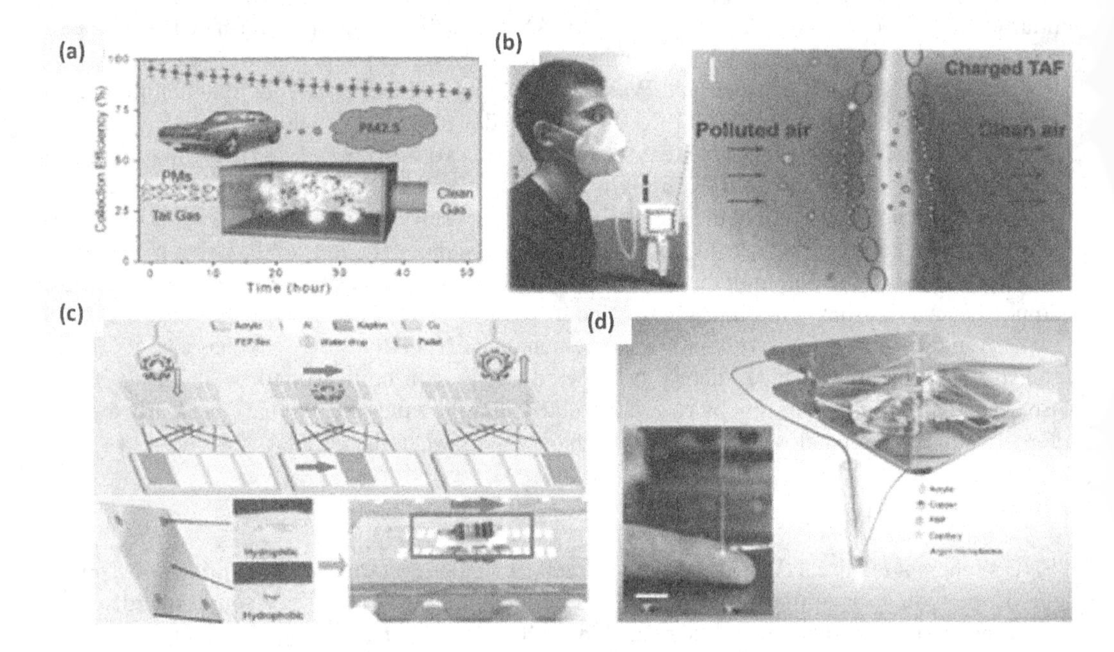

FIGURE 14.9 (a) PM extracted from exhaust gases from automobiles, (b) washable-air-filter, (c) microfluidic conveyance unit, and (d) schematic of triboelectric microplasma by combining TENG and a plasma resource. (Reprinted with permission from: (a) Han et al. 2015, (b) Bai et al. 2018, (c) Nie et al. 2018, and (d) Cheng et al. 2018a.)

14.3 APPLICATIONS OF PIEZOELECTRIC NANOGENERATORS (PENGs)

PENGs serve as a reliable alternative power source and a driving force for a variety of fashionable applications. This technology serves as new techniques for the advancement of an eco-friendly environment.

The nanoscale piezoelectric harvesting devices have used different fields such as industrial production, pharmacy, automobiles, and telecommunications.

14.3.1 Self-Powered Nano/Microsensors

The small-scale NG makes it possible to generate energy for micro and nanoscale devices, which can be stable at a place somewhere constant mechanical vibrations are accessible. Self-powered gas sensors have been developed using 1D MOs nanostructures, like ZnO, SnO_2, and In_2O_3, due to having effective features, like large (S/V) surface-to-volume ratio (Gurlo 2011; Li et al. 2007).

Deng et al. (2014) reported that design of a self-powered active biosensor using ZnO nanowire.

Nanowire generates power that acts as both a biosensing signal and a power source. Surface molecular adsorption can alter the density of free carriers on ZnO nanowire to regulate the screening effect. Due to the combined properties of piezotronic and bio-sensitive features of ZnO nanowire, this nanowire is a feasible alternative to active biomolecule detection (Zhao et al. 2014).

A self-powered, active gas H_2S sensor, based on CuO/ZnO nanoarray, has been developed by Nie et al. (2014). This piezoelectric response acts as a sensor signal and power supplier. The piezoelectric contribution of the nanoarray at room temperature reduced from 0.738 to 0.101 V when 800-ppm H_2S was applied by the device. CuS/ZnO ohmic contact was formed by the modification of CuO/ZnO p-n junction as a result of exposure of the device to H_2S gas. An additional screening effect was obtained due to the increase of electron density in the nanowire (Nie et al. 2014).

14.3.2 Self-Powered Electronics

Innovative consumer electronics have become an indispensable element of comfortable human life with the exponential development of electronics. Nano-sized harvesting systems can be used as power generation source and also foundation for the development of self-propelled units. A polyvinylidene fluoride (PVDF)/graphene-based NG can synchronize finger motions fully, and the energy produced can charged a LED for 30 s. This PVDF generator can be used as a self-triggered power resource for transportable peculiar electronic systems with low power requirements (Abolhasani, Shirvanimoghaddam, and Naebe 2017).

Hu et al. (2016) proposed an anti-peep transparent PENG that could extract energy from light finger tapping and produce an output current of 0.8 nA. When lead zirconate titanate (PZT) nanowires were aligned on polydimethylsiloxane (PDMS), the light transmittance and electromechanical properties of this transparent NG improved (Hu et al. 2016).

14.3.3 Smart Wearable Systems

By combining nanosensors (NSs) with usable fabrics, smart clothing can be woven. They are used to charged less current required sensors via mechanical energy from human body movements (Zhong et al. 2014). Figure 14.10 demonstrated a schematic of the fabric NG and its assembly. Smart cloth made of a fiber-based 3D hybrid NG (FBHNG) could be used to create a self-powered strain sensor. The rectified output of an FBHNG could also be used to create a self-powered alarm system (Li et al. 2014a).

14.3.4 Biomedical Applications

Nanostructures like nanochannels, nanotubes, nanopatterns, and various material are used in range of biomedical applications such as drug and gene delivery, and the biodetection of pathogens and proteins (Bruchez et al. 1998; Chan and Nie 1998). The exposure of protein molecules with the help of nanoparticles is illustrated in Figure 14.11.

Nanochannels and nanotubes have been used in a variety of applications, including DNA structural studies and tissue engineering (Ma et al. 2003) people magnetic nanoparticles have a broad variety of applications in medicine, including the isolation and purification of biotic particles and

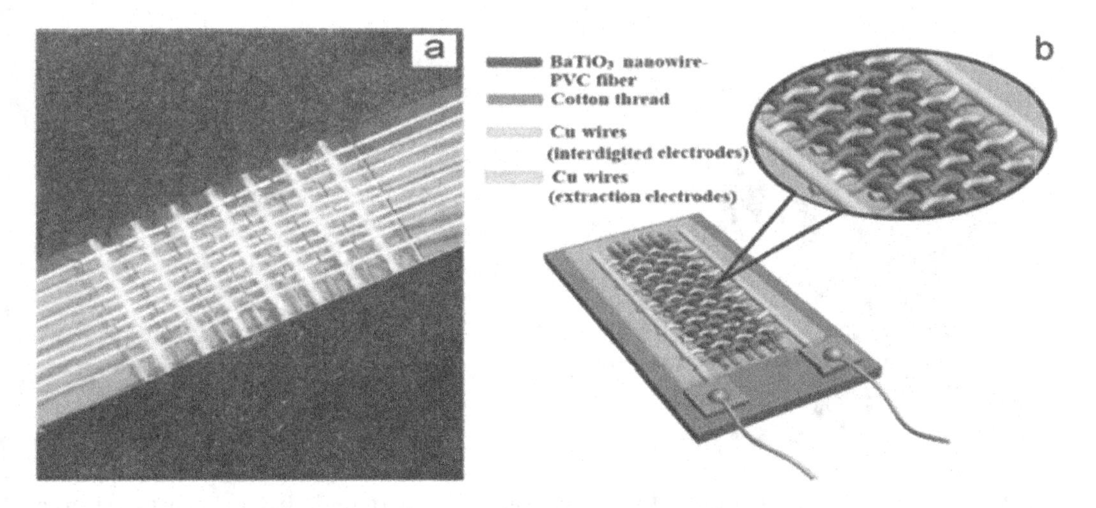

FIGURE 14.10 The photo and schematic of the fabric nanogenerator (FNG). (a) The photo of the FNG and (b) the structure of the FNG. (Reprinted with permission from Zhang et al. 2015.)

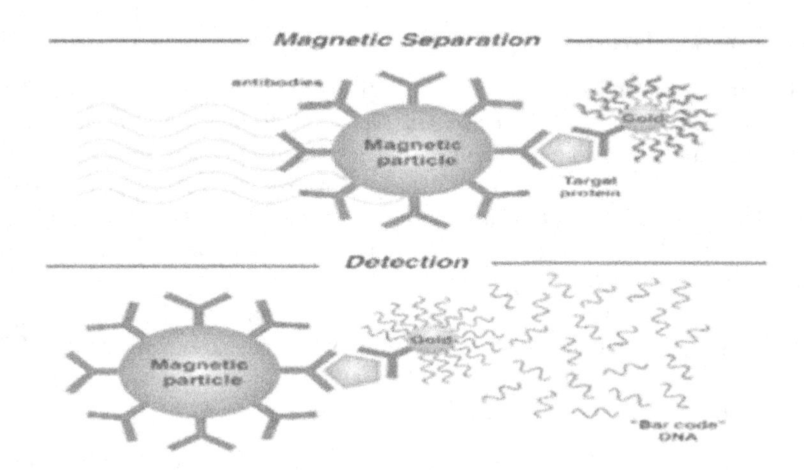

FIGURE 14.11 Protein detection using nanoparticles. (Reprinted with permission from Nam, Thaxton, and Mirkin 2003.)

cells (Molday and Mackenzie 1982). However, increased in magnetic resonance imaging (MRI) and hyperthermia (the heating-induced death of tumor cells), and several phagokinetic studies are currently being researched (Parak et al. 2003).

14.3.5 Security Systems

PENGs have been used as many types of motion sensors for transport monitoring, biomedical sensors, wireless sensors, and tactile sensors and they have much potential as energy-harvesting power sources (Lin, Hu et al. 2013; Yang, Wang et al. 2012). Keun et al. (2014) announced a hybrid energy-scavenging NG that showed high compatibility with various parts of the human body, making it a better option for portable gadgets, and robots with exceptional longevity (Alam et al. 2015). A hybrid device is depicted in Figure 14.12 (Lee et al. 2014a). They are extremely beneficial in the implementation of safety systems.

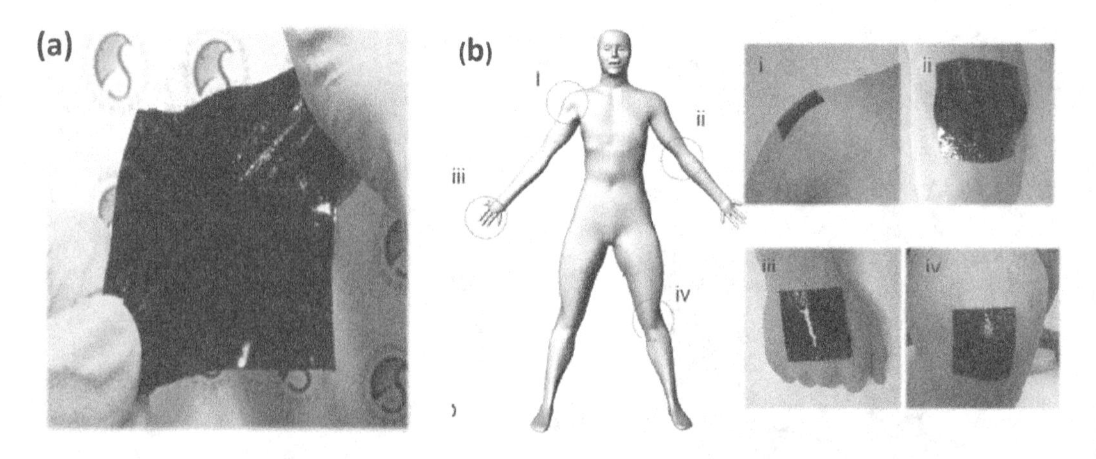

FIGURE 14.12 (a) Photo image of the HSNG, (b) photo images of the HSNG at various locations on human body, showing good compatibility of the device with various parts of body. (Reprinted with permission from Lee et al. 2014a.)

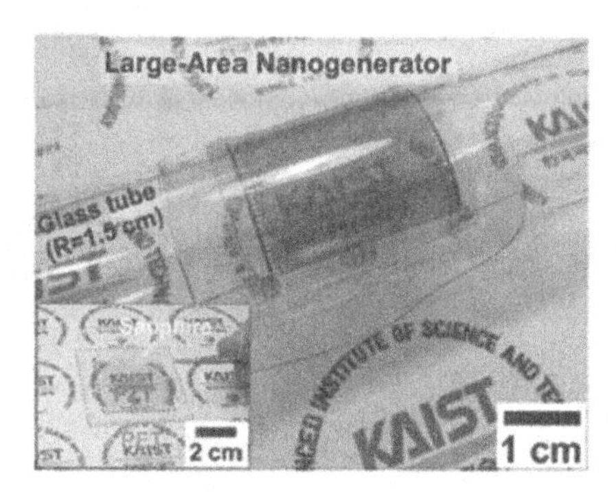

FIGURE 14.13 A large-area PZT thin film NG on a curved glass tube with a radius of 1.5 cm. The inset shows a large-area PZT thin film on PET substrate transferred from a sapphire substrate by LLO process. (Reprinted with permission from Park et al. 2014.)

14.3.6 Transparent and Flexible Devices

Thin metal or metal oxide films are used as transparent conductors in optical devices like LEDs and solar cell. Thus, due to their low elasticity, chemical instability, and high cost, these transparent metal-assisted layers have a restricted adoptability in various field (Granqvist 2007). Due to their efficient optical, mechanical, and electrical properties, CNTs are easy substitute of metal-assisted elastic films in a variety of flexible optical products (From百度文库 2013; Schlatmann et al. 1996). CNTs have been limited in their application in flexible devices.

Using a traditional sol-gel process, Kwi et al. (2014) created an effective, and versatile NGs (Park et al. 2014). The thin film was transferred from a sapphire substrate to a plastic substrate using a laser lift-off (LLO) process that caused no mechanical destruction, as shown in Figure 14.13.

Yang et al. (2009) developed a single wire generator (SWG)-assisted flexible piezoelectric zinc oxide nanowire, altered bio-mechanical energy into an appropriate electric response. Integrating four SWGs in series could produce an output voltage of up to approx. 0.1–0.15 V. Figure 14.14 shows the harvesting of energy from NG attached to a human finger or a running hamster.

The device's transparency can be improved without raising the cost by replacing the indium tin oxide (ITO) electrode with a graphene film (Choi et al. 2010). Figure 14.15 depicts the graphene-based NG. Graphene has shown to be a promising candidate for use in a variety of fields, including nanoelectronics, optoelectronics, and spintronics (Geim, Novoselov, and Blake 2006; Liu et al. 2008).

14.4 APPLICATION OF PYROELECTRIC NANOGENERATORS (PyENGs)

The applications of PyENGs in the field of energy harvesters and sensors will be the focus of this section.

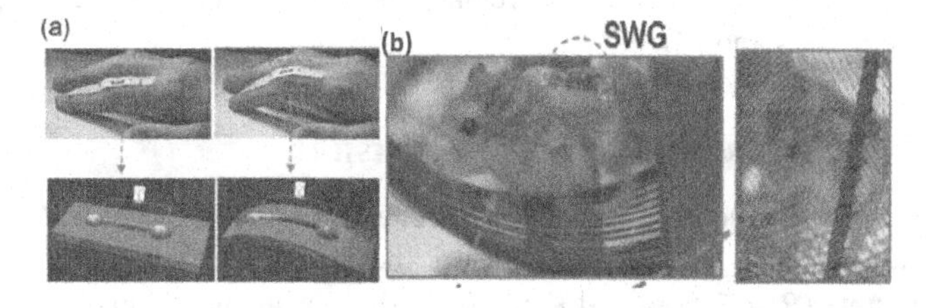

FIGURE 14.14 Harvesting energy using SWG from (a) a vibrating individual's index finger and (b) a real rabbit with running and scratching (Reprinted with permission from Yang et al. 2009.)

FIGURE 14.15 Fully rollable graphene-based nanogenerator. (Reprinted with permission from Choi et al. 2010.)

14.4.1 Flexible Energy Harvesters

Flexible energy harvesting devices have sparked a lot of interest among the reported pyroelectric energy harvesting (PyEH) devices because portable and plantable devices on the body have an almost infinite potential (Chen et al. 2017; Narita and Fox 2018; Potnuru and Tadesse 2014; Sultana et al. 2018; You et al. 2018). You et al. (2018) demonstrated a self-powered flexible hybrid (piezoelectric and pyroelectric) NG based on nonwoven nanofiber membranes (comprising PVDF polymer). A versatile PyENG based on a thin PVDF film was demonstrated by Zhang et al. (2016). Yang, Zhang, Zhu et al. (2013) revealed that versatile cell with a hybrid energy harvesting system made up of a pyro and piezo NG, and a solar cell that could harvest thermal, mechanical, and solar energies separately or concurrently. The proposed cells have many potential application including wireless sensor systems, environmental monitoring, medical diagnostics, and security technologies.

Chen et al. (2016) recorded a versatile hybrid piezo-PyENG constructed on PVDF-trifluoro ethylene (TrFE) with advanced in its electric output. Zhao et al. (2017) recently improved the output voltage and electric charge of the PVDF-assisted stretchable PyENG for self-propelled temperature monitoring applications. With a load of 100 MΩ, the output voltage and current were 9.1 V and 95 nA, respectively.

Due to their excellent pyroelectric properties, ceramic thin films with and without lead have also been investigated in addition to polymer-based energy harvesters (Chen et al. 2017; Narita and Fox 2018; Yang, Jung et al. 2012). Chen et al. presented a flexible Pb(Mg1/3Nb2/3) O3–0.3PbTiO3 (PMN-PT) ribbon-assisted piezo-pyroelectric hybrid NG for temperature regulation and scavenging mechanical movement of human organs (Figure 14.16). It can also be used to track acoustic sounds with pinpoint accuracy. Temperature-related behavior can be monitored using the implanted sensor.

Ko et al. (2016) investigated a versatile and hybrid pyro-PENG-assisted $Pb(Zr_{0.52}Ti_{0.48})O_3$ films. To make the flexible unit, using $LaNiO_3$ as a bottommost electrode, a PZT film deposited on an extremely flexible nickel-chromium metal foil substrate, allowing the film to expand at high temperatures. Yang, Jung et al. (2012) revealed versatile PyENGs made of a composite structure of lead-free $KNbO_3$ nanowires and PDMS.

14.4.2 Pyroelectric Materials in a Hybrid Harvester

The efficacy of PyEH improved using a variety of energy sources. In general, combining two or more energy transfer systems, such as pyro, piezo, thermoelectric, and photothermal, developed hybrid energy harvesters (Ryu, Yoon, and Kim 2019), thus ensuring that each of the integrated harvesting systems is self-contained. In this part, we'll look at hybrid energy harvesters that combine pyro-piezoelectric technologies. The related equations of pyroelectricity and piezoelectricity are

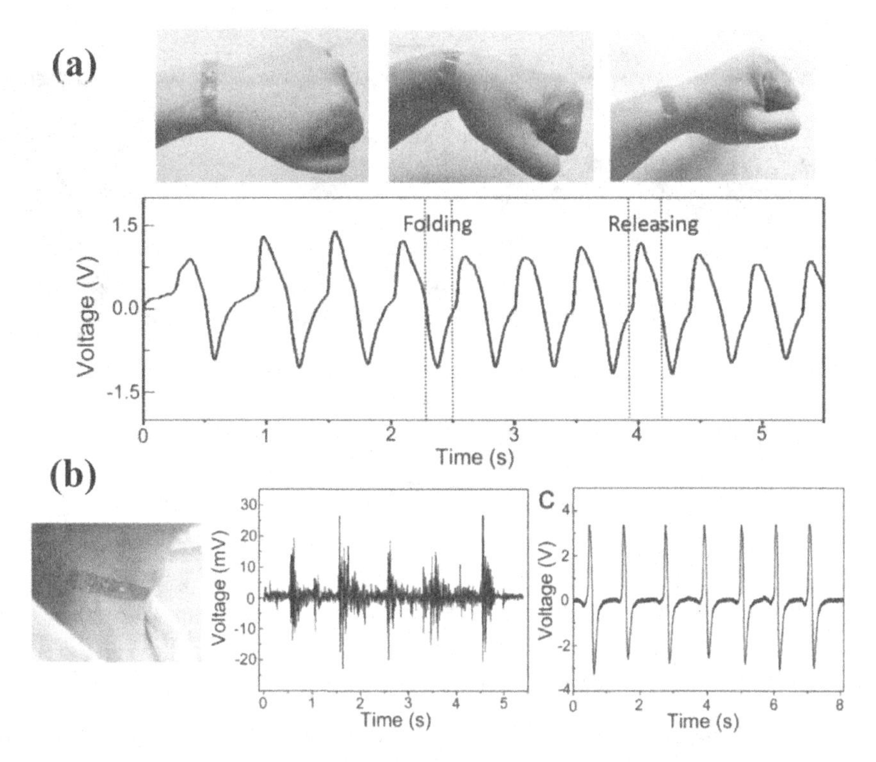

FIGURE 14.16 (a) Illustration of the PMN-PT ribbon-based unit attached to the skin of the wrist. The output voltage obtained while folding and resealing wrist behavior is shown in the plots below the picture. (b) Image of the apparatus connected to neck and a graph obtained from voltage output by coughing, and (c) the voltage-output produced by a stick striking. (Reprinted with permission from Chen et al. 2017.)

very similar since the change in temperature (ΔT) is identical to the stress parameter in piezoelectric effect. Thus, hybrid pyro-piezoelectric energy harvesting systems have received a lot of attention (Chen et al. 2016; Chen et al. 2017; Ko et al. 2016; Lee et al. 2014b; Yang, Zhang, Zhu et al. 2013; You et al. 2018; Zi, Lin et al. 2015). Changes in polarization should be positive in such hybrid systems, enhancing the energy harvester's power generation even further.

Lee et al. (2014b) developed a pyro-piezoelectric hybrid NG with fine-patterned PDMS-CNT composites, fine-patterned piezoelectric PVDF-PTFE polymer, and nano-sized sheets of graphene, as depicted in Figure 14.17. PDMS-CNTs were used at the base of the device to make it more versatile. The equation established the material's ability to harvest (ΔT) and mechanical stress (σ). The polarization accumulative conversion is as follows:

$$\Delta P = d \cdot \sigma + \lambda \cdot \Delta T, \text{ d denotes the piezoelectric coefficient.}$$

The hybrid NG, comprising a PVDF nanowire-PDMS composite/ITO configuration, was demonstrated by Wang, Wang, and Yang (2016). Erturun et al. (2014) looked into a hybrid energy harvester that used a heating lamp to heat a vibrating beam. Even though few variables limit hybrid energy harvesters' overall performance, the use of pyroelectric-piezoelectric hybrid energy harvesters can boost production capacity.

14.4.3 Flexible Sensors

Small and portable electronic devices have gotten a lot of attention in recent decades. Because of the pyroelectric effect, temperature differences between the electrode transformed into the power,

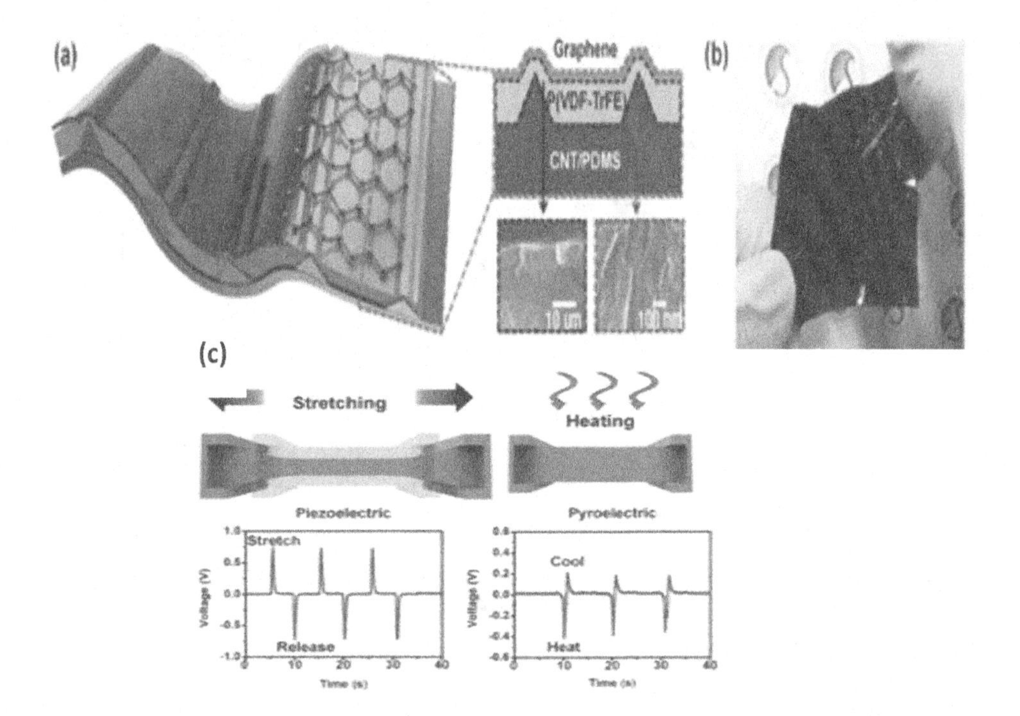

FIGURE 14.17 (a) Schematic illustration of the flexible hybrid energy harvester device, (b) photo images of the hybrid stretchable nanogenerator (HSNG), and (c) piezoelectric and pyroelectric output voltages from the HSNG under stretch-release and cool-heat conditions. (Reprinted with permission from Lee et al. 2014a.)

as PyENGs are mounted in the human body as wearable devices. People tend to wear a respirator in the winter, mainly due to cold weather or air pollution. Xue et al. (2017) revealed a self-propelled respiratory apparatus using the PyENG. Also, Pullano, Islam, and Fiorillo (2014) showed PVDF as temperature sensors.

For pyroelectric energy conversion applications, PMN-PT and PZT are used as lead-assisted ceramics that have been shown to have excellent figure of merits (FoMs) (Cuadras, Gasulla, and Ferrari 2010; Hsiao, Siao, and Ciou 2012; Kotipalli et al. 2010; Mane et al. 2011; McKinley and Pilon 2013; Ravindran et al. 2011; Sebald et al. 2006; Yang et al. 2012; Yang, Wang et al. 2012). Similarly, several lead-free ceramic pyroelectric materials with excellent FoMs have been published, including bismuth sodium titanate-barium titanante (BNT-BT) (Sun et al. 2014), barium strontium titanate (BST) (Huang et al. 2012), and potassium sodium niobite (KNN) (Kumar et al. 2014; Sharma et al. 2015). Changes in the chemical and physical composition of ceramics, such as doping or the introduction of pores into the ceramics, have been studied by researchers (Ohji and Fukushima 2012; Patel, Chauhan, and Vaish 2016; Zhang et al. 2009). $Pb(Zr_{0.965}Ti_{0.035})O_3$ and 1 wt.% Nb_2O_5, $BaSn_{0.05}Ti_{0.95}O_3$, $Ba_{0.67}Sr_{0.33}TiO_3$, and $PbZr_{0.45}Ti_{0.55}O_3$ have all used pore formers, such as poly(methyl methacrylate), carbon nanotubes, and ethyl cellulose (Jiang et al. 2015; Suyal and Setter 2004; Zhang et al. 2009; Zhang et al. 2015; Zhang, Jiang, and Kajiyoshi 2010). These studies showed that increasing the porosity of the ceramics improved the FoMs (Suyal and Setter 2004; Zhang et al. 2005).

14.5 CONCLUSIONS

This chapter discussed thoroughly types and large-scale application of NGs. As move to mobility, and efficiency, it revealed that harvesting ambient mechanical energy is sustainable for electrical as well as electronics sectors. It has vastly adopted in various areas, including conveyance, sensors, recuperation systems like smooth sensors, and so on. NGs have some inherent properties such as tiny size, fabricated with low-cost materials, effortlessly designed, and, most importantly, produce commendable performance, making them an encouraging technology for long-term feasibility. NGs can also be used as substitute for powered batteries, hence effectively acceptable in the field of self-propelled sensors. NGs output voltages are frequently high when compared to output current. According to this chapter, the TENGs are proficient of producing higher voltage and power density than the other two types of NGs. Voltage transformers and rectifiers support to raise the electric current while lowering the voltage, can help with this high-voltage, low-current problem.

The output power features can be improved using transformers, inductors, and electronic logic control switches. TENG has a higher power density. In the field of blue energy harvesting, this is required in to identify how well the system works with the power management module. One of the key issues is the durability of NGs in blue energy harvesting. The progress of long-established NGs is still difficult for researchers. To boost overall power conversion efficiency, NGs are combining with several other energy-harvesting systems. Using PyENGs, the concern of high-temperature control in solar cells is resolved. In the past decade, NG research for real-world applications has developed at an exponential rate, suggesting that this technology may widely commercialized.

REFERENCES

Abolhasani, Mohammad Mahdi, Kamyar Shirvanimoghaddam, and Minoo Naebe. 2017. "PVDF/Graphene Composite Nanofibers with Enhanced Piezoelectric Performance for Development of Robust Nanogenerators." *Composites Science and Technology* 138: 49–56.

Alam, Md Mehebub, Sujoy Kumar Ghosh, Ayesha Sultana, and Dipankar Mandal. 2015. "Lead-Free $ZnSnO_3$/MWCNTs-Based Self-Poled Flexible Hybrid Nanogenerator for Piezoelectric Power Generation." *Nanotechnology* 26 (16): 165403–08.

Askari, Hassan, Amir Khajepour, Mir Behrad Khamesee, Zia Saadatnia, and Zhong Lin Wang. 2018. "Piezoelectric and Triboelectric Nanogenerators: Trends and Impacts." *Nano Today* 22: 10–3.

Atzori, Luigi, Antonio Iera, and Giacomo Morabito. 2010. "The Internet of Things: A Survey." *Computer Networks* 54 (15): 2787–805.

Bae, Jihyun, Jeongsu Lee, Seongmin Kim, Jaewook Ha, Byoung Sun Lee, Youngjun Park, Chweelin Choong, et al. 2014. "Flutter-Driven Triboelectrification for Harvesting Wind Energy." *Nature Communications* 5: 4929–37.

Bai, Yu, Chang Bao Han, Chuan He, Guang Qin Gu, Jin Hui Nie, Jia Jia Shao, Tian Xiao Xiao, Chao Ran Deng, and Zhong Lin Wang. 2018. "Washable Multilayer Triboelectric Air Filter for Efficient Particulate Matter PM2.5 Removal." *Advanced Functional Materials* 28 (15): 1706680–8.

Batra, Ashok K., and Almuatasim Alomari. 2017. "Power Harvesting via Smart Materials." *Power Harvesting via Smart Materials* PM277: 1–15.

Beeby, S. P., M. J. Tudor, and N. M. White. 2006. "Energy Harvesting Vibration Sources for Microsystems Applications." *Measurement Science and Technology* 17 (12) : 120175–95.

BP Energy Outlook. 2018. "2018 BP Energy Outlook 2018 BP Energy Outlook," 125.

Bruchez, Marcel, Mario Moronne, Peter Gin, Shimon Weiss, and A. Paul Alivisatos. 1998. "Semiconductor Nanocrystals as Fluorescent Biological Labels." *Science* 281 (5385): 2013–6.

Chan, Warren C.W., and Shuming Nie. 1998. "Quantum Dot Bioconjugates for Ultrasensitive Nonisotopic Detection." *Science* 281 (5385): 2016–8.

Chandrasekhar, Arunkumar, Nagamalleswara Rao Alluri, Venkateswaran Vivekananthan, Yuvasree Purusothaman, and Sang Jae Kim. 2017. "A Sustainable Freestanding Biomechanical Energy Harvesting Smart Backpack as a Portable-Wearable Power Source." *Journal of Materials Chemistry C* 5 (6): 1488–93.

Chen, Bo, Ya Yang, and Zhong Lin Wang. 2018. "Scavenging Wind Energy by Triboelectric Nanogenerators." *Advanced Energy Materials* 8 (10): 1–13.

Chen, Jun, and Zhong Lin Wang. 2017. "Reviving Vibration Energy Harvesting and Self-Powered Sensing by a Triboelectric Nanogenerator." *Joule* 1 (3): 480–521.

Chen, Jun, Jin Yang, Zhaoling Li, Xing Fan, Yunlong Zi, Qingshen Jing, Hengyu Guo, et al. 2015. "Networks of Triboelectric Nanogenerators for Harvesting Water Wave Energy: A Potential Approach toward Blue Energy." *ACS Nano* 9 (3): 3324–31.

Chen, Xiangyu, Tao Jiang, Yanyan Yao, Liang Xu, Zhenfu Zhao, and Zhong Lin Wang. 2016. "Stimulating Acrylic Elastomers by a Triboelectric Nanogenerator – Toward Self-Powered Electronic Skin and Artificial Muscle." *Advanced Functional Materials* 26 (27): 4906–13.

Chen, Xiaoliang, Jinyou Shao, Xiangming Li, and Hongmiao Tian. 2016. "A Flexible Piezoelectric-Pyroelectric Hybrid Nanogenerator Based on P(VDF-TrFE) Nanowire Array." *IEEE Transactions on Nanotechnology* 15 (2): 295–302.

Chen, Xuexian, Liming Miao, Hang Guo, Haotian Chen, Yu Song, Zongming Su, and Haixia Zhang. 2018. "Waterproof and Stretchable Triboelectric Nanogenerator for Biomechanical Energy Harvesting and Self-Powered Sensing." *Applied Physics Letters* 112 (20): 1–6.

Chen, Yan, Yang Zhang, Feifei Yuan, Fei Ding, and Oliver G. Schmidt. 2017. "A Flexible PMN-PT Ribbon-Based Piezoelectric-Pyroelectric Hybrid Generator for Human-Activity Energy Harvesting and Monitoring." *Advanced Electronic Materials* 3 (3): 1600540–6.

Cheng, Jia, Wenbo Ding, Yunlong Zi, Yijia Lu, Linhong Ji, Fan Liu, Changsheng Wu, and Zhong Lin Wang. 2018a. "Triboelectric Microplasma Powered by Mechanical Stimuli." *Nature Communications* 9 (1): 1–11.

Choi, Dukhyun, Min Yeol Choi, Won Mook Choi, Hyeon Jin Shin, Hyun Kyu Park, Ju K. Seo, Jongbong Park, et al. 2010. "Fully Rollable Transparent Nanogenerators Based on Graphene Electrodes." *Advanced Materials* 22 (19): 2187–92.

Cuadras, A., M. Gasulla, and V. Ferrari. 2010. "Thermal Energy Harvesting through Pyroelectricity." *Sensors and Actuators, A: Physical* 158 (1): 132–9.

Davies, D. K. 1969. "Charge Generation on Dielectric Surfaces." *Journal of Physics D: Applied Physics* 2 (11): 1533–7.

Deng, Jianan, Xiao Kuang, Ruiyuan Liu, Wenbo Ding, Aurelia C. Wang, Ying Chih Lai, Kai Dong, et al. 2018. "Vitrimer Elastomer-Based Jigsaw Puzzle-Like Healable Triboelectric Nanogenerator for Self-Powered Wearable Electronics." *Advanced Materials* 30 (14): 1–10.

Dong, Kai, Jianan Deng, Yunlong Zi, Yi Cheng Wang, Cheng Xu, Haiyang Zou, Wenbo Ding, et al. 2017. "3D Orthogonal Woven Triboelectric Nanogenerator for Effective Biomechanical Energy Harvesting and as Self-Powered Active Motion Sensors." *Advanced Materials* 29 (38): 1702648–58.

Erturun, Ugur, Christopher Green, Matthew L. Richeson, and Karla Mossi. 2014. "Experimental Analysis of Radiation Heat-Based Energy Harvesting through Pyroelectricity." *Journal of Intelligent Material Systems and Structures* 25 (14): 1838–49.

Fan, Xing, Jun Chen, Jin Yang, Peng Bai, Zhaoling Li, and Zhong Lin Wang. 2015. "Triboelectric Nanogenerator for Acoustic Energy Harvesting and a Self Powered Sound Recording." *ACS Nano* 9 (4): 4236–43.

Fan, Feng Ru, Zhong Qun Tian, and Zhong Lin Wang. 2012. "Flexible Triboelectric Generator." *Nano Energy* 1 (2): 328–34.

From百度文库. 2013. "済無No Title No Title." *Journal of Chemical Information and Modeling* 53 (9): 1689–99.

Gao, Fengxian, Wanwan Li, Xiaoqian Wang, Xiaodong Fang, and Mingming Ma. 2016. "A Self-Sustaining Pyroelectric Nanogenerator Driven by Water Vapor." *Nano Energy* 22: 19–26.

Gao, Yifan, and Zhong Lin Wang. 2007. "Electrostatic Potential in a Bent Piezoelectric Nanowire. The Fundamental Theory of Nanogenerator and Nanopiezotronics." *Nano Letters* 7 (8): 2499–505.

Geim, Andre K., Konstantin Novoselov, and Peter Blake. 2006. "Graphene Spin Valve Devices." *IEEE Transactions on Magnetics* 42 (10): 2694–6.

Granqvist, Claes G. 2007. "Transparent Conductors as Solar Energy Materials: A Panoramic Review." *Solar Energy Materials and Solar Cells* 91 (17): 1529–98.

Gu, Guang Qin, Chang Bao Han, Cun Xin Lu, Chuan He, Tao Jiang, Zhen Liang Gao, Cong Ju Li, and Zhong Lin Wang. 2017. "Triboelectric Nanogenerator Enhanced Nanofiber Air Filters for Efficient Particulate Matter Removal." *ACS Nano* 11 (6): 6211–7.

Guo, Hengyu, Xianjie Pu, Jie Chen, Yan Meng, Min-Hsin Yeh, Guanlin Liu, Qian Tang, et al. 2018. "A Highly Sensitive, Self-Powered Triboelectric Auditory Sensor for Social Robotics and Hearing Aids." *Science Robotics* 3 (20): eaat2516.

Gurlo, Aleksander. 2011. "Nanosensors: Towards Morphological Control of Gas Sensing Activity. SnO_2, In_2O_3, ZnO and WO_3 Case Studies." *Nanoscale* 3 (1): 154–65.

Han, Chang Bao, Tao Jiang, Chi Zhang, Xiaohui Li, Chaoying Zhang, Xia Cao, and Zhong Lin Wang. 2015. "Removal of Particulate Matter Emissions from a Vehicle Using a Self-Powered Triboelectric Filter." *ACS Nano* 9 (12): 12552–61.

He, Chuan, and Zhong Lin Wang. 2018. "Triboelectric Nanogenerator as a New Technology for Effective PM2.5 Removing with Zero Ozone Emission." *Progress in Natural Science: Materials International* 28 (2): 99–112.

He, Xu, Yunlong Zi, Hengyu Guo, Haiwu Zheng, Yi Xi, Changsheng Wu, Jie Wang, Wei Zhang, Canhui Lu, and Zhong Lin Wang. 2017. "A Highly Stretchable Fiber-Based Triboelectric Nanogenerator for Self-Powered Wearable Electronics." *Advanced Functional Materials* 27 (4): 1604378–84.

Hsiao, Chun Ching, An Shen Siao, and Jing Chih Ciou. 2012. "Improvement of Pyroelectric Cells for Thermal Energy Harvesting." *Sensors* 12 (1): 534–48.

Hu, Caixia, Li Cheng, Zhe Wang, Youbin Zheng, Suo Bai, and Yong Qin. 2016. "A Transparent Antipeep Piezoelectric Nanogenerator to Harvest Tapping Energy on Screen." *Small* 12 (10): 1315–21.

Hu, Youfan, and Zhong Lin Wang. 2014. "Recent Progress in Piezoelectric Nanogenerators as a Sustainable Power Source in Self-Powered Systems and Active Sensors." *Nano Energy* 14: 3–14.

Huang, K., J. B. Wang, X. L. Zhong, B. L. Liu, T. Chen, and Y. C. Zhou. 2012. "Significant Polarization Variation Near Room Temperature of $Ba_{0.65}Sr_{0.35}TiO_3$ Thin Films for Pyroelectric Energy Harvesting." *Sensors and Actuators, B: Chemical* 169 (July): 208–12.

Indira, Sridhar Sripadmanabhan, Chockalingam Aravind Vaithilingam, Kameswara Satya Prakash Oruganti, Faizal Mohd, and SaidurRahman. 2019. "Nanogenerators as a Sustainable Power Source: State of Art, Applications, and Challenges." *Nanomaterials* 9: 773–807.

JHLee, Ju Hyuck, Keun Young Lee, Manoj Kumar Gupta, Tae Yun Kim, Dae Yeong Lee, Junho Oh, Changkook Ryu, et al. 2014a. "Highly Stretchable Piezoelectric-Pyroelectric Hybrid Nanogenerator." *Advanced Materials* 26 (5): 765–69.

Jiang, Shenglin, Pin Liu, Xiaoshan Zhang, Ling Zhang, Qi Li, Junlong Yao, Yike Zeng, Qing Wang, and Guangzu Zhang. 2015. "Enhanced Pyroelectric Properties of Porous $Ba_{0.67}Sr_{0.33}TiO_3$ Ceramics Fabricated with Carbon Nanotubes." *Journal of Alloys and Compounds* 636 (July): 93–6.

Jiang, Tao, Yanyan Yao, Liang Xu, Limin Zhang, Tianxiao Xiao, and Zhong Lin Wang. 2017. "Spring-Assisted Triboelectric Nanogenerator for Efficiently Harvesting Water Wave Energy." *Nano Energy* 31 (December 2016): 560–7.

Kagan, Cherie R., Laura E. Fernandez, Yury Gogotsi, Paula T. Hammond, Mark C. Hersam, André E. Nel, Reginald M. Penner, C. Grant Willson, and Paul S. Weiss. 2016. "Nano Day: Celebrating the Next Decade of Nanoscience and Nanotechnology." *ACS Nano* 10 (10): 9093–103.

Ko, Young Joon, Dong Yeong Kim, Sung Sik Won, Chang Won Ahn, Ill Won Kim, Angus I. Kingon, Seung Hyun Kim, Jae Hyeon Ko, and Jong Hoon Jung. 2016. "Flexible $Pb(Zr_{0.52}Ti_{0.48})O_3$ Films for a Hybrid Piezoelectric-Pyroelectric Nanogenerator under Harsh Environments." *ACS Applied Materials and Interfaces* 8 (10): 6504–11.

Kotipalli, Venu, Zhongcheng Gong, Pushparaj Pathak, Tianhua Zhang, Yuan He, Shashi Yadav, and Long Que. 2010. "Light and Thermal Energy Cell Based on Carbon Nanotube Films." *Applied Physics Letters* 97 (12): 124102–4.

Kumar, Anuruddh, Anshul Sharma, Rajeev Kumar, Rahul Vaish, and Vishal S. Chauhan. 2014. "Finite Element Analysis of Vibration Energy Harvesting Using Lead-Free Piezoelectric Materials: A Comparative Study." *Journal of Asian Ceramic Societies* 2 (2): 138–43.

Kumar, Brijesh, and Sang Woo Kim. 2012. "Energy Harvesting Based on Semiconducting Piezoelectric ZnO Nanostructures." *Nano Energy* 1: 342–55.

Kwi Il Park, Jung Hwan Son, Geon Tae Hwang, Chang Kyu Jeong, Jungho Ryu, Min Koo, Insung Choi, et al. 2014. "Highly-Efficient, Flexible Piezoelectric PZT Thin Film Nanogenerator on Plastic Substrates." *Advanced Materials* 26 (16): 2514–20.

Lee, Ju Hyuck, Keun Young Lee, Manoj Kumar Gupta, Tae Yun Kim, Dae Yeong Lee, Junho Oh, Changkook Ryu, et al. 2014a. "Highly Stretchable Piezoelectric-Pyroelectric Hybrid Nanogenerator." *Advanced Materials* 26 (5): 765–9.

Li, Anyin, Yunlong Zi, Hengyu Guo, Zhong Lin Wang, and Facundo M. Fernández. 2017. "Triboelectric Nanogenerators for Sensitive Nano-Coulomb Molecular Mass Spectrometry." *Nature Nanotechnology* 12 (5): 481–7.

Li, C. C., Z. F. Du, L. M. Li, H. C. Yu, Q. Wan, and T. H. Wang. 2007. "Surface-Depletion Controlled Gas Sensing of ZnO Nanorods Grown at Room Temperature." *Applied Physics Letters* 91 (3): 2005–8.

Li, Shengming, Wenbo Peng, Jie Wang, Long Lin, Yunlong Zi, Gong Zhang, and Zhong Lin Wang. 2016. "All-Elastomer-Based Triboelectric Nanogenerator as a Keyboard Cover to Harvest Typing Energy." *ACS Nano* 10 (8): 7973–81.

Li, Xiuhan, Zong Hong Lin, Gang Cheng, Xiaonan Wen, Ying Liu, Simiao Niu, and Zhong Lin Wang. 2014a. "3D Fiber-Based Hybrid Nanogenerator for Energy Harvesting and as a Self-Powered Pressure Sensor." *ACS Nano* 8 (10): 10674–81.

Li, Zhaoling, Jun Chen, Jin Yang, Yuanjie Su, Xing Fan, Ying Wu, Chongwen Yu, and Zhong Lin Wang. 2015. "β-Cyclodextrin Enhanced Triboelectrification for Self-Powered Phenol Detection and Electrochemical Degradation." *Energy and Environmental Science* 8 (3): 887–96.

Lin, Long, Youfan Hu, Chen Xu, Yan Zhang, Rui Zhang, Xiaonan Wen, and Zhong Lin Wang. 2013. "Transparent Flexible Nanogenerator as Self-Powered Sensor for Transportation Monitoring." *Nano Energy* 2 (1): 75–81.

Lin, Long, Yannan Xie, Sihong Wang, Wenzhuo Wu, Simiao Niu, Xiaonan Wen, and Zhong Lin Wang. 2013. "Triboelectric Active Sensor Array for Self-Powered Static and Dynamic Pressure Detection and Tactile Imaging." *ACS Nano* 7 (9): 8266–74.

Lin, Zhiming, Jun Chen, and Jin Yang. 2016. "Recent Progress in Triboelectric Nanogenerators as a Renewable and Sustainable Power Source." *Journal of Nanomaterials* 2016: 1–24.

Liu, Ruiyuan, Xiao Kuang, Jianan Deng, Yi Cheng Wang, Aurelia C. Wang, Wenbo Ding, Ying Chih Lai, et al. 2018. "Shape Memory Polymers for Body Motion Energy Harvesting and Self-Powered Mechanosensing." *Advanced Materials* 30 (8): 1705195–202.

Liu, Zunfeng, Qian Liu, Yi Huang, Yanfeng Ma, Shougen Yin, Xiaoyan Zhang, Wei Sun, and Yongsheng Chen. 2008. "Organic Photovoltaic Devices Based on a Novel Acceptor Material: Graphene." *Advanced Materials* 20 (20): 3924–30.

Lu, Ming Pei, Jinhui Song, Ming Yen Lu, Min Teng Chen, Yifan Gao, Lih Juann Chen, and Zhong Lin Wang. 2009. "Piezoelectric Nanogenerator Using P-Type ZnO Nanowire Arrays." *Nano Letters* 9 (3): 1223–7.

Ma, J., Huifen Wong, L. B. Kong, and K. W. Peng. 2003. "Biomimetic Processing of Nanocrystallite Bioactive Apatite Coating on Titanium." *Nanotechnology* 14 (6): 619–23.

Mane, Poorna, Jingsi Xie, Kam K. Leang, and Karla Mossi. 2011. "Cyclic Energy Harvesting from Pyroelectric Materials." *IEEE Transactions on Ultrasonics, Ferroelectrics, and Frequency Control* 58 (1): 10–7.

McKinley, Ian M., and Laurent Pilon. 2013. "Phase Transitions and Thermal Expansion in Pyroelectric Energy Conversion." *Applied Physics Letters* 102 (2): 023906–10.

Meng, Bo, Xiaoliang Cheng, Xiaosheng Zhang, Mengdi Han, Wen Liu, and Haixia Zhang. 2014. "Single-Friction-Surface Triboelectric Generator with Human Body Conduit." *Applied Physics Letters* 104 (10): 103904–8.

MM Alam, Md Mehebub, Sujoy Kumar Ghosh, Ayesha Sultana, and Dipankar Mandal. 2015. "Lead-Free ZnSnO3/MWCNTs-Based Self-Poled Flexible Hybrid Nanogenerator for Piezoelectric Power Generation." *Nanotechnology* 26 (16). IOP Publishing.

Molday, Robert S., and Donald Mackenzie. 1982. "Immunospecific Ferromagnetic Iron-Dextran Reagents for the Labeling and Magnetic Separation of Cells." *Journal of Immunological Methods* 52 (3): 353–67.

Monrose, Fabian, and Aviel Rubin. 2009. "Keystroke Dynamics as a Biometric." *University of Southampton* 16: 93.

Nam, Jwa-Min, C. Shad Thaxton, and Chad A. Mirkin. 2003. "Nanoparticle-Based Bio – Bar Codes for the Ultrasensitive." *Science* 301 (September): 1884–7.

Narita, Fumio, and Marina Fox. 2018. "A Review on Piezoelectric, Magnetostrictive, and Magnetoelectric Materials and Device Technologies for Energy Harvesting Applications." *Advanced Engineering Materials* 20: 1700743–64.

Nechibvute, Action, Albert Chawanda, and Pearson Luhanga. 2012. "Piezoelectric Energy Harvesting Devices: An Alternative Energy Source for Wireless Sensors." *Smart Materials Research* 2012 (May): 1–13.

Nie, Jinhui, Tao Jiang, Jiajia Shao, Zewei Ren, Yu Bai, Mitsumasa Iwamoto, Xiangyu Chen, and Zhong Lin Wang. 2018. "Motion Behavior of Water Droplets Driven by Triboelectric Nanogenerator." *Applied Physics Letters* 112 (18): 183701–5.

Nie, Yuxin, Ping Deng, Yayu Zhao, Penglei Wang, Lili Xing, Yan Zhang, and Xinyu Xue. 2014. "The Conversion of PN-Junction Influencing the Piezoelectric Output of a CuO/ZnO Nanoarray Nanogenerator and Its Application as a Room-Temperature Self-Powered Active H_2S Sensor." *Nanotechnology* 25 (26): 265501–9.

Niu, Simiao, and Zhong Lin Wang. 2014. "Theoretical Systems of Triboelectric Nanogenerators." *Nano Energy* 14: 161–92.

Noliac. n.d. "Piezo Basics – Tutorial." 1–19.

Ohji, T., and M. Fukushima. 2012. "Macro-Porous Ceramics: Processing and Properties." *International Materials Reviews* 57 (2): 115–31.

Ornes, Stephen. 2016. "The Internet of Things and the Explosion of Interconnectivity." *Proceedings of the National Academy of Sciences of the United States of America* 113 (40): 11059–60.

Parak, Wolfgang J., Daniele Gerion, Teresa Pellegrino, Daniela Zanchet, Christine Micheel, Shara C. Williams, Rosanne Boudreau, Mark A. Le Gros, Carolyn A. Larabell, and A. Paul Alivisatos. 2003. "Biological Applications of Colloidal Nanocrystals." *Nanotechnology* 14 (7): 070015–27.

Park, Jiwon, A. Young Choi, Chang Jun Lee, Dogyun Kim, and Youn Tae Kim. 2017. "Highly Stretchable Fiber-Based Single-Electrode Triboelectric Nanogenerator for Wearable Devices." *RSC Advances* 7 (86): 54829–34.

Park, Jiwon, Dogyun Kim, A. Young Choi, and Youn Tae Kim. 2018. "Flexible Single-Strand Fiber-Based Woven-Structured Triboelectric Nanogenerator for Self-Powered Electronics." *APL Materials* 6 (10): 101106–12.

Park, Kwi Il, Jung Hwan Son, Geon Tae Hwang, Chang Kyu Jeong, Jungho Ryu, Min Koo, Insung Choi, et al. 2014. "Highly-Efficient, Flexible Piezoelectric PZT Thin Film Nanogenerator on Plastic Substrates." *Advanced Materials* 26 (16): 2514–20.

Patel, Satyanarayan, Aditya Chauhan, and Rahul Vaish. 2016. "Large Pyroelectric Figure of Merits for Sr-Modified $Ba_{0.85}Ca_{0.15}Zr_{0.1}Ti_{0.9}O_3$ Ceramics." *Solid State Sciences* 52: 10–8.

Potnuru, Akshay, and Yonas Tadesse. 2014. "Characterization of Pyroelectric Materials for Energy Harvesting from Human Body." *Integrated Ferroelectrics* 150 (1): 23–50.

Pu, Xianjie, Hengyu Guo, Jie Chen, Xue Wang, Yi Xi, Chenguo Hu, and Zhong Lin Wang. 2017. "Eye Motion Triggered Self-Powered Mechnosensational Communication System Using Triboelectric Nanogenerator." *Science Advances* 3 (7): 1–8.

Pullano, Salvatore Andrea, Syed Kamrul Islam, and Antonino S. Fiorillo. 2014. "Pyroelectric Sensor for Temperature Monitoring of Biological Fluids in Microchannel Devices." *IEEE Sensors Journal* 14 (8): 2725–30.

Rathore, Saurabh, Shailendra Sharma, Bibhu P. Swain, and Ranjan Kr Ghadai. 2018. "A Critical Review on Triboelectric Nanogenerator." *IOP Conference Series: Materials Science and Engineering* 377 (1): 012186–201.

Ravindran, S. K. T., T. Huesgen, M. Kroener, and P. Woias. 2011. "A Self-Sustaining Micro Thermomechanic-Pyroelectric Generator." *Applied Physics Letters* 99 (10): 1–4.

Roji, Ani Melfa M., G. Jiji, and Ajith Bosco T. Raj. 2017. "A Retrospect on the Role of Piezoelectric Nanogenerators in the Development of the Green World." *RSC Advances* 7: 33642–70.

Ryu, Hanjun, Hong Joon Yoon, and Sang Woo Kim. 2019. "Hybrid Energy Harvesters: Toward Sustainable Energy Harvesting." *Advanced Materials* 31 (34): 1–19.

Schlatmann, A. R., D. Wilms Floet, A. Hilberer, F. Garten, P. J. M. Smulders, T. M. Klapwijk, and G. Hadziioannou. 1996. "Indium Contamination from the Indium-Tin-Oxide Electrode in Polymer Light-Emitting Diodes." *Applied Physics Letters* 69 (12): 1764–6.

Sebald, Gael, Laurence Seveyrat, Daniel Guyomar, Laurent Lebrun, Benoit Guiffard, and Sebastien Pruvost. 2006. "Electrocaloric and Pyroelectric Properties of $0.75Pb(Mg_{1/3}Nb_{2/3})O_3$-$0.25PbTiO_3$ Single Crystals." *Journal of Applied Physics* 100 (12): 124112–7.

Sharma, Anshul, Rajeev Kumar, Rahul Vaish, and Vishal S. Chauhan. 2015. "Performance of $K_{0.5}Na_{0.5}NbO_3$ (KNN)-Based Lead-Free Piezoelectric Materials in Active Vibration Control." *International Journal of Applied Ceramic Technology* 12 (S1): E64–72.

Su, Yuanjie, Jun Chen, Zhiming Wu, and Yadong Jiang. 2015. "Low Temperature Dependence of Triboelectric Effect for Energy Harvesting and Self-Powered Active Sensing." *Applied Physics Letters* 106 (1): 013114–8.

Sultana, Ayesha, Md Mehebub Alam, Tapas Ranjan Middya, and Dipankar Mandal. 2018. "A Pyroelectric Generator as a Self-Powered Temperature Sensor for Sustainable Thermal Energy Harvesting from Waste Heat and Human Body Heat." *Applied Energy* 221 (March): 299–307.

Sun, Renbing, Jinzhi Wang, Fang Wang, Tangfu Feng, Yanlong Li, Zhenhua Chi, Xiangyong Zhao, and Haosu Luo. 2014. "Pyroelectric Properties of Mn-Doped $94.6Na_{0.5}Bi_{0.5}TiO_3$-$5.4BaTiO_3$ Lead-Free Single Crystals." *Journal of Applied Physics* 115 (7): 74101–4.

Suyal, G., and N. Setter. 2004. "Enhanced Performance of Pyroelectric Microsensors through the Introduction of Nanoporosity." *Journal of the European Ceramic Society* 24 (2): 247–51.

Wang, Jiyu, Wenbo Ding, Lun Pan, Changsheng Wu, Hua Yu, Lijun Yang, Ruijin Liao, and Zhong Lin Wang. 2018. "Self-Powered Wind Sensor System for Detecting Wind Speed and Direction Based on a Triboelectric Nanogenerator." *ACS Nano* 12 (4): 3954–63.

Wang, Shuhua, Zhong Lin Wang, and Ya Yang. 2016. "A One-Structure-Based Hybridized Nanogenerator for Scavenging Mechanical and Thermal Energies by Triboelectric-Piezoelectric-Pyroelectric Effects." *Advanced Materials* 28 (15): 2881–7.

Wang, Xiaofeng, Simiao Niu, Yajiang Yin, Fang Yi, Zheng You, and Zhong Lin Wang. 2015. "Triboelectric Nanogenerator Based on Fully Enclosed Rolling Spherical Structure for Harvesting Low-Frequency Water Wave Energy." *Advanced Energy Materials* 5 (24): 1–9.

Wang, Xudong. 2012. "Piezoelectric Nanogenerators-Harvesting Ambient Mechanical Energy at the Nanometer Scale." *Nano Energy* 1 (1): 13–24.

Wang, Zhong Lin. 2013. "Triboelectric Nanogenerators as New Energy Technology for Self-Powered Systems and as Active Mechanical and Chemical Sensors." *ACS Nano* 7 (11): 9533–57.

Wang, Zhong Lin, Jun Chen, and Long Lin. 2015. "Progress in Triboelectric Nanogenerators as a New Energy Technology and Self-Powered Sensors." *Energy and Environmental Science* 8 (8): 2250–82.

Wang, Zhong Lin, Tao Jiang, and Liang Xu. 2017. "Toward the Blue Energy Dream by Triboelectric Nanogenerator Networks." *Nano Energy* 39 (June): 9–23.

Wang, Zhong Lin, and Jinhui Song. 2006. "Piezoelectric Nanogenerators Based on Zinc Oxide Nanowire Arrays." *Science* 312 (5771): 242–6.

Wen, Xiaonan, Weiqing Yang, Qingshen Jing, and Zhong Lin Wang. 2014. "Harvesting Broadband Kinetic Impact Energy from Mechanical Triggering/Vibration and Water Waves." *ACS Nano* 8 (7): 7405–12.

Wen, Zhen, Min Hsin Yeh, Hengyu Guo, Jie Wang, Yunlong Zi, Weidong Xu, Jianan Deng, et al. 2016. "Self-Powered Textile for Wearable Electronics by Hybridizing Fiber-Shaped Nanogenerators, Solar Cells, and Supercapacitors." *Science Advances* 2 (10): 1600097–104.

Wu, Changsheng, Wenbo Ding, Ruiyuan Liu, Jiyu Wang, Aurelia C. Wang, Jie Wang, Shengming Li, Yunlong Zi, and Zhong Lin Wang. 2018. "Keystroke Dynamics Enabled Authentication and Identification Using Triboelectric Nanogenerator Array." *Materials Today* 21 (3): 216–22.

Wu, Changsheng, Ruiyuan Liu, Jie Wang, Yunlong Zi, Long Lin, and Zhong Lin Wang. 2017. "A Spring-Based Resonance Coupling for Hugely Enhancing the Performance of Triboelectric Nanogenerators for Harvesting Low-Frequency Vibration Energy." *Nano Energy* 32 (December 2016): 287–93.

Wu, Changsheng, Aurelia C. Wang, Wenbo Ding, Hengyu Guo, and Zhong Lin Wang. 2019. "Triboelectric Nanogenerator: A Foundation of the Energy for the New Era." *Advanced Energy Materials* 9 (1): 1–25.

Wu, Changsheng, Xin Wang, Long Lin, Hengyu Guo, and Zhong Lin Wang. 2016. "Paper-Based Triboelectric Nanogenerators Made of Stretchable Interlocking Kirigami Patterns." *ACS Nano* 10 (4): 4652–9.

Wu, Fan, Congju Li, Yingying Yin, Ran Cao, Hui Li, Xiuling Zhang, Shuyu Zhao, et al. 2019. "A Flexible, Lightweight, and Wearable Triboelectric Nanogenerator for Energy Harvesting and Self-Powered Sensing." *Advanced Materials Technologies* 4 (1): 1–7.

Xiong, Jiaqing, Hongsheng Luo, Dace Gao, Xinran Zhou, Peng Cui, Gurunathan Thangavel, Kaushik Parida, and Pooi See Lee. 2019. "Self-Restoring, Waterproof, Tunable Microstructural Shape Memory Triboelectric Nanogenerator for Self-Powered Water Temperature Sensor." *Nano Energy* 61 (July): 584–93.

Xiuhan Li, Zong Hong Lin, Gang Cheng, Xiaonan Wen, Ying Liu, Simiao Niu, and Zhong Lin Wang. 2014. "3D Fiber-Based Hybrid Nanogenerator for Energy Harvesting and as a Self-Powered Pressure Sensor." ACS Nano 8 (10): 10674–81

Xu, Liang, Tao Jiang, Pei Lin, Jia Jia Shao, Chuan He, Wei Zhong, Xiang Yu Chen, and Zhong Lin Wang. 2018. "Coupled Triboelectric Nanogenerator Networks for Efficient Water Wave Energy Harvesting." *ACS Nano* 12 (2): 1849–58.

Xue, Hao, Quan Yang, Dingyi Wang, Weijian Luo, Wenqian Wang, Mushun Lin, Dingli Liang, and Qiming Luo. 2017. "A Wearable Pyroelectric Nanogenerator and Self-Powered Breathing Sensor." *Nano Energy* 38 (August): 147–54.

Yang, Jin, Jun Chen, Ying Liu, Weiqing Yang, Yuanjie Su, and Zhong Lin Wang. 2014. "Triboelectrification-Based Organic Film Nanogenerator for Acoustic Energy Harvesting and Self-Powered Active Acoustic Sensing." *ACS Nano* 8 (3): 2649–57.

Yang, Rusen, Yong Qin, Cheng Li, Guang Zhu, and Zhong Lin Wang. 2009. "Converting Biomechanical Energy into Electricity by a Muscle-Movement-Driven Nanogenerator." *Nano Letters* 9 (3): 1201–5.

Yang, Ya, Wenxi Guo, Ken C. Pradel, Guang Zhu, Yusheng Zhou, Yan Zhang, Youfan Hu, Long Lin, and Zhong Lin Wang. 2012. "Pyroelectric Nanogenerators for Harvesting Thermoelectric Energy." *Nano Letters* 12 (6): 2833–8.

Yang, Ya, Jong Hoon Jung, Byung Kil Yun, Fang Zhang, Ken C. Pradel, Wenxi Guo, and Zhong Lin Wang. 2012. "Flexible Pyroelectric Nanogenerators Using a Composite Structure of Lead-Free $KNbO_3$ Nanowires." *Advanced Materials* 24 (39): 5357–62.

Yang, Ya, Sihong Wang, Yan Zhang, and Zhong Lin Wang. 2012. "Pyroelectric Nanogenerators for Driving Wireless Sensors." *Nano Letters* 12 (12): 6408–13.

Yang, Ya, Hulin Zhang, Ruoyu Liu, Xiaonan Wen, Te Chien Hou, and Zhong Lin Wang. 2013. "Fully Enclosed Triboelectric Nanogenerators for Applications in Water and Harsh Environments." *Advanced Energy Materials* 3 (12): 1563–8.

Yang, Ya, Hulin Zhang, Guang Zhu, Sangmin Lee, Zong Hong Lin, and Zhong Lin Wang. 2013. "Flexible Hybrid Energy Cell for Simultaneously Harvesting Thermal, Mechanical, and Solar Energies." *ACS Nano* 7 (1): 785–90.

Yang, Ya, Yusheng Zhou, Jyh Ming Wu, and Zhong Lin Wang. 2012. "Micro, Single, Nanowire Pyroelectric Nanogenerators, and Self-Powered Temperature Sensors." *Single Micro/Nanowire Pyroelectric* 9: 8456–61.

Yang, Ya, Guang Zhu, Hulin Zhang, Jun Chen, Xiandai Zhong, Zong-Hong Lin, Yuanjie Su, and Peng Bai. 2013. "Triboelectric Nanogenerator for Harvesting Wind Energy and as Self-Powered Wind Vector Sensor System." *ACS Nano* 7 (10): 9461–8.

Yayu Zhao, Ping Deng, Yuxin Nie, Penglei Wang, Yan Zhang, Lili Xing, and Xinyu Xue. 2014. "Biomolecule-Adsorption-Dependent Piezoelectric Output of ZnO Nanowire Nanogenerator and Its Application as Self-Powered Active Biosensor." *Biosensors and Bioelectronics* 57. Elsevier: 269–75.

Yeh, Min-Hsin, Long Lin, Po-Kang Yang, and Zhong Lin Wang. 2015. "Motion-Driven Electrochromic Reactions for Self-Powered Smart Window System." *ACS Nano* 9: 4757–65.

Yi, Fang, Xiaofeng Wang, Simiao Niu, Shengming Li, Yajiang Yin, Keren Dai, Guangjie Zhang, et al. 2016. "A Highly Shape-Adaptive, Stretchable Design Based on Conductive Liquid for Energy Harvesting and Self-Powered Biomechanical Monitoring." *Science Advances* 2 (6): 1501624–33.

You, Ming Hao, Xiao Xiong Wang, Xu Yan, Jun Zhang, Wei Zhi Song, Miao Yu, Zhi Yong Fan, Seeram Ramakrishna, and Yun Ze Long. 2018. "A Self-Powered Flexible Hybrid Piezoelectric-Pyroelectric Nanogenerator Based on Non-Woven Nanofiber Membranes." *Journal of Materials Chemistry A* 6 (8): 3500–9.

Yu, Aifang, Ming Song, Yan Zhang, Yang Zhang, Libo Chen, Junyi Zhai, and Zhong Lin Wang. 2015. "Self-Powered Acoustic Source Locator in Underwater Environment Based on Organic Film Triboelectric Nanogenerator." *Nano Research* 8 (3): 765–73.

Zhang, Guangzu, Shenglin Jiang, Yike Zeng, Yangyang Zhang, Qingfeng Zhang, and Yan Yu. 2009. "High Pyroelectric Properties of Porous $Ba_{0.67}Sr_{0.33}TiO_3$ for Uncooled Infrared Detectors." *Journal of the American Ceramic Society* 92 (12): 3132–4.

Zhang, Haibo, Shenglin Jiang, and Koji Kajiyoshi. 2010. "Enhanced Pyroelectric and Piezoelectric Figure of Merit of Porous $Bi_{0.5}(Na_{0.82}K_{0.18})_{0.5}TiO_3$ Lead-Free Ferroelectric Thick Films." *Journal of the American Ceramic Society* 93 (7): 1957–64.

Zhang, Hulin, Yuhang Xie, Xiaomei Li, Zhenlong Huang, Shangjie Zhang, Yuanjie Su, Bo Wu, Long He, Weiqing Yang, and Yuan Lin. 2016. "Flexible Pyroelectric Generators for Scavenging Ambient Thermal Energy and as Self-Powered Thermosensors." *Energy* 101 (April): 202–10.

Zhang, Min, Tao Gao, Jianshu Wang, Jianjun Liao, Yingqiang Qiu, Quan Yang, Hao Xue, et al. 2015. "A Hybrid Fibers Based Wearable Fabric Piezoelectric Nanogenerator for Energy Harvesting Application." *Nano Energy* 13: 298–305.

Zhang, Q., S. Corkovic, C. P. Shaw, Z. Huang, and R. W. Whatmore. 2005. "Effect of Porosity on the Ferroelectric Properties of Sol-Gel Prepared Lead Zirconate Titanate Thin Films." *Thin Solid Films* 488 (1–2): 258–64.

Zhang, Yan, Yinxiang Bao, Dou Zhang, and Chris R. Bowen. 2015. "Porous PZT Ceramics with Aligned Pore Channels for Energy Harvesting Applications." *Journal of the American Ceramic Society* 98 (10): 2980–3.

Zhao, Kun, Zhong Lin Wang, and Ya Yang. 2016. "Self-Powered Wireless Smart Sensor Node Enabled by an Ultrastable, Highly Efficient, and Superhydrophobic-Surface-Based Triboelectric Nanogenerator."*ACS Nano* 10 (9): 9044–52.

Zhao, Tingting, Weitao Jiang, Dong Niu, Hongzhong Liu, Bangdao Chen, Yongsheng Shi, Lei Yin, and Bingheng Lu. 2017. "Flexible Pyroelectric Device for Scavenging Thermal Energy from Chemical Process and as Self-Powered Temperature Monitor." *Applied Energy* 195: 754–60.

Zhao, Yayu, Ping Deng, Yuxin Nie, Penglei Wang, Yan Zhang, Lili Xing, and Xinyu Xue. 2014. "Biomolecule-Adsorption-Dependent Piezoelectric Output of ZnO Nanowire Nanogenerator and Its Application as Self-Powered Active Biosensor." *Biosensors and Bioelectronics* 57: 269–75.

Zheng, Li, Zong Hong Lin, Gang Cheng, Wenzhuo Wu, Xiaonan Wen, Sangmin Lee, and Zhong Lin Wang.2014. "Silicon-Based Hybrid Cell for Harvesting Solar Energy and Raindrop Electrostatic Energy." *Nano Energy* 9: 291–300.

Zheng, Li, Yali Wu, Xiangyu Chen, Aifang Yu, Liang Xu, Yongsheng Liu, Hexing Li, and Zhong Lin Wang. 2017. "Self-Powered Electrostatic Actuation Systems for Manipulating the Movement of Both

Microfluid and Solid Objects by Using Triboelectric Nanogenerator." *Advanced Functional Materials* 27 (16): 1606408–17.

Zheng, Qiang, Yang Zou, Yalan Zhang, Zhuo Liu, Bojing Shi, Xinxin Wang, Yiming Jin, Han Ouyang, Zhou Li, and Zhong Lin Wang. 2016. "Biodegradable Triboelectric Nanogenerator as a Life-Time Designed Implantable Power Source." *Science Advances* 2 (3): 1–10.

Zhong, Junwen, Yan Zhang, Qize Zhong, Qiyi Hu, Bin Hu, Zhong Lin Wang, and Jun Zhou. 2014. "Fiber-Based Generator for Wearable Electronics and Mobile Medication." *ACS Nano* 8 (6): 6273–80.

Zhou, Linglin, Di Liu, Jie Wang, and Zhong Lin Wang. 2020. "Triboelectric Nanogenerators: Fundamental Physics and Potential Applications." *Friction* 8: 481–506.

Zhou, Tao, Limin Zhang, Fei Xue, Wei Tang, Chi Zhang, and Zhong Lin Wang. 2016. "Multilayered Electret Films Based Triboelectric Nanogenerator." *Nano Research* 9 (5): 1442–51.

Zhou, Yu Sheng, Guang Zhu, Simiao Niu, Ying Liu, Peng Bai, Qingsheng Jing, and Zhong Lin Wang. 2014. "Nanometer Resolution Self-Powered Static and Dynamic Motion Sensor Based on Micro-Grated Triboelectrification." *Advanced Materials* 26 (11): 1719–24.

Zhu, Guang, Jun Chen, Ying Liu, Peng Bai, Yu Sheng Zhou, Qingshen Jing, Caofeng Pan, and Zhong Lin Wang. 2013. "Linear-Grating Triboelectric Generator Based on Sliding Electrification." *Nano Letters* 13 (5): 2282–9.

Zhu, Guang, Zong Hong Lin, Qingshen Jing, Peng Bai, Caofeng Pan, Ya Yang, Yusheng Zhou, and Zhong Lin Wang. 2013. "Toward Large-Scale Energy Harvesting by a Nanoparticle-Enhanced Triboelectric Nanogenerator." *Nano Letters* 13 (2): 847–53.

Zhu, Guang, Wei Qing Yang, Tiejun Zhang, Qingshen Jing, Jun Chen, Yu Sheng Zhou, Peng Bai, and Zhong Lin Wang. 2014. "Self-Powered, Ultrasensitive, Flexible Tactile Sensors Based on Contact Electrification." *Nano Letters* 14 (6): 3208–13.

Zhu, Huarui, Ying Xu, Yu Han, Shuwen Chen, Tao Zhou, Magnus Willander, Xia Cao, and Zhonglin Wang. 2015. "Self-Powered Electrochemical Anodic Oxidation: A New Method for Preparation of Mesoporous Al_2O_3 without Applying Electricity." *Nano Research* 8 (11): 3604–11.

Zi, Yunlong, Hengyu Guo, Zhen Wen, Min Hsin Yeh, Chenguo Hu, and Zhong Lin Wang. 2016. "Harvesting Low-Frequency (<5 Hz) Irregular Mechanical Energy: A Possible Killer Application of Triboelectric Nanogenerator." *ACS Nano* 10 (4): 4797–805.

Zi, Yunlong, Long Lin, Jie Wang, Sihong Wang, Jun Chen, Xing Fan, Po Kang Yang, Fang Yi, and Zhong Lin Wang. 2015. "Triboelectric-Pyroelectric-Piezoelectric Hybrid Cell for High-Efficiency Energy-Harvesting and Self-Powered Sensing." *Advanced Materials* 27 (14): 2340–7.

Zi, Yunlong, Simiao Niu, Jie Wang, Zhen Wen, Wei Tang, and Zhong Lin Wang. 2015. "Standards and Figure-of-Merits for Quantifying the Performance of Triboelectric Nanogenerators." *Nature Communications* 6 (September): 8376–83.

Zi, Yunlong, and Zhong Lin Wang. 2017. "Nanogenerators: An Emerging Technology towards Nanoenergy." *APL Materials* 5 (7): 074103–15.

15 Electroactive Polymers and Their Carbon Nanocomposites for Energy Harvesting

Ramanujam B.T.S., Reshma Haridass, Pranesh Muralidharan, Ashok Kumar Nanjundan, Deepak Dubal, and Pratheep K. Annamalai

CONTENTS

15.1 INTRODUCTION

The global energy demand is constantly increasing due to the growing population and emerging advancements in technologies and is estimated to rise by 50% in the year 2050 [1]. On the other hand, the depletion of conventional sources of energy like fossil fuels, coal, etc., has raised alarms. Hence, there is a need to invest in clean energy technologies. In this regard, the use of advanced materials for energy harvesting and energy storage applications is explored as the probable solution to address the hiking energy demands. Recently, advanced functional materials have been synthesized to design and fabricate new devices or technological products that can harvest energies from light, heat, vibrations, the flow of fluids, etc. This is achieved through designing devices to convert ambient energies in the form of utilizable electrical power through piezoelectric, thermoelectric, and pyroelectric effects. In such applications, the choice of the material plays a crucial

DOI: 10.1201/9781003187615-15

role in determining the efficiency of energy harvesting devices. Conventional piezoceramics such as barium titanate and lead zirconate titanate are less ductile and tend to crack. Furthermore, the processing of those ceramic materials in the form of a thin film increases the cost. Hence, electroactive polymers (EAPs) and polymer nanocomposites, which are flexible, easily processible, and cost-effective, have become the focus of intense research [2]. Energy harvesting can be effectively achieved using EAPs by enhancing their properties such as piezoelectricity [3], electrical conductivity [4], etc. For the development of smart, portable, and wearable electronics, harnessing energy from nonconventional sources using EAPs has become a preferable option to provide an extended life for the energy harvester [5].

Insulating polymers such as polyvinylidene fluoride (PVDF) and its copolymers (polyvinylidene fluoride-trifluoroethylene [PVDF-TrFE] and polyvinylidene fluoride-hexafluoropropylene [PVDF-HFP]) are extensively being used for mechanical energy harvesting application due to their flexibility, enhanced piezoelectric performance, mechanical strength, chemical resistance, ease of processing, non-toxicity, etc. The significance of those polymers and their nanocomposites for energy harvesting applications can be gauged from the statistical data of increasing research publications every year as depicted in Figure 15.1(a) (simple search words of 'PVDF-based mechanical energy harvesting' in sciencedirect.com yields a number of research papers published in different years, which are used for plotting). PVDF exists in five different chain conformations to form α, β, γ, δ, and ε crystalline phases. The electroactive β and γ phases exhibit piezoelectric, ferroelectric, and pyroelectric properties [6]. As reported in the literature, the effective solution to enhance the efficiency of polymer-based mechanical energy harvesters is to enhance their piezoelectric performance [7]. The piezoelectric properties, electroactive β phase of PVDF, and the power generation from the PVDF-based nanocomposites are correlated to each other. Hence, various studies have been focused on the maximum conversion of nonpolar α phase to polar β phase to enhance the piezoelectric performance of the PVDF and its copolymers [8]. Incorporating carbon nanostructures into the PVDF matrix, functionalizing the fillers, inclusion of metal complexes in PVDF matrix, annealing, electrospinning, electrical poling, drawing, etc. are the suggested methods for enhancing β phase of the PVDF and PVDF-based polymer nanocomposites [9]. Among various fillers, the inclusion of carbon nanofillers in PVDF matrix is preferred to enhance the β phase of PVDF-based polymer nanocomposites due to their higher conductivity, lower density, ductility, and surface area [10]. Incorporation of the carbon nanofillers such as expanded graphite [11, 12], carbon nanofiber (CNF) [13], graphene oxide (GO) [14], carbon nanotube (CNT) [15], etc. in the base polymer matrix such as PVDF not only enhances the electrical conductivity of the composites but also improves the crystallization of β phase and piezoelectric properties of the polymer nanocomposites [16].

In a similar way, the mechanical energy can also be harnessed in the form of triboelectricity using the EAP-carbon nanofiller-based nanocomposites. The triboelectric performance of the composites is directly proportional to the surface charge density of composites, which in turn depends on surface roughness, temperature, strain, etc. [17]. In the case of PVDF-based nanocomposites, increasing the fraction of β phase not only enhances the piezoelectric performance but also triboelectric performance of the nanocomposite systems. As mentioned earlier, the inclusion of carbon nanofillers usually improves the β-phase content in the PVDF matrix, which in turn increases not only the dielectric constant of the composite but also the surface charge density in the composites. Although the triboelectric performance can be improved by increasing the dielectric constant of the composite systems to obtain high output power, the percolation threshold of the nanofillers is one of the limiting factors. The dispersion of the nanofillers in the PVDF matrix must be enhanced to prevent agglomeration of the nanofillers, and hence, triboelectric properties can be improved [18].

Similarly, power generation using waste heat is an active area of research. Materials capable of converting waste heat into electricity are known as thermoelectric (TE) materials. The field of TE materials is dominated by inorganic conducting materials such as bismuth telluride (Bi_2Te_3), lead (IV) telluride ($PbTe_2$), and antimony telluride (Sb_2Te_3) due to their efficiency at obtaining large figure of merit ($ZT > 1$) (parameter that is directly proportional to electrical conductivity (σ) square

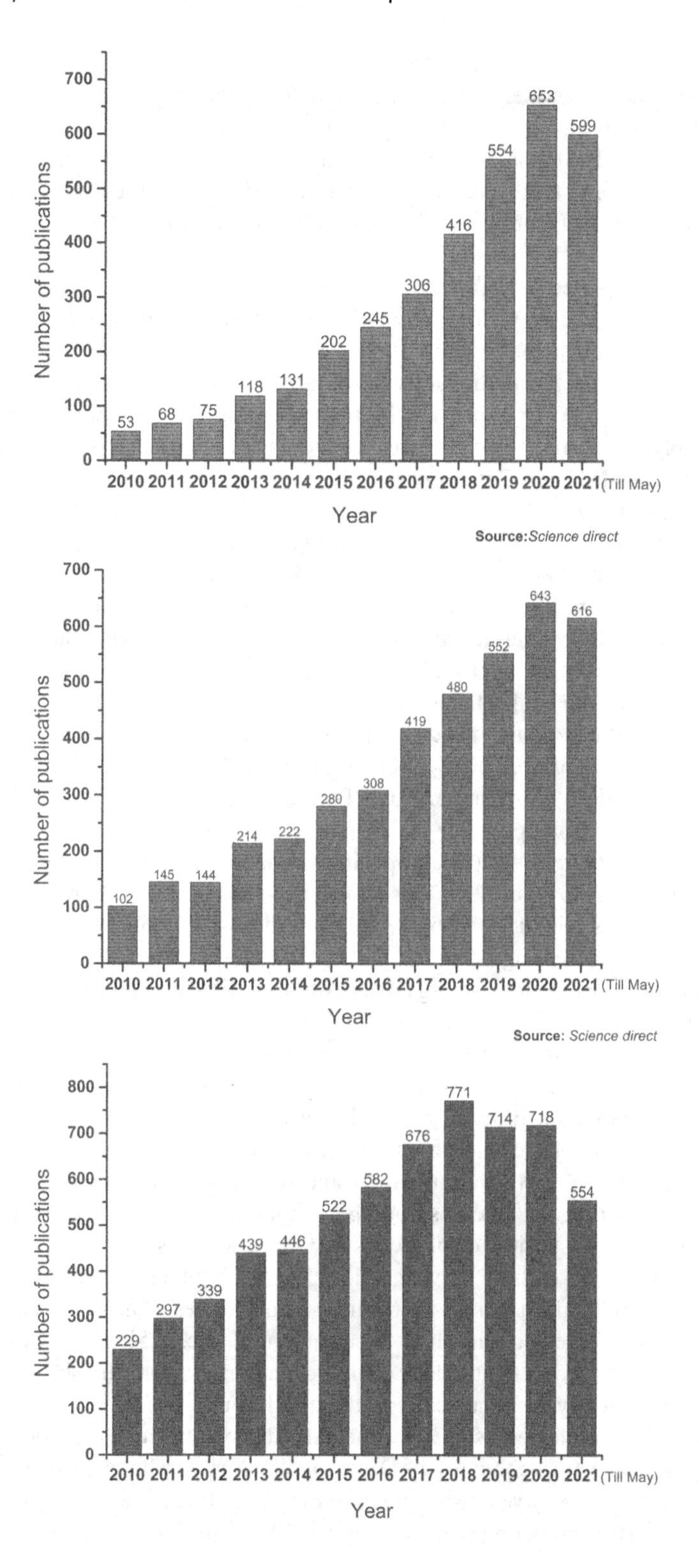

FIGURE 15.1 Plot of number of publications vs. years for (a) electroactive polymer-based mechanical energy harvesters, (b) electroactive polymer-based thermoelectric energy harvesters, and (c) electroactive polymer-based dye-sensitized solar cells. (sciencedirect.com.)

of Seebeck coefficient (α), and inversely proportional to thermal conductivity (κ)). But the high cost, scarceness of the raw materials, intrinsic defects, difficulty in mass production, etc. pose impediments for their usage in different applications [19]. Intrinsically conducting polymers (ICPs) are a class of EAPs that can offer significant advantages over the conventional inorganic conducting materials due to their mechanical flexibility, lightweight, ease of processing, remarkably low thermal conductivity, ability to be thermoelectrically active at ambient temperature, etc. [20]. The research based on ICP-based TE materials have been increasing due to the abovementioned advantages as depicted in Figure 15.1(b). (Search words of 'Conducting polymer-based TE materials' in Sciencedirect.com yields a number of papers published in different years. This data is used for plotting). Though the power factor (PF) (parameter that is the product of square of Seebeck coefficient and electrical conductivity) obtained using conducting polymers is relatively low, there is plenty of scopes to improve it by incorporating various carbon nanostructures into the conducting polymer matrix. The PF must be increased to enhance the efficiency of TE performance of the materials. The TE properties of ICPs can be improved by various techniques like doping, the addition of conducting carbonaceous nanofillers, modifying the molecular arrangement of polymer chains, controlling the crystallinity of polymers during the process, incorporation of inorganic fillers, thermal annealing, etc. [21]. However, the interdependence between electrical, thermal conductivities and Seebeck coefficient is a huge hindrance in achieving higher PF. Usually, increasing the carrier concentration to enhance electrical conductivity of the materials results in the decrease of Seebeck coefficient. Hence, fine control over electrical conductivity is essential to maximize the PF [22]. The EAPs like polyaniline (PANI), polypyrrole (PPy), etc. are deemed as potential candidates for developing efficient TE energy harvesters owing to their low thermal conductivity and possibility of enhancing the electrical conductivity by the incorporation of carbon nanofillers like graphene, CNTs, etc. Usually when these conducting nanofillers are incorporated in porous polymer matrices, due to phonon scattering and energy filtering effects, the thermal conductivity of the composite is reduced, whereas the electrical conductivity is retained at an appreciable level [23]. This synergism in properties of conducting polymer and carbon nanofillers makes conducting polymer-carbon nanofiller composites a versatile material for efficient TE energy harvesting.

EAPs can also be used for solar energy harvesting application. The most dominant and conventional solar energy harvesters in the market are silicon-based inorganic solar cells. The first-generation solar cells are wholly based on silicon. Though silicon-based solar energy harvesters exhibit an efficiency of around 15%–20% and longevity, they have shortcomings such as rigid, overpriced, tedious fabrication process, and a decrease in the efficiency at higher temperatures. The second generation of solar cells is based on amorphous silicon, thin films of copper indium gallium selenide (CIGS), etc. [24]. Both the first- and second-generation solar cells are made from opaque materials demanding specific positioning to capture the solar energy. The third-generation solar cells, namely, the dye-sensitized solar cells (DSSCs), serve as a better alternative to the first- and second-generation solar cells to scavenge clean energy. To fabricate a highly efficient DSSC, the counter electrode (CE) must possess low charge transfer resistance, superior electrocatalytic activity, higher electrical conductivity, higher surface area, etc. [25]. DSSCs predominantly use platinum (Pt) as CE for its excellent electrocatalytic properties. But, the scarceness of platinum, high price, and the difficulty in large-scale production limit its use. Hence, conducting polymers like PANI, PPy, etc. are preferred as CEs for their flexibility, low cost, easy processing, etc. [26]. The importance of conducting polymers in DSSC application can be gauged from the number of research papers published every year as depicted in Figure 15.1(c). (The data for the plot is obtained using the search words 'Conducting polymer-based DSSCs' in Sciencedirect.com, which yields a number of papers published in different years.) To enhance the electrical conductivity and facilitate vigorous charge transfer for reducing the oxidized species of electrolyte to maintain effective dye regeneration, carbon nanofillers like SWCNT, multiwalled carbon nanotube (MWCNT), graphene, etc. are incorporated into conducting polymer matrices [27]. The superior electrical conductivity and the higher surface area of these carbon nanofillers coupled with the mechanical stability of the conducting polymers shall serve as a promising polymer composite CE for DSSC.

Hence, in this chapter, recent advancements and developments in the areas of mechanical energy harvesting, TE, and solar energy harvesting based on carbon nanofillers such as graphene, CNT, graphite nanoplatelets, reduced graphene oxide (rGO), and GO incorporated EAPs are discussed in detail.

15.2 BASICS OF ENERGY HARVESTING: ELECTRICITY FROM SOLAR AND AMBIENT MICRO-ENERGIES

15.2.1 Piezoelectricity

The generation of electric charges across the surfaces of solid materials when mechanical stress is applied is known as piezoelectric effect. The piezoelectric effect is of two types: the direct and inverse piezoelectric effects. The direct piezoelectric effect is related to producing electric charge by applying mechanical stress to the material, which is important for energy harvesting and sensing applications. The converse (applying an electric field to deform the material) is the inverse piezoelectric effect exploited for actuator application. In crystalline/semicrystalline materials, there exists charge neutrality between the positive and negative charges. On applying stress to these materials, the charge neutrality is disrupted and dipoles (equal and opposite charges separated by a distance in a material) are oriented uniformly in the direction of electric field (polarization) as shown in Figure 15.2. Due to polarization, the charges are accumulated across the surface of the crystal, which can be collected through electrodes resulting in current. The formation of electric dipoles in a piezoelectric material determines its piezoelectric performance. Usually, polycrystalline materials that lack centrosymmetry (center of inversion is absent) exhibit piezoelectric effect [28]. From a total of 32-point groups, 21 are non-centrosymmetric and 20 among them can result in piezoelectric response.

PVDF exhibits five polymorphs such as α, β, γ, δ, and ε, out of which polar β and γ phases exhibit finite dipole moment. The dipole moment of β and γ phases are 8×10^{-30} cm and 4×10^{-30} cm, respectively [29]. The larger dipole moment of β phase is due to the arrangement of fluorine and hydrogen atoms as depicted in Figure 15.3. The complete conversion of the nonpolar α (*trans*-gauche-*trans*-gauche′ [TGTG′]) phase to polar β (*trans-trans-trans-trans* [TTTT]) phase is essential to achieve enhanced piezoelectricity. The β phase of PVDF is particularly focused due to its high piezoelectric nature. Processes like poling the PVDF film (or PVDF-based composite film) at high electric field orient the dipoles uniformly retaining the electroactive β phase even after the removal of the applied electric field. The polarization can be adjusted by modifying the poling conditions to derive better piezoelectric effect.

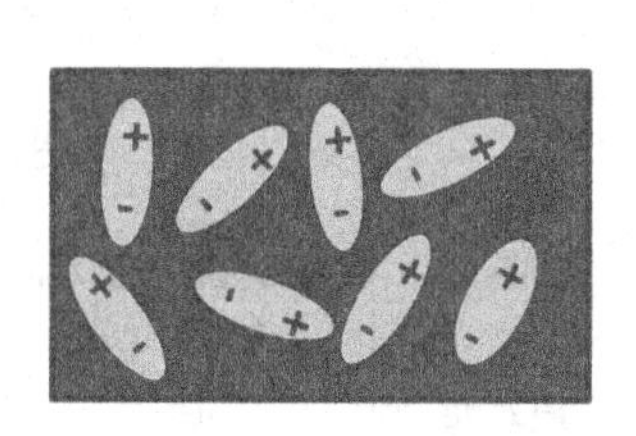

(a) Randomly oriented dipoles in the polymer

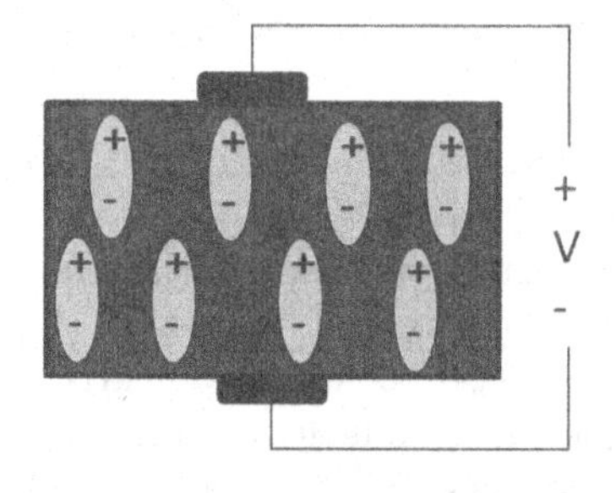

(b) Orientation of dipoles during poling in the direction of electric field

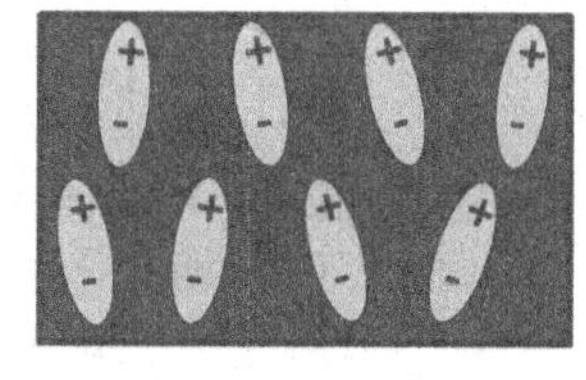

(c) Oriented dipoles are mostly retained even after removal of the electrical field

FIGURE 15.2 Possible arrangements of electric dipoles in a piezoelectric polymer: (a) randomly oriented dipoles in the absence of external electric field, (b) orientation of dipoles during poling in the direction of electric field, and (c) orientation of dipoles after removal of electric field.

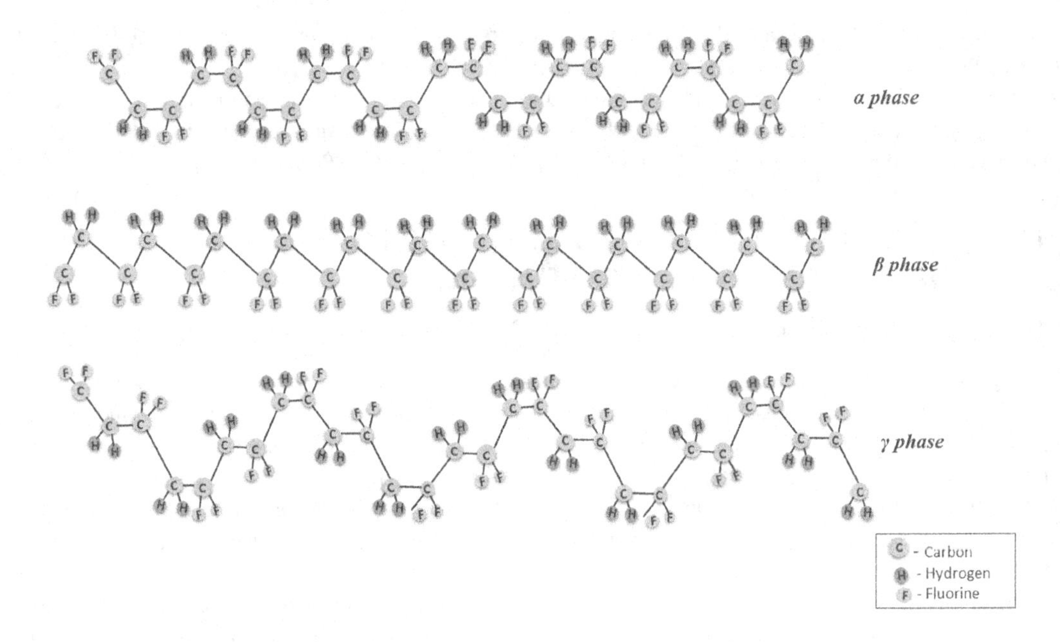

FIGURE 15.3 Electroactive phases of PVDF.

The piezoelectric constants are used to determine the piezoelectric response of the materials. The piezoelectric coefficient, d_{ij} is related to polarization as

$$P = d_{ij}\sigma \text{ (Direct piezoelectric effect)} \tag{15.1}$$

$$\varepsilon = d_{ij}E \text{ (Indirect piezoelectric effect)} \tag{15.2}$$

where P is the polarization, σ is the applied stress, ε is the strain, and E is the applied electric field. The subscripts of piezoelectric constant d_{ij}, i and j, stand for the direction of dipole orientation and applied force, respectively. The most commonly used piezoelectric constants are d_{33} and d_{31} (where '1' stands for the axis in the direction of stretch of the material and '3' stands for the polarization axis of the material) and their unit is pC/N or pm/V. The constituent equations of piezoelectric effects are:

$$\text{Direct piezoelectric effect, } D_i = e_{ij}^{\sigma}E^j + d_{im}^{d}\sigma_m \tag{15.3}$$

$$\text{Inverse piezoelectric effect, } \varepsilon_k = d_{jk}^{c}E_j + S_{km}^{E}\sigma_m \tag{15.4}$$

where D_i is the electric displacement, ε_k is a mechanical strain vector, E_j is the applied electric field vector, e_{ij}^{σ} is dielectric permittivity, σ_m is a mechanical stress vector, d_{im}^{d} and d_{jk}^{c} are piezoelectric coefficients, and S_{km}^{E} is the elastic compliance matrix. On applying stress/strain (force) perpendicular to the surface of piezoelectric materials, their electromechanical responses can be calculated using the following parameters:

- Short-circuit current,

$$I_{sc} = \frac{Q}{\Delta t} = \left(\frac{d_{33}\ \Delta F}{\Delta t}\right) \tag{15.5}$$

- Open-circuit voltage,

$$V_{oc} = \frac{Q}{C} = \left(\frac{d_{33}}{\varepsilon_o \varepsilon_r}\right)\sigma Yh = \left(\frac{d_{33}}{\varepsilon_o \varepsilon_r}\right)h\frac{\Delta F}{A} \tag{15.6}$$

- Power density,

$$P_d = \frac{V^2}{A \times R_L} \tag{15.7}$$

where Q is the total charge at the surface of the material, C is the capacitance, ε_r is the relative permittivity, ε_o is the permittivity of free space, h is the thickness, Y is Young's modulus, V is the output voltage, $\Delta F/\Delta t$ is the rate of change of applied force, σ is the strain, R_L is the load resistance, and $\Delta F/A$ is the change in force per area where the force is applied. The higher value of these parameters (I_{sc}, V_{oc}, and P_d) indicates an enhancement in the piezoelectric performance of the material [30].

15.2.2 Triboelectricity

Triboelectricity is produced due to the frictional contact between materials. When two different materials come in contact with each other due to an external mechanical stimulus (pressing, bending, tapping, etc.), they become charged after the materials are separated. On releasing the contact between the two materials due to the uneven distribution of charges, an electrostatic field results. When the electrodes are connected through the external load, current is registered. This is known as triboelectric effect. When these triboelectric materials are brought into contact again, the current flows in a reverse direction. On repeating the cycle, a continuous alternating current is generated as an output. After certain number of cycles, the charges transferred will stay on the surface of materials for a long time (lasts from hours to days). The triboelectric effect is based on contact electrification. The contact electrification is due to the static polarized charges on the surface of the material, i.e., on separating the materials after a contact, the materials turn out to be electrically charged due to the transfer of charges between them [31]. Understanding the mechanism of charging the materials by the physical contact at a nanoscale is challenging and is being actively investigated.

The triboelectric nanogenerator (TENG), made up of triboelectrically positive and negative materials, is packed with conducting materials as electrodes on either side. It can be operated in four different modes such as vertical contact-separation mode, single electrode mode, lateral sliding mode, and freestanding triboelectric layer mode as explained elsewhere [32].

15.2.3 Thermoelectricity

Materials capable of directly converting the waste heat into electricity are known as TE materials. The direct conversion of heat to electricity and vice versa can be realized by the Seebeck and Peltier effects. TE generators consist of both n-type and p-type materials. The TE generator can be cascaded both in series and parallel. On supplying heat energy, the temperature difference across the two electrically conducting dissimilar materials (one p-type and other n-type) results in voltage that can be harnessed as energy. This phenomenon is known as the Seebeck effect. The temperature difference drives the electrons from higher to lower temperature regions in the material. The reverse phenomenon, i.e., the temperature difference is produced by applying an electric potential across the materials, is known as the Peltier effect. The schematic representation of these effects is shown in Figure 15.4.

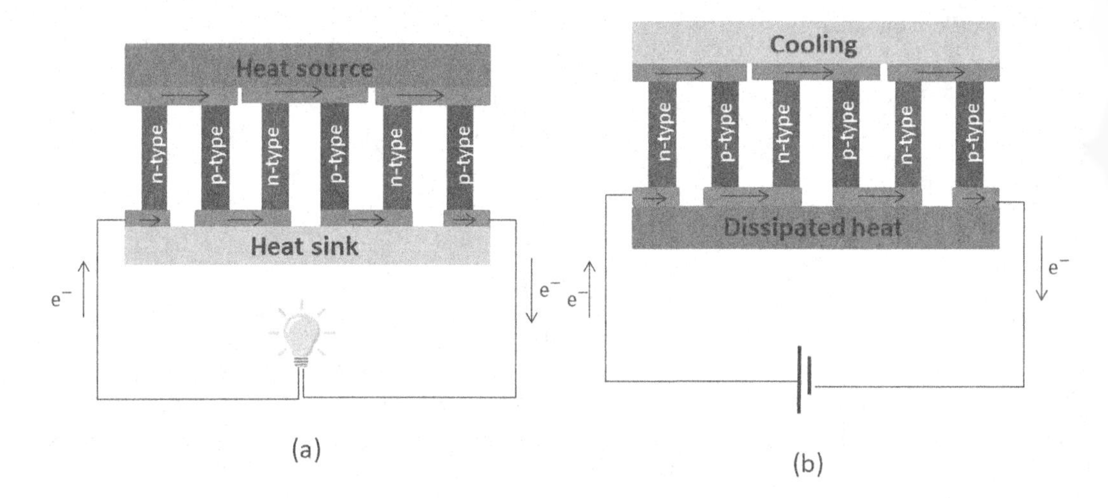

FIGURE 15.4 Schematic representation of (a) Seebeck effect and (b) Peltier effect.

The three important parameters that determine the efficiency of TE materials are described as follows:

- **Seebeck coefficient (α):** It is the measure of the magnitude of induced TE voltage with respect to the difference in the temperature across the material.

$$\alpha = \frac{\Delta V}{\Delta T} \tag{15.8}$$

  where ΔV is the voltage difference and ΔT is the temperature difference.
- **Figure of merit (ZT):** It is a dimensionless quantity relating electrical conductivity, Seebeck coefficient, and thermal conductivity as given in the following equation:

$$ZT = \frac{\alpha^2 \sigma T}{\kappa_e + \kappa_l} \tag{15.9}$$

  where α, σ, T, κ_e, and κ_l represent Seebeck coefficient, electrical conductivity, temperature, thermal conductivity due to electron, and thermal conductivity due to lattice, respectively [33]. To improve the overall figure of merit, the lattice thermal conductivity must be reduced effectively.
- ***PF*:** It is the parameter used to characterize TE materials, which is mathematically expressed as

$$PF = \alpha^2 \sigma \tag{15.10}$$

To enhance the effectiveness of TE materials, the Seebeck coefficient and electrical conductivity must be tuned to be high and thermal conductivity must be reduced. The interdependence of these three parameters poses a great challenge to develop an ideal TE material. The PF is dependent on square of Seebeck coefficient and the electrical conductivity of the material. Various processing routes employed to improve these parameters for developing an efficient TE material are described in this chapter.

15.2.4 Dye-Sensitized Solar Cells

The DSSCs work on the principle of photo-electrochemical process. On irradiating the cell using sunlight, current is produced. The basic components of a typical DSSC are dye, electrolyte, working electrode, and the CE.

- **Transparent, conductive substrates:** The working electrode and CEs are made by coating the active materials onto these conductive substrates. The substrates must be at least 80% transparent to allow maximum sunlight to pass through them into the cell and they should be highly electrically conductive for effective transfer of charges. FTO (fluorine-doped tin oxide), with a transmittance greater than 80% and sheet resistance of 18 Ω/cm^2, and ITO (indium-doped tin oxide), with a transmittance approximately 75% and sheet resistance of 8.5 Ω/cm^2, are mostly used as conductive substrates in DSSC.
- **Working electrode:** Generally, wide bandgap semiconductors like titanium dioxide (TiO_2), zinc oxide (ZnO), tin(IV) oxide (SnO_2), nickel oxide (NiO), etc. are used. TiO_2 is mostly used as a working electrode. Specifically, anatase phase is the most effective allotropic form of TiO_2 used as working electrodes in DSSC due to their wide bandgap of 3.2 eV.
- **Dye:** Since the semiconducting material such as TiO_2 in the working electrode is only capable of absorbing a small fraction of light in the ultraviolet (UV) region, the working electrode is soaked in dye to absorb maximum sunlight. The absorption spectra of the dye must be in the range of UV-visible to infrared region. Organometallic compounds are used as dyes that act as sensitizers to absorb more light. Usually, the working electrodes will be dipped into dye solution. The absorbed light excites the electron, knocks it off, and transfers it to the conduction band of the semiconducting material coated onto FTO.
- **Electrolyte:** Iodide/triiodide$\left(I^- / I_3^-\right)$, bromide/bromine (Br^- / Br_2), cobalt redox mediators (CO(II)/CO(III)), etc. are commonly used electrolytes in DSSCs [34]. Most commonly used and the efficient electrolyte is I^- / I_3^-. The reduction of I_3^- to $3I^-$ occurs on interaction with the CE. The electrolyte acts as a medium to return these electrons to oxidized dye molecules to regenerate the dye.
- CE – Generally, platinum, conducting polymer-based composites, etc. are used as CEs. The CE is a significant part of the DSSC as it facilitates the reduction of oxidized species in the electrolyte (e.g., iodide/triiodide system) and provides electrons to regenerate the dye. To produce a highly efficient DSSC, the CE must possess the following characteristics – low charge transfer resistance (R_{ct}), high concentration of electron donor, high surface area, porosity, superior conductivity, and electrochemical stability.

  The efficiency of a DSSC depends on (1) the energy gap between the HOMO (highest occupied molecular orbital) and LUMO (lowest unoccupied molecular orbital) of the dye, (2) Fermi level of the scaffold material, and (3) redox potential of the electrolyte. The conduction band of wide bandgap semiconducting material such as TiO_2 should be below the LUMO of the dye molecules, which ensures the flow of electrons from the excited dye molecule to the semiconducting material and then to the external circuit through the transparent conducting electrodes. The flow of electrons from the photoanode to the CE generates electricity.

The working of DSSC is based on four steps – light absorption, electron injection, transport of carriers, and collection of current as shown in Figure 15.5. These steps are explained as follows:

- When the DSSC is exposed to the sunlight, the dye absorbs the light that is incident on the working electrode. The electrons in the dye get excited.
- The life span of excited electrons is usually in nanoseconds. The excited electrons move to the conduction band of a working electrode, which must be lower than that of the excited state of dye. Thus, the dye molecule becomes oxidized (S^+).

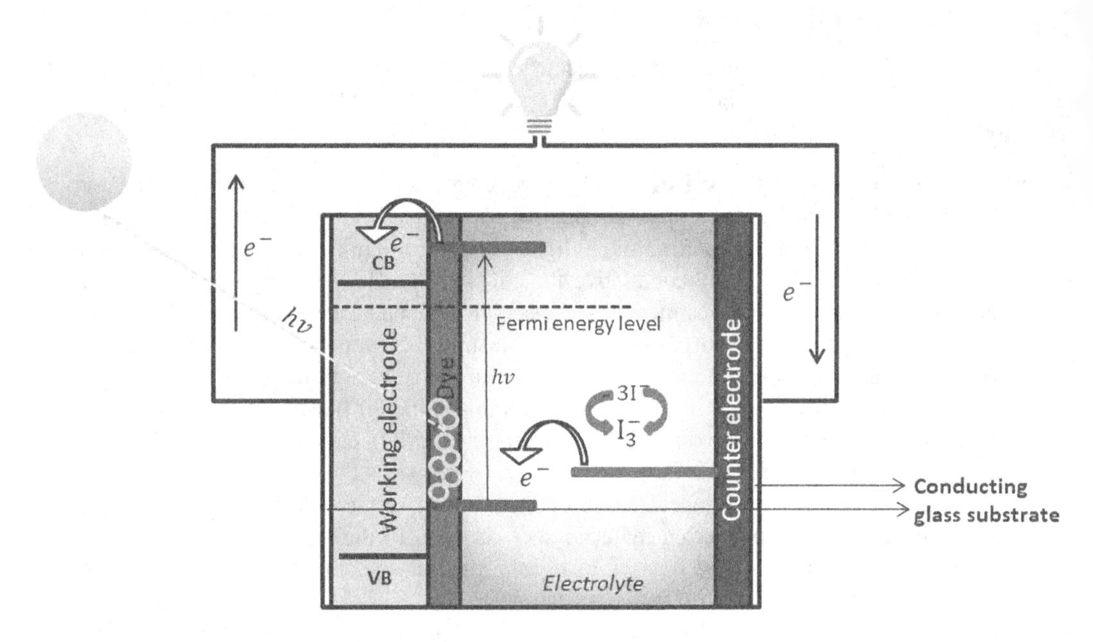

FIGURE 15.5 Schematic representation of a DSSC.

- The electron travels from the working electrode through the circuit and reaches the CE. The CE acts as a catalyst and reduces I_3^- to I^- by transferring electrons. The electrolyte consists of an iodine molecule (I_2) and iodide ion (I^-) normally combined as triiodide ion (I_3^-). This triiodide ion accepts electron from the CE and gets reduced to $3I^-$. These iodide ions get oxidized to become I_3^- by donating electron to the dye.
- On accepting electrons from I^- ion redox mediator, the regeneration of dye back to its ground state and oxidation of I^- to I_3^- occurs. This oxidized I_3^- ion moves to the CE and gets reduced to I^-. This cycle is repeated till the light is shed on the cell. The charges traveling through the external circuit can be utilized to do work.

The working steps of photo-electrochemical process are described in terms of the equation as below:

$(TiO_2/S) + h\nu \rightarrow (TiO_2/S^*)$ (Excitation of dye molecule 'S' after absorption of light so that it becomes S^*)

$(TiO_2/S^*) \rightarrow (TiO_2/S^+) + e^-$ (Injection of electron into the conduction band of TiO_2. Dye molecule become oxidized as S^+)

$(TiO_2/S^+) + 3I^- \rightarrow (TiO_2/S) + I_3^-$ (Regeneration of the dye)

$I_3^- + 2e^- \rightarrow 3I^-$ (Reduction of the oxidised species of the electrolyte)

where 'S' is the ground state of the dye and S^* is the excited state of dye in the DSSC [35].

The DSSC is connected to a source meter and illuminated to analyze the photovoltaic performance.

The photovoltaic performance of the DSSC can be evaluated using the following parameters:

- The current when applied voltage is zero is known as short-circuit current, I_{sc}.
- Current density is the maximum current per unit area that can be drawn from the DSSC.
- Open-circuit voltage (V_{oc}) is calculated when the current is zero between the positive and negative electrodes.

- Maximum power generated per unit area (P_{max}) is the product of the maximum voltage (V_{max}) and maximum current density (J_{max}) produced:

$$P_{max} = V_{max} J_{max} \tag{15.11}$$

- Fill factor is the ratio of maximum power generated to the theoretical power calculated:

$$\text{Fill Factor, } FF = \frac{J_{mp} V_{mp}}{J_{sc} V_{oc}} \tag{15.12}$$

$$\text{Efficiency, } \eta = \frac{P_{max}}{P_{in}} \times 100\ (\%) \tag{15.13}$$

where J_{mp} and V_{mp} are the current density and voltage values, respectively, obtained when power per unit area is maximum. P_{in} is the input power per unit area.

15.3 EAP NANOCOMPOSITES FOR MECHANICAL ENERGY HARVESTING

Research based on insulating polymers such as PVDF and polyimides (PI) [36] for mechanical energy harvesting application are gaining prominence in recent years due to their flexibility, etc. as briefed in Section 15.1. Out of other polymers, PVDF is focused extensively due to the possibility of inducing electroactive β and γ phases. The formation of electric dipoles in PVDF is the key for the mechanical energy harvesting application.

Incorporation of carbon nanofillers like GO, graphene, CNT, etc. at an optimal loading will effectively induce electroactive phases in PVDF and enhance the piezoelectric properties of the composite films. Adopting appropriate processing techniques, optimizing parameters of the process also play a significant role in developing an efficient polymer-based energy harvester. Hence, this section briefs about the recent advancements in the area of mechanical energy harvesting based on carbon nanofillers incorporated in PVDF nanocomposites.

15.3.1 PVDF-Carbon Nanostructure Nanocomposite-Based Mechanical Energy Harvesters

Since the most stable α phase of PVDF is nonpolar, the electroactive phases of PVDF should be induced by a suitable process or incorporation of carbon nanostructures for better mechanical energy harvesting. Inclusion of carbon nanofillers like graphene, single-walled carbon nanotube (SWCNT), etc. into PVDF matrix results in the crystallization of the electroactive β polymorph of the polymer. The process of electrospinning at a strong electric field directly generates the piezoelectric β phase in PVDF-based composites without any additional poling process. Yang et al. [37] synthesized and investigated efficient piezoelectric nanogenerators (PENGs) using PVDF-GO and PVDF-rGO nanofiber mats. The PVDF/GO nanofiber mats were synthesized by the electrospinning method. The nanogenerator was made using the nanofiber mats that were glued with copper foils on both sides and then packed with polydimethylsiloxane (PDMS) substrates on either side. Vertical force was applied to the PENG by a vibrator exciter. Incorporation of GO and rGO in PVDF resulted in decreased diameter of composite nanofiber. The authors claimed that the decrease in the diameter of the composite fibers was attributed to the formation of the Taylor cone. The GO and rGO fillers could easily be charged at higher voltages (usually in kV) resulting in enhanced Coulombic force. The electrostatic repulsion between the charged filler particles eventually decreased the diameter of the fiber. PVDF with 2 wt% GO or rGO had the highest β-phase fraction of 87%. The increase in β-phase content in the composites was attributed to the specific interaction between PVDF and graphene layers. The short-circuit current at 2 wt% rGO loading in PVDF was observed to be 700 nA and current signal was stronger than what can be observed in PVDF-GO nanocomposites.

The conducting network formation between the filler especially rGO resulted in the decrease of the internal resistance of PENG. The open-circuit voltage of PVDF-2 wt% rGO PENG has been measured to be 16 V, which was 20 times higher than that of pure PVDF. On finger pressing mode, the power density of PVDF-2 wt% rGO PENG obtained was 2800 nW/cm^2. This value was approximately 150 times the power density of the pure PVDF (15.75 nW/cm^2).

Yu et al. [38] have reported an enhanced piezoelectric performance of electrospun PVDF-MWCNT nanofiber mats. The authors claim that PVDF-5 wt% MWCNT nanofiber mat exhibit an increased fraction of the polar β phase (68.4%) when compared to that of pure PVDF nanofiber mat (58.4%). During electrospinning, in the presence of electric field, the incorporated conductive MWCNTs can produce inductive charges on the surface of PVDF leading to increased Coulomb force that promotes the crystallization of β phase in PVDF-5 wt% MWCNT nanofiber mat. PVDF-5 wt% MWCNT nanofiber mat while bending produced a maximum power output voltage of 6 V and average output power of 81.8 nW. Gebrekrstos et al. [39] have studied the effect of GO and functionalized graphene oxide (FGO) on the structure and piezoelectric performance of PVDF. The GO was functionalized in two different ways such that carboxylated graphene oxide (GOCOOH) and fluorinated graphene oxide (GF) were synthesized. Electrospinning technique was employed to synthesize the nanocomposite fiber. The highest β-phase content was observed in PVDF-1 wt% GF nanocomposite fiber (89%) when compared to that of what could be obtained in PVDF-1 wt% GOCOOH (79%), PVDF-1 wt% GO (69%), and pure PVDF (38%). The higher conversion of α–β phase was attributed to the nucleating effect of the graphene-based fillers and uniaxial stretching of the nanocomposite fiber during the electrospinning process. The highest β-phase fraction was observed in PVDF-GF samples due to the electronegativity of fluorine atoms in GF that interacts with PVDF chain causing β conformation of PVDF. The piezoelectric coefficient d_{33} has been reported to be higher for PVDF-1 wt% GF (63 pC/N) nanocomposite when compared to that of PVDF-1 wt% GO (40 pC/N) and PVDF-1 wt% GOCOOH (46 pC/N) nanocomposites, respectively.

Garain et al. [40] developed a flexible, ultrasensitive acoustic nanogenerators (ANGs) using electrospun PVDF-Ce^{3+}-doped graphene nanofibers. The delocalized π electrons in the graphene interact well with $-CH_2$ dipoles of PVDF resulting in 99% β polymorph in PVDF-Ce^{3+}-doped 1 wt% graphene nanofibers. The ANG developed using PVDF-Ce^{3+}-doped 1 wt% graphene nanofibers could sense even a low-pressure sound wave (2 Pa). The authors reported that when 6.6-kPa pressure was applied to the ANG, voltage of 11 V and a maximum power of 6.8 μW were obtained. On applying 88 dB to the ANG, an output voltage of 3 V was obtained. The study by Ahn et al. on PVDF-MWCNT nanocomposites resulted in a high degree of β-phase conversion from the nonpolar α phase of PVDF at 0.2 wt% loading of MWCNT when drawing and poling of electrospun composites have been carried out [41]. The electrospinning method of synthesis of nanocomposites enhanced the dispersion of MWCNTs in the polymer matrix resulting in the charge accumulation at the interface of PVDF and MWCNT. This interaction enhanced the β phase formation as revealed through large remnant polarization values obtained for PVDF-1 wt% MWCNT composite membranes (60 mC/cm^2), when compared to that of PVDF membranes (55 mC/cm^2) synthesized by the same processing method.

Abolhasani et al. [42] have reported the enhancement of piezoelectric properties in PVDF-0.1 wt% graphene nanoplatelets nanocomposite synthesized by electrospinning technique. The Fourier transform infrared spectroscopy (FTIR) and wide-angle X-ray diffraction (WAXD) studies confirmed the presence of both α and β phases in pure PVDF and the crystalline β-phase fraction has been increased from 77% in pure PVDF to 83% in PVDF-0.1 wt% graphene nanoplatelet nanocomposites. The process of electrospinning along with the nucleating effect of graphene nanoplatelets has led to the enhanced crystallization of β polymorph at an exceptionally low loading of the filler (0.1 wt%). Open-circuit voltage of 3.8 V in pure PVDF nanofibers has been increased to 7.9 V when 0.1 wt% graphene nanoplatelets were incorporated in PVDF due to enhanced piezoelectric properties of the nanocomposite nanofibers. Tiwari et al. [43] have developed a highly efficient energy harvester using PVDF-FGO nanocomposites to harness energy from waste mechanical resources. Electrospinning parameters are optimized to yield high-quality nanofibers. Incorporation of sulfonated graphene oxide (GS) (sulfonated in the presence of ammonium sulfate) in PVDF resulted in very thin fibers due to the interaction of polar sulfate groups of GS with PVDF molecular chains. This specific interaction has resulted in the crystallization of the β phase as confirmed by FTIR, differential scanning calorimetry (DSC), and Raman spectroscopy analyses. The FTIR studies show an increase in β-phase content in PVDF-1 wt% GS nanocomposite (88%) when compared to that of pure PVDF (59%) and PVDF-1 wt% GO nanocomposite (76%), respectively. The authors have reported the highest electromechanical response recorded from human motions when PVDF-1 wt% GS nanocomposite is used. PVDF-1 wt% GS nanocomposite resulted in the output voltage of 62 V and power density of 48.13 μW/cm^2 compared to that of pure PVDF (20 V and 23.4 μW/cm^2) and PVDF-1 wt% GO nanocomposite (45 V and 33.7 μW/cm^2) synthesized by electrospinning process.

Abbasipour et al. [44] have developed polymer nanocomposites with three different nanofillers such as graphene (Gr), GO, and halloysite (HNT) incorporated in PVDF matrix to produce an efficient PENG. The composites were electrospun. PVDF-1.6 wt% GO resulted in the highest output voltage of 2 V when compared to that of PVDF (1 V), PVDF-1.6 wt% Gr (1.3 V), and PVDF-1.6 wt% HNT (1.5 V) nanocomposites, respectively, on bending mode. Similarly, the incorporation of GO in PVDF resulted in enhanced piezoelectric properties and thermal stability of the nanocomposite. The enhanced interfacial adhesion between the GO and PVDF molecular chains was reported to be responsible for the improved piezoelectric performance of the nanocomposite. The piezoelectric constant of 24 pC/N reported for PVDF-1.6 wt% GO nanocomposite was higher than that of pure PVDF (17 pC/N), PVDF-1.6 wt% Gr (18 pC/N), and PVDF-1.6 wt% HNT (22.7 pC/N) nanocomposites due to enhanced β-phase content in the polymer.

There are various other processing routes by which an improved crystallization of polar β phase can be achieved to enhance the efficiency of the PVDF-based nanocomposite energy harvesters as described later. The study by Guo et al. [45] proves enhancement in the piezoelectric performance of PVDF-graphene aerogel (GA) composite. The PVDF-GA composites were synthesized by vacuum-assisted impregnation process followed by corona polarization. The composite films were made with different weight fractions of PVDF dissolved in dimethyl sulfoxide (DMSO). The synthesized GA was made up of interconnecting rGO sheets in the form a honeycomb structure. The authors claim that there was a transition from β to γ phase with an increase in the concentration of PVDF. At lower concentrations of PVDF (lesser than or equal to 60 mg/ml), the interaction between –CF_2 groups of PVDF and C=O of rGO results in the formation of β phase. On increasing the concentration of PVDF above 60 mg/ml, the GA was completely coated by PVDF, which hindered the excessive PVDF from interacting with rGO sheets and thus resulted in the formation of γ phase. The transition from β to γ phase on increasing the concentration of PVDF was supported by X-ray diffraction (XRD) and FTIR data. The value of d_{33} decreased with the increase in PVDF content due to the formation of γ phase. Out of all the samples, at PVDF concentration of 80 mg/ml, the PVDF-GA composites exhibited enhanced piezoelectric properties with d_{33} value of 16 pC/N that was lesser than that of β phase of PVDF (d_{33} = 28 pC/N) [46]. The average peak output current of the same composite was reported to be 0.51 μA. Even under repeated pressing, the 80-mg/ml PVDF-GA composites exhibited good cycling stability. On increasing the frequency of pressing and reducing the force, output current value decreased to 0.22 μA.

Alluri et al. [47] have developed PVDF-activated carbon PENG as an approach to unconventional energy harvesting from waste mechanical energy. The PVDF-activated carbon films were synthesized by ultra-sonication of activated carbon dispersed in *N,N*-dimethylformamide (DMF) and PVDF solution prepared by dissolving PVDF in DMF and acetone mixture under sonication process using a probe sonicator. The composite mixture was poured in a Petri dish and heated at 70°C overnight to remove the remnant solvents. The PENG was made by placing the composite film between aluminum foil electrodes and external connections were made using copper wires attached to aluminum foils with the silver paste. The entire assembly was kept in between PDMS layers to protect the films against external mechanical stresses. The authors demonstrated composite nanogenerator device as a self-powered acceleration sensor. Specific attraction of $-CH_2$ dipoles toward delocalized π electrons in the activated carbon resulted in the generation of electroactive phases of PVDF. Apart from existence of β phase, the authors had reported traces of electroactive γ phase also. Unpoled PVDF PENG exhibited a high peak to peak output voltage and current of 37.7 V and 299 nA, respectively. No significant change was observed in the output voltage and current of both poled and unpoled PVDF PENG. However, the unpoled PVDF-activated carbon PENG (30V/V%-PVDF solution: activated carbon in DMF solution) resulted in an output voltage and current of 37.87 V and 0.831 μA, respectively. Poling of this composite film drastically increased the output voltage and the current by 30% and 96%, respectively when compared to that of unpoled films. Power density of PVDF-activated carbon PENG (30V/V%) was approximately 63.07 mW/m^2. Yang et al. [48] achieved exceptional piezoelectric performance by topological structure modulation of PVDF-3D (three-dimensional)-CNTs (MWCNTs) nanohybrid composites. The 3D-CNT nanohybrids were made up of CNTs as the main structure with manganese dioxide (MnO_2) and graphene as its branches. The MnO_2 was used as a shielding layer to improve the breakdown strength of the composites. CM23 (23 wt% with lower amount of MnO_2 contents) and CM66 (66 wt% with higher amount of MnO_2 contents) were the two different types of hybrid CNT fillers used. The hybrid CNT fillers were made by one-pot synthesis method (successive chemical reactions took place in a single reactor at a stretch to obtain a final product). The sandwich structure of nanocomposite was formed by stacking of the layers – L_x ($x = 2, 4$) and H_y ($y = 0, 1, 2$) where 'L' is CM23-PVDF, 'H' is CM66-PVDF, x, and y are 10 times the weight fractions of CM23 and CM66, respectively, incorporated into the PVDF matrix. The nanocomposites were synthesized by a three-step process such as solution casting, hot pressing, and rolling. The authors explicitly proved that rolling process resulted in enhanced β-phase content in the polymer even in the three-layered sandwiched structure that was attributed to the enhanced piezoelectric characteristics of the sandwiched structures. Further rolling process was efficient in enhancing crystallinity of the nanocomposites. The conductive nature of CM23 fillers resulted in highly oriented β chain conformation in the sandwiched nanocomposites. So, the L_2 and L_4 layers had better polarization when compared with H_0 or H_1 or H_2 layers. Hence, a combination of L_x and H_y layers resulted in the improved piezoelectric performance of the nanocomposites. The superior piezoelectric behavior was observed in the sandwiched nanocomposite, L_2-H_1-L_2 with a high piezoelectric constant (d_{33}) of 48 pC/N at an electric field of 80 MV/m, while poling compared to other layers. The piezoelectric performance of sandwiched nanocomposites was observed to be stable (using different combinations of both fillers in PVDF matrix) over the temperature range of 10°C–60°C making them promising candidates for sensing and energy harvesting applications.

An inexpensive route to induce enhanced piezoelectric performance in PVDF-GO nanosheets composite for energy harvesting applications was reported by El Achaby et al. [49]. Flexible PVDF-GO films were synthesized by solution casting method using DMF as solvent. Unlike in electrospinning method, where in situ polarization occurred under a strong electric field to yield a high fraction of β phase of PVDF, the simplest process of solution casting resulted in the formation of pure β phase at an extremely low loading of 0.1 wt% GO in PVDF without any additional poling, The α–β phase transition was primarily due to the interaction between the molecular chains of PVDF and functional groups on the surface of GO. Im et al. [50] obtained superior piezoelectric properties in PVDF-MWCNT composites by macroscopic deformation. The process of drawing and poling increased the fraction of β phase over 90% in PVDF-x wt% MWCNT ($x = 0$–1 in steps of 0.1) films as well as membranes indicating the possibility of easily achieving pure β phase in the composite using this process. The authors synthesized the nanocomposites by non-solvent precipitation route followed by hot pressing for forming films and electrospinning technique for the making membranes. In both the synthesis routes, the authors obtained enhanced electroactive β after drawing and poling.

15.3.2 PVDF Copolymers-Carbon Nanostructure Nanocomposites for Mechanical Energy Harvesting

Several copolymers of PVDF such as P(VDF-TrFE), PVDF-HFP, polyvinylidene fluoride tetrafluoroethylene (PVDF-TFE), etc. have been synthesized and used for mechanical energy harvesting applications. The copolymers are high-performance polymers with properties superior to that of homopolymer itself. P(VDF-TrFE) can directly crystallize polar β phase from solution or melt without any additional poling process. The greater number of fluorine atoms in P(VDF-TrFE) creates hindrance in the molecular structure to form the α phase (TGTG′) and makes the formation of β phase easier. Bhavanasi et al. [51] reported effective approach of enhancing the energy harvesting efficiency of P(VDF-TrFE)-GO films by forming bilayers. The P(VDF-TrFE)-GO bilayer films were prepared by drop casting GO solution onto the polymer film already formed. Prior to the formation of bilayer film, the polymer film was annealed at 135°C for 2 hours and cooled to room temperature to enhance the ferroelectric properties. Followed by annealing, pure polymer films were gold coated on both sides by sputtering and electrically poled (30 MV/m) at 105°C. Complex interface was formed in the bilayered film when materials such as GO (due to the existence of the functional groups) and ferroelectric polymer (PVDF-TrFE) were mixed together. The interaction between hydrogen of P(VDF-TrFE) matrix and functional groups of GO resulted in enhanced piezoelectric properties of P(VDF-TrFE)-GO bilayer films. As a result, an output voltage and current of 4 V and 1.88 μA, respectively, were obtained. These values were found to be greater than that of poled pure P(VDF-TrFE) films (1.9 V and 0.96 μA). The power density of 4.41 μW/cm^2 obtained for P(VDF-TrFE)-GO bilayer film was greater than that of poled P(VDF-TrFE) films (1.77 μW/cm^2).

Bhunia et al. [52] have synthesized P(VDF-TrFE)-rGO nanocomposites with different loadings of GO by sol-gel method for PENG and hybrid piezoelectric-triboelectric nanogenerator (HPTENG) applications. The authors reported that on incorporation of rGO fillers in PVDF-TrFE matrix, the fraction of crystalline β phase of the polymer was enhanced due to the interaction of carbonyl group of rGO with –CH_2-CF_2 dipoles of P(VDF-TrFE). During the interaction, the centrosymmetry of rGO was broken enhancing the piezoelectric properties of the P(VDF-TrFE)-rGO composites. The composites were spin-coated onto the ITO/PET sheets using MEK (methyl ethyl ketone) solvent. The PENG device was made by thermally depositing the aluminum above P(VDF-TrFE)-rGO composite film and ITO at the bottom as the contact surface. Similarly, the HPTENG device was made by stacking P(VDF-TrFE)-rGO, ITO, and PET in multiple layers as shown in Figure 15.6. The ITO layer of each stack was used as contact surface. The PET layer of ITO/PET sheet and the composite layer of P(VDF-TrFE)/ITO/PET were used as working layers of TENG. The potential was generated in the HPTENG device when the layers of PET and P(VDF-TrFE) film or P(VDF-TrFE)-rGO film came in contact with each other. The positive charge of PET was transferred to P(VDF-TrFE) or the composite film, and the negative charge of the film to PET. This charge separation disrupted the charge neutrality of piezoelectric layer (P(VDF-TrFE) or the composite film, which resulted in potential difference across the electrodes and when these layers were separated, they retained their charge neutrality. The author claimed that among all the samples, P(VDF-TrFE)-0.5 wt% rGO composite film exhibited outstanding ferroelectric and piezoelectric properties with the highest fraction of β phase. The piezoelectric coefficient (d_{33}) value of poled P(VDF-TrFE)-0.5 wt% rGO composite was reported to be 80 pm/V. PENG based on P(VDF-TrFE)-0.5 wt% rGO composite film exhibited a high piezoelectric output voltage of 90 V and power density of 0.34 mW/cm^2. HPTENG device using P(VDF-TrFE)-1 wt% rGO composite film produced the highest output voltage and power density values of 227 V and 0.287 W/cm^2, respectively.

Silibin et al. [53] have investigated piezoelectric response, dielectric permittivity, and mechanical properties of PVDF-HFP-GO spin-coated films. The film thickness varied from 400 nm to 500 nm for various compositions. The load-displacement curve at indentation depth of less than 200 nm (0.3-mN maximum loading) showed linear elastic loading and unloading behavior and the Young's modulus of PVDF-HFP composite film was increased from 32.9 GPa to 34 GPa for 1 wt%

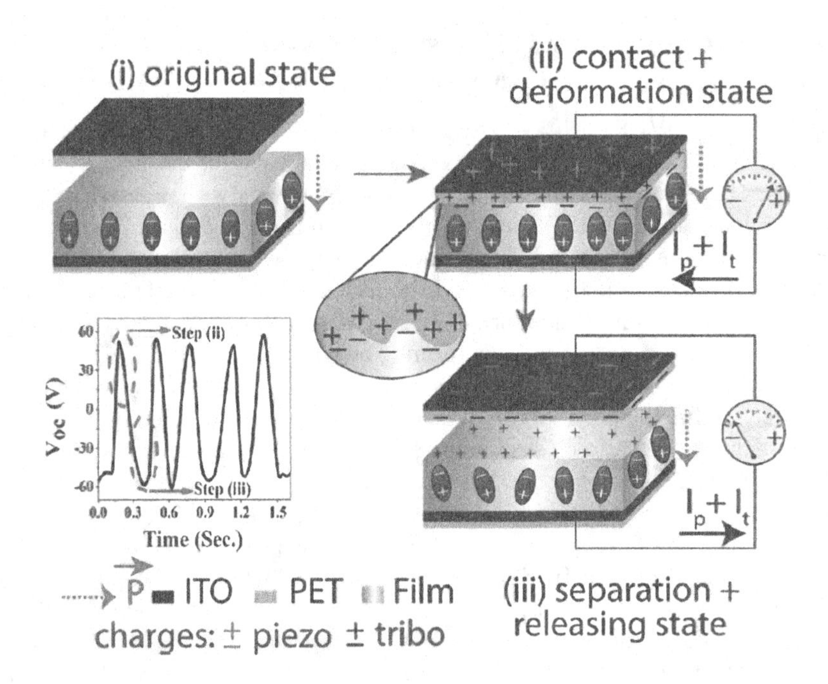

FIGURE 15.6 The contact-separation mechanism in P(VDF-TrFE)-GO/ITO/PET HPTENG. (Adapted from Ref. [52].)

GO incorporation. In order to understand the mechanism of piezoresponse of the nanocomposite film, the authors computed piezoelectric coefficients by constructing models of PVDF-GO composites. The three variants considered were (1) H-side connected from PVDF to GO, (2) fluorine-side connected from PVDF to GO, and (3) GO/PVDF on both sides (sandwich type). First, the model without externally applied electric field to determine the initial optimal parameters of PVDF chain height was considered. Then with the application of electric field, the authors tried to search for the new geometry for new atomic configuration. Due to the changes of the main parameters such as PVDF chain height in its central part, the authors could calculate d_{33} coefficient and it decreased with GO incorporation (single side with eight OH groups) with a value –14.6 pm/V when compared to that of pure PVDF (–38.5 pm/V). In case of double-sided GO model, the d_{33} coefficient got increased to –29.8 pm/V. The qualitative agreement between the experimental and calculated values of piezoelectric coefficient suggested that the reduction in the piezo signal for 1 wt% GO loading in the polymer could be correlated to statistical disorientation of GO and PVDF layers. However, modeling the sandwich structure of PVDF-GO and the piezoresponse force microscopy (PFM) analysis of the GO grains suggests enhanced piezoelectric property of the multilayer structure, which can be witnessed.

Zhao et al. [54] have developed highly stable, sensitive, and efficient nanogenerators using electrospun P(VDF-TrFE)-MWCNT nanofiber membranes. Nanogenerator was prepared by packing the P(VDF-TrFE)-MWCNT membranes with copper foil electrodes along with the (polyethylene terephthalate) PET substrates. External connections are made through copper wires connected to copper electrodes. The addition of MWCNTs at an optimal loading (3 wt%) in the polymer matrix accumulates charges at the interface increasing the β phase of the polymer in the nanocomposites, which results in better piezoelectric properties. The nanogenerator made using P(VDF-TrFE)-3 wt% MWCNT exhibited a high peak output voltage of 35.45 V and current of 3.04 μA when compared to that of pure P(VDF-TrFE) (approximate values obtained from graph are 8 V and 9 μA). A maximum power density of 6.53 μW/cm^2 at load resistance 10 MΩ was achieved.

Wu et al. [55] have reported drastic enhancement in power generation when 0.15 wt% graphene was incorporated into P(VDF-TrFE) matrix. The composites were synthesized as a combination of solution casting, stretching, heat treatment, and electric poling. The poling was done in the electric field range of 10 MV/m–60 MV/m with the increment of 10 MV/m at each step. DMF was used as a solvent. The stretching of the composites had a major influence on crystallization of β phase of the composite films. Scanning electron microscopy (SEM) images clearly revealed the alignment of P(VDF-TrFE) molecular chains in the direction of stretching. The stretching of polymer films enhanced the piezoelectric performance of the composite. Maximum open-circuit voltage of 12.43 V was obtained at an optimal loading of 0.15 wt% graphene, which was twice that of the value obtained for pure P(VDF-TrFE) (6.10 V). A maximum power of 148.06 W/m^3 was harvested at 0.15 wt% graphene in the polymer at a load resistance of 16.92 MΩ. Under identical conditions, the power value obtained for P(VDF-TrFE) film was 36.77 W/m^3.

Cherumannil Karumuthil et al. [56] have investigated the piezoelectric properties of P(VDF-TrFE)-1%ZnO-0.01% exfoliated graphene oxide (EGO) hybrid composites synthesized by electrospinning process. The polymer was dissolved in DMF and MEK mixture (3:1). The nanogenerator was made using electrospun P(VDF-TrFE)-ZnO-EGO composite fiber packed with flexible PDMS substrates on either side. Inclusion of EGO prevented agglomeration of ZnO by uniformly distributing it in the polymer matrix and facilitated more nucleating sites to improve the crystallization of the β phase in the P(VDF-TrFE). Output voltage of 0.40 V was obtained on finger tapping mode and maximum value of 2.50 V on mechanized tapping from hybrid PNEG (polymer nanocomposite energy generator). The current generated using pristine P(VDF-TrFE) was increased from 0.12 nA to 0.23 nA on the addition of ZnO-EGO fillers to P(VDF-TrFE) at the loading mentioned earlier. Xu et al. [57] have developed a high-efficiency beam-structured energy harvester using P(VDF-TrFE)-SWCNT composite with appropriate boundary conditions. The boundary conditions imposed were tilting the beam-structured composite at different angles and using a mechanical pin stop for the beam to bend as shown in Figure 15.7. The composites were solution blended and cast. The beam structure was placed on a 3D printing testing platform to conveniently tilt the samples. DSC and FTIR studies proved that incorporation of SWCNT enhanced crystallization of the β phase. As depicted in Figure 15.7, on tilting the beam-structured composite at 45° and using a mechanical stop (pin) perpendicular to the beam structure, the beam structure was bent to its maximum to yield the highest peak voltage and current. The output voltage and current were improved by 160% and 200%, respectively, when compared to that of films with no such boundary conditions. The power output from the P(VDF-TrFE)-0.1 wt% SWCNT beam structure with boundary condition of tilting the beam structure at 45° was increased to 122%. The output voltage and the current values of 2.2 V and 23.3 nA, respectively, were obtained for the same condition. Wang et al. [58] have achieved enhanced piezoelectric properties using transparent and ultrathin P(VDF-TrFE)-GO films (<200 nm) for nanogenerator application. The P(VDF-TrFE)-GO films were synthesized by spin coating on ITO substrates. The generators were made using ITO glass substrates onto which P(VDF-TrFE)-GO film was coated, and copper foil electrodes were linked according to the configurations of the generators. The generators were subjected to dynamical compression pressure and the output voltage measured. The authors investigated the output voltage of the generators at a compression pressure of 3.06 kPa both for in-plane and out-of-plane configurations. An increased β-phase content was obtained on the

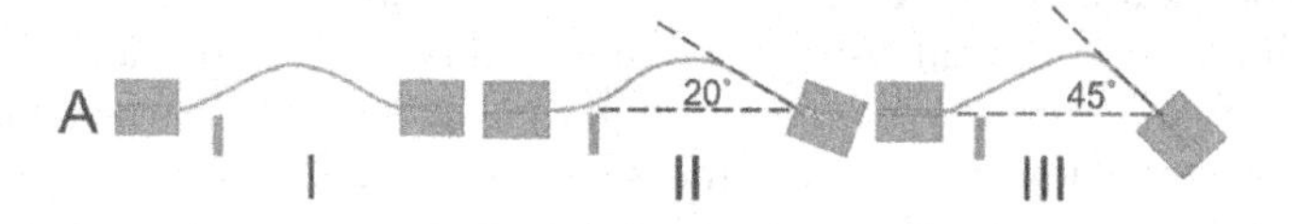

FIGURE 15.7 The maximum bending of P(VDF-TrFE)/SWCNT beam structure composite with mechanical pin (green bar) and tilting at 45°. (Adapted from Ref. [57].)

inclusion of GO fillers into the P(VDF-TrFE) matrix. XRD pattern of P(VDF-TrFE)-0.1 wt% GO films proved the fact that the complete transformation of α–β phase could be realized upon the incorporation of GO in the polymer matrix as only 20.1° peak in 2θ (corresponding to β phase) was observed. Enhanced piezoelectric properties were obtained using P(VDF-TrFE)-0.1 wt% GO films as revealed through the output voltage obtained for in-plane (copper electrodes are placed on either end of the film) and out of plane (copper electrode is placed on top of the film and sealing it) configurations (27 mV and 49 mV). The accumulation of charges at the interface of GO and P(VDF-TrFE) on incorporating GO in the P(VDF-TrFE) matrix enhanced electroactive β phase in nanocomposites and resulted in high-voltage output for nanocomposite film when compared to that of a pure polymer film.

Similarly, Wang et al. [59] investigated in detail the piezoelectric properties of P(VDF-TrFE)-CNT nanocomposites. Dense CNT array was grown on silicon wafer by chemical vapor deposition. P(VDF-TrFE) solution was then deposited over these arrays and dried at room temperature. This work reports that using random CNT film as the sandwiched structure on either side of P(VDF-TrFE) resulted in higher piezoelectric performance as the CNT orients more β crystals of the polymer. Output voltages of approximately 35 mV and 25 mV were obtained in CNT-P(VDF-TrFE)-CNT composite films on stretch-release and bending modes, respectively, when compared to that of P(VDF-TrFE) film (approximately 15 mV, 10 mV). The remnant polarization was increased in CNT-P(VDF-TrFE)-CNT sandwiched structure films when compared to that of P(VDF-TrFE) film supporting the enhancement in β-phase fraction of the polymer.

Apart from P(VDF-TrFE), PVDF-HFP is also extensively used for mechanical energy harvesting applications. PVDF-HFP is preferred due to its increased hydrophobicity, superior mechanical strength, etc. than PVDF. Cai et al. [60] have developed PVDF-HFP-carbon black (CB)-few-layered graphene hybrid nanocomposite film by solution casting followed by uniaxial stretching and stepwise poling methods. The thickness of the prepared film varied from 85 μm to 110 μm for different loadings of fillers. Incorporation of few-layered graphene (FLG) and CB resulted in a high fraction of β phase, which was evident from FTIR studies. The process of stretching increased the β phase of PVDF-HFP from 27% to 79%. The PVDF-HFP-0.3 wt%, CB-0 wt%, FLG and PVDF-HFP-3 wt%, CB-0.02 wt%, and FLG composites resulted in 97% (before stretching 71%) and 95% (before stretching 74%) of β phase of the polymer, respectively, on stretching. The process of poling had a much lesser impact on the formation of β phase. The enhancement in the piezoelectric properties was revealed through enhanced open-circuit voltage. The open-circuit voltage obtained in PVDF-HFP-0.3 wt%, CB-0.02 wt%, FLG and PVDF-HFP-0.5 wt%, CB-0.02 wt%, and FLG were 79% and 81%, respectively, higher than that of what was obtained for pure PVDF-HFP film (2.29 V). Li et al. [61] have developed a conducting, breathable, washable, and an efficient piezoelectric material from PVDF-HFP-MWCNT nanofibrous mat synthesized by electrospinning method followed by thermally induced selective welding at specific cross points of MWCNTs. The electrospun nanofiber of the polymer was collected by a copper plate floating in a tank containing 4% MWCNT solution. The PVDF-HFP nanofibrous mat coated with MWCNT finally annealed at 120°C and 130°C, respectively, for 10 minutes. Contact resistance of nanofiber was reduced by selective welding. The increase in β-phase content of the polymer has been proved through DSC, XRD, and FTIR studies. The fraction of β phase obtained in PVDF-HFP nanofibers was 88% and enhanced to 92.6% and 93% in PVDF-HFP-MWCNT nanofibrous mat on annealing them at 120°C and 130°C for about 10 minutes. The orientation of the PVDF-HFP chains increased with the increase in annealing temperature facilitating improved β-phase content. Enhancement in piezoelectric properties in the nanofibrous mat resulted in maximum output voltage of 0.62 V when a pressure of 15 N was applied by pressing with finger. The authors also investigated the response of PVDF-HFP-MWCNT nanofibrous mat to the finger joint bending and wrist bending to prove the bending stability of the mat.

TABLE 15.1
Comparison of Piezoelectric Parameters of Various Electroactive Polymer/Carbon Nanostructure Composite-Based Mechanical Energy Harvesters

Materials (Ref.)	Processing Route	β Phase %	d_{33}	Power Density (J/cm^2)/Output Voltage/Output Current
PVDF-rGO [37]	Electrospinning	87	–	2800 nW/cm^2, 16 V
PVDF-CNT [38]	Electrospinning	68.4	–	6 V, 81.8 nW
PVDF-GF(GO/GOCOOH) [39]	Electrospinning	69–89	40–63 pC/N	–
PVDF-Ce^{3+}-doped graphene [40]	Electrospinning	99	–	3 V
PVDF-GNP [42]	Electrospinning	77–83	–	3.8–7.9 V
PVDF-GS [43]	Electrospinning	76–88	–	25–45 V, 23.4–33.7 $\mu W/cm^2$
PVDF-Gr (or HNT or GO) [44]	Electrospinning	–	18–24 pC/N	1.3–2 V
PVDF-graphene aerogel [45]	Vacuum-assisted impregnation	–	16 pC/N	0.51 μA
PVDF-activated carbon [46]	Ultrasonication and heat treatment	–	–	37.7 V, 299 nA, 63.07 mW/m^2
PVDF-3D CNTs [47]	Solution casting, hot pressing, and rolling	–	48 pC/N	–
PVDF-GO [48]	Solution casting	100	–	–
PVDF-MWCNT [49]	Drawing	90	–	–
PVDF-TrFE/GO [50]	Drop casting	–	–	4V, 1.88 μA, 4,41 $\mu W/cm^2$
P(VDF-TrFE)-GO [51]	Sol-gel process	–	–80 pm/V	90–227 V, 0.34 mW/cm^2-0.287W/cm^2
P(VDF-TrFE)-GO [52]	Spin coating	–	–29.8 pC/N	–
P(VDF-TrFE)-MWCNT [53]	Electrospinning	–	–	35.45V, 3.04 μA, 6.53 $\mu W/cm^2$
P(VDF-TrFE)-graphene [54]	Solution casting, stretching, heat treatment, and electric poling	–	–	12.43 V, 148.06 W/cm^2
P(VDF-TrFE)-ZnO-Gr [55]	Electrospinning	–	–	0.40–2.5 V, 0.23 nA
P(VDF-TrFE)-SWCNT [56]	Solution casting	–	–	2.2 V, 23.3 Na
P(VDF-TrFE)-GO [57]	Spin coating	–	–	27–49 mV
P(VDF-TrFE)-CNT [58]	Templating effect, annealing	–	–	25–35 mV
P(VDF-HFP)-CB/FLG [59]	Uniaxial stretching, stepwise poling	95–97	–	51.9 W/m^2, 4.14 V
P(VDF-HFP)-MWCNT [61]	Electrospinning and thermally induced selective welding	92–93	–	0.62 V

The piezoelectric performance of various composites of PVDF and its copolymers incorporated with carbon nanostructures and their processing routes have been summarized in Table 15.1. The piezoelectric property of the composites depends on the electroactive β or γ phase generated in the chosen polymer matrix. It can be realized that most of the works have adopted various processing techniques to achieve maximum transition of α to β phase in the composites. Selecting effective processing routes and optimizing the processing conditions to improve the molecular interaction between PVDF and carbon nanofillers will enhance the β-phase content at minimal loading of fillers. With enhanced β-phase content, mechanical energy harvesting can be accomplished using polymer nanocomposite film or mat.

15.4 TE ENERGY HARVESTING USING ICP-BASED NANOCOMPOSITES

Carbon nanostructures such as graphene, CNT, and graphite nanoplatelets have been incorporated in various inherently conducting polymers such as PANI, PPy, etc. to synthesize conducting polymer nanocomposites for TE application. The higher electrical conductivity and surface area of the carbon nanostructures are the important factors to enhance TE properties. Thus, in this section, recent advancements in the synthesis and properties of carbon nanostructures that incorporated PANI and PPy conducting nanocomposites will be discussed with reference to TE applications.

15.4.1 PANI-Carbon Nanostructure-Based TE Nanocomposites

15.4.1.1 PANI-Graphene/CNT TE Nanocomposites

PANI due to its easy processability, low cost, good environmental stability, and excellent electrical conductivity has received huge attention for TE applications. The electrical conductivity of PANI can be fine-tuned by doping with suitable dopants like camphor sulfonic acid (CSA), dodecyl benzene sulfonic acid (DBSA), etc. To achieve enhanced electrical conductivity better than inorganic conductors (bismuth telluride (Bi_2Te_3), lead telluride (PbTe), silicon germanium (SiGe), etc., carbon nanostructures such as graphene, CNTs, GNP (graphene nanoplatelets), etc. are incorporated into the polymer matrix. Graphene is one of the most widely used carbon nanofiller due to its higher electrical conductivity and mechanical strength. In a study by Amirabad et al. [62], PANI-graphene nanosheets (GNS)-coated fabric was prepared at varying filler concentrations. The polyester fabric was soaked in polymerized aniline/GNS suspension in 1-M HCl after the slow addition of required amount of ammonium persulfate (APS) dissolved in 1-M HCl. The reaction was carried out under nitrogen atmosphere and the reaction temperature was maintained between 0°C and 4°C. The study revealed that the addition of GNS filler increased the electrical conductivity and Seebeck coefficient of the nanocomposite on increasing the filler content (0.5 wt%, 2.5 wt%, 5 wt%, and 10 wt%). The enhancement in electrical conductivity on incorporating GNS was attributed to the π-π interactions between GNS and PANI, which facilitated better charge transport. The temperature-dependent variation of Seebeck coefficient for PANI-2.5 wt%, GNS, and PANI-5 wt% GNS-coated fabrics is depicted in Figure 15.8. The upward progression of Seebeck coefficient with respect to temperature for the PANI-GNS nanocomposite suggested the p-type nature of the nanocomposite. The PANI-2.5 wt% GNS had the highest PF value of ~0.0032 $\mu W/(m\ K^2)$ at 338°C. Functionalization of graphene is another viable technique to improve the overall TE performance of the composite by obtaining enhanced dispersion of nanofillers in the chosen polymer matrix.

In a study by Lin et al. [63], enhancement of the TE properties on incorporating *p*-phenylenediamine-modified graphene (PDG) in PANI matrix has been reported. The temperature during the synthesis of PDG was varied between 30°C and 80°C to understand the effect of temperature on the modification of graphene. Modified graphene when incorporated in PANI matrix resulted in the formation of semi-interpenetrating networks (SIPNs) due to chemical bond between linear PANI chains and the graphene sheets. The SIPNs forming conductive pathways facilitate smooth transfer of charge carriers and hence reduce the overall resistance during the charge transport improving TE performance of the composites. PDG prepared using higher modification temperatures resulted in higher electrical and thermal conductivities with decreased Seebeck coefficient. To understand the effect of modification temperature on the properties of the PANI-PDG composite, the concentration of PDG was fixed at 1 wt% in the composite and synthesized at various modification temperatures (30°C, 45°C, 60°C, and 80°C). The authors have shown that PANI-1 wt% PDG composite prepared at modification temperature of 80°C exhibited the highest electrical conductivity of 37 S/cm due to the formation of SIPNs between PANI and PDG. The effect of PDG filler concentrations (1 wt%–5 wt%) in PANI on the TE properties of the composite was also studied, at modification temperatures of 30°C and 80°C, respectively. In general, the electrical conductivity was higher for the PANI-PDG composites synthesized at modification temperature of 80°C (37 S/cm–68.5 S/cm) compared to

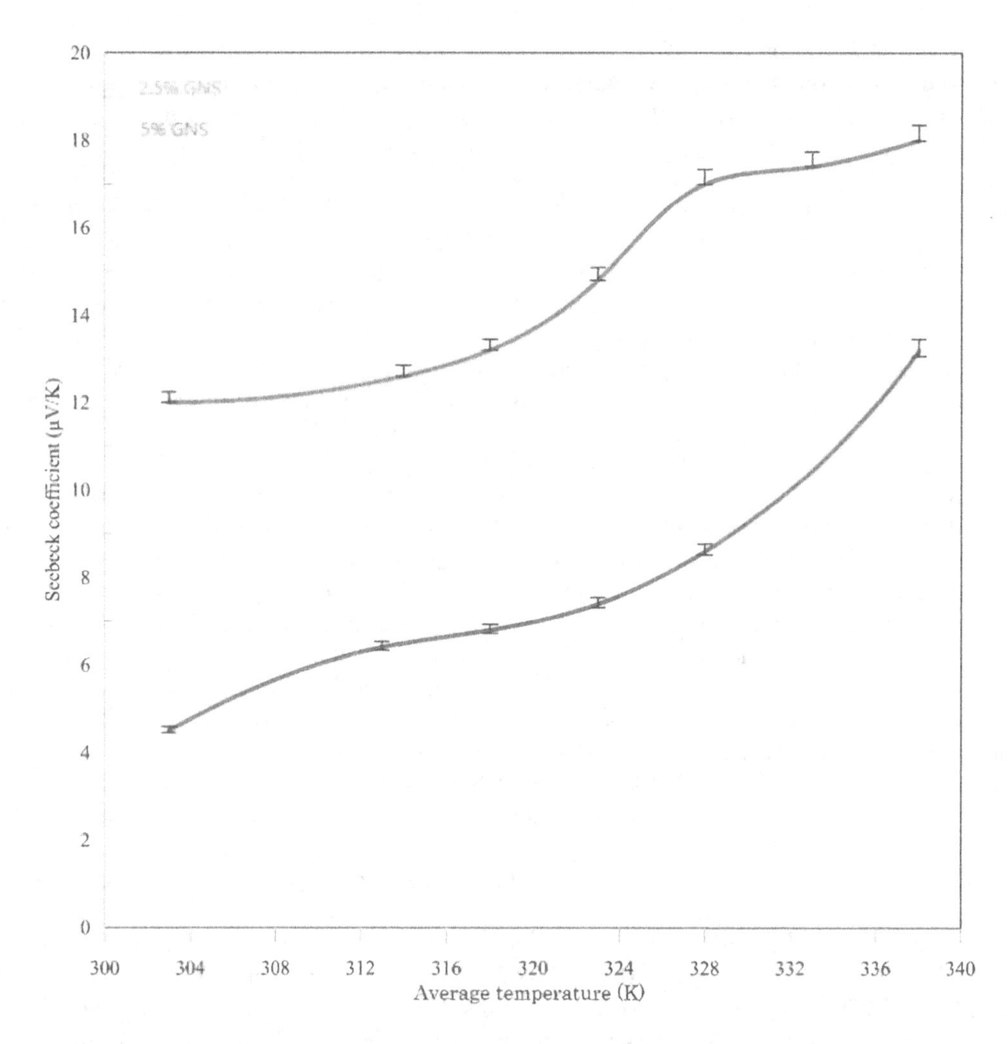

FIGURE 15.8 Temperature dependent Seebeck coefficient for PANI-2.5 wt% GNS and PANI-5 wt% GNS coated fabric. (Adapted from Ref. [62].)

those synthesized at 30°C (14.5 S/cm–20.7 S/cm). The highest electrical conductivity of 68.5 S/cm was reported for PANI-15 wt% PDG synthesized at 80°C. The enhancement in electrical conductivity was attributed to the increased crosslinks formed between PANI and PDG at the increased modification temperature of 80°C and concentration of fillers used. The minimal amount of PDG fillers (1 wt%–3 wt%) formed SIPNs as a result of which enhancement in electrical conductivity and reduction in the thermal conductivity of the composites were realized eventually leading to higher ZT values. The highest ZT values 0.74 and 0.68 were obtained for PANI-3 wt% PDG at modification temperature of 30°C and PANI-1 wt% PDG composites at a modification temperature of 80°C, respectively.

An alternative route to enhance the TE properties of PANI-graphene composites has been adopted in a study by Hsieh et al. [64]. The author used a 3D graphene (3DG) scaffold as a matrix and PANI as filler. The porous nature and interconnected graphene flakes in the 3DG scaffold are reported to increase the dispersion and conductivity of the PANI-graphene composites. The PANI is incorporated into the 3DG scaffold by in situ electrochemical polymerization. At first, the pristine 3DG was soaked in a mixture of 0.5-M H_2SO_4 and 0.05-M aniline aqueous solution. The monomer molecules were uniformly adsorbed onto the graphene flakes. The galvanostatic electrodeposition was carried out with as prepared 3DG as the working electrode and platinum as the CE. The amount of PANI deposited was controlled by manipulating the time of electrodeposition. The PANI content of 45 wt%, 64 wt%, 80 wt%, and 93 wt% were deposited onto 3DG. Furthermore, this technique resulted in a uniform dispersion of PANI in a 3DG matrix that effectively increased the electrical conductivity of the composite. The composite with 80 wt% PANI deposited onto 3DG exhibited the highest PF and electrical conductivity of 81.9 μW/(m K^2) and 394.1 S/cm, respectively. Bending tests performed on 80 wt% PANI-3DG sample showed that the PF saturated after 1500 cycles with a value of 57 μW/(m K^2). Enhancement in Seebeck coefficient was observed up to 80 wt% PANI-3DG sample due to strong π-π interactions between PANI and graphene and the electron energy filtering effect at the interfaces of graphene and PANI. By preventing low-energy carriers with energies below the barrier height from participating in the conduction process, the average energy per carrier transported can be enhanced. This phenomenon is known as electron energy filtering effect. Beyond 80 wt% PANI, overcoating of PANI molecules onto graphene-reduced electron energy filtering effect resulting in the decrease of Seebeck coefficient.

Anisotropy in the TE properties of PANI-CNT composites synthesized by in situ polymerization method followed by cold pressing has been studied by Chen et al. [65]. The PANI-CNT composites were cold pressed in the form of cylindrical pellets. The authors have used two different fillers, SWCNT and MWCNT, to analyze their effect on the TE properties of the composite. The TE properties measured in the perpendicular direction of the cold press axis were considerably higher than that of the TE properties measured in the parallel direction of cold press axis. This was due to the better alignment of CNTs in the perpendicular direction. The TE parameters were evaluated in the perpendicular direction to cold press axis to study the effect of SWCNT and MWCNT concentrations on the TE performance of the composites. The maximum PF of 0.2 μW/(cm K^2) and 0.03 μW/(cm K^2) were reported for PANI-40 wt% SWCNT and PANI-40 wt% MWCNT, respectively, at 400K when measured in the direction perpendicular to that of cold pressing pressure axis. The electrical conductivity obtained in the perpendicular direction to the cold pressing pressure axis was found to be the highest for 40 wt% CNT loading. The electrical conductivity of the composites reduced further by increasing the filler content due to agglomeration of CNTs. Overall, the electrical conductivity of PANI-SWCNT nanocomposites was observed to be higher than that of PANI-MWCNT composites. In general, PANI-SWCNT composites exhibit better TE properties when compared to PANI-MWCNT composites. The largest PF ratios ($PF_{perpendicular}/PF_{parallel}$) of 1.7 and 1.5 were obtained for PANI-10 wt% SWCNT and PANI-30 wt% MWCNT composites, respectively.

Wang et al. [66] investigated the TE properties of PANI-MWCNT composite nanofiber prepared by a combination of in situ polymerization and electrospinning processes. Aniline monomer was polymerized with APS in the presence of different weight fractions of MWCNT. First aniline monomer and MWCNT were mixed and 1-M HCl was added followed by dropwise addition of APS in 1-M HCl in order to polymerize aniline. The polymerization reaction was carried out at 0°C in an ice bath for 6 hours. The nanocomposite powder obtained finally was mixed with CSA at a mol ratio of 1:0.5 (emeraldine base:CSA). The composite powder obtained through in situ polymerization of aniline-MWCNT mixture, pure PANI, direct mixing of PANI powder, and MWCNT were electrospun with chloroform used as the solvent. The thick fiber mats were pressed into pellets and used for various characterizations. The fiber mats were cut into strips along two different directions: parallel and perpendicular to the fiber axis and pressed to form pellets. The electrical conductivity (σ) and Seebeck coefficient (S) of PANI-40 wt% MWCNT were reported to be 17.1 S/cm and 10 μV/K, respectively, when in situ polymerization approach was utilized for the synthesis of composite nanofibers. The previous values of electrical conductivity and Seebeck coefficient were 10 times greater than that of pure PANI synthesized under identical conditions. In the in situ polymerization approach, arrangement of PANI chains on MWCNT was greatly improved due to π-π interaction that resulted in the enhanced mobility of charge carriers and hence increase in σ and S were realized. Furthermore, anisotropies in σ and S along parallel and perpendicular to the fiber direction were realized. For direct mixed PANI powder and MWCNT composite followed by electrospinning, isotropic behaviors of σ and S were obtained. It was reported that both σ and S were higher when measured along parallel direction when compared to the values obtained along perpendicular direction for the composite nanofiber synthesized by in situ polymerization followed by electrospinning techniques. The enhancement in the TE properties was realized due to the high level of orientation of polymer chains. Thus, the effect of processing route greatly affected the TE properties of the PANI-MWCNT nanofibers.

Besides the usage of different fillers with and without functionalization, polymer matrix properties also have a significant effect on the TE properties of composites. Wu et al. [67] synthesized PANI by chemical oxidative polymerization method to understand its effect on the TE properties. An amount of 1.25-mmol ammonium persulfate in 100-ml 1-M HCl was added to 0.25-mmol aniline in 100-ml HCl during initial synthesis of PANI by oxidative polymerization. Then PANI was later doped with CSA to prepare PANI-SWCNT solution cast films. The polymerization of aniline in the presence of required amount of SWCNT was carried out for 5 hours. PANI-60 wt% SWCNT exhibited enhanced electrical conductivity and PF of 4000 S/cm and 56 μW/(m K^2), respectively, at room temperature. The π-π interactions between SWCNT and PANI resulted in a well-ordered arrangement of PANI chains at the interface enhancing the electrical conductivity of the PANI-SWCNT composite. However, at remarkably high filler loading (greater than 90 wt%) of SWCNT, the formation of interface layer was reduced due to improper coating of PANI over SWCNT resulting in the decrease of electrical conductivity and Seebeck coefficient of the composites. Polymerization duration had a remarkable impact on the TE properties of pure PANI films due to the variation in molecular weight of PANI. However, the TE properties of PANI-60 wt% SWCNT composite films with 5 hours of polymerization duration for the synthesis of PANI have been observed to be superior to that of the same composite synthesized with different polymerization durations of aniline. Thus, the polymerization duration for the synthesis of PANI had a major influence on the TE properties of the composite films. The authors had also studied the transport properties of the composite film through temperature-dependent TE properties in the range of 300K–410K. The composition of CNT was varied from 30 wt% to 70 wt% in PANI. For all the compositions, the electrical conductivity was found to decrease with an increase in temperature suggesting metallic conducing nature of the film. The maximum PF reported by the authors for PANI-60 wt% SWCNT at 410K was 100 μW/(m K^2).

15.4.1.2 PPy-Based TE Nanocomposites

PPy is another extensively used conducting polymer for TE application due to several attractive properties such as good electrical conductivity (after doping with appropriate dopants), low thermal conductivity, environmental stability, etc.. PPy is also extensively focused on applications such as EMI shielding, drug delivery, supercapacitors, etc. [68, 69]. Jana Chatterjee et al. [70] have reported enhancement in TE properties of PPy-5 wt% SWCNT nanocomposite. PPy-SWCNT nanocomposites with varying loadings of the filler were prepared by chemical oxidative in situ polymerization technique with 5-sulfosalicylic acid (SSA) as the dopant. The incorporation of SWCNT in PPy was found to enhance the TE properties by enhancing the electrical conductivity and PF through network formation and energy filtering effect at the interfaces, respectively. The PPy-5 wt% composite exhibited higher electrical conductivity of 26.2 S/cm and a PF of 19.1 μV/K at room temperature when compared to that of pure PPy pellets (12.5 S/cm, 3.25 μV/K). The figure of merit of the composite was low with a value of nearly 0.002. However, it was found to be about 55 times higher than that of PPy pellets (0.00004). Also, TE properties of PPy-5 wt% SWCNT pellet were significantly improved when compared with that of pure PPy pellets.

Apart from binary conducting nanocomposites, ternary nanocomposites have also shown promise in enhancing the TE properties. TE properties of PANI-PPy-SWCNT ternary nanocomposite at different mass ratios were synthesized by oxidative in situ polymerization method as reported by Wang et al. [71]. APS was used as oxidant for the polymerization of aniline and pyrrole monomer.

The ternary composites were further doped with 0.1-M $FeCl_3$ to improve the electrical conductivity. After adding $FeCl_3$, the undoped nitrogen atoms of PANI and PPy in the ternary composite tend to oxidize and form new pairs of anions and cations that dissociated and formed free charge carriers. This resulted in higher carrier concentration and improved electrical conductivity. The room temperature electrical conductivity and PF of the doped PANI-PPy-SWCNT composite (with a mass ratio of aniline:pyrrole:SWCNT equal to 2:1:3) were 2.9 and 2 times higher than that of undoped PANI-PPy-SWCNT composite. A reduction in electrical conductivity was observed with increasing PPy concentration due to the agglomeration of PPy at the interface of SWCNT. In a study by Baghdadi et al. [72], PPy-MWCNT 1D nanocomposites were synthesized using SDS (sodium dodecyl sulfate) as surfactant by in situ polymerization of pyrrole with different filler loadings. Addition of surfactant was found to improve the coating of PPy onto the MWCNT walls preventing agglomeration of MWCNTs. Due to the perfect coating of PPy on MWCNT, higher electrical conductivity and PF of 4000 S/m and 1.3 $\mu W/(m\ K^2)$, respectively, were obtained for PPy-14.6 wt% MWCNT composite. In another study by Wang et al. [73], a similar ternary PANI-PPy composite was synthesized with graphene as the conducting filler. The TE properties were measured using cold-pressed PANI-PPy-graphene nanocomposites. The in situ polymerization route utilized for the synthesis of PPy-graphene nanocomposites was found to increase the packing of PPy chains onto graphene resulting in better dispersion of graphene. PANI-PPy-32 wt% graphene ternary composite was synthesized by ultrasonic mixing of 24 wt% PPy, 32 wt% graphene, and 44 wt% of PANI nanorods. PANI-PPy-graphene composites exhibited excellent electrical conductivity and Seebeck coefficient owing to π-π interactions between the PANI and PPy-graphene. Furthermore, hydrogen bonds were formed between nitrogen and hydrogen atoms in PANI-PPy-graphene composites. These interactions resulted in better arrangement of PANI on the surface of PPy-graphene. The presence of interfaces in ternary composites also enhanced the Seebeck coefficient due to the energy filtering effect. The ternary composite resulted in a PF value of 52.5 $\mu W/(m\ K^2)$ that is significantly higher than that of the PFs obtained for PANI-32 wt% graphene and PPy-32 wt% graphene nanocomposites (14.6 $\mu W/(m\ K^2)$ and 19.4 $\mu W/(m\ K^2)$) at 90°C, respectively.

In a study by Du et al. [74], the impact of surface morphology modifications of PPy on the TE properties of the PPy-graphene composites has been reported. The study involved the synthesis of PPy-graphene composites by chemical oxidative polymerization with three different morphologies of PPy such as nanowires, particles, and nanotubes. The PPy nanowires were prepared by using APS as an oxidant for the polymerization of pyrrole in the presence of cetyl trimethyl ammonium bromide (CTAB) that was used as a surfactant for the better dispersion of PPy nanowires in the solution. A similar procedure was used for the synthesis of PPy-nanowires-graphene nanocomposites in which graphene contents were varied (5 wt%, 10 wt%, 15 wt%, and 20 wt%). The PPy nanotubes and PPy particles were synthesized by using $FeCl_3 \cdot 6H_2O$ for the polymerization of pyrrole monomer. Among all three morphologies, PPy-nanowires-graphene composites exhibited higher PF of nearly 0.1 $\mu W/(m\ K^2)$ at 300K compared to the same composite with other morphologies of PPy. The enhanced TE performance of PPy nanowires employed nanocomposites was attributed to the electrostatic attraction and π-π interaction between the graphene and pyrrole monomer. The pyrrole monomer was adsorbed onto the graphene surface. After adding APS, pyrrole monomer was polymerized into PPy nanowire adhering to the surface of graphene resulting in an improved interaction between them. The electrical conductivity of the pure PPy nanowires was 15.7 S/cm and increased to 32.3 S/cm on inclusion of 20 wt% graphene due to the highly conducting nature of graphene and the network formation between PPy-nanowires and graphene in nanocomposites. Similarly, the Seebeck coefficient and PF were observed to be enhanced with increased graphene loading. The PF of 0.1 $\mu W/(m\ K^2)$ in PPy nanowires was reported to be increased to 0.6 $\mu W/(m\ K^2)$ on adding 20 wt% graphene to the PPy nanowires at 300K. The highest PF of 1.01 $\mu W/(m\ K^2)$ at 380K was reported for PPy nanowires-20 wt% graphene nanocomposite.

The TE performance of various ICP-based nanocomposite systems discussed in this section has been summarized in Table 15.2. The low electrical conductivity of ICP-based nanocomposites

TABLE 15.2
Comparison of Various TE Parameters Reported in the Literature Related to Electroactive Polymer/Carbon Nanostructure Composite-Based Thermoelectric Energy Harvesters

Materials (Ref)	Process	ZT	Power Factor (μW/mk²)	Electrical Conductivity	Seebeck Coefficient (μV/K)
PANI-graphene nanosheets [62]	Ultra-sonication, annealing	–	0.0032	0.10 S/cm	12–18
PANI-PDG [63]	One-step chemical route	0.68–0.74	–	37–68 S/cm	0.27
PANI-graphene [64]	In situ electrochemical polymerization	–	57–81.9	394.1 S/cm	–
PANI-CNT [65]	In situ polymerization, cold pressing	–	0.03–0.2	–	–
PANI-MWCNT [66]	In situ polymerization, electrospinning	–	–	17.1 S/cm	10
PANI-CSA-doped SWCNT [67]	Oxidative polymerization	–	–	3000 S/cm	56
PPy-SSA-doped SWCNT [68]	In situ oxidative polymerization	0.002	19.1	26.2 S/cm	–
PPy-SWCNT [71]	In situ oxidative polymerization	–	16–20	–	–
PPy-MWCNT [72]	In situ polymerization	–	1.3	4000 S/m	–
PANI-PPy-graphene [73]	In situ polymerization	–	52.5	–	–
PPy-graphene [74]	Chemical oxidative polymerization	–	1.01	15–32.3 S/cm	–

restrains the performance of the TE devices made out of them. It is also important to identify a suitable process for the synthesis of a selected polymer nanocomposite system that will result in a better figure of merit for commercial applications. Hence, there exists plenty of opportunities to work toward the development of an ideal ICP matrix-filler combination, suitable processing methods, etc. so that better TE devices with enhanced performance at low cost can be fabricated.

15.5 SOLAR ENERGY HARVESTING BASED ON ICP-CARBON NANOSTRUCTURES NANOCOMPOSITES

The third-generation photovoltaic cells, namely, the DSSCs, have turned out to be a clean energy option for harvesting ambient solar energy from the environment and converting it into electricity. Platinum (Pt) is usually used as a CE in the DSSCs due to its superior electrocatalytic properties and low charge transfer resistance. But owing to its high cost, mass production, etc., inherently conducting polymers like PANI and PPys are being focused to develop low-cost CEs with superior electrocatalytic properties. To achieve high power conversion efficiency (PCE), it is important to enhance the charge transfer by CEs for reducing the redox species in the electrolyte. PANI is preferred as the CE material for its excellent redox behavior and low cost. To improve the electrical conductivity and charge transfer for redox reaction, carbon nanofillers like CNT, graphene,

MWCNT, etc. are incorporated in inherently conducting polymers. Mehmood et al. [75] have developed a highly efficient and electrochemically stable CE, using PANI-graphene composites. The PANI-graphene nanocomposite was made by chemical oxidative polymerization. The graphene was added to hydrochloric acid (HCl) and sonicated to prepare graphene dispersion. Aniline in HCl was added to the graphene dispersion in HCl and was polymerized in the presence of oxidant, APS dissolved in HCl for 24 hours at 0°C–5°C. The PANI-graphene nanocomposite thus obtained was dispersed in water/isopropanol at a ratio of 1:1 (by volume). The graphene content was varied from 3 wt% to 15 wt% in the nanocomposites. The CE was prepared by spray coating the PANI-graphene nanocomposite dispersion on FTO substrate and TiO_2-coated FTO was used as the working electrode. TiO_2-coated FTO was sensitized using N3 dye (*cis*-bis(isothiocyanato)bis (2,2′-bipyridyl-4,4′-dicarboxylato ruthenium (II)). The electrolyte contained I^-/I_3^- redox couple. The incorporation of graphene improved the electrocatalytic properties of PANI by increasing the charge transfer at the interface, hence enhancing the PCE to 7.45% at a loading of 9 wt% graphene, which was almost close to that of standard Pt CE (7.63%). Current density of about 15.504 mA/cm^2 was obtained for PANI-9 wt% graphene composite-based CE.

An effective strategy of layer-by-layer self-assembly was employed to create a multilayered CE $(\text{PANI-graphene (G)-GO})_n$ (*n* indicates no of layers) for rapid charge transfer by Wang et al. [76]. Aniline-graphene complex was synthesized first by reflux method (heating the blended solution in a controlled manner using condenser to prevent loss of solvent at a constant temperature), which resulted in a vigorous charge transfer. Aqueous solution consisting of 1-M HCl and APS was added to aniline-graphene complex at 0°C. The reaction was carried out at 4°C for 10 hours to form PANI-graphene complexes. In order to coat multilayers of (PANI-G-GO), first FTO was dipped in PANI-G aqueous solution for 10 minutes and then dipped in GO aqueous solution. By repeating this process, bilayers can be increased and hence $(\text{PANI-G-GO})_n$ multilayers can be deposited on FTO (*n* refers to deposition cycle). Increasing the number of bilayers enhanced the electron delocalization in PANI-conjugated structures and thus improved the electrical conductivity and electrocatalytic activity of the multilayered CE. TiO_2-based working electrode and N719 dye (di-tetrabutylammonium *cis*-bis(isothiocyanato)bis(2,2′-bipyridyl-4,4′-dicarboxylato) ruthenium (II)) were used in the DSSC. The redox electrolyte consisted of 100-mM tetraethyl ammonium iodide, 100-mM tetramethylammonium iodide, 100-mM tetrabutylammonium iodide, 100-mM NaI (sodium iodide), 100-mM LiI (lithium iodide), 50-mM I_2 (iodine), and 500 mM of 4-*tert*-butylpyridine in 50-ml acetonitrile. High PCE of 7.88% and current density of 16.79 mA/cm^2 using $(\text{PANI-10 wt\% G-GO})_{10}$ CEs with 10 bilayers were obtained.

Bora et al. [77] investigated PANI nanotube-CB nanocomposites synthesized by in situ chemical oxidative polymerization route as CEs in DSSC. Two separate solutions of aniline monomer and APS in aqueous reaction medium of 0.4-mol/l acetic acid and 1-mol/l methanol were prepared and mixed at 0°C–5°C. The mixture was left undisturbed overnight to form PANI nanotubes. Different concentration of CB was mixed with aniline solution and in situ polymerized to form PANI nanotube-CB nanocomposites. The nanocomposite was made as a paste and coated onto FTO glass substrate. N719 dye was used, and the liquid electrolyte comprised LiI, I_2, 0.5-mol/l TBP (*tert*-butylpyridine), and 0.6-mol/l MPI (1-methyl-3-propylimidazolium iodide) in a mixed solvent of NMP and acetonitrile (1:4 volume ratio). A comparative study on the photovoltaic performance of the DSSC was done using a liquid and polymer gel electrolyte. Poly(methyl methacrylate) (PMMA) and PANI nanotubes dispersed in a mixture of NMP (*N*-methylpyrrolidone)-acetonitrile solvent was used as polymer gel electrolyte. Effect of varying thickness of the CE on the efficiency of the photovoltaic device was also studied. Increasing the thickness of 0.75 wt% CB-PANI nanotube nanocomposite coating from 0.78 μm to 10.58 μm considerably increased the PCE due to the enhanced reduction of I_3^-. The authors reported that the addition of CB to PANI nanotubes enhanced catalytic surface area and improved the triiodide reduction ($I_3^- \rightarrow I^-$). On using PANI nanotube-0.75 wt% CB as CE with a liquid electrolyte, a PCE of 6.62% was obtained. The PCE was reduced to 4.82% on using the same PANI nanotube-0.75 wt% CB along with a polymer gel electrolyte due to the hindrance caused by polymer network for diffusion of ions.

Seema et al. [78] have investigated the effectiveness of graphene-incorporated PANI as an efficient alternative to Pt CE. The PANI-SDS-H_2SO_4-rGO (PANI-rGO) nanocomposite was synthesized by in situ emulsion polymerization technique using GO as the precursor. TiO_2 was used as the working electrode and immersed in 0.5-mM N719 dye solution while fabricating the DSSC. The electrochemical performances of PANI, rGO, PANI-rGO (40 wt%) nanocomposite, and Pt CE were understood through cyclic voltammetry studies in a three-electrode cell using Ag/AgCl and Pt as the reference and CEs, respectively. The electrolyte used was 0.01-M LiI, 0.1-M $LiClO_4$ (lithium perchlorate), and 1-mM I_2 and the CV was recorded at a scan rate of 50 mV/s. The redox peaks could be clearly observed and well separated. The magnitude of peak current separation (potential difference) is a measure of electrocatalytic performance of the chosen material used as the CE. The potential difference corresponding to redox peaks in PANI-40 wt% RGO (159 mV) was observed to be lesser than that of pure PANI (203 mV) or rGO (297 mV) with respect to reference Pt-based CE. The lower peak separation for the nanocomposite CE was attributed to enhanced charge transfer kinetics of I^-/I_3^- redox reaction that occurs at a much faster rate at PANI-rGO surface. In a similar fashion, the authors proved a decrease in the charge transfer resistance for PANI-rGO nanocomposite CE using the same electrolyte by electrochemical impedance spectroscopy analysis. The PCE of 3.98% was obtained using PANI-40 wt% RGO CE-based DSSC device under AM 1.5G (100 mW/cm^2) illumination. This value was lesser than that of Pt CE (4.75%) used in the DSSC. Current density obtained for the abovementioned composition when used as CE was 12.58 mA/cm^2 greater than that of what could be obtained for PANI CE (11.03 mA/cm^2). The enhanced electrocatalytic activity was due to a large surface area of PANI-40 wt% rGO as proved by the authors through SEM analysis.

Wu et al. [79] have investigated PANI-MWCNT modified by transition metals to produce an effective CE. Nickel (Ni), cobalt (Co), manganese (Mn), and copper (Cu) were used as dopants in that work. The authors first made 10 mg of MWCNT dispersed in isopropanol to form paste and deposited onto FTO glass followed by sintering at 400°C. Then electropolymerization of aniline monomers in the presence of various transition metal ions onto the MWCNT-coated FTO glass was performed using cyclic voltammetry in the potential range of 0.2 V–1.2 V vs. saturated calomel electrode at a sweep rate of 20 mV/s using an electrochemical analyzer. Typically, the electrolyte contained 0.2-M aniline monomers, 0.5-M HCl, and 0.1-M transition metal ions. DSSC device was fabricated with photoanode (8-μm thick TiO_2 film sensitized with N719 dye). The electrolyte contained 0.1-M LiI, 0.6-M 1-propyl-3-methylimidazolium iodide, 0.07-M I_2, 0.5-M 4-TBP, and 0.1-M guanidinium thiocyanate in 3-methoxypropionitrile. Out of different transition metals used, modifying PANI-MWCNT CE with Ni^{2+} and Co^{2+} exhibited higher catalytic activity in the composites as they were active dopants, whereas Mn^{2+} was neutral dopant with absolutely no effect on the catalytic activity and Cu^{2+} was a negative dopant that deteriorated the catalytic activity in the PANI-MWCNT CE. The varying doping mechanism for different metal ions depends on the empty d orbital of the metal. Ni^{2+} can form octahedral structure with ammonium cationic ligand. Mn^{2+} can form tetrahedral configuration with ammonium-cationic ligands that favored electron transfer like that of octahedral structure. Cu^{2+} ligand attained planar structure and hence the electron transport was retarded. Through EIS and CV studies, the authors proved that suitable modification of PANI-MWCNT with metal ions such as Ni^{2+} and Co^{2+} decreased the charge transfer resistance. The PCE was improved to 6% and 5.75%, respectively, on modifying PANI-MWCNT composites with Ni^{2+} and Co^{2+} in comparison with undoped PANI-MWCNT CE (4.58%). The current densities obtained were 14.25 mA/cm^2 and 13.75 mA/cm^2 for PANI-MWCNT CEs modified with Ni^{2+} and Co^{2+}, respectively. In a study by He et al. [80], enhanced PCE was achieved by using PANI-graphene complex CE in DSSC. Aniline-graphene complex (with varying concentrations of graphene) was refluxed in inert atmosphere using N_2 gas for about 6 hours at 184°C. The aniline-graphene complex was then polymerized (onto FTO substrate) in acidic medium (1-M HCl) with oxidant APS for about 10 hours at 4°C. It was then vacuum desiccated at 50°C for 24 hours. Working electrode, TiO_2-coated FTO, was sensitized by soaking it in a solution of 0.5-mM N 719 dissolved in ethanol. A combination of 100-mM tetraethylammonium iodide, 100-mM tetramethylammonium iodide, 100-mM tetrabutylammonium iodide, 100-mM NaI, 100-mM KI, 100-mM LiI, 50-mM I_2, and 500-mM 4-*tert*-butylpyridine in 50 mL of acetonitrile was used as the electrolyte. The active area of DSSC was 0.25 cm^2. Among all the CE with varying concentration of graphene, the authors reported that PANI-8 wt% graphene complex CE exhibited the highest electrocatalytic activity for reducing triiodide. This enhancement in electrocatalytic activity of PANI-8 wt% graphene complex CE was attributed to its porous structure and increased number of bonding sites formed between PANI and graphene (due to the covalent bond formed between the imine group (–NH–) of PANI and carbon atoms (–C=)of graphene), which improved the transfer of charges within the PANI-graphene complex. Due to this reason, the charge transfer resistance was reduced to 4.96 $\Omega\ cm^2$ compared to its pristine form (188.8 $\Omega\ cm^2$). Also, a maximum PCE of 7.78% and current density of 14.85 mA/cm^2 were obtained in PANI-8 wt% graphene complex CE.

Thuy et al. [81] studied the photovoltaic performance of PPy-graphene-coated aluminum oxide (GCA) composite thin film deposited onto FTO as CE in DSSC. The PPy-GCA composites were synthesized by electrochemical deposition process. The GCA nanoparticles were prepared by coating graphene on Al_2O_3. Different loadings (1.5 wt%, 3 wt%, 4.5 wt%, 6 wt%) of GCA with respect to pyrrole monomer (0.1 M) were dispersed in four drops of perchloric acid in 50 ml of deionized water and 0.1-M pyrrole monomer and then electropolymerized to form PPy-GCA composite film on FTO. DSSC device was fabricated with thin TiO_2 layer deposited onto FTO to form photoanode and immersed in 0.3-M N719 dye in ethanol (sensitizer). The electrolyte used was 0.05-M I_2, 0.2-M LiI, 0.7-M 3-propyl-1-methyl imidazoliumiodide, and 0.5-M *t*-butylpyridine in acetonitrile. The authors reported that the coating of graphene over Al_2O_3 enhanced the interaction between the PPy and GCA nanoparticles by preventing the aggregation of graphene sheets and improved the charge collection and transport within the PPy matrix. Furthermore, the interaction between the PPy and graphene improved the electrocatalytic properties of PPy-3 wt% GCA nanocomposite CE. A PCE of 7.33% almost similar to that of Pt CE (7.57%) was obtained for PPy-3 wt% GCA composites. The current density of 16.54 mA/cm^2 was also obtained for the same composite. The authors also proved the enhanced catalytic properties of PPy-GCA composite CE through CV studies as the peak current increased when GCA was incorporated in PPy in comparison to that of pure PPy under the same scan rate. Also, the peak-to-peak (first oxidation and first reduction peak current) separation potential was reduced for PPy-GCA CE when compared to that of pure PPy CE.

Jin et al. [82] have demonstrated the use of PANI-4 wt% SWCNT and PPy-2 wt% SWCNT complexes synthesized by reflux method as both CE and conducting gel electrolyte in a quasi-solid-state DSSC. The PANI-SWCNT complex was synthesized at 0°C by adding 50-ml aqueous solution of 1-M HCl and $(NH_4)_2S_2O_8$ to 50 ml of aqueous mixture from 1-M HCl and 0.325-M aniline-CNT and PPy-CNT complex by polymerizing pyrrole-CNT with $FeCl_3$ for about 10 hours at 4°C. The samples were then vacuum-desiccated at 50°C–60°C for 24 hours. After solution polymerization, PANI-CNT/PPy-CNT complexes were deposited onto FTO glass substrates and also incorporated into poly (acrylic acid)/poly (ethylene glycol) (PAA/PEG) matrix to form conducting gel electrolytes. Quasi-solid state DSSC was fabricated by sandwiching gel electrolyte between a dye-sensitized TiO_2 photoanode and Pt-free CE. The dye used was N719. The solar cell architectures fabricated were N719-TiO_2/PANI-CNT gel electrolyte/PANI-CNT CE, N719-TiO_2/PPy-CNT gel electrolyte/PPy-CNT CE, and N719-TiO_2/PAA-PEG gel electrolyte/Pt CE. PCE was significantly enhanced to 8.23% on using PANI-CNT complex as CE and conducting gel, i.e., PANI-CNT complex incorporated into PAA/PEG matrix in the same DSSC when compared to the DSSC fabricated using Pt CE and poly (acrylic acid)/poly (ethylene glycol) (PAA/PEG) electrolyte (5.74%). The salient feature of PAA/PEG 3D structures is the enhanced absorption capacity of aqueous or organic solutions. Hence, the absorption of liquid electrolyte containing I^-/I_3^- redox couple in the presence of PANI-CNT and PPy-CNT complexes was enhanced. Therefore, the photoconversion efficiency of conducting gel

TABLE 15.3
Comparison of Parameters Describing the Performance of DSSCs Fabricated with Electroactive Polymer/Carbon Nanostructure Nanocomposites for Solar Energy Harvesting

Counter Electrode Materials (Ref.)	Processing Route	Working Electrode	Dye	Power Conversion Efficiency (%)	Current Density (mA/cm²)
PANI-Graphene [75]	Spray coating	TiO_2	N 3	7.45	15.504
(PANI-Graphene/GO)$_n$ [76]	Layer-by-layer self assembly	TiO_2	N 719	7.88	16.79
PANI nanotube-CB [77]	In situ polymerization	TiO_2	N 719	6.62	9.257
PANI-rGO [78]	In situ emulsion polymerization	TiO_2	N 719	3.98	12.58
PANI-MWCNT-(Ni^{2+}/CO^{2+}) [79]	Electrochemical deposition	TiO_2	N 719	5.75-6	13.75–14.25
PANI-MWCNT complex [80]	Reflux method, oxidative polymerization	TiO_2	N 719	7.78	14.85
PPy-GCA [81]	Electrochemical polymerization	TiO_2	N 719	7.33	6.54
PANI-CNT and PPy-CNT [82]	Reflux method	TiO_2	N 719	7.69–8.23	–
PPy-MWCNT [83]	Three-step oxidative process	TiO_2	N 719	4.9	9.85
PPy-MWCNT [84]	Cyclic voltammetry electrodeposition	TiO_2	N 719	7.15	17.56
PPy-Graphene [85]	In situ emulsion polymerization	TiO_2	N 719	3.58	10.638

electrolyte employed DSSC showed remarkable enhancement as reported by the authors. Similarly, on using PPy-CNT complex as CE in DSSC along with conducting gel electrolyte with PPy-CNT complex, the PCE of 7.69% was obtained, higher than that of traditional quasi-solid state DSSC with Pt as CE with PAA/PEG gel electrolyte (5.74%). The authors further carried out CV and EIS analyses to prove enhanced charge transfer when conducting gel electrolytes were used.

The photovoltaic performances of different ICP-based nanocomposites, their CEs, and the processing routes as described in this section are summarized and provided in Table 15.3. In most of the works, various methods have been adopted to obtain better photovoltaic performance in ICP-based nanocomposite materials. However, the PCE achieved is not on par with the inorganic DSSCs. Employing optimized processing methods to improve the interaction between the polymer matrix and various carbon nanostructures will facilitate enhanced charge transport at the CE to achieve maximum PCE.

15.6 FUTURE PERSPECTIVE

The pivotal role of incorporating carbon nanostructures in the EAPs for harnessing power from mechanical vibrations, thermal energy, and solar energy has been discussed briefly in this chapter. For an efficient mechanical energy harvesting, complete conversion of nonpolar α phase to polar β phase in PVDF and its copolymers is particularly important. Appropriate processing techniques and optimization of the synthesis parameters of the composites will induce the electroactive phase in

PVDF and its copolymer matrices. By achieving maximum β-phase fraction, it is possible to develop an ultrasensitive, effective power generator from the ambient mechanical vibrations. The dispersion of carbon nanostructures in those polymer matrices is the key to induce electroactive phases of the polymer. Hence, more research activities in improving the dispersion of carbon nanostructures in different EAP matrices are still required. With regard to TE materials, one of the biggest challenges faced by the researchers is to achieve reduced thermal conductivity and enhanced electrical conductivity to improve the TE performance of polymer-based nanocomposites. Though the energy filtering mechanism shall enhance the TE performance of the EAP-carbon nanostructure composites, still the dispersion of nanofillers in polymer matrices, identification of suitable synthesis routes for the preparation of nanocomposites poses a great challenge. Thus, there exists plenty of room to improve the TE properties of ICP-carbon nanostructure-based nanocomposite materials. Also, suitable model development in predicting the TE properties of polymer nanocomposites is an open area of research.

Solar energy harvesting using DSSCs is an eco-friendly and relatively cheaper energy harnessing technology. In order to develop an efficient DSSC using EAP-carbon nanostructure-based nanocomposite as CE, the charge transfer resistance must be reduced as much as possible. The interaction between polymer matrices and the chosen fillers must be enhanced so that vigorous charge transport can result. The stability of the DSSC must also be improved in order to avoid frequent replacements. Development of theoretical models to predict the ideal parameters for obtaining higher efficiency in DSSCs is still an open area of research. Thus, the EAP-carbon nanostructure-based nanocomposites offer plenty of scopes to improve the performance of the devices made out of them as well as their commercialization.

15.7 CONCLUSION

The recent advancements in mechanical energy harvesting, TE energy harvesting, and solar energy harvesting using carbon nanostructures incorporated EAP-based nanocomposites have been reviewed in this chapter. With the increasing global demand for energy and deterioration of conventional energy sources, EAP-carbon nanostructure-based nanocomposites can serve as potential advanced material to scavenge the energy from mechanical vibrations, heat, and solar radiations. They are used in developing eco-friendly, low-cost, flexible, and portable power generators. Despite their effectiveness and their potential to be utilized in various ranges of applications, their efficiency in the respective domains of applications is limited. However, adopting appropriate processing techniques, optimizing the processing parameters, and improving the dispersion of the carbon nanostructures in the selected polymer matrices shall improve the performance and efficiency of energy harvesting devices made from EAP-carbon nanostructures nanocomposites.

ABBREVIATIONS

Bi_2Te_3	bismuth telluride
CE	counter electrode
CNF	carbon nanofiber
CNT	carbon nanotube
CO(II)/CO(III)	cobalt redox mediators
CSA	camphor sulfonic acid
DBSA	dodecyl benzene sulfonic acid
DMF	*N,N*-dimethylformamide
DSC	differential scanning calorimetry
DSSCs	dye-sensitized solar cells
EGO	exfoliated graphene oxide
FTIR	Fourier transform infrared spectroscopy
FTO	fluorine-doped tin oxide

GA	graphene aerogel
GF	fluorinated graphene oxide
GOCOOH	carboxylated graphene oxide
GO	graphene oxide
GS	sulfonated graphene oxide
HNT	halloysite
HPTENG	hybrid piezoelectric-triboelectric nanogenerator
ICPs	intrinsically conducting polymers
ITO	indium-doped tin oxide
MEK	methyl ethyl ketone
MPI	1-methyl-3-propylimidazolium iodide
MWCNT	multiwalled carbon nanotube
N3	*cis*-bis(isothiocyanato)bis (2,2′-bipyridyl-4,4′-dicarboxylato ruthenium (II)
N719	di-tetrabutylammonium *cis*-bis(isothiocyanato)bis(2,2′-bipyridyl-4,4′-dicarboxylato) ruthenium (II)
NiO	nickel oxide
P(VDF-TrFE)	polyvinylidene fluoride-trifluoroethylene
PANI	polyaniline
$PbTe_2$	lead (IV) telluride
PDG	phenediamino-modified graphene
PDMS	polydimethylsiloxane
PENGs	piezoelectric nanogenerators
PET	polyethylene Terephthalate
PF	power factor
PPy	polypyrrole
PVDF	polyvinylidene fluoride
PVDF-HFP	polyvinylidene fluoride-hexafluoropropylene
rGO	reduced graphene oxide
Sb_2Te_3	antimony telluride
SEM	scanning electron microscopy
SnO_2	tin(IV) oxide
SWCNT	single-walled carbon nanotube
TBP	*tert*-butylpyridine
TE	thermoelectric
TENG	triboelectric nanogenerator
TiO_2	titanium dioxide
UV	ultraviolet
WAXD	wide angle X-ray diffraction
XRD	X-ray diffraction
ZnO	zinc oxide
ZT	figure of merit
α	Seebeck coefficient
σ	electrical conductivity
κ	thermal conductivity

REFERENCES

1. https://www.iea.org/reports/world-energy-outlook-2020
2. Colonnelli, Stefania, Giuseppe Saccomandi, and Giuseppe Zurlo. 2015. "The Role of Material Behavior in the Performances of Electroactive Polymer Energy Harvesters." *J. Polym. Sci., B Polym. Phys.* 53(18):1303–14.

3. Anton, Steven R, and Henry A Sodano. 2007. "A Review of Power Harvesting Using Piezoelectric Materials-2003-2006." *Smart Mater. Struct.* 16(3):R1–21.
4. Junior, O H, A L O Maran, and N C Henao. 2018. "A Review of the Development and Applications of Thermoelectric Microgenerators for Energy Harvesting." *Renew. Sustain. Energy Rev.* 91:376–93.
5. Jean-Mistral, C, S Basrour, and J-J Chaillout. 2010. "Comparison of Electroactive Polymers for Energy Scavenging Applications." *Smart Mater. Struct.* 19(8):085012.
6. Lovinger, Andrew J. 1983. "Ferroelectric Polymers." *Science* 220(4602):1115–21.
7. Yuan, Xuan, Shuai Changgeng, Gao Yan, and Zhao Zhenghong. n.d. "Application Review of Dielectric Electroactive Polymers (DEAPs) and Piezoelectric Materials for Vibration Energy Harvesting." *J. Phys. Conf. Ser.* 744:012077.
8. Jeong, Changyoon, Chanwoo Joung, Seonghwan Lee, Maria Q Feng, and Young-Bin Park. 2020. "Carbon Nanocomposite Based Mechanical Sensing and Energy Harvesting." *Int. J. Precis. Eng. Manuf.-Green Technol.* 7:247–67.
9. Safaei, M, H A Sodano, and S R Anton. 2019. "A Review of Energy Harvesting Using Piezoelectric Materials: State-of-the-Art a Decade Later (2008-2018)." *Smart Mater. Struct.* 28(11):3001–41.
10. Wan, Chaoying, and Christopher Rhys Bowen. 2017. "Multiscale-Structuring of Polyvinylidene Fluoride for Energy Harvesting: The Impact of Molecular-, Micro- and Macro-Structure." *J. Mater. Chem. A* 5(7):3091–128.
11. Ramanujam, B T S, and Pratheep K Annamalai. 2017. "Conducting Polymer–Graphite Binary and Hybrid Composites: Structure, Properties, and Applications." *Hybrid Polymer Composite Materials.* Amsterdam, The Netherlands: Elsevier Ltd. 1–34.
12. Sachin, M, Reshma Haridass, and B T S Ramanujam. 2020. "Polyvinylidene Fluoride (PVDF)-Poly(Methyl Methacrylate) (PMMA)-Expanded Graphite (ExGr) Conducting Polymer Blends: Analysis of Electrical and Thermal Behavior." *Mater. Today: Proc.* 28(1):103–7.
13. Ramanujam, B T S, Parag V Adhyapak, S Radhakrishnan, and R Marimuthu. 2021. "Effect of Casting Solvent on the Structure Development, Electrical, Thermal Behavior of Polyvinylidene Fluoride-Carbon Nanofiber Conducting Binary and Hybrid Nanocomposites." *Polym. Bull.* 78:1735–51.
14. Batabyal, Sudip Kumar, K Y Yasoda, S Kumar, M S Kumar, and K Ghosh. 2021. "Fabrication of MnS/GO/PANI Nanocomposites on a Highly Conducting Graphite Electrode for Supercapacitor Application." *Mater. Today: Chem.* 19:100394–6.
15. Ramanujam, B T S, and S Radhakrishnan. 2017. "Electrical Properties of Conducting Polymer-MWCNT Binary and Hybrid Nanocomposites."*Trends and Applications in Advanced Polymeric Materials.* Hoboken, NJ: Wiley. 127–43.
16. Sankar, Nitin, Mamilla Nagarjun Reddy, and R Krishna Prasad. 2016. "Carbon Nanotubes Dispersed Polymer Nanocomposites: Mechanical, Electrical, Thermal Properties and Surface Morphology." *Bull. Mater. Sci.* 39(1):47–55.
17. Wang, J, C Wu, Y Dai. et al. 2017. "Achieving Ultrahigh Triboelectric Charge Density for Efficient Energy Harvesting." *Nat. Commun.* 8(88):1–8.
18. Annamalai, P K, A K Nanjundan, D P Dubal, and J-B Baek. 2021. "An Overview of Cellulose-Based Nanogenerators." *Adv. Mater. Technol.* 6(3):2001164.
19. Yan, Haiyan, and Kaichang Kou. 2014. "Enhanced Thermoelectric Properties in Polyaniline Composites with Polyaniline-Coated Carbon Nanotubes." *J. Mater. Sci.* 49:1222–8.
20. Bierschenk, James L. 2009. "Optimized Thermoelectrics for Energy Harvesting Applications." *Energy Harvesting Technologies.* Salmon Tower Building, NY: Springer. 337–51.
21. Zhang, Yinhang, and Soo Jin Park. 2019. "Flexible Organic Thermoelectric Materials and Devices for Wearable Green Energy Harvesting." *Polymers* 11(5):909.
22. Erden, F, H Li, X Wang, F Wang, and C He. 2018. "High-Performance Thermoelectric Materials Based on Ternary TiO_2/CNT/PANI Composites." *Phys. Chem. Chem. Phys.* 20(14): 9411–8.
23. Gao, Caiyan, and Guangming Chen. 2016. "Conducting Polymer/Carbon Particle Thermoelectric Composites: Emerging Green Energy Materials." *Compos. Sci. Technol.* 124:52–70.
24. Ahmed, Usman, Mahdi Alizadeh, Nasrudin Abd, Syed Shahabuddin, Muhammad Shakeel, and A K Pandey. 2018. "A Comprehensive Review on Counter Electrodes for Dye Sensitized Solar Cells : A Special Focus on Pt-TCO Free Counter Electrodes." *Solar Energy* 174:1097–125.
25. He, B, Q Tang, T Liang, and Q Li. 2014. "Efficient Dye-Sensitized Solar Cells From Polyaniline–Single Wall Carbon Nanotube Complex Counter Electrodes." *J. Mater. Chem. A* 2(9):3119–26.
26. Li, Qinghua, Jihuai Wu, Qunwei Tang, Zhang Lan, Pinjiang Li, Jianming Lin, and Leqing Fan. 2008. "Electrochemistry Communications Application of Microporous Polyaniline Counter Electrode for Dye-Sensitized Solar Cells." *J. Electrochem.* 10:1299–302.

27. Zhang, Xueni, Jing Zhang, Yanzheng Cui, Jiangwei Feng, and Yuejin Zhu. 2012. "Carbon/Polymer Composite Counter-Electrode Application in Dye-Sensitized Solar Cells." *J. Appl. Polym. Sci.* 128(1):75–9.
28. Kholkin, A L, N A Pertsev, and A V Goltsev. 2008. "Piezoelectricity and Crystal Symmetry." *Piezoelectric and Acoustic Materials for Transducer Applications*. Salmon Tower Building, NY: Springer. 17–38.
29. Abdalla, S, A Obaid, and F M Al-Marzouki. 2016. "Preparation and Characterization of Poly(Vinylidene Fluoride): A High Dielectric Performance Nano-composite for Electrical Storage." *Results Phys.* 6:617–26.
30. Mishra, Suvrajyoti, Lakshmi Unnikrishnan, Sanjay Kumar Nayak, and Smita Mohanty. 2019. "Advances in Piezoelectric Polymer Composites for Energy Harvesting Applications: A Systematic Review." *Macromol. Mater. Eng.* 304(1):1–25.
31. Wang, Jie, Linglin Zhou, Chunlei Zhang, and Zhong Lin Wang. 2020. "Small-Scale Energy Harvesting from Environment by Triboelectric Nanogenerators." *A Guide to Small-Scale Energy Harvesting Techniques*. London, UK: Intech. 83703.
32. Wu, Changsheng, Aurelia C Wang, Wenbo Ding, Hengyu Guo, and Zhong Lin Wang. 2019. "Triboelectric Nanogenerator: A Foundation of the Energy for the New Era." *Adv. Energy Mater.* 9(1):1–25.
33. Snyder, G J. 2009. "Thermoelectric Energy Harvesting." *Energy Harvesting Technologies*. Boston, MA, USA: Springer. 325–36.
34. Taylor, Ajay Jena, Shyama Prasad Mohanty, Pragyensh Kumar, Johns Naduvath, P Lekha, Jaykrushna Das, et al. 2012: "Dye Sensitized Solar Cells: A Review." *Trans. Indian Ceram. Soc.* 71:1–16.
35. Saranya, K, Md Rameez, and A Subramania. 2015. "Developments in Conducting Polymer Based Counter Electrodes for Dye-Sensitized Solar Cells – An Overview." *Eur. Polym. J.* 66:207–27.
36. Kim, Heung Soo, Joo Hyong Kim, and Jaehwan Kim. 2011. "A Review of Piezoelectric Energy Harvesting Based on Vibration." *Int. J. Precis. Eng. Manuf.* 12(6):1129–41.
37. Yang, Jie, Yihe Zhang, Yanan Li, Zhihao Wang, Wenjiang Wang, Qi An, and Wangshu Tong. 2020. "Piezoelectric Nanogenerators Based on Graphene Oxide/PVDF Electrospun Nanofiber with Enhanced Performances by In-Situ Reduction." *Mater. Today: Commun.* 26:1–19.
38. Yu, H, T Huang, M Lu, M Mao, Q Zhang, and H Wang. 2013. "Enhanced Power Output of an Electrospun PVDF/MWCNTs-Based Nanogenerator by Tuning Its Conductivity." *Nanotechnology* 24(40):405401.
39. Gebrekrstos, Amanuel, Giridhar Madras, and Suryasarathi Bose. 2018. "Piezoelectric Response in Electrospun Poly(Vinylidene Fluoride) Fibers Containing Fluoro-Doped Graphene Derivatives." *ACS Omega* 3(5):5317–26.
40. Garain, Samiran, Santanu Jana, Tridib Kumar Sinha, and Dipankar Mandal. 2016. "Design of in Situ Poled Ce3+-Doped Electrospun PVDF/Graphene Composite Nanofibers for Fabrication of Nanopressure Sensor and Ultrasensitive Acoustic Nanogenerator." *ACS Appl. Mater. Interfaces* 8(7):4532–40.
41. Ahn, Yongjin, Jun Young Lim, Soon Man Hong, Jaerock Lee, Jongwook Ha, Hyoung Jin Choi, and Yongsok Seo. 2013. "Enhanced Piezoelectric Properties of Electrospun Poly(Vinylidene Fluoride)/Multiwalled Carbon Nanotube Composites Due to High β-Phase Formation in Poly(Vinylidene Fluoride)." *J. Phys. Chem. C* 117(22):11791–9.
42. Abolhasani, Mohammad Mahdi, Kamyar Shirvanimoghaddam, and Minoo Naebe. 2017. "PVDF/Graphene Composite Nanofibers with Enhanced Piezoelectric Performance for Development of Robust Nanogenerators." *Compos. Sci. Technol.* 138:49–56.
43. Tiwari, Shivam, Anupama Gaur, Chandan Kumar, and Pralay Maiti. 2020. "Electrospun Hybrid Nanofibers of Poly(Vinylidene Fluoride) and Functionalized Graphene Oxide as a Piezoelectric Energy Harvester." *Sustain. Energy Fuels RSC* 4(5):2469–79.
44. Abbasipour, Mina, Ramin Khajavi, Ali Akbar Yousefi, Mohammad Esmail Yazdanshenas, Farhad Razaghian, and Abdolhamid Akbarzadeh. 2019. "Improving Piezoelectric and Pyroelectric Properties of Electrospun PVDF Nanofibers Using Nanofillers for Energy Harvesting Application." *Polym. Adv. Technol.* 30(2):279–91.
45. Guo, Yiting, Jie Xu, Wanli Wu, Shuhang Liu, Jia Zhao, Emilia Pawlikowska, Mikołaj Szafran, and Feng Gao. 2020. "Ultralight Graphene Aerogel/PVDF Composites for Flexible Piezoelectric Nanogenerators." *Compos. Commun.* 22:100542.
46. Martins, P, A C Lopes, and S Lanceros-Mendez. 2014. "Electroactive Phases of Poly(Vinylidene Fluoride): Determination, Processing and Applications." *Prog. Polym. Sci.* 39(4):683–706.
47. Alluri, Nagamalleswara Rao, Arunkumar Chandrasekhar, Ji Hyun Jeong, and Sang Jae Kim. 2017. "Enhanced Electroactive β-Phase of the Sonication-Process-Derived PVDF-Activated Carbon Composite Film for Efficient Energy Conversion and a Battery-Free Acceleration Sensor." *J. Mater. Chem. C* 5(20):4833–44.

48. Yang, Lu, Qiuying Zhao, Ying Hou, Rujie Sun, Meng Cheng, Mingxia Shen, Shaohua Zeng, Hongli Ji, and Jinhao Qiu. 2018. "High Breakdown Strength and Outstanding Piezoelectric Performance in Flexible PVDF Based Percolative Nanocomposites through the Synergistic Effect of Topological-Structure and Composition Modulations." *Composites, A. Appl. Sci. Manuf.* 114:13–20.
49. El Achaby, M, F Z Arrakhiz, S Vaudreuil, E M Essassi, and A Qaiss. 2012. "Piezoelectric β-Polymorph Formation and Properties Enhancement in Graphene Oxide – PVDF Nanocomposite Films." *Appl. Surf. Sci.* 258(19):7668–77.
50. Im, Joon Young, Sehyun Kim, Hyoung Jin Choi, and Yongsok Seo. 2014. "Effect of Elongational Deformation on the β-Phase Formation of Poly(Vinylidene Fluoride)/Multiwalled Carbon Nanotube Composites and Their Piezoelectric Properties." *Macromol. Symp.* 346(1):7–13.
51. Bhavanasi, Venkateswarlu, Vipin Kumar, Kaushik Parida, Jiangxin Wang, and Pooi See Lee. 2016. "Enhanced Piezoelectric Energy Harvesting Performance of Flexible PVDF-TrFE Bilayer Films with Graphene Oxide." *ACS Appl. Mater. Interfaces* 8(1):5219.
52. Bhunia, Ritamay, Shashikant Gupta, Bushara Fatma, Prateek, Raju Kumar Gupta, and Ashish Garg. 2019. "Milli-Watt Power Harvesting from Dual Triboelectric and Piezoelectric Effects of Multifunctional Green and Robust Reduced Graphene Oxide/P(VDF-TrFE) Composite Flexible Films." *ACS Appl. Mater. Interfaces* 11(41):38177–89.
53. Silibin, M V, V S Bystrov, D V Karpinsky, N Nasani, G Goncalves, I M Gavrilin, A V Solnyshkin, P A A P Marques, Budhendra Singh, and I K Bdikin. 2017. "Local Mechanical and Electromechanical Properties of the P(VDF-TrFE)-Graphene Oxide Thin Films." *Appl. Surf. Sci.* 421:42–51.
54. Zhao, Chaoxian, Jin Niu, Yangyang Zhang, Cong Li, and Penghao Hu. 2019. "Coaxially Aligned MWCNTs Improve Performance of Electrospun P(VDF-TrFE)-Based Fibrous Membrane Applied in Wearable Piezoelectric Nanogenerator." *Composites, B. Eng.* 178:107447.
55. Wu, Liangke, Min Jing, Yaolu Liu, Huiming Ning, Xuyang Liu, Shifeng Liu, Liyang Lin, Ning Hu, and Liangbing Liu. 2019. "Power Generation by PVDF-TrFE/Graphene Nanocomposite Films." *Composites B. Eng.* 164:703–9.
56. Cherumannil Karumuthil, Subash, Sreenidhi Prabha Rajeev, Uvais Valiyaneerilakkal, Sujith Athiyanathil, and Soney Varghese. 2019. "Electrospun Poly(Vinylidene Fluoride-Trifluoroethylene)-Based Polymer Nanocomposite Fibers for Piezoelectric Nanogenerators." *ACS Appl. Mater. Interfaces* 11(43):40180–8.
57. Xu, Zhe, Yin Liu, Lin Dong, Andrew B Closson, Nanjing Hao, Meagan Oglesby, Gladys Patricia Escobar, et al. 2018. "Tunable Buckled Beams with Mesoporous PVDF-TrFE/SWCNT Composite Film for Energy Harvesting." *ACS Appl. Mater. Interfaces* 10(39):33516–22.
58. Wang, Yunqiu, Minjie Fang, Bobo Tian, Pinghua Xiang, Ni Zhong, Hechun Lin, Chunhua Luo, Hui Peng, and Chun Gang Duan. 2017. "Transparent PVDF-TrFE/Graphene Oxide Ultrathin Films with Enhanced Energy Harvesting Performance." *Chem. Sel.* 2(26):7951–5.
59. Wang, Min, Lingdong Li, Shenglin Zhou, Rujun Tang, Zhaohui Yang, and Xiaohua Zhang. 2018. "Influence of CNTs on the Crystalline Microstructure and Ferroelectric Behavior of P(VDF-TrFE)." *Langmuir* 34(36):10702–10.
60. Cai, Jing, Ning Hu, Liangke Wu, Yuhang Liu, Yuan Li, Huiming Ning, Xuyang Liu, and Liyang Lin. 2019. "Preparing Carbon Black/Graphene/PVDF-HFP Hybrid Composite Films of High Piezoelectricity for Energy Harvesting Technology." *Composites, A. Appl. Sci. Manuf.* 121:223–31.
61. Li, Haoxuan, Wenxin Zhang, Qian Ding, Xiangyu Jin, Qinfei Ke, Zhaoling Li, Dong Wang, and Chen Huang. 2019. "Facile Strategy for Fabrication of Flexible, Breathable, and Washable Piezoelectric Sensors via Welding of Nanofibers with Multiwalled Carbon Nanotubes (MWCNTs)." *ACS Appl. Mater. Interfaces* 11(41):38023–30.
62. Amirabad, Reza, Ahmad Ramazani Saadatabadi, and M Hossein Siadati. 2020. "Preparation of Polyaniline/Graphene Coated Wearable Thermoelectric Fabric Using Ultrasonic-Assisted Dip-Coating Method." *Mater. Renew. Sustain. Energy* 9(4):1–12.
63. Lin, Yen Hao, Tsung Chi Lee, Yu Sheng Hsiao, Wei Keng Lin, Wha Tzong Whang, and Chun Hua Chen. 2018. "Facile Synthesis of Diamino-Modified Graphene/Polyaniline Semi-Interpenetrating Networks with Practical High Thermoelectric Performance." *ACS Appl. Mater. Interfaces* 10(5):4946–52.
64. Hsieh, Yu Yun, Yu Zhang, Lu Zhang, Yanbo Fang, Sathya Narayan Kanakaraaj, Je Hyeong Bahk, and Vesselin Shanov. 2019. "High Thermoelectric Power-Factor Composites Based on Flexible Three-Dimensional Graphene and Polyaniline." *Nanoscale* 11(14):6552–60.
65. Chen, Jikun, Liming Wang, Dudi Ren, Yanhui Chu, Yong Wu, Kangkang Meng, Jun Miao, Xiaoguang Xu, and Yong Jiang. 2018. "Revealing the Anisotropy in Thermoelectric Transport Performances in CNT/PANI Composites." *Synth. Met.* 239:13–21.

66. Wang, Q, Q Yao, J Chang, and L Chen. 2012. "Enhanced Thermoelectric Properties of CNT/PANI Composite Nanofibers by Highly Orienting the Arrangement of Polymer Chains." *J. Mater. Chem.* 22(34):17612.
67. Wu, Ruili, Haocheng Yuan, Chan Liu, Jin Le Lan, Xiaoping Yang, and Yuan Hua Lin. 2018. "Flexible PANI/SWCNT Thermoelectric Films with Ultrahigh Electrical Conductivity." *RSC Adv.* 8(46):26011–9.
68. Yan, Jing, Ying Huang, Xudong Liu, Xiaoxiao Zhao, Tiehu Li, Yang Zhao, and Panbo Liu. 2021. "Polypyrrole Based Composite Materials for Electromagnetic Wave Absorption." *Polym. Rev.* 61(1):325–47.
69. Afzal, Adeel, Faraj A Abuilaiwi, Amir Habib, Muhammad Awais, Samaila B Waje, and Muataz A Atieh. 2017. "Polypyrrole/Carbon Nanotube Supercapacitors Technological Advances and Challenges." *J. Power Sources* 352:174–86.
70. Jana Chatterjee, Mukulika, Mousumi Mitra, and Dipali Banerjee. 2018. "Thermoelectric Performance of Polypyrrole and Single Walled Carbon Nanotube Composite." *Mater. Today* 5(3):9743–8.
71. Wang, Shichao, Yan Zhou, Yijia Liu, Lei Wang, and Chunmei Gao. 2020. "Enhanced Thermoelectric Properties of Polyaniline/Polypyrrole/Carbon Nanotube Ternary Composites by Treatment with a Secondary Dopant Using Ferric Chloride." *J. Mater. Chem. C* 8(2):528–35.
72. Baghdadi, Neazar, M Sh Zoromba, M H Abdel-Aziz, A F Al-Hossainy, M Bassyouni, and Numan Salah. 2021. *Polymers* 13(2):1–16.
73. Wang, Yihan, Jie Yang, Lingyu Wang, Kai Du, Qiang Yin, and Qinjian Yin. 2017. "Polypyrrole/Graphene/Polyaniline Ternary Nanocomposite with High Thermoelectric Power Factor." *ACS Appl. Mater. Interfaces* 9(23):20124–31.
74. Du, Yong, Hao Niu, Jun Li, Yunchen Dou, Shirley Z Shen, Runping Jia, and Jiayue Xu. 2018. "Morphologies Tuning of Polypyrrole and Thermoelectric Properties of Polypyrrole Nanowire/Graphene Composites." *Polymers* 10(10):6–13.
75. Mehmood, Umer, Nayab Abdul Karim, Hafiza Fizza Zahid, Tahira Asif, and Muhammad Younas. 2019. "Polyaniline/Graphene Nanocomposites as Counter Electrode Materials for Platinum Free Dye-Sensitized Solar Cells (DSSCSs)." *Mater. Lett.* 256:126651.
76. Wang, Min, Qunwei Tang, Peipei Xu, Benlin He, Lin Lin, and Haiyan Chen. 2014. "Counter Electrodes from Polyaniline-Graphene Complex/Graphene Oxide Multilayers for Dye-Sensitized Solar Cells." *Electrochim. Acta* 137:175–82.
77. Bora, Anindita, Kiranjyoti Mohan, Palash Phukan, and Swapan Kumar Dolui. 2018. "A Low Cost Carbon Black/Polyaniline Nanotube Composite as Efficient Electro-Catalyst for Triiodide Reduction in Dye Sensitized Solar Cells." *Electrochim. Acta* 259:233–44.
78. Seema, Humaira, Zaiba Zafar, and Ayesha Samreen. 2020. "Evaluation of Solution Processable Polymer Reduced Graphene Oxide Transparent Films as Counter Electrodes for Dye-Sensitized Solar Cells." *Arab. J. Chem.* 13(4):4978–86.
79. Wu, Kezhong, Lei Chen, Chongyuan Duan, Jing Gao, and Mingxing Wu. 2016. "Effect of Ion Doping on Catalytic Activity of MWCNT-Polyaniline Counter Electrodes in Dye-Sensitized Solar Cells." *Mater. Des.* 104:298–302.
80. He, B, Q Tang, M Wang, H Chen, and S Yuan. 2014. "Robust Polyaniline–Graphene Complex Counter Electrodes for Efficient Dye-Sensitized Solar Cells." *ACS Appl. Mater. Interfaces* 6(11):8230–6.
81. Thuy, Chau Thi Thanh, Joo Hei Jung, Suresh Thogiti, Woo Sik Jung, Kwang Soon Ahn, and Jae Hong Kim. 2016. "Graphene Coated Alumina-Modified Polypyrrole Composite Films as an Efficient Pt-Free Counter Electrode for Dye-Sensitized Solar Cells." *Electrochim. Acta* 205:170–7.
82. Jin, Xiao, Lai You, Zhongping Chen, and Qinghua Li. 2018. "High-Efficiency Platinum-Free Quasi-Solid-State Dye-Sensitized Solar Cells from Polyaniline (Polypyrrole)-Carbon Nanotube Complex Tailored Conducting Gel Electrolytes and Counter Electrodes." *Electrochim. Acta* 260:905–11.
83. Gemeiner, Pavol, Jaroslav Kuliček, Milan Mikula, Michal Hatala, Ľubomír Švorc, Lenka Hlavatá, Matej Mičušík, and Mária Omastová. 2015. "Polypyrrole-Coated Multi-Walled Carbon Nanotubes for the Simple Preparation of Counter Electrodes in Dye-Sensitized Solar Cells." *Synth. Met.* 210:323–31.
84. Hou, Wenjing, Yaoming Xiao, Gaoyi Han, and Haihan Zhou. 2016. "Electro-Polymerization of Polypyrrole/Multi-Wall Carbon Nanotube Counter Electrodes for Use in Platinum-Free Dye-Sensitized Solar Cells." *Electrochim. Acta* 190:720–8.
85. Shahid, Muhammad Umair, Norani Muti Mohamed, Ali Samer Muhsan, Robabeh Bashiri, Adel Eskandar Shamsudin, and Siti Nur Azella Zaine. 2019. "Few-Layer Graphene Supported Polyaniline (PANI) Film as a Transparent Counter Electrode for Dye-Sensitized Solar Cells." *Diamond Relat. Mater.* 94:242–51.

16 Polymer Nanogenerators for Biomedical Applications

Jaison Jeevanandam, Sharadwata Pan, and Michael K. Danquah

CONTENTS

16.1 INTRODUCTION

Generators transform mechanical or thermal energy synthesized via a physical change into electric power or electricity (Coxeter and Moser 2013). When nanosized materials are developed as a generator, they are termed as nanogenerators, as they are distinct from the bulk generators in characteristics and properties (Wang et al. 2007). Nanogenerators represent a domain that exploits displacement current as the principal facilitator for the operational conversion of mechanical or thermal energy into electrical signal or power, via the assistance of a nanomaterial (Wang 2020). The conventional generators are usually used in applications such as wind turbines (Petersson 2005), bicycle dynamos (Valdès and Darque 2003), and certain next-generation hydroelectricity converters (Bakis 2007). However, these generators cannot be utilized as a power source for small equipment or as an energy source to provide electricity to implantable or wearable technologies for biomedical applications (Pietrzyk et al. 2016). Thus, nanogenerators are conceived with energy harnessing nanomaterials to be implanted as an energy source for the efficient performance of wearable biomedical appliances (Song et al. 2018).

These nanogenerators are widely categorized into piezoelectric, pyroelectric, and triboelectric generators, depending on the type of energy conversion (Zi et al. 2015). It can be noted that while the triboelectric and piezoelectric nanogenerators possess ability to transform motive power (mechanical energy) into electric power (electricity) (Wang et al. 2017), the pyroelectric nanogenerators can yield thermal energy via a temporal vacillation in the temperature (Yang, Guo, et al. 2012). These nanogenerators are recently proposed to be useful to facilitate muscle movement by heartbeat (Yang et al. 2009) or lung inhalation (Lin, Wuu, et al. 2013). Furthermore, the nanogenerators have proven potential in getting employed toward the fabrication of self-powered nano- or microdevices (Zhao et al. 2019), smart wearable systems (Gunawardhana, Wanasekara, and Dharmasena 2020),

DOI: 10.1201/9781003187615-16

transparent flexible devices (Lin, Hu, et al. 2013), and implantable telemetric energy (Yoon and Kim 2020). Metal, metal oxides, carbon, and polymer nanomaterials are extensively used for the fabrication of nanogenerators (Sriphan et al. 2020). However, it has been revealed in recent studies that polymer-based nanocomposites can be beneficial in elevating the efficiency of nanogenerators for biomedical applications (Singh, Singh, and Khare 2018). The present chapter lays an overview of various synthesis procedures to fabricate different types of polymer-based nanogenerators. In addition, the biomedical applications of polymer-based nanogenerators, such as wearable smart devices, micro or nano transparent devices, and as an energy source to power up implantable biomedical appliances, are also discussed.

16.2 SYNTHESIS OF POLYMER NANOGENERATORS

The types of precursor polymer materials, synthesis approach, and functionalization procedure are all critical toward regulating the class of nanogenerator fabrication.

16.2.1 Piezoelectric Nanogenerators

In general, polymeric nanomaterials are widely used for the synthesis of piezoelectric nanogenerators due to their robust and flexible nature and being less disposed to a mechanical breakdown that is crucial from the perspectives of the vibrational energy reapers. It has been revealed that the nanowires of polymers synthesized using polyvinylidene fluoride (PVDF), as well as its copolymers, such as polylactic acid and nylon-11, are beneficial in the formation of polymer-ceramic nanocomposite nanogenerators, for enhancing the energy harvester design (Jing and Kar-Narayan 2018). Recently, Cherumannil Karumuthil et al. (2019) fabricated a novel 'poly(vinylidene fluoride-trifluoroethylene)-dependent nanocomposite fibers' via the electrospinning approach for the formation of piezoelectric nanogenerators. In this study, self-poled nanocomposite fibers were formed with piezoelectric polymer, zinc oxide, and exfoliated graphene oxide to exhibit energy generation efficiency for energy harvesting applications (Cherumannil Karumuthil et al. 2019). Furthermore, Mao et al. (2014) synthesized an exclusive sponge-like polymer film using PVDF for the scalable and integration-worthy production of piezoelectric nanogenerators, in addition to the self-executing or automatic electronic schemes. The study showed arguments in favor of a straightforward integration of these nanogenerators to an electronic device for the generation of electricity via effective transformation of mechanical energy from surrounding surface oscillations, which benefits or advances the development of self-powered electronic gadgets (Mao et al. 2014). Furthermore, Ponnamma et al. (2019) synthesized original, smart, robust, and self-directing electronic textiles by using piezoelectric polymer nanocomposites that are formed via electrospinning approach. In the study, PVDF hexafluoropropylene is used as a piezoelectric polymer, which is blended with the electrospun uniform membranes of cellulose nanocrystals and iron-doped zinc oxide nanoparticles, to form the composites. The study showed that the polymer composite with ferroelectric properties can exhibit a maximum of 12 V as the highest, 'peak-to-peak output voltage' with 1.9 $\mu A/cm^2$, which is 2.3 times higher related to polymeric fibers (Ponnamma et al. 2019).

In a recent study, Ponnamma et al. (2020) fabricated a novel PVDF-based nanocomposites with core-shell nanofibers, to be useful as exclusive piezoelectric nanogenerators. These flexible nanogenerators are synthesized via the coaxial electrospinning method by using polyvinylidene hexafluoropropylene (PVDHFP) as a copolymer with two semiconducting metal oxide nanostructures, such as titanium dioxide and zinc oxide in the nanotube and nanoflower morphologies, respectively. The study showed that the lightweight of these flexible nanogenerators in fiber mat morphology possess enhanced flexibility and mechanical strength to produce a maximum of 14 V as output voltage, depending on the motorized pulsations at explicit regularities in reaction to human movements (Ponnamma et al. 2020). Likewise, Jin et al. (2020) also fabricated adaptable piezoelectric nanogenerators via utilizing zinc oxide-PVDF films, which are metal-doped. In this study,

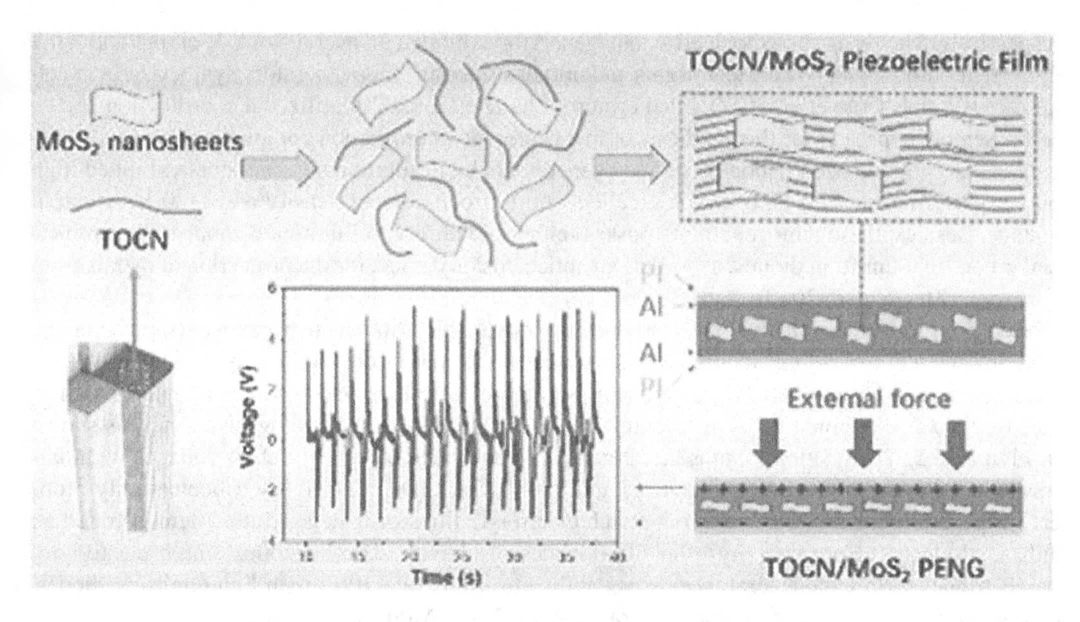

FIGURE 16.1 Polymer-based piezoelectric nanogenerators using cellulose nanofibers. (Reproduced with permission from Wu et al. (2021). Copyright 2021, Elsevier.)

cobalt, silver, sodium, and lithium were doped with zinc oxide and blended with PVDF to synthesize the composites. The study revealed that the nanogenerator-based device possesses the ability to harvest energy of ~2 Hz from the tapping of fingers, which may induce capacitor charging with 0.45 W/cm^3 of power density productivity or yield to illuminate an UV light-emitting diode (Jin et al. 2020). Similarly, Wu et al. (2021) synthesized novel flexible piezoelectric nanogenerator composite films using natural polymeric cellulose nanofibrils and molybdenum disulfide nanosheets, as shown in Figure 16.1. The study identified that the resultant nanogenerators possess excellent mechanical properties, including 8.2 GPa of enhanced Young's modulus, and 307 MPa of tensile strength. Furthermore, decent highest output or yield voltage (4.1 V) as well as short-circuit current (0.21 μA) were generated, nanogenerator-rectified, and eventually transformed to DC (Wu et al. 2021). It can be noted that the electrospinning approach is widely utilized for the fabrication of piezoelectric nanogenerators.

16.2.2 Triboelectric Nanogenerators

The triboelectric nanogenerators generally operate under the collective influences of both electrostatic induction and triboelectrification, where the polymer surface charge density sets the basis of their productivity (Yu and Wang 2016). Yu et al. (2015) used the 'sequential infiltration synthesis approach' toward the development of efficient triboelectric nanogenerator via polymer films (doped) with amendable electrical characteristics. The authors used 'atomic layer deposition' to manufacture thin films of polymers, such as polyimide, polydimethylsiloxane, and poly(methyl methacrylate), which are doped with aluminum oxide. The study showed that the sequential infiltration can lead to introduction of aluminum oxide molecules into polymers for about 3-μm deep to tune their electrical properties. The resultant metal-oxide-doped polymer synthesized via the sequential infiltration approach elevated the electronic and dielectric accomplishments of triboelectric nanogenerators for an effective reaping of mechanical energy (Yu et al. 2015). Furthermore, Yao et al. (2017) built a novel triboelectric nanogenerator using natural cellulose nanofibrils with nitro and methyl groups, as well as fluorinated ethylene propylene. The nitro-cellulose and the methyl cellulose nanofibrils possess a negative surface charge density of 85.8 μC/m^2 and a surface charge density of 62.5 μC/m^2, respectively. The resultant triboelectric nanogenerator exhibited an average of 8-V output and 9 μA of current output, which represent higher magnitudes than the nanogenerators synthesized via cellulosic materials (Yao et al. 2017). Furthermore, Zhang et al. (2020) utilized a coordination-driven self-assembly approach for the synthesis of metal-organic coordination complex with organosulfonate counter-anion-based triboelectric nanogenerator. The resultant nanogenerators exhibited high output performance with 98.6 μA of excellent short-circuit current density and 1180 V of output voltage. Besides, these nanogenerators possess enhanced ability to illuminate about 1488 commercial green-light-emitting diodes, as well as an anticorrosion device, for the protection of metals from corrosion (Zhang et al. 2020).

Recently, Ccorahua et al. (2019) produced a novel 'bio-triboelectric nanogenerator' via the 'cleanroom-free processing' technique, using electrolytes based on starch polymer. In this study, a triboelectric nanogenerator fabricated using starch-polytetrafluoroethylene (PTFE) and starch as electrolyte was combined to form calcium chloride-PTFE as corresponding dielectrics, as shown in Figure 16.2. The resultant nanogenerator showed enhanced electrical output voltages with narrow film thickness, improved frequencies and loads, which is similar to the triboelectricity property of other biopolymers. In addition, starch electrolyte films of nanogenerators demonstrated an enhanced electrical productivity after 5000 cycles of activity. These Ecoflex starch electrolyte-based triboelectric nanogenerators possess ability to illuminate 100 green-light-emitting diodes, compared to the synthetic polymers (Ccorahua et al. 2019). Moreover, Hwang et al. (2018) utilized an extremely distortable, mesoporous, polymer composite targeted toward the fabrication of triboelectric nanogenerator without gap, using a 'single-step metal oxidation' procedure. The authors employed a nano-flake morphology bearing aluminum oxide-polydimethylsiloxane mesoporous composites, with 71.35% porosity. The results showed that the fabricated nanogenerator film possesses around 1.8 times higher capacitance and 6.67 times higher output performance (Hwang et al. 2018). All these studies show that exclusive synthesis methods are necessary for the fabrication of triboelectric nanogenerators.

16.2.3 Pyroelectric Nanogenerators

Numerous synthesis methods are widely utilized for the fabrication of pyroelectric nanogenerators. Gao et al. (2016) synthesized a water-vapor-facilitated, self-enduring pyroelectric nanogenerator with ubiquitous low-speed air flow velocity of 1–2 m/s. In this study, conventional ferroelectric polymers, for instance, polyvinylidene difluoride with copolymer of trifluoroethylene and vinylidene difluoride, were used to fabricate a sizeable area malleable slender nanogenerators with an enhanced pyroelectric property. The resultant nanogenerator possessed an enhanced ability to produce an open-circuit voltage of 145 V, 0.12 $\mu A/cm^2$, 1.47 mW/cm^3 of peak power density by volume, and 4.12 $\mu W/cm^2$ by area (Gao et al. 2016). Furthermore, Li et al. (2021) synthesized a novel vibration-free pyroelectric nanogenerator via local hot charge density regulation, for the effective enhancement of catalysts and *in situ* monitoring of Raman scattering on the surface. In this study, the pyroelectric property of the plasmonic graphene-silver nanowire layers with the polymer PVDF film is easily improved by the simulated sunlight illumination (Li et al. 2021). Furthermore, Sultana et al. (2019) synthesized a flexible piezoelectric-pyroelectric nanogenerator (see Figure 16.3) utilizing the poly(vinylidene fluoride) nanofibers, incorporated with methylammonium lead iodide, via electrospinning approach. The resultant pyroelectric nanogenerator generated ~220 mV of output voltage at 4 Hz of frequency through periodic compressive contact force. The results showed that the nanogenerator possess ~19.7 pC/N of piezoelectric coefficient (d_{33}) with 60,000 cycles of high durability and ~1 millisecond of quick response time. In addition, the nanogenerator also exhibited 18.2 pA of pyroelectric output current, 41.78 mV of voltage, 1.14 s of rapid response time, 1.25 s of reset time, and ~44 pC/m^2 of pyroelectric coefficient after cyclic hearing, including a cooling exposure at 38K of temperature (Sultana et al. 2019).

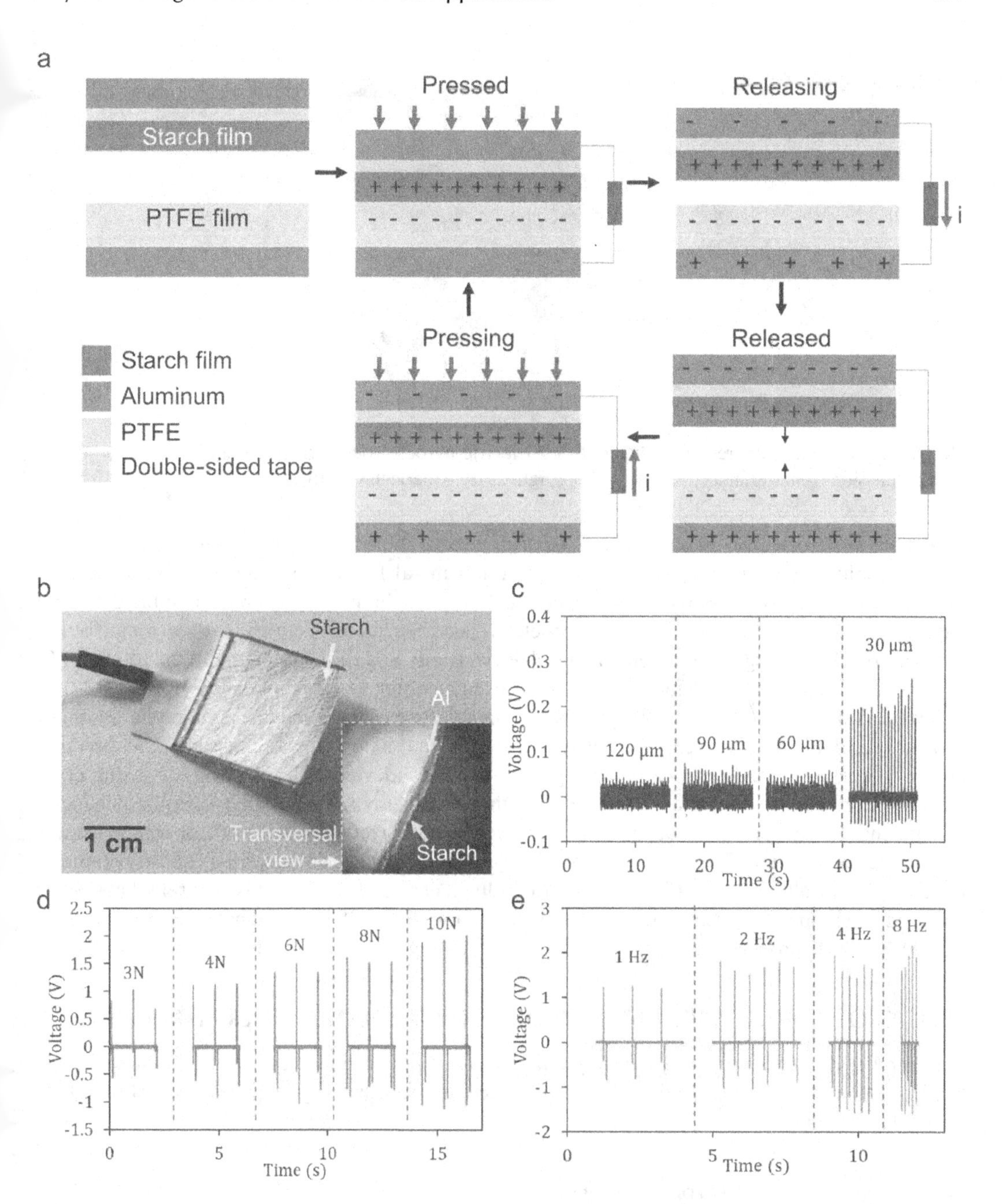

FIGURE 16.2 Dielectric starch film and polymer used for triboelectric nanogenerator: (a) starch-based nanogenerator mechanism, (b) aluminum-doped starch dielectric film, (c) film thickness dependent on electrical output, (d) loads, and (e) frequency (Ccorahua et al. 2019). (Reproduced with permission from Ccorahua et al. (2019). Copyright 2019, Elsevier.)

Recently, Roy et al. (2019) developed a pyroelectric breathing sensor using a graphene oxide interfaced poly(vinylidene fluoride) nanofibers via electrospinning approach. The sensor possesses 4.3 V/kPa of high sensitivity to identify considerably low pressure (10 Pa), where the pyroelectric nanogenerator generates a maximum of ~1.2 nW/m^2 output power density to be beneficial for the fabrication of wearable sensors (Roy et al. 2019). Similarly, Ko et al. (2016) produced a hybrid

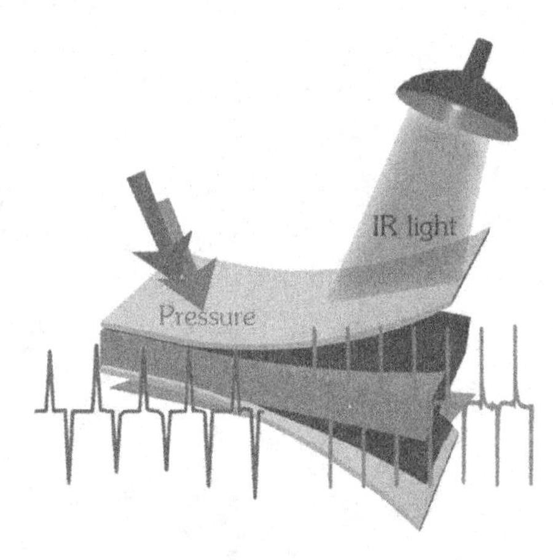

FIGURE 16.3 Schematic representation of a piezoelectric-pyroelectric hybrid nanogenerators. (Reproduced with permission from Sultana et al. (2019). Copyright 2019, American Chemical Society.)

piezoelectric-pyroelectric nanogenerator using a flexible film, made up of perovskite oxide of lead, zirconium, and titanium with nickel-chromium metal foil flexible substrate alongside lithium niobite (conductive), as the bottom electrode. The resultant nanogenerator exhibited 50 nC/cm^{2}K and 140 pC/N of pyroelectric and piezoelectric coefficient, respectively, at room temperature. Furthermore, the nanogenerator generated stable electricity at a raised temperature of 100°C, 70% RH, and a stronger resistance for base and water even at a high Curie temperature (Ko et al. 2016). Likewise, Yang, Jung, et al. (2012) synthesized tunable pyroelectric nanogenerators via an amalgamated configuration of potassium niobate (devoid of lead) nanowires. In this study, the direction poling process was used for manufacturing the nanowires (devoid of lead) alongside a sing crystalline structure, and formed as a composite with polydimethylsiloxane polymer to be beneficial as a potential pyroelectric nanogenerator. The study emphasized that the resultant nanogenerator can utilize electric field for controlling the voltage or the current output, which can be eventually increased by the alterations in temperature (Yang, Jung, et al. 2012). Thus, it is apparent that several novel synthesis approaches are utilized for the fabrication of pyroelectric and hybrid nanogenerators with pyroelectric properties.

16.3 BIOMEDICAL APPLICATIONS OF POLYMER NANOGENERATORS

Polymer-based nanogenerators are widely utilized to self-powered medical devices, wearable electronics, transparent flexible devices, and implantable telemetric energy devices for exclusive biomedical applications.

16.3.1 Self-Powered Medical Devices

In recent times, there are several studies that showed that the polymer-based nanogenerators are widely useful to fabricate self-powered medical devices. Guan et al. (2020) fabricated flexible piezoelectric nanogenerators using polydopamine-modified barium titanate with electrospun poly(vinylidene fluoride-trifluoroethylene) nanocomposite fiber mats via a hierarchical architecture. These flexible nanogenerators are used for an effective detection of human body movements as self-powered sensors. Such a nanogenerator exhibited 40–68% of output voltage with ~8.78 mW/m^2 of maximum power density, compared to other nanocomposite membranes (Guan, Xu, and Gong 2020). Furthermore, Kim et al. (2020) synthesized innovative, nanofiber-dependent, *in situ* poled tunable piezoelectric nanogenerators toward an effective self-directed censoring of human body movements. The authors employed tunable, hybrid PVDF-graphene oxide piezoelectric nanofibers,

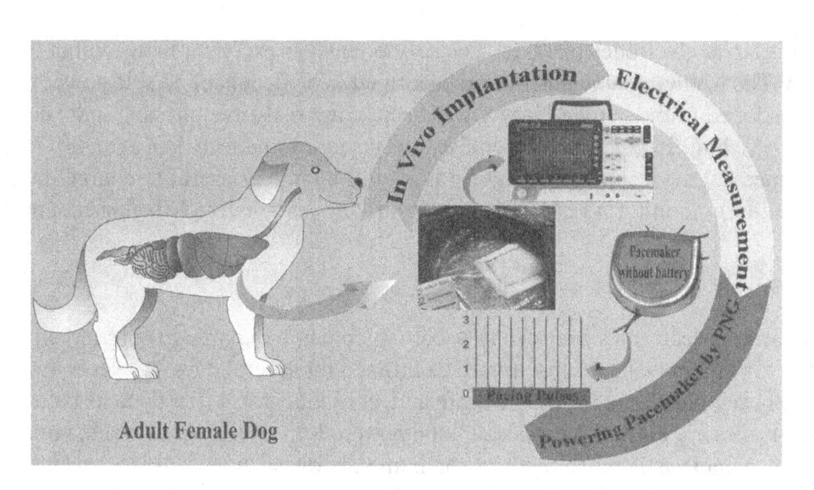

FIGURE 16.4 Schematic representation of nanogenerator-dependent self-effected cardiac pacemaker. (Reproduced with permission from Azimi et al. (2021). Copyright 2021, Elsevier.)

which can generate 2.1 V of an output voltage at a 12 N uniform force, as nanogenerators to be beneficial as locomotion and self-powered motion sensors (Kim et al. 2020). Furthermore, Azimi et al. (2021) advanced a cardiac pacemaker utilizing piezoelectric polymer nanogenerator and employing them as a potential self-powered implant. In this study, the nanogenerator was synthesized via PVDF-trifluoroethylene-based films as effective cardiac energy harvesters. The study showed that the *in vivo* (the 'dog animal model') implanted nanogenerator-based battery-free heart pacemaker harvested 0.487 μJ from every heartbeat, especially after implanting in the left ventricle cardiac motions, as shown in Figure 16.4 (Azimi et al. 2021). Moreover, Sun et al. (2021) fabricated scalable graphite-polydimethylsiloxane composite films in a hierarchical structure, to be used as triboelectric nanogenerators for self-powered tactile sensors with large area. The polymer composite-film-based triboelectric nanogenerators generated the highest power of 2.4-mW power corresponding to an external load of 3 MΩ. This nanogenerator has also been identified to possess ability to charge a capacitor of 2.2 μf, 1.5 V within 2 s, an electronic watch and power up 30 commercial green LEDs. Furthermore, the nanogenerator was incorporated in a self-powered tactile sensor and included in a rubber glove for the detection of progress in grabbing the objects (Sun et al. 2021).

Recently, Qian et al. (2020) synthesized mechano-informed biomimetic scaffolds using polymers by the incorporation of self-directed nanogenerators (zinc oxide-dependent) for the augmentation of neural function and motor recovery. In this study, poly(vinylidene difluoride-trifluoroethylene)-based polycaprolactone composite scaffolds were combined with piezoelectric zinc oxide nanogenerators. The study showed that the nanogenerator helps in the recovery of nerve defects in rats within 18 weeks. Additionally, it generates electricity, which could be beneficial in smart implantable medical devices for neuroengineering applications (Qian et al. 2020). Likewise, Maity and Mandal (2018) developed self-powered, multifunctional sensors via multilayer-assembled nanofiber, mat-based, all-organic, piezoelectric nanogenerator synthesized via electrospinning approach, to exhibit high performance. In this study, the nanogenerator was fabricated using nanofiber mats made of poly(vinylidene fluoride), where poly(3,4-ethylenedioxythiophene) polymerized via vapor-phase was coated on the nanofiber to be used as an electrode, and free nanofibers were used as an active compound. The nanogenerators demonstrated excellent ultrasensitivity toward the body movements of human, such as walking and foot strikes with enhanced robustness to be beneficial as a wearable harvester of mechanical energy (Maity and Mandal 2018). Besides, Chen et al. (2017) developed a transparent and stretchable nanocomposite-based nanogenerator for the self-driven scrutinizing of physiological alterations. This work was embodied by piezoelectric barium titanate and polydimethylsiloxane polymer-based nanocomposites in sandwich arrangement as the sensor layer, and transparent electrodes (sprayed silver nanowires) as excellent input-output linearity, with 60% of strain. The resultant nanogenerator is identified to be beneficial to self-power wearable sensors that are ascribed to the human body toward censoring real-time physical indicators, including pronunciation, eye blinking, radial artery pulse, and arm movements (Chen et al. 2017). Thus, these self-powered medical devices with nanogenerators will be highly useful to charge the implantable devices for long-term monitoring of health conditions and personalized diagnosis of diseases.

16.3.2 Wearable Electronics

Currently, wearable electronics are widely used, which can be charged using an external electric source, and storing them using a capacitor, to be utilized for several days. However, overuse of such electronics will eventually lead to battery-drain and, consequently, will reduce its battery life. Thus, wearable electronics are currently proposed to be powered-up by nanogenerators, i.e., to charge the electronics via characteristic movements of the human body, such as walking or physical exercise. Patnam et al. (2020) developed hybrid piezo-triboelectric nanogenerators using composites of solid-state technique synthesized calcium-doped barium zirconate titanate and polydimethylsiloxane films. The resultant nanogenerators exhibited an open-circuit voltage of 550 V, a short-circuit current of 34 μA, and a power density of 23.6 W/m^2, to act like the reservoir of power for transferable and lightweight electronic gadgets (Patnam et al. 2020). Furthermore, Su et al. (2020) demonstrated the fabrication of a wearable, simple structure bearing triboelectric nanogenerator, based on a hybrid polymer and carbon nanotube layer composition. In this study, the nanogenerator was fabricated by combining frictional silk material and liquid carbon nanotube as an electrode material, where the liquidized nanogenerator was attached to a glove to generate electric field based of bodily motions to charge-up wearable electronics. The resultant nanogenerator-based glove exhibited 75,600 cycle of stable continuous operation conditions with 262 V of voltage, 8.73 μA of current and 285.91 μW/cm^2 of power density with 16.7% of voltage reduction (Su, Brugger, and Kim 2020). Furthermore, Mule et al. (2019) established an innovative triboelectric 'single-electrode-mode' wearable nanogenerator for self-powered electronics using conductive polymer-coated textiles. In this study, a polypyrrole-based wearable and flexible nanogenerator was developed with support from interlaced microfibrous mesh of cotton fabric via the chemical polymerization process. The cotton textile coated with polypyrrole was used as an electrode in the nanogenerator, where soft-imprint lithography was used to develop a layer of sandpaper-assisted microtextured polydimethylsiloxane on top of the electrode. This novel nanogenerator device effectively converted mechanical energy into electrical energy via continuous contacts with counter-friction objects, such as the human skin, and cellulose membrane was used for dialysis (Mule et al. 2019). However, extensive research is required in the field of nanogenerators to utilize them for charging up commercial wearable devices.

16.3.3 Transparent Flexible Devices

Recently, several polymer-based nanogenerators are widely employed toward the manufacturing of transparent supple gadgets. Fatma et al. (2019) fabricated novel noncontact, magneto-, flexible, and transparent triboelectric nanogenerators using PVDF-maghemite nanocomposites. In this study, nanogenerators were synthesized using iron oxide (maghemite) fillers incorporated in PVDF with polyethylene terephthalate matrix. The resultant nanogenerators exhibited an average of 250-V open-circuit voltage, short-circuit current of 5 μA, 0.117 W/m^2 of power density, and 0.17 mW of maximum power output. This transparent nanogenerator was attached to the footwear sole to harvest electricity via biomechanical foot motion to facilitate the charging of a 1-μF value of capacitor to ~30 V in just 90 s (Fatma et al. 2019). Likewise, Lee et al. (2018) demonstrated the synthesis of a flexible and see-through high-power triboelectric nanogenerator with tribo-negative conducting polymer embedded with metallic nanowires. In this study, poly(3,4-ethylenedioxythiophene) (conducting nature): polymer [poly(styrene sulfonate)] characterized by tribo-negativity beneath

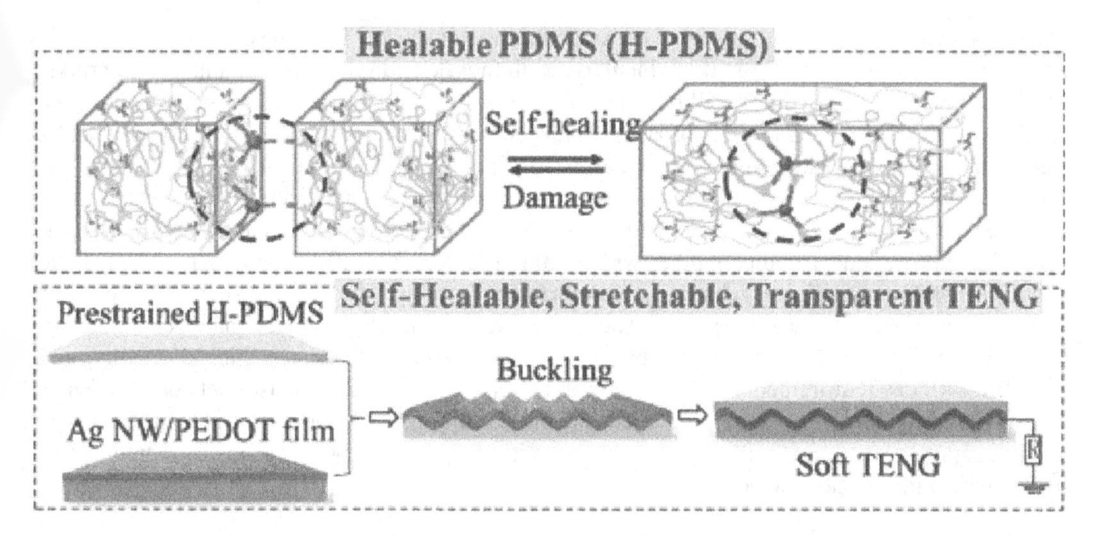

FIGURE 16.5 Schematic of self-healable transparent polymer-based triboelectric nanogenerator. (Reproduced with permission from Sun et al. (2018). Copyright 2018, American Physical Society.)

the layer of silver nanowire was utilized with increased surface roughness and conductivity. The nanogenerator possesses abilities to produce 160 V of output voltage, > 50 μA with > 1.5 mW/cm^2 of instant power density. The resultant nanogenerator with stretchable substrate possesses superior transparency, including an excellent output performance for instant touch visualization via self-powered system, by fixing them on an arbitrary fabric or the human skin surface (Lee et al. 2018). Similarly, Sun et al. (2018) developed original soft power foundations using stretchable, self-curable, and see-through triboelectric nanogenerators, as shown in Figure 16.5. The nanosized generators were formed by buckling up silver nanowires with composites of poly(3,4-ethylenedioxythiophene) as electrodes, which are sandwiched with self-correctable elastomers of poly(dimethylsiloxane) at room temperature. The resultant nanogenerators were identified to possess ~73% of transmittance, ~50% of stretchability, 100% of healing efficiency with ~100 V, and 327 mW/m^2 of electricity generation, even after accidental cutting of the polymeric generator (Sun et al. 2018). However, research endeavors concerning the polymer-nanogenerators for transparent flexible devices are still in their infancy, which necessitates extensive investigations to realize their optimum applications in commercial devices.

16.3.4 Telemetry Energy for Implantable Devices

Recently, polymer-based nanogenerators are widely used for the generation of telemetry energy for implantable devices, which will be able to harness electricity via wireless approach. He et al. (2019) fabricated a self-driven, glove-dependent, instinctual interface for differentiated real or cyber space control applications. In this study, triboelectric nanogenerators made up of poly(styrene sulfonate) (PEDOT: PSS): poly(3,4-ethylenedioxythiophene) were used to process the signals from human-machine interfaces. The nanogenerator was attached to a glove by coating them on the textile strip via stitching process, whereas a slender film of silicone rubber was also layered on the glove. This (along with the nanogenerator) was identified to successfully control car, drone, minigame, virtual reality device and pointer for cyber spending via wireless approach, and alphabet writing with an intuitive and a straightforward maneuver procedure (He et al. 2019). Furthermore, Vera Anaya et al. (2020) fabricated an eye motion sensor, which is self-powered, using a novel triboelectric nanogenerator that works via 'electrostatic induction' (near-field) toward enhanced wearable rehabilitative know-hows. In this study, PEDOT: PSS-conducting polymer was used for the fabrication of a triboelectric nanogenerator to form a flexible motion transducer. The transducer generates electricity upon sensing movement in the orbicularis oculi muscle and monitors involuntary eye blinks. The device with this transducer will be beneficial to assist mobility-impaired people to control the computer cursor via a hands-free approach, as well as to monitor driving behaviors, control hands-free drones and remote cars (Vera Anaya et al. 2020). In addition, Mishra et al. (2015) evaluated the efficiency of piezoelectric nanogenerators to be incorporated into microelectromechanical systems (MEMS) to fabricate self-powered implants for soldiers. The study reported that lead zirconate titanate with multiwalled carbon nanotubes with 5–20 nm of diameter, 10 μm of length and synthetic polydimethylsiloxane silicone rubber in chloroform and tetrahydrofuran as solvents were used for the synthesis of flexible nanogenerators. The study revealed that these flexible nanogenerators are highly beneficial to power up implantable devices with peak performance, high durability, and shelf-life with low requirement of power for mobile devices and wireless sensor networks (Mishra, Jain, and Prasad 2015). Currently, the telemetry energy production property of nanogenerators is gaining significant attraction among researchers as it can assist and boost the endowment of electricity toward implantable devices.

16.4 FUTURE PERSPECTIVES

In general, it is worthy to note from the literature that the fabrication of nanogenerators to harvest electricity using human body motion is not a proposed concept anymore and can be customized for several biomedical applications. However, most reports revealed the possibility of synthesizing nanogenerators in the laboratory scale. Thus, research efforts investigating large-scale and commercial production potential, as well as biomedical applications of nanogenerators, are still in their infancy (Zhu et al. 2016, Wang et al. 2020). The major limitations of polymer-based nanogenerators are the biocompatibility and biodegradability in the live animal models, since synthetic polymers may not be compatible with human skin or may lead to certain complications, while being utilized as self-powered implants (Li et al. 2018). In addition, synthetic polymers, especially nanosized polymers, in nanogenerators may also lead to cytotoxicity, which can limit their usage for potential biomedical applications (Miederer et al. 2004, Silva et al. 2019). Hence, biodegradable polymers were widely used for the fabrication of nanogenerators, particularly for biomedical applications (Pan et al. 2018). However, it may be noted that the stability of these biodegradable polymers is low, compared to their synthetic counterparts (Orts et al. 2007). Furthermore, the conversion of biomechanical energy to electricity will generate heat and most of the biopolymers are noncompatible to heat transfers, which may even damage the nanogenerators (Markarian 2008). This challenge can be avoided by fabricating a nanogenerator with the composite of nickel-gold or nickel-titanium (nitinol) alloy and biopolymers. The biopolymer will have the property of flexibility and electrical conductivity, whereas the nickel-gold or nitinol alloy will have heat resistance property at a low weight (Silva et al. 2020). Besides, the efficiency to harvest electrical energy from biomechanical motion should also be improved by fabricating polymeric composite-based nanogenerators, with distinct combinations of metal, metal oxides, carbon, and quantum dot nanoparticles with less toxicity, high biocompatibility, biodegradability, and energy harvesting capability, in the future.

16.5 CONCLUSION

An overview of various synthesis procedures to fabricate different types of polymer-based nanogenerators is provided in this chapter. In addition, the biomedical applications of polymer-based nanogenerators, such as wearable smart devices, micro or nano transparent devices, and as an energy source to power up implantable biomedical appliances, are also discussed. Based on past studies, it is apparent that polymer-based nanogenerators will replace current generators and conventional implantable devices in the future, catering to their self-powering capability derived from human bodily movements. However, duly acknowledging the upsurge of the recent advancements, there is an urge to develop even more original and sustainable synthesis strategies to fabricate nanogenerators that are less toxic and biocompatible to the human cells.

REFERENCES

Azimi, Sara, Allahyar Golabchi, Abdolhossein Nekookar, Shahram Rabbani, Morteza H. Amiri, Kamal Asadi, and Mohammad M. Abolhasani. 2021. "Self-powered cardiac pacemaker by piezoelectric polymer nanogenerator implant." *Nano Energy* 83:105781. https://doi.org/10.1016/j.nanoen.2021.105781.

Bakis, R. 2007. "The current status and future opportunities of hydroelectricity." *Energy Sources, Part B* 2 (3):259–266.

Ccorahua, Robert, Juan Huaroto, Clemente Luyo, Maria Quintana, and Emir A. Vela. 2019. "Enhanced-performance bio-triboelectric nanogenerator based on starch polymer electrolyte obtained by a clean-room-free processing method." *Nano Energy* 59:610–618. https://doi.org/10.1016/j.nanoen.2019.03.018.

Chen, Xiaoliang, Kaushik Parida, Jiangxin Wang, Jiaqing Xiong, Meng-Fang Lin, Jinyou Shao, and Pooi S. Lee. 2017. "A stretchable and transparent nanocomposite nanogenerator for self-powered physiological monitoring." *ACS Applied Materials & Interfaces* 9 (48):42200–42209. https://doi.org/10.1021/acsami.7b13767.

Cherumannil Karumuthil, Subash, Sreenidhi Prabha Rajeev, Uvais Valiyaneerilakkal, Sujith Athiyanathil, and Soney Varghese. 2019. "Electrospun poly(vinylidene fluoride-trifluoroethylene)-based polymer nanocomposite fibers for piezoelectric nanogenerators." *ACS Applied Materials & Interfaces* 11 (43):40180–40188. https://doi.org/10.1021/acsami.9b17788.

Coxeter, Harold S.M., and Moser William O.J. 2013. *Generators and relations for discrete groups*. Vol. 14: Springer Science & Business Media, Berlin, Germany.

Fatma, Bushara, Ritamay Bhunia, Shashikant Gupta, Amit Verma, Vivek Verma, and Ashish Garg. 2019. "Maghemite/polyvinylidene fluoride nanocomposite for transparent, flexible triboelectric nanogenerator and noncontact magneto-triboelectric nanogenerator." *ACS Sustainable Chemistry & Engineering* 7 (17):14856–14866. https://doi.org/10.1021/acssuschemeng.9b02953.

Gao, Fengxian, Wanwan Li, Xiaoqian Wang, Xiaodong Fang, and Mingming Ma. 2016. "A self-sustaining pyroelectric nanogenerator driven by water vapor." *Nano Energy* 22:19–26. doi: https://doi.org/10.1016/j.nanoen.2016.02.011.

Guan, Xiaoyang, Bingang Xu, and Jianliang Gong. 2020. "Hierarchically architected polydopamine modified BaTiO3@P(VDF-TrFE) nanocomposite fiber mats for flexible piezoelectric nanogenerators and self-powered sensors." *Nano Energy* 70:104516. doi: https://doi.org/10.1016/j.nanoen.2020.104516.

Gunawardhana, K.R.S.D., Nandula D. Wanasekara, and R.D.I.G. Dharmasena. 2020. "Towards truly wearable systems: optimising and scaling up wearable triboelectric nanogenerators." *Iscience* 23 (8):101360.

He, Tianyiyi, Zhongda Sun, Qiongfeng Shi, Minglu Zhu, David V. Anaya, Mengya Xu, Tao Chen, Yuce, Aaron V.-Y. Thean, and Chengkuo Lee. 2019. "Self-powered glove-based intuitive interface for diversified control applications in real/cyber space." *Nano Energy* 58:641–651. https://doi.org/10.1016/j.nanoen.2019.01.091.

Hwang, Hee J., Younghoon Lee, Choongyeop Lee, Youngsuk Nam, Jinhyoung Park, Dukhyun Choi, and Dongseob Kim. 2018. "Mesoporous highly-deformable Mehmet Rasit composite polymer for a gapless triboelectric nanogenerator via a one-step metal oxidation process." *Micromachines* 9 (12). https://doi.org/10.3390/mi9120656.

Jin, Congran, Nanjing Hao, Zhe Xu, Ian Trase, Yuan Nie, Lin Dong, Andrew Closson, Zi Chen, and John X.J. Zhang. 2020. "Flexible piezoelectric nanogenerators using metal-doped ZnO-PVDF films." *Sensors and Actuators A: Physical* 305:111912. https://doi.org/10.1016/j.sna.2020.111912.

Jing, Qingshen, and Sohini Kar-Narayan. 2018. "Nanostructured polymer-based piezoelectric and triboelectric materials and devices for energy harvesting applications." *Journal of Physics D: Applied Physics* 51 (30):303001. https://doi.org/10.1088/1361-6463/aac827.

Kim, Minjung, Vignesh K. Kaliannagounder, Afeesh R. Unnithan, Chan H. Park, Cheol S. Kim, and Arathyram Ramachandra, Kurup Sasikala. 2020. "Development of in-situ poled nanofiber based flexible piezoelectric nanogenerators for self-powered motion monitoring." *Applied Sciences* 10 (10). https://doi.org/10.3390/app10103493.

Ko, Young J., Dong Y. Kim, Sung S. Won, Chang W. Ahn, Ill W. Kim, Angus I. Kingon, Seung-Hyun Kim, Jae-Hyeon Ko, and Jong H. Jung. 2016. "Flexible Pb(Zr0.52Ti0.48)O3 films for a hybrid piezoelectric-pyroelectric nanogenerator under harsh environments." *ACS Applied Materials & Interfaces* 8 (10):6504–6511. https://doi.org/10.1021/acsami.6b00054.

Lee, Bo-Yeon, Se-Um Kim, Sujie Kang, and Sin-Doo Lee. 2018. "Transparent and flexible high power triboelectric nanogenerator with metallic nanowire-embedded tribonegative conducting polymer." *Nano Energy* 53:152–159. https://doi.org/10.1016/j.nanoen.2018.08.048.

Li, Chonghui, Shicai Xu, Jing Yu, Zhen Li, Weifeng Li, Jihua Wang, Aihua Liu, Baoyuan Man, Shikuan Yang, and Chao Zhang. 2021. "Local hot charge density regulation: vibration-free pyroelectric nanogenerator for effectively enhancing catalysis and in-situ surface enhanced Raman scattering monitoring." *Nano Energy* 81:105585. https://doi.org/10.1016/j.nanoen.2020.105585.

Li, Jun, Lei Kang, Yanhao Yu, Yin Long, Justin J. Jeffery, Weibo Cai, and Xudong Wang. 2018. "Study of long-term biocompatibility and bio-safety of implantable nanogenerators." *Nano Energy* 51:728–735.

Lin, Hung-I., Dong-Sing Wuu, Kun-Ching Shen, and Ray-Hua Horng. 2013. "Fabrication of an ultra-flexible ZnO nanogenerator for harvesting energy from respiration." *ECS Journal of Solid State Science and Technology* 2 (9):P400.

Lin, Long, Youfan Hu, Chen Xu, Yan Zhang, Rui Zhang, Xiaonan Wen, and Zhong Lin Wang. 2013. "Transparent flexible nanogenerator as self-powered sensor for transportation monitoring." *Nano Energy* 2 (1):75–81.

Maity, Kuntal, and Dipankar Mandal. 2018. "All-organic high-performance piezoelectric nanogenerator with multilayer assembled electrospun nanofiber mats for self-powered multifunctional sensors." *ACS Applied Materials & Interfaces* 10 (21):18257–18269. https://doi.org/10.1021/acsami.8b01862.

Mao, Yanchao, Ping Zhao, Geoffrey McConohy, Hao Yang, Yexiang Tong, and Xudong Wang. 2014. "Sponge-like piezoelectric polymer films for scalable and integratable nanogenerators and self-powered electronic systems." *Advanced Energy Materials* 4 (7):1301624. doi: https://doi.org/10.1002/aenm.201301624.

Markarian, Jennifer. 2008. "Biopolymers present new market opportunities for additives in packaging." *Plastics, Additives and Compounding* 10 (3):22–25.

Miederer, Matthias, Michael R. McDevitt, George Sgouros, Kim Kramer, Nai-Kong V. Cheung, and David A. Scheinberg. 2004. "Pharmacokinetics, dosimetry, and toxicity of the targetable atomic generator, 225Ac-HuM195, in nonhuman primates." *Journal of Nuclear Medicine* 45 (1):129–137.

Mishra, Ritendra, Shruti Jain, and C.D. Prasad. 2015. "A review on piezoelectric material as a source of generating electricity and its possibility to fabricate devices for daily uses of army personnel." *International Journal of Systems, Control and Communications* 6 (3):212–221.

Mule, Anki R., Bhaskar Dudem, Harishkumarreddy Patnam, Sontyana A. Graham, and Jae S. Yu. 2019. "Wearable single-electrode-mode triboelectric nanogenerator via conductive polymer-coated textiles for self-power electronics." *ACS Sustainable Chemistry & Engineering* 7 (19):16450–16458. https://doi.org/10.1021/acssuschemeng.9b03629.

Orts, William J., Aicardo Roa-Espinosa, Robert E. Sojka, Gregory M. Glenn, Syed H. Imam, Kurt Erlacher, and Jan S. Pedersen. 2007. "Use of synthetic polymers and biopolymers for soil stabilization in agricultural, construction, and military applications." *Journal of Materials in Civil Engineering* 19 (1):58–66.

Pan, Ruizheng, Weipeng Xuan, Jinkai Chen, Shurong Dong, Hao Jin, Xiaozhi Wang, Honglang Li, and Jikui Luo. 2018. "Fully biodegradable triboelectric nanogenerators based on electrospun polylactic acid and nanostructured gelatin films." *Nano Energy* 45:193–202.

Patnam, Harishkumarreddy, Bhaskar Dudem, Nagamalleswara R. Alluri, Anki R. Mule, Sontyana Adonijah Graham, Sang-Jae Kim, and Jae S. Yu. 2020. "Piezo/triboelectric hybrid nanogenerators based on Ca-doped barium zirconate titanate embedded composite polymers for wearable electronics." *Composites Science and Technology* 188:107963. https://doi.org/10.1016/j.compscitech.2019.107963.

Petersson, Andreas. 2005. *Analysis, modeling and control of doubly-fed induction generators for wind turbines*: Chalmers University of Technology, Sweden.

Pietrzyk, Kyle, Joseph Soares, Brandon Ohara, and Hohyun Lee. 2016. "Power generation modeling for a wearable thermoelectric energy harvester with practical limitations." *Applied Energy* 183:218–228.

Ponnamma, Deepalekshmi, Mariem M. Chamakh, Abdulrhman M. Alahzm, Nisa Salim, Nishar Hameed, and Mariam A. AlMaadeed. 2020. "Core-shell nanofibers of polyvinylidene fluoride-based nanocomposites as piezoelectric nanogenerators." *Polymers* 12 (10). https://doi.org/10.3390/polym12102344.

Ponnamma, Deepalekshmi, Hemalatha Parangusan, Aisha Tanvir, and Mariam A.A. AlMa'adeed. 2019. "Smart and robust electrospun fabrics of piezoelectric polymer nanocomposite for self-powering electronic textiles." *Materials & Design* 184:108176. https://doi.org/10.1016/j.matdes.2019.108176.

Qian, Yun, Yuan Cheng, Jialin Song, Yang Xu, Wei-En Yuan, Cunyi Fan, and Xianyou Zheng. 2020. "Mechano-informed biomimetic polymer scaffolds by incorporating self-powered zinc oxide nanogenerators enhance motor recovery and neural function." *Small* 16 (32):2000796. doi: https://doi.org/10.1002/smll.202000796.

Roy, Krittish, Sujoy K. Ghosh, Ayesha Sultana, Samiran Garain, Mengying Xie, Christopher R. Bowen, Karsten Henkel, Dieter Schmeiβer, and Dipankar Mandal. 2019. "A self-powered wearable pressure sensor and pyroelectric breathing sensor based on GO interfaced PVDF nanofibers." *ACS Applied Nano Materials* 2 (4):2013–2025. https://doi.org/10.1021/acsanm.9b00033.

Silva, Amélia M., Helen L. Alvarado, Guadalupe Abrego, Carlos Martins-Gomes, Maria L. Garduño-Ramirez, María L. García, Ana C. Calpena, and Eliana B. Souto. 2019. "In vitro cytotoxicity of oleanolic/ursolic acids-loaded in PLGA nanoparticles in different cell lines." *Pharmaceutics* 11 (8):362.

Silva, Emmanuel J.N.L., Mayara Zanon, Fernanda Hecksher, Felipe G. Belladonna, Rafaela A. de Vasconcelos, and Tatiana K.da Silva Fidalgo. 2020. "Influence of autoclave sterilization procedures on the cyclic fatigue resistance of heat-treated nickel-titanium instruments: a systematic review." *Restorative Dentistry & Endodontics* 45 (2): e25.

Singh, Huidrom H., Simrjit Singh, and Neeraj Khare. 2018. "Enhanced β-phase in PVDF polymer nanocomposite and its application for nanogenerator." *Polymers for Advanced Technologies* 29 (1):143–150.

Song, Peiyi, Guang Yang, Tingting Lang, and Ken-Tye Yong. 2018. "Nanogenerators for wearable bioelectronics and biodevices." *Journal of Physics D* 52 (2):023002.

Sriphan, Saichon, Thitirat Charoonsuk, Tosapol Maluangnont, Phakkhananan Pakawanit, Catleya Rojviriya, and Naratip Vittayakorn. 2020. "Multifunctional nanomaterials modification of cellulose paper for efficient triboelectric nanogenerators." *Advanced Materials Technologies* 5 (5):2000001.

Su, Meng, Juergen Brugger, and Beomjoon Kim. 2020. "Simply structured wearable triboelectric nanogenerator based on a hybrid composition of carbon nanotubes and polymer layer." *International Journal of Precision Engineering and Manufacturing-Green Technology* 7 (3):683–698. https://doi.org/10.1007/s40684-020-00212-8.

Sultana, Ayesha, Sujoy K. Ghosh, Md M. Alam, Priyabrata Sadhukhan, Krittish Roy, Mengying Xie, Chris R. Bowen, Subrata Sarkar, Sachindranath Das, Tapas R. Middya, and Dipankar Mandal. 2019. "Methylammonium lead iodide incorporated poly(vinylidene fluoride) nanofibers for flexible piezoelectric–pyroelectric nanogenerator." *ACS Applied Materials & Interfaces* 11 (30):27279–27287. https://doi.org/10.1021/acsami.9b04812.

Sun, Jiangman, Xiong Pu, Mengmeng Liu, Aifang Yu, Chunhua Du, Junyi Zhai, Weiguo Hu, and Zhong L. Wang. 2018. "Self-healable, stretchable, transparent triboelectric nanogenerators as soft power sources." *ACS Nano* 12 (6):6147–6155. https://doi.org/10.1021/acsnano.8b02479.

Sun, Qi-Jun, Yanqiang Lei, Xin-Hua Zhao, Jing Han, Ran Cao, Jintao Zhang, Wei Wu, Hadi Heidari, Wen-Jung Li, Qijun Sun, and Vellaisamy A.L. Roy. 2021. "Scalable fabrication of hierarchically structured graphite/polydimethylsiloxane composite films for large-area triboelectric nanogenerators and self-powered tactile sensing." *Nano Energy* 80:105521. doi: https://doi.org/10.1016/j.nanoen.2020.105521.

Valdès, L.C., and J. Darque. 2003. "Design of wind-driven generator made up of dynamos assembling." *Renewable Energy* 28 (3):345–362.

Vera Anaya, David, Tianyiyi He, Chengkuo Lee, and Mehmet R. Yuce. 2020. "Self-powered eye motion sensor based on triboelectric interaction and near-field electrostatic induction for wearable assistive technologies." *Nano Energy* 72:104675. https://doi.org/10.1016/j.nanoen.2020.104675.

Wang, Xudong, Jinhui Song, Jin Liu, and Zhong L. Wang. 2007. "Direct-current nanogenerator driven by ultrasonic waves." *Science* 316 (5821):102–105.

Wang, Xingzhao, Bin Yang, Jingquan Liu, and Chunsheng Yang. 2017. "A transparent and biocompatible single-friction-surface triboelectric and piezoelectric generator and body movement sensor." *Journal of Materials Chemistry A* 5 (3):1176–1183.

Wang, Zhong L. 2020. "On the first principle theory of nanogenerators from Maxwell's equations." *Nano Energy* 68:104272.

Wang, Wei, Aifang Yu, Xia Liu, Yudong Liu, Yang Zhang, Yaxing Zhu, Ying Lei, Mengmeng Jia, Junyi Zhai, and Zhong L. Wang. 2020. "Large-scale fabrication of robust textile triboelectric nanogenerators." *Nano Energy* 71:104605.

Wu, Tao, Yiheng Song, Zhuqun Shi, Dongning Liu, Siling Chen, Chuanxi Xiong, and Quanling Yang. 2021. "High-performance nanogenerators based on flexible cellulose nanofibril/MoS2 nanosheet composite piezoelectric films for energy harvesting." *Nano Energy* 80:105541. https://doi.org/10.1016/j.nanoen.2020.105541.

Yang, Rusen, Yong Qin, Cheng Li, Guang Zhu, and Zhong L. Wang. 2009. "Converting biomechanical energy into electricity by a muscle-movement-driven nanogenerator." *Nano Letters* 9 (3):1201–1205.

Yang, Ya, Jong H. Jung, Byung K. Yun, Fang Zhang, Ken C. Pradel, Wenxi Guo, and Zhong L. Wang. 2012. "Flexible pyroelectric nanogenerators using a composite structure of lead-free KNbO3 nanowires." *Advanced Materials* 24 (39):5357–5362. https://doi.org/10.1002/adma.201201414.

Yang, Ya, Wenxi Guo, Ken C. Pradel, Guang Zhu, Yusheng Zhou, Yan Zhang, Youfan Hu, Long Lin, and Zhong L. Wang. 2012. "Pyroelectric nanogenerators for harvesting thermoelectric energy." *Nano Letters* 12 (6):2833–2838.

Yao, Chunhua, Xin Yin, Yanhao Yu, Zhiyong Cai, and Xudong Wang. 2017. "Chemically functionalized natural cellulose materials for effective triboelectric nanogenerator development." *Advanced Functional Materials* 27 (30):1700794. https://doi.org/10.1002/adfm.201700794.

Yoon, Hong-Joon, and Sang-Woo Kim. 2020. "Nanogenerators to Power Implantable Medical Systems." *Joule* 4 (7):1398–1407.

Yu, Yanhao, Zhaodong Li, Yunming Wang, Shaoqin Gong, and Xudong Wang. 2015. "Sequential infiltration synthesis of doped polymer films with tunable electrical properties for efficient triboelectric nanogenerator development." *Advanced Materials* 27 (33):4938–4944. https://doi.org/10.1002/adma.201502546.

Yu, Yanhao, and Xudong Wang. 2016. "Chemical modification of polymer surfaces for advanced triboelectric nanogenerator development." *Extreme Mechanics Letters* 9:514–530. https://doi.org/10.1016/j.eml.2016.02.019.

Zhang, Yingying, Jiarui Wu, Siwen Cui, Wutao Wei, Weihua Chen, Rui Pang, Zijie Wu, and Liwei Mi. 2020. "Organosulfonate counteranions—a trapped coordination polymer as a high-output triboelectric nanogenerator material for self-powered anticorrosion." *Chemistry – A European Journal* 26 (3):584–591. https://doi.org/10.1002/chem.201904873.

Zhao, Gengrui, Yawen Zhang, Nan Shi, Zhirong Liu, Xiaodi Zhang, Mengqi Wu, Caofeng Pan, Hongliang Liu, Linlin Li, and Zhong L. Wang. 2019. "Transparent and stretchable triboelectric nanogenerator for self-powered tactile sensing." *Nano Energy* 59:302–310.

Zhu, Minshen, Yang Huang, Wing S. Ng, Junyi Liu, Zifeng Wang, Zhengyue Wang, Hong Hu, and Chunyi Zhi. 2016. "3D spacer fabric based multifunctional triboelectric nanogenerator with great feasibility for mechanized large-scale production." *Nano Energy* 27:439–446.

Zi, Yunlong, Long Lin, Jie Wang, Sihong Wang, Jun Chen, Xing Fan, Po-Kang Yang, Fang Yi, and Zhong L. Wang. 2015. "Triboelectric-pyroelectric-piezoelectric hybrid cell for high-efficiency energy-harvesting and self-powered sensing." *Advanced Materials* 27 (14):2340–2347.

Index

Note: Locators in *italics* represent figures and **bold** indicate tables in the text.

G

H

I

K

L

V

W

X

Y

Z

Printed in the United States
by Baker & Taylor Publisher Services